Bats in the Anthropocene: Conservation of Bats in a Changing World

Christian C. Voigt · Tigga Kingston
Editors

Bats in the Anthropocene: Conservation of Bats in a Changing World

Springer Open

Editors
Christian C. Voigt
Berlin
Germany

Tigga Kingston
Lubbock, TX
USA

ISBN 978-3-319-25218-6 ISBN 978-3-319-25220-9 (eBook)
DOI 10.1007/978-3-319-25220-9

Library of Congress Control Number: 2015950865

Springer Cham Heidelberg New York Dordrecht London

Printed on acid-free paper

Springer International Publishing AG Switzerland is part of Springer Science+Business Media (www.springer.com)

For Thomas H. Kunz and Otto von Helversen for sharing with us their passion for bats.

For Silke, Philippa and Florian (CCV) and for Danny (TK) for their inspiration and patience.

Contents

Chapter 1
Bats in the Anthropocene

Christian C. Voigt and Tigga Kingston

Abstract Humans have inadvertently changed global ecosystems and triggered the dawn of a new geological epoch, the Anthropocene. While some organisms can tolerate human activities and even flourish in anthropogenic habitats, the vast majority are experiencing dramatic population declines, pushing our planet into a sixth mass extinction. Bats are particularly susceptible to anthropogenic changes because of their low reproductive rate, longevity, and high metabolic rates. Fifteen percent of bat species are listed as threatened by the IUCN, i.e., they are considered Critically Endangered, Endangered or Vulnerable. About 18 % of species are Data Deficient, highlighting the paucity of ecological studies that can support conservation status assessments. This book summarizes major topics related to the conservation of bats organized into sections that address: the response of bats to land use changes; how the emergence of viral and fungal diseases has changed bat populations; our perception of bats; and drivers of human–bat conflicts and possible resolutions and mitigation. The book ends with approaches that might advance bat conservation through conservation networks and a better understanding of human behavior and behavioral change.

C.C. Voigt (✉)
Leibniz Institute for Zoo and Wildlife Research, Alfred-Kowalke-Str. 17,
10315 Berlin, Germany
e-mail: voigt@izw-berlin.de

C.C. Voigt
Institute for Biology, Freie Universität Berlin, Takustr. 6,
14195 Berlin, Germany

T. Kingston (✉)
Department of Biological Sciences, Texas Tech University, Lubbock, TX, USA
e-mail: tigga.kingston@ttu.edu

C.C. Voigt and T. Kingston (eds.), *Bats in the Anthropocene: Conservation of Bats in a Changing World*, DOI 10.1007/978-3-319-25220-9_1

1.1 The Emergence of a New Geological Epoch: The Anthropocene

The world in which we live is fragile; a small layer of organismic activity covers the planet like a microbial film on top of a large boulder. Nonetheless, humans treat the Earth as if anthropogenic impacts on this delicate biological layer may be absorbed by unfailing natural buffers. Yet, convergent and overwhelming evidence from all over the world underlines that mankind has already changed and continues changing the face of our planet. Among the many transformations humans imposed on our planet, some of the most severe appear to be (1) the addition of more than 550 billion metric tons of carbon to the atmosphere which are the main drivers of global climate change and ocean acidification (Gray 2007; Ciasi and Sabine 2013), (2) the alteration of the global nitrogen cycle by the use of artificial fertilizers (Canfield et al. 2010), (3) the routing of more than one third of global primary production to human consumption (Krausmann et al. 2013), (4) the ongoing mass extinction of species (Barnosky et al. 2011), and (5) the globalization of transport which has resulted in the spread of invasive species and pathogens (Lewis and Maslin 2015). It is now widely recognized that global ecosystem services may be inadvertently suffering from human action, because human-induced environmental impacts are overriding natural process that have dominated our planet for millions of years (Steffen et al. 2011).

In the face of lasting human impacts on the Earth's geological conditions and processes, many scientists, beginning with Paul Crutzen and Eugene Stoermer in 2000, now posit that our actions have brought us to the dawn of a new geological epoch—the Anthropocene. The pros and cons regarding this definition, which literally means "Human Epoch" and would succeed the Holocene, are still heavily debated (Monastersky 2015). Yet skeptics are declining in number, and much of the current debate focuses on the exact beginning of the Anthropocene, generally considered to be c. 1800. The Anthropocene working group of the Subcommission on Quaternary Stratigraphy reports to the International Commission on Stratigraphy with a proposal to formalize the Anthropocene in 2016. For the purpose of this book, we do not refer to an exact starting point of the Anthropocene, but merely acknowledge the fact that humans have an impact on virtually all global ecosystems and that wildlife species such as bats (order Chiroptera) have adjusted to these changes, experienced substantial population declines, or gone extinct.

1.2 Bats in the Anthropocene: The Conservation of a Nocturnal Taxon

Bats (order Chiroptera) include more than 1300 extant species, forming the second largest mammalian order, and are unique among mammals in their evolution of powered flight. Although the common ancestor of living bats dates back to the K/T boundary (c. 70 mya), the most rapid radiation of any mammalian order resulted in all 18 extant families by the end of the Eocene c. 37 mya (Teeling et al.2005). Moreover, although the majority of bat species are insectivorous, trophic diversity is extraordinary for a single order, with frugivores, nectarivores, piscivores, sanguinivores, and carnivores represented. Bats currently inhabit all continents except Antarctica, and in many parts of the world, especially the tropics, are the most species-rich mammalian group at a given locality, with alpha diversity reaching about 70 species in the Paleotropics (Kingston et al. 2010) and over 100 in the Neotropics (Voss and Emmons 1996; Rex et al. 2008). From any perspective, bats are an evolutionary and ecological success story. Nonetheless, bat populations are under severe threat in many regions of the world (Racey and Entwistle 2003). The last recorded case of a bat species driven to extinction is that of the Christmas Island pipistrelle, *Pipistrellus murrayi* (Lumsden and Schulz 2009; Lumsden 2009; Martin et al. 2012), yet this species is most likely not the last one to vanish from our planet.

The IUCN Bat Specialist Group is in the process of reassessing the Red List status of bat species, with the current assessments of 1150 species mostly completed in 2008, with 34 species assessed since. From these assessments, five species were assessed as Extinct (giant vampire bat (*Desmodus draculae*), dusky flying fox (*Pteropus brunneus*), large Pelew flying fox (*P. pilosus*), dark flying fox (*P. subniger*), and Guam flying fox (*P. tokudae*)). The giant vampire bat is known only from the fossil and subfossil records, and the causes of its extinction are unknown. However, the four island *Pteropus* spp. are all victims of the Anthropocene, with hunting and habitat loss as the main drivers of extinction. Fifteen percent of bat species are listed in the threatened categories [Critically Endangered (CE), Endangered (EN), and Vulnerable (VU)] and 7 % are Near Threatened (Fig. 1.1). Around 18 % of species are Data Deficient (DD), and there have been a wealth of new species discovered since the last assessment. The pattern of vulnerability is fairly consistent across families (Fig. 1.2), with the notable exception of the Pteropodidae with 36 % of species extinct or threatened, probably because of their size, their appeal as bushmeat and for traditional medicine,

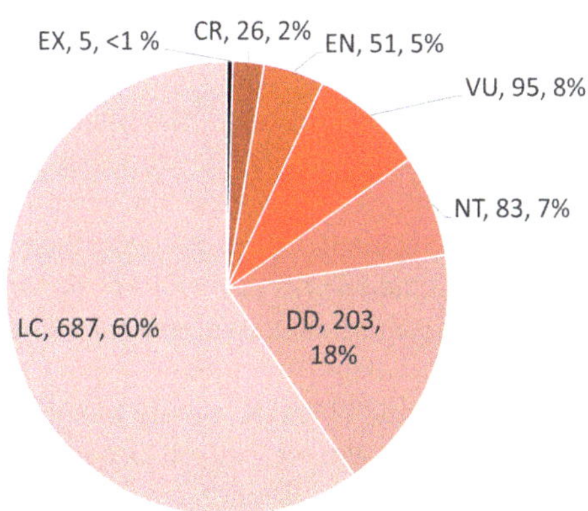

Fig. 1.1 *Red* List status of the 1150 bat species assessed 2008–2014 (IUCN 2015). IUCN categories are *EX* Extinct, *CR* Critically Endangered, *EN* Endangered, *VU* Vulnerable, *NT* Near Threatened, *DD* Data Deficient, *LC* Least Concern. Number of species and percentage of all species given as labels

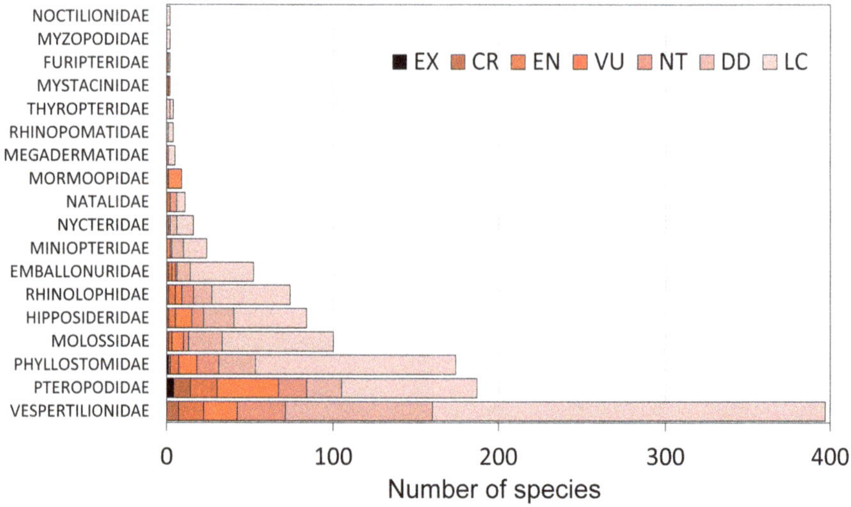

Fig. 1.2 *Red* List status of bats by family. Abbreviations as for Fig. 1.1

and because many form susceptible island populations. Even this depicts only part of the picture; populations are only considered stable in 21 % of all species and increasing in less than 1 %. Of the remaining species, populations are decreasing (23 %) or the trend is unknown (55 %). Moreover, of the 687 species assessed as Least Concern (LC), current specific threats were identified for about 27 % of species. Declining populations and identified threats suggest a bleak future, and it is probable that more species will satisfy the rigorous criteria of the threatened categories in the coming years.

Globally, the major threats to bat species identified by IUCN assessments are land use change (logging, non-timber crops, livestock farming and ranching, wood and pulp plantations, and fire), urbanization, hunting and persecution, quarrying and general human intrusions on bat habitats (Fig. 1.3). Bats are particularly susceptible to these human-induced perturbations of habitats because of their distinct life history. Bats are on the slow side of the slow-fast continuum of life histories (Barclay and Harder 2003). For example, they reproduce at a low rate (Barclay et al. 2004) and are long-lived mammals (Munshi-South and Wilkinson 2010; Wilkinson and South 2002). Thus, bat populations recover slowly from increased mortality rates. Despite their low reproductive rate and longevity, bats have relatively high metabolic rates owing to their small size which leads to relatively high food requirements (Thomas and Speakman 2003).

Lastly, bats are nocturnal animals with often cryptic habits. Even though they are present in many larger cities of the temperate zone, they often go unnoticed by their human neighbors. It is quite likely that perceptions of bats would be very different if *Homo sapiens* evolved as a nocturnal hominid. Or to put it in the words of Rich and Longcore: What if we woke up one morning and realize that we missed

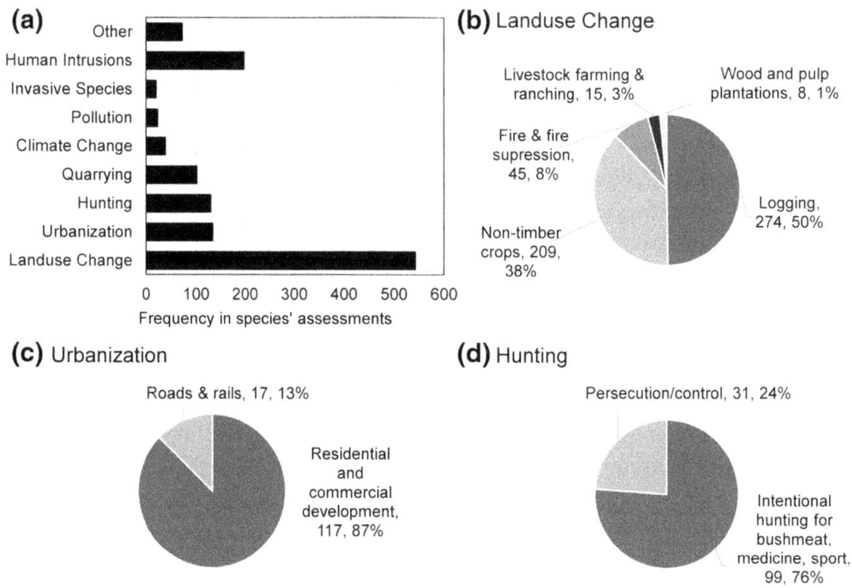

Fig. 1.3 Frequency of threats listed in the IUCN assessments of bat species. **a** Distribution of major threats across assessments. Land use changes, urbanization. and hunting are aggregations of IUCN listed threats given in **b**–**d**. Frequency of threat and percentage contribution are given

half of the story in our conservation efforts, namely the night part? (modified after Rich and Longcore 2004, p. 1). This brings up an important question: Do nocturnal animals benefit less from legal protection than diurnal animals? Are we more concerned about animals that we see and interact with during daytime? Do human societies perceive and evaluate, for example, fatalities of birds of prey at wind turbines in a different way than bat fatalities when both ought to benefit from the same level of protection? Do we consider recommendations to reduce light pollution for the sake of nocturnal animals such as bats, or does the expansion of the human temporal niche into the night come at high costs for all nocturnal animals? In summary, we speculate that bats as nocturnal animals might be particularly exposed to human-induced ecological perturbations because we are driven by our visual system and therefore tend to neglect the dark side of conservation, i.e., the protection of nocturnal animals.

1.3 Why Care About Bat Conservation?

The reasoning for the conservation of nature can be manifold, reaching from purely moral to monetary arguments and legal requirements. It may also vary according to the scale of the conservation approach, i.e., whether it is driven by local, national, or international perspectives. Indeed, ethical considerations for the protection of species—although quite often neglected in modern civilization—should be the primary motivation; i.e., the obligation of humans to conserve nature for the simple reason of its existence and for the more selfish reason to make the diversity of biological life accessible and useable to following generations of humans. Lately, economic arguments for the conservation of nature are increasingly used, e.g., the importance of protecting water catchment areas to provide potable water or irrigation in agriculture. So-called ecosystem services of nature are highly valued in modern societies and therefore benefit from increasing protection.

Recent attempts to critically review the ecosystem services provided by bats have revealed that many species offer unique and large-scale monetary benefits to agricultural industry (Kunz et al. 2011; Ghanem and Voigt 2012; Maas et al. 2015). For example, flowers of the Durian tree are only effectively pollinated by the Dawn bat, *Eonycteris spelaea*, in Southeast Asia (Bumrungsri et al. 2009). Durian is a highly valued fruit in Asia with Thailand producing a market value of durians of almost 600 million US$ annually (Ghanem and Voigt 2012). Other bats consume large amounts of pest insects, thereby offering services that could save millions of US$ for national industries (Boyles et al. 2011; Wanger et al. 2014). However, the monetary approach for protecting bat species is a double-edged sword, since bat species without apparent use for human economy may not benefit from protection compared to those that provide some ecosystem services. Moreover, arguments based on economic or utilitarian values of wildlife may appeal to self-interest motivations and suppress environmental concern (Kingston 2016). In this context, it is important to note that we have just started to understand the ecological role bats fill in natural ecosystems. For example, bats have been recently documented as top-down regulators of insect populations in forest habitats of the tropics and temperate zone (Kalka et al. 2008; Boehm et al. 2011) and also in subtropical coffee and cacao plantations (Williams-Guillen et al. 2008; Maas et al. 2013). Finally, bats are protected by law in some countries. For example, they are covered by the Habitat Directive of the European Union and thus strictly protected in E.U. countries. Also, migratory bats benefit from some level of protection because they are covered by the UN Convention for the Protection of Migratory Species. Countries that have signed this convention are obliged to support conservation actions that are beneficial for migratory species. CITES (The Convention on International Trade in Endangered Species of Wild Fauna and Flora) protects threatened species through controls of international trade in specimens. The precarious conservation status of the flying foxes is apparent. Currently, *Acerodon jubatus* and ten *Pteropus* spp are on CITES Appendix I, with trade only permitted in exceptional circumstances, and the remaining *Acerodon* and *Pteropus* species on Appendix II, by which trade is controlled to avoid utilization incompatible with their survival.

1.4 About This Book

The idea to publish a book about bat conservation was stimulated by the "3rd International Berlin Bat Meeting: Bats in the Anthropocene" in 2013. The overall goal is to provide a summary of the major threats bats are facing in a rapidly changing world. The book is organized in four major sections: (1) bats in anthropogenically-shaped landscapes, (2) emerging diseases, (3) human–bat conflicts, and (4) conservation approaches. The basic concept of chapters in all of these sections is to review the literature that is available in peer-reviewed journals. We are aware that many topics related to bat conservation have also been addressed in brochures or books published by non-governmental or governmental organizations. Sometimes these sources have been cited in the corresponding chapters, yet in most cases authors of this book have focused on the aforementioned sources of information.

From our editorial perspective, the chapters cover the majority of relevant topics in bat conservation. However, we acknowledge that at least three topics are missing in this book. First, this book misses a chapter on "bats and global climate change," because Jones and Rebelo (2013) published a recent review on this topic and the body of literature about this topic has not largely increased since then. Second, we did not commission a chapter on "Bats and chemical pollutants," as current knowledge of heavy metals was recently synthesized by Zukal et al. (2015) and information for other pollutants is sparse. That said, the subject is referenced in several chapters (Williams-Guillen et al. 2015; Korine et al. 2015; Voigt et al. 2016). Third, we did not include a chapter on "island bats," although many of them are endangered and some even are threatened by extinction, as Fleming and Racey (2010) provide a detailed overview of this topic in their recent book. Finally, authors integrate successful interventions into their accounts and make specific recommendations for future research, but additional evidence-based evaluations of the success of conservation interventions per se are found in Berthinussen et al. (2014).

The Anthropocene has gained momentum. It is a geological epoch that is not in equilibrium but is constantly changing by the action of mankind. For a handful of bat species anthropogenic changes may prove beneficial, but for the vast majority our actions precipitate drastic population declines that must be slowed if we are to conserve the extraordinary diversity of this unique mammalian order. We hope that this book will stimulate new directions for research and support conservation interventions that will keep the night sky alive with bats in the Human Epoch.

Acknowledgements We would like to acknowledge the financial support provided by the Leibniz Institute for Zoo and Wildlife Research in Berlin, Germany, by EUROBATS and a National Science Foundation grant to the Southeast Asian Bat Conservation Research Unit (NSF Grant No. 1051363) that enabled us to publish this book as an open-access electronic book. We thank Mark Brigham, Anne Brooke, Justin Boyles, Gabor Csorba, Brock Fenton, Jorge Galindo-González, Chris Hein, Carmi Korine, Allen Kurta, Pia Lentin, Herman Limpens, Lindy Lumsden, Jörg Müller, Alison Peel, Paul Racey, Hugo Rebelo, DeeAnn Reeder, Scott Reynolds, Danilo Russo, Armando Rodríguez-Durán, N. Singaravelan, Vikash Tatayah, Peter Taylor, and numerous anonymous reviewers for providing constructive comments on chapters of this book.

References

Barclay RMR, Harder LD (2003) Life histories of bats: life in the slow land. In: Kunz TH, Fenton MB (eds) Bat ecology. University of Chicago Press, Chicago and London

Barclay RMR, Ulmer J, MacKenzie CJA, Thompson MS, Olson L, McCool J, Cropley E, Poll G (2004) Variation in the reproductive rate of bats. Can J Zool 82:688–693

Barnosky AD, Matzke N, Tomiya S, Wogan GOU, Swartz B, Quental TB, Marshall C, McGuire JL, Lindseay EL, Maguire KC, Mersey B, Ferrer EA (2011) Has the Earth's sixth mass extinction already arrived. Nature 471:51–57

Berthinussen A, Richardson OC, Altringham JD (2014) Bat conservation: global evidence for the effects of interventions. Pelagic Publishing, Exeter

Boehm SM, Wells K, Kalko EKV (2011) Top-down control of herbivory by birds and bats in the canopy of temperate Broad-Leaved Oaks (*Quercus robur*). PLoS One 6(4):e17857. doi:10.1371/journal.pone.0017857

Boyles JG, Cryan PM, McCracke GF, Kunz TH (2011) Economic importance of bats in agriculture. Science 332:41–42

Bumrungsri S, Sripaoraya E, Chongiri T, Sridith JK, Racey PA (2009) The pollination ecology of durian (*Durio hibethinus*, Bombacaceae) in southern Thailand. J Trop Ecol 25:85–92

Canfield DE, Glazer AN, Falkowski PG (2010) The evolution and future of Earth's nitrogen cycle. Science 330:192–196

Ciasi P, Sabine C (2013) Chapter 6. In: Stocker TF, Qin D, Plattner G-K, Tignor MMB, Allen SK, Boschung H, Nauels A, Xia Y, Bex V, Midgley PM (eds) Climate change 2013: the physical science basis. Contribution of working group I to the fifth assessment report of the intergovernmental panel on climate change, pp 465–570

Fleming TH, Racey PA (2010) Island bats: evolution, ecology and conservation. University of Chicago Press, Chicago, p 592

Ghanem SH, Voigt CC (2012) Increasing awareness of ecosystem services provided by bats. Adv Study Behav 44:279–302

Gray V (2007) Climate change 2007: the physical science basis summary for policymakers. Energy Environ 18:433–440

IUCN (2015) The IUCN Red List of Threatened Species. Version 2015-3.http://www.iucnredlist.org. Downloaded on 9 September 2015

Jones G, Rebelo H (2013) Responses of bats to climate change: learning from the past and predicting the future. In: Adams RA, Pedersen SC (eds) Bat evolution, ecology, and conservation. Springer, New York, Berlin, pp 457–478

Kalka MB, Smith AR, Kalko EKV (2008) Bats limit arthropods and herbivory in a tropical forest. Science 320:71. doi:10.1126/science.1153352

Kingston T (2010) Research priorities for bat conservation in Southeast Asia: a consensusapproach. Biodivers Conserv 19:471–484

Kingston T (2016) Cute, Creepy, or Crispy - How values, attitudes, and norms shape human behavior toward bats. In: Voigt CC, Kingston T (eds) Bats in the Anthropocene: conservation of bats in a changing world. Springer International AG, Cham, pp 571–588

Korine C, Adams R, Russo D, Fisher-Phelps M, Jacobs D (2015) Bats and water: anthropogenic alterations threaten global bat populations. In: Voigt CC, Kingston T (eds) Bats in the Anthropocene: conservation of bats in a changing world. Springer International AG, Cham, pp 215–233

Krausmann F, Erb K-H, Gingrich S, Haberla H, Bondeau A, Gaube V, Lauk C, Plutzar C, Searchinger TD (2013) Global human appropriation of net primary production doubled in the 20th century. Proc Natl Acad Sci USA 110:10324–10329

Kunz TH, Braun de Torrez E, Bauer DM, Lobova TA, Fleming TH (2011) Ecosystem services provided by bats. In: Ostfeld RA, Schlesinger WH (eds) The year in ecolgoy and conservation 2011: annals of the New York academy of sciences. Wiley, New York, USA, pp 1–38
Lewis SL, Maslin MA (2015) Defining the Anthropocene. Nature 519:171–180
Lumsden L (2009) The extinction of the Christmas Island Pipistrelle. Aus Bat Soc Newsl 33:21–25
Lumsden L, Schulz M (2009) Captive breeding and future in-situ management of the Christmas Island Pipistrelle Pipistrellus murrayi. A report to the Director of National Parks. Arthur Rylah Institute, Department of Sustainability and Environment, Heidelberg, Victoria
Maas B, Clough Y, Tscharntke T (2013) Bats and birds increase crop yield in tropical agroforesty landscapes. Ecol Lett 16:14801
Maas B, Karp DS, Bumrungstri S, Darras K, Gonthier D, Huang JCC, Lindell CA, Maine JJ, Mestre L, Michel NL, Morrison EB, Perfecto I, Philpott SM, Sekergioglu CH, Silva RM, Taylor PJ, Tscharntke T, Van Bael SA, Whelan CH, Williams-Guillen K (2015) Bird and bat predation services in tropical forests and agroforestry landscapes. Biol Rev. doi:10.1111/brv.12211
Martin TG, Nally S, Burbidge AA, Arnall S, Garnett ST, Hayward MW, Lumsden L, Menhhorst P, McDonald-Madden E, Possingham HP (2012) Acting fast helps avoid extinction. Conserv Lett 5:274–280
Monastersky R (2015) Anthropocene: the human age—editorial. Nature 519:144–147
Munshi-South J, Wilkinson GS (2010) Bats and birds: exceptional longevity despite high metabolic rates. Ageing Res Rev 9:12–19
Racey PA, Entwistle AC (2003) Conservation ecology of bats. In: Kunz TH, Fenton MB (eds) Bat ecology. University of Chicago Press, Chicago, pp 680–743
Rex K, Kelm DH, Wiesner K, Kunz TH, Voigt CC (2008) Species richness and structure of three Neotropical bat assemblages. Biol J Linn Soc 94:617–629
Rich C, Longcore T (2004) Ecological consequences of artificial night lighting. Island Press, Washington
Steffen W, Grinevald J, Crutzen P, McNeill J (2011) The anthropocene: conceptual and historical perspectives. Philos Trans R Soc A 369:842–867
Teeling EC, Springer MS, Madsen O, Bates P, O'Brien SJ, Murphy WJ (2005) A molecular phylogeny for bats illuminates biogeography and the fossil record. Science 307:580–584
Thomas D, Speakman JR (2003) Physiological ecology and energetics of bats. In: Kunz TH, Fenton MB (eds) Bat ecology, University of Chicago Press. Chicago and London, pp 430–490
Voss RS, Emmons LH (1996) Mammalian diversity in Neotropical lowland rainforests: a preliminary assessment. Bull Am Mus Nat Hist 230:1–115
Voigt CC, Phelps KL, Aguirre L, Schoeman MC, Vanitharani J, Zubaid A (2016) Bats and buildings: the conservation of synanthropic bats. In: Voigt CC, Kingston T (eds) Bats in the Anthropocene: conservation of bats in a changing world. Springer International AG, Cham, pp 427–453
Wanger TC, Darras K, Bumrungsri S, Tscharntke T, Klein AM (2014). Bat pest control contributes to food security in Thailand. Biol Conserv 171:220–223
Wilkinson GS, South JM (2002) Life history, ecology and longevity in bats. Aging Cell 1:124–131
Williams-Guillen K, Perfecto I, Vandermeer J (2008) Bats limit insects in a neotropical agroforestry system. Science 320:70. doi:10.1126/science.1152944
Williams Guillen et al. (2016) Bats in the anthropogenic matrix: challenges and opportunities for the conservation of Chiroptera and their ecosystem services in agricultural landscapes. In: Voigt CC, Kingston T (eds) Bats in the Anthropocene: conservation of bats in a changing world. Springer International AG, Cham, pp 151–178
Zukal J, Pikula J, Bandouchova H (2015) Bats as bioindicators of heavy metal pollution: history and prospect. Mamm Biol-Z Saugertierkd 80:220–227

Part I
Bats in Anthropogenically Changed Landscapes

Chapter 2
Urbanisation and Its Effects on Bats—A Global Meta-Analysis

Kirsten Jung and Caragh G. Threlfall

Abstract Urbanisation is viewed as the most ecologically damaging change to land use worldwide, posing significant threats to global biodiversity. However, studies from around the world suggest that the impacts of urbanisation are not always negative and can differ between geographic regions and taxa. Bats are a highly diverse group of mammals that occur worldwide, and many species persist in cities. In this chapter, we synthesise current knowledge of bats in urban environments. In addition, we use a meta-analysis approach to test if the general response of bats depends on the intensity of urbanisation. We further investigate if phylogenetic relatedness or functional ecology determines adaptability of species to urban landscapes and if determining factors for urban adaptability are consistent worldwide. Our meta-analysis revealed that, in general, habitat use of bats decreases in urban areas in comparison to natural areas. A high degree of urbanisation had a stronger negative effect on habitat use compared to an intermediate degree of urbanisation. Neither phylogenetic relatedness nor functional ecology alone explained species persistence in urban environments; however, our analysis did indicate differences in the response of bats to urban development at the family level. Bats in the families Rhinolophidae and Mormoopidae exhibited a negative association with urban development, while responses in all other families were highly heterogeneous. Furthermore, our analysis of insectivorous bats

K. Jung (✉)
Institute of Evolutionary Ecology and Conservation Genomics, University of Ulm, Ulm, Germany
e-mail: kirsten.jung@uni-ulm.de

C.G. Threlfall (✉)
School of Ecosystem and Forest Sciences, University of Melbourne, Melbourne, Australia
e-mail: caragh.threlfall@unimelb.edu.au

C.C. Voigt and T. Kingston (eds.), *Bats in the Anthropocene: Conservation of Bats in a Changing World*, DOI 10.1007/978-3-319-25220-9_2

revealed that the adaptability of individual families, e.g. Emballonuridae and Vespertilionidae, to urbanisation is not consistent worldwide. These results suggest that behavioural and/or morphological traits of individual species may better determine species' adaptability to urban areas, rather than phylogenetic or functional classifications, and that driving factors for species adaptability to urban areas might be regionally divergent. We thus argue that future research should focus on behavioural and morphological traits of bats, to assess if these determine urban adaptability in this species-rich group of mammals.

2.1 Introduction

2.1.1 The Urban Context

Urbanisation results in extreme forms of land use alteration (Shochat et al. 2006; Grimm et al. 2008). In the last century, the human population has undergone a transition in which the majority of people now live in urban rather than rural areas (UNPD 2012). The rate of change at which urban areas are evolving due to natural population growth is dramatic, including significant rural-to-urban migration and spatial expansion (Grimm et al. 2008; Montgomery 2008; UN 2012; McDonnell and Hahs 2013). In the last 50 years, the global human population in urban areas increased from 2.53 to 6.97 billion people (UNPD 2012). Yet human pressure resulting from urbanisation is not uniformly distributed on the planet. While urbanisation in the developed countries is slowing down slightly, it is increasing rapidly in developing countries of Asia, Africa, Latin America and the Caribbean, many of which harbour hotspots of biodiversity (Myers et al. 2000). In addition, over half of the urban population growth is projected to occur in smaller towns and cities (UN 2012). This implies that urbanisation is not a locally concentrated event, it is rather a fundamentally dispersed process and a happening worldwide (McDonald 2008).

The ecological footprint of cities reaches far beyond their boundaries (McGranahan and Satterthwaite 2003; McDonald and Marcotullio 2013). Effects of cities operate from local (e.g. through urban sprawl) to global scales (e.g. through greenhouse gas emission) (McDonald et al. 2008), and act both directly, through expansion of urban areas, and indirectly through growth in infrastructure and changes in consumption and pollution (McIntyre et al. 2000; Pickett et al. 2001). Apart from the obvious loss in natural area, expansion of cities also impacts the surrounding rural and natural habitats through increased fragmentation, and edge effects with increasing temperature and noise levels, which together introduce new anthropogenic stressors on fringe ecosystems (Grimm et al. 2008) and nearby protected areas (McDonald et al. 2008; McDonald and Marcotullio 2013). However, despite the radical land transformation and habitat loss incurred through urbanisation, many species (native and introduced) can still persist in urban environments and some even experience population increases (McKinney 2006). This suggests that urban landscapes can actually provide suitable habitat for a variety of species, albeit an anthropogenically altered habitat. Nevertheless, our understanding of what constitutes a suitable habitat in urban areas and what determines a species' adaptability to an urban environment is currently very limited.

Generally, urban areas are characterised by high quantities of impervious surfaces (McKinney 2002). There are however many additional physical and chemical changes incurred via the process of urbanisation (McDonnell and Pickett 1990), such as increased pollution, eutrophication, increased waste generation, altered hydrology (Vitousek et al. 1997; Grimm et al. 2008), increased urban noise (e.g. Slabbekoorn and Ripmeester 2008) and artificial light (Longcore and Rich 2004). Urban areas can provide a more thermally stable environment via the urban heat island effect (e.g. Zhao et al. 2006); less radiation is reflected during the day and more heat is trapped at night, which can increase minimum temperatures in cities (Grimm et al. 2008). The changed climate profile of cities can benefit some species by making the area more inhabitable year round. In addition, the planting of attractive introduced and native plant species throughout the suburbs and along city roads also changes the resources available to fauna, for example by providing nectar or fruits throughout the year. Altogether these changes can impact local species assemblages within cities and regional biodiversity beyond the municipal boundaries (Grimm et al. 2008).

Anthropogenic changes in urban ecosystems typically occur at rates drastically faster than long-lived organisms are capable of adapting to, and thus disrupt ecological processes that historically governed community structure (Duchamp and Swihart 2008). However, generalisations about the negative effects of urbanisation should not overlook biologically meaningful differences in how taxa respond to human land use (Dixon 2012). Some wildlife species are able to adjust to a life in urban areas. Among vertebrates, a range of birds are relatively abundant in urban environments and bird species richness may peak at intermediate levels of urbanisation because of increased heterogeneity of edge habitats (Blair 2001; McKinney 2002) and changes in resource availability due to provision of artificial feeding stations (Sewell and Catterall 1998). In contrast, only a few mammals have been documented as successful species in urban areas (Macdonald and Newdick 1982; Septon et al. 1995; Luniak 2004). For example, the grey-headed flying fox (*Pteropus poliocephalus*) has established a year-round camp in urban Melbourne, Australia, an area outside of its normal climatic range. Warmer temperatures from the urban heat effect, enhanced precipitation from local irrigation and year-round food resources appear to have facilitated the colony's arrival and persistence (Parris and Hazell 2005). Many animals, however, disappear from cities because they depend on habitat features that no longer exist (Gilbert 1989; McKinney 2002; Luniak 2004; Haupt et al. 2006; McDonnell and Hahs 2008). Declining species often suffer from increased habitat isolation, or face competition from invasive and more dominant species (McDonald and Marcotullio 2013). Some species in urban areas also suffer from additional stress (Isaksson 2010), increased infection and parasitism rates (Giraudeau et al. 2014) and reductions in potential reproductive success (Chamberlain et al. 2009). Urbanisation can also trigger a change in behaviour (Ditchkoff et al. 2006; Grimm et al. 2008). For example, urban noise alters the pitch at which some birds call (Slabbekoorn and Peet 2003), and affects activity patterns of larger vertebrates (Ditchkoff et al. 2006). Furthermore, increased artificial lighting can potentially disturb the circadian rhythms of nocturnal animals and interfere with the navigation of migrating species (Longcore and Rich 2004; Hölker et al. 2010; see Rowse et al., Chap. 7 this volume).

2.1.2 Urban Wildlife

Persistence of wildlife in urban environments may be linked to opportunism and a high degree of ecological and behavioural plasticity (Luniak 2004). In contrast, species that decline in response to urbanisation are often habitat and resource specialists (McKinney and Lockwood 1999; Jokimäki et al. 2011). Typically this results in altered assemblage structures in urban environments, often with a few highly abundant species, which account for a much higher proportion of the whole community in urban environments than in surrounding wild lands (Shochat et al. 2006). In addition, many native species are replaced by non-native, weedy or pest species (McKinney 2002). The resulting mix of introduced and native species in urban areas can lead to novel species interactions and altered ecosystem functioning (Hobbs et al. 2006). Often these non-native and introduced species are the same species across cities throughout the world. Thus, the flora and fauna of cities are becoming increasingly homogeneous (Hobbs et al. 2006; Grimm et al. 2008), however recent evidence suggests that many cities still retain several endemic species (Aronson et al. 2014).

Multi-scaled and multi-taxa investigations are required to provide detailed information about urban biodiversity (Clergeau et al. 2006). To date, urban ecologists have focused on few taxa, examining the response of conspicuous species to an urbanisation gradient (McDonnell and Hahs 2008). Population- and assemblage-level responses to urbanisation have been examined most prolifically for highly diverse and mobile bird taxa (McKinney 2002; McDonnell and Hahs 2008). Unfortunately, our understanding of how other wildlife, including bats, respond to the complex process of urbanisation is still limited (Barclay and Harder 2003). Research conducted to date provides a general indication that many bats may be declining due to urbanisation, however an understanding of the processes driving these patterns remains largely unknown.

2.1.3 Bats in Urban Environments

Bats likely form the most diverse group of mammals remaining in urban areas (van der Ree and McCarthy 2005; Jung and Kalko 2011). Of the studies conducted in urban landscapes to date, many show that overall bat activity and species richness are greatest in more natural areas, and decreases with increasing urban influence (Kurta and Teramino 1992; Walsh and Harris 1996; Gaisler et al. 1998; Legakis et al. 2000; Lesiñski et al. 2000). However, certain bat species may better be able to adapt to urban landscapes (Avila-Flores and Fenton 2005; Duchamp and Swihart 2008). Coleman and Barclay (2011), however, cautioned that most researchers have worked in forested regions directing less attention to other biomes, including grasslands. They argue that because urban tree cover is fairly constant (<30 %) in all cities (McKinney 2002), urbanisation in tree-rich regions implies deforestation and thus reduced tree cover may cause the negative effect of urbanisation. In contrast, urban areas within grassland regions might enhance structural heterogeneity and thus benefit species richness and relative abundance patterns (see Coleman and Barclay 2011 for more details). This is in accordance with the results of Gehrt and Chelsvig (2003, 2004) investigating the response of bats in and around the highly populated city of Chicago, USA. Here species diversity and occurrence were higher in habitat fragments within urban areas than in similar fragments in rural areas (Gehrt and Chelsvig 2003, 2004). However the large, forested parks in the region may offset the habitat loss caused by urbanisation, and hence they mitigate any negative impacts to bats at the regional scale.

The majority of studies on bats in urban environments come from the temperate regions of Europe and North America. Many studies focus on the response of bats to differently structured areas within the urban environment including historic and newly built city districts (Gaisler et al. 1998; Legakis et al. 2000; Guest et al. 2002; Dixon 2012; Hale et al. 2012; Pearce and Walters 2012), illuminated and non-illuminated areas (Bartonicka and Zukal 2003), industrial areas (Gaisler et al. 1998) small and larger parklands (Kurta and Teramino 1992; Fabianek et al. 2011; Park et al. 2012) and areas that receive waste water (Kalcounis-Rueppell et al. 2007). Most of these studies report relatively high bat activity and species richness in areas with remaining vegetation such as older residential areas, riverine habitats or parklands. Certain bat species appear to thrive in these urban environments, and success has been linked to species-specific traits (Duchamp and Swihart 2008). In particular, bat species with high wing loadings and aspect ratios, so presumed to forage in open areas (Norberg and Rayner 1987), which also roost primarily in human structures appeared to adjust to urban environments, provided that there is sufficient tree cover (Dixon 2012). Many of these studies imply that protecting and establishing tree networks may improve the resilience of some bat populations to urbanisation (Hale et al. 2012). Population- and assemblage-level responses along gradients of urbanisation reveal that generally foraging activity of bats seems to be higher in rural and forested areas than in urban areas (Geggie and Fenton 1985; Kurta and Teramino 1992; Lesiñski et al. 2000). However, it is important to note that some species might be highly flexible in their habitat use. The European bat *Eptesicus nilsonii*, for example, spends a much higher proportion of its foraging time in urban areas after birth of the juveniles than before (Haupt et al. 2006). This raises the importance of repeat observations during different seasons when investigating the response of bats to urbanisation.

In the Neotropics, most studies concerning bats and environmental disturbance have concentrated on fragmentation effects due to logging or agricultural land use (e.g. García-Morales et al. 2013). Persistence of bats in fragmented landscapes has been associated with edge tolerance and mobility in phyllostomids (Meyer and Kalko 2008), and the predominant use of open space as foraging habitat for aerial insectivorous bats (Estrada-Villegas et al. 2010). Of the few studies focusing on urban areas, most report an overall decrease in species richness and relative abundance of bats in urban areas (Avila-Flores and Fenton 2005; Siles et al. 2005; Pacheco et al. 2010; Jung and Kalko 2011) compared to forested areas. Predominantly, insectivorous bats seem to remain in large urban environments (Bredt and Uieda 1996; Filho (2011). Of these, it is typically members of the Molossidae, which are known to forage in the open spaces above the tree canopy that seem to tolerate and potentially profit from highly urbanised areas (Avila-Flores and Fenton 2005; Pacheco et al. 2010; Jung and Kalko 2011). In addition, many buildings in cities provide potential roost sites that resemble natural crevices (Burnett et al. 2001; Avila-Flores and Fenton 2005) and are known to be readily occupied by molossid bats (Kössl et al. 1999; Scales and Wilkins 2007). In a smaller urban setting in Panama, where mature forest meets very restricted urban development, a high diversity of bats occurs within the town and bats frequently forage around street lights (Jung and Kalko 2010). Nevertheless, even in such a low impact urban setting some species of the bat assemblage such as *Centronycteris centralis* revealed high sensitivity and were never recorded within the town, albeit foraging frequently in the nearby mature forest (Jung and Kalko 2010).

Recent investigations from large metropolitan urban centres in Australia show suburban areas can provide foraging habitat for bats (Rhodes and Catterall 2008; Threlfall et al. 2012a), and support greater bat activity and diversity than more urban and even forested areas (Hourigan et al. 2010; Basham et al. 2011; Threlfall et al. 2011, 2012b; Luck et al. 2013). However, studies from regional urban centres in Australia suggest that any urban land cover, even if low-density residential, can decrease bat activity and species richness (Hourigan et al. 2006; Gonsalves et al. 2013; Luck et al. 2013), and can deter some species of clutter-tolerant bats altogether (Gonsalves et al. 2013; Luck et al. 2013). Evidence also suggests that species adapted to open spaces and edges, such as those within the molossid family, do not display the same response to urbanisation in small regional versus large metropolitan urban centres, indicating subtle behavioural differences among species with similar ecomorphology (Luck et al. 2013; McConville et al. 2013a, b). The few studies that have investigated species-specific foraging and roosting requirements, suggest that although bats display high roost site fidelity within urban areas (Rhodes and Wardell-Johnson 2006; Rhodes et al. 2006; Threlfall et al. 2013a), species differ in their ability to forage successfully on aggregations of insects across the urban matrix, reflecting variation in flight characteristics and sensitivity to artificial night lighting (Hourigan et al. 2006; Scanlon and Petit 2008; Threlfall et al. 2013b).

Asian bat assemblages comprise a variety of frugivore and insectivore bat species; however, there is limited information on urban impacts to bats in this region of the world. Many roosting and foraging resources for frugivore species such as *Cynopterus* and *Pteropus* species are provided by exotic trees that grow easily in urban centres in Asia, for example *Ficus*, *Livistona* and *Syzygium* species, which have been studied in Hong Kong (Corlett 2005, 2006), India (Caughlin et al. 2012) and Japan (Nakamoto et al. 2007). Frugivore species in these systems provide critical seed dispersal services and can play a role in regeneration and pollination of some tree species (Mahmood-ul-Hassan et al. 2010; Caughlin et al. 2012). Radio-tracking studies show that some bat species roost in forested areas (Nakamoto et al. 2012) or in-built structures (Nadeem et al. 2013), however many frugivore species appear to profit from the density of planted exotic vegetation and both frugivore and insectivore bats can benefit from increased foraging resources in urban areas (Corlett 2005; Nakamoto et al. 2007; Utthammachai et al. 2008; Caughlin et al. 2012; Nakamoto et al. 2012). However, it appears that Asian bats, particularly large pteropodids, are also under threat from direct human impacts via hunting (Thomas et al. 2013), in addition to human land use alteration, and hence, any impact of urbanisation may be confounded by direct human impacts. However, increasing land use change and growing urban populations have been stated as a likely cause of dramatic declines of many bat species (including pteropodids) in Singapore (Pottie et al. 2005; Lane et al. 2006), where it is suggested the reported declines may reflect the declining status of bats in Southeast Asia more broadly (Lane et al. 2006). The only study to our knowledge that has examined bat species distribution in relation to increasing urbanisation was conducted in Pakistan, where greater bat capture success was recorded in urban areas in comparison to suburban and rural areas (Nadeem et al. 2013), and in line with other studies worldwide, the urban bat assemblage was dominated by a few common species. However, it is unclear whether these results were influenced by trapping success, and as such, should be interpreted cautiously.

The co-location of biodiversity and high human population densities raises the importance of conservation-related studies in urban areas where anthropogenic growth directly interacts with the highest levels of biodiversity (Rompré et al. 2008). In these landscapes, it is especially important to identify the underlying mechanisms determining the potential of different species to adjust to urban environments. Currently, our general understanding of what influences a species success and details of urban foraging and roosting habitat selection is incomplete. Yet, arguably the conservation of species such as bats in urban areas dependents upon this knowledge (Fenton 1997).

2.2 Evidence-Based Evaluation of the Effect of Urbanisation on Bats Worldwide Using a Meta-Analysis

Within this book chapter, we were in particular interested in the general conclusions concerning the potential of bats to adjust to urban environments. We thus synthesised pre-existing data of published literature with a focus on bats in urban versus natural environments in a worldwide meta-analysis. Meta-analysis has been previously used in ecology and conservation because results can lead to evidence-based environmental policies.

Here, we investigated the general response of bats to urbanisation and tested whether this is consistent across cities differing in the intensity of urban development. In addition, we address the question of whether adaptability of species to urban landscapes correlates with phylogeny or rather functional ecology. Functional ecology of species can be linked to species traits, where traits refer to morphological, behavioural or physiological attributes of species (Violle et al. 2007). Using such functional traits can improve understanding of and help predict how species respond to environmental change (Didham et al. 1996; Flynn et al. 2009), such as increasing urbanisation. A key challenge is to develop frameworks that can predict how the environment acts as a filter by advantaging or disadvantaging species with certain traits. Urbanisation has been demonstrated to select for, or against, species with specific response traits within flora and fauna communities, including remnant grasslands (Williams et al. 2005), bat communities (Threlfall et al. 2011) and bird communities (Evans et al. 2011). To more fully understand and predict the impact of increasing urban land cover on urban bat communities, the identification and investigation of traits across a variety of studies in urban landscapes worldwide may prove useful. To do this, we investigated the response of bats to urbanisation using a functional ecology approach and further investigated if these mechanisms are consistent worldwide and thus separately analysed the compiled literature for America (North and South America combined) versus Europe, Asia and Australia. Based on previous studies in urban and other human disturbed landscapes, we expected that predominant food item (fruits, nectar and insects), foraging mode (aerial, gleaning) and foraging space (narrow, edge and open, following Schnitzler and Kalko (2001)) may impact upon a species ability to adapt to urban environments, as suggested by (e.g. Avila-Flores and Fenton 2005; Jung and Kalko 2011; Threlfall et al. 2011)

2.2.1 Data Acquisition and Meta-Analysis

We used the Web of Knowledge (Thomson Reuter) to search for publications containing the following key words "bats" AND, "urban", "urbanis(z)ation", AND "gradient", "community", "assemblage", "species composition". This resulted in 99 studies reporting bat responses to urbanisation. In addition, we searched the reference list of these publications for further relevant articles. We compiled all studies focusing on bats in urban areas in our primary dataset. This selection also including different bat inventory methods such as acoustic monitoring, mist net and harp trap sampling as well as visual observations and roost surveys. In many of these articles however, quantitative data on bats were missing, sampling effort was not standardised, or studies did not reciprocally sample bats in urban versus natural areas. We excluded all of these studies from our final dataset, as it was impossible to calculate a standard effect size of urbanisation. We thus only included studies into our final meta-analysis that reported species-specific data on capture success, roosting individuals, occurrence counts or activity per sampling time in both urban and natural areas (Table 2.1). In a few cases, we extracted data from graphs. We considered all of these measures as indicators of the relative intensity of habitat use and thus assumed comparability of these datasets and hence eligibility to be combined in a meta-analysis. Our final data set for the meta-analysis consisted of 23 articles (Table 2.1) and 96 bat species. Within this dataset we discriminated between studies with high ($N = 5$) and intermediate intensity ($N = 5$) of urbanisation following the individual authors' statements in their articles (Table 2.1). Our designation of 'high' and 'intermediate' was qualitative and based on descriptions of the urban study area from the original papers. For example, Avila-Flores and Fenton (2005) state that their study area of Mexico City is one of the "largest and most populated cities in the world", hence we assigned this study a 'high' urban intensity. Gonsalves et al. (2013) state that no quantification of urban intensity was made in their study, however they suggest that housing density in their study area was low and could be classified as suburban, hence we assigned this study an 'intermediate' urban intensity. This classification is by no means comprehensive, however we believe for comparative purposes these two classifications give some indication and context of the intensity of urban development in the study area for each study used. Some articles ($N = 13$) reported the response of bats to multiple intensities of urbanisation; here we extracted data on the highest, the lowest and the intermediate degrees of urbanisation. Data from urban parks, suburbia or small towns we considered as intermediate degrees of urbanisation.

For each species reported in an article we compared the relative intensity of habitat use in urban (treatment group) versus natural areas (control group) and calculated the log odds ratio as a standardised effect size (Rosenberg et al. 2000). A positive log odds ratio > 0 indicated species that showed a higher intensity of habitat use in urban areas, while a negative log odds ratio < 0 indicated higher intensity of habitat use in natural areas. For multiple reports on a species' response to urbanisation in distinct articles we averaged the log odds ratios to avoid pseudoreplication. Species with incomplete identifications were deleted from the dataset, except for *Mormopterus species 2* (Australia) which has not yet been formally named (Adams et al. 1988) and *Eumops* sp. (Panama) which most likely includes the two species *Eumops glaucinus* and *Eumops auripendulus* (Jung and Kalko 2011). For our analysis we thus considered each bat species ($N = 96$) as a study case for our final meta-analysis models. For all statistical analysis, we used the statistical software package R Version 2.1.4. (R Development Core Team 2011), package "metafor" (Viechtbauer 2013) (version 1.6-0).

In a first approach, we focused on the general response of bats to urbanisation and investigated if the overall response of bats depends on the degree of urbanisation. Hereby we distinguished between high and intermediate intensity of urbanisation (see above) and calculated log odds ratios for the respective contrast to natural areas. We then conducted a random effect model meta-analysis for the

Table 2.1 List of publications included in the meta-analyses

Reference	Country	Urban intensity	Study type	N species urban	N species suburban	N species forest	Survey method	Considered habitat types
Avila-Flores and Fenton (2005)	Mexico	High	Urban gradient	2	3	4	Acoustic monitoring	Residential areas, large parks, forest
Basham et al. (2010)	Australia	Intermediate	Urban/forest	NA	11	13	Acoustic monitoring	Backyards, natural bushland
Bihari (2004)	Hungary	High	Urban/forest	1	NA	1	Roost survey	Residential area, forested park
Chirichella (2004)	Italy	High	Urban gradient	1	1	1	Public survey	Urban, suburban, forest
Duchamp et al. (2004)	USA	High	Urban/forest	2	NA	2	Captures/ telemetry	Urban, woods
Fabianek et al. (2011)	Canada	Intermediate	Urban gradient	NA	3	3	Acoustic monitoring	Urban parks < 100 ha, urban parks > 100 ha
Gaisler et al. (1998)	Czech Republic	High	Urban gradient	2	2	2	Acoustic monitoring	Historical city centre, old suburbs, outskirts
Gehrt and Chelsvig (2004)	USA	High	Urban gradient	5	NA	4	Acoustic monitoring	Urban index: 0 (urban) urban index: −5 (rural)
Gonsalves et al. (2013)	Australia	Intermediate	Urban/forest	NA	9	13	Acoustic monitoring	Small urban, forest
Hale et al. (2013)	United Kingdom	High	Urban gradient	2	2	2	Acoustic monitoring	Dense urban, suburban, rural
Haupt et al. (2006)	Germany	High	Urban/forest	1	NA	1	Captures/ telemetry	Urban areas, forest (before and after birth of juveniles
Hourigan et al. (2006)	Australia	Intermediate	Urban gradient	NA	8	8	Acoustic monitoring	Intermediate suburbs, woodland
Hourigan et al. (2010)	Australia	High	Urban gradient	13	13	13	Acoustic monitoring	High and low-density residential, bush land

(continued)

Table 2.1 (continued)

Reference	Country	Urban intensity	Study type	N species urban	N species suburban	N species forest	Survey method	Considered habitat types
Jung and Kalko (2010)	Panama	Intermediate	Urban/forest	NA	21	22	Acoustic monitoring	Small town, forest
Jung and Kalko (2011)	Panama	High	Urban gradient	16	21	25	Acoustic monitoring	Urban, small town, forest
Kurta and Teramino (1992)	USA	High	Urban/forest	4	NA	5	Captures	Urban/rural
Lesinski et al. (2000)	Poland	High	Urban gradient	3	3	3	Acoustic monitoring	Central, suburban(III), suburban(V)
Nadeem et al. (2013)	Pakistan	High	Urban gradient	4	4	3	Roost surveys	Urban, suburban, rural dwelling
Pottie et al. (2005)	Malaysia/ Singapore	High	Urban gradient	5	5	12	Roost surveys/captures/ acoustic monitoring	City/urban, suburban, primary forest
Silva et al. (2005)	Brazil	High	Urban gradient	7	8	9	Acoustic monitoring	Small farm, campus Funcesi, forest
Threlfall et al. (2011, 2012)	Australia	High	Urban gradient	10	15	7	Acoustic monitoring	Urban, suburban, forest
Utthammachai et al. (2008)	Thailand	High	Urban gradient	1	1	1	Acoustic monitoring	Forest patch, village/ others, urban
Walters et al. (2007)	USA	High	Urban/forest	1	1	1	Captures	Woodlots, low-density residential, commercial lands

Given are the reference, the country where the study was carried out, the urban intensity (large or small) the study type (gradient, paired design), the number of bat species reported by each author for each habitat type, the survey methodology of the original data and the considered habitat types included in the meta-analysis. NA indicates that the corresponding study had no record of bats for the specified habitat type effect of high and intermediate urban development, respectively. Random effect models provide an unconditional inference of a larger set of studies from which only a few are included in the meta-analysis and assumed to be a random sample (Viechtbauer 2010). We compared both models based on the reported effect size and assessed the proportion of heterogeneity of bat responses between high and intermediate urban development (τ^2 highly urban- τ^2 small urban/τ^2 highly urban).

In a second approach, we pooled data from high and intermediate urbanisation categories to investigate if the potential of bats to adjust to urban environments is determined by phylogeny or rather functional ecology using a mixed model meta-analysis. For this analysis we classified bats according to their taxonomic family and genus, their predominant food item (fruits, nectar and insects), foraging mode (aerial, gleaning) and foraging space (narrow, edge and open, following Schnitzler and Kalko (2001)) and included these classifications as moderators in our mixed model meta-analysis. We further investigated in detail how each of the categorical moderators influences effect size. Further, focusing on aerial insectivores, the majority of study cases in our dataset, we then investigated if moderators influencing the adaptability to urban areas are consistent between North and South America versus Europe, Asia and Australia. P-levels for all models were assessed using a permutation test with 1000 randomizations. In none of our models did the funnel plot technique (Viechtbauer 2013) reveal any significant publication bias or asymmetry in our dataset (function: regtest, package metaphor).

2.2.2 High Versus Lower Levels of Urbanisation

Our random effect meta-analysis revealed that in general, urbanisation negatively affects bats, and areas with high (deviance = 453.14, z-value = −3.9, $p < 0.001$) and intermediate (deviance = 439.73; z-value = −2.4, $p < 0.05$) degrees of urban development reveal significantly lower intensity of habitat use across species compared to natural areas (Fig. 2.1). A high degree of urbanisation had a stronger negative effect on the general intensity of habitat use (estimate: −1.47) than an intermediate degree of urban development (estimate: −0.79). However, in both high and intermediate urban development, we found significant variation in the

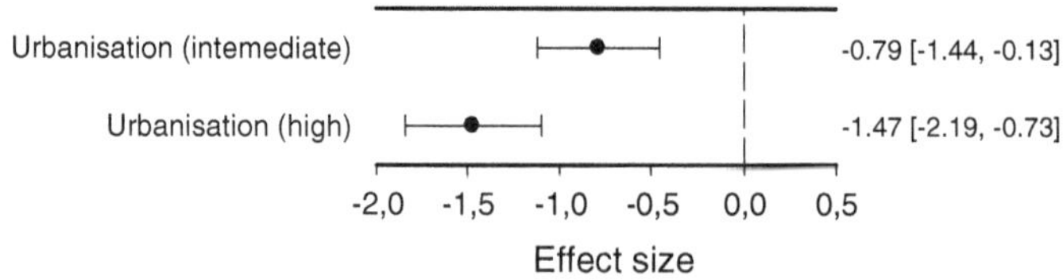

Fig. 2.1 Effect sizes of relative intensity of habitat use by bats in high and intermediate urban development, compared to natural areas. Solid symbols indicate the mean effect size (log odds ratio) and whiskers indicate the estimated standard error. Values of the estimated effect size, including the 95 % confidence intervals are listed on the right side of the figure

effect sizes (high urban development: $Q_{(df=84)} = 641.2$, $p < 0.0001$; intermediate urban development $Q_{(df=85)} = 989.9$, $p < 0.0001$), indicating a high variability in the response of bat species to urbanisation. This species-specific variability in the intermediate degree of urbanisation ($\tau^2 = 7.74$) accounted for 21 % of the variability in the areas with high urban development ($\tau^2 = 9.80$). This suggests that although intermediate urban development clearly has a negative influence on bats it still permits the use of this habitat by more species showing fewer extremes in the species-specific response to urbanisation, compared to high urban development.

2.2.3 Phylogeny Versus Functional Ecology

Neither phylogeny ($Q_{M(df=3)} = 11.57$, $p > 0.05$) nor functional ecology ($Q_{M(df=3)} = 12.18$, $p > 0.05$) explained the heterogeneity in bat response to urbanisation. However, a different pattern emerged when investigating the effect of single moderators in detail. Response to urbanisation differed between families ($Q_{M(df=10)} = 32.4$, $p = 0.05$) with bat species in the Rhinolophidae being negatively affected by urban development ($p < 0.01$). In addition, bat species in the Mormoopidae tended to respond negatively towards urbanisation, as the 95 % confidence interval did not overlap with zero. All other families revealed a high heterogeneity in the response to urbanisation. Effect size was neither genera—($Q_{M(df=46)} = 81.4$, $p > 0.05$) nor species-specific ($Q_{M(df=86)} = 99.7$, $p > 0.05$).

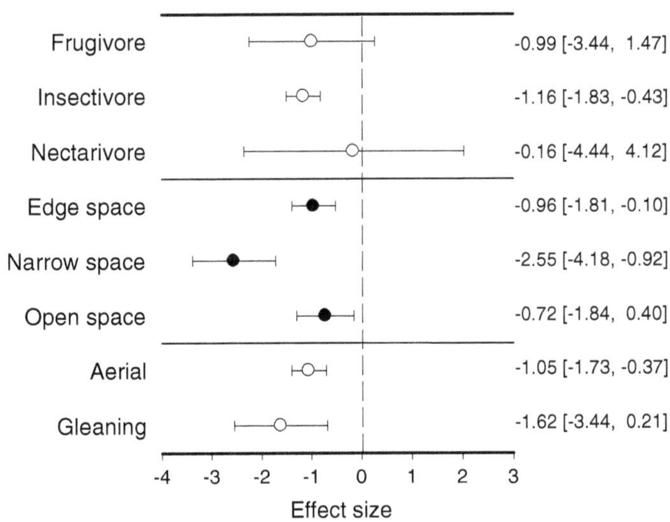

Fig. 2.2 Effect of urbanisation (log odds ratio and the estimated standard error) on relative intensity of habitat use in relation to the predominant food item (**a**), foraging space (**b**), and foraging mode (**c**). Solid symbols indicate the mean effect size (log odds ratio) and whiskers indicate the estimated standard error. Values of the estimated effect size, including the 95 % confidence intervals are listed on the right side of the figure

None of the functional classifications, food item, foraging mode and foraging space, revealed a significant association with the persistence of bats in urban areas. However narrow space foragers (estimate -2.55 ± 0.83, $p = 0.06$) revealed a tendency to be associated with natural areas (Fig. 2.2).

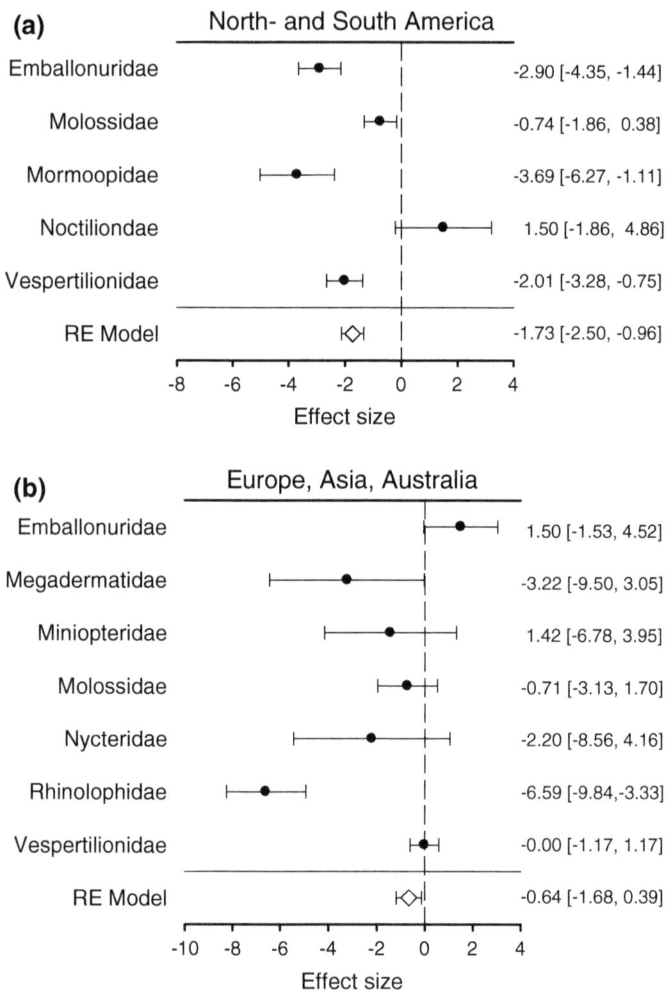

Fig. 2.3 Response of insectivorous bat families to urbanisation in **a** North and South America and **b** Europe, Asia and Australia. A negative effect size reflects a higher association with natural areas, a positive effect size an association with urban areas. Depicted are the mean effect sizes (log odds ratio) and the estimated standard errors by family. Values of the estimated effect size, including the 95 % confidence intervals are listed on the right side of the figure. The overall effect of urbanisation on insectivorous bats, based on the random effect model (RE Model), is given at the bottom of the respective figure

2.2.4 *Contrasting the Effects between North and South America and Europe, Asia and Australia Focusing on Insectivores*

The general response of insectivorous bats differed between the Americas and Europe, Asia and Australia. While insectivorous bats in the Americas revealed a significant negative response to urbanisation (deviance = 171.18, *z*-value = −4.4, $p < 0.001$) the overall response of insectivorous bats to urbanisation in Europe, Asia and Australia was insignificant (deviance = 258.9, *z*-value = −1.2, $p > 0.05$, Fig. 2.3a, b).

However, in both the Americas ($Q_{M(df=5)} = 35.1$, $p < 0.05$) and Europe, Asia and Australia ($Q_{M(df=7)} = 18.7$, $p < 0.05$) the response to urbanisation differed significantly across families. Interestingly this family-level response was inconsistent between the Neo- and Paleotropics. While Neotropical bats in the Emballonuridae showed a strong tendency to be associated with natural areas (estimate: -2.9 ± 0.7, $p = 0.06$), emballonurids in the Paleotropics (estimate: 1.5 ± 1.5, $p > 0.05$) occurred frequently in urban areas. We found a similar trend in the globally distributed family of Vespertilionidae, which showed a higher association with natural areas in the Americas (estimate: -2.0 ± 0.6, $p > 0.05$) but did not reveal any clear association in Europe, Asia and Australia (estimate: -0.0 ± 0.6, $p > 0.05$) (Fig. 2.3a, b).

2.3 Adaptability of Species to Urban Areas: General Trends, Species-Specific Differences and Future Research

Urban areas can provide suitable habitat for a variety of species, albeit an anthropogenically altered habitat (McKinney 2006). However, our general understanding of what influences a species' success in urban environments is limited. Arguably the conservation of species such as bats in urban areas is dependent upon this knowledge (Fenton 1997). Within this book chapter, we reviewed the existing literature on bats in urban areas. In addition, we combined published data in a meta-analysis to evaluate and derive general patterns in the response of bats to urban development.

Our meta-analysis revealed that, in general, habitat use of bats decreases in urban areas. A high degree of urbanisation had a stronger negative effect on overall habitat use of bats compared to an intermediate degree of urban development. However, habitat use in intermediate urban development was much lower compared with natural areas. This is alarming, as it is generally thought that small towns and suburban landscapes could potentially provide suitable habitat for a wide range of species (McKinney 2006), including bats. The combination of habitats with different complexity in smaller urban developments should lead to greater complementarity at a local scale and should favour species diversity and abundance. Some of the publications in our meta-analysis dataset indeed report a higher bat diversity, activity (Hourigan et al. 2010; Threlfall et al. 2011, 2012b) and feeding activity (Jung and Kalko 2011; Threlfall et al. 2012a) at intermediate levels of disturbance compared to natural or urban habitats. Other studies reported that any urban land cover, even if low-density residential, can decrease bat activity and species richness (Hourigan et al. 2006; Gonsalves et al. 2013; Luck et al. 2013), and even deter individual species (Jung and Kalko 2010; Gonsalves et al. 2013; Luck et al. 2013). Altogether, this strongly suggests regional differences in the intensity of urban development and points towards an interacting effect of the surrounding landscape (see Coleman and Barclay 2011).

Results from recent urban bat studies suggest that bats of some families (e.g. molossids Jung and Kalko 2011) are better pre-adapted for life in an urban environment compared to others (e.g. rhinolophids Stone et al. 2009; Threlfall et al. 2011). Our analysis also indicated a family-specific effect of urbanisation and confirmed the negative response of Rhinolophidae to urban development across the Old World. However, the responses of Molossidae and Vespertilionidae, which are known to frequently roost in man-made structures in North and South America, did not reveal consistent associations with either urban or natural areas across continents. This might be due to the high morphological and behavioural heterogeneity within these families. We believe that the likely explanation for our results is that the response to urbanisation is dictated by the behavioural and morphological traits of species, regardless of geographic region or phylogeny. In particular, species foraging in open space seem to persist in urban areas, as due to their wing morphology (high aspect ratio and wing loading) they might be able to commute large distances between roosting sites and feeding areas (Jung and Kalko 2011). Thus traits predicting species mobility have been associated with urban tolerance (Jung and Kalko 2011; Threlfall et al. 2012a), and the ability to forage around street lights (see Rowse et al., Chap. 7 this volume). In addition, traits that allow for flexible roost and foraging strategies confer an advantage for urban-tolerant species. Our current results support these findings and thus suggest that adaptability of bats to urban environments (or disturbance in general) might be correlated with, and reflected by, species behavioural flexibility. Advancement of knowledge in this area will assist with conservation efforts of bat species globally, and potentially allow development of a predictive framework for assessing the impacts of urban development on bats.

References

Adams M, Reardon T, Baverstock P, Watts C (1988) Electrophoretic resolution of species boundaries in Australian microchiroptera. IV. The Molossidae (Chiroptera). Aust J Biol Sci 41(3):315–326

Aronson MFJ, La Sorte FA, Nilon CH, Katti M, Goddard MA, Lepczyk CA, Warren PS, Williams NSG, Cilliers S, Clarkson B, Dobbs C, Dolan R, Hedblom M, Klotz S, Kooijmans JL, Kühn I, MacGregor-Fors I, McDonnell M, Mörtberg U, Pyšek P, Siebert S, Sushinsky J, Werner P, Winter M (2014) A global analysis of the impacts of urbanization on bird and plant diversity reveals key anthropogenic drivers. Proc R Soc B Biol Scie 281(1780):20133330

Avila-Flores R, Fenton B (2005) Use of spatial features by foraging insectivorous bats in a large urban landscape. J Mammal 86(6):1193–1204

Barclay RMR, Harder LD (2003) Life histories of bats: life in the slow lane. In: Kunz TH, Fenton B (eds) Bat ecology. The University of Chicago Press, Chicago, pp 209–253

Bartonicka T, Zukal J (2003) Flight activity and habitat use of four bat species in a small town revealed by bat detectors. Folia Zool 52(2):155–166

Basham R, Law B, Banks PB (2011) Microbats in a 'leafy' Sydney urban landscape: are they persisting, and what factors influence their presence? Austral Ecol (in press)

Blair RB (2001) Birds and butterflies along urban gradients in two ecoregions of the U.S. In: Lockwood J, McKinney L (eds) Biotic homogenization. Kluwer, Norwell, pp 33–56

Bredt A, Uieda W (1996) Bats from urban and rural environments of the Distrito Federal, mid-western Brazil. Chiroptera Neotropical 2(2):54–57

Burnett SE, Jennings JB, Rainey JC, Best TL (2001) Molossus bondae. Mamm Species 668:1–5

Caughlin TT, Ganesh T, Lowman MD (2012) Sacred fig trees promote frugivore visitation and tree seedling abundance in South India. Curr Sci 102(6):918–922

Chamberlain DE, Cannon AR, Toms MP, Leech DI, Hatchwell BJ, Gaston KJ (2009) Avian productivity in urban landscapes: a review and meta-analysis. Ibis 151(1):1–18

Clergeau P, Jokimäki J, Snep R (2006) Using hierarchical levels for urban ecology. Trends Ecol Evol 21(12):660–661

Coleman JL, Barclay RMR (2011) Influence of urbanization on demography of little brown bats (*Myotis lucifugus*) in the prairies of North America. PLoS ONE 6(5):e20483

Corlett RT (2005) Interactions between birds, fruit bats and exotic plants in urban Hong Kong, South China. Urban Ecosyst 8(3):275–283

Corlett RT (2006) Figs (Ficus, Moraceae) in Urban Hong Kong, South China. Biotropica 38(1):116–121

Didham RK, Ghazoul J, Stork NE, Davis AJ (1996) Insects in fragmented forests: a functional approach. Trends Ecol Evol 11(6):255–260

Ditchkoff SS, Saalfeld ST, Gibson CJ (2006) Animal behavior in urban ecosystems: modifications due to human-induced stress. Urban Ecosyst 9(1):5–12

Dixon MD (2012) Relationship between land cover and insectivorous bat activity in an urban landscape. Urban Ecosyst 15(3):683–695

Duchamp J, Swihart R (2008) Shifts in bat community structure related to evolved traits and features of human-altered landscapes. Landscape Ecol 23(7):849

Estrada-Villegas S, Meyer CFJ, Kalko EKV (2010) Effects of tropical forest fragmentation on aerial insectivorous bats in a land-bridge island system. Conserv Biol 143:597–608

Evans KL, Chamberlain DE, Hatchwell BJ, Gregory RD, Gaston KJ (2011) What makes an urban bird? Glob Change Biol 17(1):32–44

Fabianek F, Gagnon D, Delorme M (2011) Bat distribution and activity in Montréal Island green spaces: responses to multi-scale habitat effects in a densely urbanized area1. Ecoscience 18(1):9–17

Fenton B (1997) Science and the conservation of bats. J Mammal 78(1):1–14
Filho HO (2011) Urban bats: aspects of ecology and health. In: Robinson WH, de Carvalho Campos AE (eds) Proceedings of the seventh international conference on urban pests, Instituto Biológico, São Paulo, Brazil
Flynn DFB, Gogol-Prokurat M, Nogeire T, Molinari N, Richers BT, Lin BB, Simpson N, Mayfield MM, DeClerck F (2009) Loss of functional diversity under land use intensification across multiple taxa. Ecol Lett 12(1):22–33
Gaisler J, Zukal J, Rehak Z, Homolka M (1998) Habitat preference and flight activity of bats in a city. J Zool 244(3):439–445
García-Morales R, Badano EI, Moreno CE (2013) Response of Neotropical bat assemblages to human land use. Conserv Biol 27(5):1096–1106
Geggie J, Fenton B (1985) A comparison of foraging by Eptesicus fuscus (Chiroptera: Vespertilionidae) in urban and rural environments. Can J Zool 83:263–265
Gehrt SD, Chelsvig JE (2003) Bat activity in an urban landscape: patterns at the landscape and microhabitat scale. Ecol Appl 13(4):939–950
Gehrt SD, Chelsvig JE (2004) Species-specific patterns of bat activity in an urban landscape. Ecol Appl 14:625–635
Gilbert OL (1989) The ecology of urban habitats. Chapman & Hall, London
Giraudeau M, Mousel M, Earl S, McGraw K (2014) Parasites in the city: degree of urbanization predicts poxvirus and coccidian infections in house finches (*Haemorhous mexicanus*). PLoS ONE 9(2):e86747
Gonsalves L, Lamb S, Webb C, Law B, Monamy V (2013) Do mosquitoes influence bat activity in coastal habitats? Wildlife Research 40(1):10–24
Grimm NB, Faeth SH, Golubiewski NE, Redman CL, Wu J, Bai X, Briggs JM (2008) Global change and the ecology of cities. Science 319(5864):756–760
Guest P, Jones KE, Tovey J (2002) Bats in greater London: unique evidence of a decline over 15 years. Br Wildl 14:1–5
Hale JD, Fairbrass AJ, Matthews TJ, Sadler JP (2012) Habitat composition and connectivity predicts bat presence and activity at foraging sites in a large UK conurbation. PLoS ONE 7(3):e33300
Haupt M, Menzler S, Schmidt S (2006) Flexibility of habitat use in *Eptesicus nilssonii*: does the species profit from anthropogenically altered habitats? J Mammal 87(2):351–361
Hobbs RJ, Arico S, Aronson J, Baron JS, Bridgewater P, Cramer VA, Epstein PR, Ewel JJ, Klink CA, Lugo AE, Norton D, Ojima D, Richardson DM, Sanderson EW, Valladares F, Vilà M, Zamora R, Zobel M (2006) Novel ecosystems: theoretical and management aspects of the new ecological world order. Glob Ecol Biogeogr 15(1):1–7
Hölker F, Wolter C, Perkin EK, Tockner K (2010) Light pollution as a biodiversity threat. Trends Ecol Evol 25(12):681–682
Hourigan CH, Johnson C, Robson SKA (2006) The structure of a micro-bat community in relation to gradients of environmental variation in a tropical urban area. Urban Ecosyst 9:67–82
Hourigan CL, Catterall CP, Jones D, Rhodes M (2010) The diversity of insectivorous bat assemblages among habitats within a subtropical urban landscape. Austral Ecol 35:849–857
Isaksson C (2010) Pollution and its impact on wild animals: a meta-analysis on oxidative stress. EcoHealth 7(3):342–350
Jokimäki J, Kaisanlahti-Jokimäki M-L, Suhonen J, Clergeau P, Pautasso M, Fernández-Juricic E (2011) Merging wildlife community ecology with animal behavioral ecology for a better urban landscape planning. Landscape Urban Plann 100(4):383–385
Jung K, Kalko EKV (2010) Where forest meets urbanization: foraging plasticity of aerial insectivorous bats in an anthropogenically altered environment. J Mammal 91(1):144–153
Jung K, Kalko EKV (2011) Adaptability and vulnerability of high flying Neotropical aerial insectivorous bats to urbanization. Divers Distrib 17(2):262–274

Kalcounis-Rueppell MC, Payne VH, Huff SR, Boyko AL (2007) Effects of wastewater treatment plant effluent on bat foraging ecology in an urban stream system. Biol Conserv 138(1–2):120–130
Kössl M, Mora E, Coro F, Vater M (1999) Two-toned echolocation calls from *Molossus molossus* in Cuba. J Mammal 80(3):929–932
Kurta A, Teramino JA (1992) Bat community structure in an urban park. Ecography 15:257–261
Lane DJW, Kingston T, Lee BPYH (2006) Dramatic decline in bat species richness in Singapore, with implications for Southeast Asia. Biol Conserv 131(4):584–593
Legakis A, Papdimitriou C, Gaetglich M, Lazaris D (2000) Survey of bats of the Athens metropolitan area. Myotis 38:41–46
Lesiñski G, Eb Fuszara, Kowalski M (2000) Foraging areas and relative density of bats (Chiroptera) in differently human transformed landscapes. Z Säugetierkunde 65:129–137
Longcore T, Rich C (2004) Ecological light pollution. Front Ecol Environ 2(4):191–198
Luck GW, Smallbone L, Threlfall C, Law B (2013) Patterns in bat functional guilds across multiple urban centres in south-eastern Australia. Landscape Ecol 28:455–469
Luniak M (2004) Synurbization—adaptation of animal wildlife to urban development. In: Shaw WW, Harris LK, VanDruff L (eds) Proceedings 4th international urban wildlife symposium. University of Arizona, Tucson, USA, pp 50–55
Macdonald DW, Newdick MT (1982) The distribution and ecology of foxes, *Vulpes vulpes* (L.) in urban areas. In: Bornkamm R, Lee JA, Seaward MRD (eds) Urban ecology. Oxford University Press, Oxford, pp 123–138
Mahmood-ul-Hassan M, Gulraiz TL, Rana SA, Javid A (2010) The diet of Indian flying-foxes (*Pteropus giganteus*) in urban habitats of Pakistan. Acta Chiropterologica 12(2):341–347
McConville A, Law B, Penman T, Mahony M (2013a) Contrasting habitat use of morphologically similar bat species with differing conservation status in south-eastern Australia. Austral Ecol 39(1):83–94
McConville A, Law BS, Mahony MJ (2013b) Are regional habitat models useful at a local-scale? A case study of threatened and common insectivorous bats in south-eastern Australia. PLoS ONE 8(8):1–10
McDonald RI (2008) Global urbanization: can ecologists identify a sustainable way forward? Front Ecol Environ 6(2):99–104
McDonald R, Marcotullio P (2013) Global effects of urbanization on ecosystem services. In: Elmqvist T, Fragkias M, Goodness J et al (eds) Urbanization, biodiversity and ecosystem services: challenges and opportunities. Springer, Dordrecht
McDonald RI, Kareiva P, Forman RTT (2008) The implications of current and future urbanization for global protected areas and biodiversity conservation. Biol Conserv 141(6):1695–1703
McDonnell MJ, Hahs AK (2008) The use of gradient analysis studies in advancing our understanding of the ecology of urbanizing landscapes: current status and future directions. Landscape Ecol 23(10):1143–1155
McDonnell M, Hahs A (2013) The future of urban biodiversity research: moving beyond the 'low-hanging fruit'. Urban Ecosyst 16(3):397–409
McDonnell MJ, Pickett STA (1990) Ecosystem structure and function along urban-rural gradients: an unexploited opportunity for ecology. Ecology 71(4):1232–1237
McGranahan G, Satterthwaite D (2003) Urban centers: an assessment of sustainability. Annu Rev Environ Resour 28:243–274
McIntyre NE, Knowles-Yánez K, Hope D (2000) Urban ecology as an interdisciplinary field: differences in the use of "urban" between the social and natural sciences. Urban Ecosyst 4(1):5–24
McKinney ML (2002) Urbanization, biodiversity, and conservation. Bioscience 52(10):883–890
McKinney ML (2006) Urbanization as a major cause of biotic homogenization. Biol Conserv 127(3):247–260

McKinney ML, Lockwood JL (1999) Biotic homogenization: a few winners replacing many losers in the next mass extinction. Trends Ecol Evol 14(11):450–453

Meyer CFJ, Kalko EKV (2008) Assemblage-level responses of phyllostomid bats to tropical forest fragmentation: land-bridge islands as a model system. J Biogeogr 35(9):1711–1726

Montgomery MR (2008) The urban transformation of the developing world. Science 319(5864):761–764

Myers N, Mittermeier RA, Mittermeier CG, da Fonseca GAB, Kent J (2000) Biodiversity hotspots for conservation priorities. Nature 403:853–858

Nadeem MS, Zafar S, Kayani AR, Mushtaq M, Beg MA, Nasir MF (2013) Distribution and roosting habitats of some Microchiropteran bats in Rawalpindi District, Pakistan. Pak J Zool 45(2):565–569

Nakamoto A, Kinjo K, Izawa M (2007) Food habits of Orii's flying-fox, *Pteropus dasymallus inopinatus*, in relation to food availability in an urban area of Okinawa-jima Island, the Ryukyu Archipelago, Japan. Acta Chiropterologica 9(1):237–249

Nakamoto A, Kinjo K, Izawa M (2012) Ranging patterns and habitat use of a solitary flying fox (*Pteropus dasymallus*) on Okinawa-jima Island, Japan. Acta Chiropterologica 14(2):387–399

Norberg UM, Rayner JMV (1987) Ecological morphology and flight in bats (Mammalia; Chiroptera): wing adaptations, flight performance, foraging strategy and echolocation. Philos Trans R Soc Lond B Biol Sci 316(1179):335–427

Pacheco SM, Sodré M, Gama AR, Bredt A, Cavallini EM, Sanches Marques RV, Guimarães MM, Bianconi G (2010) Morcegos Urbanos: Status do Conhecimento e Plano de Ação para a Conservação no Brasil. Chiroptera Neotropical 16(1):629–647

Park KJ, Mochar F, Fuentes-Montemayor E (2012) Urban biodiversity: successes and challenges: bat activity in urban green space. Glasg Naturalist 25(4)

Parris KM, Hazell DL (2005) Biotic effects of climate change in urban environments: the case of the grey-headed flying-fox (*Pteropus poliocephalus*) in Melbourne, Australia. Biol Conserv 124(2):267–276

Pearce H, Walters CL (2012) Do green roofs provide habitat for bats in urban areas? Acta Chiropterologica 14(2):469–478

Pickett STA, Cadenasso ML, Grove JM, Nilon CH, Pouyat RV, Zipperer WC, Costanza R (2001) Urban ecological systems: linking terrestrial ecological, physical, and socioeconomic components of metropolitan areas. Ann Rev Ecol Syst 32:127–157

Pottie SA, Lane DJW, Kingston T, Lee BPYH (2005) The microchiropteran bat fauna of Singapore. Acta Chiropterologica 7(2):237–247

Rhodes M, Catterall C (2008) Spatial foraging behaviour and use of an urban landscape by a fast-flying bat, the Molossid *Tadarida australis*. J Mammal 89(1):34–42

Rhodes M, Wardell-Johnson G (2006) Roost tree characteristics determine use by the white-striped freetail bat (*Tadarida australis*, Chiroptera: Molossidae) in suburban subtropical Brisbane, Australia. Austral Ecol 31:228–239

Rhodes M, Wardell-Johnson GW, Rhodes MP, Raymond BEN (2006) Applying network analysis to the conservation of habitat trees in urban environments: a case study from Brisbane. Australia. Conservation Biology 20(3):861–870

Rompré G, Robinson WD, Desrochers A (2008) Causes of habitat loss in a neotropical landscape: the Panama Canal corridor. Landscape and Urban Planning 87(2):129–139

Rosenberg MS, Adams DC, Gurevitch J (2000) Metawin: manual of statistical software for meta-analysis. Sinauer, Sunderland

Scales J, Wilkins KT (2007) Seasonality and fidelity in roost use of the mexican free tailed bat *Tadarida brasiliensis* in an urban setting. Western North American Naturalist 67(3):402–408

Scanlon AT, Petit S (2008) Effects of site, time, weather and light on urban bat activity and richness: considerations for survey effort. Wildlife Research 35(8):821–834

Schnitzler H-U, Kalko EKV (2001) Echolocation by insect-eating bats. Bioscience 51(7):557–569

Septon G, Marks JB, Ellestad T (1995) A preliminary assessment of peregrine Falcon *Falco peregrinus* recovery in the Midwestern North America. Acta Ornithologica 30:65–68

Sewell SR, Catterall CP (1998) Bushland modification and styles of urban development: their effects on birds in south-east Queensland. Wildlife Research 25(1):41–63

Shochat E, Warren PS, Faeth SH, McIntyre NE, Hope D (2006) From patterns to emerging processes in mechanistic urban ecology. Trends Ecol Evol 21(4):186–191

Siles L, Peñaranda D, Pérez-Zubieta JC, Barboza K (2005) Los murciélagos de la ciudad de Cochabamba. Revista Boliviana Ecologia 18:51–64

Slabbekoorn H, Peet M (2003) Birds sing at a higher pitch in urban noise. Nature 424(6946):267

Slabbekoorn H, Ripmeester EAP (2008) Birdsong and anthropogenic noise: implications and applications for conservation. Mol Ecol 17:72–83

Stone EL, Jones G, Harris S (2009) Street lighting disturbs commuting bats. Current Biology 19:1123–1127

Thomas NM, Duckworth J, Douangboubpha B, Williams M, Francis CM (2013) A checklist of bats (Mammalia: Chiroptera) from Lao PDR. Acta Chiropterologica 15(1):193–260

Threlfall CG, Law B, Penman T, Banks PB (2011) Ecological processes in urban landscapes: mechanisms influencing the distribution and activity of insectivorous bats. Ecography 34(5):814–826

Threlfall CG, Law B, Banks PB (2012a) Influence of landscape structure and human modifications on insect biomass and bat foraging activity in an urban landscape. PLoS ONE 7(6):e38800

Threlfall CG, Law B, Banks PB (2012b) Sensitivity of insectivorous bats to urbanization: Implications for suburban conservation planning. Biol Conserv 146:41–52

Threlfall CG, Law B, Banks PB (2013a) Roost selection in suburban bushland by the urban sensitive bat *Nyctophilus gouldi*. J Mammal 94(2):307–319

Threlfall CG, Law B, Banks PB (2013b) The urban matrix and artificial light restricts the nightly ranging behaviour of Gould's long-eared bat (*Nyctophilus gouldi*). Austral Ecol 38(8):921–930

UN (2012) System Task Team on the post-2015 UN development agenda: sustainable urbanization. Thematic think piece

UNPD (2012) World urbanization prospects: the 2011 revision. Highlights. United Nations Population Division, New York, USA

Utthammachai K, Bumrungsri S, Chimchome V, Russ J, Mackie I (2008) The Habitat Use and Feeding Activity of *Tadarida plicata* in Thailand. Thai J For 27(2):21–27

van der Ree R, McCarthy MA (2005) Inferring persistence of indigenous mammals in response to urbanisation. Anim Conserv 8(3):309–319

Viechtbauer W (2010) Conducting meta-analysis in R with the metafor package. J Statistical Softw 36(3):1–48

Viechtbauer W (2013) The metafor package: a meta-analysis package for R. http://www.metafor-project.org

Violle C, Navas M-L, Vile D, Kazakou E, Fortunel C, Hummel I, Garnier E (2007) Let the concept of trait be functional! Oikos 116(5):882–892

Vitousek PM, Mooney HA, Lubchenco J, Melillo JM (1997) Human domination of earth's ecosystems. Science 277(5325):494–499

Walsh AL, Harris S (1996) Factors determining the abundance of Vespertilionid bats in Britain: geographical, land class and local habitat relationships. J Appl Ecol 33(3):519–529

Williams NSG, Morgan JW, McDonnell MJ, McCarthy MA (2005) Plant traits and local extinctions in natural grasslands along an urban–rural gradient. J Ecol 93(6):1203–1213

Zhao S, Da L, Tang Z, Fang H, Song K, Fang J (2006) Ecological consequences of rapid urban expansion: Shanghai, China. Front Ecol Environ 4(7):341–346

Chapter 3
Bats and Roads

John Altringham and Gerald Kerth

Abstract The effects of roads on bats have been largely neglected until recently, despite growing evidence for profound effects on other wildlife. Roads destroy, fragment and degrade habitat, are sources of light, noise and chemical pollution and can kill directly through collision with traffic. The negative effects of roads on wildlife cannot be refuted but at the same time road building and upgrading are seen as important economic drivers. As a consequence, infrastructure projects and protection of bats are often in conflict with each other. There is now growing evidence that fragmentation caused by roads reduces access to important habitat, leading to lower reproductive output in bats. This barrier effect is associated with reduced foraging activity and species diversity in proximity to motorways and other major roads. The effects of light and noise pollution may add to this effect in the immediate vicinity of roads and also make bats even more reluctant to approach and cross roads. Several studies show that vehicles kill a wide range of bat species and in some situations roadkill may be high enough to lead directly to population decline. Current mitigation efforts against these effects are often ineffective, or remain largely untested. The limited information available suggests that underpasses to take bats under roads may be the most effective means of increasing the safety and permeability of roads. However, underpass design needs further study and alternative methods need to be developed and assessed.

J. Altringham (✉)
School of Biology, Faculty of Biological Sciences, University of Leeds, Leeds, UK
e-mail: J.D.Altringham@leeds.ac.uk

G. Kerth
Applied Zoology and Conservation, Zoological Institute and Museum,
University of Greifswald, Greifswald, Germany
e-mail: gerald.kerth@uni-greifswald.de

C.C. Voigt and T. Kingston (eds.), *Bats in the Anthropocene: Conservation of Bats in a Changing World*, DOI 10.1007/978-3-319-25220-9_3

3.1 Introduction

The global road network gets longer, wider, faster and more complex as existing road systems are upgraded and new roads are built. Despite the widely acknowledged need to reduce our dependence on fossil fuel and growing concerns about the environmental impact of roads, improved communication by road, and even the act of road-building itself, are often seen as essential economic drivers. As road networks expand, traffic volumes increase and congestion remains a problem. A few statistics highlight the pervasive nature of our road networks: only 2 % of Germany is made up of landscape fragments greater than 100 km^2 (Jaeger et al. 2007) and only 17 % of the US landscape is more than 1 km from a road (Riiters and Wickham 2003). In 2012, the UK had 395,000 km of roads, of which over 50,000 km are major roads and 3700 km motorways (Defra 2013). Major roads account for only 13 % of all UK roads, but carry 65 % of the traffic. 50 % of all traffic is on motorways and other major roads in rural areas. Almost 20 % of major road length is dual carriageway. Over 3200 km have been added to the UK network in the last decade and many more have been upgraded.

Roads have several negative impacts on animals. First, building roads and their ancillary structures destroys habitat directly. Secondly, the resulting road network fragments the landscape, potentially restricting animal movements, thereby blocking their access to the remaining habitat. Thirdly, roads are also sources of light, noise and chemical pollution, and so degrade the habitat around them. Moreover, the increased human access provided by roads usually accelerates urban, commercial and agricultural development and increases human disturbance in many ways, e.g. through increased recreational pressure and the introduction of non-native predators and other invasive species. Finally, fast moving traffic kills animals directly. Broad reviews of the effects of roads on vertebrates include Bennett (1991), Forman and Alexander (1998), Trombulak and Frissell (2000), Coffin (2007), Fahrig and Rytwinski (2009), Laurance et al. (2009), Benítez-López et al. (2010), and Rytwinski and Fahrig (2012). Surprisingly, despite the many ways in which roads can impact on wildlife, it is only in the last 20 years that significant attention has been given to what is now often referred to as 'road ecology' (Forman et al. 2003). Little of this attention was directed at bats. Moreover, the few existing studies on the impact of roads on bats have all been carried out in North America and Europe.

Globally many bat species are endangered (Racey and Entwistle 2003; Jones et al. 2009), including regions with a dense infrastructure such as North America and Europe (Safi and Kerth 2004). As a consequence, in Europe, for example, bats are of high priority for conservation and all bat species have been strictly protected for two decades by European law (CMS 1994). Despite the importance of bats in conservation, rigorous, peer-reviewed studies on the impact of roads on bats have only begun to be published in the last few years. Only over the last decade it has been widely accepted that roads must have an effect on bats. As a result, mitigation against these effects is becoming increasingly integrated in the road building process and practical mitigation guidelines have been published in a number of countries (e.g. Highways Agency 2001, 2006; Limpens et al. 2005). However, the precise nature and scale of the effects of roads on bats were mostly unknown, and as a consequence mitigation has often been poorly monitored and therefore rarely informed by sound evidence (Altringham 2008; O'Connor et al. 2011).

This review describes the ways in which roads do or may affect bats, discusses the available evidence in relation to each, and where appropriate suggests action for the future, in terms of both research and conservation action. Because work on the impacts of roads on bats is still scarce and biased towards the temperate zone, some work on other animals will be discussed, in particular birds, to help fill important gaps. Roads can affect bats in many ways, and because the mitigation solutions will to some extent be unique to each, the mechanisms will be discussed separately. However, there is considerable interaction between them and the impacts in many cases are cumulative, so some topics will appear under more than one heading.

To our knowledge almost no studies have been published yet that investigated the effects of railways on bats (but see Vandevelde et al. 2014). However, as linear development features, they have the potential to disrupt bats and will be discussed briefly at the end of the review.

3.1.1 Bat Life History

In order to assess the impact of roads on bats, an important consideration is of course the biology of the bats themselves. Bats are small mammals with the life history strategy of very much larger species (e.g. Barclay and Harder 2003; Altringham 2011). They have taken the low fecundity, long life option, often producing only a single pup each year, but frequently living for more than 10 years and not unusually 20 or more (e.g. Barclay and Harder 2003; Altringham 2011). Any external factors that reduce reproductive success, increase mortality, or both, can lead to severe population declines—and recovery will be slow (e.g. Sendor and Simon 2003; Papadatou et al. 2011). Furthermore, bats typically have large summer home ranges compared to other similar sized mammals and many bats migrate over considerable distances between winter and summer roosts (Altringham 2011). Finally, bats are highly gregarious (Kerth 2008). As a result, negative impacts of roads on local bat colonies can affect large numbers of individuals simultaneously. Because of their particular life history, bats are susceptible to a wider range of environmental disturbances than many other small mammals.

3.1.2 Bat Conservation Status

A substantial number of the more than 1200 extant bat species are considered to be endangered (Racey and Entwistle 2003; Jones et al. 2009). Reasons for the decline of bats include habitat loss, pollution, direct persecution and diseases (Jones et al.

2009). Several of these threat factors are also relevant during the construction and maintenance of roads. In Europe, all bats are strictly protected, as all are listed in Annex 4 of the Habitats Directive, and several species have designated protected areas because they are also listed in the Annex 2 of the Habitats Directive (Council Directive 92/43/EEC). As a consequence, whenever bat populations are likely to be adversely affected by the construction of roads, environmental assessments are required and mitigation often becomes a necessity. Thus assessments of bats have been carried out during many recent infrastructure projects (e.g. Kerth and Melber 2009) and this process will continue to be important in the future.

3.2 The Effects of Roads on Bats—Habitat Destruction, Fragmentation, Degradation and Collision Mortality

3.2.1 Loss of Habitat

Road development frequently involves the removal of trees and buildings that hold potential or actual bat roosts. The removal of trees, hedges, scrub, water bodies and unimproved ('natural') grassland also reduces available foraging habitat. The road surface alone destroys significant areas of habitat: 7 ha for every 10 km of 7 m wide, two-lane road. Roadside hard shoulders, verges, junctions, service areas and other structures remove yet more potential habitat. As a result, road construction leads to the permanent loss of habitats for bats and thus is likely to reduce population sizes directly.

3.2.2 The Barrier Effect

Roads are potential barriers to flight between roosts and foraging sites and between summer, mating and winter roosts. They could therefore reduce the available home range size and quality and may restrict migration, which could increase mortality and reduce reproductive potential. Roads may act as barriers because they interrupt existing linear flight lines, because some species are reluctant to cross open ground, because some species avoid lit areas (road and vehicle lights) and, at least initially, because they represent sudden changes in the bats' familiar landscape. Roads may therefore fragment habitat, decreasing its accessible area and quality. Since habitat area and quality are major determinants of population size, then habitat fragmentation will lower the sustainable population size.

Barriers such as roads may also limit the flow of individuals between populations with two major consequences. First, barriers may slow the recovery from local population declines since recruitment of individuals from neighbouring populations ("rescue effect") will be reduced and this will further increase the probability of local extinction. Secondly, barriers may also reduce gene flow between populations and increase inbreeding, reducing individual fitness and increasing the risk of local extinction. Genetic isolation such as this can only occur with very low levels of dispersal. These factors may only be significant for rare bat species that already have small and fragmented populations. Of course it may be that they are rare because of their susceptibility to these and other anthropogenic pressures.

Genetic isolation as a direct result of roads has not been studied in bats. In several other mammal species an effect of roads on genetic population structure has been found (Frantz et al. 2012). For example, Gerlach and Musolf (2000) have shown that populations of bank vole are genetically different either side of a four-lane highway. However, even in bat species such as Bechstein's bat, *Myotis bechsteinii*, for which barrier effects of motorways haven been shown to occur in the summer habitat (Kerth and Melber 2009), local populations living in an area with several motorways show only weak genetic differentiation (Kerth et al. 2002; Kerth and Petit 2005). In accordance with the findings on Bechstein's bats, population genetic studies on other temperate zone bats typically found no or very little evidence for genetic isolation on the regional scale (Moussy et al. 2013), despite the dense road network in Europe and North America. This suggests that in the temperate zone roads probably have no significant effect on gene flow in most bat species. For tropical bats much less data on population genetic structures are available but the situation may be different from the temperate zone. In general, mammal and bird species living in tropical rainforests are often particularly reluctant to cross open areas (Laurance et al. 2009). Moreover, unlike most bats in Europe and North America, tropical bats often mate close to or at the breeding sites of the females. Both features make tropical bats likely to suffer more from fragmentation by roads by means of restricted gene flow than temperate zone bat species. Clearly, further studies are needed to test this.

There is considerable evidence to suggest that roads act as barriers to bats during foraging and movements between different day roosts (roost switching) in the summer habitat. Bats have been shown to make major detours to avoid roads or to find appropriate crossing points (e.g. Kerth and Melber 2009). This behaviour could lead to longer journeys that consume time and energy or even deny bats access to parts of their habitat. In the study by Kerth and Melber (2009) of 32 radiotracked, female Bechstein's bats, only three individuals, belonging to two different maternity colonies, crossed a four-lane motorway cutting through a German forest to forage (Fig. 3.1). All three bats used an underpass to cross the motorway. Other bats from four nearby colonies did not cross the motorway. Moreover, during roost switching none of the colonies crossed the motorway. In addition, foraging areas of females were smaller in those colonies whose home range was bounded by the motorway, relative to those bounded by more natural forest edges. Importantly, females in colonies bounded by the motorway had lower reproductive success than other females, persuasive evidence for the adverse effects on reproductive output. In the same study, six barbastelle bats, *Barbastella barbastellus*, belonging to one maternity colony, were also tracked and five made several flights

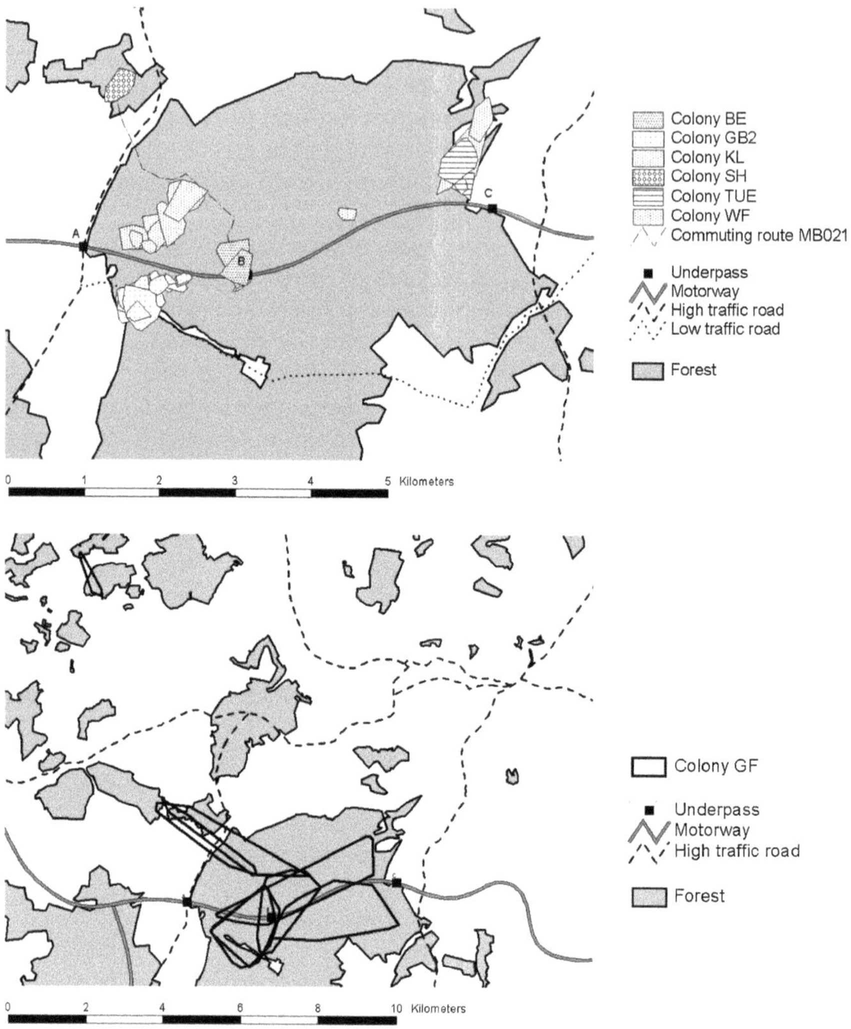

Fig. 3.1 Home range use of two forest bat species living close to a motorway in Germany. The *upper picture* shows the polygons depicting the individual foraging areas of 32 Bechstein's bats belonging to six different colonies living in a German forest that is cut by a motorway. The *lower picture* shows the polygons depicting individual foraging areas of six barbastelle bats belonging to one colony living in the same forest as the Bechstein's bat colonies. From Kerth and Melber (2009)

over the road itself (Fig. 3.1). Moreover, the barbastelle bat colony used roosts on both sides of the motorway. These findings highlight the fact that the effects of roads are species-specific, as will be discussed in more detail later. Berthinussen and Altringham (2012a) observed only three bats flying over a six-lane motorway, all belonging to *Nyctalus* species, at heights above 20 m. *Nyctalus* species are known to fly high and to forage in open spaces (e.g. Jones 1995), behaviour that is likely to make them less susceptible to the barrier effects of roads and to collision mortality. The absence of other species of bat flying over the road in this study suggests that the severance of linear elements by the road may have caused the abandonment of previous flight lines.

Roads may be perceived as barriers by bats for several reasons: open spaces and artificial light expose them to predation, and moving traffic and noise may be seen as threats. Small gaps (<5 m) in cover along flight routes can interrupt commuting bats (e.g. Bennett and Zurcher 2013), but many species will cross open spaces, even those adapted to forage in woodland (e.g. Kerth and Melber 2009; Abbott 2012; Abbott et al. 2012a; Berthinussen and Altringham 2012b), although they will typically do so close to the ground (e.g. Russell et al. 2009; Abbott 2012; Abbott et al. 2012a; Berthinussen and Altringham 2012b). Abbott et al. (2012a) observed low-flying species crossing at sites where mature hedgerows had been severed by the road, even when the gap was >50 m. However, Abbott (2012) found that the rate of bat crossing decreased with increasing distance between mature hedgerows on opposite sides of the road, suggesting a greater barrier effect. Russell et al. (2009) reported that reduced cover at the roadside reduced the number of crossing bats.

That some bats will cross roads is not an indicator that open roads are not a problem—the proportion of bats that do cross may be very small and they are at risk of collision with traffic. The presence of traffic does appear to have a direct effect on the likelihood of crossing, since Indiana bats, *Myotis sodalis*, reverse their flight paths and exhibit anti-predator avoidance behaviour in response to approaching vehicles (Zurcher et al. 2010; Bennett and Zurcher 2013). No specific study has been made of crossing behaviour in relation to traffic volume and road width but anecdotal evidence suggests that it matters. For example, in the study of Kerth and Melber (2009) an individual Bechstein's bat that flew over a two-lane road did only cross a four-lane highway through an underpass. Light and noise are discussed below.

Evidence for a barrier effect is seen in other studies. Berthinussen and Altringham (2012a) found that total bat activity, the activity of the most abundant species (*Pipistrellus pipistrellus*) and the number of species, were all positively correlated with distance from a 40 year-old, six-lane, unlit motorway in rural north-west England (30–40,000 vehicles/day). Total activity increased more than threefold between 0 and 1600 m from the road. These effects were consistent over the two years of study and similar results were obtained on a rural motorway in south-west England (25–90,000 vehicles/day) (Berthinussen 2013). Unpublished work (A. Berthinussen and J.D. Altringham, in preparation) shows that this effect can extend to single carriageway (two-lane) roads. The most likely explanation for this spatially extensive reduction in bat activity is a long-term barrier effect, possibly in combination with increased mortality, driving colonies away from the road, and this is discussed further below.

3.2.3 Roadkill

Bats that attempt to cross roads risk collision, and hotspots for mortality have been found where flyways cross roads and where there is favourable habitat for bats on both sides of a road (e.g. Lesiński 2007; Russell et al. 2009; Medinas et al. 2013). Although agile and manoeuvrable in flight, most bat species fly at low speeds (<20 km/h) and many fly close to the ground (0–4 m: e.g. Russell et al. 2009; Berthinussen and Altringham 2012b), particularly when crossing open spaces. In contrast to the majority of birds, most bats also spend most of the time they are out of the roost in flight. They make extensive use of linear landscape features, such as woodland edges and hedgerows along roads, for foraging and as navigational aids when commuting and several recent studies have shown how important these linear elements are to bats (e.g. Boughey et al. 2012; Frey-Ehrenbold et al. 2013; Bellamy et al. 2013). Flying close to such edges may also reduce predation risk. In combination, these behavioural traits make bats highly vulnerable to moving vehicles when either foraging along roads or when attempting to cross roads on commuting flights. Being small, bats can probably be pulled easily into the slipstream of passing vehicles. Russell et al. (2009) watched over 26,000 bat crossings (primarily little brown bats, *Myotis lucifugus*) on a highway in the USA. Bats approached the road using tree canopy cover and fewer bats were recorded crossing where cover was absent. The lower the cover, the lower the bats crossed the road. Where bats were forced to cross an open field on leaving the roost most did so at a height of less than 2 m. Berthinussen and Altringham (2012b) recorded bats of four or more species crossing roads at mean heights well below 5 m (Fig. 3.2).

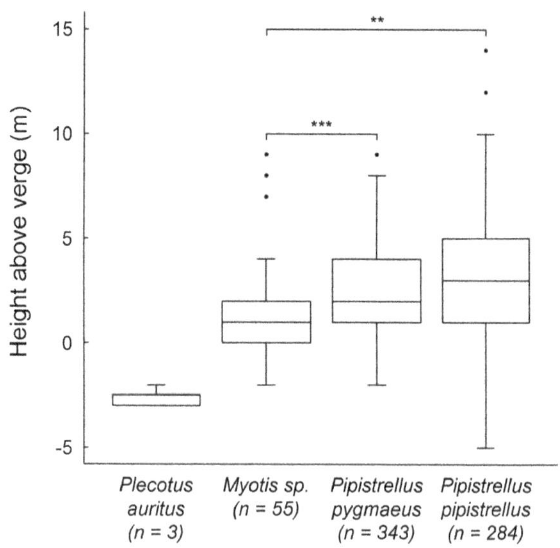

Fig. 3.2 Boxplot of flight height above verge height of identified crossing bats. Median with upper and lower quartiles. Significant differences shown for *Myotis* and *Pipistrellus* species $^{**}P < 0.0005$,$^{***}P < 0.0001$. Verges are elevated on either side of the road and are above road height, therefore negative values indicate bats flying across the road below the height of the verge. From Berthinussen and Altringham (2012b)

Lesinski (2007) recorded bat casualties on an 8 km section of two-lane highway by weekly searches for carcasses over four summers. Casualties ranged from 0.3 bats/km/year in built-up areas to 6.8 bats/km/year where roads were bordered by trees. However, a study by Slater (2002) of the rate of removal of 'carcasses' (small pieces of chicken!) by scavengers on Welsh roads, suggests that a census of this kind may underestimate wildlife road kills as much as 12–16 fold, since dawn scavengers typically removed small carcasses within 30 min. More recently Santos et al. (2011) have also shown that bat carcasses persist on roads in Portugal for a similarly brief period due to scavenging. Teixeria et al. (2013) studied roads in Brazil and found that roadkill estimates increased 2–40 fold when scavenging and low detectability were accounted for. This wide variation was due to taxonomic differences and bats would be at the high end of this range. In addition, small bat carcasses are difficult to spot and many will be thrown clear of the road or carried some distance on the vehicle, suggesting that underestimates will be even greater. Arnett (2006) found that humans (in the absence of scavengers) were able to find only 14 and 42 % of bat carcasses placed at two wind farm sites and Mathews et al. (2013) reported that humans found only 20 % of bat carcasses at wind farms, relative to 73 % found by dogs. Road mortality studies will therefore inevitably under-estimate true mortality rates.

A significant proportion of European bat species, occupying a range of ecological niches, have been documented as roadkill (e.g. Billington 2001–2006; Lesiński 2007; Lesiński et al. 2010). Woodland-adapted species should be most affected due to their characteristic low and slow flight, but this prediction was not supported by Lesiński et al. (2010), as noctules (*Nyctalus noctula*) were killed in significant numbers. Clearly other factors can play an important role locally. Forman et al. (2003, pp 120–122) show that wildlife collisions increase as vehicle speed and traffic volume increase, and with proximity to wildlife habitat and wildlife movement corridors. There are no data on bats relating mortality to speed and traffic volume, but there is no reason to believe they will be different from that of other taxa. There are data from bats to show that roadkill is greater in good habitat and at natural crossing points (Lesiński et al. 2010; Medinas et al. 2013). The effects of traffic speed and volume, road width and height, habitat characteristics, and bat species on rates of roadkill should be explored in greater depth to help us understand how best to mitigate against the effects of roads.

Collection of roadkill carcasses by Russell et al. (2009) led to a conservative estimate of an annual mortality of 5 % of the bats in local roosts. Altringham (2008) arrived at a similar estimate, based on conservative calculations for a road in the UK crossed by lesser horseshoe bats from a large roost (data from Billington 2001–2006). Theoretical studies (e.g. Lande 1987; With and King 1999; Carr and Fahrig 2001) show that populations of animal species with low reproductive rates and high intrinsic mobility, such as bats, are more susceptible to decline and ultimately extinction by the additional mortality caused by roads.

3.2.4 Habitat Degradation—Light, Noise and Chemical Pollution

Light Several studies (e.g. Rydell 1992; Blake et al. 1994; Stone et al. 2009, 2012) have shown that road lighting deters many bat species, notably slow-flying, woodland-adapted species such as members of the genera *Rhinolophus, Myotis* and *Plecotus*, from approaching the road. Lighting will probably exacerbate the barrier effect of roads, since those species reluctant to cross open spaces are also those most likely to avoid light. Both high-pressure sodium and white LED light deter woodland-adapted species, even at low intensity (Stone et al. 2009, 2012). Because light intensity drops rapidly away from the source and will often be blocked by vegetation, the effects of isolated sources are not likely to be far reaching in the landscape, but large arrays of high intensity lights will have a significant effect close to roads.

Light can also attract some bat species, in particular open air foragers such as *Nyctalus* and generalists like *Pipistrellus* (e.g. Rydell 1992; Blake et al. 1994), since short wavelength light attracts insect prey, concentrating them around lights and increasing bat foraging efficiency. This may be not be all good news, since bats exploiting insect swarms around lights may be at greater risk of collision with traffic.

As discussed above, many woodland-adapted bats avoid all forms of visible light, so insects around lights are not available to them. Many insects may indeed be drawn out of woodland towards lights, reducing prey availability to woodland specialists. This could effectively enhance the edge effect around woodland. This has yet to be demonstrated but is worth investigation. The chapter by Rowse et al. discusses the detrimental and beneficial effects of artificial lights on bats in detail.

Noise Most insectivorous bats rely on hearing the returning echoes of their ultrasonic echolocation calls to orientate, detect prey and even communicate. Some species locate and capture prey by listening for sounds generated by their prey, such as wing movements or mating calls. Traffic noise may mask prey-generated sounds and the lower frequency components of echolocation calls. During indoor flight room experiments, simulated traffic noise reduced the feeding efficiency of the greater mouse-eared bat, *Myotis myotis*, which typically hunts by listening for sounds made by its prey on the ground (Siemers and Schaub 2011). It is likely that habitats adjacent to noisy roads would therefore be unattractive as feeding areas for this and other species that glean their prey from the ground or vegetation by listening to rustling noises. Vehicle noise may also exacerbate the barrier effect: bats become less likely to fly across a road as traffic noise increases (Bennett and Zurcher 2013). Currently, there are no published field studies that have assessed the effect of traffic noise on bat diversity, abundance or breeding success. However, as described below, traffic noise, like light, is only likely to have a significant effect over relatively short distances.

Pollution Chemical pollution is another significant factor potentially affecting bats close to roads: transport is the fastest growing source of greenhouse gases. In the USA, over 50 % of domestic CO_2 emissions come from cars, putting 1.7 billion tonnes into the atmosphere every year—a major contributor to climate change. In addition there are the local effects of other chemical pollutants. Automobile exhaust gases close to a road have been shown to be associated with a decline in arthropod diversity and abundance (Przybylski 1979). Motto et al. (1970) and Muskett and Jones (1980) found significant effects on invertebrates of lead and other metals from cars up to 30 m from roads.

3.2.5 Species-Specific Effects

Body size, wing form, echolocation call structure and feeding and roosting ecology all determine how bats fly and use the landscape. Thus, it is not surprising that the effects of roads on bats are to a significant extent species-specific. Larger, fast-flying species, adapted to foraging in the open, appear from most studies to be less affected by roads (e.g. Kerth and Melber 2009; Abbott et al. 2012a; Berthinussen and Altringham 2012a), as they typically fly high above the ground. Their greater flight efficiency and speed relative to woodland-adapted species mean that even if they are forced to make long diversions to find safe crossing points or to avoid roads altogether, the consequences are likely to be less important. Smaller, slower flying, woodland-adapted species are more manoeuvrable and typically capable of gleaning and hovering but this necessarily makes them less efficient flyers (Altringham 2011). Woodland species are also more reluctant to fly in the open and tend to commute along linear features in the landscape such as treelines, waterways, and woodland edges. These features provide protection from weather and predators, are sources of insect prey, and provide conspicuous acoustic and visual landmarks for orientation. Figure 3.3 shows schematically the main patterns of flight and habitat use by insectivorous bats. It is unfortunate that the species most likely to be affected by roads, the slow-flying, woodland-adapted bats, such as *Rhinolophus* and some *Myotis* species, are also those that have suffered most from human activity in Europe and North-America and are at highest risk of extinction there (Safi and Kerth 2004).

3.2.6 Road Class and Speed

The greater width of motorways may make them more effective barriers (Berthinussen and Altringham 2012a) than most other roads. However, traffic density may be equally important (Russell et al. 2009; Zurcher et al. 2010; Bennett and Zurcher 2013) and many major non-motorway roads carry similar or greater traffic volumes, at comparable speed, to rural motorways.

Even minor roads are avoided by many bat species. In a habitat suitability modelling (HSM) study in northern England based on extensive acoustic surveys,

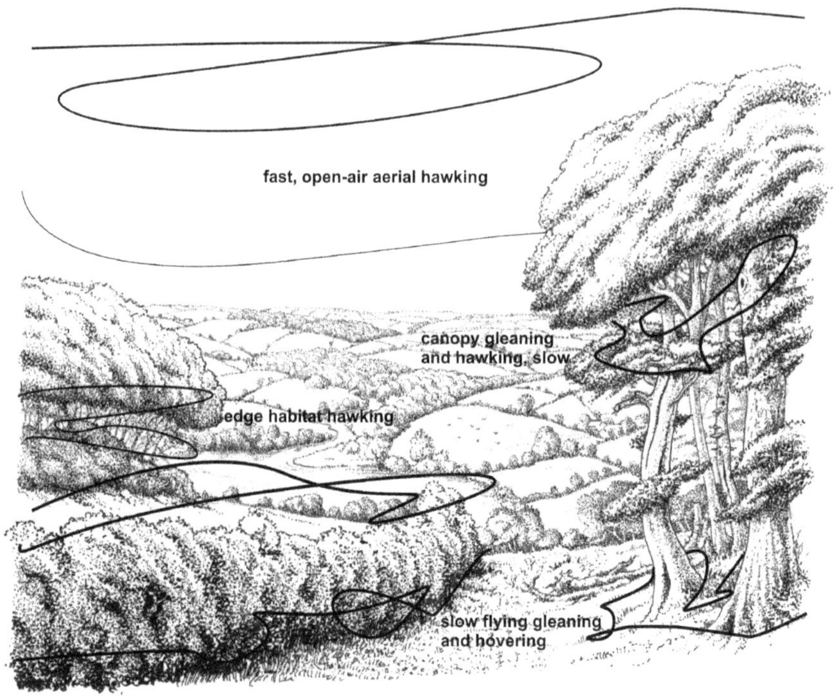

Fig. 3.3 Flight style and habitat use by insectivorous bats. Drawing by Tom McOwat

Bellamy et al. (2013) found that only *Nyctalus* and *Pipistrellus* species showed a positive association with roads and then only when roads were at low densities and in close proximity to woodland. This association is likely due to the use by bats of hedgerows along roads that connect to woodland. Other species, particularly woodland specialists, such as *Myotis* and *Plecotus* species, avoided roads and all species avoided roads when they became dense around settlements. All road classes were combined in this study, but minor roads predominate in the region, so the effects of major roads were probably underestimated. Studies of birds support these conclusions: Develey and Stouffer (2001) and Laurance et al. (2004) have shown that even narrow, unpaved forest roads can act as barriers to tropical forest birds.

In the absence of further work on bats we can look at other animals. Forman et al. (2003) demonstrated that roads act as significant barriers to a variety of mammals from voles to grizzly bears, that primary roads are significantly more effective barriers than secondary roads, and the barrier effect increases with increasing traffic volume. The effects in some cases are severe. Gerlach and Musolf (2000) have shown that populations of bank vole are genetically distinct either side of a busy four-lane highway (50 m wide, 30,000 vehicles/day), but not either side of a two-lane country road (10 m, 5000 vehicles/day) or a railway. Highways can be major genetic barriers even to large and mobile animals such as coyotes and lynx (Riley et al. 2006) or red deer (Frantz et al. 2012).

3.2.7 *Cumulative Effects, Extinction Debt and the Importance of Scale*

Most of the factors discussed above will be cumulative. The effects of each individually need not therefore be great for the combination to have a profound effect on a bat population. Furthermore, in many cases there will be a lag, known as the extinction debt, between cause and effect (e.g. Tilman et al. 1994; Loehle and Li 1996). This is illustrated in Fig. 3.4.

The effects of habitat loss and reduced habitat quality on the distribution of flying bats may be seen quickly, as bats alter their foraging and commuting behaviour to adapt as best they can to the altered landscape. Collision mortality, unless very high, may not have a significant and detectable effect for several generations. The barrier effect may take several more generations to show itself, since it is likely to involve the decline and/or relocation of nursery and other roosts, but it too may be rapid, for example when bats are completely excluded from key foraging areas. Although no data exist for bats, a study of the effects of roads on wetland biodiversity (birds, mammals, reptiles, amphibian and plants) suggests that the full effects may not be seen for several decades (Findlay and Bourdages 2000). This has important implications for monitoring the effects of roads and assessing the effectiveness of mitigation, as discussed later.

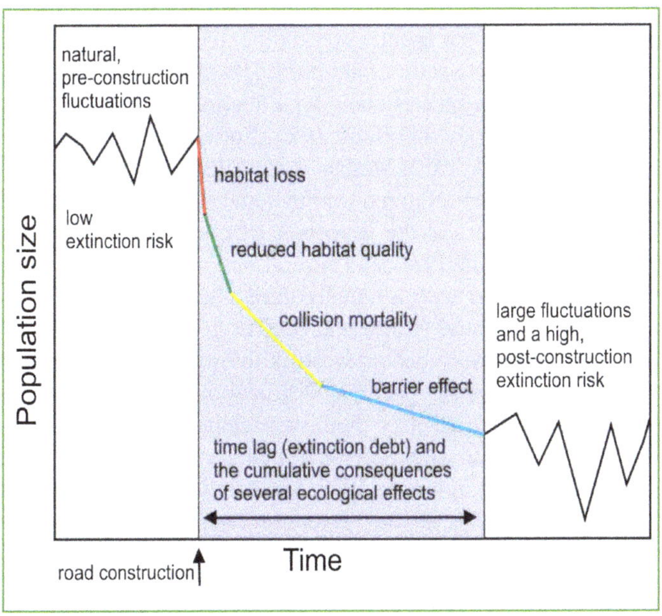

Fig. 3.4 The multiple causes of bat population reduction by roads and the delayed response (extinction debt). Adapted from Forman et al. (2003)

Berthinussen and Altringham (2012a) found that the decline in diversity and abundance of bats extended to at least 1.6 km from a motorway. Which of the above mechanisms contribute to this extensive effect? Low activity and diversity close to the road may be due to most or all of the factors identified: habitat degradation resulting from light, noise and chemical pollution, a barrier effect, or increased mortality due to roadkill. Noise pollution can contribute only to short-range effects, since noise levels in the study fell rapidly over the first 200 m and were close to ambient thereafter. Lab studies on the gleaning greater mouse-eared bat *Myotis myotis* (Schaub et al. 2008; Siemers and Schaub 2011) show that even species that hunt by listening for prey-generated noise are not likely to be affected by roads more than 60 m away. Light pollution was not considered by Berthinussen and Altringham, since the road sections studied were unlit. However, any effect of light pollution from road and vehicle lights is also likely to operate over relatively short distances, due to the inverse square relationship between distance and light intensity. In addition vegetation alongside of roads will further reduce the effect of light and noise pollution quickly. Road developments can disrupt local hydrology and polluted run-off may degrade wetland foraging habitats (Highways Agency 2001), but the scale of such effects will be very variable. As discussed above, chemical pollution is likely to be a factor only over relatively short distances unless dispersion is facilitated by drainage. The many processes that may be degrading roadside habitats need further study, but none of those discussed are likely to explain changes in bat activity over 1.6 km.

Reduced activity over long distances can however be explained by the combination of a barrier effect and increased mortality due to roadkill. The home ranges of temperate insectivorous bat species typically extend 0.5–5 km from their roost (e.g. Bontadina et al. 2002; Senior et al. 2005; Davidson-Watts et al. 2006; Smith and Racey 2008), and most species show high fidelity to roosts, foraging sites and commuting routes (e.g. Racey and Swift 1985; Entwistle et al. 2000; Senior et al. 2005; Kerth and van Schaik 2012; Melber et al. 2013). A major road built close to a nursery roost has the potential to reduce the home range area of a colony through both destruction of habitat and the severance of commuting routes that reduces access to foraging areas. The bats have several options. One is to continue to use the roosts close to the road with a reduced foraging area, reduced resources and reduced reproductive potential (Kerth and Melber 2009). The colony is therefore likely to decline. Alternatively bats may cross the road to maintain their original home range area. Local habitat loss and degradation and increased roadkill will compromise the colony, which may therefore decline. Mortality from roadkill is likely to be high since most species cross at heights that put them in the paths of vehicles (e.g. Verboom and Spoelstra 1999; Gaisler et al. 2009; Russell et al. 2009; Berthinussen and Altringham 2012b). Bats may waste time and energy by commuting greater distances, either away from the road to find new foraging sites, or to find 'safe' crossing points along the road to commute to their original foraging sites. All of these outcomes will reduce the reproductive output of nursery colonies (e.g. Tuttle 1976; Kerth and Melber 2009). Alternatively the colonies may relocate away from the road, into habitat that is presumably already fully exploited by other colonies. All 'solutions' will lead to a fall in bat density near to the road. The overall fall in habitat quality will most likely lead to reduced reproductive success and increased adult mortality and in long-lived bats these will have a profound effect on local colony size and overall population size (Sendor and Simon 2003; Papadatou et al. 2011).

Given the magnitude and spatial scale of the effects on bat activity and diversity observed by Berthinussen and Altringham (2012a), it is likely that barrier and edge effects, together with increased roadkill are having a strong negative effect on the demographics and distribution of local bat populations in proximity to major roads. Similar effects have been found in other vertebrates. Reijnen and Foppen (1994) showed that a decreased density of willow warblers up to 200 m from a major highway was due to the negative influence of the road on population sizes, with reduced breeding success and increased emigration of territorial males. Studies on breeding grassland birds revealed a decrease in density of seven out of 12 species, with disturbance distances up to 3500 m from the busiest roads (50,000 vehicles per day), with collision mortality being a major contributor (Reijnen et al. 1996). A meta-analysis of 49 studies that between them investigated 234 bird and mammal species, found that bird population densities declined up to 1 km, and mammal population densities declined up to 5 km from roads (Benítez-López et al. 2010).

3.2.8 Secondary Effects—Infill and Increased Urban and Industrial Development

Bypasses are frequently built in the countryside to divert traffic around rather than through population centres, to reduce congestion and improve the environment for people in the town or village. In addition to the direct effects of the road itself, there are frequently other consequences. The typically narrow strip of land between the settlement and the new road may be too small to support viable bat populations. This land is also frequently taken over by residential and industrial/commercial development and indeed this development is often part of the initial plan. This leads to further loss and degradation of habitat and a direct increase in traffic. Many of the secondary effects of roads are more severe in the tropics (Laurance et al. 2009), where roads allow people easy access to the remaining undisturbed habitats, which as a consequence suffer further degradation and an increase in the hunting pressure for bush meat, including bats.

3.3 Can Roads Benefit Bats?

Although the balance of the impact of roads on bats is clearly strongly negative, there are potential benefits.

Roosts Some of the ancillary structures built with roads, in particular bridges (e.g. Keeley and Tuttle 1999), can provide roosts for bats. Road bridges over water or wooded valleys are the most likely to be used, those over busy roads much less so. Old stone road bridges over water are widely used by bats, most notably by Daubenton's bat in Europe, but also other *Myotis* species and by *Nyctalus* species (e.g. Senior et al. 2005; Celuch and Sevcik 2008; Angell et al. 2013). In North America bridges are widely used by Brazilian free-tailed bats, *Tadarida brasiliensis* (e.g. Allen et al. 2011) and some other species (e.g. Bennett et al. 2008). Effective mitigation and compensation for the loss of roosting and foraging sites will make the environment close to a road more attractive to bats, but may do so at the expense of greater risk of collision with traffic.

Light Artificial light, particularly short-wavelength light such as mercury-vapour (not most LED lights) attract insects that are common prey to bats. Insect swarms around lights are exploited by open-air foraging bats such as *Pipistrellus* and *Nyctalus* (Rydell 1992; Blake et al. 1994; Stone et al. 2009, 2012). One consequence of this is that bats feeding around lights on busy roads may be at significantly greater risk of mortality from collision with traffic. The balance between the positive and negative effects will be dependent on species, topography, the position of lights, etc. and further study would be useful. A very thorough discussion of the positive and negative effects of artificial light can be found in the chapter by Rowse et al.

Flight corridors In rural environments roads are often bounded by hedgerows or treelines. The wide verges often associated with hedges in landscapes managed for wildlife can be among the most species-rich habitats in some agricultural areas. Minor roads in particular can therefore be both foraging sites and commuting routes, but even major roads are used by some species (e.g. *Nyctalus leisleri*, Waters et al. 1999) where they are bounded by suitable habitat such as a woodland edge. Depending upon structure, this habitat could be used by a wide range of species. However, Bellamy et al. (2013) found that even low road densities had a negative effect on most species of bats, most noticeably the woodland-adapted species *Myotis* and *Plecotus*. Only the distributions of common pipistrelles and noctules had a positive association with roads at low to moderate densities and only when in close proximity (<100 m) to woodland. A similar result was found for railway verges (Vandevelde et al. 2014). As road density increased above moderate levels, the probability of presence of all species declined. The effects of roads of different classes have yet to be investigated in depth—the roads in this study were predominantly minor and rural.

3.4 Conservation in Principle: Avoidance, Mitigation, Compensation and Enhancement

In many countries, legislation has been passed stating that infrastructural development should be carried out in such a way as to minimise the impact of development on the environment, and on protected species such as bats in particular. In principle, there should be no net loss to the environment. In the European Union this is formalised in the Habitats Directive (Council Directive 92/43/EEC). In practice, the system is usually flawed, sometimes severely, due to a lack of knowledge, resources and commercial and political will. Poor goal-setting, planning and execution contribute to either failure, or the absence of any evidence for success, for all wildlife (Tischew et al. 2010) and bats in particular (Altringham 2008; Berthinussen and Altringham 2012b; Stone et al. 2013). As in many other areas of conservation a more scientifically robust, evidence-based approach is urgently needed. European policy and practice also involve a hierarchal approach, starting with avoidance of environmental damage, moving to mitigation when damage is deemed to be unavoidable, then compensation when mitigation is not possible or only partial. Finally, there is an increasing expectation that replacing like with like is not enough, particularly given the uncertainty of success in mitigation and the continued loss of biodiversity. When habitat is lost or degraded, some level of habitat enhancement must accompany development so that in principle, the habitat is better than it was before development. The reality is less than perfect.

The first step in a conservation strategy to minimise the impact of a new road should be to select a route that avoids important bat habitat. To be effective this requires an understanding of the behaviour and ecology of the affected species and detailed knowledge of their distribution. Our knowledge in both areas is growing but far from complete. One approach that can deliver detailed, site-specific information relatively quickly is GIS-based HSM, which can be based on existing data sets, such as those held by museums and record centres (e.g. Jaberg and Guisan 2001; Bellamy and Altringham 2015) or data collected specifically for the purpose, for example by acoustic survey (e.g. Bellamy et al. 2013). This approach yields fine scale distribution maps of probability of occurrence for each species with an estimate of reliability, providing a useful practical tool. However, the route that best avoids bats may not meet human social and economic criteria, particularly if conservation is undervalued. The next step is therefore to build the road in such a way as to mitigate against its effects—that is remove or minimise the many detrimental effects described above. In principle, mitigation under European legislation (Habitats Directive, Council Directive 92/43/EEC) reduces 'damage' to a minimum that is consistent with maintaining bat populations in favourable conservation status.

Where significant loss cannot be avoided, it is expected that compensation will provide alternative roosting and foraging habitat to at least make good the loss. The expectation now is that there is in fact habitat enhancement, to allow for uncertainties in mitigation and to promote long-term habitat improvement.

In practice, avoidance and mitigation are compromised by competing operational and financial constraints. Furthermore, for practical and economic reasons, habitat restoration and creation are long-term processes and it may be many years before these sites are useful to bats, by which time a disturbed bat colony may have been lost. As we will show in the following section, the absence of adequate and well-planned survey and monitoring means that the consequences of road-building and the effectiveness of current avoidance, mitigation, compensation and enhancement practices are all largely unknown (Altringham 2008; O'Connor et al. 2011). In some cases, they have even been shown to be ineffective (Berthinussen and Altringham 2012b).

3.5 Conservation in Practice

We are not aware of any cases in which proposed roads have been rerouted to avoid key bat habitat. Almost all work in this area concerns attempts to remove or minimise the damaging effects of roads. This has usually involved building structures that aim to guide bats safely under or over roads to reduce both the barrier effect and roadkill. The structures built may be multifunctional, for example underpasses for people and wildlife, and use by bats has often been an incidental and unanticipated use of structures built for other purposes, such a drainage culverts. Additional features include tree and hedge planting to guide bats towards crossing points, modified lighting schemes to achieve the same ends or deter bats from crossing at dangerous locations and a wide range of more general 'enhancements' to improve roosting or foraging opportunities.

3.5.1 Over-the-Road Methods: Gantries, Green Bridges, Hop-Overs and Adapted Road/Foot Bridges

Bat bridges or 'bat gantries' have been built on many UK and continental European roads in recent years. However, the most widely used design (Fig. 3.5) in the UK does not help bats to cross the road safely, even when on the line of pre-construction flyways and after up to nine years in situ as shown in Fig. 3.6 (Berthinussen and Altringham 2012b). Other designs have yet to be tested effectively. Berthinussen and Altringham (2012b) found that only a very small proportion of bats that approached gantries 'used' them (i.e. flew in close proximity to them) and for those that did, their flight paths were not raised above the traffic collision zone (Fig. 3.6). This failure of a widespread design highlights the need for effective monitoring and assessment to be an integral part of mitigation practice.

Overpasses built to carry minor roads or footpaths appear to be largely ineffective (Bach et al. 2004; Abbott et al. 2012a) and certainly less effective than underpasses as crossing points (Bach et al. 2004; Abbott et al. 2012a). Most of the structures evaluated have been no more than footbridges and road bridges, with no adaptations to encourage bats, such as tree or shrub planting or careful design of lighting. To date studies have assessed only use, not effectiveness, in that the criterion for success in most studies has been use by an unspecified proportion of bats. A more useful approach would be to assess what proportion of bats crossing a road do so with the aid of crossing structure (Berthinussen and Altringham 2012b).

Fig. 3.5 The most common bat gantry design in the UK—steel wires with plastic spheres at intervals that are intended to be acoustic guides for bats

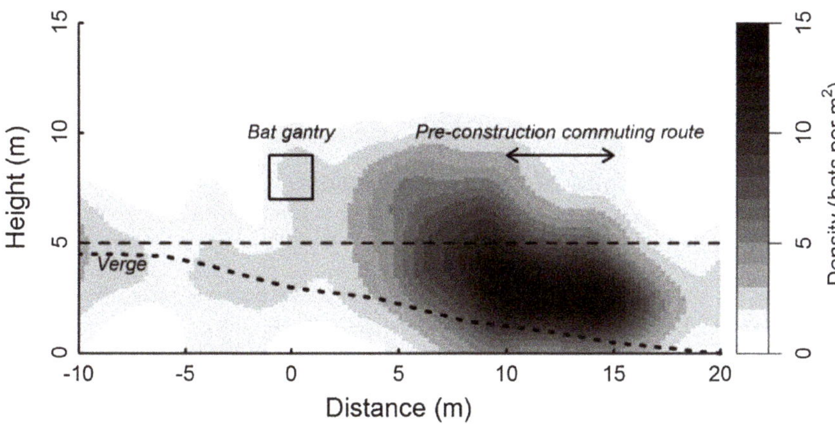

Fig. 3.6 Bat crossing activity at a 'bat gantry' that had been in place for nine years. Gaussian kernel and bandwidth of 1 m used ($n = 1078$). The gantry is located at distance 0 m on the x-axis, with distance from the gantry increasing to the *left* and *right*. The height of the gantry is marked by the *square* at 0 m, and the pre-construction commuting route is 10–15 m to the *right*. 'Unsafe' crossing heights are located below the *dashed line*, which is the maximum vehicle height in Europe. The *dotted line* marked verge shows the decrease in verge height above the road from *left* to *right*. From Berthinussen and Altringham (2012b)

Land or green bridges have been designed and built specifically for other wildlife, and if planted with tall vegetation and linked to existing bat flyways, they have obvious potential as bat crossing structures. As yet, few have been assessed, but bats have been shown to use one land bridge in Germany. Stephan and Bettendorf (2011) found that only a small proportion of woodland-adapted bats crossed a busy motorway using a new land bridge: most crossed the road itself at other locations. It will be interesting to see if bats adapt to it over time. Specific features of the design and connectivity to surrounding habitat of green bridges are probably critical factors for bat use—as they will be for other structures. Further research is required before conclusions can be drawn, but several features are likely to be positively related to use: their strategic location on known flightlines, connectivity to treelines, mature vegetation on the bridge, and bridge width.

'Hop-overs' (Limpens et al. 2005) have been put forward as a relatively low cost and unobtrusive way to encourage bats to cross roads at safe heights. These consist of close planting of trees up to the road edge on both sides of the road, with tall vegetation in the central reservation of wide roads. Branches should overhang the carriageway, ideally giving continuous canopy cover over the road. Safety concerns arising from overhanging branches may have led to reluctance to adopt hop-overs and even to remove trees from road margins. However, many roads have overhanging trees along their margins, so this is an illogical or at least inconsistent objection. The effectiveness of hop-overs has yet to be assessed. Russell et al. (2009) observed that bat flights across a 20 m road gap were at greater heights where bats approached the road along flight routes with taller roadside vegetation and Berthinussen and Altringham (2012b) found a positive correlation between road-crossing height and the height of the roadside embankment.

3.5.2 *Under-the-Road Methods: Underpasses, Culverts and Other 'Tunnels'*

Many studies show that a wide range of bat species use underpasses to fly beneath roads (e.g. Bach et al. 2004; Kerth and Melber 2009; Boonman 2011; Abbott et al. 2012a; Berthinussen and Altringham 2012b). However, most of these studies report only that a small number of bats of particular species were seen to fly through an underpass. In some cases not reported here underpasses were monitored using automated bat detectors with no guarantee that detected bats actually flew through the underpass. For an underpass (or indeed any other mitigation structure) to be effective it must help to maintain bats in favourable conservation status. That is, it must protect the population, not a few individuals, by making a road permeable and safe to cross. Assessing abundance, let alone changes in abundance, is very difficult without considerable survey effort. It is also difficult to measure changes in the permeability of a road to bats without monitoring a very large proportion of the bats in the vicinity of a newly built or upgraded road. Ideally, we would need data before the construction of the road and compare them with data after the road had been built. However, it is possible to determine whether the majority of bats at a location use an underpass (or bridge, gantry, etc.) to cross a road safely. Despite the existence of three underpasses within a 5 km stretch of motorway bisecting a forest, resident Bechstein's bats rarely used them and lost access to important roosting and feeding habitat (Kerth and Melber 2009). Lesser horseshoe bats made frequent use of three underpasses along a 1 km stretch of motorway, but 30 % still crossed directly over the road at traffic height

(Abbott et al. 2012b). Some bats have been recorded making extensive detours to avoid crossing roads (e.g. Kerth and Melber 2009 and references cited in Bach et al. 2004), but we do not know how prevalent this behaviour is: many bat species appear reluctant to deviate from their original flight paths after road severance (Kerth and Melber 2009; Abbott 2012; Berthinussen and Altringham 2012b). Where a road cuts through a dense network of flight routes it may not be straightforward providing a population with an adequate number of safe crossing points. Efforts to re-route bat flight paths, for example by planting new hedgerows linking old routes with new underpasses, should be undertaken well in advance of road clearance, and ideally tested for effectiveness before road opening. Bats were not diverted effectively to underpasses studied by Berthinussen and Altringham (2012b): the great majority of bats flew over the road, near to the original commuting routes. In the same study, one underpass on a known flightline was used by 96 % of the bats on the commuting route.

Underpasses are more likely to be used if they are well connected to the landscape by treelines, hedges or watercourses (Boonman 2011; Abbott 2012), but there is scope for further study in this area. Where possible, they should be located on pre-construction flight routes and tall enough to allow bats to pass without changing flight height or direction (Berthinussen and Altringham 2012b). Even with these precautions, a high proportion of bats may ignore the underpass and fly over the road above it, particularly if the underpass is too small. Underpass height, more than width, was the critical dimension determining the number of bats flying through underpasses in studies in Ireland (Abbott 2012; Abbott et al. 2012a, b). Required heights of underpasses will generally be lower for woodland-adapted species (~3 m) compared to generalist edge-adapted species (~6 m), and open-air species are more likely to fly high above roads. For small gleaning bat species, such as some *Myotis* species, which generally have small home ranges, it may be beneficial to build a higher number of small underpasses (Fig. 3.7) along a road instead of a few large underpasses, which then would by located outside of the home range of most individuals. Mitigation practice would benefit greatly from objective testing and reporting to determine if underpasses are actually providing safe passage for a high enough proportion of bats to protect a local population.

Bats can potentially make use of underpasses that are used by people during the day but have little use at night, such as pedestrian underpasses, minor roads, railways and forestry or agricultural tracks. Use could be maximised by restricting lighting in and around these underpasses, placing them on tree and hedge lines, and making smaller wildlife underpasses or drainage culverts larger to accommodate woodland-adapted bat species. Provision of well-placed, numerous and spacious underpasses should be integral to the overall design of road mitigation, particularly near major roosts. Roads built on embankments are likely to be particularly dangerous to bats, particularly when they sever treelines, since bats appear to maintain flight height on leaving the treeline, bringing them into collision risk over raised road sections. These sites are ideal candidates for underpasses, since they can be built relatively cheaply.

Fig. 3.7 A bat of the genus *Myotis* using a small underpass (about 2 m in diameter) to cross a motorway in Germany. Above the underpass, a wall was built to prevent bats from flying directly into the traffic. Similar walling/fencing has been used in the UK but has not yet been shown to be effective (e.g. Billington 2001–2006)

3.5.3 Light Avoidance

To reduce the potential for disturbance of roosts, flight routes and feeding sites lighting is often directed down toward the road surface, and light spill into the surroundings is minimised. However, since the most vulnerable bats, such as *Rhinolophus* species, fly close to the ground, downward pointing lighting may still have a significant impact on their behaviour. Restricting lighting in crossing structures such as pedestrian underpasses could increase their use by bats. In addition to choosing the intensity, wavelength and direction of lighting, it could also be controlled be timers and motion sensors. Lighting at river and stream crossings should always be avoided, as these are particularly important foraging areas and commuting routes for bats.

Conversely, light may be used to purposely deflect bats away from a dangerous flight route toward a safe crossing point. This has been done, but has not yet been tested for effectiveness and may exacerbate any barrier effect. This assessment is important not only to protect bats, but other wildlife too, since many species avoid light.

3.5.4 The Importance of Connectivity and the Maintenance of Existing Flightlines

An important consideration that is frequently referred to is the need to maintain existing flightlines. There is evidence to support this and it is clearly a sensible precaution. As discussed above, Berthinussen and Altringham (2012b) found that an underpass on a pre-existing flightline was used by 96 % of the bats crossing the road, but attempts to deflect bats to two other underpasses displaced from known routes were not successful.

An extension of this is the general recommendation to maintain and enhance a 'connected' landscape, i.e. a landscape with a broad range and high density of interconnecting linear features such as hedgerows and treelines. This would not only increase the value of the landscape for foraging and commuting, but may give bats more flexibility in how they adapt to a changing landscape and in particular the appearance of barriers in the form of roads. This makes intuitive sense, given the known behaviour of many bat species, and there is a growing body of evidence based on spatial analysis to support it (e.g. Boughey et al. 2012; Bellamy et al. 2013; Frey-Ehrenbold et al. 2013; Bellamy and Altringham 2015). These studies highlight, using different approaches, the importance of these features to bats, and also reveal species differences: woodland-adapted species (e.g. *Myotis*, *Plecotus, Rhinolophus*) and small generalists (e.g. *Pipistrellus*) make more use of (and are more dependent upon) these features than larger open-air species (e.g. *Nyctalus*, *Eptesicus*).

3.5.5 Habitat Improvement and Effective Landscape-Scale Planning

Some general forms of mitigation not specifically related to roads are also relevant, such as the planting of trees and the creation of ponds to replace lost habitat or enhance existing habitat as compensation for damage done by roads. Berthinussen and Altringham (2012a) have shown that the effects of major roads are less easily detected in high quality habitat. This is not a reason to build roads in high quality habitat, since a greater number of bats will still be affected than alongside a road through poor habitat, and the species affected may be more vulnerable. However, it is a reason to attempt to mitigate and compensate using habitat improvement, when a road is built in good habitat. Improvements must not increase roadkill or the costs may outweigh the benefits, so habitat design will be an interesting challenge.

Habitat improvement methods have not been tested effectively, so the scale of the benefits is generally unknown. Habitat improvement and creation obviously have the potential to be beneficial if done on an appropriate scale, but are unlikely to be effective in the short or even medium term, since new woodland and wetland take many years to become established. Over the time taken for habitat to mature, bat colonies may be lost, so long-term planning is needed. Considerable financial incentives may be needed to persuade landowners to undertake habitat improvement. Woodland and wetland creation are more likely to be used for compensation and enhancement than direct mitigation.

As discussed earlier, the Habitats Directive stipulates that in preparing development plans, the avoidance of damage is the preferred option. Mitigation and compensation should only be considered when alternative sites, routes or methods are unavailable and the avoidance of damage is not possible. There must also be over-riding social, economic or safety reasons for development. The planning of new road and rail routes now makes extensive use of GIS-based techniques to assist in the evaluation of the many factors involved. However, the environmental components of these analyses often rely on limited and biased data and do not take full advantage of the developing GIS and modelling techniques described earlier. GIS-based HSM is becoming widely used in ecology. HSM uses the detailed relationships between bat presence and habitat variables to build detailed and accurate distribution maps from relatively small datasets. Bellamy et al. (2013) and Bellamy and Altringham (2015) have used HSM to produce high resolution, accurate predictive maps of the distribution of eight bat species in the Lake District National Park. Similar maps have been, and are being, prepared for other protected areas. These techniques determine the associations between bats and their habitat over multiple spatial scales to give greater accuracy and ecological insight. As our knowledge of bat distributions improves, we will be in a better position to identify those routes that will have minimum impact on bats, and better able to devise appropriate mitigation strategies.

3.5.6 Rail

The effects of rail systems on both bats and other wildlife are even less well understood than those of roads. However, intuitively they have characteristics that may reduce their impact on wildlife. Rail systems are often (but not always) narrower than roads, giving them a smaller footprint and potentially creating a less-effective barrier to animal movement. Trains pass a given point on a network much less frequently than vehicles on roads, which are often continuous. On the busy East Coast line in northern England train noise was detectable for only 8 min/h and this noise decreased to background levels over very much shorter distances than road noise (Altringham 2012). It is nevertheless important that the effects of railways are assessed objectively, particularly in view of the proposed new HS2 line in England, on which trains will travel faster and more frequently. In a study on bat activity of railway verges, Vandevelde et al. (2014) found that bat of the genus *Myotis* seem to avoid the vicinity of railways whereas species foraging in more open space such as pipistrelle and noctule bats use railway verges as foraging habitat.

References

Abbott IM (2012) Assessment of the effectiveness of mitigation measures employed on Irish national road schemes for the conservation of bats. Ph.D. thesis, University College Cork, Ireland

Abbott IM, Butler F, Harrison S (2012a) When flyways meet highways—the relative permeability of different motorway crossing sites to functionally diverse bat species. Landscape Urban Plan 106:293–302

Abbott IM, Harrison S, Butler F (2012b) Clutter-adaptation of bat species predicts their use of under-motorway passageways of contrasting sizes—a natural experiment. J Zool (Lond) 287:124–132

Allen LC, Turmelle AS, Widmaier EP et al (2011) Variation in physiological stress between bridge- and cave-roosting Brazilian free-tailed bats. Cons Biol 25:374–381

Altringham JD (2008) Bat ecology and mitigation; proof of evidence; public enquiry into the A350 Westbury bypass. White Horse Alliance, Neston

Altringham JD (2011) Bats: from evolution to conservation, 2nd edn. Oxford University Press, Oxford

Altringham AE (2012) Train noise mapping: sound in the landscape and implications for wildlife. M.Sc. dissertation, University of Edinburgh

Angell RL, Butlin RK, Altringham JD (2013) Sexual segregation and flexible mating patterns in bats. PLoS ONE 8:e54194

Arnett EB (2006) A preliminary evaluation on the use of dogs to recover bat fatalities at wind energy facilities. Wildl Soc Bull 34:1440–1445

Bach L, Burkhard P, Limpens HJGA (2004) Tunnels as a possibility to connect bat habitats. Mammalia 68:411–420

Barclay RMR, Harder LD (2003) Life histories of bats: life in the slow lane. In: Kunz TH, Fenton MB (eds) Bat ecology. University of Chicago Press, Chicago, pp 209–253

Bellamy CC, Altringham JD (2015) Predicting species distributions using record centre data: multi-scale modelling of habitat suitability for bat roosts. PLoS ONE 10:e0128440

Bellamy C, Scott C, Altringham J (2013) Multiscale, presence-only habitat suitability models: fine-resolution maps for eight bat species. J Appl Ecol 50:892–901

Benítez-López A, Alkemade R, Verweij PA (2010) The impacts of roads and other infrastructure on mammal and bird populations: a meta-analysis. Biol Cons 143:1307–1316

Bennett AF (1991) Roads, roadsides and wildlife conservation: a review. In: Saunders DA, Hobbs RJ (eds) Nature conservation 2: the role of corridors. Surrey Beatty & Sons, Chipping Norton

Bennett VJ, Zurcher AA (2013) When corridors collide: road-related disturbance in commuting bats. J Wildl Man 77:93–101

Bennett FM, Loeb SC, Bunch MS et al (2008) Use and selection of bridges as day roosts by Rafinesque's big-eared bats. Amer Mid Nat 160:386–399

Berthinussen A (2013) The effects of roads on bats in the UK: a model for evidence based conservation. Ph.D. dissertation, University of Leeds

Berthinussen A, Altringham J (2012a) The effect of a major road on bat activity and diversity. J Appl Ecol 49:82–89

Berthinussen A, Altringham J (2012b) Do bat gantries and underpasses help bats cross roads safely? PLoS ONE 8:e38775

Billington G (2001–2006) A487 Llanwnda to South Llanllyfni improvement. Bat surveys. Greena Ecological Consultancy, Devon UK

Blake D, Huston AM, Racey PA, Rydell J, Speakman JR (1994) Use of lamplit roads by foraging bats in Southern England. J Zool 234:453–462

Bontadina F, Schofield H Naef, Naef-Daenzer B (2002) Radio-tracking reveals that lesser horseshoe bats (*Rhinolophus hipposideros*) forage in woodland. J Zool (Lond) 258:281–290

Boonman M (2011) Factors determining the use of culverts underneath highways and railway tracks by bats in lowland areas. Lutra 54:3–16
Boughey K, Lake I, Haysom K et al (2012) Improving the biodiversity benefits of hedgerows: how physical characteristics and the proximity of foraging habitat affect the use of linear features by bats. Biol Cons 144:1790–1798
Carr LW, Fahrig L (2001) Impact of road traffic on two amphibian species of different vagility. Cons Biol 15:1071–1078
Celuch M, Sevcik M (2008) Road bridges as roosts for noctules (*Nyctalus noctula*) and other bat species in Slovakia. Lynx 39:47–54
CMS (1994) Agreement on the conservation of populations of European Bats. Under the convention on the conservation of migratory species of wild animals. Bonn, Germany. Now part of the United Nations Environment Programme
Coffin AW (2007) From road kill to road ecology: a review of the ecological effects of roads. J Trans Stud 15:396–406
Davidson-Watts I, Walls S, Jones G (2006) Differential habitat selection by *Pipistrellus pipistrellus* and *Pipistrellus pygmaeus* identifies distinct conservation needs for cryptic species of echolocating bats. Biol Cons 133:118–127
Defra (2013) Road length statistics. Statistical release June 2013. Defra
Develey PF, Stouffer PC (2001) Effects of roads on movements of understorey birds in mixed-species flocks in central Amazonian Brazil. Cons Biol 15:1416–1422
Entwistle AC, Racey PA, Speakman JR (2000) Social and population structure of a gleaning bat, *Plecotus auritus*. J Zool (Lond) 252:11–17
Fahrig L, Rytwinski T (2009) Effects of roads on animal abundance: an empirical review and synthesis. Ecol Soc 14:21
Findlay CS, Bourdages J (2000) Response time of wetland biodiversity to road construction on adjacent lands. Conserv Biol 14:86–94
Forman RTT, Alexander LE (1998) Roads and their major ecological effects. Ann Rev Ecol Sys 8:629–644
Forman RTT, Sperling D, Bissonette JA et al (2003) Road ecology: science and solutions. Island Press, Washington
Frantz AC, Berttouille S, Eloy MC et al (2012) Comparative landscape genetic analyses show a Belgian motorway to be a gene flow barrier for red deer (*Cervus elaphus*), but not wild boars (*Sus scrofa*). Molec Ecol 21:3445–3457
Frey-Ehrenbold A, Bontadina F, Arlettaz R et al (2013) Landscape connectivity, habitat structure and activity of bat guilds in farmland-dominated matrices. J Appl Ecol 50:252–261
Gaisler J, Rehak Z, Bartonicka T (2009) Bat casualties by road traffic (Brno-Vienna). Acta Theriol 54:147–155
Gerlach G, Musolf K (2000) Fragmentation of landscape as a cause for genetic subdivision in bank voles. Cons Biol 14:1066–1074
Highways Agency (2001) Nature conservation advice in relation to bats: design manual for roads and bridges. Volume 10, environmental design. Section 4, nature conservation. Part 3, HA80/99. Highways Agency UK
Highways Agency (2006) Best practice in enhancement of highway design for bats. Highways Agency and Bat Conservation Trust, UK
Jaberg C, Guisan A (2001) Modelling the distribution of bats in relation to landscape structure in a temperate mountain environment. J Appl Ecol 38:1169–1181
Jaeger JAG, Schwarz-von Raumer HG et al (2007) Time series of landscape fragmentation caused by transportation infrastructure and urban development: a case study from Baden-Württemberg (Germany). Ecol Soc 12:22
Jones G (1995) Flight performance, echolocation and foraging behaviour in noctule bats, *Nyctalus noctula*. J Zool (Lond) 237:303–312
Jones G, Jacobs DS, Kunz TH et al (2009) Carpe noctem: the importance of bats as bioindicators. Endangered Species Res 8:93–115

Keeley BW, Tuttle MD (1999) Bats in American bridges. Bat Conservation International, Austin, Texas
Kerth G (2008) Causes and consequences of sociality in bats. Bioscience 58:737–755
Kerth G, Melber M (2009) Species-specific barrier effects of a motorway on the habitat use of two threatened forest-living bat species. Biol Cons 142:270–279
Kerth G, Petit E (2005) Colonization and dispersal in a social species, the Bechstein's bat (*Myotis bechsteinii*) Mol Ecol 14:3943–3950
Kerth G, van Schaik J (2012) Causes and consequences of living in closed societies: lessons from a long-term socio-genetic study on Bechstein's bats. Mol Ecol 21:633–646
Kerth G, Mayer F, Petit E (2002) Extreme sex-biased dispersal in the communally breeding, non-migratory Bechstein's bat (*Myotis bechsteinii*) Mol Ecol 11:1491–1498
Lande R (1987) Extinction thresholds in demographic models of territorial populations. Am Nat 130:624–635
Laurance SGW, Stouffer PC, Laurance WF (2004) Effects of road clearings on movement patterns of understory rainforest birds in central Amazonia. Cons Biol 18:1099–1109
Laurance WF, Goosem M, Laurance SG (2009) Impacts of roads and linear clearings on tropical forests. TREE 24:659–669
Lesiński G (2007) Bat road casualties and factors determining their number. Mammalia 2:138–142
Lesiński G, Sikora A, Olszewski A (2010) Bat casualties on a road crossing a mosaic landscape. Eur J Wildl Res 57:217–223
Limpens HJGA, Twisk P, Veenbaas G (2005) Bats and road construction. Dutch Ministry of Transport, Public Works and Water Management Directorate-General for Public Works and Water Management, Road and Hydraulic Engineering Institute, Delft, the Netherlands and the Association for the Study and Conservation of Mammals, Arnhem, the Netherlands
Loehle C, Li BL (1996) Habitat destruction and the extinction debt revisited. Ecol Inter 6:784–789
Mathews F, Swindells M, Goodhead R et al (2013) Effectiveness of search dogs compared with human observers in locating bat carcasses at wind-turbine sites: a blinded randomised trial. Wildl Soc Bull: Early View
Medinas D, Tiago MJ, Mira A (2013) Assessing road effects on bats: the role of landscape, road features, and bat activity on road kills. Ecol Res 28:227–237
Melber M, Fleischmann D, Kerth G (2013) Female Bechstein's bats share foraging sites with maternal kin but do not forage together with them—results from a long-term study. Ethol 119:793–801
Motto HL, Daines RH, Chilko DM et al (1970) Lead in soils and plants: its relation to traffic volume and proximity to highways. Env Sci Tech 4:231–237
Moussy C, Hosken DJ, Mathews F et al (2013) Migration and dispersal patterns of bats and their influence on genetic structure. Mamm Rev 43:183–195
Muskett CJ, Jones MP (1980) The dispersal of lead, cadmium and nickel from motor vehicles and effects on roadside invertebrate macrofauna. Env Poll 23:231–242
O'Connor G, Green R, Wilson S (2011) A review of bat mitigation in relation to highway severance. Highways Agency, UK
Papadatou E, Ibáñez C, Pradel R et al (2011) Assessing survival in a multi-population system: a case study on bat populations. Oecologia 165:925–933
Przybylski Z (1979) The effects of automobile exhaust gases on the arthropods of cultivated plants, meadows and orchards. Env Poll 19:157–161
Racey PA, Entwistle AC (2003) Conservation ecology of bats. In: Kunz TH, Fenton MB (eds) Bat ecology. University of Chicago Press, pp 680–743
Racey PA, Swift SM (1985) Feeding ecology of *Pipistrellus pipistrellus* (Chiroptera: Vespertilionidae) during pregnancy and lactation. I Foraging Behaviour. J Anim Ecol 54:205–215
Reijnen R, Foppen R (1994) The effects of car traffic on breeding bird populations in woodland. 1. Evidence of reduced habitat quality for willow warblers (*Phylloscopus trochilus*) breeding close to a highway. J Appl Ecol 31:85–94

Reijnen R, Foppen R, Meeuwsen H (1996) The effects of traffic on the density of breeding birds in Dutch agricultural grasslands. Biol Cons 75:255–260
Riitters KH, Wickham JD (2003) How far to the nearest road? Front Ecol Env 1:125–129
Riley SPD, Pollinger JP, Sauvajot RM et al (2006) A southern California freeway is a physical and social barrier to gene flow in carnivores. Mol Ecol 15:1733–1741
Russell AL, Butchkoski CM, Saidak L et al (2009) Roadkilled bats, highway design, and the commuting ecology of bats. Endangered Species Res 8:49–60
Rydell J (1992) Exploitation of insects around streetlamps by bats in Sweden. Funct Ecol 6:744–750
Rytwinski T, Fahrig L (2012) Do species life history traits explain population responses to roads? A meta-analysis. Biol Cons 147:87–98
Safi K, Kerth G (2004) A comparative analysis of specialisation and extinction risk in temperate-zone bats. Cons Biol 18:1293–1303
Santos SM, Carvalho F, Mira A (2011) How long do the dead survive on the road? Carcass persistence probability and implications for roadkill monitoring surveys. PLoS ONE 6:e25383
Schaub A, Ostwald J, Siemers BM (2008) Foraging bats avoid noise. J Exp Biol 211:3174–3180
Sendor T, Simon M (2003) Population dynamics of the pipistrelle bat: effects of sex, age and winter weather on seasonal arrival. J Anim Ecol 72:308–320
Senior P, Butlin RK, Altringham JD (2005) Sex and segregation in temperate bats. Proc Roy Soc Lond B 272:2467–2473
Siemers BM, Schaub A (2011) Hunting at the highway: traffic noise reduces foraging efficiency in acoustic predators. Proc Roy Soc Lond B 278:1646–1652
Slater FM (2002) An assessment of wildlife road casualties—the potential discrepancy between numbers counted and numbers killed. Web Ecol 3:33–42
Smith PG, Racey PA (2008) Natterer's bats prefer foraging in broad-leaved woodlands and river corridors. J Zool (Lond) 272:314–322
Stephan S, Bettendorf J (2011) Home ranges of Bechstein's bats overlapping a motorway. In: Paper presented at the European Bat Research Symposium, Vilnius, Lithuania
Stone EL, Jones G, Harris S (2009) Street lighting disturbs commuting bats. Curr Biol 19:1123–1127
Stone EL, Jones G, Harris S (2012) Conserving energy at a cost to biodiversity? Impacts of LED lighting on bats. Global Change Biol 19:2458–2465
Stone EL, Harris S, Jones G (2013) Mitigating the effect of development on bats in England with derogation licensing. Cons Biol 27:1324–1334
Teixeria FZ, Coelho AVP, Esperandio B et al (2013) Vertebrate road mortality estimates: effects of sampling methods and carcass removal. Biol Cons 157:317–323
Tilman D, May RM, Lehman CL, Nowak MA (1994) Habitat destruction and the extinction debt. Nature 371:65–66
Tischew S, Baasch A, Conrad MK et al (2010) Evaluating restoration success of frequently implemented compensation measures: results and demands for control procedures. Restor Ecol 18:467–480
Trombulak SC, Frissell CA (2000) Review of ecological effects of roads on terrestrial and aquatic communities. Cons Biol 14:18–30
Tuttle MD (1976) Population ecology of gray bat (*Myotis grisescens*)—factors influencing growth and survival of newly volant young. Ecology 57:587–595
Vandevelde JC, Bouhours A, Julien JF, Couvet C, Kerbiriou C (2014) Activity of European common bats along railway verges. Ecol Eng 64:49–56
Verboom B, Spoelstra K (1999) Effects of food abundance and wind on the use of tree lines by an insectivorous bat. Can J Zool 77:1393–1401
Waters D, Jones G, Furlong M (1999) Foraging ecology of Leisler's bat (*Nyctalus leisleri*) at two sites in southern Britain. J Zool (Lond) 249:173–180
With KA, King AW (1999) Extinction thresholds for species in fractal landscapes. Cons Biol 13:314–326
Zurcher AA, Sparks DW, Bennett VJ (2010) Why the bat did not cross the road? Acta Chirop 12:337–340

Chapter 4
Responses of Tropical Bats to Habitat Fragmentation, Logging, and Deforestation

Christoph F.J. Meyer, Matthew J. Struebig and Michael R. Willig

Abstract Land-use change is a key driver of the global biodiversity crisis and a particularly serious threat to tropical biodiversity. Throughout the tropics, the staggering pace of deforestation, logging, and conversion of forested habitat to other land uses has created highly fragmented landscapes that are increasingly dominated by human-modified habitats and degraded forests. In this chapter, we review the responses of tropical bats to a range of land-use change scenarios, focusing on the effects of habitat fragmentation, logging, and conversion of tropical forest to various forms of agricultural production. Recent landscape-scale studies have considerably advanced our understanding of how tropical bats respond to habitat fragmentation and disturbance at the population, ensemble, and assemblage level. This research emphasizes that responses of bats are often species and ensemble specific, sensitive to spatial scale, and strongly molded by the characteristics of the prevailing landscape matrix. Nonetheless, substantial knowledge gaps exist concerning other types of response by bats. Few studies have assessed responses at the genetic, behavioral, or physiological level, with regard to disease prevalence, or

C.F.J. Meyer (✉)
Centre for Ecology, Evolution, and Environmental Changes, Faculty of Sciences, University of Lisbon, Lisbon, Portugal
e-mail: cmeyer@fc.ul.pt

M.J. Struebig
Durrell Institute of Conservation and Ecology, School of Anthropology and Conservation, University of Kent, Canterbury, UK
e-mail: m.j.struebig@kent.ac.uk

M.R. Willig
Center for Environmental Sciences and Engineering and Department of Ecology and Evolutionary Biology, University of Connecticut, Storrs, CT, USA
e-mail: michael.willig@uconn.edu

C.C. Voigt and T. Kingston (eds.), *Bats in the Anthropocene: Conservation of Bats in a Changing World*, DOI 10.1007/978-3-319-25220-9_4

the extent to which human disturbance erodes the capacity of tropical bats to provide key ecosystem services. A strong geographic bias, with Asia and, most notably, Africa, being strongly understudied, precludes a comprehensive understanding of the effects of fragmentation and disturbance on tropical bats. We strongly encourage increased research in the Paleotropics and emphasize the need for long-term studies, approaches designed to integrate multiple scales, and answering questions that are key to conserving tropical bats in an era of environmental change and dominance of modified habitats (i.e., the Anthropocene).

4.1 Habitat Conversion: A Key Aspect of Global Change

Bats are valuable indicators of biodiversity and ecosystem health, and respond to a range of stressors related to environmental change (Jones et al. 2009). Alteration in land use is one of the principal aspects of global environmental change and a key driver of biodiversity loss in terrestrial ecosystems. Indeed, biodiversity impacts of land-use change are generally considered to be more immediate than those from climate change (Sala et al. 2000; Jetz et al. 2007; Pereira et al. 2010). However, the effects of land-use change on tropical species could exacerbate those of changing climate, leading to challenges for long-term conservation efforts (Struebig et al. 2015), including those for bats. Over the last decades, human transformation of much of the Earth's natural ecosystems has greatly accelerated, and the twenty-first century will herald profound changes in land use, particularly in developing tropical countries (Lee and Jetz 2008). The most recent quantification of global forest change revealed an overall increasing trend in annual forest loss across the tropics between 2000 and 2012 (Hansen et al. 2013), highlighting the continued prevalence of tropical deforestation.

Drivers of tropical deforestation have shifted from being promoted mostly by government policies for rural development toward urban population growth and industrial-scale, export-oriented agricultural production (DeFries et al. 2010). Fueled by unabated human population growth, global food demand is escalating, and the current trajectory of agricultural expansion will have serious negative long-term consequences for the preservation of the planet's biodiversity (Tilman et al. 2011; Laurance et al. 2014). In tropical countries, conversion of natural habitats to agricultural and pastoral land is one of the greatest threats to biodiversity (Phalan et al. 2013), as cropland expansion in recent decades has largely come at the expense of intact old-growth forest (Gibbs et al. 2010). Rampant commercial logging is also a major force of tropical forest destruction and degradation, with around 20 % of such forests subjected to some level of timber harvesting (Asner et al. 2009).

Loss of habitat as a result of extensive land conversion and associated fragmentation are ubiquitous throughout the tropics. Resulting landscapes typically comprise a mosaic of human-modified habitats that include agroforests, agricultural land, and tree plantations, as well as remnants of old-growth, logged forest, and secondary forests regenerating from clearance or burning (Gardner et al. 2009; Chazdon 2014).

Indeed, except for large areas of tropical forest in Papua New Guinea and in the Amazon and Congo basins, such a description accurately characterizes most tropical landscapes (Melo et al. 2013). Anthropogenic activities in many tropical countries have resulted in the creation of fragmented landscapes that are dominated by small (often < 50 ha), isolated, and irregularly shaped forest patches. These patches are highly prone to edge effects (Broadbent et al. 2008; Ribeiro et al. 2009), defined as systematic changes in abiotic and biotic variables at the boundary between adjacent land-use types. Although deforestation and degradation of old-growth forests are the dominant forms of land-use alteration, forest regeneration and the expansion of secondary forests are the second most important type of land-use change occurring across the tropics (Asner et al. 2009; Dent and Wright 2009). These recovering forest habitats could potentially mitigate, or even reverse, current trends of forest loss and degradation as well as concomitant biodiversity loss (Wright and Muller-Landau 2006; Dent and Wright 2009; Chazdon 2014). A pan-tropical meta-analysis of land-use change studies points to the irreplaceable value of old-growth forests, but also highlights the high species diversity found in regenerating logged forests compared to secondary forests (Gibson et al. 2011). Although the long-term conservation value of regenerating forests has been questioned (Melo et al. 2013), biodiversity representation clearly varies among logged and secondary habitats, and so not all recovering forests should be treated equally.

4.2 Tropical Bats in a Changing World

Bats exhibit the general mammalian pattern of greatest diversity in the tropics, from both a taxonomic and a functional perspective (Willig et al. 2003). Bats also provide ecosystem services that are critically important in tropical ecosystems—as pollinators and seed dispersers for hundreds of plant species and as agents of suppression of arthropod herbivores and insect pest species (Muscarella and Fleming 2007; Kalka et al. 2008; Williams-Guillén et al. 2008; Kunz et al. 2011; Maas et al. 2013). Nonetheless, many tropical bat species face an uncertain future and show declining population trends due to many of the threats outlined previously (e.g., Kingston 2013).

How do tropical bats fare in the Anthropocene, in which they are exposed to increasing levels of land-use change, potentially exacerbated by climate change (Struebig et al. 2015), and the synergistic effects of both processes? Simple pan-tropical meta-analyses suggest that the impacts of land-use change on mammal diversity, particularly on bats, are somewhat less severe than for other animal groups (Gibson et al. 2011). Nevertheless, such studies can potentially miss subtle, yet important, responses in assemblage structure. In this chapter, we summarize the accumulated knowledge on the responses of tropical bats to human-induced habitat fragmentation and forest disturbance. By providing a synthetic overview of the topic, we hope to shed light on the conservation value of anthropogenically modified habitats for bats across the major tropical regions and identify future research priorities.

4.3 Review Methodology

We followed a systematic review methodology (Pullin and Stewart 2006) to synthesize information about tropical bat responses to habitat fragmentation, logging, and deforestation. Studies were identified through a comprehensive search in the ISI Web of Science online database (accessed in September 2013), performing a topic search using the string "bat? AND *tropic* AND (fragment* OR logg* OR deforest* OR disturb*)," without restriction on publication year. The use of this combination of key words allowed for the identification of an inclusive set of studies on the effects of fragmentation, logging, and disturbance on tropical bats. The search identified 248 publications that were subsequently screened for suitability for the review based on the article's title, abstract, and, when necessary, text. We excluded review articles and studies that were conducted in urban landscapes (see Chap. 2). As our purpose here was to review evidence for the effects of *anthropogenic* habitat modification on tropical bats, we also excluded studies that were conducted in naturally fragmented landscapes (e.g., forest islands embedded in savannah, oceanic islands). Our review thus focuses on a range of human-modified matrix types of varying structural complexity and contrast—from relatively low-contrast secondary forests, agroforests, and plantation forests, to high-contrast agricultural fields and water matrices resulting from dam construction.

From the 248 studies, 93 met our criteria. In addition, we extended our search using the same key word combinations in Google Scholar through which we identified an additional eight relevant studies within the first 100 records. Sixteen additional publications were found based on a search of our own literature databases, thus bringing the total number of studies considered in our synthesis to 117. Each article was characterized according to geographic region, taxonomic focus, response type, and disturbance type. Response types included (a) population- and assemblage-level responses, (b) genetic effects, (c) behavioral responses, (d) physiological responses, parasite and disease prevalence, and (e) effects on the provisioning of ecosystem services. Disturbance type included the following broad categories: (a) habitat fragmentation, (b) logging, (c) secondary forests and succession, (d) agroforestry systems, (e) tree plantations, and (f) agriculture.

4.4 Biases in Our Understanding of Responses of Tropical Bats to Habitat Alteration

The collated literature revealed substantial geographic and taxonomic biases in the current understanding of tropical bat responses to anthropogenic disturbance. Studies covered 34 distinct study landscapes in 21 countries. Despite a general increase in the number of studies over the last 20 years (Fig. 4.1), most research has been undertaken in the New World tropics (96 studies), with research in Southeast Asia and Australasia lagging far behind (19 studies) and studies in

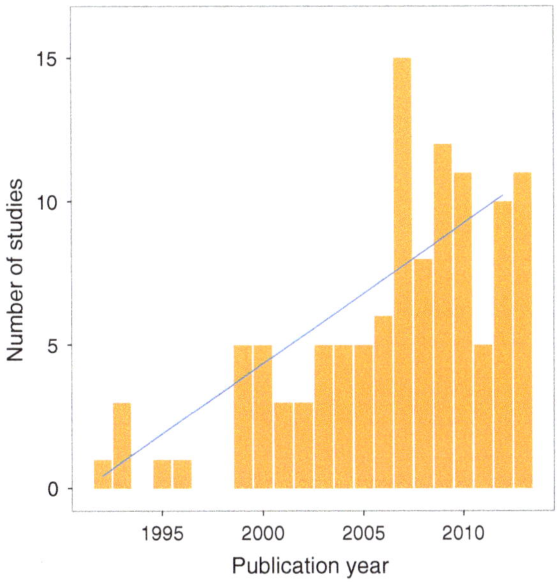

Fig. 4.1 Number of publications on the effects of fragmentation, logging, or disturbance on tropical bats based on a systematic search of the literature. There is a general increase in publications over the last 20 years (linear model fit, $R^2_{\text{adj}} = 0.55$, $p < 0.001$). Data for 2013 represent an underestimate as the literature search did not include the entire year, and therefore, they were not considered in the model fit

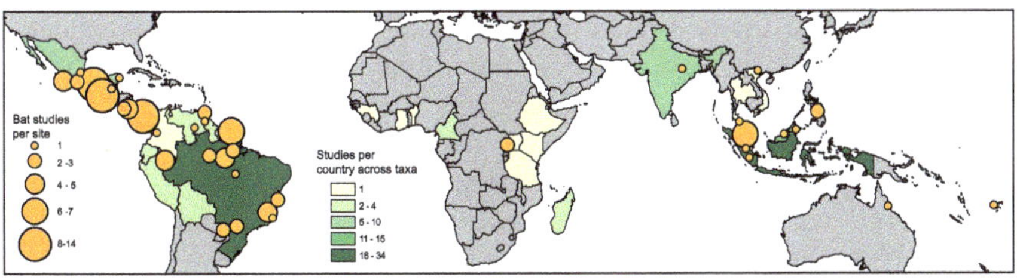

Fig. 4.2 Map illustrating the geographic distribution of research effort based on 117 studies of bats in anthropogenically modified landscapes. Sizes of *orange circles* represent the number of studies per site, where a site is defined as a particular study landscape. Colors of tropical countries represent the number of studies based on the pan-tropical analysis of the impact of disturbance and land conversion on birds, mammals, arthropods, and plants by Gibson et al. (2011)

Africa being rare (2 studies; Fig. 4.2). Geographic variation in this research effort (Fig. 4.2) broadly parallels the pattern reported for multiple taxa across the tropics (Gibson et al. 2011). A few notable differences include a disproportionately high number of bat studies in Mexico and low number of studies in Indonesia compared to other taxa. A large taxonomic bias therefore characterizes our understanding of disturbance effects on tropical bats as a consequence of the prevalence of studies in the Neotropics. With a few exceptions (Estrada et al. 2004; Estrada Villegas

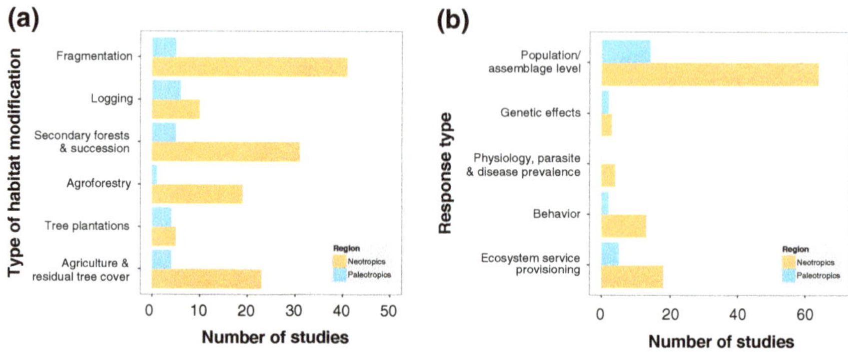

Fig. 4.3 Number of studies by region (Neotropics [n = 96 studies] vs. Paleotropics [n = 21 studies]) based on **a** type of disturbance or habitat modification and **b** type of response. Studies in many cases, especially for (**a**), matched more than one of the broad categories and were counted multiple times

et al. 2010; Williams-Guillén and Perfecto 2011), New World studies focused on the species-rich Phyllostomidae, in turn largely reflecting the use of mist nets to capture bats. Phyllostomids are easily sampled with mist nets and dominate studies. In contrast, non-phyllostomids are underrepresented in samples based on mist netting. Although acoustic methods hold much promise for sampling non-phyllostomid and non-pteropodid bats, considerable difficulties remain in the wider implementation of these techniques in tropical countries, including the lack of call libraries, taxonomic uncertainty, and practical challenges of tropical climates (Harrison et al. 2012). As a result, acoustic sampling has not yet been employed intensively in landscape-scale studies of tropical bats (see also Cunto and Bernard 2012). Finally, a considerable bias exists with respect to studied aspects of fragmentation and disturbance. Comparatively few studies have targeted bat responses to logging or agroforestry (Fig. 4.3a). The vast majority of studies evaluated responses at the population or assemblage level. Far fewer have examined the consequences of anthropogenic disturbance for the provision of ecosystem services by bats. Genetic, physiological, and behavioral effects remain poorly explored, as do effects on disease dynamics associated with bat hosts (Fig. 4.3b).

4.5 Responses at the Population and Assemblage Level

4.5.1 Habitat Fragmentation

Habitat fragmentation has become a major research theme in conservation biology, as reflected in the burgeoning literature on the subject (Fahrig 2003; Ewers and Didham 2006a; Lindenmayer and Fischer 2006; Fischer and Lindenmayer 2007; Collinge 2009). Although the exact definition of "habitat fragmentation" is contentious (Fahrig 2003; Ewers and Didham 2007; Fischer and Lindenmayer 2007), we follow a widely used definition—the landscape-scale process by which habitat loss results in the subdivision of continuous habitat into smaller patches that are isolated from each other by a matrix of modified habitat (Didham 2010).

4.5.1.1 General Patterns

Despite numerous and increasing attempts to detect consistent responses of tropical bats to habitat fragmentation, studies to date suggest relatively few generalizations. At the population level, many studies have documented that abundance responses to fragmentation are highly species and ensemble specific. For instance, in the Neotropics, abundances of gleaning animalivorous bats (Pons and Cosson 2002; Meyer et al. 2008; Meyer and Kalko 2008a) and certain forest-dependent aerial insectivores (Estrada Villegas et al. 2010) decline in response to fragmentation, whereas frugivorous and nectarivorous bats often increase (Sampaio et al. 2003; Delaval and Charles-Dominique 2006; Meyer and Kalko 2008a). In the Paleotropics, insectivorous bat species that roost in tree cavities or foliage are more vulnerable to fragmentation than are cave-roosting species (Struebig et al. 2008, 2009). At the assemblage level, studies that have compared fragmented and continuous forest in terms of species richness, diversity, and composition demonstrate inconsistent responses (Cosson et al. 1999; Schulze et al. 2000; Estrada and Coates-Estrada 2002; Faria 2006). Differences among sites with regard to fragmentation history and structural contrast between fragments and the surrounding matrix complicate the detection of general patterns. This may be a more important issue for the study of tropical bats compared to other taxonomic groups because of the wide range of dispersal abilities exhibited by chiropteran species.

4.5.1.2 Area and Isolation Effects

Early fragmentation studies generally emphasized the effects of area and isolation, reflecting the pervasive influence of island biogeographic theory (IBT, MacArthur and Wilson 1967) in ecology, while ignoring influences of the surrounding landscape matrix. This same pattern is also apparent within the fragmentation literature on tropical bats. Studies have found evidence for effects of both fragment area (Cosson et al. 1999; Struebig et al. 2008, 2011) and isolation (Estrada et al. 1993a; Meyer and Kalko 2008a, b) on population- and assemblage-level responses, whereas effects were weak or absent in others (Faria 2006; Pardini et al. 2009). Moreover, bat ensembles and species often respond differentially to fragment area or isolation, with responses of some taxa being particularly strong (Struebig et al. 2008; Estrada Villegas et al. 2010).

The relative importance of isolation versus area in shaping bat responses to fragmentation is governed by three main factors: the range of fragment sizes relative to isolation in the landscape, the history of landscape change (time since isolation, rate of change), and, probably most importantly, the type and quality of matrix habitats in which fragments are embedded. For instance, the high explanatory power of area relative to isolation reported by Struebig et al. (2008) likely reflects the low structural contrast between fragments and matrix (mostly rubber and oil palm plantations), limited range of isolation distances compared to area in the study system, and a possible time lag in the realization of isolation effects due to landscape change being fairly recent. In contrast, isolation rather than island area best predicted bat species richness and composition on Neotropical land-bridge islands (Meyer and Kalko 2008a) where fragments were surrounded by water.

The simplified dichotomous view of landscapes underlying IBT, albeit applicable in special cases (e.g., land-bridge islands), often fails to capture the influence that other land-cover types in the surrounding matrix can have and so may not be broadly applicable to most anthropogenically modified landscapes (Kupfer et al. 2006; Laurance 2008). After more than 40 years of research beyond the origins of IBT, it is now clear that for most animal taxa, including tropical bats, the majority of terrestrial habitat fragments are not islands in a homogeneous sea of inhospitable habitat. Indeed, island ecosystems support tropical bat biodiversity in fundamentally different ways compared to complex agricultural mosaic landscapes, the former adhering to IBT predictions of species loss, while countryside ecosystems are capable of maintaining high levels of species richness, evenness, and compositionally novel assemblages in human-made habitats (Mendenhall et al. 2014).

4.5.1.3 Responses to Landscape Structure

Fragmentation studies have increasingly shifted their focus from being largely patch-centered toward taking a broader landscape-scale approach, thus acknowledging the overriding importance of the matrix and the existence of gradients of habitat conditions and quality as crucial determinants of species responses (Kupfer et al. 2006; Driscoll et al. 2013; Cisneros et al. 2015). Such gradients are provided, for example, by mosaics of old-growth forest, successional habitat, and different forms of agriculture.

This paradigm shift is to some degree reflected within the more recent bat literature, as a growing number of studies have adopted matrix-inclusive approaches to studying fragmentation, although overall the number of studies is still small. In the broader literature, empirical evidence suggests widespread negative effects of habitat loss on many taxa (i.e., reduced abundance or density), whereas the effects of fragmentation per se are generally much weaker and may vary strongly in magnitude and direction of response (Fahrig 2003). In agreement with this, forest cover is a better predictor of bat assemblage characteristics (species richness or composition) than are measures of landscape configuration in Neotropical land-bridge island systems (Meyer and Kalko 2008a; Henry et al. 2010). On the other hand, consistent responses to landscape composition or configuration at the assemblage level were harder to identify in studies conducted in fragmented Neotropical rain forest landscapes in which the matrix was a mix of anthropogenic land uses (Gorresen and Willig 2004; Klingbeil and Willig 2009, 2010; Cisneros et al. 2015). A difficulty facing bat fragmentation studies is that responses tend to be highly species specific, which is often overlooked by diversity metrics applied at the assemblage level (Klingbeil and Willig 2009). This might be more important in low-contrast systems, in which the quality of matrix habitats likely mitigates some of the negative effects of fragmentation on biological communities.

At the population level, available evidence suggests that tropical bats respond in complex ways to landscape composition (i.e., the amount of suitable habitat available across the patch types represented in the landscape) and configuration (Gorresen and Willig 2004; Henry et al. 2007b; Klingbeil and Willig 2009, 2010). For instance, Klingbeil and Willig (2009, 2010) found that, apart from being scale dependent (see Sect. 4.5.1.4), abundance responses by phyllostomid bats to landscape structure in the Amazon were highly species and ensemble specific, and differed between seasons. In the dry season, abundances of frugivores responded primarily to changes in forest cover (i.e., landscape composition), whereas configurational metrics elicited the strongest response in the wet season. Gleaning animalivores showed the opposite pattern, responding to landscape configuration in the dry season and to landscape composition in the wet season. Such divergent responses suggest an important role of spatiotemporal variation in the abundance and diversity of food resources (Klingbeil and Willig 2010; Cisneros et al. 2015). Together with seasonal differences in time and energy budgets linked to reproduction, these will affect species' foraging and movement behavior, and could lead to seasonal shifts in diet composition (Durant et al. 2013; Cisneros et al. 2015). Such links remain little explored, yet future research in this regard may prove highly informative.

4.5.1.4 Spatial and Temporal Scale Dependence in Responses to Fragmentation

The scale at which bat species perceive their environment in fragmented landscapes is likely influenced by spatiotemporal variation in the distribution of resources, as well as by species-specific differences in ecological traits such as diet, wing morphology, and movement behavior. For example, in a low-contrast fragmented system in Malaysia, the provision of large cave systems in the landscape provided clear population subsidies for cave-roosting bats, but also potentially masked the impact of forest fragmentation on this ensemble (Struebig et al. 2009). Consequently, single-scale assessments may be inadequate for capturing the complex interactions between species' ecology and landscape patterns (Gorresen and Willig 2004). While there is accumulating evidence of the diverse ways by which tropical bats respond to landscape structure, equally important is the increased recognition that the detection of such responses is also sensitive to the spatial scale at which the system is examined (Gorresen et al. 2005).

Recent studies provide evidence for widespread scale dependence in associations between landscape metrics and bat responses at the assemblage, population, ensemble, and species levels (Gorresen and Willig 2004; Meyer and Kalko 2008a; Pinto and Keitt 2008; Klingbeil and Willig 2009, 2010; Henry et al. 2010; Cisneros et al. 2015). Pinto and Keitt (2008) quantified forest cover at a range of scales (buffers with radii from 50 to 2000 m) and found positive associations with bat abundance, whereby the scale that elicited the strongest response was species specific. Differential species responses to forest cover in this case were best explained by interspecific variation in diet, body size, and home range size. Similarly, multiple species- and ensemble-specific abundance responses of phyllostomid bats to landscape characteristics at multiple focal scales (buffers with 1, 3, and 5 km radii) have been reported from moderately fragmented, lowland Amazonian forest (Klingbeil and Willig 2009) and highly fragmented Atlantic forest in Paraguay (Gorresen and Willig 2004). In both studies, species were demonstrated to interact with their environment simultaneously at a range of spatial scales. In the Amazon, a change in the focal scale of response occurred between dry and wet seasons, a finding which is likely linked to seasonal differences in food abundance and diversity as well as energetic constraints associated with reproduction (Klingbeil and Willig 2010; Cisneros et al. 2015). Scale dependence in response patterns has also been observed in landscapes with an aquatic matrix (Meyer and Kalko 2008a; Henry et al. 2010), suggesting that scale effects are ubiquitous and operate in fragmented landscapes across a broad range of matrix types.

Overall, such findings emphasize that multiscale approaches to determining the effects of landscape structure on tropical bats are essential. In agreement with recent findings for tropical birds (Banks-Leite et al. 2013), the available evidence suggests, however, that the extremely idiosyncratic responses of tropical bats to landscape structure make it difficult to identify any particular landscape predictor or spatial scale that performs best at predicting responses at the assemblage level.

Despite the general importance of a landscape-level perspective in the study of habitat fragmentation, patch characteristics remain important for patch-dependent species (Driscoll et al. 2013). However, fragmentation studies on tropical bats that have jointly assessed the relative contribution of patch- and landscape-scale variables for explaining response patterns are scarce. Meyer and Kalko (2008a) found that the relative importance of local- versus landscape-scale characteristics in explaining species richness and compositional patterns of phyllostomids on Panamanian land-bridge islands varied with spatial scale. At the patch scale, isolation distance from the mainland was the strongest predictor, whereas the proportion of forest cover in the surrounding landscape was the most prominent descriptor explaining variation in assemblage attributes at larger scales.

Although the importance of spatial scale and spatial variation in matrix quality have received some attention in the bat fragmentation literature, we know little about how species responses to fragmentation vary over time or how they are mediated by changes to the matrix. Across many human-modified landscapes in the tropics, secondary forest regrowth may reclaim once deforested land, for instance in response to the abandonment of agriculturally unproductive areas (Bobrowiec and Gribel 2010; Chazdon 2014). Matrix recovery following disturbance can alter responses of fragment biota that may be driven by temporal changes in resource availability and of permeability of the matrix to dispersal (Bissonette and Storch 2007; Driscoll et al. 2013). In this context, research at the Biological Dynamics of Forest Fragments Project in the Brazilian Amazon indicates strong divergence in phyllostomid bat assemblage structure, high levels of species turnover, and marked reorganization in the rank order of the most abundant species in response to changes in matrix quality and composition over 15 years (Meyer et al., unpublished data).

Time lags in the manifestation of species responses to fragmentation are ubiquitous and constitute an important temporal aspect to consider when studying fragmentation impacts (Ewers and Didham 2006a; Bissonette and Storch 2007), but so far have been rarely investigated in tropical bat studies. Notable exceptions are a series of studies conducted in the St. Eugène land-bridge island system in French Guiana, in which fragmentation effects prior to, and for several years after, fragmentation provided clear evidence for time lags in species loss (Cosson et al. 1999; Pons and Cosson 2002; Henry et al. 2010). These time lags occurred gradually over the course of ca. 10 years.

Future assessments of tropical bat responses to fragmentation (and other types of anthropogenic disturbance) should therefore address not only the spatial but also the temporal dimension of human impacts. This is particularly notable as long-term studies in intact habitats reveal tropical bat assemblages to be highly dynamic in space and time (Pech-Canche et al. 2011; Kingston 2013).

4.5.1.5 Edge Effects

Recent reviews concur that edge effects critically affect biodiversity in habitat fragments (Ewers and Didham 2006a; Fischer and Lindenmayer 2007; Laurance et al. 2011). However, responses of tropical bats to habitat edges remain understudied, particularly in the Paleotropics. Current evidence from the Neotropics suggests that responses vary according to matrix contrast and land-use history, and are ensemble and species specific.

Several studies have modeled bat responses in relation to the amount and complexity of edge habitat, revealing that some tropical bats are sensitive to habitat edges (Gorresen and Willig 2004; Meyer and Kalko 2008a; Klingbeil and Willig 2009, 2010; Henry et al. 2010). While significant associations between species richness or composition with edge density have been found in fragmented systems with a water matrix (Meyer and Kalko 2008a), studies conducted in a low-contrast landscape did not detect significant edge-related responses at the assemblage level (Gorresen and Willig 2004; Klingbeil and Willig 2009, 2010). This again underlines the importance of matrix contrast in affecting species' edge sensitivity and also shows that, at least in landscapes with low-contrast edges, composite community measures such as species richness may fail to capture edge responses that may otherwise be evident at the species or ensemble level (Klingbeil and Willig 2009). At the population level, abundances of six frugivorous and gleaning animalivorous phyllostomid bat species in the Peruvian Amazon were positively related to edge density, whereby responses varied depending on spatial scale (Klingbeil and Willig 2009) and season (Klingbeil and Willig 2010). In contrast, in fragmented Atlantic forest, two frugivorous species exhibited negative responses to edge density (Gorresen and Willig 2004). The discrepancy in the direction of response may be explained by differences in the prevailing patterns of land conversion (small- vs. large-scale deforestation). A strong negative response of gleaning animalivores to edge cover was also found by Henry et al. (2010) in a land-bridge island system in French Guiana.

These studies indicate the sensitivity of phyllostomid bats to edges driven by changes in landscape configuration. However, quantifying the strength of edge effects requires explicit consideration of two distinct aspects: edge extent and edge magnitude. Edge extent is the distance over which a change in the response variable can be detected, and edge magnitude is the amplitude of the effect (Harper et al. 2005; Ewers and Didham 2006b). The few studies that have examined the magnitude of edge effects on tropical bats by comparing interior sites of large, mature forest stands and forest edges reported declines in phyllostomid richness, in landscape matrices of high (water; Meyer and Kalko 2008a) and low structural contrast (secondary forest and shade cacao plantations; Faria 2006). The pattern of reduced species richness at edges in the low-contrast system was mainly attributable to the decline of gleaning animalivorous species (Faria 2006; Pardini et al. 2009). Even though species composition did not significantly change between forest edge and interior, Meyer and Kalko (2008a) found that gleaning animalivorous bats exhibited a strong negative numerical response toward edges. In fact, edge sensitivity was identified as the species trait that best explained species vulnerability to fragmentation (Meyer et al. 2008). Similar to phyllostomids, aerial insectivorous bats in the same land-bridge island system had significantly lower species richness at edges compared to interiors. The two functional groups of narrow-space foragers and open-space bats responded differently to forest edges. Open-space foragers had higher abundance counts at edges, whereas those of forest species were not significantly altered (Estrada Villegas et al. 2010). Comparing general bat activity, Estrada et al. (2004) did not detect significant differences between continuous forest interiors and forest–pasture edges.

Only one study to date has tried to quantify the distance of edge influence for tropical bats. Delaval and Charles-Dominique (2006) captured phyllostomid bats along 3-km transects perpendicular to the edges of a road traversing primary forest in French Guiana. Capture rates along the transects were more than seven times higher than those at a control site, 150 km inside the primary forest block. Moreover, along the transects abundances decreased with increasing distance from the road edge, a pattern attributable to the proliferation of opportunistic frugivores such as *Carollia perspicillata* and *Artibeus jamaicensis* that exploit abundant food resources provided by young regrowth along road margins. Species richness decreased significantly with distance from the road edge, probably related to an influx of species from the open habitat into the edges. Species richness at edges was, however, not significantly greater than that in the control site that harbored seven species not present at road edges or along transects. Differences in rank abundance patterns between transects and control site provided further evidence that even narrow road clearings can alter bat assemblage structure over distances of at least 3 km into forest interiors.

Key research needs:

- Studies that try to disentangle the relative importance of habitat amount and habitat configuration in shaping species responses, in particular studies that identify portions of the gradient in habitat amount within which the effects of spatial arrangement become important, i.e., explicit tests of the "habitat threshold hypothesis" (Fahrig 2003).
- Research that addresses the relative tolerance of different species to changes in habitat configuration (see Villard and Metzger 2014).
- Studies that jointly assess the relative contribution of patch- and landscape-scale variables to explaining response patterns.
- Long-term investigations that address the effects of matrix transformation on bat species responses over time.
- More studies that quantify edge effects in terms of both magnitude and extent.
- Further research investigating how consistently species respond to habitat edges across a broad range of edge types to identify ecological traits correlated with and potentially driving edge sensitivity (Ries and Sisk 2010).
- Studies that try to disentangle edge and area effects (Fletcher et al. 2007; Banks-Leite et al. 2010).

4.5.2 Logging

Rain forests are selectively logged at 20 times the rate at which they are cleared (Asner et al. 2009), and large expanses (403 million ha) are officially designated for timber extraction (Blaser et al. 2011). Selective logging exposes vast areas to potentially detrimental edge effects (Broadbent et al. 2008) and may often be the precursor to complete deforestation (Asner et al. 2006). Yet, the impacts of selective logging on biodiversity depend critically on the harvest intensity (Asner et al. 2013; Burivalova et al. 2014) as well as the extraction techniques (Bicknell et al. 2014). Selective harvesting methods range from large-scale conventional extraction that can cause substantial loss in canopy cover and associated mortality of non-harvested trees, to reduced-impact logging (RIL), in which collateral damage is reduced as a result of improved planning and control of harvesting activities (Putz et al. 2008; Asner et al. 2013).

Recent meta-analyses indicate that selectively logged forests can retain a large proportion of the diversity of old-growth forest for a variety of taxa (Gibson et al. 2011; Putz et al. 2012) and the available evidence, though scant due to the low number of studies (Fig. 4.3a), largely supports this notion for tropical bats (Bicknell et al. 2014). At the assemblage level, selective logging appears to have little or no effect on bat species richness in the Neotropics (Ochoa 2000; Clarke et al. 2005a, b; Castro-Arellano et al. 2007). In contrast, compositional or structural differences between bat assemblages in logged and unlogged sites are more common, which suggests that if forests are unable to recover from logging disturbance, species losses may be detected in the long term (i.e., similar to time lags for fragmentation effects, see Sect. 4.5.1). Structural differences between bat assemblages in unlogged and logged forests are evident from changes in the proportional abundance of species within ensembles (Clarke et al. 2005a, b; Peters et al. 2006) and shifts in species rank distributions and dominance (Castro-Arellano et al. 2007). A consistent pattern emerging from Neotropical studies is that, similar to habitat fragmentation (see Sect. 4.5.1), selective logging appears to adversely affect the abundance of gleaning animalivorous phyllostomids, whereas frugivorous and nectarivorous species tend to increase in abundance (Ochoa 2000; Clarke et al. 2005a, b; Peters et al. 2006; Presley et al. 2008).

In a study in Trinidad, Clarke et al. (2005a) found that the magnitude of change in species composition is linked to the intensity of timber harvesting. Comparing a continuous logging system with few harvest controls (open range [OR] system) to a polycyclic, selective system that incorporated stricter controls on felling (periodic block [PB] system), the study demonstrated that PB-managed sites resembled undisturbed primary forest much more closely in bat species composition and abundance than did OR forest. Despite structural changes associated with PB management, bat assemblages in such well-managed forest stands had great potential for recovery to near predisturbance levels (Clarke et al. 2005b). The number of years post-logging was positively correlated with the number and abundance of species of gleaning animalivores but not frugivores, whereas the proportional abundance of the dominant frugivore decreased with forest recovery. Together, these findings suggest that PB or similar low-intensity selective management systems may be compatible with the conservation of bat diversity. Unfortunately, similar studies that evaluate responses of tropical bats to different management systems or across a series of logged sites of different ages within the same general study landscape are lacking.

Short-term population-level responses of phyllostomid bats to RIL in Amazonia were idiosyncratic (Castro-Arellano et al. 2007) and RIL sites had reduced species richness, linked to the local absence of rare species from logged forest, whereas the populations of common species remained unaffected (Presley et al. 2008). As argued by Presley et al. (2008), landscape context may be important in mediating the effects of RIL on bats, and for this harvesting practice to be sustainable, it may be essential that RIL blocks be located in close proximity to undisturbed forest to facilitate rescue effects that can mitigate the negative impacts of RIL on rare species. Furthermore, due to the short post-harvest interval (<42 months) in both studies, the observed responses may be short term (Castro-Arellano et al. 2007; Presley et al. 2008), stressing the necessity for longer-term evaluations of logging impacts.

In the only logging effect study on African bats, Monadjem et al. (2010), using acoustic sampling, found no significant differences in activity levels between primary and logged forests in Uganda for the insectivorous *Neoromicia nana*. Elsewhere in the Paleotropics, early studies reported higher species richness, diversity, and abundance in unlogged compared to selectively logged forest in Malaysia (Zubaid 1993) and profound changes in species composition due to logging in Sumatra (Danielsen and Heegaard 1995). However, in addition to having small sample sizes, these studies employed only mist nets, which are ineffective at capturing the numerous insectivorous species that dominate Paleotropical bat assemblages (Kingston 2013). Conclusions based on these studies alone should therefore be interpreted with caution. More recent studies in Southeast Asia have employed larger sampling effort and harp traps, which are adequate for sampling forest interior insectivores. In peninsular Malaysia, a comparison of forest reserves and adjacent logged-over forests >30 years post-extraction showed little overall difference in assemblage composition (Christine et al. 2013). In nearly all site comparisons, species richness and abundances were higher in logged forest. However, certain tree- or foliage-roosting species were only captured inside forest reserves, suggesting that forest reserves embedded in a matrix of production forest could play an important role as reservoirs to restock logged forest and to maintain populations of disturbance-sensitive species (Christine et al. 2013).

Logging effects may multiply spatially and temporally as a result of multiple harvesting cycles (Lindenmayer and Laurance 2012). However, only recently have researchers examined the impacts of multiple rounds of extraction. One such study examined bat assemblages on Borneo across a disturbance gradient ranging from old-growth to twice-logged to repeatedly logged forest (Struebig et al. 2013). Logging had little effect on bat species richness, even in heavily degraded forest that had been logged multiple times, corroborating research on other taxa in the region (Edwards et al. 2011). Changes in insectivorous bat assemblage structure and abundance between old-growth and repeatedly logged forest were nonetheless evident and degraded sites that were characterized by a low, open canopy harbored a depauperate bat fauna. Canopy height was an important determinant of assemblage change across the disturbance gradient, as was the availability of tree cavities for forest-roosting taxa. By quantifying microhabitat over the gradient, the study revealed that post-logging recovery of assemblages could be enhanced via restoration investments in canopy cover and tree cavity availability. Moreover, cave-dwelling hipposiderid and rhinolophid bats were less abundant in repeatedly logged sites, in line with findings from a study in Vietnamese karst forests in which these taxa were also less abundant in logged than in primary forest (Furey et al. 2010).

A key theme emerging from the recent logging effect literature is the potential confounding issue of spatial pseudoreplication in study design, a problem whereby study sites in continuous forest stands are inappropriately treated as independent replicates (Ramage et al. 2013). The most effective way to overcome these problems is to sample the same forest sites before and after logging. The only bat-logging study to have implemented such a robust Before–After–Control–Impact

(BACI) design to date was undertaken in RIL forests in Guyana (Bicknell et al. 2015). Differences in bat assemblage structure before and after logging were relatively weak and varied substantially across study sites. Although three species were classified as indicators of disturbed or undisturbed forest, there were no clear changes in bat assemblages at control sites, indicating that overall responses could not be reliably attributed to logging.

In conclusion, given the paucity of studies available, it remains difficult to ascertain definitive responses of tropical bats to logging. The short-term effects appear to be relatively benign, especially in low-intensity extraction systems. Reported effects vary, largely owing to differences among studies with regard to the type of forest management system, and spatial and temporal variability in disturbance attributes, including time post-harvest.

Key research needs:

- Studies comparing bat responses between different forest management systems and across a range of spatial and temporal scales.
- More studies implementing BACI designs, as exemplified by Bicknell et al. (2015).
- Integration of logging disturbance into studies of forest fragmentation in order to distinguish true fragmentation responses from those of forest degradation.

4.5.3 Secondary Forests and Succession

The future of tropical biodiversity will critically depend on our ability to manage the large expanses of regenerating secondary forests (Chazdon et al. 2009; Chazdon 2014) that account for approximately half of the remaining area of tropical moist forests (Asner et al. 2009). Studies that have examined the conservation value of secondary forests for tropical bats are largely in line with assessments with regard to other tropical taxa (Barlow et al. 2007; Gardner et al. 2010) by suggesting that regenerating forests act as important repositories of bat biodiversity. Secondary forests are effective at conserving a subset of primary forest bat species richness (Louzada et al. 2010), but usually host assemblages that differ in structure and composition from those in mature forest (Faria 2006; Barlow et al. 2007).

Secondary successional vegetation in Neotropical humid forests represents important habitat for many frugivorous and nectarivorous phyllostomids (e.g., *Carollia* spp., *Sturnira* spp., *Glossophaga* spp.). These taxa become numerically dominant in secondary forests representing early to intermediate stages (Brosset et al. 1996; Castro-Luna et al. 2007a, b; Willig et al. 2007; de la Peña-Cuéllar et al. 2012; Vleut et al. 2013). This pattern is likely attributable to an increase in the abundance, diversity, or quality of fruit and flower resources associated with early successional vegetation and emphasizes the fundamental importance of phyllostomid bats in the regeneration of tropical forests (Muscarella and Fleming 2007). In contrast, the abundance of frugivores was not elevated in earlier successional stages of tropical dry forest in Mexico (Avila-Cabadilla et al. 2009). This likely reflects distinct differences in the composition of early successional vegetation, and consequently resource scarcity, in tropical dry compared to wet forests. Pinto and Keitt (2008) found that the abundances of *Sturnira* spp. were positively associated with secondary forest cover, reflecting the species' preference for early successional vegetation. Conversely, *Carollia* spp. responded to forest cover that included both primary and secondary forests, implying that habitat connectivity may be more important than successional stage for populations in this genus. As with logged forests, these findings suggest species-specific responses to secondary vegetation linked to interspecific differences in diet, home range size, and body size. Contrary to the flexible responses observed for many frugivores and nectarivores, a large body of empirical evidence indicates that gleaning animalivorous phyllostomines are sensitive to forest degradation, as they are absent or occur at low abundance in secondary regrowth (Fenton et al. 1992; Brosset et al. 1996; Medellín et al. 2000; Faria 2006; Castro-Luna et al. 2007a, b; Mancina et al. 2007; Willig et al. 2007; Pardini et al. 2009; Bobrowiec and Gribel 2010; de la Peña-Cuéllar et al. 2012; Vleut et al. 2012, 2013).

Some studies have detected a clear pattern of species richness increasing across successional gradients (Avila-Cabadilla et al. 2009; de la Peña-Cuéllar et al. 2012), but this pattern has not been evident in others (Castro-Luna et al. 2007a; Mancina et al. 2007). Nonetheless, for Neotropical wet and dry forests, floristically more diverse and structurally more complex habitats harbor greater taxonomic and functional richness than do early or intermediate stages of succession. Here, vegetation complexity appears to be an important factor shaping assemblage composition (Medellín et al. 2000; Avila-Cabadilla et al. 2009; Bobrowiec and Gribel 2010; Avila-Cabadilla et al. 2012; de la Peña-Cuéllar et al. 2012). Late successional forest stands often host many bat species not found in earlier stages, in particular rare taxa, and through succession, the number of species and ensembles increases for frugivorous, nectarivorous, and gleaning animalivorous taxa (Avila-Cabadilla et al. 2009, 2012; de la Peña-Cuéllar et al. 2012). In tropical wet forest in Mexico, abundances of the most common bat species were associated positively or negatively with variation in canopy cover across successional stages, rather than with landscape attributes (Castro-Luna et al. 2007a). In contrast, a study in Mexican tropical dry forest found evidence for an important role of local (vegetation complexity) and landscape attributes (area and cover of different vegetation types) as determinants of variation in abundance, which were ensemble specific and scale dependent (Avila-Cabadilla et al. 2012). In Central Amazonia, gleaning animalivorous phyllostomid bats exhibited greater abundance and richness in *Cecropia*-dominated regrowth, whereas stenodermatine frugivores were more abundant in abandoned pastures and *Vismia*-dominated regrowth, demonstrating that different successional trajectories result from differences in land-use history (cutting versus cutting and burning) that lead to distinct differences in bat assemblage composition (Bobrowiec and Gribel 2010).

Despite the recovery potential of Neotropical bat assemblages during succession, the conservation value of secondary forests for bats critically hinges on landscape context and is maximized in mosaic landscapes in which patches of forest at different successional stages are located close to old-growth forest (Bobrowiec and Gribel 2010; Vleut et al. 2012).

Key research needs:

- Comprehensive assessments of the conservation value of secondary forests for bats in the Paleotropics, which are essentially lacking (but see Fukuda et al. 2009).
- Studies addressing the recovery potential of Paleotropical bat assemblages during secondary succession.

4.5.4 Agroforestry Systems

As agriculture and associated biodiversity losses continue to rise across the tropics, agroforestry systems have been advocated as biodiversity-friendly alternatives, capable of conserving biodiversity while enhancing rural livelihoods (Perfecto and Vandermeer 2008; Clough et al. 2011). Coffee (*Coffea arabica*, *Coffea canephora*) and cacao (*Theobroma cacao*) are the principal cash crops of many tropical countries (Donald 2004; Tscharntke et al. 2011) and are the primary examples in the bat literature (but see bat inventories of Sumatran rubber agroforests in Prasetyo et al. 2011). In traditional coffee and cacao agroforestry, these crops are commonly grown under a stratified canopy layer of a more or less diverse range of native shade tree species. Much of their potential for conservation derives from the fact that such traditional agroforestry systems resemble natural forest habitat in many structural aspects (Perfecto and Vandermeer 2008).

Empirical studies that have assessed the value of agroforests for tropical bats to date come almost exclusively from the Neotropics (Fig. 4.3a). Pineda et al. (2005) compared the bat fauna of Mexican cloud forest fragments and shade coffee plantations and found that both habitats had very similar species richness and composition, although there were changes in the species' rank order between habitats. Large frugivorous phyllostomids (*Artibeus* spp.) reached higher abundance in shade coffee than in the natural habitat, possibly as a result of increased food availability due to the cultivation of important fruit tree species alongside coffee, a management strategy that also favored the abundance and richness of fruit- and nectar-eating bats in coffee plantations elsewhere in Mexico (Castro-Luna and Galindo-González 2012a). Contrasting abundance responses for large *Artibeus* were found in another study in Mexico (Saldaña-Vázquez et al. 2010). Here, shade coffee plantations and disturbed cloud forest fragments did not differ in abundance levels and also had similar availability of food plants. On the other hand, abundances of *Sturnira* spp. were higher in forest fragments, probably linked to a decline in food resources for these small frugivores in the coffee plantations. This reduction in resources resulted from the pruning of understory vegetation and was reinforced by the effects of a resource-poor pasture matrix surrounding the forest fragments.

Williams-Guillén and Perfecto (2010, 2011) investigated how bat diversity patterns in coffee agroforestry change with increasing management intensity. Phyllostomid bats maintained similar richness across management regimes, but showed significant declines in abundance across the intensification gradient, from forest fragments through low-management shade polyculture and commercial polyculture to high-management coffee monocultures (Williams-Guillén and Perfecto 2010). Compositional similarity differed significantly between fragments and coffee plantations of all management intensities, and between high-shade polycultures and low-shade monocultures. The proportions of large frugivores increased with management intensity, in line with Pineda et al.'s (2005) findings. Conversely, those of nectarivorous and gleaning animalivorous bats decreased, the latter being absent from intensively managed coffee monocultures. Both forest fragments and the diverse and structurally complex shade polyculture systems may provide adequate roosting and food resources to sustain high levels of phyllostomid diversity. This contrasts strongly with the situation in low-shade monocultures, which offer reduced feeding and roosting opportunities, and may consequently serve more as commuting than foraging habitat. This was also suggested in a study on non-phyllostomid aerial insectivorous bats in the same landscape, which reported reduced foraging activity in the most intensively managed monocultures (Williams-Guillén and Perfecto 2011). Both of the functional groups of aerial insectivores, forest and open-space foragers, had similar species richness across habitat types. The two groups, however, showed opposite responses with respect to activity levels and compositional similarity. Forest-adapted species differed in ensemble composition across the management gradient and responded negatively to agricultural intensification in terms of activity. For open-space foragers, reductions in shade tree diversity and cover did not manifest in compositional changes, but were associated with increased levels of overall activity, albeit not feeding activity.

Collectively, these studies demonstrate the high conservation value of structurally diverse shade coffee for bats, but less so of intensively managed systems. The former constitutes a permeable high-quality matrix, while intensive coffee monocultures represent poor matrix habitat (Numa et al. 2005). Landscape context, in particular the dominant matrix type, is an important modulator of how bat assemblages respond to agroforest management intensity. Forest fragments harbored significantly greater phyllostomid richness than did management systems when the landscape matrix was dominated by sun coffee, whereas richness was similar among habitats in a shade coffee matrix (Numa et al. 2005).

For cacao, studies show results similar to those for coffee, supporting the notion that traditional, structurally complex shade cacao plantations sustain high levels of bat diversity. Insights come from a series of studies conducted in the Atlantic forest region of Una, Brazil. Cacao agroforests in this region provide foraging and roosting habitat for members of all feeding ensembles, including forest-dependent gleaning animalivorous species (Pardini et al. 2009), primarily because of the structural complexity retained compared to intact forest (Faria et al. 2006). In fact, bat assemblages in shade cacao showed greater richness, diversity, and abundance than did those in nearby mature or secondary forest (Faria 2006; Faria and Baumgarten 2007; Pardini et al. 2009). However, shade cacao plantations per se may not provide adequate habitat conditions for forest-dwelling bats, as the proximity of shade cacao to forest remnants was a key determinant of species persistence. Bat assemblages in plantations isolated by more than 1 km from forest were characterized by low richness and diversity, with clear shifts in species dominance, suggesting a crucial role of native forest remnants as population sources (Faria and Baumgarten 2007). Isolating distance to forest was also an important factor influencing species richness and abundance in Mexican shade plantations (Estrada et al. 1993a). These plantations maintained diverse and structurally similar bat assemblages to those in remnants of native forest (Medellín et al. 2000; Estrada and Coates-Estrada 2001b). As for coffee (Numa et al. 2005), landscapes dominated by cacao agroforests and comprising reduced native forest cover may harbor impoverished bat assemblages (Faria et al. 2006; 2007), highlighting that landscape context generally plays a crucial role in determining bat species responses in tropical agroforestry landscapes, as it does for fragmented forest systems.

In conclusion, both coffee and cacao, when grown under a traditional shade regime, comprise a high-quality matrix that offers suitable conditions for maintaining diverse phyllostomid assemblages. These agroecosystems, in turn, benefit from pest control services provided by bats as has been shown for agroforests in the Neotropics (Williams-Guillén et al. 2008) and Southeast Asia (Maas et al. 2013) (see Chap. 6). Studies in cacao agroforestry at least in some cases entailed comparison between large tracts of mature forest and the agricultural system (Medellín et al. 2000; Faria 2006), but these important baseline data are lacking for studies in coffee agroforests.

Key research needs:

- Studies that assess response patterns for non-phyllostomid bats.
- Assessments of bat responses to cacao agroforestry intensification, especially in view of globally increasing levels of conversion of shade cacao systems into unshaded monocultures (Tscharntke et al. 2011).
- Linkages between levels of bat biodiversity and crop yields.

4.5.5 Tree Plantations

Given the extent to which forested land is being converted to tree plantations across much of the tropics (Gibbs et al. 2010), there have been surprisingly few studies investigating the value of these habitats for bats. Three systems dominate tree plantation mosaics in the tropics: fast-growing timbers for the paper/pulp industry (e.g., *Acacia*, *Eucalyptus*), rubber (*Hevea brasiliensis*), and, increasingly, oil palm (*Elaeis guineensis*).

In a multitaxon assessment in Brazil, Barlow et al. (2007) found similar numbers of bat species in *Eucalyptus* plantations and secondary forests recovering from burning, but both habitats supported much lower richness than did unlogged forests. Bat assemblages in plantations were nested subsets of those in forests; approximately 11 % of all species were shared between plantations and primary forest, 4 % were shared with secondary forest, and 39 % found in all habitats (Louzada et al. 2010). Nevertheless, three species (ca. 6 % of total) were captured exclusively in *Eucalyptus* plantations.

A study in Brazilian Cerrado found lower species richness, diversity, and evenness of bat assemblages in *Eucalyptus* monocultures than in fragments of native Cerrado vegetation (Pina et al. 2013). Gleaning animalivorous phyllostomid bats were not captured in plantation forests. An earlier comparative study in Sumatra documented a distinct shift in bat assemblage structure in rubber and oil palm plantations, which supported only 13–25 % of the bat species richness found in forest (Danielsen and Heegaard 1995). However, more recent surveys have revealed additional species utilizing rubber plantations, especially those grown as agroforests or close to forest areas (Prasetyo et al. 2011). These studies point to an adverse response by bats to plantation development in both the New and Old World tropics. However, the extent to which these findings reflect true bat declines versus sampling bias (i.e., difficulties in capturing bats in open plantation habitats) is open to question. Tree plantations present a much more open habitat compared to forests, but can provide canopy structure similar to that in forest. This may present difficulties for capturing bats in these habitats, particularly in the Paleotropics, where much of the insectivorous bat fauna can only be captured in harp traps. Bat surveys in Sumatra and Borneo have resulted in extremely low capture rates for insectivorous species in oil palm plantations using mist nets and harp traps (Fukuda et al. 2009; Syamsi 2013), a finding that could reflect differential capture success in closed versus open habitats as well as true differences between habitats. Acoustic surveys could potentially contribute additional information concerning bat activity and the structure of bat assemblages in these habitats. The first insights from the Old World come from southern Thailand, where Phommexay et al. (2011) sampled bats in forest and neighboring rubber plantations using bat detectors, mist nets, and harp traps. Although diversity and overall bat activity were much lower in plantations than in forests, differences between the two habitat types were not as severe as indicated by capture-based surveys. Acoustic sampling in plantations detected less than half the number of bat species found in forest and fewer bat passes. Although bat activity was clearly reduced in plantations, a substantial number of feeding buzzes were detected, suggesting that bats were still foraging in this modified habitat.

Key research needs:

- Further studies, particularly those using acoustic methods, to accurately assess the conservation value of tree plantations for tropical bats.

4.5.6 Agriculture and Residual Tree Cover

Agricultural encroachment and cropland expansion are key threats to biodiversity in tropical countries (Phalan et al. 2013). The dominant crop will determine the permeability of the agricultural matrix, the likelihood of species persistence, and ultimately whether sustainable configurations in human-modified landscapes emerge in which biodiversity conservation and food production can be reconciled (Melo et al. 2013).

Apart from several studies in agroforestry systems (see Sect. 4.5.4) and oil palm plantations (see Sect. 4.5.5), little research has examined responses of tropical bats to forest conversion into other agricultural land uses, or the value of residual vegetation in agricultural matrices (Fig. 4.3a). By far, most of the available evidence comes from studies in Mexico and Central American tropical wet and dry forests. These studies generally suggest that human-modified landscapes comprising a heterogeneous mosaic of different land- and tree-cover types can preserve species-rich bat assemblages (Estrada et al. 1993a, b, 2004; Medellín et al. 2000; Moreno and Halffter 2001; Estrada and Coates-Estrada 2002; Harvey et al. 2006; Medina et al. 2007; Barragan et al. 2010; Mendenhall et al. 2014). For instance, in a comparison of bat diversity in forest fragments, agricultural habitats, and live fences in Mexico, agricultural habitats contained 77 % of the species recorded, whereby species richness declined with increasing distance from forest fragments (Estrada et al. 1993a). Certain frugivorous species (e.g., *Carollia* spp., *Sturnira* spp.) may become dominant in agricultural areas, whereas phyllostomine species are adversely affected by agriculture (Medellín et al. 2000). A similar pattern was found by Willig et al. (2007) in lowland Amazonian rain forest in Peru. Here, half of the frugivorous and nectarivorous species that responded consistently to habitat conversion reached highest abundances in agricultural areas, a response probably linked to the ample food resources provided by these habitats. Due to the presence of rare species not captured in forest, species richness in disturbed agricultural and early successional habitats was high compared to that in mature forest. However, the long-term persistence of most species likely still depends on the availability of forest (Willig et al. 2007). Moreover, these findings relate to small-scale habitat conversion and may not be generalizable to landscapes characterized by large-scale deforestation.

Knowledge of the conservation value of agricultural habitats for bats in the Old World is scant (see Chap. 6). In a study in Fiji (Luskin 2010), foraging densities of the Pacific flying fox, *Pteropus tonganus*, an important seed disperser were four times higher in agricultural habitats than in remnants of dry forest, illustrating a strong preference for foraging on abundant food resources in farmland. Resource subsidies provided by farmland were responsible for sustaining high abundances of the species despite severe deforestation across the region. Roosting sites, however, were restricted to native forest fragments, highlighting their importance for population persistence. Agricultural habitats provided important resources for some species of pteropodid bats in Borneo, as evidenced by high capture rates in orchards relative to forest habitats (Fukuda et al. 2009). Fukuda et al. (2009) suggest that some pteropodids in Southeast Asian dipterocarp forests, which are characterized by a supra-annual flowering and fruiting pattern, may augment food resources by feeding on cultivated plants during non-flowering periods when food supply in the forest is scarce. However, other fruit bat species were restricted to forest, suggesting that the value of agricultural land is species specific. Sedlock et al. (2008) reported that fewer species persist in mixed agricultural habitat than in tall secondary forest in the Philippines. Nevertheless, 19 of 26 species were present in agro-pastoral areas. Results from studies in the Paleotropics are thus largely congruent with those from the Neotropics in suggesting that agricultural habitats harbor considerable bat diversity and provide important foraging habitat for some fruit bat species.

Linear landscape elements (corridors of residual vegetation such as live fences or strips of riparian forest) and scattered trees, commonly found in Neotropical countryside landscapes, may enhance functional connectivity (Villard and Metzger 2014), and studies indicate that bats extensively use them (Estrada and Coates-Estrada 2001a; Galindo-González and Sosa 2003; Estrada et al. 2004; Harvey et al. 2006; Medina et al. 2007; Barragan et al. 2010). For instance, in agricultural landscapes in Nicaragua, riparian forests and live fences harbor greater bat species richness and abundance than do secondary forest and pastures with low tree cover (Harvey et al. 2006; Medina et al. 2007). Riparian forests constitute favorable habitats for foraging and roosting, particularly in tropical dry forest ecosystems, where they often have higher tree diversity and food availability compared to other types of cover (Estrada and Coates-Estrada 2001a; Harvey et al. 2006). Live fences and riparian corridors facilitate movement by bats across fragmented agricultural landscapes and may effectively reduce isolation between remnant forest patches, which, in turn, enhances species persistence at the landscape level. Similar to live fences, isolated pasture trees provide food and roosting opportunities for bats and act as important stepping stones for bat movement (Galindo-González and Sosa 2003), suggesting that they can render agro-pastoral landscapes more hospitable to bats and consequently deserve attention in conservation strategies. In contrast, studies concur that pastures are low-quality habitat for bats, likely as a consequence of resource scarcity (food, roosts) and elevated predation pressure (Estrada et al. 1993a, b, 2004; Harvey et al. 2006; Griscom et al. 2007; Medina et al. 2007).

Key research needs:

- In-depth studies in the Old World tropics that assess bat responses across a range of agricultural habitat types and landscape settings.
- Assessments of the value of residual tree cover in agricultural matrices for Paleotropical bats, particularly in Africa.
- Research addressing the effects of large-scale, commercial agriculture (e.g., cultivation of soybean, corn, sugarcane), which plays an increasingly significant role in driving deforestation in some tropical regions such as the Amazon.

4.6 Genetic Consequences

Tropical taxa are generally underrepresented in landscape genetic studies (Storfer et al. 2010). Bats are no exception, as only few studies have assessed how they are affected by anthropogenic habitat loss and fragmentation at the genetic level (Fig. 4.3b). Meyer et al. (2009) studied populations of two Neotropical bats in fragments that were isolated by a water matrix and detected significant population differentiation that matched the species' relative mobility. In contrast to the more mobile canopy frugivore, *Uroderma bilobatum*, population subdivision in the understory frugivore, *C. perspicillata*, showed a significant effect of fragmentation and isolation by distance, as well as reduced genetic diversity on islands relative to mainland populations. Also employing mitochondrial DNA sequence data, Ripperger et al. (2013) documented small-scale genetic differentiation for another small understory frugivore, *Dermanura watsoni*, in fragments embedded in a matrix dominated by agriculture. Landscape connectivity as measured by the amount of suitable habitat surrounding forest patches was most strongly correlated with genetic variation when quantified within small-scale (400 m) landscape buffers, likely reflecting the reduced mobility of this species. Importantly, empirical levels of genetic diversity in fragments were best explained by past rather than present habitat conditions. Because anthropogenic habitat fragmentation is recent on evolutionary timescales, populations may not show immediate genetic responses to fragmentation, highlighting the importance of considering time lags in these scenarios.

In a microsatellite study of three codistributed insectivorous bat species in forest fragments in peninsular Malaysia, Struebig et al. (2011) observed area-related declines in genetic diversity in *Kerivoula papillosa*, the species that was most sensitive to fragmentation based on ecological characteristics (low vagility, low population density, tree-cavity-roosting habit). Based on the genetic-area relationship observed for *K. papillosa*, the authors estimated that preserving the genetic diversity of this species at levels similar to those of intact forest would require extensive areas (>10,000 ha), several times larger than necessary to maintain comparable levels of species richness. In view of the fact that most forest patches in heavily fragmented production landscapes across Southeast Asia are much smaller, it is evident that maintaining genetic diversity of the dozens of forest specialist species that exhibit trait combinations similar to those of *K. papillosa* constitutes a substantial conservation challenge (Struebig et al. 2011). Roosting ecology and social organization may generally be important predictors of genetic structuring in insectivorous Old World bats. Rossiter et al. (2012) found that less vagile, tree-roosting species exhibit reduced gene flow, even across continuous intact rain forest, compared to more wide-ranging colonial cave-roosting species, indicating that the former should be disproportionately affected by landscape-scale habitat fragmentation.

Only weak genetic population subdivision was demonstrated for *Artibeus lituratus*, an abundant, highly mobile, and generalist frugivore, in a study in fragmented Atlantic forest (McCulloch et al. 2013). High levels of contemporary population connectivity in an abundant and widespread seed disperser like *A. lituratus* may buffer numerous plant species in Neotropical forests that rely on dispersal services of this bat species to counterbalance the negative impacts of deforestation.

In summary, the available evidence suggests, both in the New and in the Old World tropics, and irrespective of fragment–matrix contrast, that some bat species may be vulnerable to genetic erosion as a result of small-scale habitat fragmentation. Further, studies indicate that susceptibility in this context is linked to individual species traits such as mobility or roosting habit.

Key research needs:

- Increasing research on a broader range of species with different ecological and life-history traits, ideally using high-resolution genetic markers such as microsatellites or single-nucleotide polymorphisms (SNPs).
- Studies that quantify the extent to which frugivorous and nectarivorous bat species are capable of maintaining gene flow among plants in fragmented tropical landscapes.

4.7 Behavioral Responses

In addition to the direct effects on diversity and abundances, species' responses to anthropogenic habitat modification and disturbance can manifest as behavioral changes, which may include disruptions to species' dispersal, movement, activity patterns, and interspecific interactions (Fischer and Lindenmayer 2007). Few studies so far have addressed these issues for tropical bats (Fig. 3.3b).

Although a number of studies have reported movement distances and space use for a variety of tropical bat species (not reviewed here), few have explicitly addressed these phenomena in anthropogenically modified landscapes. Mark–recapture and radiotracking studies in the Neotropics suggest that in areas where landscape connectivity is relatively high, bats may regularly traverse open areas between forest fragments or between fragments and continuous forest. Evidence for interhabitat movements comes from landscapes with agricultural matrices (Estrada et al. 1993a; Estrada and Coates-Estrada 2002; Bianconi et al. 2006; Medina et al. 2007; Mendes et al. 2009; Trevelin et al. 2013) or from those with a more inhospitable aquatic matrix (Albrecht et al. 2007; Meyer and Kalko 2008a). Recapture data from a study in a fragmented landscape in Malaysia also indicate long-distance between-habitat movements for some cave-roosting species (Struebig et al. 2008). Whether a species is able to move over fragmented landscapes may be linked to the species' foraging ecology (Albrecht et al. 2007; Henry et al. 2007b). Overall, these studies were fundamental in determining the general capacity of tropical bats to move across human-modified habitats. However, they provide mostly circumstantial evidence and cannot establish whether anthropogenic disturbance elicits direct behavioral responses in bats that manifest as changes in movement distances or patterns of space use. Better insights into how habitat modification influences movement behavior can be gained through detailed radiotracking or long-term banding studies that compare movement patterns for species with different autecologies. Such studies, although difficult and costly to implement, would ideally compare continuous forest with fragments or other disturbed habitats.

Studies that have assessed behavioral changes to habitat modification in terms of effects on temporal activity patterns have followed such a rigorous approach. Disturbance-related changes in resource abundance, diversity, or predictability can be assumed to potentially alter temporal activity of species that exploit those resources (Presley et al. 2009b). Presley et al. (2009a) found no interspecific differences in activity patterns of eight abundant frugivorous bats in primary lowland Amazonian rain forest. However, for five species, activity patterns differed between primary or secondary forest and agricultural habitats, whereby bats in larger agricultural areas exhibited reduced crepuscular activity compared to those in undisturbed forest. Elsewhere in Amazonia, Castro-Arellano et al. (2009) detected no differences in activity levels for nectarivores and gleaning animalivores in response to RIL. Conversely, understory frugivores (*Carollia* spp.) decreased activity at dusk. Another study found reduced activity by some frugivores in small forest clearings created by tree removal, although the overall effects of RIL on activity patterns of frugivores were negligible (Presley et al. 2009b). In all cases, the curtailment of activity in open areas at twilight or during periods of high lunar illumination was best explained by increased predation risk (Saldaña-Vázquez and Munguía-Rosas 2013). Habitat modification and disturbance may consequently influence energy budgets of bats as they have less time available for foraging, with possible negative repercussions for their ability to meet daily energy requirements.

Human disturbance may also affect roosting behavior and roost site selection. In fragmented rain forest in Mexico, Evelyn and Stiles (2003) found that both sexes of cavity-roosting *Sturnira lilium* selected large-diameter trees in mature forest stands, as did females of the foliage-roosting *Artibeus intermedius*, whereas males of the latter species roosted in secondary forest. These findings underscore that preferences in terms of roosting and foraging habitat are not necessarily correlated and point to the importance of preserving mature forest patches in human-dominated landscapes for meeting the roosting requirements of tree-cavity-roosting species.

Key research needs:

- More studies, particularly in the Paleotropics, that assess the extent to which human-driven habitat change affects bat behavior in terms of roosting and foraging ecology.
- Research that addresses how such behavioral changes translate into fitness consequences (e.g., in terms of survival, reproductive success, physiology) that may affect long-term population persistence.

4.8 Effects on Selected Species Interactions

In recent years, bats have moved to the forefront of public attention, mostly as a result of accumulating evidence that they comprise important reservoir hosts for numerous zoonotic viruses (e.g., lyssaviruses, SARS, Ebola) that may pose a serious health risk to humans (Calisher et al. 2006; Hayman et al. 2013, Chap. 10). Recent studies have highlighted the urgency of gaining a better understanding of how habitat loss, land-use change and disturbance and an associated increase in bat–human interactions may, for instance, accelerate viral spillover (Peel et al. 2013). However, few studies to date have explored to what extent these stressors influence patterns of parasite and disease prevalence and transmission, as well as physiological stress responses in bats (Fig. 4.3b).

Cottontail et al. (2009) found that trypanosome prevalence in *A. jamaicensis* was significantly higher in fragmented sites than in continuous forest, linked to a loss of bat species richness and fragmentation-related changes in vegetation cover that may favor disease transmission. The negative relationship between trypanosome prevalence and bat species richness reflects the "dilution effect," i.e., a situation in which high host species richness reduces parasite transmission if vectors feed on multiple host species that vary in their ability to contract, amplify, or transmit the pathogen (Ostfeld and Keesing 2012). In contrast, prevalence of hemoparasitic nematodes (*Litomosoides* spp.) showed no significant difference among habitats, probably as a result of greater host specificity (Cottontail et al. 2009). In another study, fragmentation affected the physiological condition of *A. obscurus*, as evidenced by elevated hematocrit levels in forest fragments versus continuous forest, even though similar abundances in both habitats indicated a high degree of fragmentation tolerance. The opposite pattern was documented for *A. jamaicensis*, suggesting that abundance may in many instances be misleading as a metric of fragmentation sensitivity (Henry et al. 2007a). Pilosof et al. (2012) found a significant effect of anthropogenic disturbance on the abundance of ectoparasitic bat flies in three of four widespread Neotropical host bat species, whereby the direction of response differed among species. Species-specific roosting habits likely play a key role in mediating the effects of disturbance on parasite transmission. A study in Mexico found significantly lower prevalence of antirabic antibodies in non-hematophagous bats in disturbed agricultural areas (22.7 %) compared to relatively undisturbed dry forest sites (51.9 %), a pattern which may arise because of more frequent interspecies encounters in the undisturbed habitat (Salas-Rojas et al. 2004).

The important role of animalivorous, frugivorous, and nectarivorous bats in arthropod suppression, seed dispersal, and pollination in tropical ecosystems is widely acknowledged (Kunz et al. 2011). The degree to which such interactions are susceptible to habitat modification and disturbance is generally better understood for seed dispersal than for pollination or arthropod suppression. Mostly using fecal analysis or seed traps, numerous studies in various human-modified landscapes across the Neotropics have documented the quantity and diversity of seeds carried by bats (mostly *Carollia* spp., *Sturnira* spp., *Artibeus* spp.) into a diverse range of disturbed habitats including pastures, agricultural areas, coffee plantations, and secondary forests (Medellín and Gaona 1999; Galindo-González et al. 2000; García et al. 2000; Aguiar and Marinho-Filho 2007; Hanson et al. 2007; Wieland et al. 2011; Castro-Luna and Galindo-González 2012b; García-Estrada et al. 2012; García-Morales et al. 2012; Gorchov et al. 2013). Voigt et al. (2012) showed that bats of the genus *Carollia* were likely to carry seeds from mid-successional forest into adjacent primary forest, suggesting that directionality of seed transfer between disturbed areas and undisturbed forest may change with forest recovery. Isolated fig trees in abandoned pastures are attractive for many frugivores and may function as regeneration nuclei that effectively facilitate forest recovery (Guevara et al. 2004). Overall, these studies provide little evidence for major disruptions of seed dispersal mutualisms in response to habitat fragmentation and disturbance, although minor effects were detectable. For instance, small frugivorous bats disperse fewer large seeds in small, disturbed compared to large, undisturbed forest patches (Melo et al. 2009), suggesting a negative impact of disturbance on the dispersal of larger-seeded trees. Although Old World fruit bats in some areas may disperse seeds of early successional species (Hamann and Curio 1999), seed input into deforested or degraded areas tends to be low in human-modified landscapes in the Paleotropics (Duncan and Chapman 1999; Ingle 2003). Pteropodids generally play a much less significant role as dispersers of early successional plants compared to phyllostomids, but are important dispersers of late successional canopy trees (Muscarella and Fleming 2007). How habitat modification affects seed dispersal of large-seeded canopy trees by pteropodid fruit bats in Paleotropical forests requires further detailed study.

Research in fragmented Central American dry forest ecosystems found a decline in flower visitation rates, number of pollen grains deposited, and fruit set of certain bombacaceous tree species, suggesting that habitat disruption can impair the pollination services of nectarivorous phyllostomids, with negative consequences for plant reproductive success (Stoner et al. 2002; Quesada et al. 2003). However, effects were dependent on plant species (Quesada et al. 2004), making general predictions regarding the effects of habitat modification on the disruption of bat pollination difficult. Through its influence on bat foraging behavior, habitat disturbance may also limit pollen exchange between trees, leading to higher progeny relatedness in isolated trees relative to those in undisturbed forest (Quesada et al. 2013). In a fragmented landscape in tropical Australia, common blossom bats (*Syconycteris australis*) were high-quality pollinators of the rain forest tree *Syzygium cormiflorum*, as inferred based on pollen loads, visitation rates, and movement patterns (Law and Lean 1999). Nectarivorous bats often attain higher abundance in response to anthropogenic disturbance (see Sect. 4.5), suggesting that provisioning of pollination services may potentially be resistant and resilient to environmental perturbation.

Key research needs:

- Detailed studies that address the causal links between human-induced landscape change and bat physiological and immune responses, as well as disease susceptibility.
- Studies, particularly in the Paleotropics, that document the full dispersal cycle—from seed deposition through germination, seedling establishment, and recruitment—and how it is affected by habitat alteration.
- Further studies across a range of pollinator and plant species, as well as fragmented landscapes with different degrees of connectivity, to directly relate behavior and movement of pollinators with reproductive success and gene flow of trees.
- Studies that address the extent to which arthropod suppression services are affected by more intensive forms of habitat alteration and disturbance such as those associated with secondary forests, tree plantations, or cropland (see Wanger et al. 2014).

4.9 General Conclusions and Future Research Directions

As a consequence of a rapid increase in the annual number of publications over the past quarter century, ecological understanding has broadened and deepened concerning the influence of land conversion and habitat fragmentation on tropical bats at the level of populations, ensembles, and assemblages. Nonetheless, large geographic and taxonomic biases characterize current understanding.

Although many studies document that human-induced changes in land use alter bat species abundances and taxonomic dimension of biodiversity, surprisingly few studies have explored how these changes manifest with regard to genetic, behavioral, physiological, or disease-related phenomena. Similarly, little is known about the way in which land-use change affects functional or phylogenetic dimensions of biodiversity (but see Cisneros et al. 2015). Studies generally are not conducted in a spatially explicit manner (Fig. 4.4a), so multiscale (e.g., alpha, beta, and gamma diversities) or cross-scale interactions cannot be explored fully, and conclusions must be tempered in the absence of a more integrated understanding of the role of unmodified habitat in rescuing local populations from extinction. Key insights from landscape-scale studies comprise the species- and ensemble-specific nature of responses, as well as their dependence on spatial scale. The most fundamental developments include the recognition that habitat fragmentation is a complex process involving the nature of patches (i.e., landscape composition and configuration), as well as the nature of the matrix that arises as a consequence of direct, human modifications of the landscape (Fig. 4.4b). Finally, the consequences of changes in the bat fauna from habitat conversion and fragmentation have not been

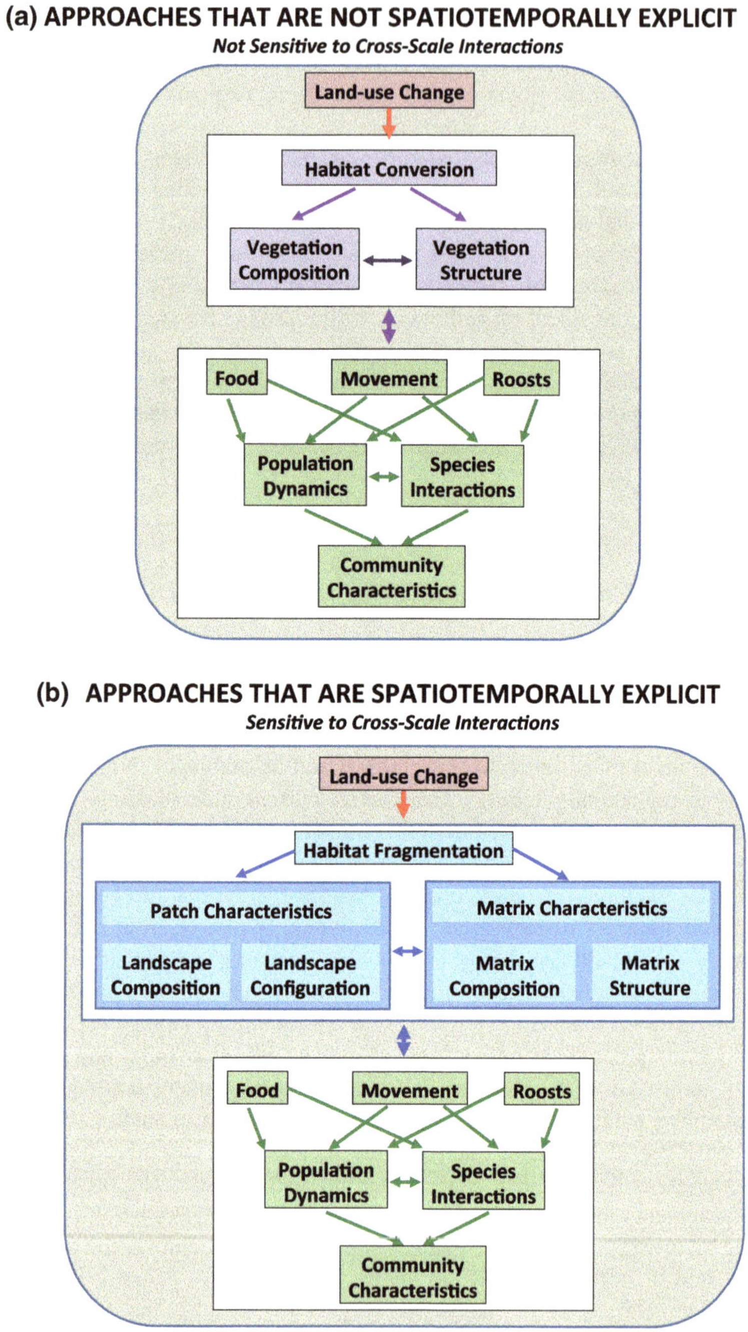
(a) APPROACHES THAT ARE NOT SPATIOTEMPORALLY EXPLICIT
Not Sensitive to Cross-Scale Interactions
Land-use Change
Habitat Conversion
Vegetation Composition
Vegetation Structure
Food
Movement
Roosts
Population Dynamics
Species Interactions
Community Characteristics
(b) APPROACHES THAT ARE SPATIOTEMPORALLY EXPLICIT
Sensitive to Cross-Scale Interactions
Land-use Change
Habitat Fragmentation
Patch Characteristics
Landscape Composition
Landscape Configuration
Matrix Characteristics
Matrix Composition
Matrix Structure
Food
Movement
Roosts
Population Dynamics
Species Interactions
Community Characteristics

◀ **Fig. 4.4** Two conceptual models that indicate the pathways whereby land-use changes affect bats in ways that **a** are not spatiotemporally explicit or **b** are spatiotemporally explicit. In both scenarios, effects of land-use change are mediated by alterations in the vegetation, but the underlying mechanisms differ (contrast the *purple boxes* with the *blue boxes*). Nonetheless, populations and assemblages of bats respond via similar mechanisms associated with feeding, roosting, and movement opportunities (*green boxes*). Generally, studies that explore the effects of habitat conversion (e.g., effects of logging or agriculture) on bats are not spatially explicit. Land-use change is reflected in habitat conversion that directly alters the composition and structure of the vegetation, with effects on the abundance and distribution of food resources or roosts, and the existence of "flyways" whereby bats navigate through the forest. In concert, these three characteristics affect the population dynamics of different bat species and the interaction likelihoods among species (e.g., bat species, other animal species, and disease-causing microorganisms). As a consequence, changes in bat species abundance distributions (e.g., richness, evenness, dominance, diversity, rarity) emerge with cascading effects on the vegetation as a consequence of altered seed dispersal, pollination, or regulation of insect herbivores. Generally, studies of habitat fragmentation are spatially explicit and explore how land-use change affects a focal habitat type (e.g., forest) by creating a network of patches embedded in a matrix of human-modified habitats. Such studies have the potential to explore how patch characteristics (e.g., landscape composition and configuration of forest patches) as well as matrix characteristics (e.g., structural or compositional attributes of the converted land) interact to affect the bat fauna. See text for additional details

quantified with regard to the maintenance of vital ecosystem processes or services. Clearly, we are still far from a comprehensive understanding of how tropical bats respond to habitat modification.

To advance ecological understanding, we have highlighted a number of more specific research needs across all themes in this chapter. We further stress the following key research directions as particularly worthy of pursuit, many of which have been summarized in different context for mammals in general (e.g., Willig 2001).

1. Geographic and taxonomic biases toward the Neotropics and a focus on just one bat family, Phyllostomidae, need to be overcome. Although research efforts in Southeast Asia are gaining momentum (Kingston 2013), Africa deserves greatly intensified research activities. As technological advances now make acoustic sampling of aerial insectivorous bats increasingly time- and cost-efficient, this ensemble should regularly be targeted in ecological research, including environmental impact assessments.
2. Research should be broadened to encompass the full spectrum of possible responses at the level of populations, ensembles, assemblages, and metacommunities. Novel mechanistic insights could be gained by studies that assess behavioral responses to particular types of habitat conversion or habitat fragmentation. Similarly, studies are needed to investigate physiological and immune responses, as well as disease susceptibility across a broad range of host and vector species. A better understanding of the genetic effects on bats from habitat modification requires integrated research on a suite of different species that explore the link between patterns of genetic variation and species' ecological and life-history traits. In general, the way in which species traits and

environmental factors interact to shape species responses to landscape change is unclear, as trait-based approaches have been rare (but see Farneda et al. 2015). Understanding how functional and phylogenetic biodiversity changes during habitat conversion and secondary succession is investigated rarely and remains poorly understood. Much also remains to be learned about how habitat disruption and modification affect the provisioning of critical ecosystem services, especially flower pollination and arthropod suppression.

3. Multiscale studies provide a more comprehensive understanding of pattern–process relationships in heterogeneous human-modified landscapes than do single-scale assessments. Future research should address bat responses to landscape change with respect to both spatial and temporal dimensions. Considerable progress in the field could be made by directing greater research effort and resources toward long-term studies that are capable of unveiling novel insights, which are hard or impossible to obtain from short-term, cross-sectional studies (cf. Lindenmayer et al. 2011). Studies currently underway at the Biological Dynamics of Forest Fragments Project (BDFFP) in Brazil (Meyer et al, unpublished data) or at the Stability of Altered Forest Ecosystems (SAFE) Project in Borneo (e.g., Struebig et al. 2013) provide examples of first efforts in this direction. The need for broader geographic coverage notwithstanding, directing more research to well-studied systems or long-term study sites, allows the responses of bats to land-use change to be compared to those of other taxa (e.g., Barlow et al. 2007; Bicknell et al. 2015; Ewers et al. 2015).
4. We stress the importance of robust study designs for assessing faunal responses to habitat alteration. Studies should have adequate replication (cf. Ramage et al. 2013) and involve controls or reference sites. Lack of controls is an important shortcoming of many of the reviewed studies, which often focused on comparisons of different types of disturbed habitats. This clearly limits their ability to ascribe observed effects to disturbance. We echo Kingston's (2013) call for studies to collect predisturbance, baseline information whenever possible, given that tropical bat assemblages exhibit considerable spatiotemporal variability even in unmodified habitats. In this context, Before–After–Control–Impact designs (e.g., Bicknell et al. 2015), in which sites affected by human disturbance are compared with undisturbed reference sites, both before and after impact, enhance inferential strength (Smith 2013), and add scientific rigor to future assessments of the effects of habitat modification on tropical bats.

Finally, an improved ecological understanding of bat responses to land-use change will be of little use to society unless it can be translated into improved management practices that ensure their long-term conservation and provision of critical ecosystem services. Across all themes in this chapter, we urge bat researchers to apply more of their science to policy and management questions. Examples of such applications include the effectiveness of specific management practices (e.g., farming intensity, cutting cycles) and mitigation measures (e.g., riparian conservation set-asides, artificial roosts).

Acknowledgments We would like to thank the editors for the invitation to contribute to this volume and Brock Fenton and Jorge Galindo-González for helpful comments on the manuscript.

References

***Studies considered in our systematic review**

Aguiar LMS, Marinho-Filho J (2007) Bat frugivory in a remnant of Southeastern Brazilian Atlantic forest. Acta Chiropt 9:251–260*

Albrecht L, Meyer CFJ, Kalko EKV (2007) Differential mobility in two small phyllostomid bats, *Artibeus watsoni* and *Micronycteris microtis*, in a fragmented Neotropical landscape. Acta Theriol 52:141–149*

Asner GP, Broadbent EN, Oliveira PJC et al (2006) Condition and fate of logged forests in the Brazilian Amazon. Proc Natl Acad Sci 103:12947–12950

Asner GP, Keller M, Lentini M et al (2013) Selective logging and its relation to deforestation. In: Keller M, Bustamante M, Gash J, Silva Dias P (eds) Amazonia and Global Change. American Geophysical Union, Washington, D.C., pp 25–42

Asner GP, Rudel TK, Aide TM et al (2009) A contemporary assessment of change in humid tropical forests. Conserv Biol 23:1386–1395

Avila-Cabadilla LD, Sanchez-Azofeifa GA, Stoner KE et al (2012) Local and landscape factors determining occurrence of phyllostomid bats in tropical secondary forests. PLoS ONE 7:e35228*

Avila-Cabadilla LD, Stoner KE, Henry M et al (2009) Composition, structure and diversity of phyllostomid bat assemblages in different successional stages of a tropical dry forest. For Ecol Manage 258:986–996*

Banks-Leite C, Ewers RM, Metzger J-P (2010) Edge effects as the principal cause of area effects on birds in fragmented secondary forest. Oikos 119:918–926

Banks-Leite C, Ewers RM, Metzger J-P (2013) The confounded effects of habitat disturbance at the local, patch and landscape scale on understorey birds of the Atlantic Forest: Implications for the development of landscape-based indicators. Ecol Indic 31:82–88

Barlow J, Gardner TA, Araujo IS et al (2007) Quantifying the biodiversity value of tropical primary, secondary and plantation forest. Proc Natl Acad Sci USA 104:18555–18560*

Barragan F, Lorenzo C, Moron A et al (2010) Bat and rodent diversity in a fragmented landscape on the Isthmus of Tehuantepec, Oaxaca, Mexico. Trop Conserv Sci 3:1–16*

Bianconi GV, Mikich SB, Pedro WA (2006) Movements of bats (Mammalia, Chiroptera) in Atlantic Forest remnants in southern Brazil. Rev Bras Zool 23:1199–1206*

Bicknell JE, Struebig MJ, Edwards DP et al (2014) Improved timber harvest techniques maintain biodiversity in tropical forests. Curr Biol 24:R1119–R1120

Bicknell JE, Struebig MJ, Davies ZG (2015) Reconciling timber extraction with biodiversity conservation in tropical forests using Reduced-Impact Logging. J Appl Ecol 52:379–388

Bissonette JA, Storch I (2007) Temporal dimensions of landscape ecology: wildlife responses to variable resources. Springer, New York

Blaser J, Sarre A, Poore D et al (2011) Status of tropical forest management. ITTO technical series no. 38. International Tropical Timber Organization, Yokohama, Japan

Bobrowiec PED, Gribel R (2010) Effects of different secondary vegetation types on bat community composition in Central Amazonia, Brasil. Anim Conserv 13:204–216*

Broadbent EN, Asner GP, Keller M et al (2008) Forest fragmentation and edge effects from deforestation and selective logging in the Brazilian Amazon. Biol Conserv 141:1745–1757

Brosset A, Charles-Dominique P, Cockle A et al (1996) Bat communities and deforestation in French Guiana. Can J Zool 74:1974–1982*
Burivalova Z, Şekercioğlu ÇH, Koh LP (2014) Thresholds of logging intensity to maintain tropical forest biodiversity. Curr Biol 24:1893–1898
Calisher CH, Childs JE, Field HE et al (2006) Bats: important reservoir hosts of emerging viruses. Clin Microbiol Rev 19:531–545
Castro-Arellano I, Presley SJ, Nelio Saldanha L et al (2007) Effects of reduced impact logging on bat biodiversity in terra firme forest of lowland Amazonia. Biol Conserv 138:269–285*
Castro-Arellano I, Presley SJ, Willig MR et al (2009) Reduced-impact logging and temporal activity of understorey bats in lowland Amazonia. Biol Conserv 142:2131–2139*
Castro-Luna AA, Galindo-González J (2012a) Enriching agroecosystems with fruit-producing tree species favors the abundance and richness of frugivorous and nectarivorous bats in Veracruz, Mexico. Mamm Biol 77:32–40*
Castro-Luna AA, Galindo-González J (2012b) Seed dispersal by phyllostomid bats in two contrasting vegetation types in a Mesoamerican reserve. Acta Chiropt 14:133–142*
Castro-Luna AA, Sosa VJ, Castillo-Campos G (2007a) Bat diversity and abundance associated with the degree of secondary succession in a tropical forest mosaic in south-eastern Mexico. Anim Conserv 10:219–228*
Castro-Luna AA, Sosa VJ, Castillo-Campos G (2007b) Quantifying phyllostomid bats at different taxonomic levels as ecological indicators in a disturbed tropical forest. Acta Chiropt 9:219–228*
Chazdon RL (2014) Second growth: The promise of tropical forest regeneration in an age of deforestation. University of Chicago Press, Chicago
Chazdon RL, Peres CA, Dent D et al (2009) The potential for species conservation in tropical secondary forests. Conserv Biol 23:1406–1417
Christine LJ, Fletcher C, Abd Rahman K (2013) Spatial effects of virgin jungle reserves (VJR) on the community of insectivorous bats in Peninsular Malaysia. J Trop For Sci 25:118–130 *
Cisneros LM, Fagan ME, Willig MR (2015) Effects of human-modified landscapes on taxonomic, functional, and phylogenetic dimensions of bat biodiversity. Divers Distrib 21:523–533 *
Clarke FM, Pio DV, Racey PA (2005a) A comparison of logging systems and bat diversity in the Neotropics. Conserv Biol 19:1194–1204*
Clarke FM, Rostant LV, Racey PA (2005b) Life after logging: post-logging recovery of a neotropical bat community. J Appl Ecol 42:409–420*
Clough Y, Barkmann J, Juhrbandt J et al (2011) Combining high biodiversity with high yields in tropical agroforests. Proc Natl Acad Sci 108:8311–8316
Collinge SK (2009) Ecology of fragmented landscapes. Johns Hopkins University Press, Baltimore
Cosson J-F, Pons J-M, Masson D (1999) Effects of forest fragmentation on frugivorous and nectarivorous bats in French Guiana. J Trop Ecol 15:515–534*
Cottontail VM, Wellinghausen N, Kalko EKV (2009) Habitat fragmentation and haemoparasites in the common fruit bat, *Artibeus jamaicensis* (Phyllostomidae) in a tropical lowland forest in Panama. Parasitology 136:1133–1145*
Cunto GC, Bernard E (2012) Neotropical bats as indicators of environmental disturbance: what is the emerging message? Acta Chiropt 14:143–151
Danielsen F, Heegaard M (1995) Impact of logging and plantation development on species diversity: a case study from Sumatra. In: Sandbukt O (ed) Management of tropical forests: towards an integrated perspective. Centre for Development and the Environment, University of Oslo*
de la Peña-Cuéllar E, Stoner KE, Avila-Cabadilla LD et al (2012) Phyllostomid bat assemblages in different successional stages of tropical rain forest in Chiapas, Mexico. Biodivers Conserv 21:1381–1397*

DeFries RS, Rudel T, Uriarte M et al (2010) Deforestation driven by urban population growth and agricultural trade in the twenty-first century. Nat Geosci 3:178–181
Delaval M, Charles-Dominique P (2006) Edge effects on frugivorous and nectarivorous bat communities in a neotropical primary forest in French Guiana. Rev Ecol-Terre Vie 61:343–352*
Dent DH, Wright SJ (2009) The future of tropical species in secondary forests: a quantitative review. Biol Conserv 142:2833–2843
Didham RK (2010) Ecological consequences of habitat fragmentation. ELS. Wiley, London
Donald PF (2004) Biodiversity impacts of some agricultural commodity production systems. Conserv Biol 18:17–38
Driscoll DA, Banks SC, Barton PS et al (2013) Conceptual domain of the matrix in fragmented landscapes. Trends Ecol Evol 28:605–613
Duncan RS, Chapman CA (1999) Seed dispersal and potential forest succession in abandoned agriculture in tropical Africa. Ecol Appl 9:998–1008*
Durant KA, Hall RW, Cisneros LM et al (2013) Reproductive phenologies of phyllostomid bats in Costa Rica. J Mammal 94:1438–1448
Edwards DP, Larsen TH, Docherty TDS et al (2011) Degraded lands worth protecting: the biological importance of Southeast Asia's repeatedly logged forests. Proc R Soc B 278:82–90
Estrada A, Coates-Estrada R (2001a) Bat species richness in live fences and in corridors of residual rain forest vegetation at Los Tuxtlas, Mexico. Ecography 24:94–102. *
Estrada A, Coates-Estrada R (2001b) Species composition and reproductive phenology of bats in a tropical landscape at Los Tuxtlas, Mexico. J Trop Ecol 17:627–646*
Estrada A, Coates-Estrada R (2002) Bats in continuous forest, forest fragments and in an agricultural mosaic habitat-island at Los Tuxtlas, Mexico. Biol Conserv 103:237–245*
Estrada A, Coates-Estrada R, Meritt J, Dennis (1993a) Bat species richness and abundance in tropical rain forest fragments and in agricultural habitats at Los Tuxtlas, Mexico. Ecography 16:309–318*
Estrada A, Coates-Estrada R, Meritt Jr. D et al (1993b) Patterns of frugivore richness and abundance in forest islands and agricultural habitats at Los Tuxlas, Mexico. Vegetatio 107/108:245–257*
Estrada A, Jiménez C, Rivera A et al (2004) General bat activity measured with an ultrasound detector in a fragmented tropical landscape in Los Tuxtlas, Mexico. Anim Biodivers Conserv 27:1–9*
Estrada Villegas S, Meyer CFJ, Kalko EKV (2010) Effects of forest fragmentation on aerial insectivorous bats in a land-bridge island system. Biol Conserv 143:597–608*
Evelyn MJ, Stiles DA (2003) Roosting requirements of two frugivorous bats (*Sturnira lilium* and *Artibeus intermedius*) in fragmented neotropical forest. Biotropica 35:405–418*
Ewers RM, Didham RK (2006a) Confounding factors in the detection of species responses to habitat fragmentation. Biol Rev 81:117–142
Ewers RM, Didham RK (2006b) Continuous response functions for quantifying the strength of edge effects. J Appl Ecol 43:527–536
Ewers RM, Didham RK (2007) Habitat fragmentation: panchreston or paradigm? Trends Ecol Evol 22:511
Ewers RM, Boyle MJ, Gleave RA et al (2015) Logging cuts the functional importance of invertebrates in tropical rainforest. Nat Commun 6:6836
Fahrig L (2003) Effects of habitat fragmentation on biodiversity. Annu Rev Ecol Evol Syst 34:487–515
Faria D (2006) Phyllostomid bats of a fragmented landscape in the north-eastern Atlantic forest, Brazil. J Trop Ecol 22:531–542*
Faria D, Baumgarten J (2007) Shade cacao plantations (*Theobroma cacao*) and bat conservation in southern Bahia, Brazil. Biodivers Conserv 16:291–312*
Faria D, Laps RR, Baumgarten J et al (2006) Bat and bird assemblages from forests and shade cacao plantations in two contrasting landscapes in the Atlantic Forest of southern Bahia, Brazil. Biodivers Conserv 15:587–612*

Faria D, Paciencia MLB, Dixo M et al (2007) Ferns, frogs, lizards, birds and bats in forest fragments and shade cacao plantations in two contrasting landscapes in the Atlantic forest, Brazil. Biodivers Conserv 16:2335–2357*

Farneda FZ, Rocha R, López-Baucells A, Groenenberg M, Silva I, Palmeirim JM, Bobrowiec PED & Meyer CFJ (2015) Trait-related responses to habitat fragmentation in Amazonian bats. J Appl Ecol 52:1381–1391*

Fenton MB, Acharya L, Audet D et al (1992) Phyllostomid bats (Chiroptera: Phyllostomidae) as indicators of habitat disruption in the Neotropics. Biotropica 24:440–446*

Fischer J, Lindenmayer DB (2007) Landscape modification and habitat fragmentation: a synthesis. Glob Ecol Biogeogr 16:265–280

Fletcher RJ Jr, Ries L, Battin J et al (2007) The role of habitat area and edge in fragmented landscapes: definitively distinct or inevitably intertwined? Can J Zool 85:1017–1030

Fukuda D, Tisen OB, Momose K et al (2009) Bat diversity in the vegetation mosaic around a lowland dipterocarp forest of Borneo. Raffles Bull Zool 57:213–221*

Furey N, Mackie I, Racey P (2010) Bat diversity in Vietnamese limestone karst areas and the implications of forest degradation. Biodivers Conserv 19:1821–1838*

Galindo-González J, Sosa VJ (2003) Frugivorous bats in isolated trees and riparian vegetation associated with human-made pastures in a fragmented tropical landscape. Southwest Nat 48:579–589*

Galindo-González J, Guevara S, Sosa VJ (2000) Bat- and bird-generated seed rains at isolated trees in pastures in a tropical rainforest. Conserv Biol 14:1693–1703*

García QS, Rezende JLP, Aguiar LMS (2000) Seed dispersal by bats in a disturbed area of Southeastern Brazil. Rev Biol Trop 48:125–128*

García-Estrada C, Damon A, Sánchez-Hernández C et al (2012) Diets of frugivorous bats in montane rain forest and coffee plantations in southeastern Chiapas, Mexico. Biotropica 44:394–401*

García-Morales R, Chapa-Vargas L, Galindo-González J et al (2012) Seed dispersal among three different vegetation communities in the Huasteca region, Mexico, analyzed from bat feces. Acta Chiropt 14:357–367*

Gardner TA, Barlow J, Chazdon R et al (2009) Prospects for tropical forest biodiversity in a human-modified world. Ecol Lett 12:561–582

Gardner TA, Barlow J, Sodhi NS et al (2010) A multi-region assessment of tropical forest biodiversity in a human-modified world. Biol Conserv 143:2293–2300

Gibbs HK, Ruesch AS, Achard F et al (2010) Tropical forests were the primary sources of new agricultural land in the 1980s and 1990s. Proc Natl Acad Sci 107:16732–16737

Gibson L, Lee TM, Koh LP et al (2011) Primary forests are irreplaceable for sustaining tropical biodiversity. Nature 478:378–381

Gorchov DL, Rondon XJ, Cornejo F et al (2013) Edge effects in recruitment of trees, and relationship to seed dispersal patterns, in cleared strips in the Peruvian Amazon. Ecol Res 28:53–65*

Gorresen PM, Willig MR (2004) Landscape responses of bats to habitat fragmentation in Atlantic forest of Paraguay. J Mammal 85:688–697*

Gorresen PM, Willig MR, Strauss RE (2005) Multivariate analysis of scale-dependent associations between bats and landscape structure. Ecol Appl 15:2126–2136*

Griscom HP, Kalko EKV, Ashton MS (2007) Frugivory by small vertebrates within a deforested, dry tropical region of Central America. Biotropica 39:278–282*

Guevara S, Laborde J, Sanchez-Rios G (2004) Rain forest regeneration beneath the canopy of fig trees isolated in pastures of Los Tuxtlas, Mexico. Biotropica 36:99–108*

Hamann A, Curio E (1999) Interactions among frugivores and fleshy fruit trees in a Philippine submontane rainforest. Conserv Biol 13:766–773*

Hansen MC, Potapov PV, Moore R et al (2013) High-resolution global maps of 21st-century forest cover change. Science 342:850–853

Hanson T, Brunsfeld S, Finegan B et al (2007) Conventional and genetic measures of seed dispersal for *Dipteryx panamensis* (Fabaceae) in continuous and fragmented Costa Rican rain forest. J Trop Ecol 23:635–642*

Harper KA, MacDonald SE, Burton PE et al (2005) Edge influence on forest structure and composition in fragmented landscapes. Conserv Biol 19:768–782
Harrison ME, Boonman A, Cheyne SM et al (2012) Biodiversity monitoring protocols for REDD+: can a one-size-fits-all approach really work? Trop Conserv Sci 5:1–11
Harvey CA, Medina A, Sanchez DM et al (2006) Patterns of animal diversity in different forms of tree cover in agricultural landscapes. Ecol Appl 16:1986–1999*
Hayman DT, Bowen RA, Cryan PM et al (2013) Ecology of zoonotic infectious diseases in bats: current knowledge and future directions. Zoonoses Public Health 60:2–21
Henry M, Cosson JF, Pons JM (2007a) Abundance may be a misleading indicator of fragmentation-sensitivity: The case of fig-eating bats. Biol Conserv 139:462–467*
Henry M, Pons J-M, Cosson J-F (2007b) Foraging behavior of a frugivorous bat helps bridge landscape connectivity and ecological processes in a fragmented rainforest. J Anim Ecol 76:801–813*
Henry M, Cosson JF, Pons JM (2010) Modelling multi-scale spatial variation in species richness from abundance data in a complex neotropical bat assemblage. Ecol Modell 221:2018–2027*
Ingle NR (2003) Seed dispersal by wind, birds, and bats between Philippine montane rainforest and successional vegetation. Oecologia 134:251–261*
Jetz W, Wilcove DS, Dobson AP (2007) Projected impacts of climate and land-use change on the global diversity of birds. PLoS Biol 5:e157
Jones G, Jacobs DS, Kunz TH et al (2009) *Carpe noctem*: the importance of bats as bioindicators. Endanger Spec Res 8:93–115
Kalka MB, Smith AR, Kalko EKV (2008) Bats limit arthropods and herbivory in a tropical forest. Science 320:71
Kingston T (2013) Response of bat diversity to forest disturbance in Southeast Asia: insights from long-term research in Malaysia. In: Adams RA, Pedersen SC (eds) Bat evolution, ecology, and conservation. Springer New York, pp 169–185
Klingbeil BT, Willig MR (2009) Guild-specific responses of bats to landscape composition and configuration in fragmented Amazonian rainforest. J Appl Ecol 46:203–213*
Klingbeil BT, Willig MR (2010) Seasonal differences in population-, ensemble- and community-level responses of bats to landscape structure in Amazonia. Oikos 119:1654–1664*
Kunz TH, de Torrez EB, Bauer D et al (2011) Ecosystem services provided by bats. Ann NY Acad Sci 1223:1–38
Kupfer JA, Malanson GP, Franklin SB (2006) Not seeing the ocean for the islands: the mediating influence of matrix-based processes on forest-fragmentation effects. Glob Ecol Biogeogr 15:8–20
Laurance WF (2008) Theory meets reality: how habitat fragmentation research has transcended island biogeographic theory. Biol Conserv 141:1731–1744
Laurance WF, Camargo JLC, Luizão RCC et al (2011) The fate of Amazonian forest fragments: A 32-year investigation. Biol Conserv 144:56–67
Laurance WF, Sayer J, Cassman KG (2014) Agricultural expansion and its impacts on tropical nature. Trends Ecol Evol 29:107–116
Law BS, Lean M (1999) Common blossom bats (*Syconycteris australis*) as pollinators in fragmented Australian tropical rainforest. Biol Conserv 91:201–212*
Lee TM, Jetz W (2008) Future battlegrounds for conservation under global change. Proc R Soc B 275:1261–1270
Lindenmayer DB, Fischer J (2006) Habitat fragmentation and landscape change. Island Press, Washington, D.C.
Lindenmayer DB, Laurance WF (2012) A history of hubris—cautionary lessons in ecologically sustainable forest management. Biol Conserv 151:11–16
Lindenmayer DB, Wood J, McBurney L et al (2011) Cross-sectional vs. longitudinal research: a case study of trees with hollows and marsupials in Australian forests. Ecol Monogr 81:557–580
Louzada J, Gardner T, Peres C et al (2010) A multi-taxa assessment of nestedness patterns across a multiple-use Amazonian forest landscape. Biol Conserv 143:1102–1109*
Luskin MS (2010) Flying foxes prefer to forage in farmland in a tropical dry forest landscape mosaic in Fiji. Biotropica 42:246–250*

Maas B, Clough Y, Tscharntke T (2013) Bats and birds increase crop yield in tropical agroforestry landscapes. Ecol Lett 16:1480–1487*
MacArthur RH, Wilson EO (1967) The theory of island biogeography. Princeton University Press, Princeton
Mancina CA, García-Rivera L, Capote RT (2007) Habitat use by phyllostomid bat assemblages in secondary forests of the 'Sierra del Rosario' biosphere reserve, Cuba. Acta Chiropt 9:203–218*
McCulloch ES, Sebastián Tello J, Whitehead A et al (2013) Fragmentation of Atlantic Forest has not affected gene flow of a widespread seed-dispersing bat. Mol Ecol 22:4619–4633*
Medellín RA, Gaona O (1999) Seed dispersal by bats and birds in forest and disturbed habitats of Chiapas, Mexico. Biotropica 31:478–485*
Medellín RA, Equihua M, Amin MA (2000) Bat diversity and abundance as indicators of disturbance in Neotropical rainforests. Conserv Biol 14:1666–1675*
Medina A, Harvey CA, Sánchez Merlo D et al (2007) Bat diversity and movement in an agricultural landscape in Matiguás, Nicaragua. Biotropica 39:120–128*
Melo FPL, Rodriguez-Herrera B, Chazdon RL et al (2009) Small tent-roosting bats promote dispersal of large-seeded plants in a Neotropical forest. Biotropica 41:737–743*
Melo FPL, Arroyo-Rodríguez V, Fahrig L et al (2013) On the hope for biodiversity-friendly tropical landscapes. Trends Ecol Evol 28:462–468
Mendenhall CD, Karp DS, Meyer CFJ et al (2014) Predicting biodiversity change and averting collapse in agricultural landscapes. Nature 509:213–217*
Mendes P, Vieira TB, Oprea M et al (2009) Long-distance movement of *Artibeus lituratus* (Chiroptera: Phyllostomidae) in the state of Espirito Santo, Brazil. Ecotropica 15:43–46*
Meyer CFJ, Kalko EKV (2008a) Assemblage-level responses of phyllostomid bats to tropical forest fragmentation: land-bridge islands as a model system. J Biogeogr 35:1711–1726*
Meyer CFJ, Kalko EKV (2008b) Bat assemblages on Neotropical landbridge islands: nested subsets and null model analyses of species co-occurrence patterns. Divers Distrib 14:644–654*
Meyer CFJ, Fründ J, Pineda Lizano W et al (2008) Ecological correlates of vulnerability to fragmentation in Neotropical bats. J Appl Ecol 45:381–391*
Meyer CFJ, Kalko EKV, Kerth G (2009) Small-scale fragmentation effects on local genetic diversity in two phyllostomid bats with different dispersal abilities in Panama. Biotropica 41:95–102*
Monadjem A, Ellstrom M, Maldonaldo C et al (2010) The activity of an insectivorous bat *Neoromicia nana* on tracks in logged and unlogged forest in tropical Africa. Afr J Ecol 48:1083–1091*
Moreno CE, Halffter G (2001) Spatial and temporal analysis of alpha, beta and gamma diversities of bats in a fragmented landscape. Biodivers Conserv 10:367–382*
Muscarella R, Fleming TH (2007) The role of frugivorous bats in tropical forest succession. Biol Rev 82:573–590
Numa C, Verdú JR, Sánchez-Palomino P (2005) Phyllostomid bat diversity in a variegated coffee landscape. Biol Conserv 122:151–158*
Ochoa GJ (2000) Effects of selective logging on the diversity of small mammals in the lowland forests of the Venezuelan Guyana region. Biotropica 32:146–164*
Ostfeld RS, Keesing F (2012) Effects of host diversity on infectious disease. Annu Rev Ecol Evol Syst 43:157–182
Pardini R, Faria D, Accacio GM et al (2009) The challenge of maintaining Atlantic forest biodiversity: a multi-taxa conservation assessment of specialist and generalist species in an agroforestry mosaic in southern Bahia. Biol Conserv 142:1178–1190*
Pech-Canche JM, Moreno CE, Halffter G (2011) Additive partitioning of phyllostomid bat richness at fine and coarse spatial and temporal scales in Yucatan, Mexico. Ecoscience 18:42–51
Peel AJ, Sargan DR, Baker KS et al (2013) Continent-wide panmixia of an African fruit bat facilitates transmission of potentially zoonotic viruses. Nat Commun 4:2770
Pereira HM, Leadley PW, Proença V et al (2010) Scenarios for global biodiversity in the 21st century. Science 330:1496–1501

Perfecto I, Vandermeer J (2008) Biodiversity conservation in tropical agroecosystems. Ann NY Acad Sci 1134:173–200
Peters SL, Malcolm JR, Zimmerman BL (2006) Effects of selective logging on bat communities in the Southeastern Amazon. Conserv Biol 20:1410–1421*
Phalan B, Bertzky M, Butchart SHM et al (2013) Crop expansion and conservation priorities in tropical countries. PLoS ONE 8:e51759
Phommexay P, Satasook C, Bates P et al (2011) The impact of rubber plantations on the diversity and activity of understorey insectivorous bats in southern Thailand. Biodivers Conserv 20:1441–1456*
Pilosof S, Dick CW, Korine C et al (2012) Effects of anthropogenic disturbance and climate on patterns of bat fly parasitism. PLoS ONE 7:e41487*
Pina SMS, Meyer CFJ, Zortéa M (2013) A comparison of habitat use by phyllostomid bats (Chiroptera: Phyllostomidae) in natural forest fragments and *Eucalyptus* plantations in the Brazilian Cerrado. Chiroptera Neotrop 19:14–30*
Pineda E, Moreno C, Escobar F et al (2005) Frog, bat, and dung beetle diversity in the cloud forest and coffee agroecosystems of Veracruz, Mexico. Conserv Biol 19:400–410*
Pinto N, Keitt TH (2008) Scale-dependent responses to forest cover displayed by frugivore bats. Oikos 117:1725–1731*
Pons J-M, Cosson J-F (2002) Use of forest fragments by animalivorous bats in French Guiana. Rev Ecol-Terre Vie 57:117–130*
Prasetyo PN, Noerfahmy S, dan Tata HL (2011) Jenis-jenis Kelelawar Agroforest Sumatera (Bat species in Sumatran agroforest). World Agroforestry Centre—ICRAF, SEA Regional Office, Bogor, Indonesia*
Presley SJ, Willig MR, Wunderle Jr. JM et al (2008) Effects of reduced-impact logging and forest physiognomy on bat populations of lowland Amazonian forest. J Appl Ecol 45:14–25*
Presley SJ, Willig MR, Castro-Arellano I et al (2009a) Effects of habitat conversion on temporal activity patterns of phyllostomid bats in lowland Amazonian rain forest. J Mammal 90:210–221*
Presley SJ, Willig MR, Saldanha LN et al (2009b) Reduced-impact logging has little effect on temporal activity of frugivorous bats (Chiroptera) in lowland Amazonia. Biotropica 41:369–378*
Pullin AS, Stewart GB (2006) Guidelines for systematic review in conservation and environmental management. Conserv Biol 20:1647–1656
Putz FE, Sist P, Fredericksen T et al (2008) Reduced-impact logging: Challenges and opportunities. For Ecol Manage 256:1427–1433
Putz FE, Zuidema PA, Synnott T et al (2012) Sustaining conservation values in selectively logged tropical forests: the attained and the attainable. Conserv Lett 5:296–303
Quesada M, Stoner KE, Rosas-Guerrero V et al (2003) Effects of habitat disruption on the activity of nectarivorous bats (Chiroptera: Phyllostomidae) in a dry tropical forest: implications for the reproductive success of the neotropical tree *Ceiba grandiflora*. Oecologia 135:400–406*
Quesada M, Stoner KE, Lobo JA et al (2004) Effects of forest fragmentation on pollinator activity and consequences for plant reproductive success and mating patterns in bat-pollinated bombacaceous trees. Biotropica 36:131–138*
Quesada M, Herrerias-Diego Y, Lobo JA et al (2013) Long-term effects of habitat fragmentation on mating patterns and gene flow of a tropical dry forest tree, Ceiba aesculifolia (Malvaceae: Bombacoideae). Am J Bot 100:1095–1101*
Ramage BS, Sheil D, Salim HMW et al (2013) Pseudoreplication in tropical forests and the resulting effects on biodiversity conservation. Conserv Biol 27:364–372
Ribeiro MC, Metzger JP, Martensen AC et al (2009) The Brazilian Atlantic Forest: how much is left, and how is the remaining forest distributed? Implications for conservation. Biol Conserv 142:1141–1153
Ries L, Sisk TD (2010) What is an edge species? The implications of sensitivity to habitat edges. Oikos 119:1636–1642

Ripperger SP, Tschapka M, Kalko EKV et al (2013) Life in a mosaic landscape: anthropogenic habitat fragmentation affects genetic population structure in a frugivorous bat species. Conserv Genet 14:925–934*

Rossiter SJ, Zubaid A, Mohd-Adnan A et al (2012) Social organization and genetic structure: insights from codistributed bat populations. Mol Ecol 21:647–661*

Sala OE, Chapin FS, Armesto JJ et al (2000) Global biodiversity scenarios for the year 2100. Science 287:1770–1774

Salas-Rojas M, Sánchez-Hernández C, Romero-Almaraz MdL et al (2004) Prevalence of rabies and LPM paramyxovirus antibody in non-hematophagous bats captured in the Central Pacific coast of Mexico. Trans R Soc Trop Med Hyg 98:577–584*

Saldaña-Vázquez R, Sosa V, Hernández-Montero J et al (2010) Abundance responses of frugivorous bats (Stenodermatinae) to coffee cultivation and selective logging practices in mountainous central Veracruz, Mexico. Biodivers Conserv 19:2111–2124*

Saldaña-Vázquez RA, Munguía-Rosas MA (2013) Lunar phobia in bats and its ecological correlates: a meta-analysis. Mamm Biol 78:216–219

Sampaio EM, Kalko EKV, Bernard E et al (2003) A biodiversity assessment of bats (Chiroptera) in a tropical lowland rainforest of Central Amazonia, including methodological and conservation considerations. Stud Neotrop Fauna Environ 38:17–31*

Schulze MD, Seavy NE, Whitacre DF (2000) A comparison of the phyllostomid bat assemblages in undisturbed neotropial forest and in forest fragments of a slash-and-burn farming mosaic in Petén, Guatemala. Biotropica 32:174–184*

Sedlock JL, Weyandt SE, Cororan L et al (2008) Bat diversity in tropical forest and agro-pastoral habitats within a protected area in the Philippines. Acta Chiropt 10:349–358*

Smith EP (2013) BACI design. Encyclopedia of environmetrics. Wiley, London

Stoner KE, Quesada M, Rosas-Guerrero V et al (2002) Effects of forest fragmentation on the Colima long-nosed bat (*Musonycteris harrisoni*) foraging in tropical dry forest of Jalisco, Mexico. Biotropica 34:462–467*

Storfer A, Murphy MA, Spear SF et al (2010) Landscape genetics: where are we now? Mol Ecol 19:3496–3514

Struebig MJ, Kingston T, Zubaid A et al (2008) Conservation value of forest fragments to Palaeotropical bats. Biol Conserv 141:2112–2126*

Struebig MJ, Kingston T, Zubaid A et al (2009) Conservation importance of limestone karst outcrops for Palaeotropical bats in a fragmented landscape. Biol Conserv 142:2089–2096*

Struebig MJ, Kingston T, Petit EJ et al (2011) Parallel declines in species and genetic diversity in tropical forest fragments. Ecol Lett 14:582–590*

Struebig MJ, Turner A, Giles E et al (2013) Quantifying the biodiversity value of repeatedly logged rainforests: gradient and comparative approaches from Borneo. In: Guy W, Eoin JOG (eds) Advances in ecological research, vol 48. Academic Press, New York, pp 183–224*

Struebig M, Wilting A, Gaveau DLA et al (2015) Targeted conservation safeguards a biodiversity hotspot from climate and land-cover change. Curr Biol 25:372–378

Syamsi F (2013) Chiroptera community in oil palm plantation. J Indonesian Nat Hist 1:49*

Tilman D, Balzer C, Hill J et al (2011) Global food demand and the sustainable intensification of agriculture. Proc Natl Acad Sci 108:20260–20264

Trevelin LC, Silveira Mc, Port-Carvalho M et al (2013) Use of space by frugivorous bats (Chiroptera: Phyllostomidae) in a restored Atlantic forest fragment in Brazil. Ecol Manage 291:136–143*

Tscharntke T, Clough Y, Bhagwat SA et al (2011) Multifunctional shade-tree management in tropical agroforestry landscapes—a review. J Appl Ecol 48:619–629

Villard M-A, Metzger JP (2014) Beyond the fragmentation debate: a conceptual model to predict when habitat configuration really matters. J Appl Ecol 51:309–318

Vleut I, Levy-Tacher SI, Galindo-González J et al (2012) Tropical rain-forest matrix quality affects bat assemblage structure in secondary forest patches. J Mammal 93:1469–1479*

Vleut I, Levy-Tacher SI, de Boer WF et al (2013) Tropical secondary forest management influences frugivorous bat composition, abundance and fruit consumption in Chiapas, Mexico. PLoS ONE 8:e77584*
Voigt CC, Voigt-Heucke SL, Kretzschmar AS (2012) Isotopic evidence for seed transfer from successional areas into forests by short-tailed fruit bats (*Carollia* spp.; Phyllostomidae). J Trop Ecol 28:181–186*
Wanger TC, Darras K, Bumrungsri S et al (2014) Bat pest control contributes to food security in Thailand. Biol Conserv 171:220–223
Wieland LM, Mesquita RCG, Bobrowiec PED et al (2011) Seed rain and advance regeneration in secondary succession in the Brazilian Amazon. Trop Conserv Sci 4:300–316*
Williams-Guillén K, Perfecto I (2010) Effects of agricultural intensification on the assemblage of leaf-nosed bats (Phyllostomidae) in a coffee landscape in Chiapas, Mexico. Biotropica 42:605–613*
Williams-Guillén K, Perfecto I (2011) Ensemble composition and activity levels of insectivorous bats in response to management intensification in coffee agroforestry systems. PLoS ONE 6:e16502*
Williams-Guillén K, Perfecto I, Vandermeer J (2008) Bats limit insects in a Neotropical agroforestry system. Science 320:70*
Willig MR (2001) Exploring biodiversity in time and space: Profitable directions for mammalogy in the 21st century. Mastozool Neotrop 8:107–109
Wright SJ, Muller-Landau H (2006) The future of tropical forest species. Biotropica 38:287–301
Willig MR, Patterson BD, Stevens RD (2003) Patterns of range size, richness, and body size in the Chiroptera. In: Kunz TH, Fenton MB (eds) Bat Ecology. University of Chicago Press, Chicago, pp 580–621
Willig MR, Presley SJ, Bloch CP et al. (2007) Phyllostomid bats of lowland Amazonia: Effects of habitat alteration on abundance. Biotropica 39:737–746*
Zubaid A (1993) A comparison of the bat fauna between a primary and fragmented secondary forest in peninsular Malaysia. Mammalia 57:201–206*

Chapter 5
Insectivorous Bats and Silviculture: Balancing Timber Production and Bat Conservation

Bradley Law, Kirsty J. Park and Michael J. Lacki

Abstract Forests are one of the most important habitats for insectivorous bats as they offer the potential for both roosting and foraging. We reviewed silvicultural literature from North America, Australia, and Europe and found that diverse research approaches have revealed commonalities in bat responses to forest silviculture. Almost all silvicultural treatments evaluated were compatible with some use by forest bats, though different bat ensembles respond in different ways. Ensemble ecomorphology was a consistent predictor of how bats respond to vegetative clutter and its dynamic changes as forests regenerate and develop a dense structure following harvesting. Sustaining high levels of bat diversity in timber production forests requires a mix of silvicultural treatments and exclusion areas staggered across the landscape, regardless of forest type or geographic region. Use of edge habitats, exclusion areas/set-asides, and riparian corridors for roosting and foraging by bats were consistent themes in the literature reviewed, and these habitat elements need to be considered in forest planning. Densities of hollow or dead trees sufficient to support large populations of roosting bats are unknown and remain a major knowledge gap, but will likely be species contingent. New paradigm shifts in forest management away from the use of even-aged systems to multi-spatial scale retention of mature forest including trees with cavities should be beneficial to bats, which are influenced by landscape-scale management. Such

B. Law (✉)
Forest Science, NSW Primary Industries, Sydney, Australia
e-mail: brad.law@dpi.nsw.gov.au

K.J. Park
Biological & Environmental Sciences, University of Stirling, Scotland, UK

M.J. Lacki
Department of Forestry, University of Kentucky, Kentucky, USA

C.C. Voigt and T. Kingston (eds.), *Bats in the Anthropocene: Conservation of Bats in a Changing World*, DOI 10.1007/978-3-319-25220-9_5

an approach is already in use in some regions, though there is a limited guidance on what constitutes a reasonable landscape threshold for retention. The effectiveness of such an approach will require long-term monitoring and research, especially with population studies which are currently lacking.

5.1 Introduction

Forests are one of the most important habitats for bats as they offer the potential for both roosting and foraging, and most species are reliant on forests for at least some part of their life cycle. Humans are also heavily reliant on the resources produced by forests, in particular timber. Consequently, forests are highly managed and modified in many areas. Understanding the effect that human manipulation of forested landscapes has on the resources required by bats is therefore of great importance to their conservation.

The use of silvicultural techniques to manipulate tree stands for timber production or biodiversity conservation goals presents several challenges. Forest bats are mobile and, as with forest birds, can use a large three-dimensional space to meet their life requisites (Kroll et al. 2012). Therefore, stand-level considerations alone are insufficient in sustaining habitat conditions for many forest bats as landscape-level needs are of equal or greater concern (Duchamp et al. 2007). Secondly, forest bats require roosting sites, high-quality foraging habitats, drinking sites, and features that provide connectivity among landscape elements. Providing all of these habitat requirements for an entire assemblage of bats simultaneously on a managed forested landscape is a difficult challenge, necessitating hierarchical approaches that assess spatial juxtaposition of habitat elements on the landscape and that implement silvicultural systems using multiple treatments applied both within and among stands.

Silvicultural practices vary greatly around the world. For example, in the northern hemisphere, clear felling typically results in cleared areas of 40–180 ha surrounded by relatively even-aged forests (Thomas 1988; Grindal and Brigham 1999; Swystun et al. 2001). In parts of Europe and North America, however, patch sizes are considerably smaller and some countries have abandoned clear felling altogether, favouring a more selective logging approach. Similarly, in parts of Australia, where broad scale clear-fall techniques are not utilised, selective logging results in a multi-aged forest (Nicholson 1999).

A key feature of insectivorous bats is their sophisticated sensory system, which enables them to navigate and forage in the dark. The foraging efficiency of echolocating bats is constrained by variations in vegetation because the echoes returning from prey need to be distinguished from background echoes returning from vegetation. These 'clutter' echoes can mask the echoes of prey making foraging inefficient in situations where vegetation is dense (Schnitzler et al.

2003). Forest bat species differ in echolocation signal design and wing morphology and this influences their flight behaviour and their tolerance to clutter, allowing classification into three broad foraging ensembles: (1) closed-space species are slow flying and highly manoeuvrable bats that can forage close to vegetation; (2) edge-space species exploit edge habitat and other linear features; and (3) open-space foragers have lower manoeuvrability and fly faster above the forest canopy or within large gaps in the forest. Changes to forest structure that influence the degree of clutter can, therefore, alter the availability of foraging habitat for each ensemble.

Our aim in this chapter was to explore how insectivorous bats respond to different silvicultural approaches used in forests around the world, incorporating studies within natural, or semi-natural, forests to intensive management within plantation forestry. We focus on three broad areas: North America, Australasia (including New Zealand), and Europe and refer the reader to Meyer et al. (2016) (Chap. 3) for tropical forests. While the majority of studies included in this review are published in scientific journals, we also include information from the grey literature (e.g. reports, conference proceedings, and unpublished theses) and some unpublished data where appropriate.

We look to highlight both commonalities and differences in the various approaches to the issue in different regions. We suggest that ecomorphology is one of the keys to understanding how bats use their environment and we use ecomorphological traits as a framework for predicting how the three broad functional ensembles of bats respond to forest logging (Hanspach et al. 2012; Luck et al. 2013). Conceptual models have been proposed previously for the relationship between the abundance of bats and key ecological resources manipulated by forest management (Fig. 5.1; Hayes and Loeb 2007). These posit the influence of thresholds for certain variables such as water availability, where further increases do not result in increased bat abundance. We assess the extent to which these models fit current data and extend them to (1) consider the time since logging as a response variable and (2) include an ecomorphological framework for the response of bats. We emphasise the importance of a long-term perspective when assessing bat responses in forests given that forests are long-lived ecosystems that undergo dynamic changes after disturbance. Finally, we consider the merits of multi-spatial scale management for bats and recommend future areas of research to advance the effective management of this diverse and functionally important group. There is some specialised terminology within this chapter that may be unfamiliar to those new to silvicultural literature, so we have provided a glossary at the end of the chapter with definitions. While the term woodland is often used to describe vegetation communities comprising trees but with a more open and lower canopy cover than forests, this definition varies by country. Here, we use the term forest to encompass the various definitions of woodland.

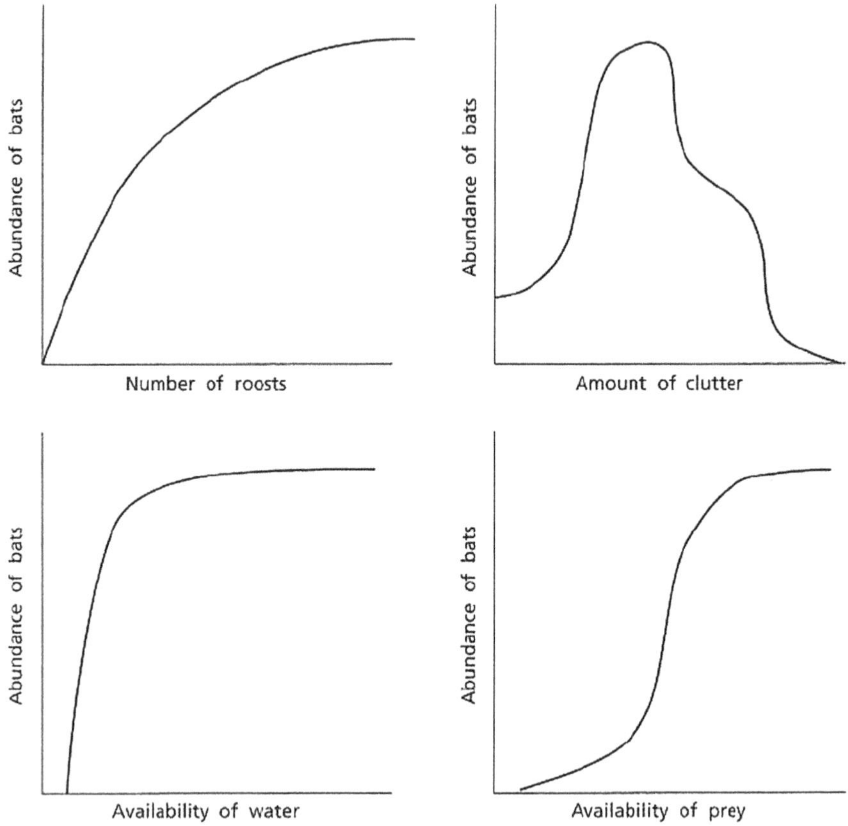

Fig. 5.1 Conceptual models illustrating hypothesised relationships among the abundance of bats and ecological resources within forests (Hayes and Loeb 2007)

5.2 Major Forest Areas

5.2.1 North America

Management of forests in North America is undergoing a renaissance, of sorts, as threats associated with habitat loss and fragmentation, climate change, increased fire frequency, and introduction of forest insect pests are leading to paradigm shifts in how forests should be managed to sustain biodiversity, increase carbon sequestration, and maintain the capacity for resource extraction (Boerner et al. 2008; Parks and Bernier 2010; Moore et al. 2012). Historically, even-aged management was practiced across the continent with clearcuts, shelterwood cuts, seed-tree cuts, and deferment cuts all used in management of forests regardless of region or forest type. These practices have reached their zenith in south-eastern pine plantations where production forestry has led to short rotation harvests of monotypic stands of loblolly pine, *Pinus taeda* (Wear and Greis 2013). More recently, silvicultural approaches have focused on mimicking natural disturbance events or ecologically based forestry (Mitchell et al. 2002; Long 2009), resulting in application of uneven-aged or multi-aged silvicultural systems (O'Hara 2002, 2009), and prescribed fires (Boerner et al. 2008), in both pine and hardwood forests. North America is >24 million km^2 in total land surface and lies entirely within the northern hemisphere. The continent supports a rich diversity of plant species across eight major forest types (Young and Geise 2003, Fig. 5.2) with each type encompassing from 1 to 8 subtypes (SAF 2010). Latitude plays a prominent role in the distribution of forest types across the continent, with a north-to-south pattern of northern coniferous, northern hardwood, central broad-leaved, oak–pine, bottomland hardwood, and tropical forests (Young and Geise 2003). Two other forest

Fig. 5.2 **a** Standing dead ponderosa pine (*Pinus ponderosa*) used as a roost tree by long-legged myotis (*Myotis volans*) in Oregon, **b** forested landscape treated using clearcut logging in Idaho with natural regeneration present, **c** stand of dead trees in California typical of habitats used by bark- and cavity-roosting bats in western coniferous forests, and **d** bottomland hardwood forest in Western Kentucky, with hollow roost tree of Rafinesque's big-eared bat (*Corynorhinus rafinesquii*) in the centre. *Photograph credits* M. Baker (**a**), M. Lacki (**b**, **c**), and J. Johnson (**d**)

types, Pacific coast and Rocky Mountain, are distributed largely in a north–south direction paralleling several mountain ranges and, thus, cross a greater expanse of latitudes. The northern coniferous and boreal forest, dominated by spruce, fir, and larch, covers the largest extent of North American land surface of any forest type, followed by Rocky Mountain and central broad-leaved forests. Rocky Mountain forests are dominated by pines across much of their range, with central broad-leaved forests supporting oaks, hickories, maple, and beech. Bottomland hardwoods, comprising gums, bald cypress, oaks, and willows, represent the smallest land area of any major forest type in North America (1.25 million ha remaining; Mississippi Museum of Natural Science 2005). Globally, North America has experienced some of the greatest forest losses with a 5.1 % decline in forest land cover from 2000 to 2005 (Hansen et al. 2010). Declines in forest cover have been greatest in the south-eastern USA, where 3.5 million ha have been lost from 1992 to 2001 (World Resources Institute 2014). Recent shifts in the region-wide approach to management of south-eastern bottomland hardwood forests, however, have brought about a reversal in the trend of loss of these forests (USDA Forest Service 2009; Miller et al. 2011).

5.2.2 *Europe*

Europe consists of 50 countries and is just over 10 million km^2 in land area. Forests cover approximately 45 % of the land area, most of which is found within the Russian Federation which comprises 40 % of the land area of Europe (FAO 2012). Europe's native forest is very diverse with 13 broad categories encompassing 74 types (EEA 2006). Boreal forest consisting primarily of spruce or pine species dominates in northerly latitudes that comprise Scandinavia (Fig. 5.3). This is replaced by hemiboreal forest and nemoral coniferous and mixed broad-leaved/coniferous forest in southern Sweden and much of eastern central Europe, with alpine coniferous forest along the mountain ranges. Moving west, mesophytic deciduous and beech forest dominates, but there is increasing amounts of plantation forest. In the southern parts of Europe coniferous (pines, firs, junipers, cypress, cedar), broadleaved (oak, chestnut) and evergreen broadleaved forests are the main wooded habitats. Parts of Europe have undergone extensive deforestation and cover has been fragmented and depleted for several centuries. While 26 % of Europe's forest area is classified as primary, this falls to <3 % excluding the Russian Federation, and approximately 52 % of all forests in Europe are now designated primarily for production (FAO 2012). In Europe, as in North America and Australia, there is growing interest in silvicultural practices that mimic natural forest ecosystem processes with the aim of developing mixed, structurally diverse stands (Lähde et al. 1999). This is a result of a move away from treating forests, particularly plantations, solely as a resource for timber, and an increased emphasis on sustainable management for multiple objectives including biodiversity conservation and recreation (Mason and Quine 1995). In practice, this has meant a

Fig. 5.3 **a** New Forest, United Kingdom: wood pasture, a historical European land management system providing shelter and forage for grazing animals as well as timber products, **b** double-leadered Corsican pines (*Pinus nigra* ssp. *laricio*) are used as roost sites by Natterer's bats (*M. nattereri*) in Tentsmuir forest in Scotland, UK; **c** wooded landscape, including olive groves, used extensively in southern Italy by *Rhinolophus euryale*; **d** typical Bechstein's bat (*Myotis bechsteinii*) foraging habitat in England, UK: a mixture of oak (*Quercus robur*) and hazel (*Corylus avellana*) woodland. *Photograph credits* J Sjolund, G Mortimer (**b**), D Russo (**c**), F Greenaway (**d**)

reduction in clear felling, although this varies greatly between countries. For example, it has been largely phased out in Switzerland and Slovenia, but is still the primary form of logging in the UK (Fries et al. 1997; Mason et al. 1999), but recent modifications include retaining stands with longer rotations where possible (Mason and Quine 1995), reducing the removal of deadwood (Humphrey and Bailey 2012), and techniques geared to mimic natural disturbance such as prescribed burning.

5.2.3 Australia

It is estimated that forests covered about a third of the Australian continent at the time of European settlement in 1788, but by the mid-2000s this had been reduced to about 19 % cover (Montreal Process Implementation Group for Australia 2008). Five million hectares of forest are classified as old growth (22 %) and over 70 % of these occur in conservation reserves. Timber harvesting on public land is now restricted to 9.4 million ha, or about 25 % of the areas potentially suitable for timber production, and much of this has been previously logged. Eucalypts dominate the forests of Australia, and they are highly diverse comprising 500–600 species (Fig. 5.4, Florence 1996). Eucalypt forests range from those with a high diversity of eucalypt species to those dominated by one or a few species, the latter most often occurring in the tall wet forests of temperate southern Australia, including Tasmania (Florence 1996). These different eucalypt species and forest communities grow on different soils, under varying climates and natural disturbance regimes that in turn influence the variety of silvicultural practices applied. Fire is

Fig. 5.4 Eucalypt forests of Australia: **a** narrow vehicle tracks through regrowth wet sclerophyll forest are used extensively by bats; **b** recently thinned regrowth forest potentially increases flight space and foraging opportunities for bats; **c** senescing crown of a Blackbutt *Eucalyptus pilularus* supports multiple hollow branches where bats, including maternity colonies, selectively roost; **d** an old-growth, spotted gum forest, *Corymbia maculata,* supports high densities of hollows and an open zone above a dense understorey/shrub layer, providing a variety of niches for foraging and roosting bats. *Photographs* B. Law

also a driving force behind the distribution and composition of eucalypt forests, and it occurs as massive wildfires that sweep across the landscape and less intensive prescribed burns that aim to reduce fuel loads and minimise damaging wildfires. To some extent, silvicultural practices aim to mimic these disturbance events and maximise regeneration after harvest.

Silviculture of Australian eucalypts is thus highly variable, although the techniques applied largely resemble those used elsewhere around the globe. For example, silviculture varies from clearcut practices in the tall wet eucalypt forests of temperate southern Australia (Tasmania and Victoria) to group selection and single tree selection in warm temperate and subtropical areas to the north. Clearcuts aim to mimic broadly the massive stand replacement events created by wildfires, which are an irregular feature of tall eucalypt forests in Australia. However, one important difference between clearcuts and wildfires is that wildfires leave legacies in the form of dead trees with hollows that can remain standing for decades. Regrowth after harvesting may take many decades to self-thin sufficiently for the forest to begin to resemble the openness of mature or unlogged forest (Florence 1996). Selective logging can occur at a range of intensities that are almost a continuum from very low levels of tree removal targeting specific size/species of trees with ~10 % of tree basal area removed to almost a seed-tree retention silviculture with >60 % of stand basal area removed. In selectively harvested forests, nominal 'rotations' are about 60–80 years though these develop from repeated logging visits to the same coupes every 10–30 years to produce a dynamic of multi-aged mosaics of even-aged regeneration cohorts (Curtin et al. 1991). Selective logging is most commonly applied to forests comprising mixed eucalypt species and uneven ages. Rainforest has a restricted occurrence in Australia, and logging of this forest type is no longer permitted.

5.3 Complexity of Bat Habitat Needs

5.3.1 Mature, Large Diameter Trees

Older age classes of trees, especially old-growth forests, have historically been viewed as important habitats for bats (Altringham 1996; Fisher and Wilkinson 2005; Hayes and Loeb 2007) and are likely to contain a greater diversity and abundance of insect prey (e.g. Fuentes-Montemayor et al. 2012; Lintott et al. 2014). Early studies demonstrated variation in bat activity across stands of different age classes, with the levels of bat activity higher in older, mature stands than young stands (Thomas 1988; Erickson and West 1996; Crampton and Barclay 1998; Law and Chidel 2002). Older forests possess canopies that are more fully developed than regenerating or early-seral forests, with complex crown architecture (Wunder and Carey 1996). Old-growth forests are also likely to contain a larger number of microhabitats which are associated with higher bat species richness and higher levels of activity in common and Nathusius pipistrelles, *Pipistrellus pipistrellus* and *P. nathusii*, in oak, *Quercus* spp., forests in southern France managed for coppice (Regnery et al. 2013a). In a parallel study, time since cutting was the best predictor of the number of tree microhabitats which were 13 times more abundant in stands >90 years post-cutting, than those <30 years in age (Regnery et al. 2013b).

Considerable research has been undertaken on roost selection since pioneering radio-tracking studies in Australia (Lunney et al. 1988; Taylor and Savva 1988). A consistent trend throughout the world is that most bats prefer to roost in larger diameter trees (>30 cm, Russo et al. 2004; ~80 cm, Baker and Lacki 2006; see also Kalcounis-Rüppell et al. 2005), often in older forest stands or mature forests (Lunney et al. 1988; Taylor and Savva 1988; Brigham et al. 1997; Betts 1998; Crampton and Barclay 1998; Sedgeley and O'Donnell 1999; Law and Anderson 2000; Lumsden et al. 2002; Mazurek and Zielinski 2004; Russo et al. 2004, 2010; Ruczyński et al. 2010). Such trees have a greater likelihood of supporting larger populations of roosting bats and persist for longer than smaller diameter dead trees (Lacki et al. 2012); thus, their identification and provision in residual patches during timber harvesting is important. Where mature forest is absent across large areas at least some species find roosts in scattered hollow trees in regrowth forest where habitat trees were not specifically retained, indicating that bats typically roost in the largest available trees. One Australian study found that the 4-g eastern forest bat, *Vespadelus pumilus*, which ranges over relatively small areas, maintains similar sizes of maternity colonies in the scarce roosts remaining within regrowth forest compared to maternity colonies in old-growth forest (Law and Anderson 2000). Russo et al. (2010) found evidence of roost selection flexibility in barbastelle bats, *Barbastella barbastellus*; dead and dying trees, a favoured roost site for this species, were six times more common in unmanaged than managed European beech, *Fagus sylvatica*, forests in central Italy. Bats, however, were able to roost within managed forest, albeit in smaller numbers by exploiting roost sites in live trees and rock crevices. Few studies have investigated roost selection in younger forest where roosts are scarce, so generalisations are difficult (although see section on Plantations below).

5.3.2 Deadwood Availability and Hollow Tree Density

Until the late twentieth century, in many parts of Europe and North America, deadwood in managed forests was removed due to concerns over forest health. While this is still common practice in some areas, the key role played by dead and decaying wood in the functioning and productivity of forest ecosystems, and its importance for biodiversity, has gained increasing recognition over the past 20 years (Humphrey 2005). In Australia, deadwood removal has been confined to plantations, though recognition of the importance of specifically retaining old trees with hollows in managed forests originated in the 1980s. A preference for roosts in dead and dying trees has been noted for *Barbastella* and *Nyctalus* species in Europe (Russo et al. 2004; Ruczyński and Bogdanowicz 2008; Hillen et al. 2010), and high densities of dead trees appear to be strongly correlated with the presence of roosts of bark and cavity-roosting bats in forested ecosystems across North America (Mattson et al. 1996; Sasse and Pekins 1996; Rabe et al. 1998; Waldien et al. 2000; Cryan et al. 2001; Bernardos et al. 2004; Broders and Forbes 2004; Miles et al. 2006; Perry and Thill 2007b; Arnett and Hayes 2009).

The importance of high roost density has also been reported in Australia. In dry Jarrah forest of Western Australia, both Gould's long-eared bat, *Nyctophilus gouldi*, and the southern forest bat, *Vespadelus regulus*, preferred roosting in older forest that contained a much higher density of trees with hollows (16–32 trees ha^{-1}) than shelterwood creation and gap release sites (8–12 trees ha^{-1}) (Webala et al. 2010). These mature forest hollow tree densities are comparable to average densities of live and dead hollow trees in roost areas used by Gould's wattled bat, *Chalinolobus gouldii*, (17 ha^{-1}) and the lesser long-eared bat, *N. geoffroyi*, (18 ha^{-1}) in a fragmented landscape in south-eastern Australia (Lumsden et al. 2002). Greater densities of hollow trees likely facilitate roost switching in bark and cavity-roosting bats or fission–fusion behaviours (Kerth and König 1999; Willis and Brigham 2004). These behaviours lead to complex patterns of use and movement among available roost trees by colonies of forest bats. The variation in numbers of roosts between core and peripheral areas of roost networks is further influenced by the density and spatial distribution of available roost trees, as demonstrated for Rafinesque's big-eared bat, *Corynorhinus rafinesquii*, in south-eastern bottomland hardwood forests of North America (Johnson et al. 2012b). Roost networks of northern long-eared bat, *Myotis septentrionalis*, in actively managed forests were scale-free and connected to a single central-node roost tree (Johnson et al. 2012a). A similar pattern was observed for the open-space foraging white-striped free-tail bat, *Tadarida australis*, in south-east Queensland (Rhodes et al. 2006). Given these patterns, we postulate that implementation of silvicultural systems, which promote retention of higher densities of dead and old living trees across forested ecosystems, should benefit bark- and cavity-roosting bats and facilitate 'natural patterns' in colony behaviours, social interactions, and the use of roost networks.

5.3.3 Understory Vegetation

The extent and composition of understory vegetation in forests strongly influences insect prey availability, the ability of bats to access the forest interior, and the microclimates available and is also likely to affect risk of predation. The degree to which understory cover affects the use of forests by bats depends greatly on their wing morphology and foraging behaviour, with some bats benefitting from a more open forest with little in the way of cover, while other species rely heavily on a well-developed dense understory (e.g. Hill and Greenaway 2008; Müller et al. 2012). Vegetation structure revealed by LiDAR in Germany indicated that while high levels of understory cover were preferred by edge-space and gleaning species, open-space foragers were more associated with relatively open forest stands (Jung et al. 2012). Foraging intensity also varies with canopy height, with the activity of open-space foragers highest above the canopy (Kalcounis et al. 1999; Müller et al. 2013), although few studies have surveyed bats at those heights. Similarly, in forest fragments in Scotland (UK), high activity levels of edge-space species, e.g. *Pipistrellus* spp., are related to low tree densities and an open understory, while closed-space gleaning species, e.g. Natterer's bat, *Myotis nattereri*, showed the opposite trend. These studies are supported by numerous species-specific studies. For example, roosts of Bechstein's bat, *Myotis bechsteinii*, and the barbastelle bat, *B. barbastellus*, are strongly associated with areas of thick understory (Greenaway and Hill 2004), and core foraging areas for brown long-eared bat, *Plecotus auritus*, a closed-space species, were associated with more cover and a well-developed understory layer more than peripheral areas (Murphy et al. 2012). An Australian study of vertical stratification (excluding above the canopy) in spotted gum forest also found the understorey to support the greatest insect abundance, although bat activity was up to 11 times greater in the canopy where there was less clutter and presumably insects were more accessible (Adams et al. 2009). There was no evidence that any one ensemble or ensemble species foraged exclusively at a particular height, although the open-space ensemble was most activity in the canopy.

5.3.4 Slope and Aspect

Slope and aspect influence roost selection in forest bats by creating variation in the amount and extent of solar heating at roosting sites due to differences in shading effects and the length of the day that roosts are in direct sunlight. Studies have demonstrated the importance of both slope position and reproductive stage in roost selection. For example, long-legged myotis, *Myotis volans*, in the northwestern USA switch between riparian bottoms and upper-slope positions during pregnancy, but select roosts in upper-slope positions during lactation, where they would be exposed to greater solar radiation (Baker and Lacki 2006). Studies of bats in south-eastern forests of North America have also observed preferences for roosting in upper-slope positions by foliage-roosting eastern red bat, *Lasiurus borealis*, and bark- and cavity-roosting bats (*Myotis* and *Eptesicus*) (Hutchinson and Lacki 2000; Lacki and Schwierjohann 2001; Perry et al. 2008), suggesting that higher slopes are important for roost selection in some forest bat species in both eastern and western parts of North America and should be accounted for in forest planning. Use of lower slope positions and riparian corridors for roosting is common in several bats in eastern and south-eastern forests, however, including bark- and cavity-roosting (Watrous et al. 2006; Perry and Thill 2008; Fleming et al. 2013) and foliage-roosting species (Perry et al. 2007a; Hein et al. 2008b; O'Keefe et al. 2009). Roosting on lower slopes was also found in a subtropical Australian forest, where lactating eastern forest bats, *V. pumilus*, roost in hollow trees in riparian zones during early summer, but shift up-slope during autumn when bats begin to mate (Law and Anderson 2000). Riparian zones may provide more buffered conditions for maternity roosts in warm, subtropical locations.

In the northern hemisphere, selection of south-eastern-facing (Willis and Brigham 2005), south-facing (Klug et al. 2012), and eastern-facing (Perry and Thill 2007a) sides of tree canopies by hoary bats, *Lasiurus cinereus*, is associated with positive energy savings and is hypothesised to facilitate rapid growth of young (Klug et al. 2012). Eastern red bat, *L. borealis*, another foliage-roosting species, was observed using the south aspect of tree canopies that were also located in south-facing slope positions (Mormann and Robbins 2007). Collectively, these behaviours suggest consideration be given to creating and maintaining edge habitats for foliage-roosting bats at the landscape scale, especially along south-facing slopes in the northern hemisphere in areas with sufficient topographic relief.

5.3.5 Forest Edge

Loss and fragmentation of forest habitat are accompanied by an increase in the ratio of forest edge to interior forest, and the response of bats to this can vary among species. Roosting ecology and edge-affinity have been identified as good predictors of the sensitivity of individual bat species to habitat fragmentation; 'forest interior' species (often tree-roosting bats) are negatively affected by fragmentation, as opposed to species which show affinity for forest edges (Meyer et al. 2008).

Edge habitats can influence roosting behaviour in bark- and cavity-roosting *Myotis* species differently. Indiana bat, *M. sodalis*, and northern long-eared bat, *M. septentrionalis*, two species with overlapping distributions in North America and similar preferences for roosting in dead trees (Foster and Kurta 1999; Lacki et al. 2009), choose roosts differently in the same forested landscapes. *M. sodalis* prefers roosts in edge habitats with low vegetative clutter and higher solar exposure of roost trees and *M. septentrionalis* selects roosts in shaded environments within intact forests (Carter and Feldhamer 2005). Russo et al. (2007) found that barbastelle bat, *B. barbastellus*, emerged later from tree roosts in more open forests, probably as a result of increased predation risks, and suggested that it was important to ensure canopy heterogeneity to provide a range of roosting conditions. Edge effects also influence foraging behaviour in forest bats although results from studies comparing bat activity at the edge compared to forest interior show contrasting results; all five species spanning the open/edge-space/closed-space spectra that were assessed in forests in Canada showed higher activity at the forest edge than in the interior (Jantzen and Fenton 2013). Bat activity was also high along coupe edges 5–8 years after clear fell in Tasmania (Law and Law 2011), partly because bats avoided the large harvested gaps in these coupes. In contrast, of three species surveyed within forest fragments on farmland in the UK, one edge-space species showed similar levels of activity at edge verses interior while the other two species (one edge-space and one closed-space) showed higher levels of activity within the forest interior (Fuentes-Montemayor et al. 2013). In Australia, harvested forests are often interspersed with old logging trails and fire trails, providing 'edge habitats' that facilitate the use of dense forest regenerating after harvest by bats that possess a range of traits (Crome and Richards 1988; Law and Chidel 2002; Webala et al. 2011). Activity on trails in regrowth forest is as high as it is in mature forest. Most importantly, foraging activity is typically much higher on forest trails than within the forest remote from trails or along narrow riparian zones (Law and Chidel 2002; Lloyd et al. 2006; Webala et al. 2011). Use of trails as linear edges in regenerating forest has also been reported in North America (Menzel et al. 2002). These observations highlight the importance of edge habitats to many bat species within each ensemble, in all the regions covered in this chapter.

5.4 Bat Responses to Silvicultural Treatments

Silviculture involves a diverse range of techniques to manipulate growth conditions, extract resources, and facilitate regeneration within forests. These influence the composition and density of tree species present, the extent and composition of the understorey vegetation and ultimately the resources available for bats. Here, we focus on the techniques for which there is at least some information on the response of bats to (1) different logging strategies, (2) thinning regimes, and (3) the use of harvest exclusion areas. We also examine the use of timber plantations by bats which, in some regions, is the focus of silvicultural activities. There is very little information on the effects of other techniques such as coppice and the use of chemical applications (e.g. herbicides to clear vegetation), and we highlight important knowledge gaps in the concluding section.

5.4.1 Logging

Historically, the strategy for logging in forest managed for timber extraction was to remove all trees within an area (clearcuts) as this is considered the most economically profitable method. In production State Forests in Australia, selective harvesting was most common before World War II, but it was subsequently recognised that this adversely affected the regeneration and growth of many of the fastest growing, commercial species, which subsequently led to increased intensity of harvests. Recent concern over the environmental (including biodiversity loss and soil erosion) and visual impacts, however, has led to increased use of more selective forms of logging including variable retention and group selection techniques, which are reviewed here.

A review of published data sets on response of forest bats to silvicultural logging indicates that there are major gaps in our understanding of relationships of bats with timber harvesting practices (Table 5.1). In particular, there is a notable

Table 5.1 Summary of bat response in activity and roost selection to silvicultural treatments referred to in this review for North America and Australasia

Treatment(s)	Treatment conditions	Forest type	Bat species	Bat response	Source
Bat activity					
Even-aged treatments					
North America					
Clearcut	30 ha	Pacific coast	*M. lucifugus*	None	Lunde and Harestad 1986
Clearcut	Not defined	Northern hardwood	*L. borealis*	Decrease	Hart et al. 1993
			L. cinereus	Increase	
			Myotis sp.	Decrease	
Clearcut	2–3 years old	Pacific coast	*E. fuscus*	Increase	Erickson and West 1996
			L. noctivagans	Increase	
			C. townsendii	Increase	
Clearcut	Not defined	Northern coniferous	Multiple	Mixed	Grindal 1996
Clearcut	Along streams	Pacific coast	*Myotis* sp.	Decrease	Hayes and Adam 1996
Clearcut	5–17 years old	Pacific coast	Multiple	Decrease	Parker et al. 1996
Clearcut and residual patches	Varied patch isolation	Northern coniferous	Multiple	Mixed	Swystun et al. 2001
Clearcut and residual patches	8–10 ha; 1.5 years old; 0.2–0.46 ha	Northern hardwood	*M. lucifugus*	Increase	Hogberg et al. 2002
			M. septentrionalis	Increase	
			L. noctivagans	None	
Clearcut	10 ha	Northern coniferous	*L. noctivagans*	Increase	Patriquin and Barclay 2003
			M. lucifugus	Increase	
			M. septentrionalis	Decrease	
Clearcut; deferment harvest	5 years old; 6–10 m^2/ha residual	Northern hardwood	*L. cinereus*	Increase	Owen et al. 2004
			L. noctivagans	Increase	
			Myotis sp.	None	
Shelterwood harvest	10 ha; 30–50 % decline in volume	Central broad-leaved	*L. borealis*	Increase	Titchenell et al. 2011
			L. noctivagans	Increase	
			E. fuscus	Increase	

(continued)

Table 5.1 (continued)

Treatment(s)	Treatment conditions	Forest type	Bat species	Bat response	Source
Seed-tree and shelterwood harvest	7.7 m^2/ha residual; 18 m^2/ha residual	Northern hardwood	Multiple	Increase	Dodd et al. 2012
Australasia					
Clearcut; post-wildfire	0–250 years old	Tall mountain ash eucalypt	Total activity	Increase with age	Brown et al. 1997
Clearcut and Variable retention	10–27 ha; 8 years old; 0.5–1 ha retention	Tall wet eucalypt forest	Multiple	Mixed	Law and Law 2011
Plantations	Non-commercial mixed; <10 and 20–25 years old	Eucalypts	Multiple	Positive, older plantations	Law and Chidel 2006
Plantations	Low rainfall monoculture; <11 years old	Eucalypts	Multiple	Neutral	Law et al. 2011
Uneven-aged treatments					
North America					
Group selection cuts	0.1–0.8 ha; ≤9 years old	Northern hardwood	Multiple	Increase	Krusic et al. 1996
Group selection cuts	60 % decline in volume	Northern coniferous	Multiple	Increase	Perdue and Steventon 1996
Small cutblocks	0.5–1.5 ha	Northern coniferous	Multiple	Increase	Grindal and Brigham 1998
Group selection cuts	0.02–0.5 ha gaps	Southern oak–pine	Multiple	Increase	Menzel et al. 2002
Canopy gaps	16–33.5 m wide	Northern hardwood	*E. fuscus*	Increase	Ford et al. 2005
			L. cinereus	Increase	
			Myotis sp.	Decrease	
Australasia					
Selective	18 % basal removal 1–6 years old	Tropical rainforest	Multiple	Mixed	Crome and Richards 1988
Selective	3 age classes	Wet sclerophyll eucalypt	Multiple	Mixed	de oliveira et al. 1999
Alternate coupe	15 ha coupes, 22 years old	Dry sclerophyll eucalypt	Multiple	Mixed	Law and Chidel 2001

(continued)

Table 5.1 (continued)

Treatment(s)	Treatment conditions	Forest type	Bat species	Bat response	Source
Group selection/ plantation/old growth	13–97 ha catchments; 16 years old; tracks versus interior	Wet sclerophyll eucalypt	Multiple	Mixed	Law and Chidel 2002
Group selection cuts	3 age classes; riparian buffers 10–50 m	Wet and dry sclerophyll eucalypt	Multiple	Mixed	Lloyd et al. 2006
Group selection cuts	Old vs young regrowth; tracks vs interior; vertical stratification	Spotted gum eucalypt	Multiple	Mixed	Adams et al. 2009
Gaps and shelterwood	3 age classes; gaps <10 ha; tracks vs interior	Dry sclerophyll eucalypt	Multiple	Mixed	Webala et al. 2011
Variable retention	10–100 % retention; 100 ha blocks	Tableland eucalypt	Guilds	Mixed	Law unpubl. data
Intermediate treatments					
North America					
Thinning	10–13 years old	Pacific coast	Multiple	Increase	Erickson and West 1996
Thinning	≥10 ha; 55 % decline in density	Pacific coast	Multiple	Increase	Humes et al. 1999
Thinning	25 % decline in density;	Northern coniferous	Multiple	None	Patriquin and Barclay 2003
Thinning	45 % decline in density	Northern pine plantation	Multiple	None	Tibbels and Kurta 2003
Thinning	18 m^2/ha residual	Southern oak–pine	*E. fuscus*	Increase	Loeb and Waldrop 2008
			L. borealis	Increase	
			P. subflavus	None	
Salvage logging	Control, moderate, and heavily logged sites × 4 replicates (12–16 ha); 1 year post-fire	Douglas, white and ponderous fir	Multiple	Positive	Hayes 2009

(continued)

Table 5.1 (continued)

Treatment(s)	Treatment conditions	Forest type	Bat species	Bat response	Source
Australasia					
Thinning	4–9 years old	Spotted gum eucalypt	Multiple	None	Adams and Law (2011)
Europe					
Salvage logging	4 stand types varying by logging & structure × 8 replicates (5 + ha each)	Norway spruce, beech and silver fir	Multiple	Varied between foraging guilds	Mehr et al. 2012
Roost selection					
Even-aged treatments					
North America					
Clearcut	7–18 ha	Northern coniferous	*M. evotis*	Positive, tree stumps	Vonhof and Barclay 1997
Cutblocks with residual patches	Not defined	Northern coniferous	*Myotis* sp.	Positive, edges	Grindal 1999
Australasia					
Clearcut	11 years old	Dry sclerophyll eucalypt	Multiple	Positive, mature forest and diameter	Taylor and Savva 1988
Plantation/regrowth versus old growth	30 years old	Wet sclerophyll forest	*V. pumilus*	Positive, gullies and diameter	Law and Anderson 2000
Plantation	Exotic; mosaic age classes	*Pinus radiata*	*C. tuberculatus*	Positive, old age classes and near water	Borkin and Parsons 2011b
Uneven-aged treatments					
North America					
Group selection and thinning	13.8 m^2/ha residual	Southern oak–pine	*M. septentrionalis*	Positive	Perry and Thill 2007b
Group selection and thinning	13.8 m^2/ha residual	Southern oak–pine	5 of 6 species	Positive	Perry et al. 2008
Australasia					
Alternate coupe	10–20 ha; 2–3 years old	Dry sclerophyll eucalypt	*N. gouldi*	Positive, gullies and diameter	Lunney et al. 1988

(continued)

Table 5.1 (continued)

Treatment(s)	Treatment conditions	Forest type	Bat species	Bat response	Source
Gaps and shelterwood	gaps <10 ha; buffers; 20–30 years old	Dry sclerophyll eucalypt	*V. regulus*	Positive, mature forest and diameter	Webala et al. 2011
			N. gouldi	Positive, retained trees & diameter	
Intermediate treatments					
North America					
Thinning	150–309 trees/ha	southern pine plantation	*L. borealis*	Positive	Elmore et al. 2004
Thinning	13.8 m^2/ha residual	Southern oak–pine	*L. borealis*	Positive	Perry et al. 2007a
			L. cinereus	Positive	

lack of long-term, longitudinal studies that track changes in bat assemblages and their forest habitat over time. Studies on bats in even-aged systems have largely focused on responses to clearcuts with limited exploration of two-age systems such as seed tree, shelterwood, or deferment harvests (Owen et al. 2004; Titchenell et al. 2011). Clearcut harvests have been used with less frequency, especially on public lands, for some time now (USDA and USDI 1994), though they still persist in cool temperate forests, such as those of Tasmania (Law 1996), and some European countries. Patterns in bat responses to clearcuts are still helpful, however, in understanding the potential effects on bats of future directions in forest management based on even-aged systems. Bat responses to uneven-aged systems, such as small cutblocks, patch cuts, or group selection harvests, have received greater attention and have been evaluated across multiple bat species and forest types, so inferences can be drawn on the efficacy of these silvicultural systems for bats. In North America, more studies have evaluated bat response to thinning than any other silvicultural treatment, with thinning often applied in combination with other treatments on the same landscape (Erickson and West 1996; Patriquin and Barclay 2003; Loeb and Waldrop 2008; Perry et al. 2008). Studies of treatment combinations are important as future directions in the management of forests in North America are emphasising multi-treatment prescriptions (Aubry et al. 2009; Harrod et al. 2009; Hessburg et al. 2010), to increase structural habitat complexity, both vertically and horizontally, while reducing the impact of insect infestations and the threats of wildfire and global climate change (Boerner et al. 2008; Parks and Bernier 2010; Duerr and Mistretta 2013). Some forest management strategies specifically target bats, though often bats are catered for under broad forest prescriptions that aim to accommodate the needs of a range of forest-dependent species in an area (Law 2004).

There is a surprising lack of European studies on the effects of any logging strategy on bats and the only study found for this review which directly related to this issue was one on the effects of salvage logging (see Sect. 4.1.4). This is especially concerning given the strict protection afforded to all bat species and particularly bat roosts under the EU Habitat Directive; this prohibits deliberate disturbance of all bats during any stage of their life cycle as well as the destruction of breeding sites or resting areas. As such, the timing of forest harvesting needs to consider whether bats may be roosting in targeted areas (e.g. Forestry Commission 2005). There are no such restrictions in Australia; though for New Zealand pine plantations, Borkin et al. (2011) recommends that harvests should be planned when bats are not heavily pregnant nor have non-volant dependents. In eastern North America, logging is currently restricted from 15 October through 31 March across the distribution of the endangered Indiana bat, *Myotis sodalis*, as this bat uses live and dead trees as maternity sites during the growing season (USFWS 2009). Restrictions are further constrained to a start date of 15 November within 16 km of known hibernacula of the species (USFWS 2009). The implications of white-nose syndrome and the extensive mortalities of cave-hibernating bats in North America (USFWS 2012) are likely to add species of forest bats to the threatened and endangered species list in the USA, leading to further restrictions on logging. Missing in all of the dialogue, however, is any direct link of impact, or mortality of bats, during logging operations and studies of these potential impacts are needed (but see Borkin et al. 2011).

5.4.1.1 Clearcut and Deferment Harvests

Response of forest bats to clearcut harvests has been mixed across forest types and species of bats (Table 5.1). For example, three studies each in different locations within the Pacific coast forest type found no response to clearcuts by little brown bats, *Myotis lucifugus*, in British Columbia (Lunde and Harestad 1986), a decrease in overall bat activity over clearcuts in south-eastern Alaskan rainforests (Parker et al. 1996), and an increase in activity of big brown bats, *Eptesicus fuscus*, silver-haired bats, *Lasionycteris noctivagans*, and Townsend's big-eared bats, *Corynorhinus townsendii*, in clearcuts in western Washington (Erickson and West 1996). Patterns in bat activity recorded in and around clearcut harvests are influenced by three factors: the number of years post-harvest when data were collected, the size and shape of cutblocks studied, and the assemblage of bat species present in the area. When reported, the age of clearcut stands in North America evaluated post-harvest ranged from 1.5 to 17 years. This range in age is wide and likely spans considerable variation in above-ground habitat structure due to differences in the amounts of regeneration present; thus, a varied response by bats across studies and geographic locations should be expected. In montane eucalypt forests of south-eastern Australia, bat activity peaked in 165-year-old wildfire regrowth rather than in younger regrowth from clear-felling operations (Brown et al. 1997). Unfortunately, the size and shape of clearcuts studied are rarely reported so an evaluation of the effects of cut size and shape on bat activity cannot be made.

Focusing on traits is likely to provide more insights into the response of bats to the large gaps created by clearcut harvests. In North America, two trends are evident. First, the creation of less obstructed flight space over clearcut stands generally leads to increased levels of activity of edge/open-space bats that possess moderate to high aspect ratios and often higher wing loadings (Lacki et al. 2007). This mix of bats includes the foliage-roosting *Lasiurus* species, along with others (*Lasionycteris* and *Eptesicus*) (Table 5.1). The length of years post-harvest at which this increase in bat activity is sustained is less clear and likely is affected by tree species composition and the speed at which regeneration proceeds in harvested stands at a particular geographic location. Second, the response to clearcut harvests between *Myotis* species varies both within and among species (Patriquin and Barclay 2003), with some increase in activity associated with linear edge habitats at the periphery of cuts but reduced activity in the centre of harvested stands, except where residual patches are left behind (Hogberg et al. 2002). As our ability to distinguish among *Myotis* species increases with technological advances in acoustic detectors and software packages (Britzke et al. 2011), resolution among the full suite of *Myotis* bats in North America should become possible allowing for a more in-depth and complete evaluation of bat response to edge effects in actively managed forests.

Data on bat responses to even-aged systems other than clearcuts are sorely lacking. A study of bat activity in deferment harvests found high levels of activity of silver-haired bats, *L. noctivagans*, in stands with 6–10 m^2/ha of basal area remaining (Owen et al. 2004), and the only study examining bat activity in shelterwood harvests (30 to 50 % reductions in basal area) observed higher levels of activity in three species of bats that have wing morphologies and echolocation call structures possessed by edge/open-space bats (Titchenell et al. 2011). Patterns of habitat use by radio-tagged northern long-eared bats, *M. septentrionalis*, a closed-space bat, showed this species spent limited time in deferment harvest stands, especially harvested sites with more open canopies and less cluttered foraging space (Owen et al. 2003).

For roosting bats, gap release and shelterwood systems retain tall and large diameter hollow-bearing trees within stands possessing less clutter than surrounding forest regenerating after harvest and these offer potential roosts for bats. However, in Western Australia, southern forest bat, *V. regulus*, avoided locating roosts in shelterwood treatments when older forest was available nearby (Webala et al. 2010). In general, remnant trees in these silvicultural treatments, including retained 'habitat trees', were not preferred as roost sites by *V. regulus*, though a second species (*N. gouldi*) frequently used such trees. One possible reason for avoiding using 'habitat trees' as roosts was the relatively low density of hollow roosts (see 3.2 Deadwood availability and hollow tree density).

5.4.1.2 Variable Retention Harvests

Variable retention has recently been proposed as an alternative to standard clearcuts, whereby old-growth elements are retained within the clearcut coupe

(Baker and Read 2011). Variable retention increases the availability of edges, for example, around retained patches (aggregates) of undisturbed forest within the clear-fell coupe and along coupe boundaries as well as increasing the area of open space. Open- and edge-space ensembles would be expected to benefit from this treatment. The 200-ha Silvicultural Systems Trial, in Tasmania, provides one of the main experimental sites in Australia for investigating responses to variable retentions. Bat activity was similar in control coupes of 45- to 60-m-tall old-growth Messmate Stringybark, *Eucalyptus obliqua*, forest, compared to variable retention coupes 5–8 years after logging (Law and Law 2011). Activity was lower above the dense young regeneration of clear-fell-burn-sow (no retention) coupes and marginally lower for dispersed tree retention coupes. This suggests that the retention of old-growth elements as aggregates or patches moderates the unsuitable young regrowth zone for total bat activity, while retention of dispersed individual trees is less effective. Surprisingly, bat activity was low at the retained aggregates themselves, both in their centre and along the edge, and it is not known to what extent bats roost in these locations. Overall the results are consistent with conceptual models (Fig. 5.1), whereby activity is predicted to be higher in areas of medium clutter levels and where hollow abundance is high. Individual bat taxa responded to treatments consistent with predictions from ecomorphology. Closed-space bats were less active in clearcuts than unharvested forest, large edge-space bats were more active in clearcuts (especially along edges), and smaller edge-space bats were less influenced by patch type and location within coupes; consistent with other studies of forest clearcuts from North America (Grindal and Brigham 1999; Menzel et al. 2002; Patriquin and Barclay 2003).

The age of regenerating forest is likely to be an important influence on how bats respond to variable retention. An unreplicated, operational scale (100-ha forest blocks) experiment established in 1984 in the temperate forests of southern New South Wales (Waratah Creek) (Kavanagh and Webb 1998) was sampled acoustically for bats after 18 years of regrowth. Treatments retained different amounts of tree canopy within four different forest blocks comprising 100 % (control), 50 % (0.5 ha patches in a chessboard pattern), 25, or 10 % tree canopy retention. Control sites supported 2–4 times more activity than logged sites, with 10 % retention supporting the lowest activity level with just 50 bat passes per night of sampling (Fig. 5.5; B. Law, unpubl. data). Thus, bat activity remained low even 18 years after logging and the amount of canopy retained within a block had little impact on activity, except for the block with the most intensive logging which supported the lowest activity level. As expected, the activity of closed-space bats was similar, though low, between the control and treatments, after 18 years. Activity of edge-space bats was three times lower in logged stands, suggesting a loss of edges and spaces between trees, especially in the treatment where logging was most intense. Logging treatments had little effect on open-space bats that forage above the canopy, except that activity was lower where logging intensity was greatest.

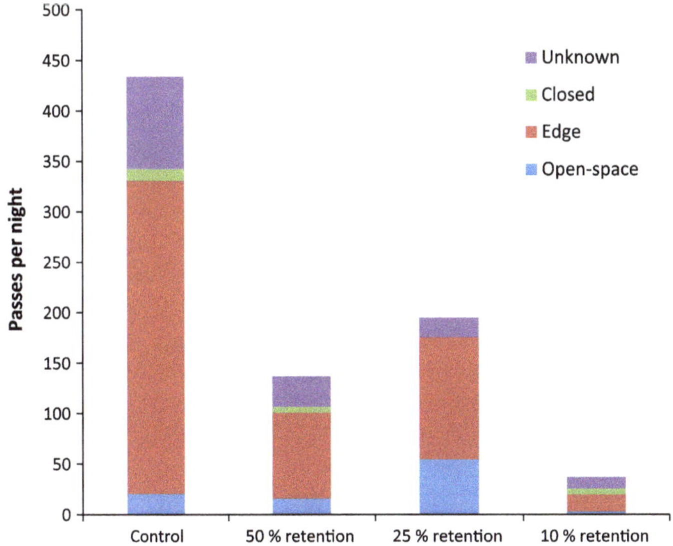

Fig. 5.5 Total bat activity (762 passes, 10 taxa) recorded 18 years after logging in an unreplicated, variable intensity logging experiment in New South Wales, Australia. Data are mean number of passes per night for two Anabat detectors deployed per forest block (~100 ha) over two entire nights of recording and exclude activity on trails (B. Law, unpubl. data). Different bat ensembles are open-space, edge-space, closed-space, and unknown

5.4.1.3 Group Selection Harvests

Changes in ensemble activity with group selection harvest are likely to depend on gap size, with an increase in edge-space activity if gaps are small and open-space activity if gaps are large. Immediately after harvest, closed-space bats are expected to decline, but we predict subsequent recovery if the retention of roost trees is catered for. All studies examining bat responses in North America to group selection harvests, canopy gaps, or small cutblocks consistently reported increases in activity of bats, primarily open/edge-space species, with the opening up of forest canopies, regardless of forest type or assemblage of bats present (Table 5.1). The one exception was a decline in activity of *Myotis* bats in canopy gaps in forests of the central Appalachian Mountains, with this drop off in use inversely correlated with increasing diameter of canopy gaps (Ford et al. 2005). In this study, the maximum gap diameter examined was 33.5 m in width, with the decline in activity with increasing gap size largely attributable to response of closed-space *Myotis* species. Studies in oak–pine forests in Arkansas have demonstrated the use of dead and live trees along gap edges for roosting by several bat species (Perry and Thill 2007b; Perry et al. 2008), demonstrating the importance of maintaining canopy gaps in managed forests. The almost universal response by bats of increased activity with canopy gap formation means this silvicultural treatment holds much promise for management of foraging habitat of bats in the short-term. Use of gaps by forest bats following a decade or more of successional change is likely to be different, however, with overall declines in activity plausible as open/edge-space species disappear or decline in abundance with increasing gap clutter. Such temporal changes need to be identified along with the optimal gap size(s) and the density of gaps required by different species of bats to permit commercially viable, sustained yield harvests while fostering high levels of bat activity and provision of roosting habitat in managed forests.

In contrast to many North American studies that have been undertaken in gaps soon after harvesting, in Australia, most bat research has focused on the use of older regrowth regenerating from group selection harvest, particularly characterising bat species by their traits in relation to the use of these dense stands. There is a general pattern of forest clutter increasing over time after group selection harvest so that old regrowth (>30 years) has significant higher clutter levels than young or older forest, which constrains use by bats to closed-space species with a low wing aspect ratio (Law and Chidel 2002; Webala et al. 2011). Less manoeuvrable edge-space species with a high wing aspect ratio tend to be scarce in regrowth forest (except on flyways provided by tracks and creeks), although their activity is greater in the subcanopy and canopy than understorey (Adams et al. 2009). Vegetation is more cluttered in regrowth at these upper heights (closer stems and less vertical space in the subcanopy), and this leads to less bat activity in such situations (Adams et al. 2009). It is not known whether open-space and low-frequency edge-space species are active above the canopy of these young forests, although this was confirmed by Müller et al. (2013) for mature forests in Europe.

5.4.1.4 Salvage Logging

Salvage logging involves the removal of dead wood after a natural disturbance (e.g. windthrow, forest fires, and insect outbreaks) and has been employed even in protected forests, provoking some controversy. To our knowledge, no research has examined the implications for roost availability of this practice, although removal of standing dead wood will inevitably reduce the abundance and diversity of roosts and would have a considerable impact when carried-out over large scales (Lindenmayer and Noss 2006). We found two studies which investigated changes in bat activity following salvage operations. In Germany, closed-space species reduced their activity in both types of forest clearing (bark beetle and logging), while the activity of open-space species slightly increased, and edge-adapted species showed a mixed response (Mehr et al. 2012). These results are similar to a study in Oregon where the highest bat activity was in the more intensely logged sites (Hayes 2009).

5.4.2 Recovery Times After Timber Harvest

Long-term studies are largely missing from assessments of the response of bats to silvicultural methods. A typical approach uses chrono-sequences or snapshots of comparisons between different silvicultural methods or logging histories and makes the assumption that the matching of treatments is equal and evenly distributed across the same environmental niche and landscape context. Most importantly, a one-year snapshot may not be representative of temporal variation and dynamism over a longer period (Recher et al. 1983; Maron et al. 2005); thus, conservation plans developed from snap-shots can have limitations. Long-term studies are ideal for tracking changes to vegetation structure as forests regenerate after harvesting and how different ensembles of bats respond to these dynamics.

One longitudinal study in Australian eucalypt forests, initiated in 1998, has been investigating alternate-coupe-integrated harvesting for woodchips and sawlogs, and although currently unpublished, a summary is presented here (B. Law and M. Chidel, unpubl. data). Alternate-coupe harvesting divides management units (e.g. 200-ha areas) into small (~15 ha) coupes that are alternately harvested in a chessboard fashion, every 20 years. In 1998, bat activity was recorded after 22 years of regrowth from the first cycle of logged coupes (Law and Chidel 2001). Bat activity in the cluttered regrowth was about half that of adjacent, more open unlogged coupes. This effect was most notable for less manoeuvrable, open- and edge-space vespertilionids that were more active in unlogged coupes.

The site was then sampled at intervals over 13 years following the second round of alternate-coupe logging (B. Law and M. Chidel, unpubl. data; Fig. 5.6). During this period, total bat activity remained low in old regrowth coupes (22 years old in 1998). Activity in unlogged controls remained similar to the initial samples taken prior to second round harvesting. Within the recently logged coupes, activity peaked soon after logging in the large gaps, but it quickly declined and remained at low levels (similar to that found in old regrowth coupes) once young regenerating eucalypts established within eight years of logging. In terms of clutter and total bat activity, these results are only partly consistent with the conceptual models of Hayes and Loeb (2007). The model predicts low bat activity when clutter is very low, yet this was not the case in this study, possibly because gaps were patchy within the 15-ha coupes due to the requirement for retention of 5 habitat trees per ha plus equivalent numbers of recruits, indicating that gap size or scale is likely to be an important issue influencing activity. High activity at intermediate clutter levels (unlogged coupes) and low activity at high clutter levels (old regrowth coupes) are consistent with the model. The response of individual species and ensembles are yet to be analysed for this study.

The lack of recovery after 36 years in old regrowth coupes is consistent with a number of other studies where low activity persisted for more than 30 years after disturbance (Brown et al. 1997; Adams et al. 2009; Webala et al. 2011), but differs from selective harvesting of wet sclerophyll forest in subtropical Queensland where recovery of bat activity was apparent in a site logged 33 years previously

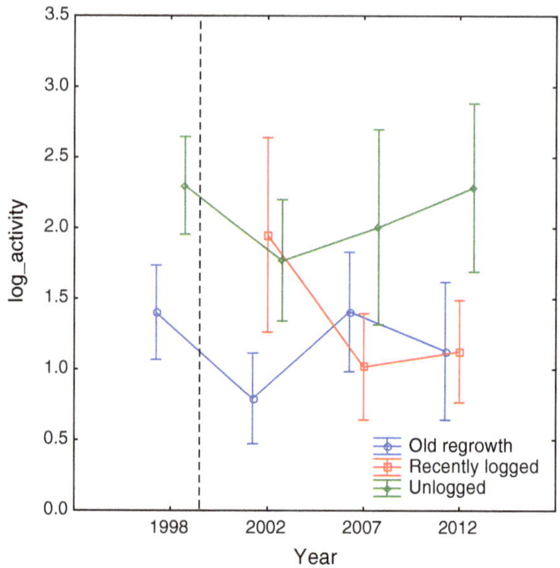

Fig. 5.6 Changes in total bat activity over 14 years in an alternate-coupe logging system in southern Australia (B. Law and M. Chidel, unpubl. data). The *dashed vertical line* indicates second round logging of the alternate unlogged coupes in 1999, which took place 23 years after the first round of logging of adjacent coupes in 1976. All but two unlogged coupes were harvested in 1999 and are thereafter referred to as recently logged coupes. Bat activity is a log transformation of the number of passes per night (±95 % confidence limits) after adjusting with mean nightly temperature as a covariate

(de Oliveira et al. 1999). It is important to note that none of these studies consider activity levels on tracks, riparian zones, or other areas of retention that potentially could ameliorate the effects of clutter from dense regrowth and loss of tree hollows.

5.4.3 Thinning Young Forests

The goal of thinning is to improve the quality and growth of the remaining trees (especially diameter) by reducing the density of trees in a stand. Reducing tree density will decrease canopy cover, at least initially, with increased light levels reaching the forest floor and thus influencing understory cover. Adams and Law (2011) reviewed the literature on thinning and bats and proposed hypotheses for testing that included: (1) activity of edge- and open-space species will increase from pretreatment levels where thinning reduces stem separation to 7 m (~200 stems per ha) but will remain at low levels where average stem separation is less than 3 m (~1100 stems per ha); (2) highly cluttered forests will have low bat activity away from flyways, regardless of the number of potential roosting sites and the abundance of insects, while bat activity in open forests will be highest where roost availability and insect abundance are high.

Consistent with the hypotheses, bat responses to silvicultural thinning have been examined across several forest types in North America with increases in bat activity associated with thinning in Pacific coast (Erickson and West 1996; Humes et al. 1999) and southern oak–pine (Loeb and Waldrop 2008) forests, but not in northern red pine, *Pinus resinosa*, plantations (Tibbels and Kurta 2003) or northern coniferous forests (Patriquin and Barclay 2003). An explanation for these differences is not readily clear, as the extent of thinning is not always reported in metrics that can be compared among study sites, and the suite of bat species present varies among locations. Further, data for bat activity within the *Myotis* genus could not be resolved to the species level with technologies used, preventing an evaluation of responses by ensemble. Patterns in roost selection of *Lasiurus* species in southern oak–pine forests indicate that thinned stands are frequently selected by these bats for roosting. Thus, as with clearcut harvests and larger-sized canopy gaps, stands thinned to basal areas <14 m^2/ha appear to be well suited to less manoeuvrable edge-space *Lasiurus* species by providing suitable roosting and foraging habitats (Perry and Thill 2007a; Perry et al. 2007a, 2008).

The response of bats to forest thinning has received little attention in Australia. A preliminary study found high variability in activity for all bats and ensembles between thinned and unthinned eucalypt stands and among vegetation layers within the forest (Adams and Law 2011). Unexpectedly, thinned regrowth had a higher percentage cover for the shrub layer, and the vertical gap between canopy and understory trees was halved, which represented an increase in clutter in the zone where bats frequently fly and this could have undermined any benefit of wider stem spacings. However, the variability in bat activity within the thinned/control treatments was too high to unequivocally state that thinning had no effect.

While thinning is a commonly employed silvicultural technique across Europe, there has been no study of its effects on bat activity, occurrence, or species richness. There are, however, a few studies which have looked at effects of tree density on bats, thereby providing indirect evidence on likely effects of thinning. For example, in one study, where tree density varied between 180 and 2500 stems per ha in mixed deciduous/coniferous fragments within agricultural landscapes in Scotland (UK), activity of the soprano pipistrelle, *Pipistrellus pygmaeus* (an edge-space forager), decreased with increased tree density. In contrast, the abundance and activity of *Myotis* spp., and the abundance of Diptera, both increased with tree density (Fuentes-Montemayor et al. 2013). This mirrors findings by Müller et al. (2012) where the activity of closed-space foragers and prey abundance increased at higher vegetation densities, while the activity of open-space foragers, and to a lesser extent, edge-space foragers declined.

5.4.4 Harvest Exclusion Areas

Given the low levels of bat activity observed in young regenerating forest after logging, mitigations are needed to ameliorate the effect of high clutter levels and lower numbers of tree hollows. Edge habitat, such as tracks and clearcut boundaries, is extensively used by a range of bat species (Sect. 3.5). In Australia, harvest exclusion areas that support naturally open, undisturbed forest constitute a much greater proportion of the forest landscape compared to forest tracks and are therefore expected to be more important at ameliorating logging impacts on bats given that they also provide roosts in the hollows of old trees. Provided attention is paid to the size and location of harvest exclusion areas these can play a vital role in landscape connectivity, acting as corridors across forested landscapes, permitting bats to reach otherwise isolated blocks of preferred habitat within landscapes where fragmentation has altered the matrix and created an abundance of suboptimal habitat blocks. As the extent of habitat fragmentation increases, so does the importance of corridors on the landscape (Duchamp et al. 2007). Indiana bats, *M. sodalis*, preferred to fly along wooded corridors and avoided open fields in Michigan, even though commuting distances increased by more than 50 % (Murray and Kurta 2004), with similar results for *Pipistrellus* spp. in the UK (Downs and Racey 2006). Activity of bats in heavily fragmented, pine plantations in South Carolina demonstrated more use by bats of edges along corridors than habitats within the corridor interior or nearby stands of timber (Hein et al. 2009a), with bat activity directly correlated with the height of the corridor overstorey.

Riparian corridors in timber production forests are often excluded from harvesting in order to ameliorate impacts of harvesting on water quality as well as providing unharvested productive habitat for biodiversity. Riparian corridors are important areas of bat foraging activity (Hayes and Adam 1996; Zimmerman and Glanz 2000; Brigham 2007), with male and female bats segregating themselves along corridor reaches in upland landscapes, with males more abundant at higher elevations (Grindal et al. 1999; Senior et al. 2005). Activity of bats along riparian corridors appears to be scale-dependent, with vegetation architecture, i.e. shrub and tree cover, influencing the use of foraging space by bats at the local, or finest spatial, scale more than landscape habitat measures or abundance of insect prey (Ober and Hayes 2008). Abundance of Lepidoptera was high in riparian corridors in Arkansas prompting the authors to hypothesise that Ozark big-eared bat, *Corynorhinus townsendii ingens*, a moth strategist (Dodd and Lacki 2007), feeds extensively in and around riparian corridors in the Ozark Mountains (Dodd et al. 2008). Use of best management practices along streamside management zones for sustaining healthy, riparian ecosystems is a well-established forest management practice in many regions of North America (Stringer and Perkins 2001; Lee et al. 2004). Regardless, data on how these practices influence habitat use by forest bats in riparian areas remain limited, with experimental studies sorely needed on the effects of habitat quality within corridors (stand age and composition) and corridor dimensions (size and width) on roosting and foraging ecology of bats. One study in Australia demonstrated that bat activity, foraging rates, and species richness in riparian corridors within selectively harvested eucalypt forest was maintained at levels similar to riparian areas in mature forest (Lloyd et al. 2006). Higher activity was recorded on larger rather than smaller order streams, a pattern also not affected by harvesting history. Such results highlight the benefits of buffers, with riparian areas effectively providing habitat for foraging and commuting bats in selectively logged forests where clutter levels are likely to be high.

Mitigating the loss of roosting habitat in hollow-bearing trees is arguably even more important than maintaining suitable foraging habitat. Forested corridors are critical habitat elements for North American foliage-roosting bats by providing both roosting and foraging opportunities. Male Seminole bats, *Lasiurus seminolus*, in south-eastern loblolly pine, *P. taeda*, plantations chose roost trees in forested corridors within harvest exclusion zones over 60 % of the time, even though corridors represented only 11 % of the landscape area (Hein et al. 2008a). Corridors were 100 to 200 m in width and comprised largely of older-aged forests in riparian and upland slope positions. Use of forested corridors for roosting has been observed in other foliage-roosting species in south-eastern forests, with tri-coloured bats, *Perimyotis subflavus*, selecting riparian corridors (O'Keefe et al. 2009), male evening bats, *Nycticeius humeralis*, choosing upland corridors of mature forest (Hein et al. 2009b), and eastern red bats, *L. borealis*, roosting in the vicinity of gated roads (O'Keefe et al. 2009). Greenbelts in riparian corridors, or unharvested inclusions of mature mixed-pine hardwoods $\geq$50 years in age, were important roosting habitats for these same species in southern oak–pine forests of Arkansas (Perry et al. 2007b; Perry and Thill 2008).

Harvest exclusion areas, especially those surrounding streams, are commonly used as roosting habitat by many tree hollow roosting Australian bats such as Gould's long-eared bat, *N. gouldi,* eastern forest bat, *V. pumilus*, and southern forest bat, *V. regulus* (Lunney et al. 1988; Law and Anderson 2000; Webala et al. 2010). A range of factors will influence the pattern of roosting close to creek-lines, but a large pool of older and mature trees in a variety of decay classes is likely to be important. Riparian areas often support a different vegetation type, with rainforest being particularly common in Australia. The specialist golden-tipped bat, *Kerivoula papuensis*, preferentially roosts in the suspended nests of small birds within riparian rainforest and such areas are excluded from harvesting (Schulz 2000; Law and Chidel 2004).

Jarrah forest in Western Australia offers one example of providing pools of mature trees using zoning. Since 2004, Fauna Habitat Zones (i.e. areas of mature forest >200 ha set 2–4 km apart within areas available for logging) have been retained for species, including bats, that rely on blocks of forest supporting mature forest attributes or characteristics (Webala et al. 2010). In some forest blocks, approximately 54 % of the total area (11,740 ha) is currently reserved from logging as conservation reserves, informal reserves (riparian buffers, diverse ecotype zones, road reserves), old-growth forest, and fauna habitat zones. Of these, about 39 % are permanently reserved, including riparian buffers, from logging in the future. Testing the effectiveness of this level of retention remains a priority for forest bat research. Collectively, these findings indicate that forested corridors are important habitat elements for roosting bats in forests across the globe.

5.4.5 Plantations

There is no internationally agreed definition of forest plantation and many very old forests we may think of as natural have been planted. However, for the purposes of this review, the term plantation is used to mean forests planted primarily for timber extraction using intensive management techniques. Timber plantations are perhaps the most extreme form of silviculture as they require replanting of typically exotic trees, with site and soil preparation required over large scales. Seedlings are planted at high densities to maximise growth and form of trees, and this has the consequence of producing high levels of clutter as the trees grow. All the silvicultural practices outlined in this section are also applicable to plantation forests. The response of bats has been documented in eucalypt plantations in Australia and pine plantations in New Zealand. As expected, bat activity in young plantations of eucalypts (<10 years) is typically low and considerably less than that found in nearby forest, and, somewhat surprisingly, activity is similar to levels over adjacent cleared farms (Law and Chidel 2006; Law et al. 2011). Bat activity is higher in older eucalypt plantations (~25 years), especially where drought and lack of maintenance leads to tree mortality and the creation of gaps (Law and Chidel 2006). Closed-space species (*Nyctophilus*) show some association with plantations as do open-space species (*Mormopterus ridei*), which presumably use the space above plantations together with adjacent open paddocks. Radio-tracked bats avoid roosting in young eucalypt plantations where tree hollows are absent, even though decorticating bark is present (Law et al. 2011).

Despite limitations in habitat quality, plantation forests provide large areas of additional habitat for threatened long-tailed bats, *Chalinolobus tuberculatus*, in New Zealand (Borkin and Parsons 2011a). Borkin and Parsons (2011b) found these bats roosting in crevices, fissures, and small hollows in the oldest stands of Monterey pine, *Pinus radiata*, plantations (25–30 years), with females choosing to roost within 150 m of waterways. In these plantations, bats selected home ranges with higher proportions of relatively old stands than available (Borkin and Parsons 2011a). Males selected edges with open unplanted areas within their home ranges, which females avoided, instead selecting older stands for foraging. Borkin et al. (2011) also documented the response to the clear-fell harvest of a pine plantation and found a pattern of declining numbers of roosts used, as well as smaller roosting areas and colony sizes. Over 3 years, 21 % of known roosts were lost with 15 % due to forestry operations and 6 % due to natural tree fall. To mitigate harvest operations, it was suggested that some suitable foraging and roosting areas should be retained within bat home ranges. Borkin et al. (2011) further suggested that priority management for this declining New Zealand bat should focus on plantation areas closest to water and harvests should be planned when bats are not heavily pregnant nor have non-volant dependents.

Pine plantations in the south-eastern USA are actively managed landscapes with extensive amounts of fragmentation and edge development. Nevertheless, these landscapes often support a diverse bat assemblage, in part due to enhanced foraging conditions along edge interfaces and to suitable foraging and roosting habitats along forested-riparian corridors (Miller 2003; Elmore et al. 2004; Hein et al. 2008b, 2009a). Experimental studies have demonstrated that activity of bats is affected by edge habitats, with highest levels of activity occurring along the edge interface regardless of echolocation call structure or wing morphology (Jantzen and Fenton 2013). Tree canopies also serve as edge interfaces in forested environments, with more manoeuvrable, high-frequency bats foraging along canopies and edges more often than less manoeuvrable, low-frequency bats (Pettit and Wilkins 2012). Relationships of age, formation, and structural characteristics of edge habitats with activity of foraging bats are complex, with newly formed, high-contrast edges supporting higher bat activity and stronger depth of edge influence, than older more developed, cantilevered edges which possess less contrast between adjacent habitats (Jantzen and Fenton 2013). Regardless, data indicate that managed forests with an abundance of edge habitat, typical of plantation forests in south-eastern North America, can support a diverse assemblage of forest bat species.

Spruce, pine, and fir species account for the largest share of the forest plantation area in Europe, with *Eucalyptus* species introduced from Australia common in the south. While eucalypt plantations appear to be avoided by some bats (Di Salvo et al. 2009), positive selection was found for the Mediterranean horseshoe bat, *Rhinolophus euryale*, in the Basque country (Aihartza et al. 2003). In Spain, *R. euryale* and Mehely's horseshoe bat, *R. mehelyi*, both closed-space foragers, were radio-tracked foraging in eucalypt plantations and dehesa (managed oak savanna) in proportion to, or greater than, their availability (Russo et al. 2005a, b). Numerous acoustic and radio-tracking studies have documented avoidance of bats from non-native coniferous plantations in Europe (e.g. Entwhistle et al. 1996; Walsh and Harris 1996). Perhaps as a consequence of this, the effects of plantation forestry practices on bat populations in Europe have been largely ignored, and surprisingly little is known about the use of timber plantations by bats. However, several long-running artificial 'bat box' schemes operated by the UK's Forestry Commission have indicated that some plantations contain large roosting bat populations (Park et al. 1998). Radio-tracking of Natterer's bat, *Myotis nattereri*, a species previously associated primarily with deciduous forests has uncovered the extensive use of areas used for commercial forestry, both for roosting and foraging (Mortimer 2006). This study conducted in a plantation in Scotland found that *M. nattereri* preferentially foraged within areas of Corsican pine, *Pinus nigra* var. *maritima*, and roosted in cavities formed from live double-leadered Corsican pine (Mortimer 2006). Given life-history parameters of the bats studied (survival, population densities) were similar or higher than those described within deciduous forests, and that double-leadered trees are usually targeted for removal by foresters as uneconomic, such findings illustrate the importance of studies in plantation forests.

A high percentage of open ground in some planted forests can benefit species that specialise on the predation of ground dwelling prey. Greater mouse-eared bat, *Myotis myotis*, for example, while often associated with deciduous forests, was found preferentially foraging in mature spruce monocultures with a high percentage of open ground in Germany, and intensively managed orchards and lowland forests with no undergrowth in Switzerland (Arlettaz 1999; Zahn et al. 2004). These studies collectively suggest that it is the forest structure that may be more important than tree species composition in many cases. Therefore, it seems clear that timber plantations have the potential to be of value to bats, but we lack an understanding of how populations of different species are affected by current silviculture practices.

5.4.6 Prey

The response of bat prey is also a critical issue when evaluating silvicultural treatments. Lepidoptera (moths–a fundamentally important prey group of bats) in temperate zone forests of North America differ little in species richness between stands regenerating after harvest and stands that remain unharvested (Burford et al. 1999; Summerville and Crist 2002; Dodd et al. 2008). Group selection logging of Australian eucalypt forests has found greater insect biomass in old regrowth Jarrah forest (>30 years since logging) than younger forest treatments (Webala et al. 2011) and a similar trend was found in spotted gum forests in eastern Australia (Adams et al. 2009). An additive effect of insect abundance and an index of vegetation openness in the spotted gum forests influenced bat activity, especially edge-space species with medium to high echolocation frequency. High values of insects and openness correlated with high levels of bat activity (Adams et al. 2009). Thus, dense clutter appears to constrain activity of some species even where insect abundance is high. This varies between bat ensembles, however, with closed-space foragers able to take advantage of the higher insect densities often associated with clutter, particularly Diptera, an important taxa for many bats (Müller et al. 2012; Fuentes-Montemayor et al. 2013; see also Sect. 4.3). While the prey base of bats can probably be sustained with application of many silvicultural systems, clearcut stands regenerating as monocultures support reduced levels of moth diversity, indicating that plant species richness is important for providing adequate populations of lepidopteran prey for insectivorous bats in managed forests (Summerville and Crist 2002; Dodd et al. 2012).

5.5 Multi-spatial Scale Forest Management

Integrating silvicultural systems into managed forested landscapes in ways that promote habitat for forest bats must account for the fact that bats are highly mobile and exhibit considerable variability in the use of habitats both spatially and temporally (Duchamp et al. 2007). Given that resource requirements differ among species and also sex, age, and reproductive classes within species (Perry et al. 2007a; Perry and Thill 2007b; Henderson et al. 2008), designing a forested-landscape matrix with a mosaic of resources that addresses the needs of all bat species in the region will likely require the application of a mix of silvicultural methods, each implemented with different objectives in mind (Guldin et al. 2007). These would include the retention of mature forest habitat at the landscape and stand scale in the form of large reserves, narrow and large strips, streamside reserves, aggregates, and clumps (Gustafsson et al. 2012). Lindenmayer and Franklin (2002) proposed a strategic landscape-scale approach with conservation measures applied at multiple spatial scales for forests. The four main strategies identified for conservation at multiple spatial scales include: (1) establishment of large ecological reserves, (2) application of landscape-level measures in off-reserve areas, (3) application of stand-level measures in off-reserve areas, and (4) monitoring and adaptive management.

There are limited data on bats for setting overall retention thresholds at the landscape scale. Gustafsson et al. (2012) suggested a strict minimum of 5–10 % retention of old-growth forest to achieve a positive ecological response for biodiversity. However, considerably higher levels are often recommended. For example, in Tasmania, 30 % is retained in some state forests (Gustafsson et al. 2012; see also Białowieża Forest in Europe ~20 %, Ruczyński et al. 2010). This retention should be spread across the landscape to facilitate dispersal. A key question is whether there are thresholds for the retention of mature forest that can optimise the trade-off between biodiversity conservation and production.

A recent study on Tasmanian bats, using both radio-tracking and acoustic detectors, assessed the response of bats to multi-spatial scale forest management (Cawthen et al. 2013). At broader scales, maternal bat colonies selected roosts in landscapes with the highest availability of hollow-bearing trees. At more fine-scales, however, maternal colonies did not exhibit strong selection for roost trees in patches with the highest availability of hollow-bearing trees. Instead, other attributes such as hollow type were important. For overall bat activity, the extent to which bats used different types of retained forest patches varied with the composition of the surrounding landscape. Large strips and small patches of wooded habitat were used by bats to a greater extent in landscapes with less mature forest in the surrounding area (<1 km radius). For small patches, this corresponded to landscapes with <22 % mature forest in the surrounding 1 km. No thresholds in bat activity were identified for large patches (370 ha) or small corridors (3 ha). Overall, these results indicate that in the landscapes sampled, activity is low in small retained patches where mature forest is readily available nearby, though these habitat elements do provide roosts and connectivity (and probably foraging habitat) where mature forest is rare or has been lost. Thus, the type, amount, and spatial arrangement of mature forest existing in the landscape need to be considered when retaining forest habitat at finer-spatial scales.

Clearly, the extent to which forest bats respond to changes at the landscape scale remains only partially understood. Studies of bat activity at stand and landscape scales in both north-western and north-eastern forests of North America demonstrated that patterns in habitat use of bats were largely determined by habitat characteristics at the local or stand level and not at landscape scales (Erickson and West 2003; Ford et al. 2006), suggesting that silvicultural systems that create a mosaic of treatments across forested landscapes with local differences in habitat structure will support a higher overall diversity of bat species (Wigley et al. 2007). This approach has been recommended in published studies (Loeb and Waldrop 2008; Perry et al. 2008); however, other sources report both stand and landscape metrics in North America and Europe to be important in selection of activity areas of bats (Loeb and O'Keefe 2006; Yates and Muzika 2006; Fuentes-Montemayor et al. 2013), with tri-coloured bats, *P. subflavus*, and eastern red bats, *L. borealis*, most affected by local stand structure, northern long-eared bats, *M. septentrionalis*, negatively affected by forest edge, and Indiana bats, *M. sodalis*, positively affected by dead tree density and non-forested land cover. Other studies corroborate that selection of roosting sites in both bark- and cavity-roosting and foliage-roosting bat species is strongly influenced by landscape-scale metrics in both eastern and western forests of North America (Limpert et al. 2007; Perry et al. 2008; Arnett and Hayes 2009; Lacki et al. 2010).

5.6 Summary and Future Possibilities

This review of the effects of silvicultural systems on forest bats demonstrated that almost all treatments evaluated were compatible with some use by forest bats, depending on the suite of species considered: closed-space species feed in intact forests, but respond to creation of small canopy gaps and less to reduced tree densities and open-edge interfaces; edge-space species exploit edge habitat along tracks, coupe edges, and other linear features such as creeks, but fare poorly within dense regrowth that often dominates soon after harvest; and, open-space foragers benefit *temporarily* from silvicultural treatments that significantly reduces cluttered air space and provides edge interfaces for roosting. These patterns were largely consistent across three different continents.

To sustain high levels of bat diversity in managed forests at the landscape scale, a balance of needs for these three groupings of bats is desirable and will likely require a mix of silvicultural treatments and exclusion areas staggered across the landscape, regardless of forest type or geographic region. Use of edge habitats, exclusion areas/set-asides, and riparian corridors for roosting and foraging by bats was a consistent theme in the literature reviewed, and these habitat elements need to be considered in forest planning. These landscape features accompany forest fragmentation, however, and it remains unclear to what extent increasing loss of the unharvested forest matrix will lead to declines in population numbers of forest bats. Unfortunately, data on densities of occupied roosts and, thus, potential for landscape-scale population estimates of bats are few (Clement and Castleberry 2013; Fleming et al. 2013). Regardless, population studies could integrate the potential benefits of multiple prescriptions at a scale over which bats themselves sample the landscape. Population studies are likely to provide the ultimate test of the effectiveness of a silvicultural regime, especially when such studies take a long-term perspective. Long-term studies on forest bats are notably lacking in the published literature.

Application of silvicultural treatments in regenerating forests to reduce tree densities and open gaps in the forest canopy shows promise for creating forested landscapes that support diverse and sustainable populations of bats. Forests with reduced tree density and vegetative clutter permit higher levels of light penetration, with this increased exposure hypothesised to enhance the suitability of live and dead trees for roosting by bark- and cavity-roosting bats in temperate climates (Boyles and Aubrey 2006). Further, LiDAR studies demonstrate that reduced clutter in the mid- and understory layers of forests is correlated with higher levels of activity by low-frequency ($\leq$34 kHz) open-space bats (Britzke et al. 2011; Dodd et al. 2013). However, closed-space bat species that glean insects from vegetation and manoeuvre well within clutter benefit from a relatively dense understorey and higher tree densities, which can act as sources of insect prey (Fuentes-Montemayor et al. 2013). Therefore, management that encourages habitat heterogeneity to fulfil the requirements of different species is needed. Bat activity is also vertically stratified, but there is a paucity of information on the effects of high canopy forest structure on bat activity (Adams et al. 2009; Müller et al. 2013), and research to address this gap would be valuable.

The quality and density of old trees in exclusion areas must not be overlooked. Roost abundance stands out as a key variable in our conceptual model (Fig. 5.1). The posited relationship is for increasing bat populations with increasing numbers of roosts, though with a threshold at the upper end of roost abundance rather than at low roost abundance. Densities of hollow trees sufficient to support populations of roosting bats are unknown and remain a major knowledge gap (Law 1996), but will likely be species contingent and based on roost switching behaviours and social dynamics within colonies (Johnson et al. 2013) and the density of other hollow-dependent fauna. Even small colonies of bats can require a large number of roosts over the active season. For example, Russo et al. (2005a, b) estimated that over a period of a month a colony of 12 female barbastelle bats, *B. barbastellus*, would require approximately 18 different trees for roosting. Although the retention and sustained recruitment of large mature trees at various stages of decay is essential in harvested forests for the future long-term maintenance of bat roosts and other hollow-dependent fauna, this might best be achieved through regular harvest exclusion areas (unharvested buffers, old-growth forest, etc.) that can maintain high local densities of potential roosts. There remains little guidance on how much undisturbed forest should be retained at a landscape scale.

Paradigm shifts in forest management away from even-aged to retention systems (Puettmann et al. 2009) are already in place in Pacific coast forests of North America and Australian eucalypt forests and are being encouraged for use in management of forests globally (Gustafsson et al. 2012; Lindenmayer et al. 2012). These systems allow for maintenance in post-harvest forests of tree species compositions, canopy structures, and ecosystem functions typical of preharvest conditions. We conclude from our review that the use of multi-scale retention systems may be a compatible approach for sustaining habitats of bats in forests. These silvicultural systems are designed to provide spatial variation in retained tree densities and distribution of residual patches of uncut forest, both of which lead to habitat complexity within stands and across landscapes. These systems intentionally mimic natural disturbance regimes and have broad biodiversity benefits across multiple taxa (Long 2009). Retention of old forest patches is likely to be most important where harvest intensity is high, such as in clearcut or heavy selection practices, or where retention of critical habitat components is low. Stand-level (site-scale) retention should be greater where old-growth forest in the surrounding landscape is scarce and where logging practices are more intense. The effectiveness of this multi-scale approach will require testing through monitoring and research tailored for different environments, multiple taxa and silvicultural practices. Monitoring the effectiveness of these strategies is an essential part of adaptive management and a fundamental part of ecological sustainable forestry and the 'social license to operate' that is increasingly required by forest certification schemes (Lindenmayer and Franklin 2002).

Acknowledgements Thanks to L. Cawthen, L. Lumsden, T. Kingston, J. Müller, D. Russo, and J. Williams for helpful comments on a draft.

Glossary

Clearcut/Clear-fell Harvest Also referred to as uniform selection and heavy group selection, it removes all trees from a large management area and allows natural regeneration to take place, resulting in even-aged regrowth with high stem density. The aim is to mimic natural stand replacing events such as wildfire or large storms

Coupe/Cutblocks A defined area of forest, which may vary in size, in which harvesting takes place usually over one year

Deferment Harvests Sometimes also referred to as a shelterwood or clearcut with reserves. A deferment harvest retains a limited number of canopy trees (reserve trees) while allowing regeneration in the understory. These two tree levels are then allowed to develop together until the end of the next rotation, whereupon other trees are retained for canopy cover

Forest Zoning Where management for multiple objectives in a forest incorporates broad exclusion areas such that logging is excluded from patches of forest deemed to be environmentally sensitive or where patches of forest are specified to allow different silvicultural practices (Florence 1996)

Gap Release Creation of canopy gaps typically <0.1 ha to allow the growth of younger, often suppressed trees

Green Tree The retention of live trees on an otherwise harvested area as part of a variable retention harvest

Group Selection Harvest Removes all trees from small patches, with the aim of using disturbance to stimulate regeneration of new trees, but simultaneously maintaining a well-connected mosaic of patches of varying size, containing varying numbers of residual mature trees

LiDAR A remote sensing technology that measures distance by illuminating a target with a laser and analyses the reflected light

Patch Cuts An area of felling smaller than a clearcut but removing a larger number of trees than a group selection harvest

Prescriptions Targeted retention that aims to mitigate the effects of logging on environmental features. Hollow tree retention and riparian exclusion zones are two common prescriptions, but can also include exclusion zones surrounding significant bat roosts

SeedTree Harvest The retention of a few residual trees in a harvested area to provide seeds for the forest to regenerate

Self-thinning Density-dependent mortality within an even-aged stand of trees as they grow in size, leading to reduced tree density

Shelterwood Harvest See deferment harvest.

Shelterwood Systems Removal of canopy trees in a series of selective harvests leaving sufficient trees for regeneration and shelter. New seedlings are left to establish before mature trees are removed

Silviculture The art and science of manipulating a stand of trees by controlling the supplies of water, nutrients, and solar radiation by altering forest structure, towards a desired future condition (Guldin et al. 2007), typically for timber production but also for biodiversity conservation goals

Single Tree Selection Removes a scattering of high value individual trees from management areas, with repeat cuts taking place at regular intervals over time. However, intensity can vary. Cumulative effects can result in reduced hollow tree density unless there is a specific retention of old trees

Stand A group of forest trees sufficiently uniform in species composition or age to be considered a management unit

Thinning Felling to decrease tree stem density within young regrowth forests to reduce competition for resources among trees and promote the growth of the stand (Florence 1996)

Variable Retention Harvests Creation of multi-aged stands in clearcut zones by retaining clumps, patches, or aggregates of old trees within the clearcut

References

Adams M, Law B (2011) The impact of forest thinning on bat activity: towards improved clutter-based hypotheses. In: Law B, Eby P, Lunney D, Lumsden L (eds) Biology and conservation of australasian bats. Royal Zoological Society of New South Wales, pp. 363–379

Adams MD, Law BS, French KO (2009) Vegetation structure influences the vertical stratification of open- and edge-space aerial-foraging bats in harvested forests. For Ecol Manage 258:2090–2100

Aihartza JR, Garin I, Goiti U et al (2003) Spring habitat selection by the mediterranean horseshoe bat (*Rhinolophus euryale*) in the Urdaibai biosphere reserve (Basque country). Mammalia 67:25–32

Altringham JD (1996) Bats: biology and behaviour. Oxford University Press, Oxford

Arlettaz R (1999) Habitat selection as a major resource partitioning mechanism between the two sympatric sibling bat species *Myotis myotis* and *Myotis blythii*. J Anim Ecol 68:460–471

Arnett EB, Hayes JP (2009) Use of conifer snags as roosts by female bats in western forests. J Wildl Manage 73:214–225

Aubry KB, Halpern CB, Peterson CE (2009) Variable-retention harvests in the Pacific Northwest: a review of short-term findings from the DEMO study. For Ecol Manage 258:398–408

Baker MD, Lacki MJ (2006) Day roosting habitat of female long-legged myotis in ponderosa pine forests. J Wildl Manage 70:207–215

Baker SC, Read SM (2011) Variable retention silviculture in Tasmania's wet forests: rationale, adaptive management and synthesis of biodiversity benefits. Aust For 74:218–232

Bernardos DA, Chambers CL, Rabe ML (2004) Selection of Gambel oak roosts by southwestern myotis in ponderosa pine-dominated forests, northern Arizona. J Wildl Manage 68:595–601

Betts BJ (1998) Roosts used by maternity colonies of silver-haired bats in northeastern Oregon. J Mammal 79:643–650

Boerner REJ, Hart SC, McIver JD (2008) The national fire and fire surrogate study: ecological consequences of alternative fuel reduction methods in seasonally dry forests. For Ecol Manage 255:3075–3080

Borkin K, Parsons S (2011a) Home range and habitat selection by a threatened bat in exotic plantation forest. For Ecol Manage 262:845–852

Borkin K, Parsons S (2011b) Sex-specific roost selection by bats in clearfell harvested plantation forest: improved knowledge advises management. Acta Chiropter 13:373–383

Borkin K, O'Donnell CFJ, Parsons S (2011) Bat colony size reduction coincides with clear-fell harvest operations and high rates of roost loss in plantation forest. Biodivers Cons 20:3537–3548

Boyles JG, Aubrey DP (2006) Managing forests with prescribed fire: implications for a cavity-dwelling bat species. For Ecol Manage 222:108–115

Brigham RM (2007) Bats in forests: what we know and what we need to learn. In: Lacki MJ, Hayes JP, Kurta A (eds) Bats in forests: conservation and management. The Johns Hopkins University Press, Baltimore, pp 1–15

Brigham RMR, Vonhof MJ, Barclay RM et al (1997) Roosting behaviour and roost-site preferences of forest-dwelling California bats (*Myotis californicus*). J Mammal 78:1231–1239

Britzke ER, Duchamp JE, Murray KL et al (2011) Acoustic identification of bats in the eastern United States: a comparison of parametric and nonparametric methods. J Wildl Manage 75:660–667

Broders HG, Forbes GJ (2004) Interspecific and intersexual variation in roost-site selection of northern long-eared and little brown bats in the Greater Fundy National Park ecosystem. J Wildl Manage 68:602–610

Brown GW, Nelson JL, Cherry KA (1997) The influence of habitat structure on insectivorous bat activity in montane ash forests of the central highlands, Victoria. Aust Forestry 60:138–146

Burford LS, Lacki MJ, Covell CV (1999) Occurrence of moths among habitats in a mixed mesophytic forest: implications for management of forest bats. For Sci 45:323–332

Carter TC, Feldhamer GA (2005) Roost tree use by maternity colonies of Indiana bats and northern long-eared bats in southern Illinois. For Ecol Manage 219:259–268
Cawthen L, Nicol S, Law B et al (2013) Effectiveness of the multi-spatial scale approach to forest management—a Tasmanian case study on bats. Abstract for the international bat conference, Costa Rica
Clement MJ, Castleberry SB (2013) Estimating density of a forest-dwelling bat: a predictive model for Rafinesque's big-eared bat. Pop Ecol 55:205–215
Crampton LH, Barclay RMR (1998) Selection of roosting and foraging habitat by bats in different-aged aspen mixedwood stands. Conserv Biol 12:1347–1358
Crome FHJ, Richards GC (1988) Bats and gaps: microchiropteran community structure in a Queensland rainforest. Ecology 69:1960–1969
Cryan PM, Bogan MA, Yanega GM (2001) Roosting habits of four bat species in the black hills of South Dakota. Acta Chiropter 3:43–52
Curtin RA, Squire RH, Mackowski CM (1991) Management of native hardwood forests of north coast New South Wales. In: McKinnell FH, Hopkins ER, Fox JED (eds) Forest management in Australia. Surrey Beatty & Sons, NSW, pp 77–107
de Oliveira MC, Smith GC, Hogan LD (1999) Current limitations in the use of bat detectors to assess the impacts of logging—a pilot study in south-east Queensland. Aust Zool 31:110–117
Di Salvo I, Russo D, Sarà M (2009) Habitat preferences of bats in a rural area of Sicily determined by acoustic surveys. It J Mamm 20:137–146
Dodd LE, Lacki MJ (2007) Prey consumed by *Corynorhinus townsendii ingens* in the Ozark mountain region. Acta Chiropter 9:451–461
Dodd LE, Lacki MJ, Rieske LK (2008) Variation in moth occurrence and implications for foraging habitat of Ozark big-eared bats. For Ecol Manage 255:3866–3872
Dodd LE, Lacki MJ, Britzke ER et al (2012) Forest structure affects trophic linkages: how silvicultural disturbance impacts bats and their insect prey. For Ecol Manage 267:262–270
Dodd LE, Skowronski NS, Dickinson MB et al (2013) Using LiDAR to link forest canopy structure with bat activity and insect occurrence: preliminary findings. Mammoth cave national park's 10th research symposium: celebrating the diversity of research in the mammoth cave region. Mammoth Cave, Kentucky, pp 50–57
Downs NC, Racey PA (2006) The use by bats of habitat features in mixed farmland in Scotland. Acta Chiropter 8:169–185
Duchamp JE, Arnett EB, Larson MA et al (2007) Ecological considerations for landscape-level management of bats. In: Lacki MJ, Hayes JP, Kurta A (eds) Bats in forests: conservation and management. The Johns Hopkins University Press, Baltimore, pp 237–261
Duerr DA, Mistretta PA (2013) Invasive pests—insects and diseases. In: Wear DN, Greis JG (eds) The southern forest futures project. USDA Forest Service, Southern Research Station, Gen. Tech. Rep. SRS-178, Asheville, North Carolina, pp 457–491
Elmore LW, Miller DA, Vilella FJ (2004) Selection of diurnal roosts by red bats (*Lasiurus borealis*) in an intensively managed pine forest in Mississippi. For Ecol Manage 199:11–20
Entwhistle AC, Racey PA, Speakman JR (1996) Habitat exploitation by a gleaning bat, *Plecotus auritus*. Phil Trans Roy Soc Lon Ser B 351:921–931
Erickson JL, West SD (1996) Managed forests in the western Cascades: the effects of seral stage on bat habitat use patterns. In: Barclay RMR, Brigham RM (eds) Bats and forests symposium, October 19–21, 1995, Victoria, British Columbia. Canadian Research Branch, BC Ministry of Forests, Victoria, BC. Working Paper 23/1996, pp 215–227
Erickson JL, West SD (2003) Associations of bats with local structure and landscape features of forested stands in western Oregon and Washington. Biol Conserv 109:95–102
European Environment Agency (2006) European forest types: categories and types for sustainable forest management reporting and policy. EEA Technical report No 9/2006. Available at: http://www.eea.europa.eu/publications/technical_report_2006_9
Fisher JT, Wilkinson L (2005) The response of mammals to forest fire and timber harvest in the North American boreal forest. Mammal Rev 35:51–81

Fleming HL, Jones JC, Belant JL et al (2013) Probability of detection and visual count error for Rafinesque's big-eared bat (*Corynorhinus rafinesquii*) and southeastern myotis (*Myotis austroriparius*) in tree cavities. Amer Midl Nat 169:56–65
Florence RG (1996) Ecology and silviculture of eucalypt forests. CSIRO Publishing, Victoria
Food and Agricultural Organisation of the United Nations (2012). State of the World's Forests, Rome. Available at: http://www.fao.org/docrep/016/i3010e/i3010e00.htm
Ford WM, Menzel MA, Rodrigue JL et al (2005) Relating bat species presence to simple habitat measures in a central Appalachian forest. Biol Conserv 126:528–539
Ford WM, Menzel JM, Menzel MA et al (2006) Presence and absence of bats across habitat scales in the upper coastal plain of South Carolina. J. Wildl Manage 70:1200–1209
Forestry Commission (2005) Woodland management for bats. Forestry Commission Publications, Wetherby UK. Available at: http://www.forestry.gov.uk/pdf/woodland-management-for-bats.pdf/$FILE/woodland-management-for-bats.pdf
Foster R, Kurta A (1999) Roosting ecology of the northern bat (*Myotis septentrionalis*) and comparisons with the endangered Indiana bat (*Myotis sodalis*). J Mammal 80:659–672
Fries C, Johansson O, Pettersson B et al (1997) Silvicultural models to maintain and restore natural stand structures in Swedish boreal forests. For Ecol Manage 94:89–103
Fuentes-Montemayor E, Goulson D, Cavin L et al (2012) Factors influencing moth assemblages in woodland fragments on farmland: implications for woodland creation and management schemes. Biol Conserv 153:265–275
Fuentes-Montemayor E, Goulson D, Cavin L et al (2013) Fragmented woodlands in agricultural landscapes: the influence of woodland character and landscape context on bats and their insect prey. Agri Ecosyst Environ 172:6–15
Greenaway F, Hill DA (2004) Woodland management advice for Bechstein's and barbastelle bat. English Nature Research Reports no 658
Grindal SD (1996) Habitat use by bats in fragmented forests. In: Barclay RMR, Brigham RM (eds) Bats and forests symposium, October 19–21, 1995, Victoria, British Columbia. Canadian Research Branch, BC Ministry of Forests, Victoria, BC. Working Paper 23/1996, pp 260–272
Grindal SD (1999) Habitat use by bats, *Myotis* spp., in western Newfoundland. Can Field-Nat 113:258–263
Grindal SD, Brigham RM (1998) Short-term effects of small-scale habitat disturbance on activity by insectivorous bats. J Wildl Manage 62:996–1003
Grindal SD, Brigham RM (1999) Impacts of forest harvesting on habitat use by foraging insectivorous bats at different spatial scales. Ecoscience 6:25–34
Grindal SD, Morissette JL, Brigham RM (1999) Concentration of bat activity in riparian habitats over an elevation gradient. Can J Zool 77:972–977
Guldin JM, Emmingham WH, Carter SA et al (2007) Silvicultural practices and management of habitat for bats. In: Lacki MJ, Hayes JP, Kurta A (eds) Bats in forests: conservation and management. The Johns Hopkins University Press, Baltimore, pp 177–205
Gustafsson L, Baker SC, Bauhus J et al (2012) Retention forestry to maintain multifunctional forests: a world perspective. Bioscience 62:633–645
Hansen MC, Stehman SV, Potapov PV (2010) Quantification of global gross forest cover loss. Proc US Nat Acad Sci 107:8650–8655
Hanspach J, Fischer J, Stagoll K et al (2012) Using trait-based filtering as a predictive framework for conservation: a case study of bats on farms in southeastern Australia. J Appl Ecol 49:842–850
Harrod RJ, Peterson DW, Povak NA et al (2009) Thinning and prescribed fire effects on overstory tree and snag structure in dry coniferous forests of the interior Pacific Northwest. For Ecol Manage 258:712–721
Hart JA, Kirkland GL Jr, Grossman SC (1993) Relative abundance and habitat use by tree bats, *Lasiurus* spp., in southcentral Pennsylvania. Can Field-Nat 107:208–212

Hayes JP (2009) Post-fire salvage logging in central Oregon: short-term response in bats, birds and small mammals. Fire Sci Brief 1–6
Hayes JP, Adam MD (1996) The influence of logging riparian areas on habitat utilization by bats in western Oregon. In: Barclay RMR, Brigham RM (eds) Bats and forests symposium, October 19–21, 1995, Victoria, British Columbia. Canadian Research Branch, BC Ministry of Forests, Victoria, BC. Working Paper 23/1996, pp 228–237
Hayes JP, Loeb SC (2007) The influences of forest management on bats in North America. In: Lacki, MJ, Hayes, JP, Kurta (eds) Bats in forests: conservation and management. The Johns Hopkins University Press, Baltimore, pp 207–235
Hein CD, Castleberry SB, Miller KV (2008a) Male Seminole bat winter roost-site selection in a managed forest. J Wildl Manage 72:1756–1764
Hein CD, Castleberry SB, Miller KV (2008b) Sex-specific summer roost-site selection by Seminole bats in response to landscape-level forest management. J Mammal 89:964–972
Hein CD, Castleberry SB, Miller KV (2009a) Site-occupancy of bats in relation to forested corridors. For Ecol Manage 257:1200–1207
Hein CD, Miller KV, Castleberry SB (2009b) Evening bat summer roost-site selection on a managed pine landscape. J Wildl Manage 73:511–517
Henderson LE, Farrow LJ, Broders HG (2008) Intra-specific effects of forest loss on the distribution of the forest-dependent northern long-eared bat (*Myotis septentrionalis*). Biol Conserv 141:1819–1828
Hessburg PF, Povak NA, Salter RB (2010) Thinning and prescribed fire effects on snag abundance and spatial pattern in an eastern Cascade range dry forest, Washington, USA. For Sci 56:74–87
Hill DA, Greenaway F (2008) Conservation of bats in British woodland. Brit Wild 161:161–169
Hillen J, Kiefer A, Veith M (2010) Interannual fidelity to roosting habitat and flight paths by female western barbastelle bats. Acta Chiropter 12:187–195
Hogberg LK, Patriquin KJ, Barclay RMR (2002) Use by bats of patches of residual trees in logged areas of the boreal forest. Amer Midl Nat 148:282–288
Humes ML, Hayes JP, Collopy MW (1999) Bat activity in thinned, unthinned, and old-growth forests in western Oregon. J Wildl Manage 63:553–561
Humphrey JW (2005) Benefits to biodiversity from developing old-growth conditions in British upland spruce plantations: a review and recommendations. Forestry 78:33–53
Humphrey JW, Bailey S (2012) Managing deadwood in forests and woodlands. Forestry Commission Practice Guide. Forestry Commission, Edinburgh. i–iv+ 1–24 pp
Hutchinson JT, Lacki MJ (2000) Selection of day roosts by red bats in mixed mesophytic forests. J Wildl Manage 64:87–94
Jantzen MK, Fenton MB (2013) The depth of edge influence among insectivorous bats at forest-field interfaces. Can J Zool 91:287–292
Johnson JB, Ford WM, Edwards JW (2012a) Roost networks of northern myotis (*Myotis septentrionalis*) in a managed landscape. For Ecol Manage 266:223–231
Johnson JS, Kropczynski JN, Lacki MJ et al (2012b) Social networks of Rafinesque's big-eared bats (*Corynorhinus rafinesquii*) in bottomland hardwood forests. J Mammal 93:1545–1558
Johnson JS, Kropczynski JN, Lacki MJ (2013) Social network analysis and the study of sociality in bats. Acta Chiropter 15:1–17
Jung K, Kaiser S, Böhm S et al (2012) Moving in three dimensions: effects of structural complexity on occurrence and activity of insectivorous bats in managed forest stands. J Appl Ecol 49:523–531
Kalcounis MC, Hobson KA, Brigham RM et al (1999) Bat activity in the boreal forest: importance of stand type and vertical strata. J Mammal 80:673–682
Kalcounis-Rüppell MC, Pysllakis JM, Brigham RM (2005) Tree roost selection by bats: an empirical synthesis using meta-analysis. Wildl Soc Bull 33:1123–1132
Kavanagh RP, Webb G (1998) Effects of variable-intensity logging on mammals, reptiles and amphibians at Waratah creek, Southeastern New South Wales. Pac Cons Biol 4:326–347

Kerth G, König B (1999) Fission, fusion and nonrandom associations in female Bechstein's bats (*Myotis bechsteinii*). Behaviour 136:1187–1202

Klug BJ, Goldsmith DA, Barclay RMR (2012) Roost selection by the solitary, foliage-roosting hoary bat (*Lasiurus cinereus*) during lactation. Can J Zool 90:329–336

Kroll AJ, Lacki MJ, Arnett EB (2012) Research needs to support management and conservation of cavity-dependent birds and bats on forested landscapes in the Pacific Northwest. Western J Appl Forest 27:128–136

Krusic RA, Yamasaki M, Neefus CD et al (1996) Bat habitat use in white Mountain National Forest. J Wildl Manage 60:625–631

Lacki MJ, Schwierjohann JH (2001) Day-roost characteristics of northern bats in mixed mesophytic forest. J Wildl Manage 65:482–488

Lacki MJ, Amelon SK, Baker MD (2007) Foraging ecology of bats in forests. In: Lacki MJ, Hayes JP, Kurta A (eds) Bats in forests: conservation and management. Johns Hopkins University Press, Baltimore, pp 83–127

Lacki MJ, Cox DR, Dickinson MB (2009) Meta-analysis of summer roosting characteristics of two species of Myotis bats. Amer Midl Nat 161:321–329

Lacki MJ, Baker MD, Johnson JS (2010) Geographic variation in roost-site selection of long-legged myotis in the Pacific Northwest. J Wildl Manage 74:1218–1228

Lacki MJ, Baker MD, Johnson JS (2012) Temporal dynamics of roost snags of long-legged myotis in the Pacific Northwest, USA. J Wildl Manage 76:1310–1316

Lähde E, Laiho O, Norokorpi Y (1999) Diversity-orientated silviculture in the Boreal zone of Europe. For Ecol Manage 118:223–243

Law BS (1996) The ecology of bats in south-east Australian forests and potential impacts of forestry practices: a review. Pac Cons Biol 2:363–374

Law BS (2004) Challenges for the management of bats in state forests of NSW. In: Lunney D (ed) The conservation of Australia's forest fauna, 2nd edn. Royal Zoological Society of NSW, Sydney, pp 748–760

Law BS, Anderson J (2000) Roost preferences and foraging ranges of the eastern forest bat *Vespadelus pumilus* under two disturbance histories on the mid-north coast of New South Wales. Aust Ecol 25:352–367

Law BS, Chidel M (2001) Bat activity 22 years after first-round intensive logging of alternate coupes near Eden, NSW. Aust Forestry 64:242–247

Law BS, Chidel M (2002) Tracks and riparian zones facilitate the use of Australian regrowth forest by insectivorous bats. J Appl Ecol 39:605–617

Law B, Chidel M (2004) Roosting and foraging ecology of the golden-tipped bat *Kerivoula papuensis* on the south coast of New South Wales. Wildl Res 31:73–82

Law B, Chidel M (2006) Eucalypt plantings on farms: use by insectivorous bats. Biol Cons 133:236–249

Law BS, Law PR (2011) Early responses of bats to alternative silvicultural treatments in wet eucalypt forests of Tasmania. Pac Cons Biol 17:36–47

Law BS, Chidel M, Penman T (2011) Do young eucalypt plantations benefit bats in an intensive agricultural landscape? Wildl Res 38:173–187

Lee P, Smyth C, Boutin S (2004) Quantitative review of riparian width buffer guidelines from Canada and the United States. J Environ Manage 70:165–180

Limpert DL, Birch DL, Scott MS et al (2007) Tree selection and landscape analysis of eastern red bat day roosts. J Wildl Manage 71:478–486

Lindenmayer DB, Franklin JF (2002) Conserving forest biodiversity: a comprehensive multi-scaled approach. Island Press, Washington

Lindenmayer DB, Noss RF (2006) Salvage logging, ecosystem processes, and biodiversity conservation. Conserv Biol 20:949–958

Lindenmayer DB, Franklin JF, Lõhmus A et al (2012) A major shift to the retention approach for forestry can help resolve some global forestry sustainability issues. Conserv Letters 5:421–431

Lintott P, Minderman J, Fuentes-Montemayor E et al (2014) Moth species richness, abundance and diversity in fragmented urban woodlands: implications for conservation and management strategies. Biodiv Conserv 23:2875–2901

Lloyd A, Law B, Goldingay R (2006) Bat activity on riparian zones and upper slopes in Australian timber production forests and the effectiveness of riparian buffers. Biol Cons 129:207–220
Loeb SC, O'Keefe JM (2006) Habitat use by forest bats in South Carolina in relation to local, stand, and landscape characteristics. J Wildl Manage 70:1210–1218
Loeb SC, Waldrop TA (2008) Bat activity in relation to fire and fire surrogate treatments in southern pine stands. For Ecol Manage 255:3185–3192
Long JN (2009) Emulating natural disturbance regimes as a basis for forest management: a North American view. For Ecol Manage 257:1868–1873
Luck GW, Smallbone L, Threlfall C et al (2013) Patterns in bat functional guilds across multiple urban centres in south-eastern Australia. Landscape Ecol 28:455–469
Lumsden LF, Bennett AF, Silins JE (2002) Location of roosts of the lesser long-eared bat *Nyctophilus geoffroyi* and Gould's wattled bat *Chalinolobus gouldii* in a fragmented landscape in south-eastern Australia. Biol Cons 106:237–249
Lunde RE, Harestad AS (1986) Activity of little brown bats in coastal forests. Northwest Sci 60:206–209
Lunney D, Barker J, Priddel D et al (1988) Roost selection by Gould's long-eared bat, *Nyctophilus gouldi* Thomas (*Chiroptera*: Vespertilionidae), in logged forest on the south coast of New South Wales. Aust Wildl Res 15:375–384
Maron M, Lill A, Watson DM et al (2005) Temporal variation in bird assemblages: how representative is a one-year snapshot? Aust Ecology 30:383–394
Mason WL, Quine CP (1995) Silvicultural possibilities for increasing structural diversity in British spruce forests: the case of Kielder. For Ecol Manage 79:13–28
Mason B, Kerrand G, Simpson J (1999) What is continuous cover forestry? Forestry commission information note 29. Forestry Commission, Edinburgh. Available at: http://www.forestry.gov.uk/pdf/fcin29.pdf/$FILE/fcin29.pdf
Mattson TA, Buskirk SW, Stanton NL (1996) Roost sites of the silver-haired bat (*Lasionycteris noctivagans*) in the Black Hills, South Dakota. Great Basin Nat 56:247–253
Mazurek MJ, Zielinski WJ (2004) Individual legacy trees influence vertebrate wildlife diversity in commercial forests. For Ecol Manage 193:321–334
Mehr M, Brandl R, Kneib T et al (2012) The effect of bark beetle infestation and salvage logging on bat activity in a national park. Biodivers Conserv 21:2775–2786
Menzel MA, Carter TC, Menzel JM et al (2002) Effects of group selection silviculture in bottomland hardwoods on the spatial activity patterns of bats. For Ecol Manage 162:209–218
Meyer CFJ, Matthew JS, Willig MR (2016) Responses of tropical bats to habitat fragmentation, logging, and deforestation. In: Voigt CC, Kingston T (eds) Bats in the Anthropocene: conservation of bats in a changing world. Springer International AG, Cham, pp. 105–141
Meyer CFJ, Fründ J, Lizano WP et al (2008) Ecological correlates of vulnerability to fragmentation in Neotropical bats. J Appl Ecol 45:381–391
Miles AC, Castleberry SB, Miller DA et al (2006) Multi-scale roost-site selection by evening bats on pine-dominated landscapes in southwestern Georgia. J Wildl Manage 70:1191–1199
Miller DA (2003) Species diversity, reproduction, and sex ratios of bats in managed pine forest landscapes of Mississippi. Southeast Nat 2:59–72
Miller DA, Stihler CW, Sasse DB et al. (2011) Conservation and management of eastern big-eared bats (*Corynorhinus* spp.). In: Loeb SC, Lacki MJ, Miller DA (eds) Conservation and management of eastern big-eared bats: a symposium. USDA, Forest Service, Southern Research Station, GTR SRS-145. Asheville, North Carolina, pp 53–61
Mississippi Museum of Natural Science (2005) Mississippi's comprehensive wildlife conservation strategy. Mississippi Department of Wildlife, Fisheries, and Parks, Jackson, Mississippi
Mitchell RJ, Palik BJ, Hunter ML Jr (2002) Natural disturbance as a guide to silviculture. For Ecol Manage 155:315–317
Montreal Process Implementation Group for Australia (2008) Australia's state of the forests report 2008. Bureau of Rural Sciences, Canberra

Moore PT, DeRose RJ, Long JN et al (2012) Using silviculture to influence carbon sequestration in southern Appalachian spruce-fir forests. Forests 3:300–316

Mormann BM, Robbins LW (2007) Winter roosting ecology of eastern red bats in southwest Missouri. J Wildl Manage 71:213–217

Mortimer G (2006) Foraging, roosting and survival of Natterer's bats, *Myotis nattereri*, in a commercial coniferous plantation. PhD thesis, St. Andrews University, UK

Müller J, Mehr M, Bässler C et al (2012) Aggregative response in bats: prey abundance versus habitat. Oecologia 169:673–684

Müller J, Brandl R, Buchner J et al (2013) From ground to above canopy—bat activity in mature forests is driven by vegetation density and height. For Ecol and Manage 360:179–184

Murphy SE, Greenaway F, Hill DA (2012) Patterns of habitat use by female brown long-eared bats presage negative impacts of woodland conservation management. J Zool (Lond) 288:177–183

Murray SW, Kurta A (2004) Nocturnal activity of the endangered Indiana bat (*Myotis sodalis*). J Zool (Lond) 262:197–206

Nicholson E (1999) Winds of change for silvicultural practice in NSW native forests. Aust For 62:223–235

O'Hara KL (2002) The historical development of uneven-aged silviculture in North America. Forestry 75:339–346

O'Hara KL (2009) Multiaged silviculture in North America. J For Sci 55:432–436

O'Keefe JM, Loeb SC, Lanham JD et al (2009) Macrohabitat factors affect day roost selection by eastern red bats and eastern pipistrelles in the southern Appalachian Mountains, USA. For Ecol Manage 257:1757–1763

Ober HK, Hayes JP (2008) Influence of vegetation on bat use of riparian areas at multiple spatial scales. J Wildl Manage 72:396–404

Owen SF, Menzel MA, Ford WM et al (2003) Home-range size and habitat used by the northern myotis (*Myotis septentrionalis*). Amer Midl Nat 150:352–359

Owen SF, Menzel MA, Edwards JW et al (2004) Bat activity in harvested and intact forest stands in the Allegheny Mountains. Northern J Appl Forest 21:154–159

Park KJ, Masters E, Altringham JD (1998) Social structure of three sympatric bat species (Vespertilionidae). J Zool (Lond) 244:379–389

Parker DI, Cook JA, Lewis SW (1996) Effects of timber harvest on bat activity in southeastern Alaska's temperate rainforests. In: Barclay RMR, Brigham RM (eds) Bats and forests symposium, October 19–21, 1995, Victoria, British Columbia. Canadian Research Branch, BC Ministry of Forests, Victoria, BC. Working Paper 23/1996, pp 277–292

Parks CG, Bernier P (2010) Adaptation of forests and forest management to changing climate with emphasis on forest health: a review of science, policies and practices. For Ecol Manage 259:657–659

Patriquin KJ, Barclay RMR (2003) Foraging by bats in cleared, thinned and unharvested boreal forest. J Appl Ecol 40:646–657

Perdue M, Steventon JD (1996) Partial cutting and bats: a pilot study. In: Barclay RMR, Brigham RM (eds) Bats and forests symposium, October 19–21, 1995, Victoria, British Columbia. Canadian Research Branch, BC Ministry of Forests, Victoria, BC. Working Paper 23/1996, pp 273–276

Perry RW, Thill RE (2007a) Roost characteristics of hoary bats in Arkansas. Amer Midl Nat 158:132–138

Perry RW, Thill RE (2007b) Roost selection by male and female northern long-eared bats in a pine dominated landscape. For Ecol Manage 247:220–226

Perry RW, Thill RE (2008) Diurnal roosts of male evening bats (*Nycticeius humeralis*) in diversely managed pine-hardwood forests. Amer Midl Nat 160:374–385

Perry RW, Thill RE, Carter SA (2007a) Sex-specific roost selection by adult red bats in a diverse forested landscape. For Ecol Manage 253:48–55

Perry RW, Thill RE, Leslie DM Jr (2007b) Selection of roosting habitat by forest bats in a diverse forest landscape. For Ecol Manage 238:156–166

Perry RW, Thill RE, Leslie DM Jr (2008) Scale-dependent effects of landscape structure and composition on diurnal roost selection by forest bats. J Wildl Manage 72:913–925
Pettit TW, Wilkins KT (2012) Canopy and edge activity of bats in a quaking aspen (*Populus tremuloides*) forest. Can J Zool 90:798–807
Puettmann KJ, Coates KD, Messier C (2009) A critique of silviculture: managing for complexity. Island Press, Washington, DC
Rabe MJ, Morrell TE, Green H et al (1998) Characterisitcs of ponderosa pine snag roosts used by reproductive bats in northern Arizona. J Wildl Manage 62:612–621
Recher HF, Gowing G, Kavanagh R et al (1983) Birds, resources and time in a tablelands forest. Proc Ecol Soc Aust 12:101–123
Regnery B, Couvet D, Kubarek L et al (2013a) Tree microhabitats as indicators of bird and bat communities in Mediterranean forests. Ecol Indic 34:221–230
Regnery B, Paillet Y, Couvet D et al (2013b) Which factors influence the occurrence and density of tree microhabitats in Meditteranean Oak forests? For Ecol Manage 295:118–125
Rhodes M, Wardell-Johnson GW, Rhodes MP et al (2006) Applying network analysis to the conservation of habitat trees in urban environments: a case study from Brisbane, Australia. Cons Biol 20:861–870
Ruczyński I, Bogdanowicz W (2008) Summer roost selection by tree-dwelling bats *Nyctalus noctula* and *N. leisleri*: a multiscale analysis. J Mammal 89:942–951
Ruczyński I, Nicholls B, MacLeod CD et al (2010) Selection of roosting habitats by *Nyctalus noctula* and *Nyctalus leisleri* in Białowieża Forest—adaptive response to forest management? For Ecol Manage 259:1633–1641
Russo D, Cistrone L, Jones G et al (2004) Roost selection in barbastelle bats (*Barbastella barbastellus*, Chiroptera: Vespertilionidae) in beech woodlands of central Italy: consequences for conservation. Biol Cons 117:73–81
Russo D, Almenar D, Aihartza J, Goiti U, Salsamendi E, Garin I (2005a) Habitat selection in sympatric *Rhinolophus mehelyi* and *R. euryale* (Mammalia: Chiroptera). J Zool Lond 266:327–332
Russo D, Cistrone L, Jones G (2005b) Spatial and temporal patterns of roost use by tree-dwelling barbastelle bats, *Barbastellus barbastellus*. Ecography 28:769–776
Russo D, Cistrone L, Jones G (2007) Emergence time in forest bats: the influence of canopy closure. Acta Oecolgica 31:119–126
Russo D, Cistrone L, Garonna AP et al (2010) Reconsidering the importance of harvested forests for the conservation of tree-dwelling bats. Biodivers Conserv 19:2501–2515
Sasse DB, Pekins PJ (1996) Summer roosting ecology of northern long-eared bats (*Myotis septentrionalis*) in the White Mountain National Forest. In: Barclay RMR, Brigham RM (eds) Bats and forests symposium, October 19–21, 1995, Victoria, British Columbia. Canadian Research Branch, BC Ministry of Forests, Victoria, BC. Working Paper 23/1996, pp 91–101
Schnitzler H-U, Moss CF, Denzinger A (2003) From spatial orientation to food acquisition in echolocating bats. Trends Ecol Evol 18:386–394
Schulz M (2000) Roosts used by the golden-tipped bat *Kerivoula papuensis* (Chiroptera: Vespertilionidae). J Zool (Lond) 250:467–478
Sedgeley JA, O'Donnell CFJ (1999) Roost selection by the long-tailed bat, *Chalinolobus tuberculatus*, in temperate New Zealand rainforest and its implications for the conservation of bats in managed forests. Biol Cons 88:261–276
Senior P, Butlin RK, Altringham JD (2005). Sex and segregation in temperate bats. Proc Roy Soc B—Biol Sci 272:2467–2473
Society of American Foresters (2010) Forest types of North America. Available at: http://www.encyclopediaofforestry.org/index.php?title=BioEco12
Stringer JW, Perkins C (2001) Kentucky forest practice guidelines for water quality management. University of Kentucky, College of Agriculture, Cooperative Extension Service, FOR-67, Lexington, Kentucky
Summerville KS, Crist TO (2002) Effects of timber harvest on forest Lepidoptera: community, guild, and species responses. Ecol Applica 12:820–835

Swystun MB, Psyllakis JM, Brigham RM (2001) The influence of residual tree patch isolation on habitat use by bats in central British Columbia. Acta Chiropter 3:197–201

Taylor RJ, Savva NM (1988) Use of roost sites by four species of bats in state forest in south-eastern Tasmania. Aust Wildl Res 15:637–645

Thomas DW (1988) The distribution of bats in different ages of Douglas-fir forests. J Wildl Manage 52:619–626

Tibbels AE, Kurta A (2003) Bat activity is low in thinned and unthinned stands of red pine. Can J For Res 33:2436–2442

Titchenell MA, Williams RA, Gehrt SD (2011) Bat responses to shelterwood harvests and forest structure in oak-hickory forests. For Ecol Manage 262:980–988

U.S. Department of Agriculture Forest Service (2009) U.S. forest resource facts and historic trends: forest facts 1952–2007 US metric revised rev072411. FS-801 M. Available at: http://fia.fs.fed.us

U.S. Department of Agriculture and U.S. Department of the Interior (1994) Record of decision for amendments to Forest Service and Bureau of Land Management planning documents within the range of the northern spotted owl. U.S. Forest Service, Portland, Oregon

U.S. Fish and Wildlife Service (2009) Range-wide Indiana bat protection and enhancement plan guidelines. Available at: http://www.fws.gov/frankfort/pdf/inbatpepguidelines

U.S. Fish and Wildlife Service (2012) North American bat death toll exceeds 5.5 million from white-nose syndrome. News Release, U.S. Fish and Wildlife Service, Arlington, Virginia

Vonhof MJ, Barclay RMR (1997) Use of tree stumps as roosts by the western long-eared bat. J Wildl Manage 61:674–684

Waldien DL, Hayes JP, Arnett EB (2000) Day-roosts of female long-eared myotis in western Oregon. J Wildl Manage 64:785–796

Walsh AL, Harris S (1996) Factors determining the abundance of vespertilionid bats in Britain: geographical, land class and local habitat relationships. J Appl Ecol 33:519–529

Watrous KS, Donovan TM, Mickey RM et al (2006) Predicting minimum habitat characteristics for the Indiana bat in the Champlain Valley. J Wildl Manage 70:1228–1237

Wear DN, Greis JG (2013) The southern forests futures project. USDA Forest Service, Southern Research Station, Gen. Tech. Rep. SRS-178, Asheville, North Carolina

Webala PW, Craig MD, Law BS et al (2010) Roost site selection by southern forest bat *Vespadelus regulus* and Gould's long-eared bat *Nyctophilus gouldi* in logged jarrah forests, south-western Australia. For Ecol Manage 260:1780–1790

Webala PW, Craig MD, Law BS et al (2011) Bat habitat use in logged jarrah eucalypt forests of south-western Australia. J Appl Ecol 48:398–406

Wigley TB, Miller DA, Yarrow GK (2007) Planning for bats on forest industry lands in North America. In: Lacki MJ, Hayes JP, Kurta A (eds) Bats in forests: conservation and management. Johns Hopkins University Press, Baltimore, pp 293–318

Willis CKR, Brigham RM (2004) Roost switching, roost sharing and social cohesion: forest-dwelling big brown bats, *Eptesicus fuscus*, conform to the fission-fusion model. Anim Behav 68:495–505

Willis CKR, Brigham RM (2005) Physiological and ecological aspects of roost selection by reproductive female hoary bats (*Lasiurus cinereus*). J Mammal 86:85–94

World Resources Institute (2014) Southern forests for the future. Available at: http://www.seesouthernforests.org

Wunder L, Carey AB (1996) Use of the forest canopy by bats. Northwest Sci 70:79–85

Yates MD, Muzika RM (2006) Effect of forest structure and fragmentation on site occupancy of bat species in Missouri Ozark forests. J Wildl Manage 70:1238–1248

Young RA, Giese RL (eds) (2003) Introduction to forest ecosystem science and management, 3rd edn. John Wiley and Sons, New York

Zahn A, Haselbach H, Güttinger R (2004) Foraging activity of central European *Myotis myotis* in a landscape dominated by spruce monocultures. Mamm Biol 70:265–270

Zimmerman GS, Glanz WE (2000) Habitat use by bats in eastern Maine. J Wildl Manage 64:1032–1040

Chapter 6
Bats in the Anthropogenic Matrix: Challenges and Opportunities for the Conservation of Chiroptera and Their Ecosystem Services in Agricultural Landscapes

Kimberly Williams-Guillén, Elissa Olimpi, Bea Maas, Peter J. Taylor and Raphaël Arlettaz

Abstract Intensification in land-use and farming practices has had largely negative effects on bats, leading to population declines and concomitant losses of ecosystem services. Current trends in land-use change suggest that agricultural areas will further expand, while production systems may either experience further intensification (particularly in developing nations) or become more environmentally friendly (especially in Europe). In this chapter, we review the existing literature

K. Williams-Guillén (✉)
Paso Pacífico and School of Natural Resources and Environment, Ventura, CA, USA
e-mail: kim@pasopacifico.org

K. Williams-Guillén
University of Michigan, Ann Arbor, MI, USA

E. Olimpi
Environmental Studies Department, University of California, Santa Cruz, CA, USA
e-mail: eolimpi@gmail.com

B. Maas
Agroecology Group, University of Gottingen, Gottingen, Germany
e-mail: beamaas@gmx.at

P.J. Taylor
South African Research Chair in Biodiversity Value and Change
and Centre for Invasion Biology, University of Venda, Thohoyandou, South Africa
e-mail: peter.taylor.univen@gmail.com

R. Arlettaz
Division of Conservation Biology, Institute of Ecology and Evolution,
University of Bern, Bern, Switzerland
e-mail: raphael.arlettaz@iee.unibe.ch

R. Arlettaz
Swiss Ornithological Institute, 6204 Sempach, Switzerland

C.C. Voigt and T. Kingston (eds.), *Bats in the Anthropocene: Conservation of Bats in a Changing World*, DOI 10.1007/978-3-319-25220-9_6

on how agricultural management affects the bat assemblages and the behavior of individual bat species, as well as the literature on provision of ecosystem services by bats (pest insect suppression and pollination) in agricultural systems. Bats show highly variable responses to habitat conversion, with no significant change in species richness or measures of activity or abundance. In contrast, intensification within agricultural systems (i.e., increased agrochemical inputs, reduction of natural structuring elements such as hedges, woods, and marshes) had more consistently negative effects on abundance and species richness. Agroforestry systems appear to mitigate negative consequences of habitat conversion and intensification, often having higher abundances and activity levels than natural areas. Across biomes, bats play key roles in limiting populations of arthropods by consuming various agricultural pests. In tropical areas, bats are key pollinators of several commercial fruit species. However, these substantial benefits may go unrecognized by farmers, who sometimes associate bats with ecosystem disservices such as crop raiding. Given the importance of bats for global food production, future agricultural management should focus on "wildlife-friendly" farming practices that allow more bats to exploit and persist in the anthropogenic matrix so as to enhance provision of ecosystem services. Pressing research topics include (1) a better understanding of how local-level versus landscape-level management practices interact to structure bat assemblages, (2) the effects of new pesticide classes and GM crops on bat populations, and (3) how increased documentation and valuation of the ecosystem services provided by bats could improve attitudes of producers toward their conservation.

6.1 Introduction

Agricultural areas cover approximately 40 % of our planet's terrestrial ecosystems (FAOSTAT 2011), with the 5 billion ha of land under farming and grazing now surpassing the extent of the world's forested areas (Robertson and Swinton 2005; Power 2010). Agricultural areas are expected to continue to expand with increasing human population growth and resultant resource use: Low- and middle-income countries will experience a 100 % increase in demand for agricultural products by 2050 (Defries et al. 2010; FAO 2011). In the face of increasing pressure on natural resources, the conservation of remaining natural areas is critical for the survival of multitudes of species. However, the ubiquity of agriculture means that farmland cannot be ignored in the context of landscape-level approaches to biodiversity conservation (Vandermeer and Perfecto 2007; Loos et al. 2014).

A growing body of research demonstrates that not only do some agricultural systems harbor high levels of biodiversity and provide a variety of ecosystem services (Tilman 1999; Foley et al. 2005; Tscharntke et al. 2005), but also that characteristics of these agricultural systems may have profound effects upon remaining natural areas (Perfecto and Vandermeer 2010). Agricultural matrices can vary drastically in their quality and permeability, impacting dispersal rates, and hence, long-term population stability of organisms found in less disturbed areas (Ricketts 2001; Laurance 2008; Perfecto and Vandermeer 2010; Tscharntke et al. 2012). On a local scale, different agricultural management approaches often coexist. Some rely on varying chemical inputs (pesticides, fertilizer), or novel plant types (e.g., genetically modified crops incorporating genes for characteristics such as insecticide functions), resulting in environmental contamination, pollution, and dissemination of toxins that could negatively impact biodiversity across multiple spatial scales (Nelson et al. 2009; Power 2010). As a consequence, agricultural management has effects not only on biodiversity, but also on human health and economies.

In the tropics, the expansion of export-oriented agriculture results from population growth and shifts in consumption patterns of developing nations, and is carried out mostly to the detriment of old growth forests and extensively managed grasslands such as pastures (Defries et al. 2010; Lambin and Meyfroidt 2011). As a consequence, croplands are still expanding dramatically, and agricultural practices are likely to further intensify in the near future (more chemical and mechanical inputs, reliance on genetically modified plants with novel manufactured traits). Short-term increases in yield will come at the cost of reduced structural and taxonomic diversity within agricultural systems (Loos et al. 2014) and concomitant loss of crucial ecosystem services.

An additional factor affecting agriculture in the Anthropocene is climate change and the need to adapt cultures to novel environmental conditions: Many areas may become unsuitable for cultivation of their current dominant crops, while extreme weather events may result in reduced yields. Resulting declines in calorie availability, particularly in the developing world (Nelson et al. 2009), will increase the need for agricultural practices that meet both productivity and sustainability goals (Tilman et al. 2002; McShane et al. 2011; Tscharntke et al. 2012). These trends portend major shifts in land-use patterns (Lambin and Meyfroidt 2011) and hence biodiversity, with agricultural intensification, forest and tree roost loss anticipated to have particularly negative effects on bat species richness, abundance, and functional diversity (Fischer et al. 2009, 2010; Jones et al. 2009).

These emerging trends pose major threats to farmland bat assemblages and populations (Jones et al. 2009; Kunz et al. 2011) and could negatively impact human populations by altering the ecosystem services that bats provide. Thus, there is a critical need to assess how agricultural management affects bat populations, and how affected bat populations will in turn affect agricultural production. In this chapter, we review the effects of agricultural land use and management on bat assemblages and the behavior and ecology of individual bat species at field, farm, and landscape scales (Vickery and Arlettaz 2012). We also review the developing literature on ecosystem services—and disservices—provided by bats in agricultural areas. Finally, we synthesize this information to suggest key management recommendations necessary to maintain bat populations in agricultural landscapes and highlight critical knowledge gaps that must be resolved in order to conserve bat diversity and ecosystem functions in a planet increasingly dominated by food production.

6.2 Methods

We used the Web of Knowledge, Google Scholar, and PubMed search engines to locate publications with the keywords "bats" AND "agriculture," "agroforestry," "farm," and "farmland." Given the potential importance of bats in provisioning ecosystem services in agricultural areas, we also searched for "bats" AND "ecosystem services," "pollination," "pest consumption," "pest control," and "pest limitation." The majority of sources stemmed from peer-reviewed publications, although we also included Master's and Ph.D. theses and published reports if results from the study in question were not available as journal articles. We also inspected the bibliographies of relevant publications. Each co-author focused on a specific geographic area (RA, assisted by Olivier Roth: Europe; BM: Australia and tropical Asia; EO: temperate North America; PT: sub-Saharan Africa; KWG: tropical Americas). Our searches were limited to publications with English language text or summaries. We focused on agriculture and animal husbandry for the production of calories for human or animal consumption, excluding forestry systems dedicated to timber or fiber production (see Law et al., Chap. 4), studies in which fallows or abandoned fields were the only agricultural systems investigated, as well as investigations that focused on fragmentation without explicit consideration of the effect of agricultural matrix (see Meyer et al., Chap. 3).

We divided results from the literature search into two broad categories of investigations: (1) How agricultural practices affect bat assemblages, ecology, behavior, and/or physiology; and (2) how bats affect agriculture through the provision of ecosystem services such as pollination and pest suppression. Within the first category, most studies addressed effects of land conversion and agricultural management on bat assemblage structure, abundance, activity levels, and behavior. We further subdivided results to consider habitat conversion to agriculture and agricultural intensification. We define agricultural intensification as consisting of at least one of the following: decreased structural complexity of native vegetation (natural and seminatural elements structuring the landscapes such as woodland patches and hedges), increased application of agrochemicals (pesticides, fertilizer), increased crop plant density, increased mechanization, or increased reliance on GM plants. We reviewed results from searches to locate studies which contrasted aspects of bat assemblage structure, abundance, activity, ranging behavior, or diet in either natural and agricultural habitat, or different agricultural systems of contrasting management.

To better quantify the responses of bats to habitat conversion and agricultural intensification across multiple disparate studies, we conducted a meta-analysis. We emphasize that this meta-analysis is based on correlational studies, rather than from controlled experiments; because assignment of treatment locations is not randomized in the majority of these studies, confounding factors could result in spurious effect sizes (Egger et al. 1998). We thus view our meta-analysis as a tool for exploring trends across a diverse suite of studies, with limited conclusive power.

We reviewed studies for the inclusion of mean values for at least one response variable in both natural and agricultural areas, or two or more agricultural areas of differing intensification; 32 studies using mist netting, harp trapping, acoustic monitoring, or a combination of these methods included appropriate data. We classify the response variable metrics into two separate categories for analyses, measures of species richness and measures of relative activity or abundance (i.e., pass rates from acoustic monitoring or capture rates from mist netting). We also consider habitat conversion and intensification responses separately.

For each pairwise comparison (natural–agricultural, or agricultural–agricultural), we calculated the effect size as the log odds ratio of the mean value from the lower intensity system divided by the mean from the higher intensity system. Thus, a positive effect size indicates higher species richness or activity/abundance in natural versus agricultural areas or lower intensity versus higher intensity agriculture. We followed García-Morales et al. (2013) and considered mean effect sizes with 95 % confidence intervals that did not include 0 as indicative of a significant effect. In the case of studies comparing multiple natural or agricultural habitats or presenting means for multiple species or species groups (i.e., producing multiple pairwise comparisons for any given combination of metric and response type), we averaged the odds ratio to avoid pseudo replication. Due to the diverse nature of the studies and a lack of clarity about numbers of replicates in some studies, we did not weight studies by sample size or replicates. For our analysis, we thus considered each study as an equally weighted case for the final model. We conducted analyses in R Version 3.0.2 (R Development Core Team 2013) using the packages lme4 and lmerTest. This diverse set of studies includes different methods (e.g., acoustic monitoring versus mist netting) from different regions with ecologically and taxonomically characteristic bat assemblages. To account for some of this variation, we included study method and continent as random effects. Fixed factors included latitudinal zone (temperate, subtropical, and tropical) and whether or not the high-intensity system comprised an agroforestry system (including monocultural orchards).

We also located several studies on ecotoxicology and demography, focusing on the effects of pesticide and GMOs use on bats. A complete review of the effects of pesticides on bats is beyond the scope of this chapter, particularly since bats and contaminants have received recent reviews (O'Shea and Johnston 2009; Bayat et al. 2014). We therefore focus on studies that explicitly link bat agrochemical exposure to changes in bat populations. Similarly, although fertilizers comprise a large portion of the chemical inputs to agriculture, their impacts on bats are indirect.

In considering the benefits of bats for agricultural production (i.e., crop yield), we focus on the provision of two ecosystem services: agricultural pest limitation by insectivorous bats and pollination by tropical bats. We did not consider their role as seed dispersers since human management of farmland vegetation limits the effect and value of bat seed dispersal. Similarly, although bat pollination is key for the unmanaged reproduction of several economically important crops, such as

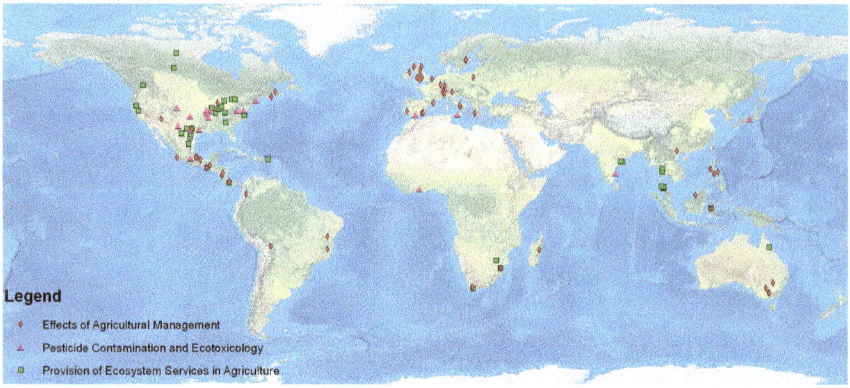

Fig. 6.1 Locations of studies on effects of habitat conversion or agricultural intensification (*red diamonds*) on bats, pesticide contamination (*pink triangles*) on bats, and ecosystem services (*green squares*) provided by bats in agriculture

bananas and agaves (Kunz et al. 2011), we did not consider these particular crops because they are mostly propagated vegetatively in such plantations. We instead focus on crops that are almost exclusively reliant on bat pollination under standard cultivation practices. Multiple investigations have characterized the diets of insectivorous bats at the order level, claiming potential consumption of pest insects. To more confidently assess consumption of insects damaging crops, we focused on studies in which known (species level identity) or probable (family level identity) agricultural pests were identified from feces of bats foraging in farms or areas dominated by agriculture. We exclude dietary studies that have sampled exclusively from natural habitats or do not describe the agricultural systems within which bats may have been foraging. We also briefly contrast these with ecosystem *disservices* of bats in agricultural areas. Bats are associated with costs to agriculturalists, particularly in the subtropics and tropics where frugivorous bats raid crops and sanguivorous bats attack domestic livestock. As with other sections, we focus on direct impacts on productive systems and do not consider the impacts of bat transmission of disease except where it directly impacts agriculture.

The majority of the nearly 140 investigations reviewed in this chapter have been conducted in temperate North America and Europe (Fig. 6.1). The bulk of studies documenting how habitat conversion or agricultural intensification affects bats has been conducted in Europe and the Neotropics (Fig. 6.1, Table 6.1). Within temperate zones, studies have focused mainly on annual cultivars and pasture, while research in tropical areas is dominated by studies on agroforestry systems, particularly coffee and cacao. Results on ecotoxicology of farmland bats come primarily from North America. Studies demonstrating the consumption of agricultural pests also derive primarily from North America, whereas studies of other ecosystem services provided by bats are limited to the tropics.

Table 6.1 Studies investigating the effects of agriculture on bat assemblage structure, ecology, or behavior across six continents

Source	Biome/life zone	Agricultural system	Bat taxa assessed	Conversion response	Intensification response
North America					
Braun de Torrez (2014)[a]	Temperate woodland savannah (mesquite-juniper)	Native and commercial pecan groves	General bat assemblage	≈ species richness, ↑ abundance, activity, habitat use	≈ species richness, abundance, activity, $\updownarrow$ by species
Farrow and Broders (2011)	Boreal forest, temperate broadleaf forest	Mixed agricultural landscape	*Perimyotis subflavus*	↓ activity levels	
Gehrt and Chelsvig (2003)[a]	Temperate prairie, woodlands, and wetlands	"Intensive" agricultural landscape	General bat assemblage	↓ activity, habitat selection	
Henderson and Broders (2008)	Boreal forest, temperate broadleaf forest	Mixed agricultural landscape	*Myotis septentrionalis*	↓ mobility	
Rambaldini and Brigham (2011)[a]	Montane forest	Vineyards	*Antrozous pallidus*	↓ activity	
Tuttle et al. (2006)	Arid desert	Rangeland with troughs	*Myotis* spp., *Antrozous pallidus*		↓ drinking efficiency
Europe					
Arlettaz and Perrin (1995), Arlettaz (1996,1999), Arlettaz et al. (1997), 2001)	Temperate agricultural landscape within European Alps	Mixed farmland	*Myotis myotis, Myotis blythii*	$\updownarrow$ foraging habitat selection	$\updownarrow$ habitat use, ↓ dietary diversity
Bontadina et al. (2002)	Temperate agricultural landscape	Mixed farmland	*Rhinolophus hipposideros*	↓ foraging habitat selection	
Boughey et al. (2011)	Temperate agricultural landscape	Conventional farmland	*Pipistrellus* spp., *Nyctalus noctula*, *Eptesicus serotinus*		$\updownarrow$ activity
Davy et al. (2007)[a]	Mediterranean landscape	Olive groves	General bat assemblage	↓ activity	↓ activity

(continued)

Table 6.1 (continued)

Source	Biome/life zone	Agricultural system	Bat taxa assessed	Conversion response	Intensification response
Dietz et al. (2013)	Temperate agricultural landscape	Traditional farmlands (orchards, meadows, pastures)	*Rhinolophus ferrumequinum*, *Myotis emarginatus*	↑ habitat selection	↓ habitat selection
Downs and Racey (2006)	Temperate agricultural landscape	Mixed farmland with woodlands and hedges	*Pipistrellus* spp., *M. daubentonii*	↓ activity	↓ activity
Drescher (2004)	Temperate agricultural landscape	Apple orchards, vineyards, pastures	*Myotis myotis*	↑ foraging activity	↓ foraging activity
Ekman and de Jong (1996)[a]	Temperate agricultural landscape	Habitat islands within crop fields	*Myotis brandti*, *Eptesicus nilssoni*, *Plecotus auritus*, *Pipistrellus pipistrellus*	↓ activity, species occurrence	
Flaquer et al. (2008)	Mediterranean landscape	Olive groves, fruit orchards, rice	*Myotis emarginatus*	↑ foraging activity (olive groves)	↓ foraging activity (orchards, rice)
Frey-Ehrenbold et al. (2013)[a]	Temperate agricultural landscape	Seminatural to intensively managed farmland	General bat assemblage		↓ activity, species richness
Fuentes-Montemayor et al. (2011)	Temperate agricultural landscape	Conventional farmland and agri-environment scheme farmlands	*Pipistrellus* spp.		↑ activity
Fuller et al. (2005)[a]	Temperate agricultural/woodland landscape	Organic and non-organic cereals	General bat assemblage		↓ species density, activity
Jones and Morton (1992)	Temperate agricultural/woodland landscape	Hay/silage, grazing	*Rhinolophus ferrumequinum*		↓ activity
de Jong (1995)	Temperate agricultural landscape	Agriculture-dominated landscape	*Myotis* spp., *Pipistrellus pipistrellus*, *Plecotus auritus*	↓ activity	
Lesiński et al. (2013)	Temperate agricultural and woodlands	Annual crops (organic and conventional)	*Eptesicus serotinus*		↓ activity

(continued)

Table 6.1 (continued)

Source	Biome/life zone	Agricultural system	Bat taxa assessed	Conversion response	Intensification response
Lisón and Calvo (2011)[a]	Semiarid Mediterranean landscape	Rainfed olive/almond, irrigated peach/citrus	General bat assemblage	≈ activity	
Lisón and Calvo (2013)	Semiarid Mediterranean landscape	Rain fed crops/xerophytic vegetation	*Pipistrellus* spp.	↕ activity (varies by species)	
Lundy and Montgomery (2010)	Temperate agricultural landscape	Improved and unimproved pasture	General bat assemblage		↓ foraging activity
Obrist et al. (2011)[a]	Temperate agricultural landscape	Managed and abandoned chestnut orchards	General bat assemblage		↑ foraging activity
Pocock and Jennings (2007)[a]	Temperate agricultural landscape	Organic/conventional farmland; hay/silage fields	General bat assemblage		↓ activity (loss of linear features), ≈ activity (agrochemical use, silage)
Rainho (2007)	Semiarid Mediterranean landscape	Cereal crops and olive groves	General bat assemblage	↓ activity	↓ activity
Russ and Montgomery (2002)	Temperate agricultural landscape	Mixed farmland with woodlands, tree lines	General bat assemblage	↓ activity	↓ activity
Russo and Jones (2003)[a]	Mediterranean landscape	Traditional farmland habitats, chestnut woodland	General bat assemblage	≈ activity	
Russo et al. (2002)	Mediterranean landscape	Olive groves, traditional farmlands	*Rhinolophus euryale*	↑ home range composition, foraging time	
Stahlschmidt et al. (2012), Stahlschmidt and Brühl (2012)[a]	Temperate agricultural/woodland landscape	Apple orchard	General bat assemblage	↓ activity (*Pipistrelle*, *Eptesicus*), ≈ activity (*Myotis*)	

(continued)

Table 6.1 (continued)

Source	Biome/life zone	Agricultural system	Bat taxa assessed	Conversion response	Intensification response
Verboom and Huitema (1997)	Temperate agricultural landscape	Linear features in traditional farmland landscapes	*Pipistrellus pipistrellus*, *Eptesicus serotinus*		↓ activity
Walsh and Harris (1996a, b)	Temperate agricultural landscape	Mixed farmland	General bat assemblage	↓ activity	↓ activity
Wickramasinghe et al. (2003, 2004)	Temperate agricultural landscape	Organic and conventional farmland	General bat assemblage		↓ activity
Australia					
Fischer et al. (2010)	Temperate woodland/agricultural landscape	Low tree density livestock pasture	General bat assemblage		↓ activity
Hanspach et al. (2012)	Temperate woodland/agricultural landscape	Pasture with varying levels of tree cover	General bat assemblage	↕ activity, species richness (peaks at intermediate tree cover)	↓ activity, species richness
Lentini et al. (2012)[a]	Temperate woodland/agricultural landscape	Cereal, canola, and pasture-dominated landscape with and without linear features	Mollossidae; Vespertilionidae; Emballonuridae		↓ activity, ≈ species richness, feeding
Lumsden et al. (2002)	Temperate woodland/agricultural landscape	Fragmented human-dominated landscape	*Nyctophilus geoffroyi*, *Chalinolobus gouldii*	↕ roost locations	↓ roost locations
Lumsden and Bennett (2005)[a]	Temperate woodland/agricultural landscape	Pasture with varying densities of trees	General bat assemblage	↓ abundance, ↑ activity	↓ abundance, activity
Neotropics					
Avila-Cabadilla et al. (2009)	Tropical dry forest	Pasture	Phyllostomids	↓ abundance, species richness	
Castro-Luna and Galindo-González (2012)[a]	Tropical montane rainforest	Diverse and simplified shade coffee, pasture	Frugivorous phyllostomids		↓ abundance, species richness

(continued)

Table 6.1 (continued)

Source	Biome/life zone	Agricultural system	Bat taxa assessed	Conversion response	Intensification response
Estrada et al. (1993)	Lowland tropical rainforest	Shaded (coffee, cacao, mixed) and unshaded (citrus, allspice) plantations, pastures	Phyllostomids, non-phyllostomids sampled with mist nets	↓ species richness	↓ species richness
Estrada and Coates-Estrada (2001)	Lowland tropical rainforest	Shaded (coffee, cacao, mixed) and unshaded (citrus, allspice) plantation	General bat assemblage	↑ abundance, ↓ expected species richness	↓ abundance, expected species richness
Estrada and Coates-Estrada (2002)[a]	Lowland tropical rainforest	Coffee, cacao, citrus, banana, pasture	General bat assemblage	↑ abundance, ≈ species richness	
Estrada et al. (2004)[a]	Lowland tropical rainforest	Fencerows, citrus, pasture	Non-phyllostomids	≈ activity	↓ activity
Faria (2006)[a]	Brazilian Atlantic forest	Shade cacao in a forest dominant landscape	General bat assemblage	↑ abundance, species richness	
Faria et al. (2006)[a]	Brazilian Atlantic forest	Shade cacao in a cacao dominant matrix	General bat assemblage	≈ species richness	
Faria and Baumgarten (2007)[a]	Brazilian Atlantic forest	Shade cacao in two contrasting landscapes	General bat assemblage		↓ abundance, species richness
García Estrada et al. (2006, 2012)	Montane tropical rainforest	Shade coffee	Phyllostomids	≈ abundance, dietary diversity, ↓ species richness	≈ abundance, dietary diversity, species richness
Harvey et al. (2006)[a]	Tropical dry forest	High and low tree cover pasture	General bat assemblage	↓ abundance, species richness (frugivores, nectarivores)	↓ abundance, species richness (frugivores), ≈ abundance (nectarivores)
Harvey and González Villalobos (2007)[a]	Tropical humid forest, premontane wet forest	Cacao agroforest, banana agroforest, plantain monoculture	General bat assemblage	↕ abundance, species richness	↓ abundance, species richness

(continued)

Table 6.1 (continued)

Source	Biome/life zone	Agricultural system	Bat taxa assessed	Conversion response	Intensification response
Medellin et al. (2000)	Montane tropical rainforest	Cacao, oldfield, and corn	General bat assemblage	↑ abundance, ↓ species richness	↓ abundance, species richness
Medina et al. (2007)[a]	Tropical moist forest	High and low tree cover pasture	General bat assemblage	↓ abundance, species richness	↓ abundance, species richness
Numa et al. (2005)[a]	Tropical montane rainforest	Sun and shade coffee in contrasting landscapes	Phyllostomids	↑ abundance, ↓ estimated species richness	≈ estimated species richness (within landscape), ↓ estimated species richness (between landscapes)
Pineda et al. (2005)[a]	Tropical montane cloud forest	Shade coffee	General bat assemblage	↑ abundance, ≈ species richness	
Saldaña Vázquez et al. (2013)	Tropical montane rainforest	Shade coffee	*Sturnira ludovici*	↓ abundance, females	
Sosa et al. (2008)	Tropical montane rainforest	Shade coffee	General bat assemblage	≈ species richness	↓ abundance
Williams-Guillén and Perfecto (2010)[a]	Tropical montane rainforest	Shade coffee	Phyllostomids	↑ abundance, ≈ species richness	↓ abundance, ≈ species richness
Williams-Guillén and Perfecto (2011)[a]	Tropical montane rainforest	Shade coffee	Non-Phyllostomids	↓ activity (cluttered space foragers), ↑ abundance (open space foragers), ≈ species richness	↓ activity (cluttered space foragers), ↑ abundance (open space foragers), ≈ species richness
Vargas Espinoza et al. (2008)	Premontane tropical rainforest	Citrus orchards	General bat assemblage	≈ abundance, species richness	
Africa					
Noer et al. (2012)	Subtropical Savanna	Sugarcane	*Chaerephon pumilus*, *Mops condylurus*	↑ foraging time	

(continued)

Table 6.1 (continued)

Source	Biome/life zone	Agricultural system	Bat taxa assessed	Conversion response	Intensification response
Randrianandrianina et al. (2006)[a]	Tropical rainforest/ mixed use landscape	Smallholder gardens	General bat assemblage	↓ species richness, ↑ activity, foraging	
Sirami et al. (2013)[a]	Subtropical grassland, fynbos	Intensive wheat, vineyards, orchards	General bat assemblage	↓ abundance, ≈ species richness	↓ abundance, ≈ species richness
Taylor et al. (2013b)[a]	Subtropical Savanna	Macadamia	General bat assemblage	≈ activity (molossids), ↓ activity (vespertilionids)	
Asia					
Fukuda et al. (2009)[a]	Fragmented tropical rainforest	Orchards, palm oil	Hipposideridae, Vespertilionidae, Pteropodidae	↑ abundance, ↓ species richness	≈ abundance, ↓ species richness
Furey et al. (2010)[a]	Limestone karst, tropical rainforest	Landscapes dominated by rice paddies and degraded forest	Insectivorous bat assemblage		↓ abundance, ≈ species richness
Graf (2010)[a]	Montane tropical rainforest	Forest and shade cacao	General bat assemblage	↑ abundance, species richness	↓ species richness
Mildenstein et al. (2005)	Tropical moist forest	Fruit orchards and hardwood tree plantations	Pteropodidae bat assemblage	↓ habitat selection	
Sedlock et al. (2008)	Montane and premontane rainforest	Mosaic of pasture, root crops, orchards, and fallows	Insectivorous bat assemblage	↓ species accumulation	
Van Weerd and Snelder (2008)[a]	Tropical moist forest	Village homegarden polycultures, shrublands used for grazing	General bat assemblage	↑ abundance, species richness	↓ abundance, ≈ species richness

[a]Studies included in meta-analysis

"Conversion response" indicates effects of agriculture versus non-anthropogenic habitat; "intensification response" indicates effects of agricultural intensification (i.e., increased amount of agrochemicals, decreased structural complexity, infrastructure construction). "↓" indicates a negative effect on specified response variable; "↕" variable effects depending on species, ensemble, or habitat contrast; "≈" no marked effects observed; "↑" positive effect on response variable

6.3 Effects of Agricultural Intensity on Bat Assemblage Structure, Behavior, and Ecology

We found 70 studies addressing the effects of habitat conversion or management on the assemblage structure, behavior, or ecology of bats. Fifty-two studies assessed bats in both natural and agricultural areas. Twenty-two studies (42 %) demonstrated negative effects of habitat conversion, twelve (23 %) showed variable responses (e.g., only some species or ensembles declined, different agricultural systems were associated with different effects), twelve (23 %) showed increased richness, activity, or abundance in agricultural areas, and six (12 %) showed little or no difference between agricultural and natural areas. Forty-five studies addressed some aspect of agricultural intensification, with 38 of these (84 %) documenting a negative effect of intensification on bats, four showing variable or neutral (9 %) responses, while three studies (7 %) documented increases in bat richness, abundance, or activity in more intensive systems.

Response variables differ in response to habitat conversion and agricultural intensification (Fig. 6.2, Table 6.2), with measures of species richness showing no significant change between treatments. In contrast, measures of relative activity and abundance show stronger responses (Fig. 6.2). Agroforestry systems are more structurally similar to the original non-anthropogenic land uses, making them less intensive than annual crops dominated by one plant species or pasture systems lacking structural complexity. This relationship presumably explains why agricultural systems that incorporate trees and other large woody perennials on farms and throughout the agricultural landscape have little effect on bat activity and abundance (Fig. 6.2). Agroforestry systems appear to mitigate negative effects on bat assemblages in cases of both habitat conversion and agricultural intensification (Table 6.2).

Several studies have considered the effects of agricultural management at landscape scales versus focusing exclusively on farm-level management practices (Estrada et al. 1993; Ekman and de Jong 1996; Verboom and Huitema 1997; Numa et al. 2005; Faria et al. 2006, 2007; Faria and Baumgarten 2007; Fuentes-Montemayor et al. 2011; Boughey et al. 2011; Maas et al. 2013). Within agricultural areas, bat activity increases with proximity to natural areas (Estrada et al. 1993; Verboom and Huitema 1997; Boughey et al. 2011) and in less fragmented landscapes (Fuentes-Montemayor et al. 2011; Frey-Ehrenbold et al. 2013) or in landscapes with more natural elements such as hedgerows and woodlots (Verboom and Huitema 1997).

Agricultural areas also serve as matrix habitat connecting fragmented non-anthropogenic habitats. Although one study has suggested that landscapes dominated by crops and open fields have a stronger negative influence on bats than water (Ekman and de Jong 1996), a recent analysis of bat responses to isolation on islands versus in forest fragments embedded in agricultural matrix suggests that the anthropogenic matrix is more permeable than water matrix (Mendenhall et al.

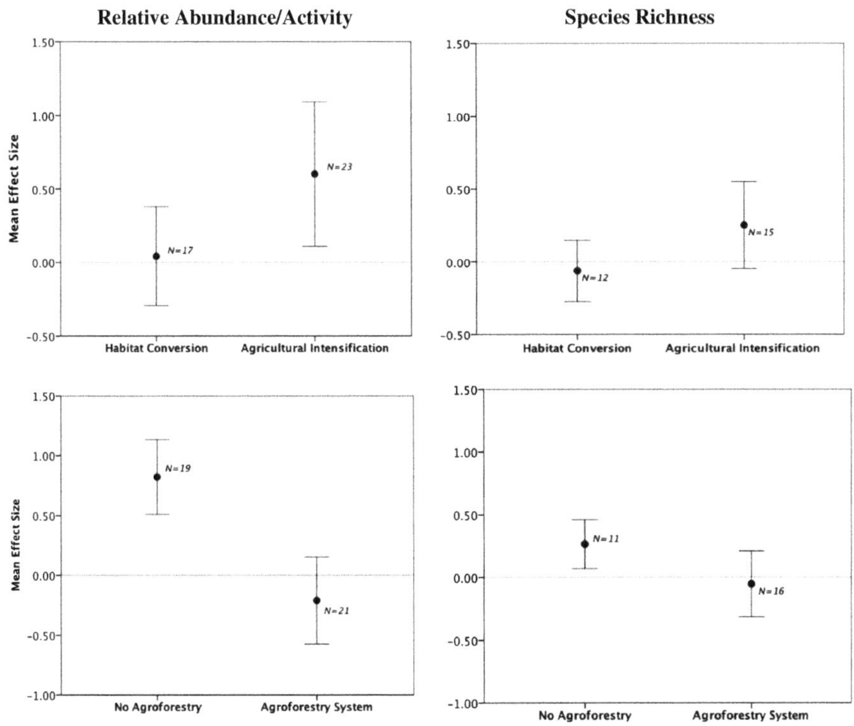

Fig. 6.2 Mean effect size (log odds ratio, circles) ±95 % CI of relative abundance and activity (*left*) and species richness (*right*) of habitat conversion versus agricultural intensification (*top row*), and of contrasts (both habitat conversion and agricultural intensification) with and without agroforestry systems (*bottom row*). Positive effect sizes indicate reductions in relative abundance and activity or species richness in response to habitat conversion and intensification

2014). Thus, agricultural intensification at the landscape level should make the matrix less permeable due to the reduction of natural resources and structural elements such as trees, affecting not only the persistence of bats in fragmented landscapes, but *also* the degree to which bat assemblages show a negative response to agriculture. A few investigations have confirmed such interactions between farm- and landscape-level intensification: Intensification in cacao matrices in Brazil (Faria et al. 2006, 2007; Faria and Baumgarten 2007) and coffee matrices in Colombia (Numa et al. 2005) resulted in reductions in the species richness and abundance of bats in diverse shade agroforests relative to forest fragments. In Europe, effects of landscape management on bat assemblage structure and ecology in temperate landscapes dedicated to the production of annual crops remain largely unexplored compared to the extensive information available at the field and farm scales.

Table 6.2 Effects of latitudinal zone and agroforestry systems on effect size (log odds ratio) for two response variable types under habitat conversion and agricultural intensification

Response variable	Land change type	Model	AIC	χ^2	*P*
Abundance/activity	Habitat conversion	Effect size ~ (Method) + (Continent)	60.7		
		Effect size ~ Agroforestry + (Method) + (Continent)	**49.7**	**13.00**	**<0.001**
		Effect size ~ Latitude + (Method) + (Continent)	62.0	0.00	1.000
		Effect size ~ Agroforestry + Latitude + (Method) + (Continent)	51.9	12.15	<0.001
	Intensification	Effect size ~ (Method) + (Continent)	52.4		
		Effect size ~ Agroforestry + (Method) + (Continent)	**49.2**	**5.22**	**0.022**
		Effect size ~ Latitude + (Method) + (Continent)	53.6	0.00	1.000
		Effect size ~ Agroforestry + Latitude + (Method) + (Continent)	50.6	4.923	0.026
Species richness	Habitat conversion	**Effect size ~ (Method) + (Continent)**	**20.7**		
		Effect size ~ Agroforestry + (Method) + (Continent)	21.7	0.99	0.319
		Effect size ~ Latitude + (Method) + (Continent)	24.0	0.00	1.000
		Effect size ~ Agroforestry + Latitude + (Method) + (Continent)	24.1	1.82	0.178
	Intensification	**Effect size ~ (Method) + (Continent)**	**22.9**		
		Effect size ~ Agroforestry + (Method) + (Continent)	24.4	0.54	0.460
		Effect size ~ Latitude + (Method) + (Continent)	26.3	0.06	0.806
		Effect size ~ Agroforestry + Latitude + (Method) + (Continent)	27.0	1.34	0.248

Parentheses indicate random effects, and *bold text* indicates best fitting model based on AIC value

6.4 Pesticide Impacts on Bat Populations

Agricultural intensification may remove potential habitat for bats and their prey; the effects of increased agrochemical inputs, such as increased exposure and changes in prey availability, may put resident bats under further pressure.

Bats may directly consume pesticides by feeding on fruits, flowers, and arthropods exposed to chemical application. Even bats foraging outside of agricultural areas can be exposed to pesticides via biomagnification as residues are incorporated into the tissues of organisms at higher trophic levels (Bayat et al. 2014).

Investigations of exposure of bats to pesticides and its effects on physiology and mortality first appeared in the 1970s, amid a wave of growing concern regarding the effects of organochlorine pesticides (e.g., DDT, DDE, dieldrin, lindane, endosulfan, aldrin) on ecosystems and observations of declining bat populations at high-profile sites such as the Carlsbad Caverns in New Mexico, USA (Clark 1988, 2001). In some cases, DDT and other organochlorines were even applied directly to bat roosts in efforts to exterminate "vermin" (Kunz et al. 1977), and declines in high-profile bat colonies were linked to organochlorine use (Clark et al. 1978; Clark 2001). Even sublethal exposure to pesticides can have negative consequences for bats, resulting in increased metabolic rates (Swanepoel et al. 1998), and ingestion of pesticide residues on arthropods may poses a potential reproductive risk to certain bat species (Stahlschmidt and Brühl 2012).

Organochlorine residues have been documented in bats in a wide variety of both agricultural and non-agricultural landscapes, although several studies have found increased contaminant loads in bats sampled near agricultural areas (Clark and Prouty 1976; White and Krynitsky 1986) or near sites of pesticide manufacture (O'Shea et al. 2001). In some cases, temporal changes in levels of different contaminants reflect shifts in local agricultural practice as farmers adopt new pesticide regimes (Miura et al. 1978; Clark et al. 1980). Organochlorines are notorious for their persistence in ecosystems, and a variety of studies demonstrate that bats continue to harbor these contaminants in their tissues 20–30 years after the use of these pesticides was banned in sampling areas (Clawson and Clark 1989; Guillén et al. 1994; Schmidt et al. 2000; Sasse 2005). In some cases, persistence may reflect the continued use of these pesticides in lower income nations, as may be the case for the migratory *Tadarida brasiliensis* (Thies and Thies 1997; Bennett and Thies 2007). Investigations in India (Senthilkumar et al. 2001) and Benin (Stechert et al. 2014) have detected levels or metabolites of organochlorines in bat samples indicative of continued recent use in these regions, especially to fight against malaria. Furthermore, pesticide standards vary between different countries, application often appears to occur non-selectively, and farmers with limited training (especially in developing countries, where agricultural expansion is greatest) are likely to be unaware of the multitude of negative nontargeted environmental impacts affecting human health and biodiversity (Tilman et al. 2001; Yadav 2010).

Despite the clear negative impacts of organochlorines on bats, the effects of agrochemical classes such as pyrethroids and neonicotinoids remain largely unknown (O'Shea and Johnston 2009; Quarles 2013; Bayat et al. 2014), although recent research demonstrates a negative impact on birds (Hallmann et al. 2014). In North America, pesticide contamination has been implicated in bat mortality associated with the fungal pathogen causing white-nose syndrome (WNS), since pesticide load can lead to immunosuppression and endocrine disruption that could make bats more vulnerable to infection (Kannan et al. 2010). "Back of the envelope" calculations suggest declines in bat populations attributed to WNS could translate into an additional 1320 metric tons of insects escaping predation each year (Quarles 2013). The trickle-down impacts on agricultural production could be substantial, although quantitative evidence is lacking. The effects of GM crops incorporating insecticidal traits have been investigated largely in the context of the provisioning of predation services (Federico et al. 2008; Lopez-Hoffman et al. 2014; see next section); however, declines in pest numbers associated with the use of these crops could result in population declines of insectivorous bats (Lopez-Hoffman et al. 2014).

6.5 Ecosystem Services Provided by Bats in Agricultural Systems

6.5.1 Insectivorous Bats and Pest Limitation

Of the potential ecosystem services provided by bats, their role in consuming insect pests has received the most attention within agricultural systems. Insectivorous bats have a global distribution and have long been identified as key suppressors of arthropod pests in agricultural systems (Kunz et al. 2011). However, surprisingly little evidence exists quantifying the impact of their predation on arthropod populations, plant damage, or its economic value (Boyles et al. 2013; Maas et al. 2013). Several studies have characterized diets of insectivorous bats (reviewed by Kunz et al. 2011), and the recent development of DNA-based methods for dietary analysis provides an unprecedented amount of detail on the composition of bat diets and allows for the identification of individual pest species. Although few studies have documented direct impacts of bat predation on agricultural pests, an increasing body of evidence documents pest consumption, impacts on arthropods, and estimates of direct economic impacts.

We review 15 studies documenting the consumption of known or probable crop pests by insectivorous bats (Table 6.3). The diets of temperate North American insectivores have received particular attention. Many bat species consume lepidopterans, and studies in North America demonstrate bat predation on devastating pests such as corn earworm (*Helicoverpa zea*) and fall armyworm (*Spodoptera frugiperda*) moths (Lee and McCracken 2005; McCracken et al. 2012). Bat species across the world feed on folivorous beetles from a variety of damaging families

Table 6.3 Dietary investigations of insectivorous bat in agricultural areas documenting consumption of pest insect families or species

Study region	Source	Bat species	Crop	Pest insects consumed
Africa (South Africa)	Taylor et al. (2012, 2013a)	Various species	Macadamia nuts	• Hemiptera: *Nezara viridula*
Africa (Swaziland)	Bohmann et al. (2011)	*Chaerephon pumilus*, *Mops condylurus*	Sugarcane	• Hemiptera: Aphidadae, Lygaeidae, Pentatomidae • Lepidoptera: *Eldana saccharina*, *Mythimna phaea*
Asia (Thailand)	Leelapaibul et al. (2005)	*Chaerephon plicatus*	Rice	• Hemiptera: *Sogatella* sp.
Europe (Switzerland)	Arlettaz and Perrin (1995, 1997, 2001)	*Myotis myotis*, *M. blythii*	Agricultural landscape with orchards, pasture	• Coleoptera: *Melolontha* sp.
Latin America (Mexico)	Williams-Guillén (unpublished data)	Various species	Shade coffee	• Coleoptera: *Hypothenemus hampeii*, *Rhabdopterus jansoni* • Orthoptera: *Idiarthron subquadratum*
North America (Canada)	Clare et al. (2011)	*Myotis lucifugus*	Agricultural landscape	• Coleoptera: *Phyllophaga* spp., *Amphimallon majale*, *Phyllobius oblongus*; Curculionidae, Chrysomelidae • Diptera: *Delia antiqua* • Hemiptera: Aphididae • Lepidoptera: *Korscheltellus lupulina*
North America (Canada)	Rambaldini and Brigham (2011)	*Antrozous pallidus*	Grapes	• Coleoptera: Curculionidae, Tenebrionidae • Orthoptera: Acrididae
North America (USA)	Braun de Torrez (2014)	Various species	Pecan	• Lepidoptera: *Acrobasis nuxvorella*

(continued)

Table 6.3 (continued)

Study region	Source	Bat species	Crop	Pest insects consumed
North America (USA)	Lee and McCracken (2005)	*Tadarida brasiliensis*	Landscape with corn and cotton	• Coleoptera: Scarabaeidae • Hemiptera: Cercopidae, Delphacidae, Pentatomidae • Lepidoptera: *Spodoptera frugiperda, Helicoverpa zea*
North America (USA)	McCracken et al. (2012)	*Tadarida brasiliensis*	Corn, cotton	• Lepidoptera: *Helicoverpa zea*
North America (USA)	Storm and Whitaker (2008)	*Eptesicus fuscus*	Agricultural landscape	• Coleoptera: Curculionidae • Hemiptera: Cicadelidae
North America (USA)	Whitaker (1995)	*Eptesicus fuscus*	Agricultural landscape	• Coleoptera: Curculionidae, Scarabaeidae • Hemiptera: Cicadellidae, Pentatomidae

and species, particularly weevils, leaf beetles, and scarab beetles. Bats may also be underappreciated predators of hemipteran pests, with many studies demonstrating consumption of leafhoppers, froghoppers, spittle bugs, and stink bugs. We emphasize that direct consumption alone is not sufficient to prove that bats are limiting insect pests: Damaging insects may comprise a small proportion of the diet, and nearly every study summarized in Table 6.3 also demonstrated consumption of the predatory arthropods that comprise part of the assemblage of natural enemies. Such intraguild predation could counteract the pest-limiting effects of bat insectivory (Brashares et al. 2010), although herbivores generally comprise the majority of diet by volume in investigations using fecal pellet dissections (Kunz et al. 2011). That the relative abundance, diets, and movements of bats may track populations of agricultural pests (Lee and McCracken 2005; McCracken et al. 2012; Taylor et al. 2013b) suggests that many species are indeed preying heavily on herbivorous insects. This has been assessed in mouse-eared bats, *Myotis* spp., that track cyclic, massive local aggregations of cockchafers known since centuries for the damages they cause to fruit trees in Central Europe (Arlettaz 1996; Arlettaz et al. 2001).

During lactation, small bat species consume 75 % to over 100 % of their body weight each night (Kurta et al. 1989; Kunz et al. 1995, 2011), and a single maternity colony of 1 million Brazilian free-tailed bats is capable of consuming over 8 tons of insects per night (Kunz et al. 2011). These numbers suggest the staggering potential for bat predation to limit pest insect populations and provide a valuable ecosystem service for agricultural production. Until recently, surprisingly little work had quantified the impact of bat predation on insect biomass (Maas et al. 2015). Exclosure studies have long been a mainstay for studying the impacts of bird predation; however, it was widely assumed that such methods would not be suitable to measure the impact of bat insectivory, due to the misconception that all insect eating bats take highly mobile, flying prey. However, bats capable of gleaning insect prey from substrates exist throughout the world, and their impacts could be monitored via exclosure studies and disentangled from those of birds. This approach has been used fruitfully in the past five years, demonstrating significant increases in arthropod density when bats are absent, in agroecosystems (Williams-Guillén et al. 2008; Maas et al. 2013), reforestation (Morrison and Lindell 2012), and natural forests (Kalka et al. 2008). In Mexican polycultural shade coffee, arthropod densities on coffee plants during the rainy season nearly doubled in the absence of bats, with marked increases in densities of hoppers, katydids, cockroaches, and beetles (Williams-Guillén et al. 2008). However, no effects on plant damage were observed in that study, perhaps as a result of the short duration of the study or release of spiders and other arthropod predators. In Indonesian shade cacao, excluding bats resulted in a 29 % increase in arthropod numbers (Maas et al. 2013). Although herbivory did not differ significantly between cacao plantations with different levels of shade or proximities to primary habitats within the landscape, exclosure of bats resulted in a significant decrease in yields, with the effects of bird and bat predation together valued at an astonishing US $730 per ha and year (bat predation was valued at US $520 per ha and year). However, the effects of bat predation on crop pests are not universal: An exclosure study in Costa Rican coffee found that excluding bats alone had virtually no effect on the density or damage caused to beans by the devastating coffee berry borer (Karp et al. 2013).

Exclosure studies are not suitable to measure the impact of high-flying insectivores, such as molossids. However, careful extrapolations taking into account bat feeding rates, population sizes, pest reproduction, and survivorship, and the costs of inputs allow for estimation of the economic impact of predation for other bats, particularly molossids forming large colonies. Cleveland et al. (2006) estimate that Mexican free-tailed bats (*T. brasiliensis*) feeding on the cotton bollworm moth in Texas provide pest limitation services worth roughly US $183 per ha and year to cotton growers. Extending these estimates to agricultural areas throughout the USA suggests that bat predation could have a value of nearly US $23 billion annually (Boyles et al. 2011). These benefits hold for both conventional and transgenic cotton (Federico et al. 2008), although the introduction of *Bt* cotton (a genetically modified organism whose tissues produce an insecticide derived from the bacterium *Bacillus thuringiensis*), coupled with reduced area in cotton cultivation, has led to a decline in the overall value of this pest limitation service (Lopez-Hoffman et al. 2014).

Valuation of bat-mediated pest suppression is limited for staple crops and for sites outside the southern USA. In northern Mexico, the impact of *T. brasiliensis* predation on avoided agricultural costs across a variety of staple and commodity crops was estimated at a far more modest \$19 per ha and year (Gándara Fierro et al. 2006). In Thailand, the value of wrinkle-lipped bat (*Tadarida plicata*) predation on a major rice pest, the white-back planthopper (*Sogatella furcifera*), was estimated to have a monetary value of \$1.2 million annually (Wanger et al. 2014). This estimate results in a seemingly paltry \$0.13 per ha and year value considered against Thailand's 8.7 million ha (Redfern et al. 2012) of rice paddies, but in this case an economic approach obscures the true value of the service: This single bat species prevents the loss of nearly 2900 metric tons of rice per year, enough to feed Thailand's entire population of 66.8 million people for a week. Such investigations underscore the potentially grave consequences for human food security should global bat populations continue declining (Kunz et al. 2011).

6.5.2 Nectarivorous Bats and Pollination Services

Pollination services to crops by bats are poorly documented. Bats are key pollinators of wild *Agave* and *Musa* spp. (Kunz et al. 2011). Although these plants are propagated vegetatively under cultivation, bat pollination plays a critical role in sustaining genetic diversity in the wild relatives of these domestic species, a key aspect of maintaining future food security (Hopkins and Maxted 2011). Within the Americas, several bat pollinated cacti are commercially important fruit species (Kunz et al. 2011). Several species of the hemiepiphytic cactus *Hylocereus* (pitahaya, dragonfruit) endemic to the Neotropics are now cultivated worldwide. In Mexico, visitation of *Hylocereus undatus* fruits by bats resulted in significantly higher fruit set than did visitation by diurnal pollinators (Valiente-Banuet et al. 2007). Although *H. undatus* is self-compatible, other species such as *H. costaricensis* (an important fruit crop in southern Mesoamerica) apparently rely on pollination by bats and sphingid moths (Weiss et al. 1994; Le Bellec et al. 2006). Nectarivorous bats, particularly the cave nectar bat (*Eonycteris spelaea*) feed on the flowers of tree beans or petai (*Parkia* spp.) (Bumrungsri et al. 2008a, b, 2013) and durian (*Durio zibethinus*) (Bumrungsri et al. 2008b), pollinating these plants in the process. The economic value of this pollination has been estimated at over US \$13 million annually in three provinces of Thailand (Petchmunee 2008).

6.6 The Issue of Ecosystem Disservices of Bats to Agricultural Production

Unfortunately, while the ecosystem services provided by bats are largely invisible, their disservices are obvious. In the Paleotropics, crop raiding by frugivorous pteropodids can cause substantial losses of commercial fruits (see Aziz et al., Chap. 12). For example, in Indian vineyards, *Cynopterus sphinx* damages up to 90 % of the crop along peripheries of plantations and may cause revenue losses of up to US $590 per ha and year (Srinivasulu and Srinivasulu 2002). In the Neotropics, sanguivorous vampire bats can cause substantial economic damage: Estimates for 1968 placed losses at $47.5 million USD for over 512,000 rabies-related cattle deaths in Latin America (Arellano-Sota 1988). Harassment by vampire bats can put cattle off their feed, resulting in annual weight losses estimated at roughly 40 kg/head and milk production loss of 261 L/head (Schmidt and Badger 1979). These estimates fail to take into account the effects of vampire bats on the medium and small domestic animals (e.g., chickens, pigs, goats) that provide critical sources of animal protein for millions of smallholder farmers across the region.

Not surprisingly, farmers with first-hand experiences of economic losses engendered by bats are more likely to have negative attitudes or report a willingness to destroy bat roosts (Reid 2013). Failure to explicitly address the negative impacts of some bat species likely reduces the efficacy of conservation messages; meanwhile, practical measures to reduce these disservices could benefit multiple bat species by reducing indiscriminate persecution. Different functional groups provide most of the ecosystem services (insectivores, nectarivores) and disservices (frugivores, sanguivores). However, local farmers may not distinguish between these groups. For example, farmers and agricultural technicians in Latin America often attempt to cull vampire bat populations by destroying bat roosts; unfortunately, the widespread belief that all bats are "*vampiros*" frequently results in the destruction of colonies of beneficial bat species (Mayen 2003; Aguiar et al. 2010). If local people perceive the ecosystem services of one bat group as offsetting the damages of another, then an ecosystem service approach could provide a framework for bat conservation more broadly. Unfortunately, the extent to which knowledge of ecosystem services changes attitudes toward bats in developing countries remains unknown.

6.7 Discussion

Our review suggests that in all biogeographic regions investigated, at least some bat species persist in and exploit agricultural areas. In many agricultural systems (e.g., tropical agroforestry or historical landscapes of Europe), bat assemblages maintain richness and may even exceed abundances observed in unmanaged areas. Nevertheless, agricultural intensification has a generally negative effect on bats and thus presumably on the ecosystem services they provide. Our analysis did not address differences between bat taxa in their sensitivity to habitat change and intensification. However, evidence from speciose assemblages suggests that forest-adapted insectivorous species are particularly sensitive to habitat conversion (Medellin et al. 2000; Faria and Baumgarten 2007; Williams-Guillén and Perfecto 2010), implying that in some regions, this valuable ecosystem service could be particularly vulnerable to loss in the face of habitat loss.

Although few investigations have considered the scale of intensification, limited information suggests that less managed systems embedded in regions dominated by intensive agriculture may show depauperate bat faunas (Numa et al. 2005; Faria et al. 2007). Declines in bat populations in agricultural regions are concerning not only from the point of view of biodiversity conservation but also regarding human well-being and food security, especially in many tropical areas where smallholder farming systems are dominant. Ongoing losses of these generalist vertebrate predators could have major impacts on insect pest limitation for a wide variety of staple and commodity crops. However, the smallholder farmers in developing nations who most depend on the ecosystem services provided by bats (due to limited access to manufactured inputs or cultivation of bat pollinated crops) may have highly negative attitudes toward these mammals as a result of visible damages caused to crops and livestock (López del Toro et al. 2009; Reid 2013), whereas beneficial impacts on crop yield productivity and the value of biodiversity (i.e., increased ecosystem resilience) are often unknown or unappreciated (Williams-Guillén, unpublished data). These results suggest a pressing need to reassess common approaches to conservation and agricultural management in the Anthropocene.

6.7.1 Sparing, Sharing, and the Devaluation of Manufactured Capital

Given the anticipated need to nearly double global food production in the twenty-first century, a vigorous debate has emerged with respect to the most viable path to increase production without degrading ecosystem services or reducing biodiversity: land sparing, which posits that increased intensification and yields will reduce pressure to convert non-agricultural lands, versus land sharing, in which agricultural areas are less intensively farmed in order to increase associated biodiversity and habitat permeability (Fischer et al. 2008). Given the vagility and critical role of bats in agricultural production, land sharing approaches might be preferable with respect to the provision of bat-dependent ecosystem services. Many sensitive bat ensembles and species (e.g., many forest-adapted and insectivorous species, e.g., from Phyllostomidae or Vespertilionidae) will require well-structured farmland, i.e., cultivated landscapes including patches of natural and seminatural features for their long-term existence. However, not only do many bat species thrive in diverse agricultural landscapes, but also their loss could affect the provision of pest suppression and pollination services and result in reduced crop productivity. Given the many disadvantages of chemical control of pests, managing agricultural landscapes to maximize the abundance and diversity of bats and other natural enemies must form a key aspect of sustainable agricultural production. However, the design and management of such systems to maximize bat diversity, activity, and ecosystem services is largely unknown, although European conservationists are at the forefront with their strategies to promote biodiversity-friendly farming.

Chemical and mechanical inputs are not the only tools of agricultural intensification. Within recent decades, genetic modification of crops (e.g., *Bt* corn and cotton) has become increasingly prevalent (James 2011). In the short term, adoption of such varieties does reduce the need to rely on bats and other predators for pest limitation (Lopez-Hoffman et al. 2014), resulting in a “devaluation” of the natural capital provided by bats, and undermines arguments for bat conservation that are based exclusively on provision of ecosystem services. However, as is the case with pesticides, insects are rapidly evolving resistance to *Bt* crops across the world, resulting in a rapid devaluation of *manufactured* capital (Lopez-Hoffman et al. 2014). While the value of bats’ natural capital may fluctuate, it likely devalues far less slowly: Bats and insects are engaged in an evolutionary arms race dating back millions of years (Conner and Corcoran 2012). Without bats to buffer the inevitable loss of efficacy of chemical inputs and GM crops, the technological advances that make agricultural intensification possible leave production vulnerable to potentially catastrophic failures to limit pest damage.

6.8 Research Priorities

6.8.1 Filling in Biogeographical Knowledge Gaps

Although the effects of habitat conversion and management have been well investigated in Europe and the Neotropics, the extent to which these processes may differ in other regions of the world remains unknown. We highlight a particular lack of knowledge from Africa and Asia; we did not find any studies from East Asia, although we suspect information exists in the Chinese language literature. Understanding the types and magnitudes of ecosystem services provided by bats in a variety of agricultural systems and regions is particularly important.

6.8.2 Linking Farm Management, Ecosystem Services, and Landscape-Level Processes

The effects of farm-level management on biodiversity and ecosystem services cannot be adequately considered without taking account of landscape-level processes (Tscharntke et al. 2005; Vickery and Arlettaz 2012). Nevertheless, the extent to which local- and landscape-level management interact to shape pest suppression or pollination services is largely uninvestigated. The effect of bats in limiting arthropod pests in agricultural areas is still poorly documented. However, the limited data that exist can demonstrate a vexing degree of divergence in results. For example, bats in Mexican shade coffee have substantial effects on herbivorous insects (Williams-Guillén et al. 2008), while bats in Costa Rican shade coffee had no significant effect on herbivores (Karp et al. 2013). In Indonesian cacao agroforestry systems, insectivorous bats strongly contribute to the suppression of many different pest insect groups and crop yield productivity across gradients of local shade-tree management and forest proximity within the agricultural landscape (Maas et al. 2013). In general, the study sites differ in landscape structure and land use, local farm history and management, habitat dynamics and conversion, intensity of farming practices, and vertebrate insectivore assemblage structure. Elucidating the factors of bat ecosystem service provision is key to managing agricultural areas to sustain bat populations and enhance food production (Maas et al. 2015).

6.8.3 Pest Suppression in the Face of Climate Change, Pesticides, and GM Crops

Not only will warming climates lead to shifts in the areas suitable for agricultural production, but it will also likely lead to range expansions of tropical pests, increases in pest numbers and damage, with a parallel risk of a drop in the efficacy of pest suppression by natural enemies that might be negatively affected by climate change (Thomson et al. 2010; Bebber et al. 2013). Such changes will make the ecosystem services provided by generalist predators like insectivorous bats more valuable than ever before. However, if agricultural adaptation to climate change relies on landscape-level intensification as a strategy, bats are likely to decline further, reducing their provision of pest suppression services. Despite the myriad negative effects of pesticides (i.e., affecting livelihoods, food security, environment, and health; reviewed by Yadav 2010), farmers across the world might turn to agrochemicals as a first response to increases in pest damage (Wilson and Tisdell 2001), with the Old World's rapid development of more environmentally friendly farming practices appearing as an exception in this general move. As reviewed in this chapter, older pesticide classes such as organochlorines have particularly detrimental effects on bat populations. However, the degree to which newer pesticide classes affect bats is largely unknown. The neonicotinoids, once touted for their low toxicity, have now been linked to major declines in bees (Van der Sluijs et al. 2013) and more recently in several species of passerines as a result of insect resource depletion (Hallmann et al. 2014). The extent to which use of next-generation pesticides and GM crops is driving and interacting with bat declines and resultant increases in pest damage is a critical research area.

6.8.4 Quantifying Impact and Value Across Crops and Biomes

Additional valuation of bats' ecosystem services could provide both guidance for bat management priorities in agricultural areas and compelling rationales for conservation. However, valuation efforts have focused almost exclusively on commodity crops quantified along the single dimension of monetary value. Most of the world's smallholder farmers focus on staple crop cultivation and may not have the means to substitute the manufactured capital of pesticides and GM crops for bat predation. As Wanger et al. (2014) demonstrate, valuation based on dollars of damage prevented misses many of the criteria most important to subsistence farmers seeking food security. There is an urgent need to better understand the importance of bat ecosystem services across a variety of crop types, regions, and management approaches. Research also highlights the importance of better quantifying the fluctuations in bat service provision across years and seasons, in relation to population fluctuations, reproductive phenology, and agricultural management (Lopez-Hoffman et al. 2014; Wanger et al. 2014; Maas et al. 2015). This level of local, nuanced knowledge is key to managing pest suppression services in such a way that they are actively used as alternatives to agrochemical inputs and GM crops, and to contribute to more biodiversity-friendly and sustainable land-use practices (Tilman et al. 2002; Maas et al. 2015).

6.8.5 Changing Attitudes and Behaviors Toward Bats in the Developing World

Although the conservation of tropical biodiversity is highly beneficial to global society (Rands et al. 2010), ultimately it is the attitudes and beliefs of farmers and other rural populations that will determine its fate (Brechin et al. 2002; Tscharntke et al. 2012). Throughout the world, bats are subject to misconceptions and poor public perceptions (see Kingston and Barlow, this volume Chap. 17). However, exposure to environmental education can significantly decrease negative attitudes toward bats (López del Toro et al. 2009; Prokop et al. 2009; Reid 2013). These results suggest that reducing bat disservices, conducting environmental education, and building local valuation of beneficial bats could work in concert to improve conservation outcomes. As much as there is a critical need to manage agricultural landscapes to conserve bats, there is a parallel need to understand the local drivers of attitudes toward bats and to develop culturally appropriate, evidence-based interventions that encourage farmers to sustainably manage bat populations and other biodiversity associated with ecosystem services and ecosystem resilience.

Acknowledgments We would like to thank the researchers who made their results available to us, particularly those who shared work in preparation or press. We are very grateful to Olivier Roth who carried out a thorough literature search on European farmland bats and Jean-Yves Humbert for feedback on the meta-analysis approach. Justin Boyle, Pia Lentin, Tigga Kingston, and an anonymous reviewer provided detailed feedback that greatly improved this manuscript. Finally, we thank the editors of the current volume for invitation to review this topic and for their patience.

References

Aguiar LMS, Brito D, Machado RB (2010) Do current vampire bat (*Desmodus rotundus*) population control practices pose a threat to Dekeysers nectar bats (*Lonchophylla dekeyseri*) long-term persistence in the Cerrado? Acta Chiropterol 12:275–282

Arellano-Sota C (1988) Vampire bat-transmitted rabies in cattle. Rev Infect Dis 10:707–709

Arlettaz R (1996) Feeding behaviour and foraging strategy of free-living mouse-eared bats, *Myotis myotis* and *Myotis blythii*. Anim Behav 51:1–11

Arlettaz R (1999) Habitat selection as a major resource partitioning mechanism between the two sympatric sibling bat species *Myotis myotis* and *Myotis blythii*. J Anim Ecol 68:460–471

Arlettaz R, Perrin N (1995) The trophic niches of sympatric sibling *Myotis myotis* and *Myotis blythii*: do mouse-eared bats select prey? Symp Zool Soc Lond vol 67, London: The Society 1960–1999, pp 361–376

Arlettaz R, Perrin N, Hausser J (1997) Trophic resource partitioning and competition between the two sibling bat species. J Anim Ecol 66:897–911

Arlettaz R, Christe P, Lugon A et al (2001) Food availability dictates the timing of parturition in insectivorous mouse-eared bats. Oikos 95:105–111

Avila-Cabadilla LD, Stoner KE, Henry M, Alvarez-Añorve MY (2009) Composition, structure and diversity of phyllostomid bat assemblages in different successional stages of a tropical dry forest. For Ecol Manage 1–11

Aziz SA, Olival KJ, Bumrungsri S, Richards GC, Racey PA (2016) The conflict between pteropodid bats and fruit growers: species, legislation and mitigation. In: Voight CC, Kingston T (eds) Bats in the Anthropocene: conservation of bats in a changing world. Springer International AG, Cham, pp. 377–420

Bayat S, Geiser F, Kristiansen P, Wilson SC (2014) Organic contaminants in bats: trends and new issues. Environ Int 63:40–52

Bebber DP, Ramotowski MAT, Gurr SJ (2013) Crop pests and pathogens move polewards in a warming world. Nat Clim Change 3:985–988
Bennett BS, Thies ML (2007) Organochlorine pesticide residues in guano of Brazilian free-tailed bats, *Tadarida brasiliensis* Saint-Hilaire, from east Texas. Bull Environ Contam Toxicol 78:191–194
Bohmann K, Monadjem A, Noer CL et al (2011) Molecular diet analysis of two African free-tailed bats (Molossidae) using high throughput sequencing. PLoS ONE 6:e21441
Bontadina F, Schofield H, Naef-Daenzer B (2002) Radio-tracking reveals that lesser horseshoe bats (*Rhinolophus hipposideros*) forage in woodland. J Zool 258:281–290
Boughey KL, Lake IR, Haysom KA, Dolman PM (2011) Improving the biodiversity benefits of hedgerows: how physical characteristics and the proximity of foraging habitat affect the use of linear features by bats. Biol Conserv 144:1790–1798
Boyles JG, Cryan PM, McCracken GF, Kunz TH (2011) Economic importance of bats in agriculture. Science 332:41–42
Boyles JG, Sole CL, Cryan PM, McCracken GF (2013) On estimating the economic value of insectivorous bats: prospects and priorities for biologists. In: Adams RA, Pedersen SC (eds) Bat evolution, ecology, and conservation. Springer, New York, pp 501–515
Brashares JS, Epps CW, Stoner CJ (2010) Ecological and conservation implications of mesopredator release. In: Terborgh J, Estes J (eds) Trophic cascades. Island Press
Braun de Torrez E (2014) Bats, insects and pecans: habitat use and ecosystem services of insectivorous bats in a pecan agroecosystem in central Texas. Boston University, Boston
Brechin SR, Wilshusen P, Fortwangler CL, West PC (2002) Beyond the square wheel: toward a more comprehensive understanding of biodiversity conservation as a social and political process. Soc Nat Resour 15:41–64
Bumrungsri S, Harbit A, Benzie C et al (2008a) The pollination ecology of two species of *Parkia* (Mimosaceae) in southern Thailand. J Trop Ecol 24:467–475
Bumrungsri S, Sripaoraya E, Chongsiri T et al (2008b) The pollination ecology of durian (*Durio zibethinus*, Bombacaceae) in southern Thailand. J Trop Ecol 25:85
Bumrungsri S, Lang D, Harrower C et al (2013) The dawn bat, *Eonycteris spelaea* Dobson (Chiroptera: Pteropodidae) feeds mainly on pollen of economically important food plants in Thailand. Acta Chiropterol 15:95–104
Castro-Luna AA, Galindo-González J (2012) Enriching agroecosystems with fruit-producing tree species favors the abundance and richness of frugivorous and nectarivorous bats in Veracruz, Mexico. Mamm Biol 77:32–40
Clare EL, Barber BR, Sweeney BW et al (2011) Eating local: influences of habitat on the diet of little brown bats (*Myotis lucifugus*). Mol Ecol 20:1772–1780
Clark DR (1988) How sensitive are bats to insecticides? Wildl Soc B 16:399–403
Clark DR (2001) DDT and the decline of free-tailed bats (*Tadarida brasiliensis*) at Carlsbad Cavern, New Mexico. Arch Environ Contam Toxicol 40:537–543
Clark DR, Prouty RM (1976) Organochlorine residues in three bat species from four localities in Maryland and West Virginia, 1973. Pestic Monit J 10:44–53
Clark DR, LaVal RK, Swineford DM (1978) Dieldrin-induced mortality in an endangered species, the gray bat (*Myotis grisescens*). Science 199:1357–1359
Clark DR, LaVal RK, Krynitsky AJ (1980) Dieldrin and heptachlor residues in dead gray bats, Franklin County, Missouri–1976 versus 1977. Pestic Monit J 13:137–140
Clawson RL, Clark DR (1989) Pesticide contamination of endangered gray bats and their food base in Boone County, Missouri, 1982. Bull Environ Contam Toxicol 42:431–437
Cleveland C, Betke M, Federico P, Frank J (2006) Economic value of the pest control service provided by Brazilian free-tailed bats in south-central Texas. Front Ecol Environ 4:238–243
Conner WE, Corcoran AJ (2012) Sound strategies: the 65-million-year-old battle between bats and insects. Annu Rev Entomol 57:21–39
Davy CM, Russo D, Fenton MB (2007) Use of native woodlands and traditional olive groves by foraging bats on a Mediterranean island: consequences for conservation. J Zool 273:397–405

de Jong J (1995) Habitat use and species richness of bats in a patchy landscape. Acta Theriol 40:237–248

Defries RS, Rudel T, Uriarte M, Hansen M (2010) Deforestation driven by urban population growth and agricultural trade in the twenty-first century. Nat Geosci 3:1–4

Dietz M, Pir JB, Hillen J (2013) Does the survival of greater horseshoe bats and Geoffroy's bats in Western Europe depend on traditional cultural landscapes? Biodivers Conserv 23:3007–3025

Downs NC, Racey PA (2006) The use by bats of habitat features in mixed farmland in Scotland. Acta Chiropterol 8:169–185

Drescher C (2004) Radiotracking of *Myotis myotis* (Chiroptera, Vespertilionidae) in South Tyrol and implications for its conservation. Mammalia 68:387–395

Egger M, Schneider M, Davey Smith G (1998) Spurious precision? Meta-analysis of observational studies. BMJ 316:140–144

Ekman M, de Jong J (1996) Local patterns of distribution and resource utilization of four bat species (*Myotis brandti, Eptesicus nilssoni, Plecotus auritus* and *Pipistrellus pipistrellus*) in patchy and continuous environments. J Zool 238:571–580

Estrada A, Coates-Estrada R (2001) Bat species richness in live fences and in corridors of residual rain forest vegetation at Los Tuxtlas, Mexico. Ecography 24:94–102

Estrada A, Coates-Estrada R (2002) Bats in continuous forest, forest fragments and in an agricultural mosaic habitat-island at Los Tuxtlas, Mexico. Biol Conserv 103:237–245

Estrada A, Coates-Estrada R, Meritt D (1993) Bat species richness and abundance in tropical rain forest fragments and in agricultural habitats at Los Tuxtlas, Mexico. Ecography 16:309–318

Estrada A, Jimenez C, Rivera A, Fuentes E (2004) General bat activity measured with an ultrasound detector in a fragmented tropical landscape in Los Tuxtlas, Mexico. Anim Biodivers Conserv 27:1–9

FAO (2011) The state of the world's land and water resources for food and agriculture: managing systems at risk, pp 1–308

FAOSTAT (2011) Food and Agriculture Organization of the United Nations. faostat. fao3. org. Accessed 13 Aug 2013

Faria D (2006) Phyllostomid bats of a fragmented landscape in the north-eastern Atlantic forest, Brazil. J Trop Ecol 22:531–542

Faria D, Baumgarten J (2007) Shade cacao plantations (*Theobroma cacao*) and bat conservation in southern Bahia, Brazil. Biodivers Conserv 16:291–312

Faria D, Laps RR, Baumgarten J, Cetra M (2006) Bat and bird assemblages from forests and shade cacao plantations in two contrasting landscapes in the Atlantic forest of southern Bahia, Brazil. Biodivers Conserv 15:587–612

Faria D, Paciencia MLB, Dixo M et al (2007) Ferns, frogs, lizards, birds and bats in forest fragments and shade cacao plantations in two contrasting landscapes in the Atlantic forest, Brazil. Biodivers Conserv 16:2335–2357

Farrow LJ, Broders HG (2011) Loss of forest cover impacts the distribution of the forest-dwelling tri-colored bat (*Perimyotis subflavus*). Mamm Biol 76:172–179

Federico P, Hallam TG, McCracken GF et al (2008) Brazilian free-tailed bats as insect pest regulators in transgenic and conventional cotton crops. Ecol Appl 18:826–837

Fischer J, Brosi B, Daily GC et al (2008) Should agricultural policies encourage land sparing or wildlife-friendly farming? Front Ecol Environ 6:380–385

Fischer J, Stott J, Zerger A et al (2009) Reversing a tree regeneration crisis in an endangered ecoregion. Proc Natl Acad Sci USA 106:10386–10391

Fischer J, Zerger A, Gibbons P et al (2010) Tree decline and the future of Australian farmland biodiversity. Proc Natl Acad Sci USA 107:19597–19602

Flaquer C, Puig-Montserrat X, Burgas A, Russo D (2008) Habitat selection by Geoffroy's bats (*Myotis emarginatus*) in a rural Mediterranean landscape: implications for conservation. Acta Chiropterol 10:61–67

Foley JA, DeFries R, Asner GP et al (2005) Global consequences of land use. Science 309:570–574

Frey-Ehrenbold A, Bontadina F, Arlettaz R, Obrist MK (2013) Landscape connectivity, habitat structure and activity of bat guilds in farmland-dominated matrices. J Appl Ecol 50:252–261
Fuentes-Montemayor E, Goulson D, Park KJ (2011) Pipistrelle bats and their prey do not benefit from four widely applied agri-environment management prescriptions. Biol Conserv 144:2233–2246
Fukuda D, Tisen OB, Momose K (2009) Bat diversity in the vegetation mosaic around a lowland dipterocarp forest of Borneo. Raffles Bull Zool 57:213–221
Fuller J, Norton LR, Feber RE et al (2005) Benefits of organic farming to biodiversity vary among taxa. Biol Lett 1:431–434
Furey NM, Mackie IJ, Racey PA (2010) Bat diversity in Vietnamese limestone karst areas and the implications of forest degradation. Biodivers Conserv 19:1821–1838
Gándara Fierro G, Correa Sandoval AN, Hernández Cienfuegos CA (2006) Valoración económica de los servicios ecológicos que prestan los murciélagos *Tadarida brasiliensis* como controladores de plagas en el norte de México, pp 1–18
García Estrada C, Damon A, Hernandez C et al (2006) Bat diversity in montane rainforest and shaded coffee under different management regimes in southeastern Chiapas, Mexico. Biol Conserv 132:351–361
García-Estrada C, Damon A, Sánchez Hernández C et al (2012) Diets of frugivorous bats in montane rain forest and coffee plantations in southeastern Chiapas, Mexico. Biotropica 44:394–401
García-Morales R, Badano EI, Moreno CE (2013) Response of Neotropical bat assemblages to human land use. Conserv Biol 27:1096–1106
Gehrt SD, Chelsvig JE (2003) Bat activity in an urban landscape: patterns at the landscape and microhabitat scale. Ecol Appl 13:939–950
Graf S (2010) Diversity and habitat use of understorey bats in forest and agroforestry systems at the Margin of Lore Lindu National Park (Central Sulawesi, Indonesia). University of Vienna, Vienna
Guillén A, Ibáñez C, Pérez JL et al (1994) Organochlorine residues in Spanish common pipistrelle bats (*Pipistrellus pipistrellus*). Bull Environ Contam Toxicol 52:231–237
Hallmann CA, Foppen RPB, van Turnhout CAM et al (2014) Declines in insectivorous birds are associated with high neonicotinoid concentrations. Nature 511:341–343
Hanspach J, Fischer J, Ikin K et al (2012) Using trait-based filtering as a predictive framework for conservation: a case study of bats on farms in southeastern Australia. J Appl Ecol 49:842–850
Harvey CA, González Villalobos JA (2007) Agroforestry systems conserve species-rich but modified assemblages of tropical birds and bats. Biodivers Conserv 16:2257–2292
Harvey CA, Medina A, Sánchez DM et al (2006) Patterns of animal diversity in different forms of tree cover in agricultural landscapes. Ecol Appl 16:1986–1999
Henderson LE, Broders HG (2008) Movements and resource selection of the northern long-eared Myotis (*Myotis septentrionalis*) in a forest–agriculture landscape. J Mammal 89:952–963
Hopkins J, Maxted N (2011) Crop wild relatives: plant conservation for food security, pp 1–58
James C (2011) Global status of commercialized biotech/GM crops: 2011. International service for the acquisition of agri-biotech applications (ISAAA), Ithaca
Jones G, Morton M (1992) Radio-tracking studies on habitat use by greater horseshoe bats (*Rhinolophus ferrumequinum*). In: 4th European conference on wildlife telemetry: Remote monitoring and tracking of animals, pp 521–537
Jones G, Jacobs DS, Kunz TH et al (2009) Carpe noctem: the importance of bats as bioindicators. Endanger Species Res 8:93–115
Kalka MB, Smith AR, Kalko EKV (2008) Bats limit arthropods and herbivory in a tropical forest. Science 320:71
Kannan K, Yun SH, Rudd RJ, Behr M (2010) High concentrations of persistent organic pollutants including PCBs, DDT, PBDEs and PFOS in little brown bats with white-nose syndrome in New York, USA. Chemosphere 80:613–618

Karp DS, Mendenhall CD, Sandí RF et al (2013) Forest bolsters bird abundance, pest control, and coffee yield. Ecol Lett 16:1339–1347
Kingston T (2016) Cute, Creepy, or Crispy—How values, attitudes, and norms shape human behavior toward bats. In: Voight CC, Kingston T (eds) Bats in the Anthropocene: conservation of bats in a changing world. Springer International AG, Cham, pp. 571–588
Kunz TH, Anthony EL, Rumage WT III (1977) Mortality of little brown bats following multiple pesticide applications. J Wildl Manage 476–483
Kunz TH, Whitaker JO Jr, Wadanoli MD (1995) Dietary energetics of the insectivorous Mexican free-tailed bat (*Tadarida brasiliensis*) during pregnancy and lactation. Oecologia 101:407–415
Kunz TH, Braun de Torrez E, Bauer D et al (2011) Ecosystem services provided by bats. Ann NY Acad Sci 1223:1–38
Kurta A, Bell GP, Nagy KA, Kunz TH (1989) Energetics of pregnancy and lactation in free ranging little brown bats (*Myotis lucifugus*). Physiol Zool 62:804–818
Lambin EF, Meyfroidt P (2011) Global land use change, economic globalization, and the looming land scarcity. Proc Natl Acad Sci USA 108:3465–3472
Laurance WF (2008) Theory meets reality: how habitat fragmentation research has transcended island biogeographic theory. Biol Conserv 141:1731–1744
Law et al. (2016) Insectivorous bats and silviculture: balancing timber production and bat conservation. In: Voight CC, Kingston T (eds) Bats in the Anthropocene: conservation of bats in a changing world. Springer International AG, Cham, pp. 105–141
Le Bellec F, Vaillant F, Imbert E (2006) Pitahaya (*Hylocereus spp.*): a new fruit crop, a market with a future. Fruits 61:237–250
Lee Y, McCracken G (2005) Dietary variation of Brazilian free-tailed bats links to migratory populations of pest insects. J Mammal 81:67–76
Leelapaibul W, Bumrungsri S, Pattanawiboon A (2005) Diet of wrinkle-lipped free-tailed bat (*Tadarida plicata* Buchannan, 1800) in central Thailand: insectivorous bats potentially act as biological pest control agents. Acta Chiropterol 7:111–119
Lentini PE, Gibbons P, Fischer J et al (2012) Bats in a farming landscape benefit from linear remnants and unimproved pastures. PLoS ONE 7:e48201
Lesiński G, Durka A, Rembiałkowska E (2013) Flight activity of serotine *Eptesicus serotinus* on organic and conventional farms in central Poland. Zool Ecol 23:93–96
Lisón F, Calvo JF (2011) The significance of water infrastructures for the conservation of bats in a semiarid Mediterranean landscape. Anim Conserv 14:533–541
Lisón F, Calvo JF (2013) Ecological niche modeling of three pipistrelle bat species in semiarid Mediterranean landscapes. Acta Oecol 47:68–73
Loos J, Abson DJ, Chappell MJ et al (2014) Putting meaning back into sustainable intensification. Front Ecol Environ 12:356–361
López del Toro P, Andresen E, Barraza L, Estrada A (2009) Attitudes and knowledge of shade-coffee farmers towards vertebrates and their ecological functions. Trop Conserv Sci 2:299–318
Lopez-Hoffman L, Wiederholt R, Sansone C et al (2014) Market forces and technological substitutes cause fluctuations in the value of bat pest-control services for cotton. PLoS ONE 9:e87912
Lumsden LF, Bennett AF (2005) Scattered trees in rural landscapes: foraging habitat for insectivorous bats in south-eastern Australia. Biol Conserv 122:205–222
Lumsden LF, Bennett AF, Silins JE (2002) Location of roosts of the lesser long-eared bat *Nyctophilus geoffroyi* and Gould's wattled bat *Chalinolobus gouldii* in a fragmented landscape in south-eastern Australia. Biol Conserv 106:237–249
Lundy M, Montgomery I (2010) Summer habitat associations of bats between riparian landscapes and within riparian areas. Eur J Wildl Res 56:385–394
Maas B, Clough Y, Tscharntke T (2013) Bats and birds increase crop yield in tropical agroforestry landscapes. Ecol Lett 16:1480–1487

Maas B, Karp DS, Bumrungsri S, et al., (2015) Bird and bat predation services in tropical forests and agroforestry landscapes. Biol Rev Jul 23. doi:10.1111/brv.12211
Mayen F (2003) Haematophagous bats in Brazil, their role in rabies transmission, impact on public health, livestock industry and alternatives to an indiscriminate reduction of bat population. J Vet Med B Infect Dis Vet Public Health 50:469–472
McCracken GF, Westbrook JK, Brown VA et al (2012) Bats track and exploit changes in insect pest populations. PLoS ONE 7:e43839
McShane TO, Hirsch PD, Trung TC et al (2011) Hard choices: making trade-offs between biodiversity conservation and human well-being. Biol Conserv 144:966–972
Medellin R, Equihua M, Amin M (2000) Bat diversity and abundance as indicators of disturbance in Neotropical rainforests. Conserv Biol 14:1666–1675
Medina A, Harvey C, Merlo D et al (2007) Bat diversity and movement in an agricultural landscape in Matiguas, Nicaragua. Biotropica 39:120–128
Mendenhall CD, Karp DS, Meyer CFJ et al (2014) Predicting biodiversity change and averting collapse in agricultural landscapes. Nature 509:213–217
Meyer C et al. (2016) Responses of tropical bats to habitat fragmentation, logging and deforestation. In: Voight CC, Kingston T (eds) Bats in the Anthropocene: conservation of bats in a changing world. Springer International AG, Cham, pp. 63–94
Mildenstein TL, Stier SC, Nuevo-Diego CE (2005) Habitat selection of endangered and endemic large flying-foxes in Subic Bay, Philippines. Biol Conserv 126:93–102
Miura T, Koyama T, Nakamura I (1978) Mercury content in museum and recent specimens of chiroptera in Japan. Bull Environ Contam Toxicol 20:696–701
Morrison EB, Lindell CA (2012) Birds and bats reduce insect biomass and leaf damage in tropical forest restoration sites. Ecol Appl 22:1526–1534
Nelson GC, Rosegrant MW, Koo J et al (2009) Climate change: impact on agriculture and costs of adaptation, pp 1–30
Noer CL, Dabelsteen T, Bohmann K, Monadjem A (2012) Molossid bats in an African agro-ecosystem select sugarcane fields as foraging habitat. Afr Zoo 47:1–11
Numa C, Verdú JR, Sánchez-Palomino P (2005) Phyllostomid bat diversity in a variegated coffee landscape. Biol Conserv 122:151–158
O'Shea TJ, Johnston JJ (2009) Environmental contaminants and bats: investigating exposure and effects. In: Kunz TH, Parsons S (eds) Ecological and behavioral methods for the study of bats. The Johns Hopkins University Press, Baltimore, pp 500–528
Obrist MK, Rathey E, Bontadina F et al (2011) Response of bat species to sylvo-pastoral abandonment. For Ecol Manage 261:789–798
O'Shea TJ, Everette AL, Ellison LE (2001) Cyclodiene insecticide, DDE, DDT, arsenic, and mercury contamination of big brown bats (*Eptesicus fuscus*) foraging at a Colorado Superfund site. Arch Environ Contam Toxicol 40:112–120
Perfecto I, Vandermeer J (2010) The agroecological matrix as alternative to the land-sparing/agriculture intensification model. Proc Natl Acad Sci USA 107:5786–5791
Petchmunee K (2008) Economic valuation and learning process construction: a case study of the cave Nectarivorous bat (*Eonycteris spelaea* Dobson), pp 1–278
Pineda E, Moreno C, Escobar F, Halffter G (2005) Frog, bat, and dung beetle diversity in the cloud forest and coffee agroecosystems of Veracruz, Mexico. Conserv Biol 19:400–410
Pocock MJO, Jennings N (2007) Testing biotic indicator taxa: the sensitivity of insectivorous mammals and their prey to the intensification of lowland agriculture. J Appl Ecol 45:151–160
Power AG (2010) Ecosystem services and agriculture: tradeoffs and synergies. Philos Trans R Soc B 365:2959–2971
Prokop P, Fančovičová J, Kubiatko M (2009) Vampires are still alive: Slovakian students' attitudes toward bats. Anthrozoos 22:19–30
Quarles W (2013) Bats, pesticides and white nose syndrome. IPM Pract 33:1–20
Rainho A (2007) Summer foraging habitats of bats in a Mediterranean region of the Iberian Peninsula. Acta Chiropterol 9:171–181

Rambaldini DA, Brigham RM (2011) Pallid bat (*Antrozous pallidus*) foraging over native and vineyard habitats in British Columbia, Canada. Can J Zool 89:816–822

Randrianandrianina F, Andriafidison D, Kofoky AF et al (2006) Habitat use and conservation of bats in rainforest and adjacent human-modified habitats in eastern Madagascar. Acta Chiropterol 8:429–437

Redfern SK, Azzu N, Binamira JS (2012) Rice in Southeast Asia: facing risks and vulnerabilities to respond to climate change. In: Building resilience for adaptation to climate change in the agriculture sector: proceedings of a Joint FAO/OECD Workshop. FAO, Rome, pp 295–314

Reid JL (2013) Restoring birds, bats, and dispersal mutualisms in a tropical agricultural landscape. University of California, Santa Cruz

Ricketts TH (2001) The matrix matters: effective isolation in fragmented landscapes. Am Nat 158:87–99

Robertson G, Swinton S (2005) Reconciling agricultural productivity and environmental integrity: a grand challenge for agriculture. Front Ecol Environ 3:38–46

Russ J, Montgomery W (2002) Habitat associations of bats in Northern Ireland: implications for conservation. Biol Conserv 107:49–58

Russo D, Jones G (2003) Use of foraging habitats by bats in a Mediterranean area determined by acoustic surveys: conservation implications. Ecography 26:197–209

Russo D, Jones G, Migliozzi A (2002) Habitat selection by the Mediterranean horseshoe bat, *Rhinolophus euryale* (Chiroptera: Rhinolophidae) in a rural area of southern Italy and implications for conservation. Biol Conserv 107:71–81

Saldaña Vázquez RA, Castro-Luna AA, Sandoval Ruiz CA et al (2013) Population composition and ectoparasite prevalence on bats (*Sturnira ludovici*; Phyllostomidae) in forest fragments and coffee plantations of central Veracruz, Mexico. Biotropica 45:351–356

Sasse DB (2005) Pesticide residues in guano of gray bats. J Ark Acad Sci 59:214–217

Schmidt KM, Badger DD (1979) Some social and economic aspects in controlling vampire bats. Proc Okla Acad Sci 59:112–114

Schmidt A, Brack V Jr, Rommé R et al (2000) Bioaccumulation of pesticides in bats from Missouri. In: Johnston JJ (ed) ACS symposium series. American Chemical Society, Washington, DC, pp 8–21

Sedlock JL, Weyandt SE, Cororan L et al (2008) Bat diversity in tropical forest and agro-pastoral habitats within a protected area in the Philippines. Acta Chiropterol 10:349–358

Senthilkumar K, Kannan K, Subramanian A, Tanabe S (2001) Accumulation of organochlorine pesticides and polychlorinated biphenyls in sediments, aquatic organisms, birds, bird eggs and bat collected from south India. Environ Sci Pollut Res 8:35–47

Sirami C, Jacobs DS, Cumming GS (2013) Artificial wetlands and surrounding habitats provide important foraging habitat for bats in agricultural landscapes in the Western Cape, South Africa. Biol Conserv 164:30–38

Sosa VJ, Hernández-Salazar E, Hernández-Conrique D, Castro-Luna AA (2008) Murciélagos. In: Manson RH, Hernández-Ortiz V, Gallina S, Mehltreter K (eds) Agroecosistemas Cafetaleros de Veracruz: Biodiverdiad, Manejo, y Conservación. INECOL & INE-SEMARNAT, México D.F, pp 181–192

Srinivasulu C, Srinivasulu B (2002) Greater short-nosed fruit bat (*Cynopterus sphinx*) foraging and damage in vineyards in India. Acta Chiropterol 4:167–171

Stahlschmidt P, Brühl CA (2012) Bats at risk? Bat activity and insecticide residue analysis of food items in an apple orchard. Environ Toxicol Chem 31:1556–1563

Stahlschmidt P, Pätzold A, Ressl L et al (2012) Constructed wetlands support bats in agricultural landscapes. Basic Appl Ecol 13:196–203

Stechert C, Kolb M, Bahadir M et al (2014) Insecticide residues in bats along a land use-gradient dominated by cotton cultivation in northern Benin. Environ Sci Pollut Res, West Africa. doi:10.1007/s11356-014-2817-8

Storm JJ, Whitaker JO Jr (2008) Prey selection of big brown bats (*Eptesicus fuscus*) during an emergence of 17-year cicadas (*Magicicada* spp.). Am Midl Nat 160:350–357

Swanepoel RE, Racey PA, Shore RF, Speakman JR (1998) Energetic effects of sublethal exposure to lindane on pipistrelle bats (*Pipistrellus pipistrellus*). Environ Pollut 104:169–177
Taylor PJ, Nemudivhiso P, Mphethe V et al (2012) Bats as biocontrol agents in macadamia orchards, Levubu Valley, Limpopo Province: results of a pilot project and future prospects. South Afr Macademia Grow Assoc Yearb 20:41–51
Taylor PJ, Bohmann K, Steyn JN et al (2013a) Bats eat pest green vegetable stink bugs (*Nezara viridula*): diet analyses of seven insectivorous species of bats roosting and foraging in macadamia orchards. South Afr Macademia Grow Assoc Yearb 21:37–43
Taylor PJ, Monadjem A, Steyn JN (2013b) Seasonal patterns of habitat use by insectivorous bats in a subtropical African agro-ecosystem dominated by macadamia orchards. Afr J Ecol 51:552–561
Thies ML, Thies KM (1997) Organochlorine residues in bats from Eckert James River Cave, Texas. Bull Environ Contam Toxicol 58:673–680
Thomson LJ, Macfadyen S, Hoffmann AA (2010) Predicting the effects of climate change on natural enemies of agricultural pests. Biol Control 52:296–306
Tilman D (1999) Global environmental impacts of agricultural expansion: the need for sustainable and efficient practices. Proc Natl Acad Sci USA 96:5995–6000
Tilman D, Fargione J, Wolff B et al (2001) Forecasting agriculturally driven global environmental change. Science 292:281–284
Tilman D, Cassman KG, Matson PA et al (2002) Agricultural sustainability and intensive production practices. Nature 418:671–677
Tscharntke T, Klein AM, Kruess A et al (2005) Landscape perspectives on agricultural intensification and biodiversity-ecosystem service management. Ecol Lett 8:857–874
Tscharntke T, Clough Y, Wanger TC et al (2012) Global food security, biodiversity conservation and the future of agricultural intensification. Biol Conserv 151:53–59
Tuttle SR, Chambers CL, Theimer TC (2006) Potential effects of livestock water-trough modifications on bats in northern Arizona. Wildl Soc B 34:602–608
Valiente-Banuet A, Santos Gally R, Arizmendi MC, Casas A (2007) Pollination biology of the hemiepiphytic cactus *Hylocereus undatus* in the Tehuacán Valley, Mexico. J Arid Environ 68:1–8
Van der Sluijs JP, Simon-Delso N, Goulson D et al (2013) Neonicotinoids, bee disorders and the sustainability of pollinator services. Curr Opin Environ Sustain 5:293–305
Van Weerd M, Snelder DJ (2008) Human-altered tree-based habitats and their value in conserving bird and bat diversity in northeast Luzon, Philippines. In: Snelder DJ, Lasco RD (eds) Smallholder tree growing for rural development and environmental services. Springer, Dordrecht, pp 347–377
Vandermeer J, Perfecto I (2007) The agricultural matrix and a future paradigm for conservation. Conserv Biol 21:274–277
Vargas Espinoza A, Aguirre LF, Galarza MI, Gareca E (2008) Ensamble de murciélagos en sitios con diferente grado de perturbación en un bosque montano del Parque Nacional Carrasco, Bolivia. Mastozool Neotrop 15:297–308
Verboom B, Huitema H (1997) The importance of linear landscape elements for the pipistrelle *Pipistrellus pipistrellus* and the serotine bat *Eptesicus serotinus*. Landsc Ecol 12:117–125
Vickery J, Arlettaz R (2012) The importance of habitat heterogeneity at multiple scales for birds in European agricultural landscapes. In: FR J (ed) Birds and habitat: relationships in changing landscapes. Cambridge University Press, Cambridge, pp 177–204
Walsh AL, Harris S (1996a) Factors determining the abundance of vespertilionid bats in Britain: geographical, land class and local habitat relationships. J Appl Ecol 33:519–529
Walsh AL, Harris S (1996b) Foraging habitat preferences of vespertilionid bats in Britain. J Appl Ecol 33:508–518
Wanger TC, Darras K, Bumrungsri S et al (2014) Bat pest control contributes to food security in Thailand. Biol Conserv 171:220–223
Weiss J, Nerd A, Mizrahi Y (1994) Flowering behavior and pollination requirements in climbing cacti with fruit crop potential. HortScience 29:1487–14982

Whitaker JO Jr (1995) Food of the big brown bat Eptesicus fuscus from materinity colonies in Indiana and Illinois. Am Midl Nat 134:346–360

White DH, Krynitsky AJ (1986) Wildlife in some areas of New Mexico and Texas accumulate elevated DDE residues, 1983. Arch Environ Contam Toxicol 15:149–157

Wickramasinghe L, Harris S, Jones G, Vaughn N (2003) Bat activity and species richness on organic and conventional farms: impact of agricultural intensification. J Appl Ecol 40:984–993

Wickramasinghe L, Harris S, Jones G, Vaughn Jenings N (2004) Abundance and species richness of nocturnal insects on organic and conventional farms: effects of agricultural intensification on bat foraging. Conserv Biol 18:1283–1292

Williams-Guillén K, Perfecto I (2010) Effects of agricultural intensification on the assemblage of leaf-nosed bats (Phyllostomidae) in a coffee landscape in Chiapas, Mexico. Biotropica 42:605–613

Williams-Guillén K, Perfecto I (2011) Ensemble composition and activity levels of insectivorous bats in response to management intensification in coffee agroforestry systems. PLoS ONE 6:e16502

Williams-Guillén K, Perfecto I, Vandermeer J (2008) Bats limit insects in a Neotropical agroforestry system. Science 320:70

Wilson C, Tisdell C (2001) Why farmers continue to use pesticides despite environmental, health and sustainability costs. Ecol Econ 39:449–462

Yadav SK (2010) Pesticide applications: threat to ecosystems. J Hum Ecol 32:37–45

Chapter 7
Dark Matters: The Effects of Artificial Lighting on Bats

E.G. Rowse, D. Lewanzik, E.L. Stone, S. Harris and G. Jones

Abstract While artificial lighting is a major component of global change, its biological impacts have only recently been recognised. Artificial lighting attracts and repels animals in taxon-specific ways and affects physiological processes. Being nocturnal, bats are likely to be strongly affected by artificial lighting. Moreover, many species of bats are insectivorous, and insects are also strongly influenced by lighting. Lighting technologies are changing rapidly, with the use of light-emitting diode (LED) lamps increasing. Impacts on bats and their prey depend on the light spectra produced by street lights; ultraviolet (UV) wavelengths attract more insects and consequently insectivorous bats. Bat responses to lighting are species-specific and reflect differences in flight morphology and performance; fast-flying aerial hawking species frequently feed around street lights, whereas relatively slow-flying bats that forage in more confined spaces are often light-averse. Both high-pressure sodium and LED lights reduce commuting activity by clutter-tolerant bats of the genera *Myotis* and *Rhinolophus*, and these bats still avoided LED lights when dimmed. Light-induced reductions in the activity of frugivorous bats may affect ecosystem services by reducing dispersal of the seeds of pioneer plants and hence reforestation. Rapid changes in street lighting offer the potential to explore

---

E.G. Rowse and D. Lewanzik: Equal contributors.

E.G. Rowse · E.L. Stone · S. Harris · G. Jones (✉)
School of Life Sciences, University of Bristol, Bristol, UK
e-mail: Gareth.Jones@bristol.ac.uk

D. Lewanzik
Department of Evolutionary Ecology, Leibniz Institute for Zoo and Wildlife Research, Berlin, Germany

D. Lewanzik
Animal Behaviour, Freie Universität Berlin, Berlin, Germany

C.C. Voigt and T. Kingston (eds.), *Bats in the Anthropocene: Conservation of Bats in a Changing World*, DOI 10.1007/978-3-319-25220-9_7

mitigation methods such as part-night lighting (PNL), dimming, directed lighting, and motion-sensitive lighting that may have beneficial consequences for light-averse bat species.

7.1 Introduction

Anthropogenic change is altering ecosystems at unprecedented rates and humans now dominate most ecosystems (Vitousek et al. 1997; McDonald 2008). Urbanisation in particular has major impacts on bat activity and abundance (Jung and Threlfall 2016), and one aspect of global change that occurs predominately, but not exclusively, in urban areas is increased artificial light at night. Almost a fifth of the global land area was affected by light pollution in 2001 (Cinzano et al. 2001). Although night-time brightness generally increased in Europe between 1995 and 2010, regional patterns are complex, with some localised declines (Bennie et al. 2014). However, the biological impacts of light pollution have only recently been recognised (Longcore and Rich 2004).

Being nocturnal, bats are likely to be affected by light pollution. In this chapter, we review the types of artificial light that bats experience, describe how light pollution has become more widespread in recent years, show how technological changes may lead to significant reductions in light pollution and describe some of the physiological consequences of light pollution that may be relevant to bats. We then discuss how artificial lighting affects the insect prey of bats, and why some bats may benefit from the growth in artificial lighting, whereas others are affected detrimentally. After highlighting some aspects of bat vision, we describe the shift from observational to experimental studies of how bats respond to lighting. Finally, we identify some of the major knowledge gaps and suggest priorities for future research on the effects of artificial lighting on bats.

7.2 Types of Artificial Light

The electromagnetic spectrum encompasses radiation with wavelengths ranging from less than a nanometre (gamma rays) to a kilometre (radio waves) (Campbell 2011). While humans perceive wavelengths between 400 and 700 nm as 'visible light' (Purves and Lotto 2003), birds, fish and invertebrates can detect light in the ultraviolet (UV) range (10–400 nm). Recent work suggests that UV sensitivity may be widespread among mammals (Douglas and Jeffery 2014), and snakes and beetles can detect spectral emissions in the infrared range (700–1000 nm) (Schmitz and Bleckmann 1998; Land and Nilsson 2012).

Artificial lighting has infiltrated all aspects of human life both indoors and outside (Gaston et al. 2012). Here, we focus on street lighting because of its universal use and potential for ecological impacts (Gaston et al. 2012). Different types of street light have distinct spectral signatures (Fig. 7.1); their primary emissions

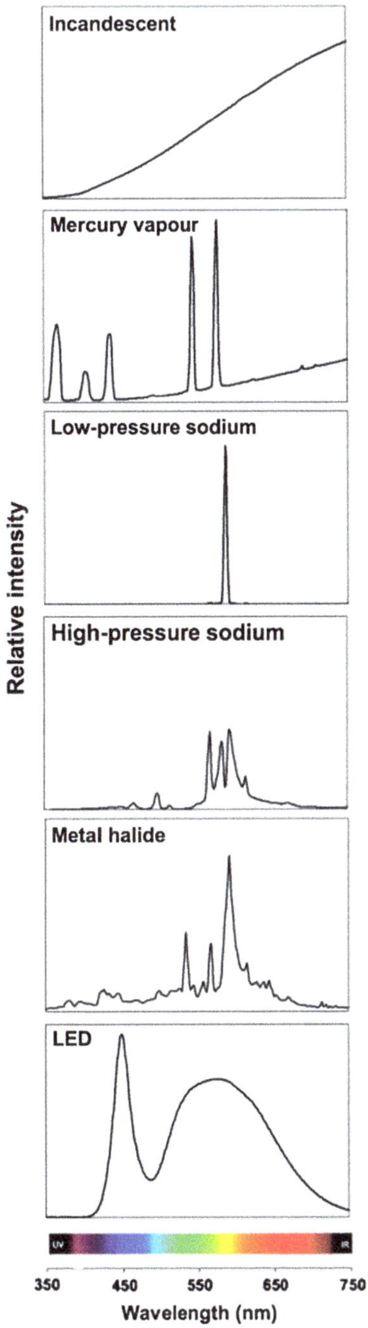

Fig. 7.1 The spectral content of different light types varies considerably. The spectral composition of common lighting technologies is shown. From Gaston et al. (2013)

depend on the type of reactive material or coating in the lamps (Buchanan 2006). Incandescent lamps, developed by Thomas Edison in 1880, mainly emit long wavelengths with a maximum intensity between 900 and 1050 nm (Elvidge et al. 2010). Despite improvements such as the quartz halogen lamp, which uses an inert gas to preserve the tungsten filament, incandescent lamps are still relatively inefficient because their emissions are predominantly near the infrared spectrum and so largely invisible to humans (Elvidge et al. 2010).

Gas discharge lamps, developed by the mid-twentieth century, produce light by passing electric arcs through gas-filled bulbs (Elvidge et al. 2010). These are further classified as low-pressure discharge and high-intensity discharge (HID) lamps (Elvidge et al. 2010). Low-pressure discharge lamps include the compact fluorescent lamp (CFL) and low-pressure sodium (LPS) lamps. Fluorescent lamps produce distinct emission peaks, which combine to emit a 'white' light (Royal Commission on Environmental Pollution 2009; Elvidge et al. 2010), whereas LPS lamps have a narrow spectral signature, emitting monochromatic orange light with a peak intensity of 589 nm (Fig. 7.1) (Rydell 2006; Elvidge et al. 2010).

HID lamps include high-pressure mercury vapour (HPMV) lamps, which produce a bluish-white light, and high-pressure sodium (HPS) and metal halide lamps that have broader spectral emissions (Fig. 7.1) (Davies et al. 2013). Emissions from HPMV lamps extend into the UV range (Rydell 2006; Elvidge et al. 2010), whereas HPS lamps emit yellow-orange light and metal halide lamps 'white' light (Royal Commission on Environmental Pollution 2009; Davies et al. 2013; Gaston et al. 2013). The colour rendering index (CRI) compares how accurately a light source replicates the full range of colours of an object viewed in natural light on a scale of 0–100, where 100 is equivalent to natural light (Schubert and Kim 2005; Elvidge et al. 2010; Davies et al. 2013). HPS lamps typically have a CRI between 7 and 32, whereas metal halide lamps have a CRI ranging from 64 to 100, reflecting their ability to render colour more suited for human vision (Elvidge et al. 2010; Gaston et al. 2012).

Gas discharge lamps replaced incandescent lamps because of their energy efficiency and improved longevity (Schubert and Kim 2005), and LPS (44 %) and HPS (41 %) lamps came to dominate street lighting in the UK (Royal Commission on Environmental Pollution 2009) and elsewhere. The luminous efficacy (LE) (amount of light produced per watt of electricity) of gas discharge lamps is five times higher than incandescent lamps (Schubert and Kim 2005; Elvidge et al. 2010). However, with pressure to reduce energy use and CO_2 emissions, the lighting industry is now turning to light-emitting diodes (LEDs) (Elvidge et al. 2010; Gaston et al. 2012). LEDs have broad spectral signatures, typically 400–700 nm, with very few emissions in the UV range (Elvidge et al. 2010). This is achieved mainly through the use of cerium-doped yttrium aluminium garnet (YAG:Ce) phosphors with a gallium nitride (GaN) which converts monochromatic blue to 'white' light. However, more recently LEDs are able to produce light by combining multiple monochromatic sources (red, green and blue), which allows for greater control over spectral emissions (Narendran et al. 2004; Gaston et al. 2012, 2013; Davies et al. 2013). LED lamps have comparable CRI scores to metal halide lamps (65–100) (Elvidge et al. 2010) but benefit from lower running costs (Gaston et al. 2012); low energy consumption (Elvidge et al. 2010); controllability of spectral, temporal and intensity of emissions; reduced CO_2 emissions (Hölker et al. 2010a); and smart lighting capabilities that enable dimming in response to weather, traffic and lunar conditions (Bennie et al. 2014).

7.3 The Growth of Light Pollution

Light pollution is defined as the changing of natural light levels in nocturnal landscapes (nightscapes) through artificial lighting sources (Falchi et al. 2011; Kyba and Hölker 2013). Here, we focus on ecological light pollution, i.e. the direct ecological effects of light as opposed to astronomical light pollution, which describes the light that disrupts viewing of stars and other celestial matter (Longcore and Rich 2004). Ecological light pollution can be caused by glare (extreme contrasts between bright and dark areas), over-illumination, light clutter (unnecessary numbers of light sources), light trespass (unwanted light) and skyglow, where artificial light is directed towards the sky, scattered by atmospheric molecules and reflected back to earth (Royal Commission on Environmental Pollution 2009; Gaston et al. 2012; Kyba and Hölker 2013).

Artificial lighting has increased as a result of urbanisation, population growth, economic development and advances in lighting technologies and provides numerous economic, commercial, recreational and security benefits (Riegel 1973; Hölker et al. 2010a; Davies et al. 2012). However, light pollution is now of global concern: the accelerated use of electric lighting, growing at 6 % per year, has escalated light pollution to threat status (Hölker et al. 2010a, b). Satellite images suggest that 19 % of the global land surface surpassed the threshold for acceptable lighting levels (Cinzano et al. 2001). However, satellites are unable to capture all illumination from light sources (Bennie et al. 2014). While light pollution is currently more apparent in developed nations (Fig. 7.2), projected increases in industrial and urban growth suggest that light pollution will become more spatially heterogeneous both locally and regionally (Cinzano et al. 2001; Gaston et al. 2012; Hölker et al. 2010b; Bennie et al. 2014).

In the UK, street lighting consumes approximately 114 Twh of energy annually (International Energy Agency 2006) and is growing at 3 % per annum (Royal Commission on Environmental Pollution 2009). The number of lighting installations is increasing (Gaston et al. 2012), and the change in emissions due to increased use of broad spectrum technologies is also likely to affect light pollution as these sources emit higher levels of blue light. This scatters more into the atmosphere than green or red light, ultimately making a bigger contribution to skyglow (Benenson et al. 2002; Falchi et al. 2011; Kyba and Hölker 2013). The growth in light pollution will be further exacerbated because, as LEDs become cheaper, non-essential uses, such as advertising and architectural lighting, may increase (Schubert and Kim 2005).

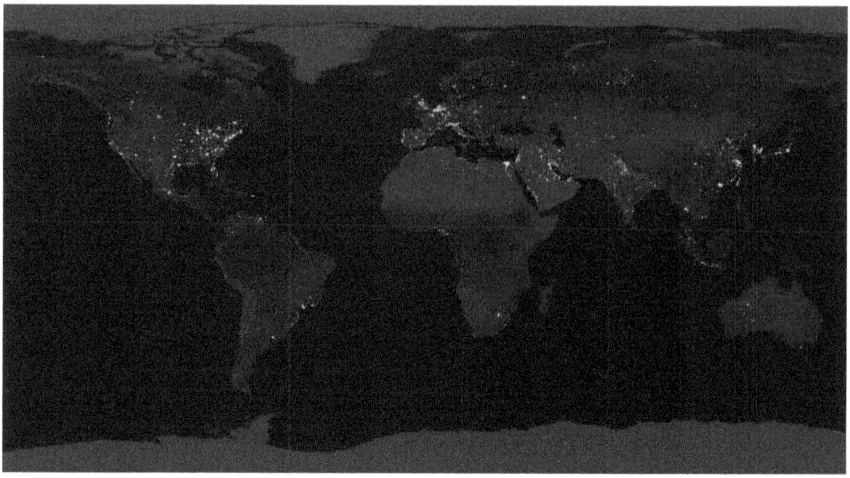

Fig. 7.2 Artificial lighting is currently most widespread in the developed world. Global use of lighting at night in 2000. From NASA Earth Observatory/NOAA NGDC (2012)

7.4 Projected Changes in Technology

International lighting policies are prioritising energy-efficient technologies to reduce costs and CO_2 emissions. The European Ecodesign Directive, for instance, encourages moves from energy-intensive technologies such as incandescent, LPS and HPMV lamps (Hölker et al. 2010a) to 'whiter' lighting with higher colour rendering capabilities (Gaston et al. 2012). This may reduce CO_2 emissions in the EU by as much as 42 Mt per year. A number of pilot studies in cities around the world (including Adelaide, Hong Kong, London, Mumbai, New York, Sydney and Toronto) have compared LED lamps against existing lighting technologies. After a three-year trial, the City of Sydney Council agreed to switch to LEDs on 6500 outdoor lights due to their reduced energy consumption, cost-effectiveness and improved illuminance (The Climate Group 2014).

Future research will focus on increasing the efficiencies of LEDs: the LE of a LED is 60–90 lm/W, compared to 80–120 lm/W for HPS lamps (California Lighting Technology Center 2010). More effective ways of producing light are also being investigated, such as combining multiple monochromatic sources as opposed to using phosphors: this will increase control over spectral emissions (Schubert and Kim 2005; Gaston et al. 2012).

7.5 The Biological Effects of Light Pollution

The number of studies revealing negative consequences of artificial night lighting on a multitude of both diurnal and nocturnal vertebrates and invertebrates is increasing rapidly (reviewed in Rich and Longcore 2006). Most negative effects are due to the disruption of natural circadian and circannual cycles, which in turn can affect a whole range of species interactions, physiological processes and behaviours.

7.5.1 Impacts of Light Pollution on Intra- and Inter-specific Competition

Light-induced changes in circadian activity patterns can alter competition both within species (e.g. for mates) and between species (e.g. interference and exploitation competition). These are best documented for birds. For instance, early singing may be a signal of male quality in songbirds and increases the rate of extra-pair copulations, which are usually higher in older males. In territories affected by artificial light, males of several songbird species start singing earlier at dawn and thereby gain access to about twice as many extra-pair mates (Kempenaers et al. 2010; Nordt and Klenke 2013; Dominoni et al. 2014). The effect of artificial light on paternity gain is even stronger in yearlings than in adults, and so street lights might result in maladaptive mate choice of females by artificially increasing the extra-pair success of yearlings (Kempenaers et al. 2010). Whether similar maladaptive effects occur with nocturnal species is less clear.

Artificial light can affect niche partitioning by extending the activity of diurnal species, bringing them into inter-specific competition with nocturnal species (Longcore and Rich 2004; Rich and Longcore 2006). The scissor-tailed flycatcher *Tyrannus forficatus*, for example, will catch insects at street lights until at least 3 h after sunset (Frey 1993); this may increase exploitation and interference competition with insectivorous bats. Light pollution may also cause inter-specific competition between bats, with light-sensitive bat species excluded from illuminated resources exploited by light-tolerant species (Arlettaz et al. 2000).

7.5.2 Effects of Artificial Light on Physiological Homeostasis

Light-induced changes in circadian rhythms may induce physiological aberrations. For instance, exposure of captive mice to light at night disrupts metabolic signals, leading to increased body mass and decreased glucose tolerance (Fonken et al. 2010). Dim night-time light can also impair learning and memory, affect stress hormone levels, compromise immune function and cause depressive-like behaviour in rodents (Bedrosian et al. 2011, 2013; Fonken et al. 2012). In humans, depression, obesity and cancer risk relate to light pollution and associated disruptions of the circadian system (Fonken and Nelson 2011; Kronfeld-Schor and Einat 2012; Haim and Portnov 2013).

Light pollution can also result in a decoupling of seasonal behaviours and physiological adaptations from the optimal time of year. So, for instance, reproduction might be desynchronised from peak food availability; even very low light levels at night advance avian reproduction (Dominoni et al. 2013) so that birds breed earlier close to street lights than in darker territories (Kempenaers et al. 2010). Light-induced decoupling can even reverse an animal's seasonal phenotype, so that it exhibits a long-day phenotype in winter and vice versa. In sheep, 1 h of light during the dark phase is enough to mimic a long-day during short-day conditions (Chemineau et al. 1992). Also in primates, artificial light at night can induce a long-day phenotype; these animals had higher core body temperatures, showed less locomotor activity during the nocturnal activity period and had fainter torpor bouts compared with short-day photoperiod acclimated animals (Le Tallec et al. 2013). Voles that experienced light interference at night showed reduced winter acclimatisation of their thermoregulatory system to such a degree that they reduced heat production and died under winter field conditions (Haim et al. 2004, 2005). Thus, light pollution may have deleterious impacts on survival when animals expend too much energy during winter (Haim et al. 2004): this may be relevant for hibernating bats.

7.5.3 Interference of Light Pollution with Nocturnal Navigation

A well-documented effect of light pollution not mediated through circadian rhythms is the impact on movement decisions of visually orienting animals. Nesting attempts of female sea turtles are disrupted by artificial light, and light attracts or confuses the hatchlings, rendering them more vulnerable to predation, exhaustion and dehydration (Salmon 2006; Perry et al. 2008; Berry et al. 2013).

Birds migrating at night often approach bright lights instead of following their normal migration route, possibly because the light interferes with their magnetic compass (Poot et al. 2008). Birds may also be trapped within the sphere of light, milling around illuminated objects until they die through collisions or exhaustion (Gauthreaux and Belser 2006; Montevecchi 2006; Spoelstra and Visser 2014). This may have relevance to bats, which also use magnetic compasses for navigation (Holland et al. 2006).

Similarly many insects, particularly moths (Lepidoptera), use artificial lights rather than the moon for orientation and die of exhaustion when circling a lamp or following a collision with the hot cover. Artificial light also provokes a 'dazzling effect': many insects become immobilised when approaching a lamp and rest on the ground or in vegetation, becoming easy prey (Eisenbeis 2006). Light pollution may even be a driver of an insect biodiversity crisis (Conrad et al. 2006). The 'vacuum cleaner' effect, i.e. the long-distance attraction of light-susceptible species to lamps, removes large numbers of insects from the ecosystem, even resulting in local extinctions. This flight-to-light behaviour strongly depends on spectral output of the lighting: white HPMV lamps have a high UV proportion of their spectrum, and so four times as many moths are captured at HPMV lights compared to yellow/orange HPS lights (Eisenbeis 2006). Warm-white and cool-white LED lights induce less flight-to-light behaviour than HPS lights (Huemer et al. 2010; Eisenbeis and Eick 2011), and the virtually monochromatic deep-orange LPS lights are least attractive to insects (Rydell 1992; Blake et al. 1994; Eisenbeis 2006; Frank 2006).

Several spiders, amphibians, reptiles, birds and bats focus their foraging on insects accumulated at street lights (Rich and Longcore 2006). For bats, this can also be advantageous because artificial light disrupts the evasive behaviour of most nocturnal Lepidoptera, rendering them more vulnerable to bat attacks (Svensson and Rydell 1998; Acharya and Fenton 1999).

7.6 Bat Vision

Vision is important in the lives of many bats; see reviews in Suthers (1970), Altringham and Fenton (2003) and Eklöf (2003). A number of species rely on vision to a large extent (Altringham 2011). Since vision is important to both bats and their predators, we briefly summarise some key recent findings relevant to bats' perception of artificial lighting.

Most pteropodids do not echolocate and use vision to locate fruit and flowers. Some echolocating bats use vision to complement auditory information when hunting (Eklöf and Jones 2003) and, if vision and echolocation provide conflicting information, visual information is used in preference (Orbach and Fenton 2010). Vision can also be more effective than echolocation over long distances (Boonman et al. 2013), and the California leaf-nosed bat *Macrotus californicus* relies more on vision when hunting prey under low levels of illumination equivalent to a moonlit night (Bell 1985).

Recent research on bat vision has focussed on the molecular evolution of light-sensitive pigments (Jones et al. 2013). As for most nocturnal mammals, bat retinas are dominated by rods: they are highly sensitive under low light and confer monochromatic vision. The opsin DNA sequences of rhodopsin (the opsin in rods) were intact in 15 bat species (Zhao et al. 2009a) and wavelengths of maximum absorbance were 497–501 nm.

Colour vision in mammals results in part from opsins in the cones that are sensitive to short and medium wavelengths. Zhao et al. (2009b) sequenced a short-wavelength sensitive opsin gene (*Sws1*) that is most sensitive to blue-violet wavelengths, and a medium-to-long-wavelength sensitive opsin gene (*M/lws*) in a range of bat species; maximum absorbance of red light wavelengths by the *M/lws* opsin was at 545–553 nm. Although many bats resemble diurnal mammals in having the potential for dichromatic vision, with both genes being intact, *Sws-1* was pseudogenised in all the rhinolophid and hipposiderid bats studied and in some pteropodids, especially cave-roosting taxa. Immunohistochemistry suggests that the primary visual cortex may not respond to stimulation by UV light in these taxa (Xuan et al. 2012a), and behavioural responses to UV were also lacking (Xuan et al. 2012b). The lesser Asiatic yellow bat *Scotophilus kuhlii* and Leschenault's rousette *Rousettus leschenaultii* showed behavioural (Xuan et al. 2012b) and immunohistochemical responses in the primary visual cortex (Xuan et al. 2012a) to UV light at 365 nm. Two phyllostomid species (Pallas's long-tongued bat *Glossophaga soricina* and Seba's short-tailed bat *Carollia perspicillata*) possess significant cone populations and express opsins that are sensitive to short and long wavelengths. The short-wavelength opsin is sensitive to UV and may be advantageous for the detections of UV-reflecting flowers (Winter et al. 2003; Müller et al. 2009). Other bat species with intact *Sws1* genes may be UV sensitive, as ancestral reconstructions suggest UV sensitivity, with maximal sensitivity close to 360 nm (Zhao et al. 2009b).

Whether differences in UV sensitivity among bat taxa affect how species with intact and pseudogenised *Sws1* genes respond to different types of lighting remains unknown. Nevertheless the findings are of interest given that the wavelengths of maximum absorbance in bat opsins lie close to some of the peak emissions of wavelengths in a range of light types (Davies et al. 2013). Moreover emerging LED lighting technologies do not emit UV wavelengths, whereas older technologies, especially HPMV lamps, emit wavelengths that extend into the UV range and so HPMV lights may have been particularly conspicuous to horseshoe bats.

7.7 Observational Studies on Bats at Street Lights

Bats have been observed foraging around lights ever since artificial lighting became pervasive (Shields and Bildstein 1979; Belwood and Fullard 1984; Barak and Yom-Tov 1989; Acharya and Fenton 1999). Artificial light attracts many positively phototactic insects (Rydell 1992; Eisenbeis 2006), and most insectivorous bats are probably opportunistic feeders. Thus, they quickly identify and exploit insect accumulations such as swarming termites (Gould 1978) and insect clusters at artificial lights (Fenton and Morris 1976; Bell 1980; de Jong and Ahlén 1991). So some insectivorous bats probably profit from street lights because resource predictability and high insect densities increase foraging efficiency (Rydell 1992, 2006). For instance, 18 of 25 Neotropical insectivorous bat species which could be detected by acoustic monitoring were observed foraging around street lights in a small settlement. While more species were recorded in mature forest, total bat activity was lowest in forest but highest around street lights (Jung and Kalko 2010).

Bats prey on relatively large insects at street lights, mostly moths (Fenton and Morris 1976; Belwood and Fullard 1984; Acharya and Fenton 1992; Acharya 1995; Hickey et al. 1996; Acharya and Fenton 1999; Jacobs 1999; Pavey 1999; Fullard 2001). While moths are the most numerous insects around artificial lights (Huemer et al. 2010; Eisenbeis and Eick 2011), their contribution to a bat's diet can be much higher than expected from their relative abundance at street lights

(Belwood and Fullard 1984). This implies that bats focus on larger moths rather than smaller prey at street lights. Although moths were only captured in 36 % of attacks, northern bats *Eptesicus nilssonii* probably gain more than twice as much energy when feeding on moths at street lights than smaller dipterans in woodlands (Rydell 1992).

Aggregations of large insects around lamps enable bats to reduce foraging time and hence energy costs while maximising energy returns (Acharya and Fenton 1999; Jung and Kalko 2010). Big brown bats *Eptesicus fuscus*, for instance, spend less than half as much time outside the roost where in habitats where they forage at street lights than where they do not use lamps for hunting (Geggie and Fenton 1985). Hence, foraging at lights might be beneficial when a high foraging efficiency compensates for the potentially higher predation risk.

Bat activity and foraging efficiency at street lights are mainly determined by the number and size of prey insects available, both of which are strongly affected by the spectral characteristics of the light (Blake et al. 1994). Thus, the type of light indirectly influences bat activity. The light's attractiveness for insects increases with its UV spectral content. Aerial-hunting long-legged myotis *Myotis volans* and California myotis *M. californicus* consistently preyed on insects clustered in the cone of experimental black (UV) lights in North America (Bell 1980). While black light is not used for street lighting, similar results are seen with street lights that produce UV emissions. Thus, bat density can be an order of magnitude higher in towns illuminated by HPMV compared with those illuminated by HPS lights and road sections illuminated by HPMV rather than deep-orange LPS lights (Rydell 1992). In Britain, mean bat activity, likely to be mainly common pipistrelles *Pipistrellus pipistrellus*, is usually equal to or lower along roads lit by LPS lights than in dark sections, whereas bat activity is higher under HPMV than LPS lights or sections with no light (Fig. 7.3; Blake et al. 1994).

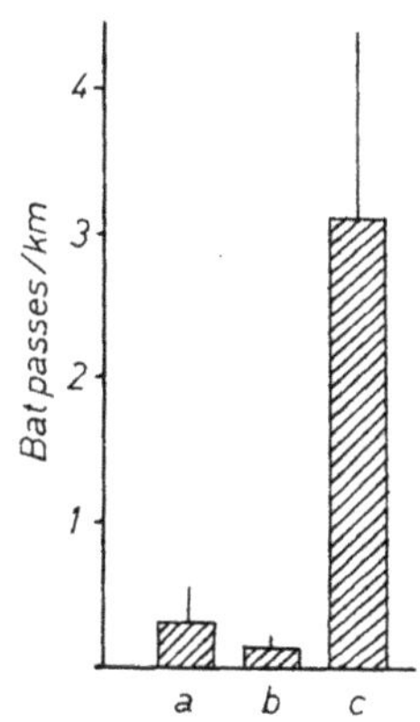

Fig. 7.3 Bat activity varies according to the type of artificial lighting. Activity of pipistrelle *Pipistrellus* spp. bats (mean and SD) along a 28 km stretch of road near Aberdeen, Scotland. *a* rural sections of the road without streetlamps, *b* village sections with sodium (*orange*) lamps and *c* a village with high-pressure mercury vapour lamps. From Rydell and Racey (1995)

7.8 Experimental Studies on Bats at Street Lights

Drawing conclusions from observational studies can be difficult, especially since confounding factors other than the presence of street lights can affect bat activity. Experimental field studies have demonstrated species-specific impacts of street lighting. Two 70 W HPS (DW Windsor Ltd, UK) lights, spaced and orientated to replicate street lights, were installed along preferred commuting routes of lesser horseshoe bats *Rhinolophus hipposideros*. The commuting activity of *R. hipposideros* (Fig. 7.4) and *Myotis* spp. was significantly reduced, and the onset of commuting delayed, on lit nights (Stone et al. 2009; Stone 2011). The following year the experiment was repeated on the same routes using white LED lights (Monaro LED, DW Windsor Ltd), at low (3.6 lux), medium (6.6 lux) and high (49.8 lux) light intensities. Activity of both *R. hipposideros* and *Myotis* spp. was significantly reduced during all lit treatments, and for *R. hipposideros,* the effect size at 49.8 lux was the same as that under HPS illumination. So both HPS and LED light disturbance caused spatial avoidance of preferred commuting routes by *R. hipposideros* and *Myotis* spp. (Stone et al. 2009), with no evidence of short-term habituation. Further work is needed to test for long-term habituation. In contrast, there was no significant change in bat activity under HPS and LED light treatments for *P. pipistrellus*, and for bats in the genera *Eptesicus* and *Nyctalus* (Fig. 7.5).

R. hipposideros and many other slow-flying species rely on linear habitat features for shelter from wind, rain and predators; acoustic orientation; and foraging

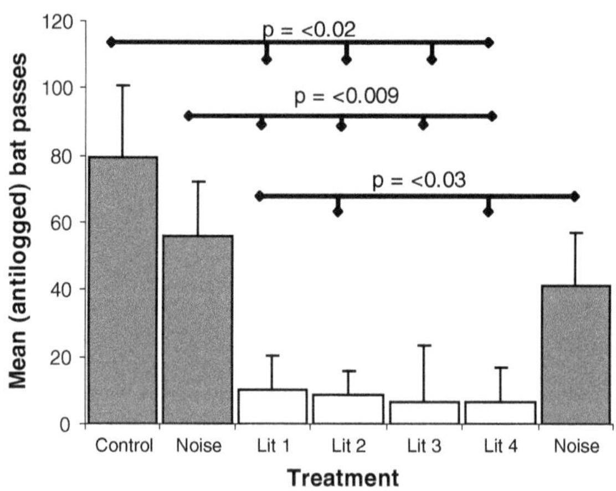

Fig. 7.4 Light-averse bat species show reduced activity along commuting routes subjected to high-pressure sodium (HPS) lighting. Activity of lesser horseshoe bats *Rhinolophus hipposideros* (mean passes and SE) in relation to lighting treatment. Significant within-subject differences with *p* values are shown. Treatments were control nights (no lighting treatment or generator), noise controls (HPS light units installed but switched off, generator running at night), 4 nights where lighting was switched on and powered by the generator (Lit 1 to Lit 4) and a final noise control. From Stone et al. (2009)

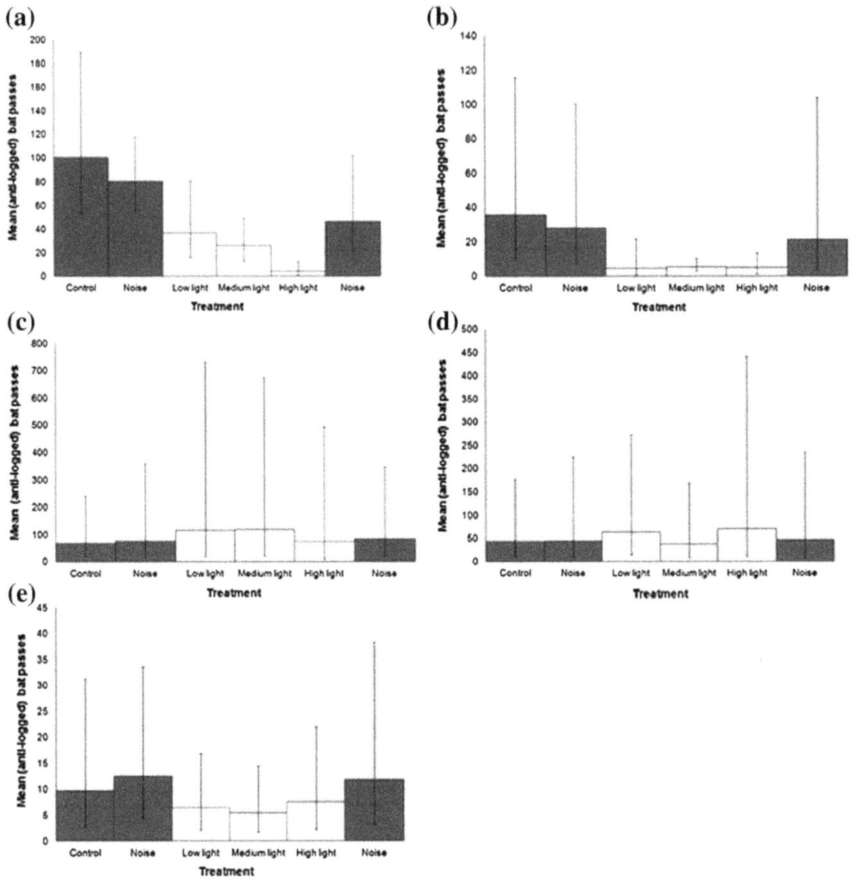

Fig. 7.5 Bats respond in different ways to LED lighting. Although the light-averse *Rhinolophus hipposideros* showed higher activity under more dimmed treatments compared with less dimmed ones, activity was still less than under unlit conditions. *Myotis* spp. showed negligible activity under all dimmed treatments. Geometric mean and confidence limits for bat passes along treatment hedges subjected to LED illumination at different light intensities are illustrated. Treatments were control nights (no lighting treatment or generator), noise controls (LED light units installed but switched off, generator running at night), 3 nights where illumination levels were modified (low light mean = 3.6 lux; medium light mean = 6.6 lux; and high light mean = 49.8 lux), and a final noise control. Bat passes were monitored on Anabat bat detectors and are shown for **a** *Rhinolophus hipposideros*, **b** *Myotis* spp., **c** common pipistrelle *Pipistrellus pipistrellus*, **d** soprano pipistrelle *Pipistrellus pygmaeus* and **e** *Nyctalus/Eptesicus*. From Stone et al. (2012)

(Verboom and Spoelstra 1999; Verboom et al. 1999). Using suboptimal routes with reduced cover to avoid artificial lighting may increase vulnerability to aerial predators and energetic costs due to increased exposure to wind and rain. So bats may have to travel further to reach foraging areas, reducing foraging time and increasing energetic losses, with consequential negative effects on reproduction rates and fitness. For example, juvenile growth rates were suppressed in the grey bat

Myotis grisescens with increased travel distance to foraging grounds (Tuttle 1976). Compensating for energetic losses by increasing foraging time may not be possible if, for instance, emergence and/or commuting is delayed by light pollution (Stone et al. 2009). Such delays also increase the risk that bats will miss the dusk peak in insect abundance, reducing the quality of foraging time. Delayed emergence could therefore affect the fitness of both individuals and the roost as whole.

Light disturbance along the commuting routes may isolate bats from their foraging grounds if the energetic costs of using alternative routes exceed the benefits. The commuting costs for *P. pipistrellus* become prohibitive when foraging areas are more than 5 km from the roost (Speakman 1991). Since bats select roosts based on the quality of surrounding habitat features, including linear connectivity (Jenkins et al. 1998; Oakeley and Jones 1998), maintaining optimal commuting routes is paramount. Whether fitness, or likely proxies of fitness, is affected by lighting needs further evaluation.

7.9 Winners and Losers: Light-Tolerant and Light-Averse Bats

Bats show variable responses to light pollution. Insectivorous bats that hunt in open spaces above the canopy (open-space foragers) or along vegetation edges such as forest edges, tree lines or hedgerows (edge foragers) are the species most tolerant of artificial lighting. They have evolved traits advantageous for foraging in sparsely structured habitats (Norberg and Rayner 1987; Neuweiler 1989) and so are preadapted to foraging in urban habitats (Rydell 2006; Jung and Kalko 2010; Jung and Threlfall 2016). Open-space foragers, such as the noctule *Nyctalus noctula*, typically have long narrow wings with a high aspect ratio, often combined with a high wing loading (weight/wing area). They have to fly fast to remain airborne and so use high-intensity, low-frequency narrowband echolocation calls that facilitate long-range detection of insects (Norberg and Rayner 1987; Rydell 2006; Kalko et al. 2008). When foraging at street lights, open-space foragers typically fly above the lamps, diving into the light cone to catch insects (Jung and Kalko 2010).

Edge foragers generally use echolocation calls with a conspicuous narrowband component, but usually also include a frequency-modulated 'broadband' component during the search phase, which is advantageous for ranging when flying close to obstacles. They comprise relatively fast-flying species with above-average aspect ratio and wing loading (e.g. *P. pipistrellus*), and species with an average aspect ratio and wing loading (e.g. *E. nilssonii*). Edge foragers tend to be more manoeuvrable than open-space foragers (Norberg and Rayner 1987; Kalko et al. 2008), and some can even conduct circuits inside the light cone when hunting insects at street lights (Jung and Kalko 2010).

Though most edge foragers fly with agility and speed (Norberg and Rayner 1987), they differ in their degree of synanthropism. While Kuhl's pipistelle

Pipistrellus kuhlii is recorded almost exclusively at street lights in southern Switzerland, *P. pipistrellus* forage to a similar extent both at lights and at least 100 m from lights (Haffner and Stutz 1985). Even within a species, foraging activity at lamps can be highly variable depending on the quantity of insects available: Geggie and Fenton (1985) never observed *E. fuscus* foraging around street lights in an urban environment, whereas in rural habitats feeding activity was greater at lights than in areas without lights. In spring and autumn, when artificial lights attract numerous insects in Sweden, *E. nilssonii* activity is about 20-fold higher in towns with street lighting than in non-illuminated towns, forest and farmland (de Jong and Ahlén 1991; Rydell 1991), with the bats flying back and forth above the street lights, regularly diving to within 1 m of the ground to catch insects.

Although fast-flying species adapted to forage in open areas, particularly bats of the genera *Eptesicus*, *Nyctalus* and *Pipistrellus*, may benefit from the increased foraging opportunities provided at lamps that attract high densities of insects, Stone et al. (2009, 2012) found no significant increases in bat activity for these 'light-tolerant' species during lit treatments. This could be due to two factors. First, HPS lights are less attractive to insects than white lights because their spectral content has less UV (Blake et al. 1994); for example, HPS street lights attracted fewer insects than white lights in Germany (Eisenbeis and Eick 2011). Second, the experimental nature of the study may have affected the results, since bats may need time to find and recognise newly installed lights as an attractive foraging source.

Though a relatively high proportion of aerial insectivorous bats may forage in suburban habitats, bat activity and the number of bat species decrease significantly towards highly urbanised areas. This is probably because both roosts and appropriate insect habitats are lacking, and those insects which are present might not aggregate at street lamps because the pervasive artificial lighting in city centres causes a dilution effect, rendering the lights less attractive for bats (Gaisler et al. 1998; Avila-Flores and Fenton 2005; Frank 2006; Rydell 2006; Jung and Kalko 2011; Jung and Threlfall 2016). In Panama, 18 of 25 insectivorous bat species frequently foraged around street lamps in a settlement bordering mature forest; the reduced vegetation cover in town constrained strictly forest-dwelling species from hunting at lamps (Jung and Kalko 2010). Yet, even some closely related and ecologically similar species may differ in their tolerance of urban habitats, and their potential to adapt to anthropologically altered habitats is best viewed from a species-specific perspective.

As compared to open-space foragers, bats at the other end of the wing shape spectrum, such as many horseshoe bats (Rhinolophidae) with their low aspect ratio wings and a low wing loading, rarely forage near artificial lights (Rydell 2006; Stone et al. 2009, 2012). They are mostly forest-dwelling and their short broad wings facilitate the high manoeuvrability needed for hawking insects in a cluttered environment (Norberg and Rayner 1987). However, their morphology only allows slow flight speeds, which might render them more vulnerable to predators when flying in a sphere of light away from protective vegetation cover (Jones and Rydell 1994; Rydell et al. 1996). Most forest-dwelling bat species emerge from their roosts relatively late in the evening, presumably to minimise predation risk from diurnal birds of prey (Jones and Rydell 1994) and so may be 'hard-wired' to be light-averse. Furthermore, slow-hawking bats use echolocation calls that are adapted for short-range prey detection among clutter (Norberg and Rayner 1987), and so these may not be suitable for orientation in semi-open habitats where most street lights are positioned.

Myotis spp. in Canada and Sweden and brown long-eared bats *Plecotus auritus* in Sweden were only recorded away from street lights (Furlonger et al. 1987; Rydell 1992). In Australia, the chocolate wattled bat *Chalinolobus morio* avoided parks when lights were switched on (Scanlon and Petit 2008). Despite having street-lit areas in their home range, they were never utilised by greater horseshoe bats *Rhinolophus ferrumequinum* (Jones and Morton 1992; Jones et al. 1995). Artificial light reduced the foraging activity of pond bats *Myotis dasycneme* over rivers in the Netherlands (Kuijper et al. 2008), and commuting activity of *R. hipposideros* and *Myotis* spp. was reduced under LED and HPS street lights (Stone et al. 2009, 2012). It is likely that the *Myotis* spp. in Stone et al.'s studies were Natterer's bats *Myotis nattereri* (Stone 2011). *M. nattereri* emerges from roosts relatively late (Jones and Rydell 1994), at median light levels (3.5 lux, Swift 1997), lower than those recorded for *R. hipposideros* (Stone et al. 2009). *M. nattereri* and *R. hipposideros* use different echolocation strategies (Parsons and Jones 2000) but have similar flight and foraging patterns. *M. nattereri* has broad wings, prefers foraging in woodlands and is slow-flying and manoeuvrable, often foraging close to vegetation to glean prey (Arlettaz 1996; Swift 1997). This suggests that light-dependent predation risk limits the ability of these bats to take advantage of illuminated areas. Nevertheless, one large-eared horseshoe bat *Rhinolophus philippinensis* was repeatedly observed traversing 200 m of open grassland to forage extensively around artificial lights in Australia. The same lights were also used by eastern horseshoe bats *Rhinolophus megaphyllus* (Pavey 1999).

Extinction risk is highest in bat species with low aspect ratios (Jones et al. 2003; Safi and Kerth 2004), which are the species that show aversion to artificial lighting. Thus, species that may suffer most from light pollution are likely to be already threatened taxa.

7.10 Effects of Light Pollution on Ecosystem Services Provided by Bats

The impacts of lighting go far beyond changing the physiology, behaviour and/or distribution of individual species. Since congeners interact with each other as well as their prey and predators, light pollution is likely to have far-reaching consequences for the entire biome and the ecosystem services that bats provide. Insectivorous bats, for instance, significantly reduce the number of insects that cause damage to flora and fauna (Ghanem and Voigt 2012). The value of insectivorous bats to the US agricultural industry by reducing insect populations was estimated to be $23 billion/year (Boyles et al. 2011).

Most studies to date have been on temperate-zone insectivorous bats. However, many tropical bats feed on nectar and fruits, thereby pollinating flowers and dispersing seeds of several hundred species of plants (Ghanem and Voigt 2012). Consequently, frugivorous bats are key for succession and maintaining plant diversity, especially in fragmented Neotropical landscapes (Medellin and Gaona 1999; Muscarella and Fleming 2007). However, very little is known about the impact of light pollution on this feeding guild. Southern long-nosed bats *Leptonycteris yerbabuenae*, a nectar- and fruit-eating species, used areas of relatively low light intensity when commuting (Lowery et al. 2009) and Oprea et al. (2009) rarely captured frugivorous bats along roads, although some were present in municipal parks. However, neither study could disentangle the influence of lighting from other factors related to urbanisation, such as altered vegetation cover or increased noise levels. Lewanzik and Voigt (2014) provided the first experimental evidence for light avoidance by frugivorous bats. They found that Sowell's short-tailed bat *Carollia sowelli*, a specialist on fruits of the genus *Piper*, harvested only about half as many fruits in a flight cage compartment lit by a sodium vapour street light than in a dark compartment, and free-ranging bats neglected ripe fruits that were experimentally illuminated (Fig. 7.6). Lewanzik and Voigt (2014) concluded that artificial light might reduce nocturnal dispersal of pioneer plant seeds. Since

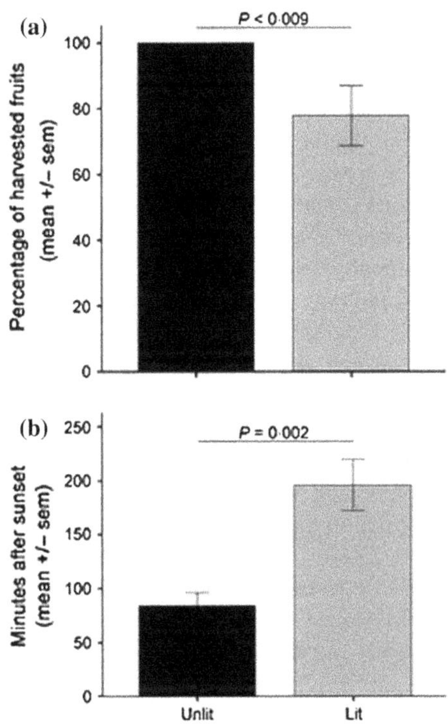

Fig. 7.6 Artificial lighting reduces and delays feeding behaviour on pepper plants by a frugivorous bat. **a** Percentage of harvested infructescences of *Piper sancti-felices* among 14 marked plants harvested by Sowell's short-tailed bats *Carollia sowelli* in non-illuminated conditions (*black*) and under conditions where plants were illuminated by a street lamp (*grey*) in the field, **b** time after sunset when infructescences were harvested. From Lewanzik and Voigt (2014)

bat-mediated seed intake is particularly important during the early stages of succession (Medellin and Gaona 1999; Muscarella and Fleming 2007), light pollution might slow down the reforestation of cleared rainforests (Lewanzik and Voigt 2014).

7.11 Knowledge Gaps, Future Challenges and Mitigation Strategies

7.11.1 Knowledge Gaps

Light pollution has only recently been acknowledged as a threat to biodiversity (Hölker et al. 2010b), and there are still many unknowns about the interactions between bat species and artificial lighting sources (Hölker et al. 2010a). Most studies have focused on specific ecological behaviours such as foraging (Rydell 1992; Blake et al. 1994), predator–prey interactions, particularly with moths (Rydell et al. 1995; Svensson and Rydell 1998), commuting routes (Stone et al. 2009, 2012) and roost emergence (Downs et al. 2003). No long-term studies have been carried out to determine whether any of these behavioural changes have fitness consequences (Beier 2006; Stone et al. 2012). The only indication of potential population-level responses has been shown in Hungary on *Myotis* species, where juveniles roosting in illuminated buildings had a lower body mass than their counterparts in unlit roosts (Boldogh et al. 2007). However, this study did not establish whether a lower body mass in these juveniles reduced their survival rate after hibernation. It is particularly important to understand higher level responses for bat species because they have low fecundity rates, usually only producing one pup per year (Dietz et al. 2009), and so populations are sensitive to sudden changes (Stone et al. 2012).

Further studies are needed to address the impact of artificial lighting at the community level (Davies et al. 2012). The current literature highlights that artificial lighting causes species-specific responses (Rydell 1992; Stone et al. 2009, 2012; Jung and Kalko 2010), which could cause light-tolerant species to exclude light-averse species (Polak et al. 2011; Stone et al. 2012). Such competitive interactions have been proposed as the driving force behind changes in bat populations in Switzerland, where decreases in photosensitive *R. hipposideros* have been linked to increases in light-tolerant *P. pipistrellus* (Stutz and Haffner 1984; Arlettaz et al. 2000). It is believed that by avoiding street lights, *R. hipposideros* are foregoing profitable prey sources exploited by *P. pipistrellus* (Arlettaz et al. 1999, 2000).

So far research has focussed largely on insectivorous bats in temperate zones. Further research in tropical ecosystems is needed. For example, the forested areas of South-east Asia contain a high diversity and abundance of horseshoe bat species that are likely to be negatively affected by light pollution, and the impact of light pollution on pollination and seed dispersal in the tropics and subtropics needs further investigation.

Research on the impacts of different light spectra in emerging technologies on bat activity and reproduction will be valuable; this is currently being investigated in the Netherlands as part of a large-scale investigation exposing a wide range of taxa to white, red and green LED lighting (see http://www.lichtopnatuur.org). With the current plans to switch to broader spectrum lighting sources, it is important to understand more about the spectral sensitivities of bats (Davies et al. 2012, 2013), especially given the recent findings on opsin genes highlighted above. Determining if there are spectral and intensity thresholds for different species would aid mitigation strategies and improve conservation initiatives (Stone et al. 2012; Gaston et al. 2013).

7.11.2 Mitigation Strategies

The most effective approach to reduce the detrimental effects of artificial lighting is to limit the growth of lighting by restricting unnecessary installations or removing them from areas already saturated with artificial lighting sources. This has the greatest potential to reduce light pollution and minimise ecological effects (Gaston et al. 2012). Turning off lights in areas commonly used by light-averse bats to forage, commute or roost during key times such as reproduction (Jones 2000) may be effective. Bats are faithful to maternity roosts due to the specific conditions they provide, and so conserving them is important for maintaining bat populations (Lewis 1995; Mann et al. 2002). However, some photosensitive bats may be disrupted even if areas were only lit for a short period of time (Boldogh et al. 2007), and switching off lighting may be challenged if it is perceived to jeopardise public safety (Lyytimäki and Rinne 2013).

Reducing the duration of illumination through part-night lighting (PNL) schemes could also help limit the adverse effects of light on nocturnal animals (Gaston et al. 2012). This has already been adopted by a number of local authorities in the UK, which switch off lights in specified areas between midnight and 05.30 to reduce CO_2 emissions and save money (Lockwood 2011). Since April 2009, lights along sections of motorways have also been switched off between these hours (Royal Commission on Environmental Pollution 2009). While this may help to reduce light pollution, it is unlikely to have significant ecological benefits since the lights remain switched on in the early part of the night, when bats and other nocturnal species undertake key activities such as foraging and commuting (Gaston et al. 2012). Intelligent lighting schemes, such as the use of motion sensors, have already been implemented in Portugal and may have more ecological benefits. The lights remain switched off unless needed and so still provide all the perceived public safety benefits (Royal Commission on Environmental Pollution 2009). However, these fluctuations in lighting levels may also be damaging to bats (Longcore and Rich 2004).

It is also important to reduce the trespass of artificial lighting to minimise the impact on bats. Newer technologies such as LEDs produce more directional light (Gaston et al. 2012), preventing the horizontal or upward emissions which contribute most to light pollution (Falchi et al. 2011). Effective luminaire design, installation of shielding fixtures and correct column height can also help focus light and avoid wasteful emissions (Royal Commission on Environmental Pollution 2009). In Lombardia, Italy, for example, 75 % of light pollution was due to poorly designed luminaires; the other 25 % was unavoidable reflection from road surfaces (Falchi 2011). Vegetation canopies such as hedgerows can also help decrease light trespass, which is crucial for many bat species that use linear features as commuting routes (Rydell 1992; Fure 2006). Diminishing trespass could create dark refuges, providing corridors for bats to forage in fragmented habitats (Longcore and Rich 2004; Stone et al. 2012; Gaston et al. 2012).

Light intensity has a significant effect on bat activity (Stone et al. 2012) and delays roost emergence (Downs et al. 2003). If bats delay foraging, they risk missing the peak abundance in insects that occurs shortly after dusk, so may not meet their energy requirements, which in turn could reduce fitness (Jones and Rydell 1994; Stone et al. 2012). In addition to implementing PNL, many local authorities are also dimming lights in specified areas (Gaston et al. 2012). This relies on local authorities already having lights such as LEDs that have the necessary centralised management system (International Energy Agency 2006). These schemes are more environmentally friendly and cost-effective (Gaston et al. 2012). However, dimming lights may not be beneficial to all bat species; Daubenton's bats *Myotis daubentonii*, for instance, only emerge from their roosts at very low light levels (less than 1 lux) (Fure 2006) and *R. hipposideros* and *Myotis* spp. avoid commuting routes illuminated to 3.6 lux (Stone et al. 2012). Since illumination levels of street lights are usually between 10 and 60 lux (Gaston et al. 2012), it may not be feasible to dim lighting to such low intensities without compromising public perceptions of safety (Stone et al. 2012; Lyytimäki and Rinne 2013).

7.11.3 Future Challenges

With a number of changes to street lighting planned in the coming years, including dimming, PNL and modifications to luminaire design to reduce light pollution, energy expenditure and greenhouse gas emissions, nightscapes could increase in heterogeneity, making it even more challenging to understand the impacts of artificial lighting on biodiversity (Gaston et al. 2012).

This is further complicated because current metrics for measuring emissions from light sources omit key biological information (Longcore and Rich 2004; Gaston et al. 2012). Illumination is measured in lux, which is defined as the brightness of a light according to human spectral sensitivities; spectral sensitivities of other taxa are often very different from ours (Peitsch et al. 1992; Briscoe and Chittka 2001). In bats, for example, many species can detect wavelengths in the UV range (Winter et al. 2003; Wang et al. 2004; Müller et al. 2009). So HPS and LPS lamps could have the same intensity of light, e.g. 50 lux, but HPS lamps emit UV wavelengths, whereas LPS lamps do not, thereby affecting both bats and their insect prey in different ways (Longcore and Rich 2004). Since lux is commonly used as a metric by lighting engineers, designers and environmental regulators, migrating from this measure may thwart interdisciplinary communication (Longcore and Rich 2004).

Another challenge is to find more effective ways of quantifying the impact of artificial lighting on bat species. Current methods use acoustic survey methods to quantify bat activity; this underestimates the activity of bats that use low-intensity echolocation calls (O'Farrell and Gannon 1999). Crucially, we also need to determine whether artificial lighting has fitness consequences (Stone et al. 2012). A decrease in bat activity may have no relevance for fitness if, for example, the bats are able to utilise equally suitable alternative sites nearby.

A transdisciplinary approach needs to be adopted to minimise the impact of light on biodiversity, reduce CO_2 emissions, increase energy efficiency and reduce costs (Hölker et al. 2010a; Gaston et al. 2012). Scientists, policymakers and engineers need to work together to implement successful strategies (Stone et al. 2012). Moreover, it is vital to find ways to broaden awareness of light pollution and its ecological impacts. Since the public plays an integral part in agreeing mitigation schemes such as dimming lights, their support is pivotal in moving forward (Hölker et al. 2010a).

Acknowledgements EGR, ELS, SH and GJ thank NERC for support. DL was supported by the Federal Ministry for Education and Research (BMBF) as part of the network project 'Loss of the Night'.

References

Acharya L (1995) Sex-biased predation on moths by insectivorous bats. Anim Behav 49:1461–1468

Acharya L, Fenton MB (1992) Echolocation behaviour of vespertilionid bats (*Lasiurus cinereus* and *Lasiurus borealis*) attacking airborne targets including arctiid moths. Can J Zool 70:1292–1298

Acharya L, Fenton MB (1999) Bat attacks and moth defensive behaviour around street lights. Can J Zool 77:27–33

Altringham JD (2011) Bats: from evolution to conservation, 2nd edn. Oxford University Press, Oxford

Altringham JD, Fenton MB (2003) Sensory ecology and communication in the Chiroptera. In: Kunz TH, Fenton MB (eds) Bat ecology. University of Chicago Press, Chicago, pp 90–127

Arlettaz R (1996) Foraging behaviour of the gleaning bat *Myotis nattereri* (Chiroptera, Vespertilionidae) in the Swiss Alps. Mammalia 60:181–186

Arlettaz R, Berthoud G, Desfayes M (1999) Tendances démographiques opposées chez deux espéces sympatriques de chauves-souris, *Rhinolophus hipposideros* et *Pipistrellus pipistrellus*: un possible lien de cause à effet? Le Rhinolophe 13:35–41

Arlettaz R, Godat S, Meyer H (2000) Competition for food by expanding pipistrelle bat populations (*Pipistrellus pipistrellus*) might contribute to the decline of lesser horseshoe bats (*Rhinolophus hipposideros*). Biol Conserv 93:55–60

Avila-Flores R, Fenton MB (2005) Use of spatial features by foraging insectivorous bats in a large urban landscape. J Mammal 86:1193–1204

Barak Y, Yom-Tov Y (1989) The advantage of group hunting in Kuhl's bat *Pipistrellus kuhli* (Microchiroptera). J Zool 219:670–675

Bedrosian TA, Fonken LK, Walton JC et al (2011) Chronic exposure to dim light at night suppresses immune responses in Siberian hamsters. Biol Lett 7:468–471

Bedrosian TA, Weil ZM, Nelson RJ (2013) Chronic dim light at night provokes reversible depression-like phenotype: possible role for TNF. Mol Psychiatry 18:930–936

Beier P (2006) Effects of artificial night lighting on terrestrial mammals. In: Rich C, Longcore T (eds) Ecological consequences of artificial night lighting. Island Press, Washington, DC, pp 19–42

Bell GP (1980) Habitat use and response to patches of prey by desert insectivorous bats. Can J Zool 58:1876–1883

Bell GP (1985) The sensory basis of prey location by the California leaf-nosed bat *Macrotus californicus* (Chiroptera: Phyllostomidae). Behav Ecol Sociobiol 16:343–347

Belwood JJ, Fullard JH (1984) Echolocation and foraging behaviour in the Hawaiian hoary bat, *Lasiurus cinereus semotus*. Can J Zool 62:2113–2120

Benenson W, Harris JW, Stocker H et al (2002) Handbook of physics. Prentice Hall, London

Bennie J, Davies TW, Duffy JP et al (2014) Contrasting trends in light pollution across Europe based on satellite observed night time lights. Sci Rep 4(3789):1–6

Berry M, Booth DT, Limpus CJ (2013) Artificial lighting and disrupted sea-finding behaviour in hatchling loggerhead turtles (*Caretta caretta*) on the Woongarra coast, south-east Queensland, Australia. Aust J Zool 61:137–145

Blake D, Hutson AM, Racey PA et al (1994) Use of lamplit roads by foraging bats in southern England. J Zool 234:453–462

Boldogh S, Dobrosi D, Samu P (2007) The effects of the illumination of buildings on house-dwelling bats and its conservation consequences. Acta Chiropterol 9:527–534

Boonman A, Bar-On Y, Cvikel N et al (2013) It's not black or white—on the range of vision and echolocation in echolocating bats. Front Physiol 4. doi:10.3389/fphys.2013.00248

Boyles JG, Cryan PM, McCracken GF et al (2011) Economic importance of bats in agriculture. Science 332:41–42

Briscoe AD, Chittka L (2001) The evolution of color vision in insects. Annu Rev Entomol 46:471–510

Buchanan BW (2006) Observed and potential effects of artificial night lighting on anuran amphibians. In: Rich C, Longcore T (eds) Ecological consequences of artificial night lighting. Island Press, Washington, DC, pp 192–220

California Lighting Technology Centre (2010) 2010 lighting technology overviews and best-practice solutions. http://cltc.ucdavis.edu/publication/2010-lighting-technology-overview. Accessed 1 June 2014

Campbell N (2011) Biology, 9th edn. Pearson Education, San Francisco

Chemineau P, Malpaux B, Delgadillo JA et al (1992) Control of sheep and goat reproduction: use of light and melatonin. Anim Reprod Sci 30:157–184

Cinzano P, Falchi F, Elvidge CD (2001) The first world atlas of the artificial night sky brightness. Mon Not R Astron Soc 328:689–707

Conrad KF, Warren MS, Fox R et al (2006) Rapid declines of common, widespread British moths provide evidence of an insect biodiversity crisis. Biol Conserv 132:279–291

Davies TW, Bennie J, Gaston KJ (2012) Street lighting changes the composition of invertebrate communities. Biol Lett 8:764–767

Davies TW, Bennie J, Inger R et al (2013) Artificial light pollution: are shifting spectral signatures changing the balance of species interactions? Glob Change Biol 19:1417–1423

de Jong J, Ahlén I (1991) Factors affecting the distribution pattern of bats in Uppland, central Sweden. Holarctic Ecol 14:92–96

Dietz C, von Helversen O, Nill D (2009) Bats of Britain. Europe and northwest Africa, Black London

Dominoni D, Quetting M, Partecke J (2013) Artificial light at night advances avian reproductive physiology. Proc R Soc Lond B 280:20123017

Dominoni DM, Carmona-Wagner EO, Hofmann M et al (2014) Individual-based measurements of light intensity provide new insights into the effects of artificial light at night on daily rhythms of urban-dwelling songbirds. J Anim Ecol 83:681–692

Douglas RH, Jeffery G (2014) The spectral transmission of ocular media suggests ultraviolet sensitivity is widespread among mammals. Proc R Soc Lond B 281:20132995

Downs NC, Beaton V, Guest J et al (2003) The effects of illuminating the roost entrance on the emergence behaviour of *Pipistrellus pygmaeus*. Biol Conserv 111:247–252

Eisenbeis G (2006) Artificial night lighting and insects: attraction of insects to streetlamps in a rural setting in Germany. In: Rich C, Longcore T (eds) Ecological consequences of artificial night lighting. Island Press, Washington, DC, pp 281–304

Eisenbeis G, Eick K (2011) Studie zur Anziehung nachtaktiver Insekten an die Straßenbeleuchtung unter Einbeziehung von LEDs. Nat Landsch 86:298–306

Eklöf J (2003) Vision in echolocating bats. Dissertation, University of Göteborg. http://www.fladdermus.net/thesis.htm. Accessed 1 June 2014

Eklöf J, Jones G (2003) Use of vision in prey detection by brown long-eared bats, *Plecotus auritus*. Anim Behav 66:949–953

Elvidge CD, Keith DM, Tuttle BT et al (2010) Spectral identification of lighting type and character. Sensors 10:3961–3988

Falchi F (2011) Campaign of sky brightness and extinction measurements using a portable CCD camera. Mon Not R Astron Soc 412:33–48

Falchi F, Cinzano P, Elvidge CD et al (2011) Limiting the impact of light pollution on human health, environment and stellar visibility. J Environ Manage 92:2714–2722

Fenton MB, Morris GK (1976) Opportunistic feeding by desert bats (*Myotis* spp.). Can J Zool 54:526–530

Fonken LK, Nelson RJ (2011) Illuminating the deleterious effects of light at night. F1000 Rep Med 3:18

Fonken LK, Workman JL, Walton JC et al (2010) Light at night increases body mass by shifting the time of food intake. Proc Natl Acad Sci USA 107:18664–18669

Fonken LK, Kitsmiller E, Smale L et al (2012) Dim nighttime light impairs cognition and provokes depressive-like responses in a diurnal rodent. J Biol Rhythm 27:319–327

Frank KD (2006) Effects of artificial night lighting on moths. In: Rich C, Longcore T (eds) Ecological consequences of artificial night lighting. Island Press, Washington, DC, pp 305–344

Frey JK (1993) Nocturnal foraging by scissor-tailed flycatchers under artificial light. Western Birds 24:200

Fullard JH (2001) Auditory sensitivity of Hawaiian moths (Lepidoptera: Noctuidae) and selective predation by the Hawaiian hoary bat (Chiroptera: *Lasiurus cinereus semotus*). Proc R Soc Lond B 268:1375–1380

Fure A (2006) Bats and lighting. Lond Nat 85:1–20

Furlonger CL, Dewar HJ, Fenton MB (1987) Habitat use by foraging insectivorous bats. Can J Zool 65:284–288

Gaisler J, Zukal J, Rehak Z et al (1998) Habitat preference and flight activity of bats in a city. J Zool 244:439–445

Gaston KJ, Davies TW, Bennie J et al (2012) Reducing the ecological consequences of nighttime light pollution: options and developments. J Appl Ecol 49:1256–1266
Gaston KJ, Bennie J, Davies TW et al (2013) The ecological impacts of nighttime light pollution: a mechanistic appraisal. Biol Rev 88:912–927
Gauthreaux SA, Belser CG (2006) Effects of artificial night lighting on migrating birds. In: Rich C, Longcore T (eds) Ecological consequences of artificial night lighting. Island Press, Washington, DC, pp 67–93
Geggie JF, Fenton MB (1985) A comparison of foraging by *Eptesicus fuscus* (Chiroptera: Vespertilionidae) in urban and rural environments. Can J Zool 63:263–267
Ghanem SJ, Voigt CC (2012) Increasing awareness of ecosystem services provided by bats. Adv Stud Behav 44:279–302
Gould E (1978) Opportunistic feeding by tropical bats. Biotropica 10:75–76
Haffner M, Stutz HP (1985) Abundance of *Pipistrellus pipistrellus* and *Pipistrellus kuhlii* foraging at street-lamps. Myotis 23:167–172
Haim A, Portnov BA (2013) Light pollution as a new risk factor for human breast and prostate cancers. Springer, Dordrecht
Haim A, Shanas U, Zisapel N et al (2004) Rodent pest control: the use of photoperiod manipulations as a tool. In: Pelz HJ, Cowan DP, Feare CJ (eds) Advances in vertebrate pest management III. Filander, Fürth, pp 29–38
Haim A, Shanas U, Zubidad AES et al (2005) Seasonality and seasons out of time—the thermoregulatory effects of light interference. Chronobiol Int 22:59–66
Hickey MBC, Acharya L, Pennington S (1996) Resource partitioning by two species of vespertilionid bats (*Lasiurus cinereus* and *Lasiurus borealis*) feeding around street lights. J Mammal 77:325–334
Hölker F, Moss T, Griefahn B et al (2010a) The dark side of light: a transdisciplinary research agenda for light pollution policy. Ecol Soc 15:13
Hölker F, Wolter C, Perkin EK et al (2010b) Light pollution as a biodiversity threat. Trends Ecol Evol 25:681–682
Holland RA, Thorup K, Vonhof MJ et al (2006) Bat orientation using Earth's magnetic field. Nature 444:702
Huemer P, Kühtreiber H, Tarmann G (2010) Anlockwirkung moderner Leuchtmittel auf nachtaktive Insekten: Ergebnisse einer Feldstudie in Tirol. Tiroler Landesumweltanwaltschaft and Tiroler Landesmuseen Betriebsgesellschaft m.b.H, Innsbruck
International Energy Agency (2006) Light's labour's lost: policies for energy-efficient lighting. http://www.iea.org/publications/freepublications/publication/name,3644,en.html. Accessed 1 June 2014
Jacobs DS (1999) The diet of the insectivorous Hawaiian hoary bat (*Lasiurus cinereus semotus*) in an open and a cluttered habitat. Can J Zool 77:1603–1608
Jenkins EV, Laine T, Morgan SE et al (1998) Roost selection in the pipistrelle bat, *Pipistrellus pipistrellus* (Chiroptera: Vespertilionidae), in northeast Scotland. Anim Behav 56:909–917
Jones J (2000) Impacts of lighting on bats. http://www.lbp.org.uk/downloads/Publications/Management/lighting_and_bats.pdf. Accessed 1 June 2014
Jones G, Morton M (1992) Radio-tracking studies and habitat use by greater horseshoe bats *Rhinolophus ferrumequinum*. In: Priede IG, Swift SM (eds) Wildlife telemetry, remote monitoring and tracking of animals. Ellis Horwood, Chichester, pp 521–537
Jones G, Rydell J (1994) Foraging strategy and predation risk as factors influencing emergence time in echolocating bats. Phil Trans R Soc Lond B 346:445–455
Jones G, Duvergé PL, Ransome RD (1995) Conservation biology of an endangered species: field studies of greater horseshoe bats. Symp Zool Soc Lond 67:309–324
Jones G, Teeling EC, Rossiter SJ (2013) From the ultrasonic to the infrared: molecular evolution and the sensory biology of bats. Front Physiol 4. doi:10.3389/fphys.2013.00117
Jones KE, Purvis A, Gittleman JL (2003) Biological correlates of extinction risk in bats. Am Nat 161:601–614

Jung K, Kalko EKV (2010) Where forest meets urbanization: foraging plasticity of aerial insectivorous bats in an anthropogenically altered environment. J Mammal 91:144–153

Jung K, Kalko EKV (2011) Adaptability and vulnerability of high flying Neotropical aerial insectivorous bats to urbanization. Divers Distrib 17:262–274

Jung K, Threlfall CG (2016) Urbanisation and its effects on bats—a global meta-analysis approach. In: Bats in the Anthropocene: conservation of bats in a changing world. Springer International AG, Cham, pp. 13–28

Kalko EKV, Villegas SE, Schmidt M et al (2008) Flying high—assessing the use of the aerosphere by bats. Integr Comp Biol 48:60–73

Kempenaers B, Borgström P, Loës P et al (2010) Artificial night lighting affects dawn song, extra-pair siring success, and lay date in songbirds. Curr Biol 20:1735–1739

Kronfeld-Schor N, Einat H (2012) Circadian rhythms and depression: human psychopathology and animal models. Neuropharmacology 62:101–114

Kuijper DPJ, Schut J, van Dullemen D et al (2008) Experimental evidence of light disturbance along the commuting routes of pond bats (*Myotis dasycneme*). Lutra 51:37–49

Kyba CCM, Hölker F (2013) Do artificially illuminated skies affect biodiversity in nocturnal landscapes? Landscape Ecol 28:1637–1640

Land MF, Nilsson D-E (2012) Animal eyes, 2nd edn. Oxford University Press, Oxford

Le Tallec T, Perret M, Théry M (2013) Light pollution modifies the expression of daily rhythms and behavior patterns in a nocturnal primate. PLoS ONE 8:e79250

Lewanzik D, Voigt CC (2014) Artificial light puts ecosystem services of frugivorous bats at risk. J Appl Ecol 51:388–394

Lewis SE (1995) Roost fidelity of bats: a review. J Mammal 76:481–496

Lockwood R (2011) A review of local authority road lighting initiatives aimed at reducing costs, carbon emissions and light pollution. https://www.google.co.uk/#q=review+of+local+authority+road+lighting+initiatives. Accessed 1 June 2014

Longcore T, Rich C (2004) Ecological light pollution. Front Ecol Environ 2:191–198

Lowery SF, Blackman ST, Abbate D (2009) Urban movement patterns of lesser long-nosed bats (*Leptonycteris curasoae*): management implications for the habitat conservation plan within the city of Tucson and the town of Marana. Arizona Game and Fish Department, Phoenix

Lyytimäki J, Rinne J (2013) Voices for the darkness: online survey on public perceptions on light pollution as an environmental problem. J Integr Environ Sci 10:127–139

Mann SL, Steidl RJ, Dalton VM (2002) Effects of cave tours on breeding *Myotis velifer*. J Wildl Manage 66:618–624

McDonald RI (2008) Global urbanization: can ecologists identify a sustainable way forward? Front Ecol Environ 6:99–104

Medellin RA, Gaona O (1999) Seed dispersal by bats and birds in forest and disturbed habitats of Chiapas, Mexico. Biotropica 31:478–485

Montevecchi WA (2006) Influences of artificial light on marine birds. In: Rich C, Longcore T (eds) Ecological consequences of artificial night lighting. Island Press, Washington, DC, pp 94–113

Müller B, Glösmann M, Peichl L et al (2009) Bat eyes have ultraviolet-sensitive cone photoreceptors. PLoS ONE 4(7):e6390

Muscarella R, Fleming TH (2007) The role of frugivorous bats in tropical forest succession. Biol Rev 82:573–590

Narendran N, Gu Y, Freyssinier JP et al (2004) Solid-state lighting: failure analysis of white LEDs. J Cryst Growth 268:449–456

NASA Earth Observatory/NOAA NGDC (2012) http://eoimages2.gsfc.nasa.gov/images/imagerecords/79000/79765/dnb_land_ocean_ice.2012.3600x1800.jpg. Accessed 1 June 2014

Neuweiler G (1989) Foraging ecology and audition in echolocating bats. Trends Ecol Evol 6:160–166

Norberg UM, Rayner JMV (1987) Ecological morphology and flight in bats (Mammalia; Chiroptera): wing adaptations, flight performance, foraging strategy and echolocation. Phil Trans R Soc Lond B 316:335–427

Nordt A, Klenke R (2013) Sleepless in town—drivers of the temporal shift in dawn song in urban European blackbirds. PLoS ONE 8(8):e71476
O'Farrell MJ, Gannon WL (1999) A comparison of acoustic versus capture techniques for the inventory of bats. J Mammal 80:24–30
Oakeley SF, Jones G (1998) Habitat around maternity roosts of the 55 kHz phonic type of pipistrelle bats (*Pipistrellus pipistrellus*). J Zool 245:222–228
Oprea M, Mendes P, Vieira TB et al (2009) Do wooded streets provide connectivity for bats in an urban landscape? Biodivers Conserv 18:2361–2371
Orbach DN, Fenton B (2010) Vision impairs the abilities of bats to avoid colliding with stationary objects. PLoS ONE 5(11):e13912
Parsons S, Jones G (2000) Acoustic identification of twelve species of echolocating bat by discriminant function analysis and artificial neural networks. J Exp Biol 203:2641–2656
Pavey CR (1999) Foraging ecology of the two taxa of large-eared horseshoe bat, *Rhinolophus philippinensis*, on Cape York Peninsula. Aust Mammal 21:135–138
Peitsch D, Fietz A, Hertel H et al (1992) The spectral input systems of hymenopteran insects and their receptor-based colour vision. J Comp Physiol A 170:23–40
Perry G, Buchanan BW, Fisher RN et al (2008) Effects of artificial night lighting on amphibians and reptiles in urban environments. Herpetol Conserv 3:239–256
Polak T, Korine C, Yair S et al (2011) Differential effects of artificial lighting on flight and foraging behaviour of two sympatric bat species in a desert. J Zool 285:21–27
Poot H, Ens BJ, de Vries H et al (2008) Green light for nocturnally migrating birds. Ecol Soc 13(2):47
Purves D, Lotto RB (2003) Why we see what we do: an empirical theory of vision. Sinauer Associates, Sunderland
Rich C, Longcore T (2006) Ecological consequences of artificial night lighting. Island Press, Washington, DC
Riegel KW (1973) Outdoor lighting is a growing threat to astronomy. Science 179:1285–1291
Royal Commission on Environmental Pollution (2009) Artificial light in the environment. Stationery Office, London
Rydell J (1991) Seasonal use of illuminated areas by foraging northern bats *Eptesicus nilssoni*. Ecography 14:203–207
Rydell J (1992) Exploitation of insects around streetlamps by bats in Sweden. Funct Ecol 6:744–750
Rydell J (2006) Bats and their insect prey at streetlights. In: Rich C, Longcore T (eds) Ecological consequences of artificial night lighting. Island Press, Washington, DC, pp 43–60
Rydell J, Racey PA (1995) Street lamps and the feeding ecology of insectivorous bats. Symp Zool Soc Lond 67:291–307
Rydell J, Jones G, Waters D (1995) Echolocating bats and hearing moths: who are the winners? Oikos 73:419–424
Rydell J, Entwistle A, Racey PA (1996) Timing of foraging flights of three species of bats in relation to insect activity and predation risk. Oikos 76:243–252
Safi K, Kerth G (2004) A comparative analysis of specialization and extinction risk in temperate-zone bats. Conserv Biol 18:1293–1303
Salmon M (2006) Protecting sea turtles from artificial night lighting at Florida's oceanic beaches. In: Rich C, Longcore T (eds) Ecological consequences of artificial night lighting. Island Press, Washington, DC, pp 141–168
Scanlon AT, Petit S (2008) Effects of site, time, weather and light on urban bat activity and richness: considerations for survey effort. Wildl Res 35:821–834
Schmitz H, Bleckmann H (1998) The photomechanic infrared receptor for the detection of forest fires in the beetle *Melanophila acuminata* (Coleoptera: Buprestidae). J Comp Physiol A 182:647–657
Schubert EF, Kim JK (2005) Solid-state light sources getting smart. Science 308:1274–1278
Shields WM, Bildstein KL (1979) Birds versus bats: behavioral interactions at a localized food source. Ecology 60:468–474

Speakman JR (1991) Why do insectivorous bats in Britain not fly in daylight more frequently? Funct Ecol 5:518–524
Spoelstra K, Visser ME (2014) The impact of artificial light on avian ecology. In: Gill D, Brumm H (eds) Avian urban ecology: behavioural and physiological adaptations. Oxford University Press, Oxford, pp 21–28
Stone EL (2011) Bats and development: with a particular focus on the impacts of artificial lighting. Dissertation, University of Bristol, Bristol
Stone EL, Jones G, Harris S (2009) Street lighting disturbs commuting bats. Curr Biol 19:1123–1127
Stone EL, Jones G, Harris S (2012) Conserving energy at a cost to biodiversity? Impacts of LED lighting on bats. Glob Change Biol 18:2458–2465
Stutz H-PB, Haffner M (1984) Summer colonies of *Vespertilio murinus* Linnaeus, 1758 (Mammalia: Chiroptera) in Switzerland. Myotis 22:109–112
Suthers RA (1970) Vision, olfaction and taste. In: Wimsatt WA (ed) Biology of bats, vol II. Academic Press, New York, pp 265–281
Svensson AM, Rydell J (1998) Mercury vapour lamps interfere with the bat defence of tympanate moths (*Operophtera* spp.; Geometridae). Anim Behav 55:223–226
Swift SM (1997) Roosting and foraging behaviour of Natterer's bats (*Myotis nattereri*) close to the northern border of their distribution. J Zool 242:375–384
The Climate Group (2014) Sydney LED trial: final report. http://www.theclimategroup.org/_assets/files/Sydney.pdf. Accessed 1 June 2014
Tuttle MD (1976) Population ecology of the gray bat (*Myotis grisescens*): factors influencing growth and survival of newly volant young. Ecology 57:587–595
Verboom B, Spoelstra K (1999) Effects of food abundance and wind on the use of tree lines by an insectivorous bat, *Pipistrellus pipistrellus*. Can J Zool 77:1393–1404
Verboom B, Boonman AM, Limpens HJGA (1999) Acoustic perception of landscape elements by the pond bat (*Myotis dasycneme*). J Zool 248:59–66
Vitousek PM, Mooney HA, Lubchenco J et al (1997) Human domination of earth's ecosystems. Science 277:494–499
Wang D, Oakley T, Mower J et al (2004) Molecular evolution of bat color vision genes. Mol Biol Evol 21:295–302
Winter Y, López J, von Helversen O (2003) Ultraviolet vision in a bat. Nature 425:612–614
Xuan F, Hu K, Zhu T et al (2012a) Immunohisochemical evidence of cone-based ultraviolet vision in divergent bat species and implications for its evolution. Comp Biochem Physiol B 161:398–403
Xuan F, Hu K, Zhu T et al (2012b) Behavioral evidence for cone-based ultraviolet vision in divergent bat species and implications for its evolution. Zoologia 29:109–114
Zhao H, Ru B, Teeling EC et al (2009a) Rhodopsin molecular evolution in mammals inhabiting low light environments. PLoS ONE 4:e8326
Zhao H, Rossiter SJ, Teeling EC et al (2009b) The evolution of color vision in nocturnal mammals. Proc Natl Acad Sci USA 106:8980–8985

Chapter 8
Bats and Water: Anthropogenic Alterations Threaten Global Bat Populations

Carmi Korine, Rick Adams, Danilo Russo, Marina Fisher-Phelps and David Jacobs

Abstract Natural bodies of open water in desert landscapes, such as springs and ephemeral pools, and the plant-life they support, are important resources for the survival of animals in hyper arid, arid and semi-arid (dryland) environments. Human-made artificial water sources, i.e. waste-water treatment ponds, catchments and reservoirs, have become equally important for wildlife in those areas. Bodies of open water are used by bats either for drinking and/or as sites over which to forage for aquatic emergent insects. Due to the scarcity of available water for replenishing water losses during roosting and flight, open bodies of water of many shapes and sizes may well be a key resource influencing the survival,

C. Korine (✉)
Mitrani Department of Desert Ecology, Jacob Blaustein Institute for Desert Research, Ben-Gurion University of the Negev, Sede Boqer Campus, Midreshet Ben-Gurion, 84990 Beersheba, Israel
e-mail: ckorine@bgu.ac.il

R. Adams
School of Biological Sciences, University of Northern Colorado, Greeley, CO 80639, USA
e-mail: battings@yahoo.com

D. Russo
Wildlife Research Unit, Laboratorio di Ecologia Applicata, Dipartimento di Agraria, Università degli Studi di Napoli Federico II, Via Università, 100, 80055 Portici, Napoli, Italy
e-mail: danrusso@unina.it

M. Fisher-Phelps
Department of Biological Sciences, Texas Tech University, Lubbock, TX 79409, USA
e-mail: m.fisher-phelps@ttu.edu

D. Jacobs
Department of Biological Sciences, University of Cape Town, 7701 Rondebosch, Republic of South Africa
e-mail: David.Jacobs@uct.ac.za

C.C. Voigt and T. Kingston (eds.), *Bats in the Anthropocene: Conservation of Bats in a Changing World*, DOI 10.1007/978-3-319-25220-9_8

activity, resource use and the distribution of insectivorous bats. In this chapter, we review the current knowledge of bats living in semi- and arid regions around the world and discuss the factors that influence their richness, behaviour and activity around bodies of water. We further present how increased anthropogenic changes in hydrology and water availability may influence the distribution of species of bats in desert environments and offer directions for future research on basic and applied aspects on bats and the water they use in these environments.

8.1 General Introduction

Dryland environments which include hyper-arid, arid and semi-arid regions can be highly complex and diverse, despite being occasionally perceived as simple ecosystems supporting low species diversity (Ayal et al. 2005). Aridity is described by ratio of precipitation to potential evapotranspiration ratio (P/ETP) (UNESCO 1979, Fig. 8.1) and dryland environments are ecosystems in which typically food availability is low, precipitation is limited and unpredictable, ambient temperature is high, humidity is low, and drinking water is scarce (Noy-Meir 1973). Consequently, there are large variations in primary production by plants that can strongly affect overall species diversity and interactions (Evenari et al. 1971). Furthermore, the distribution, abundance and persistence of several desert-dwelling mammal species is affected by water availability, especially during dry summer months, when the challenges of minimizing energy use and water losses is greatest (Calder 1984; Morton et al. 1995; Lovegrove 2000; Marom et al. 2006).

In desert environments, bats are an important component of the mammalian fauna. Carpenter (1969) asserted that, based on the number of species and abundance, bats are one of the most successful desert mammals, although they are outnumbered by rodents in the driest parts of the Sahara and the Namib Desert (Findley 1993). In the deserts of Israel, insectivorous bats are the most diverse

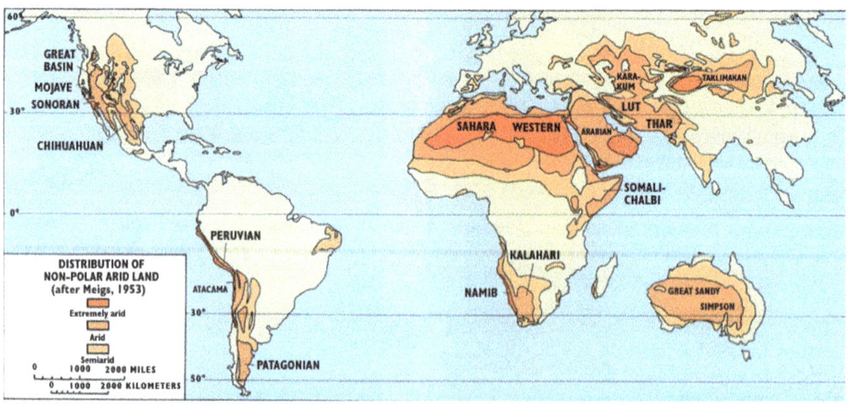

Fig. 8.1 The arid lands of the world (U.S. Geological Survey, science information services)

Fig. 8.2 A drinking event of the lesser horseshoe bat (*Rhinolophus hipposideros*) from a spring in the Dead Sea, Israel. Photo by Jens Rydell

group of mammals (Mendelssohn and Yom-Tov 1999), with 12 species recorded in the Negev Desert (Korine and Pinshow 2004) and 17 species in the Dead Sea area (Yom-Tov 1993). Benda et al. (2008) recorded 14 species of insectivorous bats in Sinai, highlighting the diversity of these mammals in desert environments. The dryland regions of South America are the most species-rich habitats of the region and have the highest number of endemic species, even when compared to the tropical lowland Amazon forest (Mares 1992; Ojeda and Tabeni 2009; Sandoval and Barquez 2013). In the Yungas dry forest of Argentina, 55 % of the bat species may be endemics (Sandoval et al. 2010). However, this area is severely under-protected and very little research has been conducted on the bat fauna (Mares 1992; Sandoval and Barquez 2013) In Mongolia, more than half of the bat species only occur in arid and semi-arid regions (Nyambayar et al. 2010).

Most bats, and in particular desert-dwelling bats, use open water sources for drinking water and/or as a foraging site (Vaughan et al. 1996; Grindal et al. 1999; Ciechanowski 2002; Campbell 2009, Fig. 8.2) with various studies reporting high levels of bat activity over open bodies of water (Rydell et al. 1994; Walsh et al. 1995; Young and Ford 2000; Mickeviciene and Mickevicius 2001; Ciechanowski 2002; Russo and Jones 2003; Korine and Pinshow 2004; Williams and Dickman 2004; Anderson et al. 2006; Davie et al. 2012; Monamy et al. 2013), making even small springs, ephemeral pools and waterholes key foraging areas for insectivorous bats worldwide (Racey 1998). Water availability was even proposed as a mechanism for elevational patterns of species richness of bats in arid mountains (McCain 2007).

In this chapter, we review our current knowledge of bats and water across regionally different semi-arid and dryland environments, and the factors that may influence their richness, behavior and activity around bodies of water. We discuss how anthropogenic development may influence water availability and thus the distribution of species of bats in desert environments. Dryland environments are also predicted to be particularly sensitive to climate change, and we will discuss patterns by which climate disruption may further reduce water availability in arid regions. Finally, we offer directions for future research on basic and applied aspects on bats and the water they use in these environments.

8.2 Ecology of Bats and Water in Drylands Environments

8.2.1 Water Sources Used by Bats

Permanent and ephemeral pools are the central characteristic of many watersheds in dry, arid and semi-arid regions. Temporary pools have largely been ignored in management programs due to their relatively small size and apparent lack of benefit for human use (Schwartz and Jenkins 2000). However, during spring and early summer, temporary pools may serve as important foraging grounds for aquatic and terrestrial species, some of which are regionally or locally rare and/or endemic (Nicolet et al. 2004). Temporary pools in the Negev Desert had equivalent levels of species richness of bats and activity to permanent pools (Razgour et al. 2010) and the activity of bats was reduced significantly when bodies of open water were dried (Korine and Pinshow 2004), highlighting the importance of pools of all shapes and sizes to desert wildlife. In the arid regions of Mongolia, even suboptimal water sources such as small human-dug wells and salty lakes are used by bats and are an important resource for their continued survival (Nyambayar et al. 2010). Conservation efforts should therefore focus on those sources offering only temporary water availability because although they support similar bat species richness and activity levels as permanent pools, they are less likely to be protected due to their ephemeral nature.

That said, the importance of permanent pools can be underestimated if landscape availability of water is not considered through time. Geluso and Geluso (2012) analyzed 34 years of data in relation to capture rates gathered at a single drinking site, which was sampled once yearly, in the San Mateo Mountains of New Mexico. They found that in non-drought years capture success was significantly lower because bats were more dispersed across the landscape. However, in drought years, capture rates at the only available water source skyrocketed, thereby indicating the importance of open-water to local species of bats.

Data gathered on foraging patterns of bats in Utah indicated a strong affinity by *Myotis* bats for riparian and edge habitats as compared to other surrounding areas (Rogers et al. 2006). Similarly, Grindal et al. (1999) showed that bat activity levels were significantly greater in riparian versus upland areas in British Columbia and capture rates were higher for females than for males indicating that female bats may be more dependent on water-driven attributes of a particular area. Williams et al. (2006) sampled across 22.5 km of the Muddy River floodplain in the Mojave Desert in Nevada, which was highly disturbed by long-standing flood control, livestock grazing, and the invasion of non-native plant species, and found that the riparian woodland habitat, which represents less than 1 % of the area, accounted for greater than 50 % of all bat activity. Areas of historically less disturbed mesquite bosque habitat maintained higher bat activity than more disturbed areas. Fortunately, restoration of habitats can increase local species richness. In Arizona, red bats (*Lasiurus blossevillii*), which had not been reported before, were captured along riparian-restoration areas of the lower Colorado River. The Arizona myotis (*Myotis occultus*), presumed extirpated, was also captured after restoration (Calvert 2012).

In Africa, there is evidence that bat activity is higher around bodies of water than in adjacent areas. For example, in two regions in southern Africa, bat abundance was higher in riverine habitat than in adjacent, dryer savannah (Rautenbach et al. 1996; Monadjem and Reside 2008). Differences in species richness and diversity between riverine and savannah habitats were not the same in the two regions. In the Kruger National Park, there was no difference in bat species richness or evenness between riverine habitat and savannah (Rautenbach et al. 1996). In contrast, at another site in Swaziland, the riverine habitat had higher species richness and diversity (Monadjem and Reside 2008). In both regions, the two assemblages differed in the relative densities of the various species, with the savannah assemblages forming a subset of the riverine assemblages (Rautenbach et al. 1996; Monadjem and Reside 2008). This reinforces the notion that bat assemblages in less mesic regions are extensions of bat assemblages in more mesic regions, but that not all species are inclined to make use of less mesic habitats when conditions are favorable. Some of them, particularly fruit eating bats (e.g. *Epomophorus crypturus*; Thomas and Fenton 1978) may be restricted to riverine habitats (Monadjem and Reside 2008).

Australian studies also indicate high levels of bat activity around bodies of water (Lumsden and Bennett 1995; Williams and Dickman 2004; Griffiths et al. 2014a). Young and Ford (2000) found that species richness of bats, abundance, and capture success in the semi-arid Idalia National Park was greatest in areas adjacent to water, with 97 % of captures occurring at sites with water. Bats in Uluru National Park and the north-eastern edge of the Simpson Desert predominantly use oasis habitats that have permanent or temporary water sources even in years with higher than average annual rainfall (Coles 1993; Williams and Dickman 2004). Multiple species of Australian insectivorous bats have even been recorded flying, foraging, and perhaps drinking over hypersaline environments (Laegdsgaard et al. 2004; Gonsalves et al. 2012; Griffiths et al. 2014a, b). *Pteropus* species in New Guinea have been recorded drinking seawater (Iudica and Bonaccorso 2003) but the prevalence of bats drinking hypersaline water in arid environments is not understood, despite natural hypersaline water bodies being common in arid and semi-arid areas in Western Australia (Halse et al. 2003; Timms 2005). In the arid regions of Mongolia, bats are mostly frequently found in association with water (Dolch et al. 2007; Davie et al. 2012).

8.2.2 Bodies of Water as a Drinking Source

Water sources that are used by bats are likely to be pools in streams, lakes, ponds, slow-flowing streams and rivers and artificial bodies of water with similar properties such as farm and urban dams (Jackrel and Matlack 2010; Sirami et al. 2013), canals (e.g. Lisón and Calvo 2011), cattle troughs, swimming pools and settling ponds at waste water treatment facilities (Vaughan et al. 1996; Abbott et al. 2009; Naidoo et al. 2013, 2014) and mines having natural seepage (Donato et al. 2007; Griffiths et al. 2014a).

Both the size and accessibility of the water source influence whether a bat can drink from it. Bats drink water by swooping over a water source while lapping at the surface (Harvey et al. 1999). Because bats drink on the wing, small and more maneuverable bats are able to drink from smaller pools, whereas less maneuverable bats need a large surface area of water to skim (Tuttle et al. 2006). In the Negev Desert, Razgour et al. (2010) found that both within and between pools, species richness of bats and activity significantly increased with pond size. Furthermore, manipulations that decreased pond size led to a significant reduction in species richness and activity and affected the bat assemblage composition. The size and situation of artificial water sources similarly affect their use by bats. In the arid Texas Panhandle, USA, bats preferentially drank water from larger livestock tanks that were full and had only light vegetation around. They tended to avoid smaller, half-full tanks with denser vegetation around them (Jackrel and Matlack 2010). Although there are many anecdotal observations (Nickerson and O'Keefe 2013) of bats drinking from swimming pools there have been no formal studies of this.

Despite the central nature of drinking and water availability for bats, there are a surprisingly small number of studies addressing this topic in Europe, even though many species do drink at open water sources regularly to rehydrate (e.g. Russo et al. 2012). Some appear more sensitive than others to water deprivation because of their stricter dependence on water habitats. For instance, in water-denial experiments Daubenton's bat, *Myotis daubentonii*, a species selectively dwelling in riparian habitat and above bodies of open water, has been found to undergo a greater body mass loss and to show signs of dehydration earlier than the brown long-eared bat, *Plecotus auritus*, a forest bat (Webb et al. 1995). Drinking sites are also of chief importance for European bats outside the semiarid Mediterranean region. In the Bavarian Forest, Germany, oligotrophic, acidic ponds are used by over a dozen species of bats for drinking (Seibold et al. 2013). Likewise, in the Italian Apennines, water cattle troughs built for traditional livestock breeding are frequently used to drink by over a dozen species of bats. Such small (often less than 15 $\times$ 1.5 m) pools of water are locally of extreme importance (Russo et al. 2010, 2012) for several threatened species (Fig. 8.3). These pools also concentrate insects, so bats occasionally forage there, but their importance for drinking is overwhelming (Russo et al. 2012). The disappearance of traditional livestock breeding due to rural depopulation in many Apennine areas has led to the abandonment of the cattle troughs, implying an unstudied yet potentially high cost for bat

Fig. 8.3 Cattle troughs used by drinking bats in the Italian Apennines. Photo by Luca Cistrone

populations (Fig. 8.3). In Italian forests, bats also drink from the small ephemeral pools which form following heavy rain and only last few days or weeks (D. Russo, pers. obs.). Eavesdropping on other drinking bats is likely to play an important role in locating such sites and this behaviour is typical of species with manoeuvrable flight such as the barbastelle bat, *Barbastella barbastellus*, and the greater horseshoe bat *Rhinolophus ferrumequinum*.

8.2.3 Bodies of Water as a Foraging Habitat

The tendency for higher insect abundance near water sources attracts bats to use water sources as foraging habitats. Furthermore, calm surface water provides a less cluttered acoustic signal return from the echolocation pulses (Mackey and Barclay 1989; Siemers et al. 2001), and there is some evidence, at least for echolocating bats, that activity over calm pools of water is higher than that over fast-flowing riffles (von Frenckell and Barclay 1987). Bat activity in a transect from dry woodland savannah to riverine habitat in southern Africa was correlated with insect abundance—both bat activity and insect abundance were higher in riverine habitat (Rautenbach et al. 1996) suggesting that bats were attracted to this habitat because of the feeding opportunities it provided.

Drought is known to reduce the abundance of insects in temperate zones (Frampton et al. 2000) and thus affect reproduction in insectivorous bats (Rhodes 2007). An eight year study by Bogan and Lytle (2011) on aquatic insects living in two study pools of a formerly perennial desert stream in the Whetstone Mountains of Arizona, USA, showed that complete water loss followed by intermittent flow caused a catastrophic regime shift in community structure that did not recover to the pre-drying configuration even after four years. Ledger et al. (2011) found significant reduction in and suppression of secondary productivity by drought that could have severe constraining effects on terrestrial vertebrate predator populations, and Love et al. (2008) found similar effects in Arkansas, USA. Furthermore, desert bats in Arizona responded to artificial-light-induced food patches (Fenton and Morris 1975) and one would presume this would be similar when small pools of water create swarms of high insect density. All of these data together suggest that small water sources with intermittent flow are vitally important as foraging sites to at least some insectivorous desert bat species.

In Europe, three species of bats are aquatic habitat specialists: Daubenton's bat, *M. daubentonii*, the long-fingered bat, *Myotis capaccinii*, and the pond bat, *Myotis dasycneme*. Besides taking insects in flight by aerial hawking, they typically forage very close to the water surface, from which prey is gaffed with their large feet or the inter-femoral membrane and transferred to the mouth while on the wing (Kalko and Schnitzler 1989; Siemers et al. 2001). Chironomidae and Trichoptera are frequent prey items of these bats (e.g. Biscardi et al. 2007; Krüger et al. 2012). *M. capaccinii* may seize adult chironomids from the water surface as they emerge from pupal casings. Trawling bats mainly forage over calm water whose surface is free from ripples (Rydell et al. 1999) as echoes from clutter interfere with prey detection (Siemers and Schnitzler 2004). On windy nights, *M. capaccinii* and *M. daubentonii* are less active (Russo and Jones 2003), presumably because wind reduces prey density and generates ripples on the water surface affecting target detection. In such circumstances, bats forage at sheltered sites where water is calmer (Lewis and Stephenson 1966; Lewis 1969).

Several other species of bats frequent riparian habitats to forage and/or drink, especially the soprano pipistrelle, *Pipistrellus pygmaeus* (e.g. Nicholls and Racey 2006), Nathusius' pipistrelle, *Pipistrellus nathusii* (Flaquer et al. 2009), and other *Pipistrellus* spp. (Scott et al. 2010), Schreiber's bat *Miniopterus schreibersii* (Serra-Cobo et al. 2000) and noctules, *Nyctalus* spp. (Rachwald 1992; Racey 1998; Vaughan et al. 1997). The stricter reliance on riparian habitats is one of the main ecological factors distinguishing *P. pygmaeus* from its sibling *P. pipistrellus* (but see Warren et al. 2000) and allowing interspecific niche partitioning and thus coexistence (Oakeley and Jones 1998; Nicholls and Racey 2006; Davidson-Watts et al. 2006; Sattler et al. 2007). However, local factors such as elevation or landscape composition may influence differences across species. At larger scales, the presence of main rivers and wetland areas are important as migratory paths and offer important stopover sites to migrating bats across Europe (Flaquer et al. 2009). Rivers and riparian vegetation also constitute important linear landscape elements used for navigation by several European bats (Serra-Cobo et al. 2000; Russo et al. 2002).

As might be expected given the above, the quality of foraging areas lacking water is influenced by their distance to water. In Portugal, proximity to a drinking water source increased foraging habitat quality for Mehely's horseshoe bat *Rhinolophus mehelyi* and *M. schreibersii* (Rainho and Palmeirim 2011). Similarly, a radio-tracking study of *R. mehelyi* in Spain showed that although this species hunted predominately in forest, the foraging areas were always within 500 m of a water source (Salsamendi et al. 2012), possibly to allow for easy rehydration between foraging bouts or perhaps to take advantage of water-emergent forest insects. In historic landscape parks of England (Glendell and Vaughan 2002) as well as in German forests (Kusch and Idelberger 2005) the relative area of available water surface is an effective proxy for levels of bat activity.

Australian bats have also been documented preferentially foraging around water sources. When compared to other habitat types in the Simpson Desert, more feedings buzzes were recorded around permanent and temporary water sources (Williams and Dickman 2004). Bats will also forage over hypersaline water bodies but more feeding buzzes are recorded over freshwater sites (Griffiths et al. 2014b). There is also evidence (e.g. Aldridge and Rautenbach 1987; Schoeman and Jacobs 2003, 2011; Naidoo et al. 2011, 2013) that insects associated with freshwater habitats (e.g. Plecoptera, Ephemeroptera and Trichoptera) occur in the diet of southern African bats.

8.2.4 Water, Roosts and Reproduction

The propensity for female bats to choose roost sites that are relatively high in ambient temperature is thought to help them save metabolic energy by allowing for continued gestation of the young during torpor (Speakman et al. 1991; Adams and Thibault 2006; Daniel et al. 2010). The cost of such a choice in roost sites in arid regions, however, is the propensity for high-levels of evaporative water loss during the diurnal roosting cycle (Webb 1995) and this is further exacerbated when females are lactating (Kurta et al. 1990). The only quantitative field study to assess the need for drinking water by lactating female bats in drylands used PIT-tagged lactating and non-reproductive females from a maternity colony of fringed myotis (*Myotis thysanodes*) in Colorado, USA. Adams and Hayes (2008) found that lactating females visited to drink an average of seven times more per night than did non-breeding adult females. In addition, lactating females visited to drink consistently night after night regardless of daily relative humidity and temperatures, whereas non-reproductive females visited more when temperatures were high and relative humidity low (Adams and Hayes 2008).

In addition, Adams (2010) synthesized 13 years of capture data from the same field sites in Colorado, USA and found that summer mean precipitation had the highest correlation with reproductive frequency followed closely by mean stream discharge rates. Of these two, the latter showed the most abrupt effect on bat reproduction. When stream discharge rates were lower than 7 m/s, the frequency of reproductively active females captured plummeted, in some years by as much as 50 %. When female reproductive condition was plotted against mean stream discharge, the frequency of lactating females tracked the amount of available water, whereas the frequency of pregnant females was not correlated. This suggests that during drought years pregnant females may give birth, but do not have access to enough drinking water to support lactation. O'Shea et al. (2010) using mark/recapture of big brown bats, *Eptesicus fuscus*, at maternity colonies in Ft. Collins, Colorado, USA found that first year survival was lowest in bats born during a drought year, although other factors were also at play.

Several species of bats have been found to roost close to bodies of water to minimize the energy expenditure required to reach important drinking or foraging sites (Racey 1998; Korine et al. 2013). The need to drink directly after emerging from the roosts may be the main factor determining the proximity of roosts to water, especially for maternity colonies (Racey 1998). *M. daubentonii*, whose foraging strictly depends on water habitat, often uses bridges over rivers, as well as buildings or cavity-bearing trees in the immediate surroundings of riparian biotopes (Racey 1998; Parsons and Jones 2003; Lučan and Radil 2010; Encarnação 2012). Several other species, such as Natterer's bat (*Myotis nattereri*), pipistrelles (*Pipistrellus* spp.) and brown long-eared bat, also tend to roost in landscapes comprising bodies of water that provide drinking and foraging opportunities (Racey 1998; Entwistle et al. 1997; Oakeley and Jones 1998). Floodplain forests of central Europe host important reproductive colonies of tree-roosting noctule bat *Nyctalus noctula* (Görföl et al. 2009). *Myotis macropus*, an Australian species, has a variable roosting behaviour but the primary force behind roost selection is proximity to waterways (Campbell 2009).

8.3 Threats to Water Sources Used by Bats

In drylands, where water resources are scarce, any loss of or degradation to open water source, such as a reduction in water quality, may create cascading affects that will be harmful to the wildlife that depends on it. When bats drink from a polluted source they ingest toxins directly and during foraging they indirectly ingest toxins that may have bio-accumulated within their insect prey. For example, if insect larvae feed on microorganisms in polluted water, they concentrate the pollutants in their bodies and when they metamorphose into adults these are consumed by bats. The effect of environmental chemical containments on bats was reviewed in 2001; most studies have occurred in Europe (~50 %) and North America (~34 %) mostly pertaining to organochlorine insecticides (58 %), metals (30 %), and polychlorinated biphenyls or PCBs (13 %) (Clark and Shore 2001). There are hardly any reports on the effect of polluted water on bat activity and richness in the drylands of North Africa, the Middle East and South America. Levels of bat activity in the Negev Desert were very high over wastewater treatment ponds (Korine and Pinshow 2004), however species richness was low and the majority of the activity was attributed to Kuhl's pipistrelle (*Pipistrellus kuhlii*). Pilosof et al. (2013) showed that sewage pollution in the Negev desert affected the immune response of Kuhl's pipistrelle and Naidoo et al. (2014) reported on DNA damage to bats that forage at wastewater treatments work.

8.3.1 Loss of Sources of Water

An estimated two-thirds of Earth's freshwater flowing to oceans is obstructed by anthropogenic development (Nilsson and Berggren 2000), with approximately 75,000 dams in the USA alone and the majority of natural wetlands having been destroyed as well. Although not the scope of this chapter, it is important to mention that for bats, wetlands provide critical foraging habitat (Johnson et al. 2008; Rainey et al. 2006) with absolute area and connectivity of wetlands being important components for foraging (Lookingbill et al. 2010).

Indeed, a recent report on total wetland loss in the USA from 2004–2009, showed a 25 % reduction from the previous reporting period. In addition, a total of 95,000 acres of saltwater wetlands and 265,720 acres of freshwater wetlands were lost (Dahl and Stedman 2013). The situation is exacerbated in the western USA, where livestock grazing has damaged at least 80 % of stream and riparian ecosystems (Belsky and Matzke 1999). The consequences for bats are illustrated by observed declines in bat activity as related to flow-reduction and drying along the San Pedro River in Arizona. Moreover, these declines corresponded to declines in insect availability at perennial sites and both bat activity and insect activity declined to imperceptible levels in areas where the river dried up (Hagen and Sabo 2012).

European rivers, lakes and wetlands are among the most seriously altered ecosystems. Human impact has caused a major structural or chemical degradation of such ecosystems with fatal repercussions for their associated biota (e.g. Abel 1996). Alteration of European rivers has often led to the loss of channel features, floodplain connectivity and structure of bank vegetation. A threatened vespertilionid, *M. capaccinii*, selects foraging sites where water is less polluted and riparian vegetation better preserved. Along with the loss or disturbance of suitable cave roosts (Papadatou et al. 2008), riparian habitat alteration poses the main threat to this bat (Biscardi et al. 2007).

Australian rivers have the highest variation in flow and flooding in the world (Williams 1981; Puckridge et al. 1988). Anthropogenic activities such as extraction and diversion of water have had adverse impacts on rivers in the arid-zone of Australia (Walker 1985; Kingsford and Thomas 1995). High natural variation in water availability coupled with anthropogenic activities and climate change has the potential to catastrophically affect arid-species that depend on water availability (Roshier et al. 2001; McKenzie et al. 2007; Saunders et al. 2013).

A major concern associated with natural rivers and lakes in urban areas is that they may be polluted by runoff from roads or other sources. When bats drink from these sources, they ingest these pollutants directly or indirectly by feeding on aquatic-emergent insects. Sources of pollution of farm and golf course dams include feces from livestock and wild animal, nitrate and phosphate in fertilizers, metals, pathogens, sediments and pesticides. Unfortunately, little research has been done on the use of polluted urban water sources by bats and the probable health impacts on bats. The little evidence that does exist suggests that at least some species of bats may not avoid polluted bodies of water in arid areas (Pilosof et al. 2013; Korine et al. 2015). In Durban, South Africa bat abundance and species richness were higher over a polluted than over an unpolluted river and bat feeding activity (measured by feeding buzzes in the echolocation sequences) was also higher at the polluted river. There was, however, no difference in insect diversity between the two rivers (Naidoo et al. 2011) and, with the exception of a single species, Rufous mouse-eared bat, *Myotis bocagii*, proportions of prey items in the diets of bats did not correspond to their proportion in the insect fauna. *M. bocagii* fed predominantly on Diptera and this was also the most abundant insect in the insect light traps (Naidoo et al. 2011).

8.3.2 Mining

Mining is a major anthropogenic source of environmental destruction and contamination globally. Toxins associated with extensive mining operations, in particular, gold mining is well documented. Cyanide used to extract gold from ore is commonly stored in open ponds, some of which are 200 acres in size. The actual numbers of bats, and other wildlife killed by drinking at these ponds is poorly understood and very difficult to track as many affected individuals either become submerged, or die from drinking contaminated water after leaving the site. Between 1980 and 1989, 34 % of all known mammals killed at cyanide ponds used for mining gold in California, Nevada, and Arizona were bats (Clark and Hothem 1991).

Other heavy metals used in mining operations such as arsenic, cadmium, chromium, copper, lead, mercury, methyl mercury, nickel, and zinc have been found in bat carcasses. In Arizona, USA where at least 20 % of bat populations are in decline (King et al. 2001), Mexican free-tailed bats (*Tadarida brasiliensis*) living 8 km from a major copper smelting mine had accumulated significant levels of atmospheric mercury in their tissues (Petit 2007). In another study in Arizona, pallid bats (*Antrozous pallidus*), western pipistrelles (*Parastrellus hesperus*), and *T. brasiliensis* had elevated mercury levels in their liver and muscles that they most likely acquired via drinking from contaminated free-water sources (Reidinger 1972; see also Syaripuddin et al. 2014).

Besides contaminated ponds, natural water flows through thousands of abandoned mines in the western USA (used by bats for hibernaculum and maternity roosts) may be highly contaminated with heavy metals. For example, at Sheep Tank Mine overlooking the Colorado River in Arizona, barium, manganese and zinc were detected in soil samples at concentrations 10 times normal levels and *E. fuscus* captured at the site had higher concentrations of these elements than those collected from three other sites (King et al. 2001). Other species included in the study had high arsenic levels as well as other contaminants (copper, lead, barium, manganese, and zinc) (King et al. 2001). Bats and other terrestrial vertebrates can also be exposed to high levels of contaminants by ingesting aquatic emergent insects living in toxic streams and High levels of bioaccumulated cadmium and zinc are known to occur as far as 381 km downstream from the pollution source, whereas lead was found to be transferred from sediments to chironomids (midges) only as far as 40 km downstream (Cain et al. 1992). Thus, large stretches of streams and rivers far from the point source of contamination pose threats to bats and other aquatic and terrestrial wildlife.

Bats are also known to fly and possibly forage/drink over gold mines in Australia (Donato and Smith 2007; Smith et al. 2008). High bat activity was recorded over gold mine water bodies containing cyanide (Griffiths et al. 2014a). Griffiths et al. (2014b) suggested that elevated salt levels in water bodies at gold mines may decrease bat activity, foraging, and drinking. Bats, including the Vulnerable (IUCN 2014) ghost bat, *Macroderma gigas*, have also been recorded around an Australian copper mine in the Great Sandy Desert, although the mine's effects on individuals or the population is unknown (Read 1998).

Africa is rich in mineral resources and this makes mining activities relatively common so likely a serious threat to water quality and therefore to bats. A matter of grave concern is that no research has been done in Africa in this regard. This situation prevails despite evidence that mining activities do pollute surface water in Africa (Olade 1987; Naicker et al. 2003).

8.3.3 Agriculture

Organochlorine pollution of streams and rivers, and other sources, is of major concern for bats (see Bayat et al. 2014 for review). Experimental testing of organochlorine insecticides such as DDT on two species widely distributed throughout the USA, found that *Myotis lucifugus* was approximately twice more sensitive than were *E. fuscus*. Furthermore, juvenile *E. fuscus* were 1.5 times more sensitive than adults (Clark et al. 1978). In addition, tests showed that individuals of *T. brasiliensis* poisoned with DDT survived for some time but later died of DDT poisoning mobilized from fat during active flight after being starved (Clark et al. 1975). Laboratory studies also show that presence of organochlorine in tissues can accelerate the catabolism of fat, causing DDE-dosed bats (*M. lucifugus*) to lose weight faster than control bats (Clark and Stafford 1981). Although banned in the USA in 1972, significant levels of DDT and DDE have been documented in tissues collected from bats foraging and drinking at the Rocky Mountain Arsenal Superfund Site (O'Shea et al. 2001). High DDT concentrations are also found in *M. lucifugus* tissues in the Eastern United States (Kannan et al. 2010). Furthermore, post-ban persistence of DDT in USA bats has been verified by sampling guano at roost sites (Clark et al. 1982; Reidinger and Cockrum 1978; Bennett and Thies 2007). DDT has also been found in bat tissues in Australia despite being banned since 1987 (Mispagel et al. 2004; Allinson et al. 2006). DDT for agricultural use was essential banned worldwide in 2001, but recent work from Africa showed that DDT is probably still being used and accumulating in the tissues of multiple species of bats (Stechert et al. 2014).

The two most common agricultural pollutants are nitrogen and phosphorus and sources of these pollutants include inorganic and organic fertilizers, leguminous crops, septic tanks, farm and municipal waste water treatment facilities, and, in the case of phosphorous, run-off from groundwater discharge and atmospheric deposition. An excess of these nutrients is the leading cause of aquatic eutrophication (Shabalala et al. 2013). Inorganic pollutants such as metals from agricultural and industrial run-off can also accumulate in these sites as well as in the tissues of insects using these bodies of water. Bats feedings on such insects are thus at risk of ingesting high levels of toxic metals such cadmium, chromium and nickel (see Naidoo et al. 2013).

8.3.4 Waste Water

European bats foraging in aquatic habitats are known to be largely exposed to toxic heavy metals which bioaccumulate in their insect food (Pikula et al. 2010). Organic pollution of rivers is also known to affect bat foraging, but its effects are variable. A British study compared the differences in bat activity found respectively upstream and downstream from sewage outputs and showed that downstream activity of pipistrelle bats decreased whereas that of *M. daubentonii* increased relative to upstream sites (Vaughan et al. 1996). The latter species is thought to benefit from the higher downstream abundance of pollution-tolerant prey such as chironomids. However, an Irish study obtained opposite results, with *P. pygmaeus* being more common downstream of sewage effluent discharges than *M. daubentonii* (Abbott et al. 2009). Park and Cristinacce (2006) compared the effects of two types of sewage treatment works for foraging bats: those with percolating filter beds, often hosting many insects potentially important for bats, and the "activated sludge" system—gradually replacing the former—in which sewage and bacteria-laden sludge are mixed and agitated so that they prove inhospitable for the invertebrate fauna. The study showed that both insect biomass and bat activity were higher at percolating filter beds and that bat activity there was comparable to that recorded at nearby natural foraging habitats. However, bats may run serious risks when foraging at such sewage treatment works: endocrine disrupting chemicals, which may alter the endocrine functions in exposed animals, have been found to concentrate in bat insect prey at percolating filter beds, with potentially harmful effects on foraging bats (Park et al. 2009).

There has been very little research in Africa on the concentration of pollutants in tissues of bats and no work on the long and short term effects of these pollutants on the health of bats. There is some evidence of the presence of the toxic metals cadmium, chromium and nickel in tissues of African bats foraging at sites downstream of waste water treatment plants (Naidoo et al. 2013). Furthermore, bats foraging over waste water treatment facilities display increased haematocrit and DNA damage and decreased antioxidant capacity in muscle tissue compared to bats that forage over unpolluted sites. Although these effects were not lethal they may result in long-term negative effects on the health of bats (Naidoo et al. 2014). These metals were probably ingested by bats via their insect prey.

There is evidence that aerial insects developing in sewage sludge and waste water at sewage treatment plants can accumulate pollutants that could disrupt endocrine functioning (Park et al. 2009). However, a similar study on the activity of the insectivorous bat, the banana bat, *Neoromicia nana*, at three urban rivers systems above and downstream of where sewage effluent enters these rivers revealed that the relative abundance and feeding activity of *N. nana* were higher at polluted sites downstream of where sewage entered the system than at the unpolluted sites upstream (Naidoo et al. 2013). In this case the bats may have been attracted by the higher abundance of dipterans over the polluted sites. Diptera were the dominant prey items in both the insect fauna at the polluted sites and in the diets of the bats (Naidoo et al. 2013). This also appeared to be the case for *M. bocagii* which also fed predominantly and opportunistically on Diptera (Naidoo et al. 2011).

The response by bats to rivers affected by waste water treatment effluent may vary both between and within species. In North America (Kalcounis-Rueppell et al. 2007) and England (Vaughan et al. 1996), some species were more active upstream from where waste water effluent entered the rivers while others were more active downstream. It appears that these differences arise from the differential effects of euthrophication on insect prey as well as on the responses of bats. Some species take advantage of eutrophication that causes an increase in the abundance of their preferred prey, and other species which apparently do not feed on insects that are affected by eutrophication, prefer to forage in less polluted habitats. Furthermore, these differences may also result from differences in the foraging behavior of the same species at different sites. For example, *N. nana* fed opportunistically on the small abundant dipterans at wastewater polluted sites, but at unpolluted river sites fed selectively on insects from other orders (Naidoo et al. 2013).

Another major anthropogenic compound found in open bodies of water in the USA is polychlorinated biphenyl or PCB, a common industrial waste product that was banned by the United States in 1979 and the United Nations in 2001. PCB poisoning in pregnant *M. lucifugus* led to stillborn young (Clark and Krynitsky 1978). Aquatic-emergent insects are key exporters of contaminants to terrestrial ecosystems (Menzie 1980; Runck 2007) and data show significant lateral transfers of PCBs to terrestrial riparian predators such as spiders, reptiles and amphibians (Walters et al. 2008). High concentrations of PCB's have been found in fat tissues of *M. lucifugus* in New York and Kentucky (Kannan et al. 2010). Along the fresh water tidal river, the Biesbosch, in the Netherlands, direct transfer from river sediments to chironomids to pond bats occurred in concentrations known to cause negative reproductive effects in mink (Reinhold et al. 1999). Frick et al. (2007) investigated the effects of an accidental chemical spill (metam sodium) on Yuma myotis (*Myotis yumanensis*) in California and found reduced female juvenile survival, but not adult female survival. The spill-affected population declined significantly during the first years of the study. Although the population increased in year four, this also coincided with an end to an extensive regional drought. Controlled experimental exposure to Lindane (an organochlorine used in wood preservatives) at sublethal levels in *P. pipstrellus* increased 24 h metabolic rates of a 7.3 g individual by 15 % and in a 6.3 g individual by 23 %, thereby posing a significant threat to survivorship of free-living individuals (Swanepoel et al. 1999) and showing that sub-lethal exposure can affect energetic balance.

8.4 Mitigation and Restoration

Both the availability and distribution of water in drylands have been drastically altered by natural processes such as decline in annual precipitation, and by anthropogenic developments such as irrigation for agriculture, over exploitation of groundwater and human-induced climate changes.

8.4.1 Restoration of Water Sources and Related Habitats

Most wetlands have been altered globally due to anthropogenic disruption, pollution, and outright destruction. In some, but too few, places, humans have begun to restore some of those wetlands. For example, in the USA, the Sierra Nevada Conservancy is working in cooperation with State Parks, the Department of Toxic Substances, California State University Chico and others, to identify mercury sources and potential remediation strategies for an abandoned hydraulic mine discharging sediment and heavy metals into the Yuba River and removing mercury from dredged sediment that have accumulated in the Combie Reservoir.

In California, restoration of the Cosumnes River floodplain re-established bat activity that broadly corresponded with flooding and an increase in aquatic emergent insects (Rainey et al. 2006). Furthermore restoration of riparian habitat, frequently damaged by cattle as well as other anthropogenic uses, and wetlands commonly destroyed by human development, is essential and is occurring in some areas, but well below necessary levels for bat conservation (Goodwin et al. 1997).

Despite some of the negative effects highlighted in the previous section concerning waste water effluent, wastewater reclamation is an important process especially in areas where water is scare (Anderson et al. 2001). Wastewater can be used to construct artificial wetlands that provide habitat for wildlife if the water is properly treated (Greenway and Simpson 1996; Fujioka et al. 1999; Greenway and Woolley 1999; Greenway 2005). Some studies have found that increased nutrient loads, such as those caused by wastewater effluent may have a positive effect on insect and bat abundance both in US and European streams (Kokurewicz 1995; Vaughan et al. 1996; Abbott et al. 2009). One US study found that bat activity and foraging levels were the same up-stream and down-stream of wastewater discharge but community structure was altered, with the riparian-specialist *Perimyotis subflavus* being more abundant (Kalcounis-Rueppell et al. 2007).

8.4.2 *Artificial Water Sources*

One way to overcome the diminishing of natural water sources in many drylands is the development of artificial catchments which are widely used for wildlife management (Krausman et al. 2006). There has long been controversy regarding the effects of catchments on local wildlife, in which critics argue that these developments do not yield expected benefits to game species and may have opposing impacts such as predation (O'Brien et al. 2006).

Small artificial ponds may be of utmost importance for wildlife (Russo et al. 2012). The large-scale expansion of intensive agriculture in semiarid Mediterranean climates has often been sustained by hydraulic engineering works, to cope with the scarcity of natural irrigation water. In southeastern Spain, Lisón and Calvo (2011) studied the effects on bats of a water transfer channel and a related network of irrigation ponds in a mixed landscape of traditional and intensive agricultural landscape. In general, artificial bodies of water had a positive effect on bat activity, but this mainly regarded common, generalist species (*P. pipistrellus* and *P. pygmaeus*) most likely because of the absence of foraging habitats suitable for more specialized species (those bearing a higher conservation value) such as riparian vegetation. In Catalonia, rice paddies sustain high bat activity, providing large amounts of insect prey. However, roost availability was the main limiting factor and installing bat boxes represents a valuable strategy to increase bat populations (Flaquer et al. 2006). In the arid Ikh Nart Nature Reserve in Mongolia, significantly more bats were caught at natural springs relative to human-made wells and no bats were captured at sites without water (Davie et al. 2012). This suggests that at least for this area, replacing lost natural water sources with artificial ones may not be as effective for preserving bat populations as conserving natural water sources.

Paradoxically, the creation of large water reservoirs may prove harmful to the entire bat community. Rebelo and Rainho (2009) looked at the effects on bats of the largest reservoir in Europe, created by construction in 2001 of the Alqueva dam, in Alentejo, Southern Portugal. The project led to the deforestation and submersion of an area of ca. 250 km^2. Consequently, bat populations were affected by the sudden disappearance of ca. 200 km of riparian habitat, together with large-scale roost loss and the replacement of important habitat with a vast homogeneous one which was not used by foraging bats. Noticeably, bat activity showed a strong decline in the submerged areas but increased in the surrounding unaffected habitat.

The expansion of Mediterranean species into surrounding arid wildlife communities may have a negative impact on local populations such as competition for the use of pools for drinking and foraging. Nine of the 12 Negev species of bats (Korine and Pinshow 2004) are associated with arid areas, and the Kuhl's pipistrelle, the European free-tailed bat (*Tadrida teniotis*), and the rare lesser horseshoe bat (*Rhinolophus hipposideros*)—are Mediterranean species that have expanded their distribution into the Negev in the twentieth century (Yom-Tov and Mendelssohn 1988). The most common bat in some desert habitats and in particular at artificial water sites in the Negev is Kuhl's pipistrelle (Korine and Pinshow 2004). The expanded distribution is probably linked to human settlements and in particular to artificial bodies of water since non-desert species of bats must drink on a daily basis and drink more frequently compared with desert-dwelling bats (Razgour 2010). Kuhl's pipistrelle competes for the use of pools for drinking and foraging, resulting in temporal and spatial partitioning between local desert bat species (Razgour et al. 2011). The documented competition between Kuhl's pipistrelle and desert-dwelling bat species (Polak et al. 2011; Razgour et al. 2011), combined with the increasing development of bodies of open water in the Negev and other drylands, may lead to further resource competition resulting in loss to the region's biodiversity. Korine et al. (2015) have shown that species richness and activity of desert dwelling bats did not differ between artificial and natural bodies of water in the Negev desert, however several species of bats drank or foraged only at natural bodies of water.

8.5 Conclusion and Future Directions

Human population growth, land use change and habitat loss have led to massive habitat alterations and destruction, particularly of water sources in arid regions. The availability of water (temporary/permanent) appears to have a strong positive influence on species of bats richness and activity. This suggests that large temporary pools are important for the conservation of bats in arid environments. A reduction in the availability of temporary pools, due to intensification of arid conditions, is expected to predominantly affect species of bats that forage over water, and will most likely increase interspecific competition for foraging space above the pools. These problems are likely be exacerbated in species of bats that are able to extend into arid areas because of their association with humans. Studies on the distribution of bats in drylands on a large scale should be the focus of future research to understand how climate change and introduction of artificial bodies of water effect species distribution, activity and richness. Studies are strongly needed in arid regions to understand the best and most efficient way to provide safe artificial water sources for bats that can mitigate increased incidences of drought due to climate change and, in some cases, the total loss of available water, especially in the more temperate arid regions with shorter growing seasons. For example, placement of artificial water sources near maternity roosts is instrumental in arid temperate areas with shorter growing seasons (Adams 2010). However, the introduction of artificial bodies of water may promote invasion by non-native species and range expansion of others, leading to resource competition. In regions of Europe likely to become water-stressed because of human induced climate changes, bats may be affected as they may lack the physiological means to cope with water limitation (Sherwin et al. 2013).

Africa, as well as other arid areas such as the Negev and the Mongolian deserts, has a high diversity of bats but compared to other areas of the world its bat fauna has been little studied. Fundamental research is most needed throughout Africa and other arid zones on how often bats need to drink and whether this varies across species, geographically and seasonally. Comparative studies on bats with distributions restricted to arid regions and species that have populations in mesic and arid regions would be particularly informative in this regard. For example, the diversity of renal capacities and habitat use amongst African species of bats of the same family (Happold and Happold 1988), and the emergence of robust family level phylogenies (e.g. Stoffberg et al. 2010) provide an excellent opportunity to study the evolution of renal form and function in African bats in an ecological context. Special focus should be placed on research determining the extent to which African bats are reliant on artificial water sources. Such research should target arid zone species of bats, especially those species that live in close association with humans because these are the species likely to be impacted by insufficient or polluted water sources.

Research is also needed on whether all water sources are used for both drinking and foraging and how bats respond to decreases in water quality as a result of pollutants. Do certain species of bats avoid drinking from low quality bodies of water as shown by Korine et al. (2015)? Would bats still use polluted bodies of water for feeding but not for drinking? If so, how do they detect low quality water, do they do so before they are adversely affected by it and do they have alternative water sources? How are desert-dwelling bats affected by pollutants in water or by waterborne toxins and pollutants in the insect fauna, and are such bats able to deal with such pollutants physiologically?

Although least is known about bats and water in sub-Sahara Africa, studies thus far in other regions of the world are in their infancy in terms of understanding the long-term effects of decreased water availability on bat and other wild populations. Due to human destruction of wetlands and riparian habitats as well as unsustainable human population growth that more and more is utilizing greater amounts of fresh water, availability of fresh water to sustain wildlife populations are reaching critically low levels, especially in areas suffering from extended droughts due to human-induced climate disruption. Because water is a key ingredient of all life, focus on this topic needs to increase and because bats act as 'canaries in a global coal mine,' studies concerning bats and water are key to better management of water resources in natural and artificial areas.

Acknowledgments We would like to acknowledge four anonymous reviewers and Dr. Tigga Kingston for their helpful comments on the chapter. RAA thanks the University of Northern Colorado, the Department of Open Space and Boulder Mountain Parks, Boulder County Department of Parks and Open Space for providing funding for research. This is paper number 871 of the Mitrani Department of Desert Ecology.

References

Abbott IM, Sleeman DP, Harrison S (2009) Bat activity affected by sewage effluent in Irish rivers. Biol Cons 142:2904–2914
Abel PD (1996) Water pollution biology. Taylor and Francis, London
Adams RA (2010) Bat reproduction declines when conditions mimic climate change projectionsfor western North America. Ecology 91:2437–2445
Adams RA, Hayes MA (2008) Water availability and successful lactation by bats as related to climate change in western regions of North America. J Anim Ecol 77:1115–1121
Adams RA, Thibault KM (2006) Temporal partitioning by bats at water holes. J Zool 270:466–472
Aldridge HDJN, Rautenbach IL (1987) Morphology, echolocation and resource partitioning in insectivorous bats. J Anim Ecol 56:763–778
Allinson G, Mispagel C, Kajiwara N, Anan Y, Hashimoto J, Laurenson L, Allinson M, Tanabe S (2006) Organochlorine and trace metal residues in adult southern bent-wing bat (*Miniopterus schreibersii bassanii*) in southeastern Australia. Chemosphere 64:1464–1471
Anderson J, Adin A, Crook J, Davis C, Hultquist R, Jimenez-Cisneros B, Kennedy W, Sheikh B, van der Merwe B (2001) Climbing the ladder: a step by step approach to international guidelines for water recycling. Water Sci Technol 43:1–8
Anderson J, Law B, Tidemann C (2006) Stream use by the large-footed myotis *Myotis macropus* in relation to environmental variables in northern New South Wales. Aust Mammal 28:15–26
Ayal Y, Polis GA, Lubin Y, Goldberg DE (2005) How can high animal diversity be supported in low productivity deserts: the role of macrodetrivory and habitat physiognomy. In: Shachak M, Gosz JR, Pickett STA, Perevolotsky A (eds) Chapter 2 in Biodiversity in Drylands: towards a unified framework. Oxford University Press, Oxford, pp. 15–29
Bayat S, Geiser F, Kristainsen P et al (2014) Organic contaminants in bats: trend and new issues. Env Inter 63:40–52
Belsky AJ, Matzke A (1999) Survey of livestock influences on stream and riparian ecosystems in the western United States. J Soil Water Cons 54:419–431
Benda P, Dietz AM et al (2008) Bats (Mammalia: Chiroptera) of the eastern mediterranean and middle east. Part 6. Bats of Sinai (Egypt) with some taxonomic, ecological and echolocation data on that fauna. Acta Soc Zool Bohem 72:1–103
Bennett BS, Thies ML (2007) Organochlorine pesticide residues in guano of Brazilian free-tailed bats, *Tadarida brasiliensis*, Saint-Hilaire, from East Texas. B Environ Contam Tox 78:191–194
Biscardi S, Russo D, Casciani V et al (2007) Foraging requirements of the endangered long-fingered bat (*Myotis capaccinii*): the influence of micro-habitat structure, water quality and prey type. J Zool 273:372–381
Bogan MT, Lytle DA (2011) Severe drought drives novel community trajectories in desert stream pools. Freshwater Biol 56:2070–2081
Cain DJ, Luoma SN, Carter JL et al (1992) Aquatic insects as bioindicators of trace element contamination in cobble-bottom rivers and streams. Can J Fish Aquat Sci 49:2141–2154
Calder WA (1984) Size, function and life-history. Harvard University Press, Cambridge, Massachusetts
Calvert AW (2012) Bats in riparian-restoration sites along the lower Colorado River, Arizona. Southwest Nat 57:340–342
Campbell S (2009) So long as it's near water: variable roosting behaviour of the large-footed myotis (*Myotis macropus*). Aust J Zool 57:89–98
Carpenter RE (1969) Structure and function of the kidney and the water balance of desert bats. Physiol Zool 42:288–302
Ciechanowski M (2002) Community structure and activity of bats (Chiroptera) over different bodies of open water. Mammal Biol 67:276–285

Clark DR Jr, Hothem RL (1991) Mammal mortality in Arizona, California, and Nevada gold mines using cyanide extraction. Calif Fish Game 77:61–69

Clark DR Jr, Krynitsky AJ (1978) Organochlorine residues and reproduction in the little brown bat, Laurel, Maryland—June 1976. Pestic Monit J 12:113–116

Clark DR Jr, Kunz TH, Kaiser TE (1978) Insecticides applied to a nursery colony of little brown bats (*Myotis lucifugus*): lethal concentrations in brain tissues. J Mammal 59:84–91

Clark DR Jr, LaVal RK, Tuttle MD (1982) Estimating pesticide burdens of bats from guano analyses. Bull Environ Contam Toxicol 29:214–220

Clark DR Jr, Martin CO, Swineford DM (1975) Organochlorine insecticide residues in the free-tailed bat (*Tadarida brasiliensis*) at Bracken Cave, Texas. J Mammal 56:429–443

Clark DR Jr, Shore RF (2001) Chiroptera. In: Shore RF, Rattner BA (eds) Ecotoxicology of wild mammals. John Wiley, Chichester, pp p159–p214

Clark DR Jr, Stafford CJ (1981) Effects of DDE and PCB (Aroclor 1260) on experimentally poisoned female little brown bats (*Myotis lucifugus*): lethal brain concentrations. J Toxicol Environ Health 7:925–934

Coles R (1993) Survey of bats of Uluru (Ayers Rock- Mount Olga) National Park. Final Report to the Australian National Parks and Wildlife Service, Canberra

Dahl TE, Stedman SM (2013) Status and trends of wetlands in the coastal watersheds of the Conterminous United States 2004 to 2009. U.S. Department of the Interior, Fish and Wildlife Service and National Oceanic and Atmospheric Administration, National Marine Fisheries Service, 46 p

Daniel S, Korine C, Pinshow B (2010) The use of torpor in reproductive female Hemprich's long-eared bats (Otonycteris hemprichii). Physiol Biochem Zool 183:142–148

Davidson-Watts I, Walls S, Jones G (2006) Differential habitat selection by *Pipistrellus pipistrellus* and *Pipistrellus pygmaeus* identifies distinct conservation needs for cryptic species of echolocating bats. Biol Cons 133:118–127

Davie HS, Murdoch JD, Naran NE, Ariunbold J, Batdorj S, Reading RP (2012) Bat diversity at Ikh Nart nature reserve, Mongolia. Mongolian J Biol Sciences 10:1–2

Dolch D, Batsaikhan N, Thiele K, Burger F, Scheffler I, Kiefer A, Mayer F et al (2007) Contributions to the Chiroptera of Mongolia with first evidences on species communities and ecological niches. Erforschung biologischer Ressourcen der Mongolischen Volksrepublik. In Schriftenreihe der Martin-Luther Universität Halle pp 407–458

Donato DB, Nichols O, Possingham H et al (2007) A critical review of the effects of gold cyanide-bearing tailings solutions on wildlife. Environ Int 33:974–984

Donato DB, Smith GB (2007) Australian centre for mine and environmental research project 58, Anglogold Ashanti Australia Sunrise Dam gold mine sponsor's report: summary of findings. Australian Centre for Mine and Environmental Research, Perth

Encarnação JA (2012) Spatiotemporal pattern of local sexual segregation in a tree-dwelling temperate bat *Myotis daubentonii*. J Ethol 30:271–278

Entwistle AC, Racey PA, Speakman JR (1997) Roost selection by the brown long-eared bat (*Plecotus auritus*). J Appl Ecol 34:399–408

Evenari M, Shanan L, Tadmor N (1971) The Negev: the challenge of a desert. Harvard University Press, Cambridge

Fenton MB, Morris GK (1975) Opportunistic feeding by desert bats (Myotis spp.). Can J Zool 54:526–530

Findley JS (1993) Bats: a community perspective. Cambridge University Press, Cambridge

Flaquer C, Puig-Montserrat X, Goiti U et al (2009) Habitat selection in Nathusius' pipistrelle (*Pipistrellus nathusii*): the importance of wetland. Acta Chiropterol 11:149–155

Flaquer C, Torre I, Ruiz-Jarillo R (2006) The value of bat-boxes in the conservation of *Pipistrelluspygmaeus* in wetland rice paddies. Biol Cons 128:223–230

Frampton GK, Van Den Brink PJ, Gould PJL (2000) Effects of spring drought and irrigation on farmland arthropods in southern Britain. J App Ecol 37:865–883

Frick WF, Rainey WE, Pierson ED (2007) Potential effects of environmental contamination on *Yuma myotis* demography and population growth. Ecol Appl 17:1213–1222

Fujioka RS, Bonilla AJ, Rijal GK (1999) The microbial quality of a wetland reclamation facility used to produce an effluent for unrestricted non-potable reuse. Water Sci Technol 40:369–374

Geluso KN, Geluso K (2012) Effects of environmental factors on capture rates of insectivorous bats, 1971–2005. J Mammal 93:161–169

Glendell M, Vaughan N (2002) Foraging activity of bats in historic landscape parks in relation to habitat composition and park management. Anim Cons 5:309–316

Gonsalves L, Law B, Webb C, Monamy V (2012) Are vegetation interfaces important to foraging insectivorous bats in endangered coastal saltmarsh on the central coast of New South Wales? Pac Conserv Biol 18:282–292

Goodwin GN, Hawkins CP, Kershner JL (1997) Riparian restoration in the western United States: overview and perspective. Restor Ecol 5:4–14

Görföl T, Dombi I, Boldogh S et al (2009) Going further south: new data on the breeding area of *Nyctalusnoctula* (Schreber, 1774) in Central Europe. Hystrix It J Mamm (n.s.) 20:37–44

Greenway M (2005) The role of constructed wetlands in secondary effluent treatment and water reuse in subtropical and arid Australia. Ecol Eng 25:501–509

Greenway M, Simpson JS (1996) Artificial wetlands for wastewater treatment, water reuse and wildlife in Queensland, Australia. Water Sci Technol 33:221–229

Greenway M, Woolley A (1999) Constructed wetlands in Queensland: performance efficiency and nutrient bioaccumulation. Ecol Eng 12:39–55

Griffiths SR, Donato DB, Coulson G, Lumsden LF (2014a) High levels of activity of bats at gold mining water bodies: implications for compliance with the international cyanide management code. Environ Sci Pollut Res. doi:10.1007/s11356-014-2651-z

Griffiths SR, Donato DB, Lumsden LF, Coulson G (2014b) Hypersalinity reduces the risk of cyanide toxicosis to insectivorous bats interacting with wastewater impoundments at gold mines. Ecotox Environ Safe 99:28–34

Grindal SD, Morissette JL, Brigham RM (1999) Concentration of bat activity in riparian habitats over an elevational gradient. Can J Zool 77:972–977

Hagen EM, Sabo JL (2012) Influence of river drying and insect availability on bat activity along the San Pedro River, Arizona (USA). J Arid Environ 84:1–8

Halse SA, Ruprecht JK, Pinder AM (2003) Salinisation and prospects for biodiversity in rivers and wetlands of south-west Western Australia. Aust J Bot 51:673–688

Happold M, Happold DCD (1988) Renal form and function in relation to the ecology of bats (Chiroptera) from Malawi, Central Africa. J Zool 215:629–655

Harvey MJ, Altenbach JS, Best TL (1999) Bats of the United States. Arkansas Game and Fish Commission

IUCN 2014. The IUCN Red List of Threatened Species. Version 2014.3. <http://www.iucnredlist.org>. Downloaded on 21st April 2015

Iudica CA, Bonaccorso FJ (2003) Anecdotal observations of seawater ingestion by flying foxes of the genus *Pteropus* (Chiroptera: Pteropodidae). Mammalia 67:455–458

Jackrel SL, Matlack RS (2010) Influence of surface area, water level and adjacent vegetation on bat use of artificial water sources. Am Midi Nat 164:74–79

Johnson JB, Gates JE, Ford WM (2008) Distribution and activity of bats at local and landscape scales within a rural-urban gradient. Urban Ecosyst 11:227–242

Kalcounis-Rueppell MC, Payne VH, Huff SR, Boyko AL (2007) Effects of wastewater treatment plant effluent on bat foraging ecology in an urban stream system. Biol Conserv 138:120–130

Kalko EKV, Schnitzler H-U (1989) The echolocation and hunting behavior of Daubenton's bat, *Myotis daubentoni*. Behav Ecol Sociobiol 24:225–238

Kannan K, Yun SH, Rudd RJ, Behr M (2010) High concentrations of persistent organic pollutants including PCBs, DDT, PBDEs and PFOS in little brown bats with white-nose syndrome in New York, USA. Chemosphere 80:613–618

King KA, Velasco AL, Record JA et al (2001) Contaminants in bats roosting in abandoned mines at Imperial National Wildlife Refuge, Arizona, 1998–1999. USFWS Region 2, Contaminants Program Report, 35 p

Kingsford RT, Thomas RF (1995) The Macquarie Marshes in arid Australia and their waterbirds: a 50-year history of decline. Environ Manage 19:867–878
Kokurewicz T (1995) Increased population of Daubenton's bat (*Myotis daubentonii* Kuhl, 1819) (Chiroptera: Vespertilionidae) in Poland. Myotis 32:155–161
Korine C, Pinshow B (2004) Guild structure, foraging space use, and distribution in a community of insectivorous bats in the Negev Desert. J Zool 262:187–196
Korine C, Daniel S, Pinshow B (2013) Roost selection by females Hemprich's long-eared bat. Behav Process 100:131–138
Korine C, Adams A, Shamir U, Gross A (2015) Effect of water quality on species richness and activity of desert-dwelling bats. Mamm Biol. doi:10.1016/j.mambio.2015.03.009
Krausman PR, Rosenstock SS, Cain JW III (2006) Developing waters for wildlife: science perception, values, and controversy. Wildlife Soc Bull 34:563–569
Krüger F, Harms I, Fichtner A et al (2012) High trophic similarity in the sympatric North European trawling bat species *Myotis daubentonii* and *Myotisdasycneme*. Acta Chiropterol 14:647–656
Kurta A, Kunz TH, Nagy KA (1990) Energetics and water flux of free-ranging big brown bats (*Eptesicus fuscus*) during pregnancy and lactation. J Mammal 71:59–65
Kusch J, Idelberger S (2005) Spatial and temporal variability of bat foraging in a western European low mountain range forest. Mammalia 69:21–33
Laegdsgaard P, McDougall A, Monamy V, Saintilan N (2004) Investigating the presence of threatened insectivorous bats on coastal NSW saltmarsh habitat. Wetlands 22:29–41
Ledger M, Edwards FK, Brown LE et al (2011) Impact of simulated drought on ecosystem biomass production: an experimental test in stream mesocosms. Glob Change Biol 17:2288–2297
Lewis T (1969) The diversity of the insect fauna in a hedgerow and neighbouring fields. J Appl Ecol 6:453–458
Lewis T, Stephenson JW (1966) The permeability of artificial windbreaks and the distribution of flying insects in the leeward zone. Ann Appl Biol 58:355–363
Lisón F, Calvo JF (2011) The significance of water infrastructures for the conservation of bats in a semi-arid Mediterranean landscape. Anim Cons 14:533–541
Lookingbill TR, Elmore AJ, Engelhardt KA et al (2010) Influence of wetland networks on bat activity in mixed-use landscapes. Biol Cons 143:974–983
Love JW, Taylor CM, Warren MP Jr (2008) Effects of summer drought on fish and macroinvertebrate assemblage properties in upland Ouachita Mountain Streams, USA. Am Mid Nat 160:265–277
Lovegrove BG (2000) The zoogeography of mammalian basal metabolic rate. Am Nat 156:201–219
Lučan RK, Radil J (2010) Variability of foraging and roosting activities in adult females of Daubenton's bat (*Myotis daubentonii*) in different seasons. Biologia 65:1072–1080
Lumsden IF, Bennett AF (1995) Bats of a semi-arid environment in south-eastern Australia: biogeography, ecology and conservation. Wildlife Res 22:217–239
Mackey RL, Barclay RMR (1989) The influence of physical clutter and noise on the activity of bats over water. Can J Zool 67:1167–1170
Mares MA (1992) Neotropical mammals and the myth of Amazonian biodiversity. Science 255:976–979
Marom S, Korine C, Wojciechowski MS et al (2006) Energy metabolism and evaporative water loss in the European free-tailed bat and Hemprich's long-eared bat (Microchiroptera): species sympatric in the Negev. Physiol Biochem Zool 79:944–956
McCain CM (2007) Could temperature and water availability drive elevational species richness patterns? A global case study for bats. Global Ecol Biogeogr 16:1–13
McKenzie NL, Burbidge AA, Baynes A, Brereton RN, Dickman CR, Gordon G, Gibson LA, Menkhorst PW, Robinson AC, Williams MR, Woinarski JCZ (2007) Analysis of factors implicated in the recent decline of Australia's mammal fauna. J Biogeogr 34:597–611

Mendelssohn H, Yom-Tov Y (1999) A report of birds and mammals which have increased their distribution and abundance in Israel due to human activity. Isr J Zool 45:35–47

Menzie CA (1980) Potential significance of insects in the removal of contaminants from aquatic systems. Water Air Soil Pollut 13:473–479

Mickeviciene I, Mickevicius E (2001) The importance of various habitat types to bats (Chiroptera: Vespertilionidae) in Lithuania during the summer period. Acta Zool Lituanica 11:3–14

Mispagel C, Allinson M, Allinson G, Iseki N, Grant C, Morita M (2004) DDT and metabolites residues in the southern bent-wing bat (*Miniopterus schreibersii bassanii*) of south-eastern Australia. Chemosphere 55:997–1003

Monadjem A, Reside A (2008) The influence of riparian vegetation on the distribution and abundance of bats in an African savannah. Acta Chiropterol 10:339–348

Monamy V, Taylor JE, Gonsalves L, Ellis MV (2013) Bassian bats in the Eyrean sub-region? Preliminary acoustic data from the Bogan and Macquarie Rivers in semi-arid NSW. Aust J Zool 36:461–469

Morton SR, Short J, Barker RD (1995) Refugia for biological diversity in arid and semi-arid Australia. Environment Australia, Biodiversity Unit, Canberra

Naicker K, Cukrowska E, McCarthy TS (2003) Acid mine drainage arising from gold mining activity in Johannesburg, South Africa and environs. Environ Pollut 122:29–40

Naidoo S, Mackey RL, Schoeman MC (2011) Foraging ecology of insectivorous bats at a polluted and an unpolluted river in an urban landscape. Durban Mus Novit 34:21–28

Naidoo S, Vosloo D, Schoeman MC (2013) Foraging at wastewater treatment works increases the potential for metal accumulation in an urban adapter, the banana bat (*Neoromicia nana*). Afri Zool 48:39–55

Naidoo S, Vosloo D, Schoeman MC (2014) Haematological and genotoxic responses in an urban adapter, the banana bat, foraging at wastewater treatment works. Ecotox Env Safe 114:304–311

Nicholls B, Racey PA (2006) Habitat selection as a mechanism of resource partitioning in two cryptic bat species *Pipistrellus pipistrellus* and *Pipistrellus pygmaeus*. Ecography 29:697–708

Nickerson Z, O'Keefe (2013) Bats and pools: assessing how bats use pools across North America. http://www.batsandpools.wordpress.com

Nicolet P, Biggs J, Fox G et al (2004) The wetland plant and macroinvertebrate assemblages of temporary ponds in England and Wales. Biol Cons 120:261–278

Nilsson C, Berggren K (2000) Alterations of riparian ecosystems caused by river regulation. BioSci 50:783–792

Noy-Meir I (1973) Desert ecosystems: environments and producers. Ann Rev Ecol Syst 4:25–51

Nyambayar B, Ariunbold J, Sukhchuluun G (2010) A contribution to the bats inhabiting arid steppe habitats in central Mongolia. Erforsch Biol Ress Mongolei 11:329–340

Oakeley SF, Jones G (1998) Habitat around maternity roosts of the 55 kHz phonic type of pipistrelle bats (*Pipistrellus pipistrellus*). J Zool 245:222–228

O'Brien CS, Waddell RB, Rosenstock SS, Rabe MJ (2006) Wildlife use of water catchments in southwestern Arizona. Wildlife Soc Bull 34:582–591

Ojeda RA, Tabeni S (2009) The mammals of the Monte Desert revisited. J Arid Environ 73:173–181

Olade MA (1987) Heavy metal pollution and the need for monitoring: illustrated for developing countries in West Africa. In: Hutchinson TC, Meema KM (eds) Lead, mercury, cadmium and arsenic in the environment. Wiley, New York, pp 335–341

O'Shea TJ, Ellison LE, Neubaum DJ et al (2010) Recruitment in a Colorado population of big brown bats: breeding probabilities, litter size, and first-year survival. J Mammal 91:418–442

O'Shea TJ, Everette AL, Ellison LE (2001) Cyclodiene insecticide, DDE, DDT, Arsenic, and Mercury contamination of big brown bats (*Eptesicus fuscus*) foraging at a Colorado Superfund site. Arch Environ Contam Toxicol 40:112–120

Papadatou E, Butlin RK, Altringham JD (2008) Seasonal roosting habits and population structure of the long-fingered bat *Myotis capaccinii* in Greece. J Mammal 89:503–512
Park KJ, Cristinacce A (2006) The use of sewage treatment works as foraging sites by insectivorous bats. Anim Cons 9:259–268
Park KJ, Muller CT, Markman S et al (2009) Detection of endocrine disrupting chemicals in aerial invertebrates at sewage treatment works. Chemosphere 77:1459–1464
Parsons KN, Jones G (2003) Dispersion and habitat use by *Myotis daubentonii* and *Myotis nattereri* during the swarming season: implications for conservation. Anim Cons 6:283–290
Petit MG (2007) A chronolocal record of environmental chemical from analysis /of stratified vertebrate excretion deposition in a sheltered environment. Env Res 6:339–343
Pikula J, Zukal J, Adam V et al (2010) Heavy metals and metallothionein in vespertilionid bats foraging over aquatic habitats in the Czech Republic. Environ Toxicol Chem 29:501–506
Pilosof S, Korine C, Moore M, Krasnov B (2013) Sewage-water contamination alters immune response in a desert bat. Mamm Biol 79:183–188
Polak T, Korine C, Yair S, Holderied MW (2011) Differential effects of artificial lighting on flight and foraging behaviour of two sympatric bat species in a desert. J Zool 285:21–27
Puckridge JT, Sheldon F, Walker KF, Boulton AJ (1998) Flow variability and the ecology of large rivers. Mar Freshwater Res 49:55–72
Racey P (1998) The importance of the riparian environment as a habitat for British bats. Symp Zool Soc Lond 71:69–91
Rachwald A (1992) Habitat preference and activity of the noctule bat *Nyctalusnoctula* in the Białowieża Primeval Forest. Acta Theriol 37:413–422
Rainey WE, Power ME, Clinton SM (2006) Temporal and spatial variation in aquatic insect emergence and bat activity in a restored floodplain wetland. Final Report to CALFED
Rainho A, Palmeirim JM (2011) The importance of distance to resources in the spatial modelling of bat foraging habitat. PLoS ONE 6(4):e19227
Rautenbach IL, Whiting MJ, Fenton MB (1996) Bats in riverine forests and woodlands: a latitudinal transect in southern Africa. Can J Zool 74:312–322
Razgour O, Korine C, Saltz D (2010) Pond characteristics as determinants of diversity and community structure in desert bats. Anim Cons 13:505–513
Razgour O, Korine C, Saltz D (2011) Does interspecific competition drive patterns of habitat use in desert bat communities? Oecologia 167:493–502
Read J (1998) Vertebrate fauna of the Nifty mine site, Great Sandy Desert, with comments on the impacts of mining and rehabilitation. West Aust Nat 22:1–22
Rebelo H, Rainho A (2009) Bat conservation and large dams: spatial changes in habitat use caused by Europe's largest reservoir. Endang Species Res 8:61–68
Reidinger, RF Jr (1972) Factors influencing Arizona bat population levels. Unpublished Ph.D. dissertation Univ. Arizona, Tucson. 172 p
Reidinger RF Jr, Cockrum EL (1978) Organochlorine residues in free-tailed bats (*Tadarida brasiliensis*) at Eagle Creek Cave, Greenlee County, Arizona. In: Olembo RJ, Castelino JB, Mutere FA (eds) Proceedings of the 4th international bat research conference, Kenya Literature Bureau, Nairobi (pp 85–96)
Reinhold JO, Hendriks AJ, Slager LK et al (1999) Transfer of microcontaminants from sediment to chironomids, and the risk for the Pond bat *Myotis dasycneme* (Chiroptera) preying on them. Aquat Ecol 33:363–376
Rhodes M (2007) Roost fidelity and fission-fusion dynamics of white-striped free-tailed bats (*Tadarida australis*). J Mammal 88:1252–1260
Rogers DS, Belk MC, González MW et al (2006) Patterns of habitat use by bats along a riparian corridor in northeastern Utah. Southwest Nat 51:52–58
Roshier DA, Whetton PH, Allan RJ, Robertson AI (2001) Distribution and persistence of temporary wetland habitats in arid Australia in relation to climate. Austral Ecol 26:371–384
Runck C (2007) Macroinvertebrate production and food web energetics in an industrially contaminated stream. Ecol Appl 17:740–753

Russo D, Cistrone L, Garonna AP et al (2010) Reconsidering the importance of harvested forests for the conservation of tree-dwelling bats. Biodivers Cons 9:2501–2515
Russo D, Cistrone L, Jones G (2012) Sensory ecology of water detection by bats: a field experiment. PLoS ONE 7(10):e48144
Russo D, Jones G (2003) Use of foraging habitats by bats in a Mediterranean area determined by acoustic surveys: conservation implications. Ecography 26:197–209
Russo D, Jones G, Migliozzi A (2002) Habitat selection by the Mediterranean horseshoe bat, *Rhinolophus euryale* (Chiroptera: Rhinolophidae) in a rural area of southern Italy and implications for conservation. Biol Cons 107:71–81
Rydell J, Bushby A, Cosgrove CC, Racey PA (1994) Habitat use by bats along rivers in north-east Scotland. Folia Zool 43:417–424
Rydell J, Miller LA, Jensen ME (1999) Echolocation constraints of Daubenton's bat foraging over water. Funct Ecol 13:247–255
Salsamendi E, Garin I, Arostegui I et al (2012) What mechanism of niche segregation allows the coexistence of sympatric sibling rhinolophid bats? Front Zool 9:30
Sandoval ML, Barquez RM (2013) The Chacoan bat fauna identity: patterns of distributional congruence and conservation implications. Revista Chilena de Historia Natural 86:75–94
Sandoval ML, Szumik CA, Barquez RM (2010) Bats and marsupials as indicators of endemism in the Yungas forest of Argentina. Dongwuxue Yanjiu 31:633–644
Sattler T, Bontadina F, Hirzel A et al (2007) Ecological niche modelling of two cryptic bat species calls for a reassessment of their conservation status. J Appl Ecol 44:1188–1199
Saunders DA, Wintle BA, Mawson PR, Dawson R (2013) Egg-laying and rainfall synchrony in an endangered bird species: Implications for conservation in a changing climate. Biol Conserv 161:1–9
Schoeman CM, Jacobs DS (2003) Support for the allotonic frequency hypothesis in an insectivorous bat community. Oecologia 134:154–162
Schoeman CM, Jacobs DS (2011) The relative influence of competition and prey defences on the trophic structure of animalivorous bat ensembles. Oecologia 166:493–506
Schwartz SS, Jenkins DG (2000) Temporary aquatic habitats: constraints and opportunities. Aquat Ecol 34:3–8
Scott SJ, McLaren G, Jones G et al (2010) The impact of riparian habitat quality on the foraging and activity of pipistrelle bats (*Pipistrellus* spp.). J Zool 280:371–378
Seibold S, Buchner J, Bässler C, Müller J (2013) Ponds in acidic mountains are more important for bats in providing drinking water than insect prey. J Zool 290:302–308
Serra-Cobo J, López-Roig M, Marques-Bonet T et al (2000) Rivers as possible land-marks in the orientation flight of *Miniopteruss chreibersii*. Acta Theriol 45:347–352
Shabalala AN, Combrinck L, McCrindle R (2013) Effect of farming activities on seasonal variation of water quality of Bonsma Dam, KwaZulu-Natal. S Afri J Sci 109:01–07
Sherwin HA, Montgomery WI, Lundy MG (2013) The impact and implications of climate change for bats. Mamm Rev 43:171–182
Siemers B, Schnitzler HU (2004) Echolocation signals reflect niche differentiation in five sympatric bat species. Nature 429:657–658
Siemers BM, Stilz P, Schnitzler H-U (2001) The acoustic advantage of hunting at low heights above water: behavioural experiments on the European 'trawling' bats *Myotis capaccinii*, *M. dasycneme* and *M. daubentonii*. J Exp Bio 204:3843–3854
Sirami C, Jacobs DS, Cumming GS (2013) Artificial wetlands and surrounding habitats provide important foraging habitat for bats in agricultural landscapes in the Western Cape, South Africa. Biol Cons 164:30–38
Smith DB, Donato DB, Griffiths SR, Gillespie CG, Rowntree J (2008) Ecology of a goldmining tailings facility in the Eastern Goldfields of Western Australia: a case study. Int J Min Reclam Environ 22:154–173
Speakman JR, Racey PA, Catto CC et al (1991) Minimum summer populations and densities of bats in N. E. Scotland, near the borders of their distributions. J Zool 225:327–345

Stechert C, Kolb M, Bahadir M, Djossa BA, Fahr J (2014) Insecticide residues in bats along a land use-gradient dominated by cotton cultivation in northern Benin, West Africa. Environ Sci Pollut R 21:8812–8821
Stoffberg SM, Jacobs DS, Mackie IJ, Matthee CA (2010) Molecular phylogenetics and historical biogeography of Rhinolophus bats. Mol Phylogenet Evol 54:1–9
Swanepoel RE, Racey PA, Shore RF et al (1999) Energetic effects of sublethal exposure to lindane on pipistrelle bats (*Pipistrellus pipistrellus*). Environ Poll 104:169–177
Syaripuddin K, Kumar A, Sing K W, Halim MRA, Nursyereen MN, Wilson JJ (2014) Mercury accumulation in bats near hydroelectric reservoirs in Peninsular Malaysia. Ecotoxicology 23:1164–1171
Thomas DW, Fenton MB (1978) Notes on the dry season roosting and foraging behaviour of *Epomophorous gambianus* and *Roussettus aegyptiacus* (Chiroptera: Pteropodidae). J Zool 186:403–406
Timms BV (2005) Salt lakes in Australia: present problems and prognosis for the future. Hydrobiologia 552:1–15
Tuttle SR, Chambers CL, Theimer TC (2006) Potential effects of livestock water-troughs modifications on bats in Northern Arizona. Wildlife Soc Bull 34:602–608
UNESCO (1979) Map of the World Distribution of Arid Regions. UNESCO, Paris
Vaughan N, Jones G, Harris S (1996) Effects of sewage effluent on the activity of bats (Chiroptera: Vespertilionidae) foraging along rivers. Biol Cons 78:337–343
Vaughan N, Jones G, Harris S (1997) Habitat use by bats (Chiroptera) assessed by means of a broad-band acoustic method. J Appl Ecol 34:716–730
von Frenckell B, Barclay RMR (1987) Bat activity over calm and turbulent water. Can J Zool 65:219–222
Walker KF (1985) A review of the ecological effects of river regulation in Australia. In: Davies BR, Walmsley RD (eds) Perspectives in Southern Hemisphere Limnology. Springer Netherlands, pp. 111–129
Walsh AL, Harris S, Hutson AM (1995) Abundance and habitat selection of foraging vespertilionid bats in Britain: a landscape-scale approach. Symp Zoo Soc Lond 67:325–344
Walters DM, Fritz KM, Otter RR (2008) The dark side of subsidies: adult stream insects export organic contaminants to riparian predators. Ecol Appl 18:1835–1841
Warren RD, Water DA, Altringham JD et al (2000) The distribution of Daubenton's bats (*Myotis daubentonii*) and pipistrelle bats (*Pipistrellu spipistrellus*) (Vespertilionidae) in relation to small-scale variation in riverine habitat. Biol Cons 92:85–91
Webb PI (1995) The comparative ecophysiology of water balance in micro-chiropteran bats. In: Racey PA Swift SM (eds) Ecology, evolution and behaviour of bats. Oxford University Press, Oxford, pp 203–218
Webb PI, Speakman JR, Racey PA (1995) Evaporative water loss in two sympatric species of vespertilionid bat, *Plecotus auritus* and *Myotis daubentoni*: relation to foraging mode and implications for roost site selection. J Zool 235:269–278
Williams AJ, Dickman CR (2004) The ecology of insectivorous bats in the Simpson Desert central Australia: habitat use. Aust Mammal 26:205–214
Williams WD (1981) Running water ecology in Australia. In: Lock MA, Dudley D (eds) Perspectives in Running Water Ecology. Springer US, pp. 367–392
Williams JA, O'Farrell MJ, Riddke BR (2006) Habitat use by bats in a riparian corridor of the Mojave desert in southern Nevada. J Mammal 87:1145–1153
Yom-Tov Y (1993) Character displacement among the insectivorous bats of the Dead Sea area. J Zool 230:347–356
Yom-Tov Y, Mendelssohn H (1988) Changes in the distribution and abundance of vertebrates in Israel during the 20th century. In: Tchernov E, Yom-Tov Y (eds) The zoogeography of Israel. Dr. W. Junk, Dordrecht, pp 515–547
Young RA, Ford GI (2000) Bat fauna of a semi-arid environment in central western Queensland, Australia. Wildlife Res 27:203–215

Part II
Emerging Diseases

Chapter 9
White-Nose Syndrome in Bats

Winifred F. Frick, Sébastien J. Puechmaille and Craig K.R. Willis

Abstract White-nose syndrome (WNS) is an infectious disease of hibernating bats that has killed millions of bats since it first emerged in eastern North America in 2006. The disease is caused by a pathogenic fungus, *Pseudogymnoascus* (formerly *Geomyces*) *destructans* that was likely introduced to North America by human trade or travel, demonstrating the serious problem of global movement of pathogens by humans in the Anthropocene. Here, we present a synthesis of the current state of knowledge on WNS, including disease mechanisms, disease ecology, global distribution and conservation and management efforts. There has been rapid research response to WNS and much about the disease is now well understood. However, critical gaps in our knowledge remain, including ways to limit spread, or effective treatment options to reduce disease mortality. There are several hibernating bat species in North America that are threatened with extinction from WNS. Protecting those species has become a race against time to find and implement creative solutions to combat the devastating impacts of this disease.

W.F. Frick (✉)
Ecology and Evolutionary Biology, University of California Santa Cruz, Santa Cruz, CA, USA
e-mail: wfrick@ucsc.edu

S.J. Puechmaille
Zoology Institute, Ernst-Moritz-Arndt University, Greifswald, Germany

S.J. Puechmaille
School of Biological and Environmental Sciences, University College Dublin, Dublin, Ireland

C.K.R. Willis
Department of Biology, University of Winnipeg, Winnipeg, MB, Canada

C.C. Voigt and T. Kingston (eds.), *Bats in the Anthropocene: Conservation of Bats in a Changing World*, DOI 10.1007/978-3-319-25220-9_9

9.1 Introduction

In late winter of 2007, biologists at the New York State Department of Environmental Conservation encountered a macabre scene during their annual winter surveys of hibernating bats in caves and mines in northern New York State: heaps of dead bats piled on cave floors (Fig. 9.1) (Veilleux 2008). Bats were also seen flying out in the middle of winter onto the snowy landscape and the number of citizen reports of dead bats found in backyards was much higher than normal. A white fuzzy growth was observed on muzzles and wings of the few remaining live bats, which led to the name white-nose syndrome (WNS) (Veilleux 2008; Reeder and Turner 2008; Turner and Reeder 2009). WNS is now recognized as one of the most devastating wildlife epidemics in recorded history and has caused the death of millions of bats in eastern North America. The research and management response to WNS has been rapid and we know much more about WNS than when those first dead bats were observed in New York, although there is still a great deal about this wildlife disease that is yet to be resolved.

The first evidence of WNS in North America is dated to a photograph taken by a caver at Howe's Cave in 2006 (Turner and Reeder 2009). Howe's Cave is a popular tourist attraction that receives hundreds of thousands of visitors each year, many of whom visit from other parts of the world. The white fuzzy growth visible on bats is caused by a pathogenic fungus, which was described as *Geomyces destructans* (Gargas et al. 2009; Blehert et al. 2009), but was recently re-named *Pseudogymnoascus destructans* after closer evaluation of its taxonomic allies (Minnis and Lindner 2013). The fungus infects the skin tissues, including the wings and tail membranes, and causes bats to arouse too frequently from torpor

Fig. 9.1 Bats that died from WNS during winter at Aeolus Cave in Vermont, USA. *Photo* by Al Hicks

Fig. 9.2 A hibernating little brown myotis (*Myotis lucifugus*) with typical WNS infection visible on skin tissues. *Photo* by Ryan von Linden

during hibernation (Lorch et al. 2011; Warnecke et al. 2012) (Fig. 9.2). Bats die before spring brings warmer weather and insects for food.

WNS has spread rapidly and by 2014 was found in 25 U.S. states and 5 Canadian Provinces (Fig. 9.3). A confirmed case of WNS is defined by the presence of cupping erosions on the skin caused by infection by *P. destructans*, which is determined by histopathological examination (Meteyer et al. 2009). There are currently seven hibernating species in North America that have been confirmed with infections characteristic of WNS, including *Myotis lucifugus, Myotis septentrionalis, Myotis sodalis, Myotis leibii, Myotis grisescens, Eptesicus fuscus and Perimyotis subflavus*. There are several additional species for which *P. destructans* has been detected on skin tissues using swab sampling and quantitative PCR methods (Muller et al. 2013), but that have not been confirmed with characteristic skin lesions that define the disease.

Two of the species confirmed with WNS (*M. sodalis, M. grisescens*) were already listed as federally endangered under the US Endangered Species Act before WNS emerged and several other species have been predicted to go globally or regionally extinct due to mortality from WNS (Frick et al. 2010; Langwig et al. 2012; Thogmartin et al. 2013). The US Fish and Wildlife Service listed *M. septentrionalis* as federally threatened in 2015 due to the risk of extinction

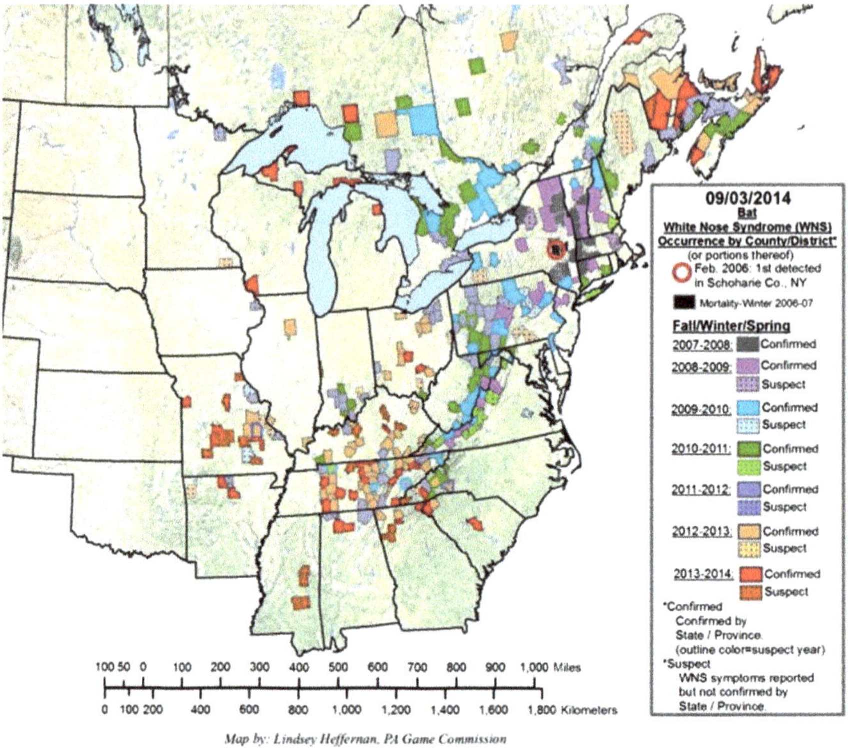

Fig. 9.3 Map of current distribution and past spread of WNS across North America. Confirmed WNS cases are those where disease has been confirmed by histological examination of tissues. Suspect cases are those that are either a molecular detection of *Pseudogymnoascus destructans* by quantitative PCR (Muller et al. 2013) or by visual signs and/or aberrant behaviour consistent with WNS disease at a site. Updated versions of this map are made publically available at whitenosesyndrome.org

from WNS-associated mortality. In addition, a status review of *M. lucifugus* is being conducted to determine whether listing as federally endangered is warranted of this once common species (Frick et al. 2010). In Canada, three species, *M. lucifugus*, *M. septentrionalis* and *P. subflavus* were listed as endangered in 2015. The rapid spread and extensive mortality associated with WNS raise serious concerns about population viability for species that are being impacted by this disease.

In this chapter, we review what is currently known about WNS, focusing on mechanisms of disease, disease ecology, global distribution patterns and conservation and management. We first explain why WNS belongs in a volume addressing bats in the Anthropocene. We review what is known about disease mechanisms, including what we currently understand about the physiology of the disease and immune response in bats. We then review what is currently known about disease ecology of WNS, including the population impacts to species, and then highlight unanswered questions about transmission dynamics. We discuss global distributions patterns, focusing on what is known about WNS in Europe. We conclude by discussing current conservation and management strategies.

Wildlife disease is increasingly recognized as a major conservation threat (Daszak 2000). Global movements of humans increase the probability and rate at which we introduce pathogens into naïve ecosystems (Cunningham et al. 2003). This human-mediated spread of pathogens has been dubbed "pathogen pollution" to highlight the role of human trade and travel in the spread of wildlife pathogens (Cunningham et al. 2003). The fungus *P. destructans* was presumably introduced to North America from Europe by people, most likely from someone who had visited caves in Europe and subsequently visited Howe's Cave with contaminated boots or gear (Puechmaille et al. 2011c; Leopardi et al. 2015). No bats are known to migrate between the Americas and other continents, implicating human trade or travel in the trans-Atlantic arrival of the fungus (Wibbelt et al. 2010). Ironically, bats are often seen as reservoirs of diseases with consequences to human health (e.g. rabies, SARS, etc.). In the case of WNS, humans were most likely the unwitting transcontinental carrier of a pathogen that has killed millions of bats and now threatens species with extinction.

The emergence of WNS has dramatically changed conservation planning and population monitoring of temperate bats in North America (Foley et al. 2011). On the positive side, this crisis prompted collaborative research efforts among bat conservationists in North America and in Europe. Although mortality from WNS is currently restricted to North America, the pathogen is a potential threat to hibernating bat populations in other parts of the globe and is a global concern for bats in the Anthropocene (Puechmaille et al. 2011c).

9.2 Disease Mechanisms

Challenge or inoculation studies (e.g. Lorch et al. 2011; Warnecke et al. 2012; Wilcox et al. 2014) and comparative studies of bats from affected versus unaffected hibernacula (Moore et al. 2011; Storm and Boyles 2011; Reeder et al. 2012; Brownlee-Bouboulis and Reeder 2013) have led to progress in our understanding of mechanisms underlying WNS. The wings of bats are physiological active tissue involved in gas exchange and fluid balance. In general, results of physiological studies are converging on a consensus that cutaneous infection of the wings accounts for the physiological and behavioural effects of WNS (Cryan et al. 2010).

Lorch et al. (2011) experimentally inoculated the wings of healthy *M. lucifugus* with *P. destructans* for comparison to sham-inoculated controls. They housed bats in temperature- and humidity-controlled incubators that maintained environmental conditions approaching natural hibernacula [82 % relative humidity (RH) at 6.5 °C]. This experiment resolved a critical question by demonstrating that experimental infection with *P. destructans* caused the defining characteristics of WNS (e.g. cupping erosions in the epidermis associated with fungal growth, Meteyer et al. 2009). They also found that *P. destructans* spread from infected to un-infected bats housed in the same cages but did not spread between cages in the same incubator confirming contact but not airborne transmission of the causal pathogen under laboratory conditions. Lorch et al. (2011) did not detect differences in survival between infected and un-infected bats possibly because the experimental duration was shorter than a typical hibernation season and/or because humidity in this experiment was lower than that of hibernacula used by *M. lucifugus* in the wild, potentially influencing hibernation patterns of both control and infected bats. Warnecke et al. (2012) repeated aspects of Lorch et al.'s (2011) experiment but increased ambient humidity to >97 % RH at 7 °C and ran the experiments for 120 days (vs. 102 days in Lorch et al. 2011). In Warnecke et al.'s (2012) experiment, all sham-inoculated bats survived four months of hibernation, while infected bats exhibited a significant increase in the frequency of periodic arousals, reduced fat reserves and reduced survival, thus confirming that infection with *P. destructans* alone causes the pathology that defines WNS, altered torpor behaviour and mortality. A field study comparing arousal frequency of bats in affected versus unaffected caves (Reeder et al. 2012) also reported a difference in arousal frequency similar to that observed by Warnecke et al. (2012). Together these findings suggest a strong role for increased arousal frequency and altered energy balance in WNS pathophysiology.

Comparisons of control and infected bats have also provided insight into immune responses (or lack of responses) of bats during and after hibernation. Hibernators generally exhibit down-regulated immune function during winter and bat species affected by WNS appear to be no exception (Meteyer et al. 2009, 2012; Moore et al. 2011). During hibernation, there is little evidence of initiation of an inflammatory response or recruitment of immune cells in bats infected by *P. destructans* based on histopathology (Meteyer et al. 2009, 2012). Despite the absence of an inflammatory response, however, variation in other aspects of cellular immunity may have a role to play. Moore et al. (2013) found differences in immunological responses of *M. lucifugus* in affected versus unaffected hibernacula, specifically higher leukocyte counts, reduced antioxidant activity and lower levels of interleukin-4 (an important precursor for differentiation of T-cells) in bats from WNS-affected caves. Although comparisons between populations of bats in different hibernacula are challenging to interpret because of the potential for underlying differences between bats independent of infection, these findings suggest that even the hardest-hit bat species attempt some, albeit weak, immune response to *P. destructans* infection. This also raises the possibility that some bats may be better equipped to resist infection than others (Puechmaille et al. 2011c) with the potential for directional selection on immune function if these differences are heritable and provide a survival advantage.

Immune responses of bats to WNS could be as much a disadvantage as an advantage. Meteyer et al. (2012) recently reported the disheartening paradox that some survivors of WNS exhibit characteristic signs of immune reconstitution inflammatory syndrome (IRIS). When infected bats emerge from hibernation and their immune function resumes, they exhibit a massive neutrophilic inflammatory response to the fungal infection. This response appears to dramatically increase tissue damage and may reflect an over-reaction to infection because euthermic body temperatures in spring would likely be sufficient to combat the fungal infection (Chaturvedi et al. 2010; Puechmaille et al. 2011b; Verant et al. 2012). The response is likely energetically expensive and the resulting wing damage could compromise flight ability and, therefore, spring energy balance by increasing healing and immunity costs, while reducing potential foraging efficiency at a time when energy balance is critical to support reproduction. Further studies of the role of IRIS in the ecology of WNS are essential for understanding the potential for populations to recover from WNS.

A down-regulated immune response in hibernating bats generally, combined with increased arousal frequency (Boyles and Willis 2010; Reeder et al. 2012; Warnecke et al. 2012) and possibly increased metabolic rate and body temperature during torpor following infection (Storm and Boyles 2011; Verant et al. 2014), appears to result in premature fat depletion and starvation. However, why fungal infection would increase arousal frequency is still not fully understood. Cryan et al. (2010) proposed the hypothesis that fungal damage to the wings of bats could lead to increased evaporative water loss (EWL) across damaged epidermis. Rates of EWL during torpor are a strong predictor of arousal frequency in hibernators (Ben-Hamo et al. 2013; Thomas and Cloutier 1992; Thomas and Geiser 1997) so an increase in EWL or fluid loss due to skin damage from infection by *P. destructans* could lead to the observed effects on arousals. Willis et al. (2011) used data on water loss and arousal frequency in healthy bats, combined with an individual-based model quantifying survival of hypothetical populations of bats, to demonstrate that even a small increase in EWL resulting from infection could cause the same patterns of arousal and mortality observed for infected bats, thus highlighting the plausibility of the dehydration hypothesis.

Two independent datasets from both captive and free-ranging bats also support a role for dehydration and fluid loss in WNS pathophysiology (Cryan et al. 2013; Warnecke et al. 2013). In addition to high hematocrit levels consistent with dramatic fluid loss, Cryan et al. (2013) and Warnecke et al. (2013) both found evidence of electrolyte depletion (with no evidence of renal pathology), consistent with hypotonic dehydration due to fluid loss across damaged wings. Presumably infected bats lose fluid containing both water and electrolytes across injured wing tissue but can only replenish or partially replenish water stores by drinking, because electrolytes are not available in hibernacula. Warnecke et al. (2013) also found preliminary evidence of a respiratory response to metabolic acidosis in infected bats which they hypothesized reflect reduced perfusion of infected tissues, localized anaerobic metabolism and acidosis, and increased respiratory rate to increase CO_2 excretion and counter acidosis. In addition to increased arousal frequency, these physiological responses also predict increased metabolic costs and elevated body temperature during torpor. To date, measurements of torpid body temperature with enough precision to test this hypothesis are unavailable but these would be valuable, especially alongside measurements of metabolism during torpor and arousal in infected versus un-infected bats.

Other physiological mechanisms could also be at play. Willis and Wilcox (2014) reviewed three (of many potential) hormone systems that could be influenced by WNS, both within individuals and via selection on traits which could favour survival. For example, the lipostat hormone leptin is strongly associated with winter energy balance and pre-hibernation fattening. Bats must enter a state of leptin resistance during fall to accumulate adequate fat stores to survive the winter. If, as the evidence suggests, WNS represents a challenge for hibernation endurance, bats with the greatest leptin resistance (and therefore potential fat stores) in autumn may be best equipped to survive increased arousals associated with WNS (Willis and Wilcox 2014). Interactions between WNS and other hormone systems important for seasonal energetics, body temperature regulation and energy and fluid balance (e.g. glucocorticoids, melatonin, thyroid hormone, vasopressin, androgens) could also play important roles in disease dynamics and evolution of remnant populations and are worth further study.

In addition to physiological research, recent studies have also examined behavioural mechanisms associated with WNS that could reflect either adaptive responses to disease or maladaptive pathological responses. Langwig et al. (2012) reported that a much greater proportion of the *M. lucifugus* surveyed in WNS-affected caves after the emergence of the disease were hibernating solitarily (i.e. without clustering) compared to bats surveyed before WNS. This could reflect a behavioural change by individuals following infection or selection by WNS for bats which tend to roost individually (Langwig et al. 2012). Wilcox et al. (2014) reported behavioural observations of bats inoculated with *P. destructans* and found evidence supporting the former hypothesis. Infected bats gradually reduced their clustering behaviour as hibernation progressed. Wilcox et al. (2014) also observed a reduction in behavioural activity during arousals, in general, for affected bats. Taken together, reduced clustering and reduced activity by infected bats could reflect general patterns known as "sickness behaviour", a coordinated response to infection characterized in part by lethargy presumably to save energy for immune responses (Adelman and Martin 2009). These behaviours could also reduce the potential for transmission among individuals in a social group within a hibernaculum. Even bats that have already been infected with *P. destructans* could benefit by reduced subsequent exposure to other infected individuals because new contacts could lead to additional areas of infection in the wings, exacerbating disease severity. On the other hand, reduced clustering behaviour could increase energy expenditure and EWL leading to negative consequences for survival. More work is needed to understand the survival consequences of a range of physiological and behavioural responses to WNS.

9.3 Disease Ecology of WNS

One of the defining characteristics of WNS is that it is a multi-host disease, meaning that *P. destructans* infects multiple bat species. Although all hibernating bat species in northeastern North America can be infected with *P. destructans* and develop the cupping erosions in their skin tissues that characterize the disease, population impacts from WNS vary widely among species (Langwig et al. 2012; Turner et al. 2011). Prior to the emergence of WNS in North America, all six hibernating bat species that occur in the northeastern United States had positive population growth trends (Frick et al. 2010; Langwig et al. 2012). With the emergence of WNS, four of these six species suffered severe population declines (*M. septentrionalis, M. lucifugus*, *M. sodalis* and *P. subflavus*) (Langwig et al. 2012). Two species (*M. leibii* and *E. fuscus*) have experienced less severe impacts from disease (Langwig et al. 2012). In addition, species of the genus *Corynorhinus* do not appear to get sick and die from WNS, despite occurring in WNS-affected caves in states in the mid-Atlantic region, such as West Virginia and Virginia. Why some species suffer higher mortality than others is an important area of current research, but there are no clear-cut answers yet. Langwig et al. (2012) showed that differences in roosting microclimates (temperature and RH) were correlated with differential impacts among sites for some species. For example, sites with warmer roosting temperatures had the highest declines for *M. lucifugus* and sites with highest RH had the highest declines for *M. sodalis*, suggesting that roosting microclimates could play an important role in WNS impacts (Langwig et al. 2012). Differences in environmental conditions as well as exposure, transmission, susceptibility, torpor physiology and immune response among species could contribute to observed differences in mortality. Future research focusing on differences in these factors among species will be critical for identifying the risks to particular species.

Understanding whether transmission is dependent on the density of hibernating populations is key to determining whether WNS will cause bats to go extinct or whether bat populations will stabilize at low numbers. For diseases where transmission is density-dependent, the probability of extinction is much lower because transmission rates decline as populations become smaller (De Castro and Bolker 2004). Langwig et al. (2012) showed that for bats that hibernate in dense clusters (e.g. *M. lucifugus* and *M. sodalis*), there was no evidence for density-dependent declines, meaning that declines from WNS were equally severe in populations that ranged from 100 to 100,000 bats. In contrast, there was evidence that declines were smaller in smaller populations for species that roost solitarily (e.g. *P. subflavus* and *M. septentrionalis*). Although the declines were density-dependent in *M. septentrionalis*, declines were not predicted to stabilize before populations went extinct in this species, suggesting that this species is at serious risk of extinction from WNS.

Determining whether a pathogen can persist in an environmental reservoir is also important for understanding disease transmission dynamics and extinction risk from disease (De Castro and Bolker 2004). Pathogens that can persist in an environmental reservoir are more likely to drive species extinct because hosts can get infected from the environment even if only a few individuals remain. Studies have shown that *P. destructans* is found in sediments and environmental substrates in hibernacula (Puechmaille et al. 2011a; Lindner et al. 2011; Lorch et al. 2013a, b). Lorch et al. (2013b) demonstrated that viable *P. destructans* can be cultured from samples taken during late summer when bats have been absent for several months, suggesting that *P. destructans* persists in the environment between hibernation seasons. An unpublished experiment conducted by Al Hicks at the New York Department of Environmental Conservation demonstrated that naïve bats that had never been exposed to *P. destructans* could contract disease and die from WNS when placed in an infected hibernaculum with no access to other infected bats (Hicks, pers. comm.). The evidence to date suggests that hibernacula are environmental reservoirs for *P. destructans*, which has potentially dire consequences if the environment proves a major source of transmission.

WNS is a seasonal disease and recent work by Langwig et al. (2015) describes how the seasonal patterns of transmission of *P. destructans* are driven by hibernation. Bats begin to become infected in the fall when they return to hibernacula during fall swarm and transmission spikes in early winter once bats begin hibernating. Infection intensity increases during hibernation and peaks in late winter at which time most bats have become infected. These seasonal patterns are similar to temporal prevalence of visual signs of *P. destructans* growth on bats at sites in Europe as described by Puechmaille et al. (2011a), where a peak of infection was also observed in late hibernation when most individuals present were infected. In Langwig et al.'s study, most bats cleared infection during summer and prevalence of infection fell to zero by late summer at maternity roosts. The seasonal timing of infection suggests that mortality occurs at a time of maximal impact for populations (before the birth pulse). However, a peak in transmission after bats begin hibernating in early winter may reduce the rate of spread among hibernacula since bats presumably move among sites less frequently once they start hibernating compared to during the fall swarm period.

9.4 Status of *P. Destructans*/WNS in Europe

In contrast to the severe impacts WNS has on North American bat species, *P. destructans* is commonly found on bats in Europe but is not associated with mass mortality (Wibbelt et al. 2010; Puechmaille et al. 2011a). Europe is a putative source of the pathogen and the pathogen likely arrived in North America by some means of human trade or travel. Ongoing studies on global distribution of *P. destructans* (S.J. Puechmaille and J.R. Hoyt, unpublished data), including surveys in temperate Asia, may reveal important insights about the global distribution of the pathogen.

Pseudogymnoascus destructans was first reported in Europe by Puechmaille et al. (2010) who sampled a hibernating *Myotis myotis* from southwestern France showing the typical powdery white fungal growth on its nose. Since then, the fungus has been morphologically and genetically confirmed in 14 countries in Europe (France, Portugal, Belgium, The United Kingdom, The Netherlands, Germany, Switzerland, Austria, Slovakia, Poland, Hungary, Ukraine and Estonia) and convincing photographic evidence further supports its presence in an additional four countries (Luxembourg, Denmark, Romania and Turkey [the European part]) (Martínková et al. 2010; Puechmaille et al. 2010, 2011a; Kubátová et al. 2011; Simonovicová et al. 2011; Mestdagh et al. 2012; Wibbelt et al. 2010, 2013; Burger et al. 2013; Paiva-Cardoso et al. 2014; Sachanowicz et al. 2014). At the continental scale, most European reports are from northeastern France through Belgium, the Netherlands, Germany and the Czech Republic, but it remains unclear whether this pattern of higher prevalence of the fungus is real or reflects sampling bias (Puechmaille et al. 2011a). Studies conducted in Italy, Slovenia and Sweden, where *P. destructans* was not detected (Voyron et al. 2010; Nilsson 2012; Mulec et al. 2013), support the hypothesis that *P. destructans* occurrence and/or prevalence varies between different geographic regions in Europe (Puechmaille et al. 2011a).

Puechmaille et al. (2011a) demonstrated that the prevalence of visible signs of *P. destructans* on bat wings and nose drastically varied through the hibernation period with the first cases appearing around mid-January. The number of cases increased to reach a peak in March and declined as bats emerged from hibernation. This pattern further complicates comparisons of prevalence of visual signs of fungal growth on bats between sites, regions or years unless surveys are carried out at the same time. Work done in the Czech Republic and Slovakia detected differences in prevalence of bats suspected to carry *P. destructans* (based on visual observations) between sub-mountain humid to mesic regions (higher prevalence) and mountainous and limestone regions (lower prevalence) (Martínková et al. 2010), supporting the idea that *P. destructans* is not equally abundant across Europe. Nevertheless, the differences in sampling strategy (spatio-temporal), sampling intensity (number of sites, number of samples), nature of the samples collected (e.g. swab from the bat vs. environment vs. guano) and analysis techniques (e.g. culture, PCR detection) between different European studies make quantification of these fine- and large-scale patterns challenging (Puechmaille et al. 2011a).

All confirmed cases of *P. destructans* infection come from fungal material collected on bats with the exception of a case from Estonia where the fungus has been isolated and cultured from the walls of the hibernation site, representing the first published isolation of viable spores from the environment in Europe or North America (Puechmaille et al. 2011a). In terms of species, available data suggest that *M. myotis* is the most commonly infected species (ca. 66 % of cases) with *P. destructans* in Europe (Martínková et al. 2010; Puechmaille et al. 2011a). The fungus is known to also infect another nine species of European *Myotis* (ranked by decreasing order of prevalence): *M. dasycneme, M. mystacinus, M. blythii, M. daubentonii, M. brandtii, M. emarginatus, M. nattereri, M. bechsteinii* and *M. escalerai*/sp. *A*. The list of species with *P. destructans* infection is likely to increase as sampling intensity increases as illustrated by the recent Zukal et al. (2014) study which reported infection of a few individuals from three more species of the family Vespertilionidae, *Eptesicus nilssonii, Plecotus auritus* and *Barbastella Barbastellus*, as well as on a single individual of *Rhinolophus hipposideros*, of the family Rhinolophidae.

Owing to the protection of bats across Europe and the absence of mass mortality, only three studies with limited to moderate numbers of samples have investigated the pathology of *P. destructans* during the hibernation period (Pikula et al. 2012; Wibbelt et al. 2013; Bandouchova et al. 2015). In Europe, *P. destructans* invasion of the wing membrane is generally restricted to the epidermis and adnexae without deep invasion into the underlying connective tissue but with occasional formation of neutrophilic pustules, contrasting with the common and extensive invasion of dermal connective tissue in bats from North America (Pikula et al. 2012; Wibbelt et al. 2013; Zukal et al. 2014; Bandouchova et al. 2015). Based on investigation of two euthanized individuals, *P. destructans* invasion in the skin of the muzzle seems to be more pronounced than invasion of the wing membrane (Pikula et al. 2012; Wibbelt et al. 2013). As damage to the skin of the muzzle may not be as physiologically important for homeostasis as damage to the wing membranes (Cryan et al. 2010; Reeder et al. 2012; Warnecke et al. 2013), we suggest that it may be important to differentiate the pathology of *P. destructans* on the wing and on the muzzle. If dehydration and fluid loss play an important role in WNS pathophysiology, quantifying wing damage consistently (e.g. following Reeder et al. 2012 or an alternative scoring system) alongside physiological measures of disease severity will be critical for a better understanding of the disease, its progression and species-specific attributes, compared to the commonly reported dichotomous presence/absence of the disease.

The term WNS was originally used to describe the symptoms associated with bats in the field before the disease was fully characterized as a cutaneous infection of skin tissues by the pathogenic fungus, *P. destructans* (Blehert et al. 2009; Meteyer et al. 2009). As such, the name 'WNS' has changed from referring to a set of symptoms, including visible fungal growth on skin surfaces, depletion of fat reserves, altered torpor patterns and aberrant winter behaviour (Blehert et al. 2009) to referring to the presence of disease as defined by the presence of cutaneous infection characterized by cupping erosions (Meteyer et al. 2009). This has led to confusion and some debate about whether the term WNS should be used to describe infections occurring in Europe, which are pathologically similar to those in North America but which do not include mass mortality or aberrant winter behaviour (Puechmaille et al. 2011a). Despite its original definition as a syndrome (Veilleux 2008; Reeder and Turner 2008; Turner and Reeder 2009), the term WNS is now routinely used to refer the cutaneous infection caused by *P. destructans*, which have been documented in Europe (Pikula et al. 2012; Wibbelt et al. 2013; Zukal et al. 2014). Some have advocated a name change to clarify a difference between a 'syndrome' and a 'disease' caused by fungal infection (Chaturvedi and Chaturvedi 2011). Inconsistency in the literature could lead to confusion but recent use of the term white-nose disease (WND; Paiva-Cardoso et al. 2014) could clarify the situation by providing terminology reminiscent enough of WNS to avoid confusion but technically consistent with the definition of a disease.

Recent work comparing colony sizes of hibernating vespertilionid bats in North America before and after the emergence of WNS, to current colony sizes in Europe, reveals an intriguing pattern. Before WNS emerged in North America, colony sizes of hibernating bats were, on average, about 10-fold larger than those of similar species in Europe (Frick et al. 2015). However, after the emergence of WNS, colony sizes in eastern North America are no longer statistically different from those in Europe (Frick et al. 2015), raising the following question: Were hibernating bat colonies in Europe once much larger prior to the emergence of WNS there? If WNS is indeed acting as a hidden force on bat populations in Europe, then small winter colony sizes in eastern North America may become the norm for species in North America that manage to persist. However, Frick et al. (2015) also show that 69 % of winter colonies of *M. septentrionalis* were entirely eliminated within 7 years of WNS detection, suggesting that this species is rapidly disappearing from the landscape. The predicted extinction of *M. septentrionalis* from WNS begs the question whether past extinctions of bat species may have also occurred in Europe.

9.5 Conservation and Management

Conservation and management strategies for WNS in North America have focused primarily on preventing spread of the pathogen to new areas through decontamination protocols as well as cave closures to limit the potential for human-mediated spread. Decontamination of gear used in hibernacula by both recreational cavers and bat researchers is an important management strategy to reduce the risk of spread of *P. destructans* by humans. *P. destructans* spores have been found on field gear after use in infected sites and therefore utmost precaution is needed to reduce the chance that researchers and cavers spread *P. destructans* to new areas. Cave closures have been controversial and have been met with some resistance by some members of the caving community. Some cave closures have subsequently been relaxed in parts of the western United States where *P. destructans* has not yet spread. Determining whether cave closures are effective can be challenging given that the absence of spread in areas is hard to measure. Bats are capable of spreading the fungus, but the primary focus of closing caves and advocating decontamination was to slow spread by people, especially to distant locations.

Finding a treatment for infected bats has proved elusive and difficult. Several studies have examined the efficacy of treating bats with anti-fungal chemicals, such as terbinafine, but none have shown any promise. There has also been interest in alternative forms of treatment, including use of naturally occurring bacteria (Fritze et al. 2012; Hoyt et al. 2015) or volatile compounds (Cornelison et al. 2014). Recent work by Cornelison et al. (2014) showed that a volatile organic compounds (VOCs) inhibited growth of *P. destructans* in vitro. Similarly, a recent study by Hoyt et al. (2015) showed that *Pseudomonas* bacteria that naturally occur on hibernating bats inhibit growth of *P. destructans* in vitro. Other strains of *Pseudomonas* found in Europe have shown similar results (Fritze et al. 2012). Research on these biological control treatment options is still in early stages and although early lab results have shown promise, experimental and field trials will need to be conducted before the efficacy of these approaches is fully evaluated. The WNS research and management community is developing standards and protocols for evaluating the safety and efficacy of biological treatment options.

Other ideas for active management have included building artificial hibernacula that can be cleaned and decontaminated each summer between hibernating seasons. An experimental artificial hibernacula was built in Tennessee and existing military bunkers have been used as artificial hibernaculum in the northeastern US. The goal of these structures is to provide a place for bats to hibernate that does not serve as an environmental source of transmission when bats re-enter the hibernaculum in fall. To date there have been no studies to determine whether bats will use these artificial hibernacula naturally and whether survival will be improved in these sites.

Given what we know about the potential role that electrolyte depletion plays in the physiology of the disease, some researchers have also explored the potential for electrolyte therapy for hibernating bats by providing access to electrolyte supplements during hibernation. Experimental trials to test this are underway. Finally, bats are very difficult to breed in captivity and, while the prospect of captive breeding and management of bats has been explored, it remains doubtful whether this approach could be useful as a management tool for bat species affected by WNS. However, if breeding programmes could be developed, they could provide a supply of animals for laboratory studies to reduce potential impacts of research on wild populations.

9.6 Conclusions

Although we have learned a great deal about WNS in the past seven years, there are still many unanswered questions about disease mechanisms, ecology, transmission dynamics, long-term impacts, global distribution patterns and potential treatment options that will be important for managing WNS and its impacts on bats. The US Fish and Wildlife Service has been pivotal in terms of coordinating meetings for information exchange among researchers and state biologists as well as directly funding much of the research on WNS in both the US and Canada. Research priorities for management and conservation of species have focused on topics such as establishing that *P. destructans* was the causative agent of infection, trying to identify potential treatment of infection, the physiology of infection and mechanisms of mortality, characterizing the environmental reservoir and understanding transmission and immunological response.

For many of us, working on WNS is a grim business. There is nothing quite like the experience of going underground and entering a chamber that was formally home to thousands of bats and seeing empty walls and a few straggling survivors covered in white fungus. However, the sense of commitment within the WNS community and the dedication of researchers and managers to try and find new ways to understand and solve this crisis provide a certain hope. We have yet to find a way to stop bats dying from WNS, but we are trying hard to do so. Whether we are able to prevent species extinctions may rely, in part, on the creativity to find solutions before it is too late and the willingness of agency biologists to implement creative solutions without clear assurances of outcomes.

References

Adelman JS, Martin LB (2009) Vertebrate sickness behaviors: adaptive and integrated neuroendocrine immune responses. Integr Comp Biol 49:202–214

Bandouchova T, Bartonicka T, Berkova H, Brichta J, Cerny J, Kovacova V, Kolarik M, Köllner B, Kulich P, Martínkova N, Rehak Z, Turner GG, Zukal J, Pikula J (2015) *Pseudogymnoascus destructans*: evidence of virulent skin invasion for bats under natural conditions, Europe. Transbound Emerg Dis 62:1–5

Ben-Hamo M, Muñoz-Garcia A, Williams JB, Korine C, Pinshow B (2013) Waking to drink: rates of evaporative water loss determine arousal frequency in hibernating bats. J Exp Biol 216:573–577

Blehert DS, Hicks AC, Behr M, Meteyer CU, Berlowski-Zier BM, Buckles EL, Coleman JTH, Darling SR, Gargas A, Niver R, Okoniewski JC, Rudd RJ, Stone WB (2009) Bat white-nose syndrome: an emerging fungal pathogen? Science 323:227

Boyles JG, Willis C (2010) Could localized warm areas inside cold caves reduce mortality of hibernating bats affected by white-nose syndrome? Front Ecol Environ 8:92–98

Brownlee-Bouboulis SA, Reeder DM (2013) White-nose syndrome-affected little brown myotis (*Myotis lucifugus*) increase grooming and other active behaviors during arousals from hibernation. J Wildl Dis 49:850–859

Burger K, Gebhardt S, Wolfhahrt G, Wibbelt G, Reiter G (2013) First confirmed records of *Geomyces destructans* (Blehert and Gargas 2009) in Austria. Ber Naturwiss-Med Ver Innsbruck 98:127–135

Chaturvedi V, Chaturvedi S (2011) What is in a name? A proposal to use geomycosis instead of white nose syndrome (WNS) to describe bat infection caused by *Geomyces destructans*. Mycopathologia 171:231–233

Chaturvedi V, Springer DJ, Behr MJ, Ramani R, Li X, Peck MK, Ren P, Bopp DK, Wood B, Samsonoff WA, Butchkoski CM, Hicks AC, Stone WB, Rudd RJ, Chaturvedi S (2010) Morphological and molecular characterizations of Psychrophilic fungus *Geomyces destructans* from New York bats with white nose syndrome (WNS). PLoS ONE 5:e10783

Cornelison CT, Gabriel KT, Barlament C, Crow SA (2014) Inhibition of *Pseudogymnoascus destructans* growth from conidia and mycelial extension by bacterially produced volatile organic compounds. Mycopathologia 177:1–10

Cryan PM, Meteyer CU, Blehert DS, Lorch JM, Reeder DM, Turner GG, Webb J, Behr M, Verant M, Russell RE, Castle KT (2013) Electrolyte depletion in white-nose syndrome bats. J Wildl Dis 49:398–402

Cryan PM, Meteyer CU, Boyles JG, Blehert DS (2010) Wing pathology of white-nose syndrome in bats suggests life-threatening disruption of physiology. BMC Biol 8:135

Cunningham AA, Daszak P, Rodriguez JP (2003) Pathogen pollution: defining a parasitological threat to biodiversity conservation. J Parasitol 89:S78–S83

Daszak P (2000) Emerging infectious diseases of wildlife—threats to biodiversity and human health. Science 287:443–449

De Castro F, Bolker B (2004) Mechanisms of disease-induced extinction. Ecol Lett 8:117–126
Foley J, Clifford D, Castle K, Cryan PM, Ostfeld RS (2011) Investigating and managing the rapid emergence of white nose syndrome, a novel, fatal, infectious disease of hibernating bats. Conserv Biol 25:223–231
Frick WF, Puechmaille SJ, Hoyt JR, Nickel BA, Langwig KE, Foster JT, Barlow KE, Bartonička T, Feller D, Haarsma AJ, Herzog C, Horáček I, van der Kooij J, Mulkens B, Petrov B, Reynolds R, Rodrigues L, Stihler CW, Turner GG, Kilpatrick AM (2015) Disease alters macroecological patterns of North American bats. Glob Ecol Biogeogr 24:741–749
Frick WF, Pollock JF, Hicks AC, Langwig KE, Reynolds DS, Turner GG, Butchkoski CM, Kunz TH (2010) An emerging disease causes regional population collapse of a common North American bat species. Science 329:679–682
Fritze M, Huong Pham TL, Irmtraut I (2012) Effekt des bodenbakteriums *Pseudomonas veronii*-like PAZ1 auf das wachstum des white-nose erregers *Geomyces destructans* in antagonistentests. Nyctalus 17:104–107
Gargas A, Trest MT, Christensen M, Volk TJ, Blehert DS (2009) *Geomyces desctructans* sp. nov. associated with bat white-nose syndrome. Mycotaxon 108:147–154
Hoyt JR, Cheng TL, Langwig KE, Hee MM, Frick WF, Kilpatrick AM (2015) Bacteria isolated from bats inhibit the growth of *Pseudogymnoascus destructans*, the causative agent of white-nose syndrome. PLoS ONE 10:e0121329
Kubátová A, Koukol O, Nováková A (2011) *Geomyces destructans*, phenotypic features of some Czech isolates. Czech Mycol 63:65–75
Langwig KE, Frick WF, Reynolds R, Parise K, Drees KP, Hoyt JR, Cheng TL, Kunz TH, Kilpatrick AM (2015) Host and pathogen ecology drive the seasonal dynamics of a fungal disease, white-nose syndrome. Proc Roy Soc Lond B 282:20142335
Langwig KE, Frick WF, Bried JT, Hicks AC, Kunz TH, Kilpatrick AM (2012) Sociality, density-dependence and microclimates determine the persistence of populations suffering from a novel fungal disease, white-nose syndrome. Ecol Lett 15:1050–1057
Leopardi S, Blake D, Puechmaille SJ (2015) White-Nose Syndrome fungus introduced from Europe to North America. Curr. Biol 25:R217–219
Lindner DL, Gargas A, Lorch JM, Banik MT, Glaeser J, Kunz TH, Blehert DS (2011) DNA-based detection of the fungal pathogen *Geomyces destructans* in soils from bat hibernacula. Mycologia 103:241–246
Lorch JM, Meteyer CU, Behr MJ, Boyles JG, Cryan PM, Hicks AC, Ballmann AE, Coleman JTH, Redell DN, Reeder DM, Blehert DS (2011) Experimental infection of bats with *Geomyces destructans* causes white-nose syndrome. Nature 480:376–378
Lorch JM, Lindner DL, Gargas A, Muller LK, Minnis AM, Blehert DS (2013a) A culture-based survey of fungi in soil from bat hibernacula in the eastern United States and its implications for detection of *Geomyces destructans*, the causal agent of bat white-nose syndrome. Mycologia 105:237–252
Lorch JM, Muller LK, Russell RE, O'Connor M, Lindner DL, Blehert DS (2013b) Distribution and environmental persistence of the causative agent of white-nose syndrome, *Geomyces destructans*, in bat hibernacula of the eastern United States. Appl Environ Microbiol 79:1293–1301
Martínková N, Bačkor P, Bartonička T, Blažková P, Cervený J, Falteisek L, Gaisler J, Hanzal V, Horáček D, Hubálek Z, Jahelková H, Kolařík M, Korytár L, Kubátová A, Lehotská B, Lehotský R, Lučan RK, Májek O, Matějů J, Rehák Z, Šafář J, Tájek P, Tkadlec E, Uhrin M, Wagner J, Weinfurtová D, Zima J, Zukal J, Horáček I (2010) Increasing incidence of *Geomyces destructans* fungus in bats from the Czech Republic and Slovakia. PLoS ONE 5:e13853
Mestdagh X, Baltus L, Hoffman L, Titeux N (2012) Découverte de chauves-souris au nez blanc au Luxembourg. Bull Soc Nat Luxemb 113:141–149
Meteyer CU, Barber D, Mandl JN (2012) Pathology in euthermic bats with white nose syndrome suggests a natural manifestation of immune reconstitution inflammatory syndrome. Virulence 3:583–588

Meteyer CU, Buckles EL, Blehert DS, Hicks AC, Green DE, Shearn-Bochsler V, Thomas NJ, Gargas A, Behr MJ (2009) Histopathologic criteria to confirm white-nose syndrome in bats. J Vet Diagn Invest 21:411–414

Minnis AM, Lindner DL (2013) Phylogenetic evaluation of *Geomyces* and allies reveals no close relatives of *Pseudogymnoascus destructans*, comb. nov., in bat hibernacula of eastern North America. Fungal Biol 117:638–649

Moore MS, Reichard JD, Murtha TD, Zahedi B, Fallier RM, Kunz TH (2011) Specific alterations in complement protein activity of little brown myotis (*Myotis lucifugus*) hibernating in white-nose syndrome affected sites. PLoS ONE 6:e27430

Moore MS, Reichard JD, Murtha TD, Nabhan ML, Pian RE, Ferreira JS, Kunz TH (2013) Hibernating little brown myotis (*Myotis lucifugus*) show variable immunological responses to white-nose syndrome. PLoS ONE 8:e58976

Mulec J, Covington E, Walochnik J (2013) Is bat guano a reservoir of *Geomyces destructans*? Open J Vet Med 03:161–167

Muller LK, Lorch JM, Lindner DL, O'Connor M, Gargas A, Blehert DS (2013) Bat white-nose syndrome: a real-time TaqMan polymerase chain reaction test targeting the intergenic spacer region of *Geomyces destructans*. Mycologia 105:253–259

Nilsson S (2012) Surveillance of *Geomyces destructans* in Swedish bats and bat hibernacula. SLU, Department of Biomedical Sciences and Veterinary Public Health, Upsalla, Sweden

Paiva-Cardoso MdN, Morinha F, Barros P, Vale-Gonçalves H, Coelho AC, Fernandes L, Travassos P, Faria AS, Bastos E, Santos M, Cabral JA (2014) First isolation of *Pseudogymnoascus destructans* in bats from Portugal. Eur J Wildl Res 60:645–649

Pikula J, Bandouchova H, Novotny L, Meteyer CU, Zukal J, Irwin NR, Zima J, Martínková N (2012) Histopathology confirms white-nose syndrome in bats in Europe. J Wildl Dis 48:207–211

Puechmaille SJ, Verdeyroux P, Fuller H, Gouilh MA, Bekaert M, Teeling EC (2010) White-nose syndrome fungus (*Geomyces destructans*) in Bat, France. Emerg Infect Dis 16:290–293

Puechmaille SJ, Wibbelt G, Korn V, Fuller H, Forget F, Mühldorfer KM, Kurth A, Bogdanowicz W, Borel C, Bosch T, Cherezy T, Drebet M, Görföl T, Haarsma AJ, Herhaus F, Hallart G, Hammer M, Jungmann C, Le Bris Y, Lutsar L, Masing M, Mulkens B, Passior K, Starrach M, Wojtaszewski A, Zöphel U, Teeling EC (2011a) Pan-European distribution of white-nose syndrome fungus (*Geomyces destructans*) not associated with mass mortality. PLoS ONE 6:e19167

Puechmaille SJ, Fuller H, Teeling EC (2011b) Effect of sample preservation methods on the viability of *Geomyces destructans*, the fungus associated with white-nose syndrome in bats. Acta Chiropterol 13:217–221

Puechmaille SJ, Frick WF, Kunz TH, Racey PA, Voigt CC, Wibbelt G, Teeling EC (2011c) White-nose syndrome: is this emerging disease a threat to European bats? Trends Ecol Evol 26:570–576

Reeder DM, Turner GG (2008) Working together to combat white nose syndrome: a report of a meeting on 9–11 June 2008, in Albany, New-York. Bat Res News 49:75–78

Reeder D, Frank CL, Turner GG, Meteyer CU (2012) Frequent arousal from hibernation linked to severity of infection and mortality in bats with white-nose syndrome. PLoS ONE 7:e38920

Sachanowicz K, Stępień A, Ciechanowski M (2014) Prevalence and phenology of white-nose syndrome fungus *Pseudogymnoascus destructans* in bats from Poland. Central Eur J Biol 9:437–443

Simonovicová A, Pangallo D, Chovanová K, Lehotská B (2011) *Geomyces destructans* associated with bat disease WNS detected in Slovakia. Biologia 66:562–564

Storm JJ, Boyles JG (2011) Body temperature and body mass of hibernating little brown bats *Myotis lucifugus* in hibernacula affected by white-nose syndrome. Acta Theriol 56:123–127

Thogmartin WE, Sanders-Reed CA, Szymanski JA, McKann PC, Pruitt L, King RA, Runge MC, Russell RE (2013) White-nose syndrome is likely to extirpate the endangered Indiana bat over large parts of its range. Biol Conserv 160:162–172

Thomas DW, Cloutier D (1992) Evaporative water loss by hibernating little brown bats, *Myotis lucifugus*. Physiol Zool 65:443–456

Thomas DW, Geiser F (1997) Periodic arousals in hibernating mammals: is evaporative water loss involved? Funct Ecol 11:585–591

Turner GG, Reeder DM (2009) Update of White Nose Syndrome in bats, September 2009. Bat Res News 50:47–53

Turner GG, Reeder DM, Coleman JTH (2011) A five-year assessment of mortality and geographic spread of white-nose syndrome in North American bats, with a look to the future. Bat Res News 52:13–27

Veilleux JP (2008) Current status of white-nose syndrome in the Northeastern United States. Bat Res News 49:15–17

Verant ML, Boyles JG, Waldrep W, Wibbelt G, Blehert DS (2012) Temperature-dependent growth of *Geomyces destructans*, the fungus that causes bat white-nose syndrome. PLoS ONE 7:e46280

Verant ML, Meteyer CU, Speakman JR, Cryan PM, Lorch JM, Blehert DS (2014) White-nose syndrome initiates a cascade of physiologic disturbances in the hibernating bat host. BMC Physiol 14(1):10

Voyron S, Lazzari A, Riccucci M, Calvini M, Varese GC (2010) First mycological investigations on Italian bats. Hystrix Italian J Mammal 22:189–197

Warnecke L, Turner JM, Bollinger TK, Lorch JM, Misra V, Cryan PM, Wibbelt G, Blehert DS, Willis CKR (2012) Inoculation of bats with European *Geomyces destructans* supports the novel pathogen hypothesis for the origin of white-nose syndrome. Proc Natl Acad Sci 109:6999–7003

Warnecke L, Turner JM, Bollinger TK, Misra V, Cryan PM, Blehert DS, Wibbelt G, Willis CKR (2013) Pathophysiology of white-nose syndrome in bats: a mechanistic model linking wing damage to mortality. Biol Lett 9:20130177

Wibbelt G, Kurth A, Hellmann D, Weishaar M, Barlow A, Veith M, Prüger J, Görföl T, Grosche L, Bontadina F, Zöphel U, Hans-Peter S, Cryan PM, Blehert DS (2010) White-nose syndrome fungus (*Geomyces destructans*) in bats Europe. Emerg Infect Dis 16:1237

Wibbelt G, Puechmaille SJ, Ohlendorf B, Mühldorfer K, Bosch T, Görföl T, Passior K, Kurth A, Lacremans D, Forget F (2013) Skin lesions in European hibernating bats associated with *Geomyces destructans*, the etiologic agent of white-nose syndrome. PLoS ONE 8:e74105

Wilcox A, Warnecke L, Turner JM, McGuire LP, Jameson JW, Misra V, Bollinger TC, Willis CKR (2014) Behaviour of hibernating little brown bats experimentally inoculated with the pathogen that causes white-nose syndrome. Anim Behav 88:157–164

Willis CKR, Menzies AK, Boyles JG, Wojciechowski MS (2011) Evaporative water loss is a plausible explanation for mortality of bats from white-nose syndrome. Integr Comp Biol 51:364–373

Willis C, Wilcox A (2014) Hormones and hibernation: possible links between hormone systems, winter energy balance and white-nose syndrome in bats. Horm Behav 66:66–73

Zukal J, Bandouchova H, Bartonička T, Berkova H, Brack V, Brichta J, Dolinay M, Jaron KS, Kovacova V, Kovarik M, Martínková N, Ondracek K, Rehák Z, Turner GG, Pikula J (2014) White-nose syndrome fungus: a generalist pathogen of hibernating bats. PLoS ONE 9:e97224

Chapter 10
Zoonotic Viruses and Conservation of Bats

Karin Schneeberger and Christian C. Voigt

Abstract Many of the recently emerging highly virulent zoonotic diseases have a likely bat origin, for example Hendra, Nipah, Ebola and diseases caused by coronaviruses. Presumably because of their long history of coevolution, most of these viruses remain subclinical in bats, but have the potential to cause severe illnesses in domestic and wildlife animals and also humans. Spillovers from bats to humans either happen directly (via contact with infected bats) or indirectly (via intermediate hosts such as domestic or wildlife animals, by consuming food items contaminated by saliva, faeces or urine of bats, or via other environmental sources). Increasing numbers of breakouts of zoonotic viral diseases among humans and livestock have mainly been accounted to human encroachment into natural habitat, as well as agricultural intensification, deforestation and bushmeat consumption. Persecution of bats, including the destruction of their roosts and culling of whole colonies, has led not only to declines of protected bat species, but also to an increase in virus prevalence in some of these populations. Educational efforts are needed in order to prevent future spillovers of bat-borne viruses to humans and livestock, and to further protect bats from unnecessary and counterproductive culling.

K. Schneeberger (✉) · C.C. Voigt
Department of Evolutionary Ecology, Leibniz Institute for Zoo and Wildlife Research, Berlin, Germany
e-mail: schneeberger@izw-berlin.de

C.C. Voigt and T. Kingston (eds.), *Bats in the Anthropocene: Conservation of Bats in a Changing World*, DOI 10.1007/978-3-319-25220-9_10

10.1 Introduction

Over the past decades, the emergence of zoonotic viruses (those that are naturally transmitted between vertebrate animals and humans) from bats has been the subject of increasing attention from both scientists and the general public (e.g. Quammen 2013). During outbreaks of diseases in humans and livestock, bats are now often the primary focus of searches for a reservoir host (Chua et al. 2002a; Leroy et al. 2005; Li et al. 2005; Halpin et al. 2007; Towner et al. 2007; Lau et al. 2010; Wibbelt et al. 2010; Memish et al. 2013). Identification of bats as natural hosts for emerging viruses has important implications for bat conservation. We review the current state of research of four important families of emerging zoonotic viruses for which bats are natural reservoir hosts and discuss direct and indirect conservation implications.

10.2 Emerging Viral Diseases: Why Bats?

Although bats have been identified as carriers of many highly virulent human pathogens (Chen et al. 2014), evidence of pathogen-related clinical signs or disease in bats is scarce, particularly for intracellular pathogens such as viruses (Brook and Dobson 2015). Post-infection survival is supported by the frequent identification of antibodies to known viruses in apparently healthy bats and long-term survival of these bats (e.g. Hayman et al. 2010). Additionally, viruses isolated or genetically detected from bat populations are highly diverse and often ancestral to related viruses in human and other mammalian species (e.g. Towner et al. 2009; Drexler et al. 2012; Baker et al. 2013a; Tong et al. 2013; Vidgen et al. 2015). Together, these findings suggest a long history of coevolution between many bat-virus relationships identified to date. Recent progress in the field of bat immunology and genomics has identified key differences in bat immunity and physiology that evolved concomitantly with the evolution of flight, resulting not only in apparently increased immunotolerance of intracellular pathogens, but also in increased longevity and decreased tumour production (Baker et al. 2013b; Zhang et al. 2013; Brook and Dobson 2015). Immunotolerance and incomplete clearance of viral infections are also likely to favour the establishment of persistent infections (Virgin et al. 2009), as proposed for a number of bat-borne viruses (Plowright et al. 2015).

Various ecological and life-history factors play a key role in the susceptibility of individuals and populations to pathogens (Allen et al. 2009; Turmelle et al. 2010; Schneeberger et al. 2013), and notable differences exist between bats and terrestrial mammals such as rodents (Luis et al. 2013). For example, the often high-population densities and the usually gregarious roosting behaviour of bats increase the likelihood of both intra- and interspecies transmission of viruses (Luis et al. 2013; Streicker et al. 2010). Large-scale movements of bats due to their ability for powered flight are also likely to facilitate viral transmission within and among species, including the exchange of novel viruses and virus variants across biomes or even continents (Calisher et al. 2006; Epstein et al. 2009; Peel et al. 2013). The extreme relative longevity of bats compared to other mammals of similar size (Wilkinson and South 2002) and the potential for persistent and/or subclinical viral infections could further increase transmission potential (Calisher et al. 2006). Reduction of body temperature associated with hibernation of temperate zone bats lowers both viral activity and the metabolism of hosts, leading to increased incubation periods and therefore reduced likelihood of epizootic fadeout (of rabies, for example; George et al. 2011). Bats are ancient mammals in evolutionary terms, and virus utilisation of highly conserved cellular receptors could facilitate transmission to other mammals (Calisher et al. 2006), for example, as has been suggested for henipaviruses (Negrete et al. 2005). Lastly, it was recently speculated that, similar to the febrile response of other mammals, the relatively high body temperature (about 38–41 °C) and metabolism of bats during flight may select for viruses tolerant to such conditions, meaning the normal febrile defence mechanism of other mammals is ineffective ("Flight as fever hypothesis", O'Shea et al. 2014), making bat-borne viruses potentially more virulent and lethal for other, non-flying mammals.

10.3 Zoonotic Viruses of Bats and Their SpillOver

10.3.1 Rhabdoviruses

Rabies virus (RABV) is the longest and best-known member of the genus *Lyssavirus* (family Rhabdoviridae) and still one of the most significant zoonoses known from bats (recent reviews include: Banyard et al. 2011; Banyard et al. 2014 and Kuzmin 2014). The genus is rapidly expanding, with 14 of the currently recognised species (plus another known from genetic material only), and all but two (Mokola and Ikoma viruses) having been isolated from bats (Table 10.1). Lyssaviruses spill over directly from bats to domestic animals, other wildlife and humans, or indirectly to humans via these other species. All lyssaviruses are potentially neurotropic, meaning that the virus infects nerve cells and replicates in the brain, resulting in clinical signs consistent with classical rabies (Schnell et al. 2009). Although isolated from a variety of tissues and body fluids in the late stages of infection, the predominant route of transmission is via saliva (mostly via biting; Kuzmin 2014).

Lyssaviruses can be divided into two distinct "phylogroups" (Badrane et al. 2001, Table 10.1), reflecting biological and genetic differences, and they are distributed globally in bats. Classical rabies virus occurs in bats across North, Central and South America (Messenger et al. 2003; Banyard et al. 2011) and was first associated with vampire bats following an outbreak in cattle in South America in 1911 (Carini 1911). It is reported most frequently in the common

Table 10.1 Known lyssaviruses and their association with different bat species (adapted from Banyard et al. 2014)

Geographical distribution	Lyssavirus species	Phylogroup	Bat species most commonly associated with lyssavirus infection	Common name	Known human cases
The Americas	Rabies virus (RABV)	I	*Eptesicus fuscus*	Big brown bat	Yes
			Tadarida brasiliensis	Mexican/ Brazilian free-tail bat	
			Lasionycteris noctivagens	Silver-haired bat	
			Perimyotis subflavus	Tri-coloured bat	
			Desmodus rotundus	Vampire bat	
Eurasia	European bat lyssavirus type 1 (EBLV-1)	I	*Eptesicus serotinus*	Serotine bat	Yes
	European bat lyssavirus type 2 (EBLV-2)	I	*Myotis daubentonii*	Daubenton's bat	Yes
	Bokeloh bat lyssavirus (BBLV)	I	*Myotis nattereri*	Natterer's bat	No
	Aravan virus (ARAV)	I	*Myotis blythi*	Lesser mouse-eared bat	No
	Irkut virus (IRKV)	I	*Murina leucogaster*	Greater tube-nosed bat	Yes
	Khujand virus (KHUV)	I	*Myotis mystacinus*	Whiskered bat	No
	West Caucasian bat virus (WCBV)	NA[a]	*Miniopterus schreibersii*	Common bent-wing bat	No
	Lleida bat lyssavirus (LLEBV)	NA[a]	*Miniopterus schreibersii*	Common bent-wing bat	No
Africa	Duvenhage virus (DUVV)	I	*Miniopterus sp?*	Undefined	Yes
			Nycteris thebaica	Egyptian slit faced bat	
	Lagos bat virus (LBV)	II	*Eidolon helvum*	Straw-coloured fruit bat	No
			Rousettus aegyptiacus	Egyptian fruit bat	
			Epomorphorus wahlbergi	Wahlberg's epauletted fruit bat	

(continued)

Table 10.1 (continued)

Geographical distribution	Lyssavirus species	Phylogroup	Bat species most commonly associated with lyssavirus infection	Common name	Known human cases
	Mokola virus (MOKV)	II	not detected		Yes
	Shimoni bat virus (SHIBV)	II	*Hipposideros commersoni*	Commerson's leaf-nosed bat	No
	Ikoma virus (IKOV)	NA[a]	not detected		No
Australasia	Australian bat lyssavirus (ABLV)	I	*Pteropus scapulatus*[b]	Little red flying fox	Yes
			Saccolaimus flaviventris	Yellow-bellied sheath-tailed bat	
			Pteropus alecto	Black flying fox	

[a]Lyssaviral phylogenies infer WCBV, LLEBV and IKOW that are more genetically distinct from other species, and they have not yet been assigned a phylogroup (Kuzmin 2014)
[b]Barrett (2004)

vampire bat (*Desmodus rotundus*; Kuzmin et al. 2011a), which has a wide distribution across Mexico, Central America, and South America. Bites from this species appear to be responsible for the majority of human and domestic animal rabies infections of bat origin in South and Central America, with increased prey availability via expansion of livestock into new areas across the region hypothesised to be contributing to increasing incidences (Schneider et al. 2009; Ruiz and Chávez 2010). In Canada and the USA, 51 cases of human rabies transmitted by non-haematophagous bats were recognised or inferred between 1951 and 2006 (mostly silver-haired bats (*Lasionycteris noctivagans*), eastern pipistrelle bats (*Perimyotis subflavus*) and Brazilian/Mexican free-tailed bats (*Tadarida brasiliensis*)) (Constantine and Blehert 2009; Banyard et al. 2011). However, across the Americas, only 15 % of human rabies cases between 1993 and 2002 were reported as resulting from encounters with bats (Belotto et al. 2005).

Reported antibody prevalences against RABV in *D. rotundus* include 3–28 % in Peru (Streicker et al. 2012) and 12 % in Brazil (Almeida et al. 2011). Depending on the year, location and species, prevalence in other bats varies from relatively low 2 % in *T. brasiliensis* in New Mexico (Steece and Altenbach 1989) and 2.5 % in the little brown bats (*Myotis lucifugus*) in New York (Trimarchi and Debbie 1977), to 58 % in Seba's short-tailed bat (*Carollia perspicillata*) in Peru (Salmón-Mulanovich et al. 2009) and 67 % in *T. brasiliensis* in Texas (Baer and Smith 1991). As with other lyssaviruses discussed below, the potential for high antibody prevalences in bat populations and infrequent reports of mortality suggest that many individuals exposed to the virus survive, contrary to the overwhelmingly lethal nature of lyssavirus infections in other mammalian species (reviewed in Banyard et al. 2011). The mechanisms for this remain unclear.

Seven bat lyssaviruses have been isolated in Eurasia (Table 10.1). **European bat lyssavirus** type 1 and type 2 (EBLV-1 and EBLV-2; Bourhy et al. 1992) are the most widely recognised and studied. Five fatal cases of human infections with EBLV have so far been reported, three from EBLV-1 (Roine et al. 1988; Selimov et al. 1989; Botvinkin et al. 2005) and two from EBLV-2 (Lumio et al. 1986; Fooks et al. 2003; Nathwani et al. 2003). Spillover of EBLV-1 into other mammals has also been observed, but rarely, with examples including zoo bats (Rønsholt et al. 1998), sheep (Tjørnehøj et al. 2006), domestic cats (Dacheux et al. 2009) and a stone marten (Müller et al. 2004). While EBLV-1 and EBLV-2 have been detected in a range of bat species (reviewed in Schatz et al. 2013), they are most frequently associated with serotine bats (*Eptesicus serotinus*) and Daubenton's bat (*Myotis daubentonii*), respectively. The dynamics of EBLV infections in their natural hosts is poorly understood, but banding and recapture data and the frequent capture of apparently healthy bats with antibodies against EBLV suggest that many bats survive infection (Serra-Cobo et al. 2002; Amengual et al. 2007; Schatz et al. 2013). In cases where bats develop clinical symptoms of EBLV infection, the affected individuals are often unable to fly, are generally weak and show abnormal behaviour, including attempts to bite (Banyard et al. 2011). Experimental studies suggest that variable development of clinical signs may be related to inoculation route and dose (reviewed in Banyard et al. 2011).

Comparatively, little is known about the remaining Eurasian bat lyssaviruses, which have each been isolated from bats only once: West Caucasian bat virus (WCBV, Botvinkin et al. 2003), Bokeloh bat lyssavirus (BBLV, Freuling et al. 2011), Aravan virus (ARAV, Kuzmin et al. 1991), Irkut virus (IRKV, Botvinkin et al. 2003) and Khujand virus (KHUV, Kuzmin et al. 2001), or is only known from partial genetic sequence data (Lleida virus, Ceballos et al. 2013, Table 10.1). Of these, only IRKV has been detected in other mammals (a human who developed rabies after a bat bite, Leonova et al. 2009). WCBV appears to have a large geographical range. It was isolated from *Miniopterus schreibersii* in Russia, but cross-reactive antibodies have also been detected in *Miniopterus* bats in Kenya (Kuzmin et al. 2008a). The relatively wide distribution and migratory behaviour of *Miniopterus* spp. may facilitate cross-continental transmission of this virus. Alternatively, given the close relationship between WCBV and Ikoma virus (IKOV), which was recently isolated in neighbouring Tanzania, the serological findings from Kenya could in fact indicate exposure to IKOV or another related lyssavirus rather than WCBV (Marston et al. 2012; Horton et al. 2014). Similarly, serological surveys have detected antibodies against ARAV virus and KHUV virus in Indian flying foxes (*Pteropus giganteus*) from Bangladesh (Kuzmin et al. 2006), and ARAV, KHUV, IRKV or Australian bat lyssavirus in Lyle's flying foxes (*P. lylei*) and dawn bats (*Eonycteris spelaea*) from Thailand (Lumlertdacha et al. 2005). Yet, given the limited lyssavirus surveillance in bats performed to date in this region and that individuals in these studies tested positive to multiple viruses, these results likely represent cross-reactivity of serological assays to unknown lyssaviruses.

Africa also hosts significant lyssavirus diversity, with five species identified, though only three of these isolated from bats to date (Table 10.1). **Duvenhage virus** (DUVV, Meredith et al. 1971) is the only phylogroup I lyssavirus in Africa and is more closely related to RABV, ABLV and the majority of the European species than other known African lyssaviruses. Since it was first isolated from a human in 1970, two more fatal human infections of DUVV have been reported, one in South Africa in 2006 (Paweska et al. 2006) and one from the Netherlands in 2007 after obtaining the infection in Kenya (van Thiel et al. 2008). DUVV has been isolated from bats twice, once from a presumed *M. schreibersii* bat in South Africa and once from an Egyptian slit-faced bat (*Nycteris thebaica*) in Zimbabwe (Schneider et al. 1985; Foggin 1988; Paweska et al. 2006). No further information is so far available on this apparently rare African lyssavirus.

In contrast, **Lagos bat virus** (LBV) is the most widely detected lyssavirus in Africa (Banyard et al. 2011). In 1956, this virus was first isolated from a straw-coloured fruit bat (*Eidolon helvum*; Boulger and Porterfield 1958). Since then, the virus has been isolated and neutralising antibodies detected in a variety of fruit bat species, one insectivorous bat species, domestic cats, domestic dogs and a water mongoose, but not in humans (reviewed in Banyard et al. 2011). *E. helvum* and *Rousettus aegyptiacus* are likely primary reservoir hosts for LBV, with seroprevalences ranging from 6 to 80 % and 29 to 46 %, respectively, depending on the region (Hayman et al. 2008, 2012; Kuzmin et al. 2008b; Dzikwi et al. 2010; Peel et al. 2013). LBV has been isolated from healthy, rabid and dead bats (reviewed in Banyard et al. 2011), but longitudinal studies in Ghana (Hayman et al. 2012) and surveys across continental Africa (Peel et al. 2010, 2013) suggest widespread exposure, no difference in survival between seropositive and seronegative *E. helvum,* and viral persistence in very small, isolated island populations. Early infection experiments with LBV suggested that LBV and other phylogroup II viruses were less pathogenic than other lyssaviruses (Boulger and Porterfield 1958; Badrane et al. 2001). However, recent experimental infections indicated the potential for comparable mortality between LBV and RABV and indicated that significant differences might instead exist between different LBV isolates (Kuzmin et al. 2010; Markotter et al. 2009).

Of the other African lyssaviruses, only Shimoni bat virus (SHIBV) has been detected in bats (Commerson's leaf-nosed bat (*Hipposideros commersoni*) in Kenya; Kuzmin et al. 2010) and only Mokola virus (MOKV) has been detected in humans (on two occasions in Nigeria, Familusi and Moore 1972; Familusi et al. 1972). MOKV has also been isolated from cats and small wild mammals, however, the natural reservoir host is unknown (Nel 2001). Ikoma virus was isolated from a rabid African civet (*Civettictis civetta*), but it is believed that the civet was a spillover host and the true reservoir host is yet to be identified (Horton et al. 2014).

The only lyssavirus detected in Australia to date—**Australian bat lyssavirus** (ABLV)—has two known lineages, one circulating in flying foxes and one in an insectivorous bat (Fraser et al. 1996; Gould et al. 2002; Warrilow 2005). In 1996, shortly after ABLV was first isolated from a black flying fox (*P. alecto*) that was unable to fly (Fraser et al. 1996), a 39-year-old woman died of clinical rabies after being bitten by a yellow-bellied sheath-tail bat (*Saccolaimus flaviventris*; Gould et al. 2002). Two subsequent human cases have been identified, a woman who died in 1998, 27 months after being bitten by a flying fox (Hanna et al. 2000), and a child who died in 2014 after being scratched by a flying fox (Francis et al. 2014). Experimental infection of grey-headed flying foxes (*P. poliocephalus*) with ABLV resulted in clinical signs of weakness, trembling and limb paralysis in three out of ten individuals (McColl et al. 2002). As with other bat lyssaviruses, a small proportion of ABLV-positive bats succumb to encephalitis-like symptoms (Hooper et al. 1997), yet serological tests show a high prevalence of antibodies in populations of surviving bats (McColl et al. 2000).

10.3.2 Paramyxoviruses

The most notable viruses from the *Paramyxoviridae* family in bats are those of the genus *Henipavirus*, which are the subject of many reviews (e.g. Halpin and Rota 2015; Smith and Wang 2013, Luby and Gurley 2012; Clayton et al. 2013; Middleton and Weingartl 2012; Field and Kung 2011). The first recognised henipavirus, **Hendra virus** (HeV), was first detected during an outbreak of infectious respiratory disease in horses and then humans in Hendra, Australia, in 1994 (Murray et al. 1995). Ultimately, 13 of 20 infected horses died or were euthanised, and of two humans working closely with horses who became infected, one died from acute pneumonia (Murray et al. 1995; Plowright et al. 2015). This spillover was preceded a month earlier by another involving two horses and one human over 800 km away in Mackay, but which went unrecognised until 1995 (Rogers et al. 1996; O'Sullivan et al. 1997). An initial serological survey of 46 wildlife species (excluding bats) failed to identify a reservoir host; however, serological evidence of HeV infection was later identified in all four species of flying foxes native to Australia (Young et al. 1996). Virus isolation (Halpin et al. 2000) and experimental studies (Halpin et al. 2011) have confirmed pteropodid bats as reservoir hosts for henipaviruses (with a lack of clinical signs), with evidence that black (*P. alecto*) and spectacled flying foxes (*P. conspicillatus*) are the main reservoir species for HeV (Smith et al. 2014; Goldspink et al. 2015).

Because HeV is frequently detected in the urine of wild flying foxes (Smith et al. 2014), the predominant transmission route to horses is likely via material recently contaminated with bat urine (e.g. pastures) or via direct transmission (Martin et al. 2015). Recognised spillover events from bats to horses occurred sporadically from 1994 to 2004 and annually since 2006, with five spillover events resulting in ongoing transmission to humans in close contact with horses (a total of seven human cases and four deaths; Field et al. 2010). Spillover events are spatiotemporally clustered, occurring year-round in the northern tropics, but seasonally clustered in winter with a peak in July in subtropical regions (Plowright et al. 2015).

The relative importance of various hypothesised drivers of HeV dynamics in bats and subsequent spillover to horses is still unclear (Plowright et al. 2015).

Nipah virus (NiV), the second henipavirus to be recognised, was first isolated in 1999 from pigs and encephalitic pig workers in Malaysia (Center of Disease Control and Prevention 1999). NiV spillover has not been observed since this time in Malaysia; however, annual seasonal outbreaks with high case fatality (average 73 %) have occurred in people in Bangladesh since 2001 (Hsu et al. 2004; Luby et al. 2009; Luby and Gurley 2012), with occasional spillover also occurring in neighbouring India (Chadha et al. 2006; Harit et al. 2006). Due to the close relatedness of HeV and NiV, fruit bats were targeted, and serological evidence quickly identified them to be the natural reservoir of NiV (Enserink 2000; Yob et al. 2001). This was subsequently supported by isolation of NiV from the urine of *P. hypomelanus* (Chua et al. 2002a), *P. vampyrus* (Rahman et al. 2010) and *P. lylei* (Reynes et al. 2005), and seroconversion in the absence of clinical signs following experimental infections in *P. vampyrus* (Halpin et al. 2011). Antibodies against NiV and NiV-related viruses have now been detected in a variety of bat species (including non-pteropid bats) across a wide geographical area (summarised in Breed et al. 2013). NiV transmission to humans appears to occur via a wider variety of routes compared with HeV. Infection of domestic animal intermediate hosts (via consumption of saliva- or urine- contaminated partially eaten fruits or raw date palm sap) has been implicated as a source of human infections in both Malaysia and Bangladesh (Chua et al. 2002b; Chowdhury et al. 2014). In Malaysia, human infections resulted from direct contacts with infected pigs (Chua et al. 1999; Paton et al. 1999; Parashar et al. 2000), whereas in Bangladesh, transmission to humans regularly occurs via consumption of contaminated date palm sap (Luby et al. 2006; Rahman et al. 2012) or directly from human to human (e.g. via nursing sick individuals or preparation for burial; Hughes et al. 2009). The risk of direct human infection with NiV from bats is considered to be lower than horizontal transmission once the virus enters the human population (Gurley et al. 2007; Luby et al. 2009; Chong et al. 2003).

A third henipavirus, **Cedar Virus** (Marsh et al. 2012), has been isolated from urine collected under a mixed *P. alecto/P. scapulatus* roost in Australia. In contrast to HeV and NiV, however, it appears to be of low pathogenicity and failed to induce clinical signs in experimentally infected laboratory animal species (Marsh et al. 2012). Serological evidence from South-East Asia and Australasia (Breed et al. 2013) and the wide diversity of paramyxovirus sequences detected in Australia (Vidgen et al. 2015) suggest more henipaviruses are yet to be found. Additionally, although henipaviruses were long thought to be restricted to Asia and Australia, antibodies cross-reactive to HeV and NiV were detected in Madagascar in 2007, suggesting a potentially wider geographical distribution of henipa-related paramyxoviruses (Iehlé et al. 2007). This was supported by serological findings and molecular detection of henipa- or henipa-like viruses in mainland Africa and its offshore islands (Hayman et al. 2008, 2012; Peel et al. 2010, 2013; Drexler et al. 2012). Indeed, a recent serological study indicates that these viruses are also occasionally transmitted to humans in Africa (Pernet et al. 2014), though no African henipavirus has been isolated to date.

Viruses from the paramyxovirus genus *Rubulavirus* (a genus which includes the human mumps virus) have also been frequently detected in bats (Barr et al. 2015). **Menangle virus** was isolated from pigs following the birth of unusually high numbers of stillborn and deformed piglets in Australia (Philbey et al. 1998). Two piggery personnel had neutralising antibodies against Menangle virus after having recovered from an unexplained febrile illness (Philbey et al. 1998). Flying fox colonies roosting in close proximity to the piggeries were a suspected source of infection for pigs, with subsequent transmission to humans (Philbey et al. 1998). This was supported by serological evidence from *P. poliocephalus, P. alecto* and *P. conspicillatus*, and recent virus isolation from *P. alecto* (Barr et al. 2012). Other isolated bat rubulaviruses with unknown or limited understanding of their zoonotic potential include Tioman virus from Malaysia (Chua et al. 2001), Tuhokovirus 1, 2 and 3 from China (Lau et al. 2010), Achimota virus 1 and 2 from Ghana (Baker et al. 2013c) and Hervey, Grove, Teviot and Yeppoon paramyxoviruses from Australia (Barr et al. 2015). Neutralising antibodies to Tioman virus and Achimota viruses have been detected in humans, suggesting previous exposure and infection with the virus (Yaiw et al. 2007; Baker et al. 2013c). Pigs experimentally infected with Tioman virus produced neutralising antibodies and excreted virus in saliva, but were either asymptomatic or developed only a fever (Yaiw et al. 2008). Undetected infection in pigs could therefore facilitate transmission to humans.

Finally, viral fragments related to rubulaviruses and the proposed genus *Jeilongvirus* have also been detected outside the range of fruit bats, in European insectivorous bat species (Kurth et al. 2012). However, nothing is yet known about the relevance of these viruses as potentially zoonotic threats to humans.

10.3.3 Coronaviruses

Bat coronaviruses were first identified from species of the genus *Miniopterus* (Poon et al. 2005), however, with unknown zoonotic potential. The most prominent coronavirus, the one causing **severe acute respiratory syndrome (SARS),** was followed by a pandemic spread in humans after the first outbreak in China in 2002 (Rota et al. 2003). Soon after the outbreak, the virus was detected in masked palm civet (*Paguma larvata*) and raccoon dogs (*Nyctereutes procyonoides*) in a market in Guangdong Province, where SARS was first reported (Guan et al. 2003). A survey of common wildlife species in the area identified bats to be the natural reservoir of SARS coronavirus, with viruses from bats showing greater genetic diversity than the ones isolated from other species, including humans (Li et al. 2005). Bats can regularly be found in markets in China, which makes direct transmission of the virus from bats to humans likely (Li et al. 2005). The followed pandemic spread with 8096 confirmed cases of which 774 were fatal can be accounted to rapid interindividual transmission of the virus once it entered the human population (World Health Organization 2003).

Outside Asia, **SARS-like coronaviruses** have been detected in the lesser horseshoe bat (*Rhinolophus hipposideros*) from Europe (Rihtarič et al. 2010), in *Chaerephon* sp. from Kenya (Tong et al. 2009) and in *Hipposideros commersoni* from Nigeria (Quan et al. 2010). Antibodies against SARS coronavirus are present in various African bat species (Müller et al. 2007). As with many newly detected viruses, their potential threat as a zoonotic disease is yet unclear.

Since the outbreak of SARS in Asia has been traced to bats as natural hosts of the virus, the same was suspected to be the case for **Middle East respiratory syndrome (MERS)**, an infection that has been occasionally spreading among humans of the Arabian peninsula since 2012 (Zaki et al. 2012). Most human infections have been traced down to close contacts with dromedary camels (*Camelus dromedarius*), which carry a virus with a similar genome organisation as human MERS (Hemida et al. 2014). There is at least one report of direct transmission of the virus from camels to humans via contact with infected animals (Memish et al. 2014). However, a small fragment of a coronavirus PCRed from an Egyptian tomb bat (*Taphozous perforatus*) showed 100 % nucleotide identity to virus from the human index case-patient of MERS, suggesting that this species may be one of the putative natural reservoirs of the virus (Memish et al. 2013). Bat-derived MERS virus has been shown to be able to use human receptors and thus could potentially infect human cells (Yang et al. 2014). However, given the generally low prevalence of MERS virus in bat populations, a direct spillover from bats to humans is unlikely, and transmission probably happens mainly via camels as intermediate hosts (Memish et al. 2013). In fact, no other bat has yet been found to carry MERS virus since the one reported by Memish and colleagues in 2013.

The intensified search for viruses in bats worldwide has led to the detection of coronaviruses other than SARS and MERS, whose potential to be or become zoonotic has yet to be investigated (Woo et al. 2006; Tang et al. 2006; Dominguez et al. 2007; Carrington et al. 2008; Brandão et al. 2008; Misra et al. 2009; Pfefferle et al. 2009; Donaldson et al. 2010; Watanabe et al. 2010; Drexler et al. 2010; Falcón et al. 2011; Annan et al. 2013; Ge et al. 2013; Anthony et al. 2013; Ithete et al. 2013). No clinical symptoms associated with infections with SARS-like and other coronaviruses have yet been described for bats.

10.3.4 Filoviruses

Ebola virus is the most prominent filovirus, causing severe haemorrhagic fever in humans with high mortality and fast spreading among African populations. The recent outbreak in 2013 in west Africa has resulted in the most severe epidemy of Ebola so far, with more than 11,000 lethal cases (as by September 2015; according to World Health Organization;http://apps.who.int/ebola/ebola-situation-reports).

All Ebola outbreaks recorded until 2004 in Gabon and the Republic of the Congo have been linked to handling of gorilla, chimpanzee or duiker carcasses, species that can carry the Ebola virus (Leroy et al. 2004; Pigott et al. 2014). It has thus became apparent that spillover from animals to humans occurs through hunting, butchering and consumption of bushmeat (Gonzalez et al. 2005; Li and Chen 2014; Chap. 12), followed by fast human-to-human transmission (World Health Organization 2014). An outbreak of Ebola in Congo in 2007 that resulted in 260 infected humans of whom 186 died has been traced to a potential direct transmission from a dead fruit bat that the first human victim bought from hunters to eat (Leroy et al. 2009). Antibodies against Ebola virus have since been detected in a total of 14 bat species, with seroprevalences of up to 44 % depending on species and location (Olival and Hayman 2014). Experimental infection of several bat species with Ebola led to high replication of the virus, but to no apparent signs of illness, suggesting that Ebola infections are subclinical in these species (Swanepoel et al. 1996). One *Eidolon helvum* has survived for at least 13 months after being tested seropositive for Ebola virus and Lagos bat virus, indicating long-term survival of an individual bat following exposure to these viruses (Hayman et al. 2010). The recent outbreak of Ebola in Guinea and neighbouring countries in 2013—countries that are at significant distance to the previous outbreaks in central Africa—has caused speculations about a possible transmission of the virus by migrating fruit bats (Bausch and Schwarz 2014; Vogel 2014). However, as the strain of the west African Ebola virus is a genetic outlier within the known Ebola viruses, it has been argued that the west African variant may have emerged from local wildlife populations rather than from migrating individuals (Gatherer 2014). Furthermore, although speculated (Saéz et al. 2015), it is yet not clear whether the spillover of Ebola virus in west Africa originated from bats.

Marburg virus is the only filovirus that has so far been directly isolated from bats (Towner et al. 2009; Amman et al. 2012; Pourrut et al. 2005). The first outbreak of the virus was caused by a spillover from laboratory monkeys to humans in Marburg, Germany, in 1967 (Jacob and Solcher 1968). In 2007, mine workers in a cave in Uganda were diagnosed with Marburg haemorrhagic fever that potentially resulted from a spillover of the virus from a colony of *Rousettus aegyptiacus*, where 5.1 % of tested individuals carried the virus (Towner et al. 2009). The high divergence of the genome sequence of Marburg in this population suggests a long-term association of the virus with the host, leading to the assumption that bats are the natural reservoir (Towner et al. 2009). However, given that no other bat species has yet tested positive for the virus (Towner et al. 2007), and seroprevalence being generally low in *R. aegyptiacus* (Pourrut et al. 2009), spillovers from bats to humans may be rare events.

The **Reston Ebolavirus** has first been detected in 1989 in crab-eating macaques (*Macaca fascicularis*) imported from the Philippines to be used for animal testing in laboratories in Reston, USA (Jahrling et al. 1990). During a second outbreak in 1990, animal handlers developed antibodies but did not get sick (Center for Disease Control and Prevention 1990). In 2008, Reston Ebolavirus was isolated from pigs in the Philippines (Marsh et al. 2011), and soon after, some sampled *R. amplexicaudatus* had antibodies against the virus, while 16 other bat species tested negative against Reston Ebolavirus (Taniguchi et al. 2011). Screening for antibodies of the Ebola virus and Reston Ebolavirus in bats in Bangladesh has found seropositive *R. leschenaultii*, suggesting that these filoviruses or related strains are distributed at a much larger geographic range than previously assumed (Olival et al. 2013).

10.4 Main Conservation Issues Related to Bat Viruses

10.4.1 Direct Effect: Viruses Killing Bats

From all the viruses described above, only a few seem to affect bats. Although experimental infection with RABV leads to mortalities between 40 and 90 % depending on the bat species (Sétien et al. 1998; Jackson et al. 2008; Turmelle et al. 2010), there are no observed mass mortalities in natural populations (Pawan 1959). The only virus that may be largely lethal for bats is the **Lloviu virus**, which is closely related to Ebola and Marburg virus, but not yet of zoonotic relevance. It was detected during investigations of a massive die-off of *Miniopterus schreibersii* in a cave in Spain (Negredo et al. 2011). However, a causal connection between the detected virus and death of the bats has not yet been confirmed, and other bat species roosting in the same caves appeared to remain unaffected (Roué and Nemoz 2004).

The lack of reports of viruses that are detrimental for bat health should not imply that viruses in general are not of importance for the conservation of bat populations. Similar to white-nose syndrome causing mass mortalities in North American bats (Frick et al. 2010), newly emerging viruses may put local populations at threat. This may be especially the case if pathogens cross geographical borders and infect naïve bat populations. *Pseudogymnoascus destructans*—the causative fungus responsible for white-nose syndrome—likely originated from Europe, where it seemingly causes no bat fatalities, in contrast to North America (Puechmaille et al. 2010; Frick et al. 2010; Frick et al. 2015, Chap. 9).

10.4.2 Indirect Effects: Biased Public Perception

Generally, the public perception of bats as aesthetically less appealing mammals as well as folklores that often associate bats with negative stigma makes bat-related conservation efforts time-consuming and demanding (Fenton 1997; Allen 2004; Knight 2008). The recent outbreaks of viral zoonotic diseases with the identification of bats as putative natural hosts have further complicated bat conservation efforts (Li et al. 2005; Knight 2008). Following numerous and often lurid reports of fatal zoonotic diseases by the media, public perception of bats is mostly skewed by fear and lack of information (Kingston 2016, Chap. 18). Therefore, it is important to highlight the context of bat-associated infections in order to provide more evidence-based information about the emergence and transmission of bat-related zoonotic diseases, which may lead to a more balanced reputation of bats. Depending on educational, cultural, legal and medial background of the targeted audience, specific aspects need to be taken into account.

In Europe and North America, rabies is, so far, the only viral disease that is associated with bats. The fact that lyssaviruses are occasionally found in temperate zone bats sometimes finds its way to the media, not always in favour of bats. Biased newspaper articles or press campaigns may result in the public misconception that bats are aggressive animals or that their mere presence can lead to human infections with these viruses. Although there are anecdotal reports of unprovoked attacks of bats on humans and dogs (Baer and Smith 1991), bats, as is the case of most mammals, usually only bite when handled or provoked. Furthermore, once bitten or scratched by a bat, immediate post-exposure vaccination can prevent a person from contracting rabies (see Sect. 10.5.2). In the case of the 37-year-old woman who died from a bat lyssavirus infection in Kenya, staff members of the health facility which the woman visited after being scratched by a bat were unaware of the possibility of rabies transmission (van Thiel et al. 2009). Likewise, two persons in Europe who worked regularly with bats and died from rabies after being bitten and scratched by bats received neither pre- nor post-exposure treatment (Roine et al. 1988; Nathwani et al. 2003). These two cases triggered a Europe-wide serological screening effort involving more than 11,000 bats, with seroprevalences varying depending on the species and location (Racey et al. 2012). EBLV-1 was most commonly detected in the serotine bat (*Eptesicus serotinus*), while EBLV-2 was very uncommon in all bat species. As a result, the public has been persuaded not to handle bats or to do so only with gloves and, in the case of bat workers, to receive pre- and/or post-exposure immunisation. Two fatal cases in which persons contracted rabies in Australia (Samaratunga et al. 1998; Hanna et al. 2000) triggered a similar campaign on this continent (Speare et al. 1997, but see Francis et al. 2014). Efficient education of medical professionals worldwide seems to be pivotal for implementing the correct treatment after scratches or bites from bats. In addition, vaccination should be mandatory for those who are frequently exposed to bats (Rupprecht and Gibbons 2004). Studies on animal models have shown that rabies vaccine also provides protection against other, although not all, lyssaviruses' variants (Brookes et al. 2005; Hanlon et al. 2005). However, there is no known case of a person developing bat-associated rabies despite having been vaccinated, neither pre- nor post-exposure. Thus, getting infected by some sort of bat-related virus is unlikely in Europe and North America and decreases virtually to zero if people who experienced bat bites and scratches are treated appropriately.

There is no case known for paramyxoviruses having spilled over to humans by direct contact with bats. An extensive serological survey among people frequently handling bats in Australia revealed no antibodies against Hendra virus (Arklay et al. 1996). The virus apparently needs horses as amplifier hosts, from where the virus can further be transmitted to persons in close contact with infected individuals. Nevertheless, the outbreak of Hendra increased the unpopularity of flying foxes in Australia, making conservation of the four native species challenging (Thiriet 2011). Unlike Hendra, Nipah virus has likely been acquired by humans via consumption of contaminated date palm sap (Luby et al. 2006; Rahman et al. 2012), followed by person-to-person transmission (Gurley et al. 2007). Although diseases associated with Hendra virus and Nipah virus have high mortality rates, the risk of infection for humans seems to be low (Chong et al. 2003), and countermeasures may be taken in order to prevent future spillover events (see Sect. 10.5.2). MERS, just as Hendra virus, apparently needs livestock as an amplifier host. In contrast to dromedaries (Hemida et al. 2014), seroprevalence of MERS seems to be low in bats (Memish et al. 2013), making direct transmission from bats to humans unlikely. As long as details on MERS infections in dromedaries and how to mitigate them are missing, it is hard to give recommendations to people who might be at risk.

In contrast to MERS, the spillover of SARS into the human populations most likely happened via the wildlife market, either directly from a bat, or from other wildlife species. Likewise, the hunting, butchering and consumption of chimpanzees, gorillas and bats seem to have been sources of Ebola spillovers from wildlife to humans. The education of local communities needs to carefully balance information about the potential risk of acquiring infectious diseases by consuming bushmeat, without implying that bats need to be eradicated in order to prevent spillovers. The recent outbreak of Ebola resulting in several thousand human victims, and with bats frequently being reported as the likely source of origin, has undoubtly led to severe loss of reputation of bats on this continent, which makes the conservation of threatened populations and species even more challenging, not only in Africa, but also worldwide.

10.4.3 Indirect Effect—Culling

The direct persecution of bats often seems to be the most effective way to deal with bat-borne diseases to members of the public. Killing of bats has long been acceptable, even if they are protected (Chap. 14). Even though culling may be officially banned and thus not supported by authorities or governmental programs, large-scale killing of bats or the destruction of roost trees may still be commonly practiced in areas where zoonotic diseases are spreading.

In Australia, for example, flying foxes are frequently harassed and killed, both legally (under permits issued by state wildlife management agencies) and illegally. This happened most prominently during periods when Hendra virus emerged in Australian flying fox populations (Roberts et al. 2012). Half of the flying fox species native to Australia have declined about 30 % in population size during the last decade, and killing of bats usually does not lead to legal measures (Booth 2005). Furthermore, large-scale culling leads to a change of movement behaviour of bats, with new, susceptible individuals being recruited from nearby colonies

(Field 2009). Instead of reducing the viral prevalence, this may therefore lead to the exact opposite (see below).

In the attempt to reduce rabies incidences, vampire bats are regularly culled in many parts of Latin America (Streicker et al. 2012). In Brazil, for example, governmental programs are in action that involve targeted campaigns against vampire bats. During these measures, vampire bats are captured and poisoned or coated with anticoagulant and released, so that allogrooming kills their conspecifics (Medellin 2003). Furthermore, bat roosts are destroyed using fire and explosives (Mayen 2003), which also leads to dramatic declines of non-target bats (Furey and Racey 2015, Chap. 15). Besides the questionable methods involved, instead of reducing viral abundance in the population, culling of wildlife can lead to an increase in viral spreading. New hosts are recruited and the dispersal probability of infected individuals increases, which results in transmission of the disease to naïve hosts (Donnelly et al. 2005; Choisy and Rohani 2006; Streicker et al. 2012). This was the case for vampire bats in Peru, where culling failed to reduce seroprevalence of rabies in bat populations, but rather had the opposite effect (Streicker et al. 2012). Therefore, persecution of bats as potential carriers of zoonotic diseases has been denounced as useless and even counterproductive by both conservationists and experts on disease transmission (Hutson and Mickleburgh 2001; Knight 2008).

10.4.4 Indirect Effect—Killing of Bats for Virus Surveys

In the scope of recently emerging zoonotic diseases, the search for new bat-borne viruses has become a well-funded field in the scientific community. While research is important to advance our understanding about the emergence of diseases and to possibly prevent further spillover events, the methods involved in these surveys are sometimes questionable from the perspective of bat conservation (Racey 2015). Some of the investigated bat species are listed as near threatened or vulnerable by the International Union for the Conservation of Nature (IUCN), with decreasing population sizes even in many species of least concern. While most surveillance studies that involve species of conservation concern use non-lethal methods such as antibody screening in blood (Hayman et al. 2008; Young et al. 1996; Lumlertdacha et al. 2005; Wacharapluesadee et al. 2005; Reynes et al. 2005), others have involved the killing of a considerable number of bats of various conservation status (e.g. in Yob et al. 2001; Kuzmin et al. 2008b, 2010, 2011b; Dzikwi et al. 2010 and Sasaki et al. 2012). In order to limit such detrimental surveys, the Food and Agriculture Organization of the United Nations (2011) has published a guideline for investigating the role of bats in emerging zoonotic diseases, including non-invasive protocols, which not only reduce the impact on bat populations, but also minimise the transmission risk of viral diseases. Such protocols have now been widely adopted, as for example by Ecohealth Alliance and other international research groups and networks.

10.5 Counter Measures in Favour of Bat Conservation

10.5.1 Preventing the Emergence of New Viral Diseases

In general, preventing the emergence of infectious diseases in wildlife populations is extremely challenging and usually underfunded, with only few practical suggestions being discussed (Daszak et al. 2000). For example, it is important that translocations of animals across geographical borders need to follow strict guidelines in order to prevent the introduction of exotic pathogens in novel areas (e.g. Woodroofe 1999). Furthermore, an integration of knowledge about disease dynamics, as well as ecological and immunological aspects of the host, may contribute to a better understanding of emerging infectious diseases in wildlife species such as bats (Daszak et al. 2000).

10.5.2 Educational Efforts

As many bat-borne viral diseases have high lethality rates for humans, preventing spillover events are of central importance. In particular, spillover by direct contact to bats, such as via bites or bat consumption, may bear severe risks to humans that could be minimised by educational programs (Kingston 2016, Chap. 18). Reducing the risk of outbreaks of zoonotic viruses may also lead to more positive attitudes towards bats, which may further be increased by highlighting their ecological importance as pollinators, seed dispersers and pest control for agriculture (Ghanem and Voigt 2012). Moreover, conservation measures that promote the preservation of bat habitats serve a dual role as they can decrease the contact zone between bats and humans, thus reducing the risk of spillover.

As aforementioned vaccination against rabies and other lyssaviruses should be mandatory for persons working with bats and recommended for other people at risk. A significant problem is that both pre- and post-exposure treatments are expensive and thus may not be readily available in developing countries, such as in Central and South America. Here, building houses in a bat-proof manner in order to avoid vampire bites during sleep and decreasing the risk of direct contact with other bats has so far been the best solution (Greenhall 1964; Voigt et al. 2016, Chap. 14).

A different issue is the transmission of Nipah viruses via consuming raw date palm sap contaminated by urine, faeces or saliva of bats (Luby et al. 2006; Rahman et al. 2012). Here, cooking the sap at temperatures above the level that viruses tolerate is an effective measure to prevent spillover (Hughes et al. 2009). Additionally, preventing bats from accessing date palms and thus contaminating the sap has been proved to be both efficient and relatively cheap (Nahar et al. 2010, 2013). The traditional "bamboo skirt" method for example uses inexpensive, recyclable bamboo to cover the part of the date palm where the sap is collected, preventing bats and

Fig. 10.1 Intact trees with colonies of *Eidolon helvum* (*left*) in Yaoundé, Cameroon, as compared to former roosting trees that have been cut (*right*) after bats were suspected to be the source of the recent Ebola outbreak in western Africa (photograph credits: Simon Ghanem)

other vertebrates from getting access. Furthermore, in contrast to bird nets, this measure is non-lethal to the bats and therefore of high conservation value to local populations. However, such protective measures are reported to be rarely used in Bangladesh (Nahar et al. 2010, 2013). This could potentially be changed by encouraging local farmers to use this method, emphasising its inexpensiveness and efficiency while highlighting the reduced risk of acquiring Nipah virus disease.

One of the key issues both for conservation and public health is the direct transmission of SARS and Ebola via wildlife markets. In South-East Asia, flying foxes are hunted regularly for the purpose of food (Mickleburgh et al. 2002; Mildenstein et al. 2016, Chap. 12), sometimes even authorised by the local Wildlife Department such as in Malaysia (Breed et al. 2006). Likewise, fruit bats are consumed regularly throughout Africa (Mickleburgh et al. 2009; Mildenstein et al. 2016, Chap. 12). Since bats are suggested as potential reservoir for the recent outbreak of Ebola, Guinea banned bats for sale from markets (Gatherer 2014). Educational efforts to reduce the threat both to public health by zoonotic diseases and to the conservation of local bat populations are challenging, as they are usually impeded by the lack of understanding of entrenched cultural behaviours and social components (Pooley et al. 2015; Kingston 2016, Chap. 18). In Ghana, for example, where the consumption of bats is part of the local culture and traditions, a survey revealed that knowledge about the ecological and economical value of bats would not make people refrain from killing and eating bats (Kamins et al.

2014). Usually, the direct economic benefit from selling hunted bats is more valuable to an individual person than the indirect, not always obvious economic value of bats, for example, for agriculture. However, about half of the hunters stated they would stop hunting bats if they could make them sick (Kamins et al. 2014). This highlights the potential effectiveness of public education, but careful consideration is needed to avoid demonising bats in the process (Pooley et al. 2015). The recent Ebola epidemic in western Africa for example has led to an increase in the persecution of bats, with roosts being destroyed and colonies being killed by communities (Fig. 10.1). Although preventing bats from being consumed may have higher priorities due to public health reasons, the culling of whole colonies as a likely result may be much more of a threat for the conservation of bats than the bushmeat trade (Pooley et al. 2015).

10.5.3 Environmental Conservation

Combining knowledge about the ecology of the host species as well as the disease dynamics of the virus may be crucial for establishing efficient disease prevention programs (e.g. Plowright et al. 2015). Here, it needs to be noted that the emergence of zoonotic diseases from bats also seems to be a consequence of anthropogenic alteration of natural environments (e.g. Daszak et al. 2001). For example, in Central and South America, the conversion of forested habitats into pastures shifted the dominant food source for vampire bats from native vertebrates to livestock. This has increased rabies transmission from vampire bats to livestock and domestic animals in many parts of Latin America (Schneider et al. 2009). Where bat habitats have been converted largely into agricultural farmland, the remaining bat populations are forced to concentrate in patches that provide them with resources they need. Flying foxes, for example, are highly sensitive to landscape modifications, as they require large forested areas for foraging. Where natural habitats are scarce, flying foxes may use fruiting or flowering trees in agricultural, suburban and urban areas, which increases the contact zone and spillover risk between bats and livestock or humans (Daszak et al. 2006; Plowright et al. 2015). Indeed, contact between bats and naïve hosts as a consequence of human landscape modification and encroachment likely sparked the transmission of Hendra viruses to horses (Epstein et al. 2006) and Nipah virus to pigs (Chua et al. 1999; Field et al. 2001).

10.5.4 Conservation of Bat Populations and Population Dynamics

Removing individuals or colonies from regional populations, either by unsustainable hunting or culling, can cause an increase in relative local resource availability, creating regional gradients along which bats from other populations may move, which may lead to an increase of virus movement (Field 2009). In Australia, for example, roosts that became empty after culling, disturbing or relocating colonies of flying foxes are usually reoccupied by immigrating individuals (Roberts et al. 2012).

Anthropogenic transformation of bat habitats in Australia has also been shown to lead to decreased migration in *Pteropus* bats, which can itself lead to a decline in population immunity (Plowright et al. 2011). This could give rise to more viral shedding after local viral reintroduction, a mechanism that may be facilitated by urban habituation of fruit bat and the resulting increased contact with human and domestic animal populations (Epstein et al. 2006; Plowright et al. 2011). In Australia, all recently emerged bat-associated viruses—Hendra, Menangle and Australian bat lyssavirus—are hypothesised to be associated with habitat loss due to deforestation and agricultural intensification (Jones et al. 2013). Therefore, protection of remaining natural habitats of bats along with farm management aiming at decreasing the contact zone between bats and livestock as well as education plans increasing awareness of environmental issues and safety may play a crucial role in the avoidance of future spillovers of bat-borne diseases to livestock and human populations, and promote further protection of local bat populations.

10.6 Conclusion

Bats harbour viruses that may become zoonotic. Circumstances facilitating spillover include direct contact with bats (bites, scratches, consumption of bats), contact with material contaminated by bat saliva, faeces or urine and amplification via intermediate hosts such as domestic animals or other wildlife species. Conservational actions are not only important to prevent spillovers, but also because emerging zoonotic viruses often lead to persecution of bats. In order to reduce the transmission risk of viruses from bats to human and livestock and to protect bat species at threat, educational efforts are needed. However, entrenched cultural and social components often act as barriers to efficient changes on how people think about and respond to bats. Whenever possible, educational efforts should be done in an informative, non-lurid way, presenting the facts rather than provoking additional fears to the already bad reputation of bats. Wherever possible, solutions should be found to enable the existence of bats in anthropogenic landscape, including the development of more affordable and readily available vaccinations (e.g. against rabies), and the reduction of potential contact between bats and humans and livestock. This however also includes that the natural habitats of bats need to be better protected to provide bat populations with sufficient space and to prevent range expansion into urban and suburban areas, where contact with humans and livestock may increase the risk of spillover events. Bat-borne viruses should be considered during bat conservation efforts, and it should be equally noticed that appropriate conservation measures may even reduce the risk of viral spillover from bat populations into human populations.

Acknowledgments We thank Alison Peel, Paul Racey and an anonymous reviewer for constructive comments that helped to improve the manuscript.

References

Allen GM (2004) Bats: biology, behavior, and folklore. Dover Publications

Allen LC, Turmelle AS, Mendonça MT, Navara KJ, Kunz TH, McCracken GF (2009) Roosting ecology and variation in adaptive and innate immune system function in the Brazilian free-tailed bat (*Tadarida brasiliensis*). J Comp Physiol B 179(3):315–323

Almeida MFD, Martorelli LFA, Sodré MM et al (2011) Rabies diagnosis and serology in bats from the State of São Paulo, Brazil. Rev Soc Bras Med Trop 44(2):140–145

Amengual B, Bourhy H, López-Roig M, Serra-Cobo J (2007) Temporal dynamics of European bat Lyssavirus type 1 and survival of *Myotis myotis* bats in natural colonies. PLoS ONE 2(6):e566

Amman BR, Carroll SA, Reed ZD et al (2012) Seasonal pulses of Marburg virus circulation in juvenile *Rousettus aegyptiacus* bats coincide with periods of increased risk of human infection. PLoS Pathog 8:e1002877

Annan A, Baldwin HJ, Corman VM et al (2013) Human betacoronavirus 2c EMC/2012–related viruses in bats, Ghana and Europe. Emerg Infect Dis 19(3):456

Anthony S, Ojeda-Flores R, Rico-Chávez O et al (2013) Coronaviruses in bats from Mexico. J Gen Virol 94(Pt 5):1028–1038

Arklay A, Selvey L, Taylor R, Gerrard J (1996) Screening of bat carers for antibodies to equine morbillivirus. Commun Dis Intell 20:477

Badrane H, Bahloul C, Perrin P, Tordo N (2001) Evidence of two Lyssavirus phylogroups with distinct pathogenicity and immunogenicity. J Virol 75(7):3268–3276

Baer GM, Smith J (1991) Rabies in nonhematophagous bats. Nat Hist Rabies 2:105–120

Baker KS, Leggett RM, Bexfield NH et al (2013a) Metagenomic study of the viruses of African straw-coloured fruit bats: detection of a chiropteran poxvirus and isolation of a novel adenovirus. Virology 441(2):95–106

Baker ML, Schountz T, Wang LF (2013b) Antiviral immune responses of bats: a review. Zoonoses Public Health 60(1):104–116

Baker KS, Todd S, Marsh GA et al (2013c) Novel, potentially zoonotic paramyxoviruses from the African straw-colored fruit bat *Eidolon helvum*. J Virol 87(3):1348–1358

Banyard AC, Hayman D, Johnson N, McElhinney L, Fooks AR (2011) Bats and lyssaviruses. Adv Virus Res 79:239–289

Banyard AC, Evans JS, Luo TR, Fooks AR (2014) Lyssaviruses and bats: emergence and zoonotic threat. Viruses 6(8):2974–2990

Barr JA, Smith C, Marsh GA, Field H, Wang LF (2012) Evidence of bat origin for Menangle virus, a zoonotic paramyxovirus first isolated from diseased. J Gen Virol 93:2590–2594

Barr J, Smith C, Smith I et al (2015) Isolation of multiple novel paramyxoviruses from pteropid bat urine. J Gen Virol 96:24–29

Barrett JL (2004) Australian bat lyssavirus. Ph.D. thesis, University of Queensland, Brisbane

Bausch DG, Schwarz L (2014) Outbreak of Ebola virus disease in Guinea: where ecology meets economy. PLoS Neglected Trop Dis 8(7):e3056

Belotto A, Leanes L, Schneider M, Tamayo H, Correa E (2005) Overview of rabies in the Americas. Virus Res 111(1):5–12

Booth C (2005) Time to stop the killing. The Australasian Bat Society Newsletter 24
Botvinkin AD, Poleschuk EM, Kuzmin IV et al (2003) Novel lyssaviruses isolated from bats in Russia. Emerg Infect Dis 9(12):1623
Botvinkin A, Selnikova O, Antonova L, Moiseeva A, Nesterenko EY, Gromashevsky L (2005) Human rabies case caused from a bat bite in Ukraine. Rabies Bull Eur 29(3):5–7
Boulger L, Porterfield J (1958) Isolation of a virus from Nigerian fruit bats. Trans R Soc Trop Med Hyg 52(5):421–424
Bourhy H, Kissi B, Lafon M, Sacramento D, Tordo N (1992) Antigenic and molecular characterization of bat rabies virus in Europe. J Clin Microbiol 30(9):2419–2426
Brandão PE, Scheffer K, Villarreal LY et al (2008) A coronavirus detected in the vampire bat *Desmodus rotundus*. Braz J Infect Dis 12(6):466–468
Breed AC, Field HE, Epstein JH, Daszak P (2006) Emerging henipaviruses and flying foxes–conservation and management perspectives. Biol Conserv 131(2):211–220
Breed AC, Meers J, Sendow I et al (2013) The distribution of Henipaviruses in Southeast Asia and Australasia: Is Wallace's line a barrier to Nipah Virus? PLoS ONE 8(4):e61316
Brook CE, Dobson AP (2015) Bats as 'special' reservoirs for emerging zoonotic pathogens. Trends Microbiol 23:172–180
Brookes S, Healy D, Fooks A (2005) Ability of rabies vaccine strains to elicit cross-neutralising antibodies. Dev Biol (Basel) 125:185–193
Calisher CH, Childs JE, Field HE, Holmes KV, Schountz T (2006) Bats: important reservoir hosts of emerging viruses. Clin Microbiol Rev 19(3):531–545
Carini A (1911) About one large epizootie of rabies. Ann Inst Pasteur 25:843–846
Carrington CV, Foster JE, Zhu HC et al (2008) Detection and phylogenetic analysis of group 1 coronaviruses in South American bats. Emerg Infect Dis 14 (12):1890
Ceballos NA, Morón SV, Berciano JM et al (2013) Novel lyssavirus in bat, Spain. Emerg Infect Dis 19(5):793
Center for Disease Control and Prevention (1990) Epidemiologic notes and reports updates: filovirus infection in animal handlers. Morb Mortal Wkly Rep 39:221
Center for Disease Control and Prevention (1999) Outbreak of Hendra-like virus–Malaysia and Singapore, 1998–1999. Morb Mortal Wkly Rep 48:265
Chadha MS, Comer JA, Lowe L et al (2006) Nipah virus-associated encephalitis outbreak, Siliguri, India. Emerg Infect Dis 12(2):235–240
Chen L, Liu B, Yang J, Jin Q (2014) DBatVir: the database of bat-associated viruses. Database:bau021. doi:10.1093/database/bau021
Choisy M, Rohani P (2006) Harvesting can increase severity of wildlife disease epidemics. Proc R Soc B: Biol Sci 273(1597):2025–2034
Chong HT, Tan C, Goh K, Lam S, Chua K (2003) The risk of human Nipah virus infection directly from bats (*Pteropus hypomelanus*) is low. Neurol J Southeast Asia 8:31–34
Chowdhury S, Khan SU, Crameri G et al (2014) Serological evidence of Henipavirus exposure in cattle, goats and pigs in Bangladesh. PLoS Neglected Trop Dis 8:e3302
Chua KB, Goh KJ, Wong KT et al (1999) Fatal encephalitis due to Nipah virus among pig-farmers in Malaysia. Lancet 354(9186):1257–1259
Chua K, Wang L, Lam SK et al (2001) Tioman virus, a novel paramyxovirus isolated from fruit bats in Malaysia. Virology 283 (2):215–229
Chua KB, Lek Koh C, Hooi PS et al (2002a) Isolation of Nipah virus from Malaysian Island flying-foxes. Microbes Infect 4(2):145–151
Chua KB, Chua BH, Wang CW (2002b) Anthropogenic deforestation, El Nino and the emergence of Nipah virus in Malaysia. Malays J Pathol 24(1):15–21
Clayton BA, Wang LF, Marsh GA (2013) Henipaviruses: an updated review focusing on the pteropid reservoir and features of transmission. Zoonoses Public Health 60(1):69–83
Constantine DG, Blehert DS (2009) Bat rabies and other lyssavirus infections. US Geological Survey
Dacheux L, Larrous F, Mailles A et al (2009) European bat lyssavirus transmission among cats, Europe. Emerg Infect Dis 15(2):280

Daszak P, Cunningham AA, Hyatt AD (2000) Emerging infectious diseases of wildlife–threats to biodiversity and human health. Science 287(5452):443–449
Daszak P, Cunningham AA, Hyatt AD (2001) Anthropogenic environmental change and the emergence of infectious diseases in wildlife. Acta Trop 78:103–116
Daszak P, Plowright R, Epstein J et al (2006) The emergence of Nipah and Hendra virus: pathogen dynamics across a wildlife-livestock-human continuum. Dis Ecol: Community Struct Pathog Dyn: 186–201
Dominguez SR, O'Shea TJ, Oko LM, Holmes KV (2007) Detection of group 1 coronaviruses in bats in North America. Emerg Infect Dis 13(9):1295
Donaldson EF, Haskew AN, Gates JE, Huynh J, Moore CJ, Frieman MB (2010) Metagenomic analysis of the viromes of three North American bat species: viral diversity among different bat species that share a common habitat. J Virol 84(24):13004–13018
Donnelly CA, Woodroffe R, Cox D et al (2005) Positive and negative effects of widespread badger culling on tuberculosis in cattle. Nature 439(7078):843–846
Drexler JF, Gloza-Rausch F, Glende J et al (2010) Genomic characterization of severe acute respiratory syndrome-related coronavirus in European bats and classification of coronaviruses based on partial RNA-dependent RNA polymerase gene sequences. J Virol 84(21):11336–11349
Drexler JF, Corman VM, Müller MA et al (2012) Bats host major mammalian paramyxoviruses. Nat Commun 3:796
Dzikwi AA, Kuzmin II, Umoh JU, Kwaga JK, Ahmad AA, Rupprecht CE (2010) Evidence of Lagos bat virus circulation among Nigerian fruit bats. J Wildl Dis 46(1):267–271
Enserink M (2000) Malaysian researchers trace Nipah virus outbreak to bats. Science 289(5479):518–519
Epstein JH, Field HE, Luby S, Pulliam JR, Daszak P (2006) Nipah virus: impact, origins, and causes of emergence. Curr Infect Dis Rep 8(1):59–65
Epstein JH, Olival KJ, Pulliam JRC et al (2009) *Pteropus vampyrus*, a hunted migratory species with a multinational home-range and a need for ro regional management. J Appl Ecol 46:991–1002
Falcón A, Vázquez-Morón S, Casas I et al (2011) Detection of alpha and betacoronaviruses in multiple Iberian bat species. Arch Virol 156(10):1883–1890
Familusi JB, Moore DL (1972) Isolation of a rabies related virus from the cerebrospinal fluid of a child with 'aseptic meningitis'. Afr J Med Sci 3(1):93–96
Familusi JB, Osunkoya BO, Moore DL, Kemp GE, Fabiyi A (1972) A fatal human infection with Mokola virus. Am J Trop Med Hyg 21(6):959–963
Fenton MB (1997) Science and the conservation of bats. J Mammal: 1–14
Field H, Young P, Yob JM, Mills J, Hall L, Mackenzie J (2001) The natural history of Hendra and Nipah viruses. Microbes Infect 3(4):307–314
Field HE (2009) Bats and emerging zoonoses: Henipaviruses and SARS. Zoonoses Public Health 56:278–284
Field H, Schaaf K, Kung N et al (2010) Hendra virus outbreak with novel clinical features, Australia. Emerg Infect Dis 16(2):338
Field H, Kung N (2011) Henipaviruses—unanswered questions of lethal zoonoses. Curr Opin Virol 1(6):658–661
Foggin C (1988) Rabies and rabies-related viruses in Zimbabwe: historical, virological and ecological aspects. University of Zimbabwe
Food and Agricultural Organisation of the United Nations (2011) Investigating the role of bats in emerging zoonoses: balancing ecology, conservation and public health interests. In: Newman SH, Field HE, de Jong CE, Epstein JH (eds) FAO animal production and health manual no 12, Rome
Fooks AR, McElhinney LM, Pounder DJ et al (2003) Case report: isolation of a European bat lyssavirus type 2a from a fatal human case of rabies encephalitis. J Med Virol 71(2):281–289

Francis JR, Nourse C, Vaska VL et al (2014) Australian bat lyssavirus in a child: the first reported case. Pediatrics 133(4):e1063–e1067

Fraser GC, Hooper PT, Lunt RA et al (1996) Encephalitis caused by a Lyssavirus in fruit bats in Australia. Emerg Infect Dis 2(4):327

Freuling CM, Beer M, Conraths FJ et al (2011) Novel lyssavirus in Natterer's bat, Germany. Emerg Infect Dis 17(8):1519–1522

Frick WF, Pollock JF, Hicks AC et al (2010) An emerging disease causes regional population collapse of a common North American bat species. Science 329:679–682

Frick WF, Puechmaille SJ, Willis CKR (2016) White-nose syndrome in bats. In: Voigt CC, Kingston T (eds) Bats in the Anthropocene: conservation of bats in a changing world. Springer International AG, Cham, pp. 245–259

Furey N, Racey P (2016) Conservation ecology of cave bats. In: Voigt, CC, Kingston, T (eds) Bats in the Anthropocene: conservation of bats in a changing world. Springer International AG, Cham, pp. 463–492

Gatherer D (2014) The 2014 Ebola virus disease outbreak in West Africa. J Gen Virol: vir 0.067199-067190

Ge X-Y, Li J-L, Yang X-L et al (2013) Isolation and characterization of a bat SARS-like coronavirus that uses the ACE2 receptor. Nature 503:535–538

George DB, Webb CT, Farnsworth ML et al (2011) Host and viral ecology determine bat rabies seasonality and maintenance. Proc Natl Acad Sci 108(25):10208–10213

Ghanem SJ, Voigt CC (2012) Increasing awareness of ecosystem services provided by bats. Adv Study Behav 44:279–302

Gonzalez J-P, Herbreteau V, Morvan J, Leroy EM (2005) Ebola virus circulation in Africa: a balance between clinical expression and epidemiological silence. Bull de la Société de Pathol Exotique 98(3):210–217

Gould AR, Kattenbelt JA, Gumley SG, Lunt RA (2002) Characterisation of an Australian bat lyssavirus variant isolated from an insectivorous bat. Virus Res 89(1):1–28

Goldspink LK, Edson D, Vidgen M, Bingham J, Field H, Smith C (2015) Natural Hendra virus infection in flying-foxes, tissue tropism and risk factors. PLoS ONE 10(6):e0128835

Greenhall AM (1964) Bats: their public health importance and control with special reference to Trinidad. In: Proceedings of the 2nd vertebrate pest control conference, p. 18

Guan Y, Zheng B, He Y et al (2003) Isolation and characterization of viruses related to the SARS coronavirus from animals in southern China. Science 302(5643):276–278

Gurley ES, Montgomery JM, Hossain MJ et al (2007) Person-to-person transmission of Nipah virus in a Bangladeshi community. Emerg Infect Dis 13(7):1031

Halpin K, Young P, Field H, Mackenzie J (2000) Isolation of Hendra virus from pteropid bats: a natural reservoir of Hendra virus. J Gen Virol 81(8):1927–1932

Halpin K, Hyatt AD, Plowright RK et al (2007) Emerging viruses: coming in ona wrinkled wing and a prayer. Clin Infect Dis 44:711–717

Halpin K, Hyatt AD, Fogarty R et al (2011) Pteropid bats are confirmed as the reservoir hosts of henipaviruses: a comprehensive experimental study of virus transmission. Am J Trop Med Hyg 85(5):946–951

Halpin K, Rota P (2015) A review of Hendra virus and Nipah virus infections in man and other animals. In: Zoonoses-infections affecting humans and animals. Springer Netherlands, pp. 997–1012

Hanlon CA, Kuzmin IV, Blanton JD, Weldon WC, Manangan JS, Rupprecht CE (2005) Efficacy of rabies biologics against new lyssaviruses from Eurasia. Virus Res 111(1):44–54

Hanna JN, Carney IK, Smith GA et al (2000) Australian bat lyssavirus infection: a second human case, with a long incubation period. Med J Aust 172(12):597–599

Harit AK, Ichhpujani RL, Gupta S et al (2006) Nipah/Hendra virus outbreak in Siliguri, West Bengal, India in 2001. Indian J Med Res 123(4):553–560

Hayman DT, Fooks AR, Horton D et al (2008) Antibodies against Lagos bat virus in megachiroptera from West Africa. Emerg Infect Dis 14(6):926

Hayman DT, Emmerich P, Yu M et al (2010) Long-term survival of an urban fruit bat seropositive for Ebola and Lagos bat viruses. PLoS ONE 5(8):e11978

Hayman DT, Fooks AR, Rowcliffe JM et al (2012) Endemic Lagos bat virus infection in *Eidolon helvum*. Epidemiol Infect 140(12):2163–2171
Hemida MG, Chu D, Poon L, Perera R, Alhammadi M, Ng H (2014) MERS coronavirus in dromedary camel herd, Saudi Arabia. Emerg Infect Dis 20(7):1231–1234
Hooper P, Lunt R, Gould A et al (1997) A new lyssavirus—fhe first endemic-related virus recognized in Australia. Bull de l'Institut Pasteur 95(4):209–218
Horton DL, Banyard AC, Marston DA et al (2014) Antigenic and genetic characterization of a divergent African virus, Ikoma lyssavirus. J Gen Virol 95:1025–1032
Hsu VP, Hossain MJ, Parashar UD et al (2004) Nipah virus encephalitis reemergence, Bangladesh. Emerg Infect Dis 10(12):2082–2087
Hughes JM, Wilson ME, Luby SP, Gurley ES, Hossain MJ (2009) Transmission of human infection with Nipah virus. Clin Infect Dis 49(11):1743–1748
Hutson AM, Mickleburgh SP (2001) Microchiropteran bats: global status survey and conservation action plan, vol 56. IUCN
Iehlé C, Razafitrimo G, Razainirina J (2007) Henipavirus and Tioman virus antibodies in pteropodid bats, Madagascar. Emerg Infect Dis 13:159
Ithete NL, Stoffberg S, Corman VM et al (2013) Close relative of human Middle East respiratory syndrome coronavirus in bat, South Africa. Emerg Infect Dis 19(10):1697
Jackson FR, Turmelle AS, Farino DM, Franka R, McCracken GF, Rupprecht CE (2008) Experimental rabies virus infection of big brown bats (*Eptesicus fuscus*). J Wildl Dis 44(3):612–621
Jacob H, Solcher H (1968) An infectious disease transmitted by *Cercopithecus aethiops* ("marburg disease") with glial nodule encephalitis. Acta Neuropathol 11(1):29
Jahrling P, Geisbert T, Johnson E, Peters C, Dalgard D, Hall W (1990) Preliminary report: isolation of Ebola virus from monkeys imported to USA. Lancet 335(8688):502–505
Jones BA, Grace D, Kock R et al (2013) Zoonosis emergence linked to agricultural intensification and environmental change. Proc Natl Acad Sci 110(21):8399–8404
Kamins AO, Rowcliffe JM, Ntiamoa-Baidu Y, Cunningham AA, Wood JL, Restif O (2014) Characteristics and risk perceptions of Ghanaians potentially exposed to bat-borne zoonoses through bushmeat. EcoHealth:1–17
Kingston T (2016) Cute, Creepy, or Crispy–how values, attitudes and norms shape human behavior toward bats. In: Voigt, CC, Kingston, T (eds) Bats in the anthropocene: conservation of bats in a changing world. Springer International AG, Cham, pp. 571–588
Knight AJ (2008) "Bats, snakes and spiders, Oh my!" How aesthetic and negativistic attitudes, and other concepts predict support for species protection. J Environ Psychol 28(1):94–103
Kurth A, Kohl C, Brinkmann A et al (2012) Novel paramyxoviruses in free-ranging European bats. PLoS ONE 7(6):e38688
Kuzmin IV, Botvinkin AD, Rybin SN, Baialiev AB (1991) A lyssavirus with an unusual antigenic structure isolated from a bat in southern Kyrgyzstan. Vopr Virusol 37(5–6):256–259
Kuzmin IV, Botvinkin AD, Khabilov TK (2001) The lyssavirus was isolated from a whiskered bat in northern Tajikistan. Plecotus et al 4:75–81
Kuzmin IV, Niezgoda M, Carroll DS et al (2006) Lyssavirus surveillance in bats, Bangladesh. Emerg Infect Dis 12(3):486
Kuzmin IV, Niezgoda M, Franka R et al (2008a) Possible emergence of West Caucasian bat virus in Africa. Emerg Infect Dis 14(12):1887
Kuzmin IV, Niezgoda M, Franka R et al (2008b) Lagos bat virus in Kenya. J Clin Microbiol 46(4):1451–1461
Kuzmin IV, Mayer AE, Niezgoda M et al (2010) Shimoni bat virus, a new representative of the Lyssavirus genus. Virus Res 149(2):197–210
Kuzmin IV, Bozick B, Guagliardo SA et al (2011a) Bats, emerging infectious diseases, and the rabies paradigm revisited. Emerg Health Threats J 4
Kuzmin IV, Turmelle AS, Agwanda B et al (2011b) Commerson's leaf-nosed bat (*Hipposideros commersoni*) is the likely reservoir of Shimoni bat virus. Vector-Borne Zoonotic Dis 11(11):1465–1470
Kuzmin IV (2014) Basic facts about Lyssaviruses. Curr Lab Tech Rabies Diagn Res Prev 1:3

Lau SK, Woo PC, Wong BH et al (2010) Identification and complete genome analysis of three novel paramyxoviruses, Tuhoko virus 1, 2 and 3, in fruit bats from China. Virology 404(1):106–116
Leonova GN, Belikov SI, Kondratov IG et al (2009) A fatal case of bat lyssavirus infection in Primorye territory of the Russian far East. Rabies Bull Europe 33(4):5–8
Leroy EM, Rouquet P, Formenty P et al (2004) Multiple Ebola virus transmission events and rapid decline of central African wildlife. Science 303(5656):387–390
Leroy EM, Kumulungui B, Pourrut X et al (2005) Fruit bats as reservoirs of Ebola virus. Nature 438(7068):575–576
Leroy EM, Epelboin A, Mondonge V et al (2009) Human Ebola outbreak resulting from direct exposure to fruit bats in Luebo, democratic Republic of Congo, 2007. Vector-Borne Zoonotic Dis 9(6):723–728
Li W, Shi Z, Yu M et al (2005) Bats are natural reservoirs of SARS-like coronaviruses. Science 310(5748):676–679
Li Y, Chen S (2014) Evolutionary history of Ebola virus. Epidemiol Infect 142(06):1138–1145
Luby SP, Rahman M, Hossain MJ et al (2006) Foodborne transmission of Nipah virus, Bangladesh. Emerg Infect Dis 12(12):1888
Luby SP, Hossain MJ, Gurley ES et al (2009) Recurrent zoonotic transmission of Nipah virus into humans, Bangladesh, 2001–2007. Emerg Infect Dis 15(8):1229
Luby SP, Gurley ES (2012) Epidemiology of henipavirus disease in humans. In: Henipavirus. Springer Berlin Heidelberg, pp. 25–40
Luis AD, Hayman DT, O'Shea TJ et al (2013) A comparison of bats and rodents as reservoirs of zoonotic viruses: are bats special? Proc R Soc B: Biol Sci 280(1756):20122753
Lumio J, Hillbom M, Roine R et al (1986) Human rabies of bat origin in Europe. Lancet 327(8477):378
Lumlertdacha B, Boongird K, Wanghongsa S et al (2005) Survey for bat lyssaviruses, Thailand. Emerg Infect Dis 11(2):232
Markotter W, Kuzmin IV, Rupprecht CE, Nel LH (2009) Lagos bat virus virulence in mice inoculated by the peripheral route. Epidemiol Infect 137(8):1155
Marsh GA, Haining J, Robinson R et al (2011) Ebola Reston virus infection of: clinical significance and transmission potential. J Infect Dis 204(Suppl 3):S804–S809
Marsh GA, De Jong C, Barr JA et al (2012) Cedar virus: a novel henipavirus isolated from Australian bats. PLoS Pathog 8(8):e1002836
Marston DA, Horton DL, Ngeleja C et al (2012) Ikoma lyssavirus, highly divergent novel lyssavirus in an African civet. Emerg Infect Dis 18(4):664–667
Martin G, Plowright R, Chen C, Kault D, Selleck P, Skerratt L (2015) Hendra virus survival does not explain spillover patterns and implicates relatively direct transmission routes from flying foxes to horses. J Gen Virol 96(6):1229–1237
Mayen F (2003) Haematophagous bats in Brazil, their role in rabies transmission, impact on public health, livestock industry and alternatives to an indiscriminate reduction of bat population. J Vet Med Ser B 50(10):469–472
McColl KA, Tordo N, Aguilar SA (2000) Bat lyssavirus infections. Revue scientifique et technique (Int Office Epizootics) 19(1):177–196
McColl K, Chamberlain T, Lunt R, Newberry K, Middleton D, Westbury H (2002) Pathogenesis studies with Australian bat lyssavirus in grey-headed flying foxes (*Pteropus poliocephalus*). Aust Vet J 80(10):636–641
Medellin RA (2003) Diversity and conservation of bats in Mexico: research priorities, strategies and Actions. Wildl Soc Bull 31:87–97
Memish ZA, Mishra N, Olival KJ et al (2013) Middle East respiratory syndrome coronavirus in bats, Saudi Arabia. Emerg Infect Dis 19(11):1819
Memish ZA, Cotten M, Meyer B et al (2014) Human infection with MERS coronavirus after exposure to infected camels, Saudi Arabia, 2013. Emerg Infect Dis 20(6):1012
Meredith C, Prossouw A, Koch H (1971) An unusual case of human rabies thought to be of chiropteran origin. South African Med J=Suid-Afrikaanse tydskrif vir geneeskd 45(28):767

Messenger SL, Rupprecht CE, Smith JS (2003) Bats, emerging virus infections, and the rabies paradigm. In: Kunz TH, Fenton MB (eds) Bat ecology. The University of Chicago Press, Chicago, pp 622–679
Mickleburgh SP, Hutson AM, Racey PA (2002) A review of the global conservation status of bats. Oryx 36(01):18–34
Mickleburgh S, Waylen K, Racey P (2009) Bats as bushmeat: a global review. Oryx 43(02):217–234
Middleton DJ, Weingartl HM (2012) Henipaviruses in their natural animal hosts. In: Henipavirus. Springer Berlin Heidelberg, pp 105–121
Mildenstein T, Tanshi I, Racey PA (2016) Exploitation of bats for bushmeat and medicine. In: Voigt, CC, Kingston, T (eds) Bats in the Anthropocene: conservation of bats in a changing world. Springer International AG, Cham, pp 325–376
Misra V, Dumonceaux T, Dubois J et al (2009) Detection of polyoma and corona viruses in bats of Canada. J Gen Virol 90(8):2015–2022
Müller T, Cox J, Peter W et al (2004) Spill-over of European bat Lyssavirus type 1 into a Stone Marten (*Martes foina*) in Germany. J Vet Med Ser B 51(2):49–54
Müller MA, Paweska JT, Leman PA et al (2007) Coronavirus antibodies in African bat species. Emerg Infect Dis 13(9):1367
Murray K, Selleck P, Hooper P et al (1995) A morbillivirus that caused fatal disease in horses and humans. Science 268(5207):94–97
Nahar N, Sultana R, Gurley ES, Hossain MJ, Luby SP (2010) Date palm sap collection: exploring opportunities to prevent Nipah transmission. EcoHealth 7(2):196–203
Nahar N, Mondal UK, Sultana R et al (2013) Piloting the use of indigenous methods to prevent Nipah virus infection by interrupting bats' access to date palm sap in Bangladesh. Health Promot Int 28(3):378–386
Nathwani D, McIntyre P, White K et al (2003) Fatal human rabies caused by European bat Lyssavirus type 2a infection in Scotland. Clin Infect Dis 37(4):598–601
Negredo A, Palacios G, Vázquez-Morón S et al (2011) Discovery of an ebolavirus-like filovirus in europe. PLoS Pathog 7:e1002304
Negrete OA, Levroney EL, Aguilar HC et al (2005) EphrinB2 is the entry receptor for Nipah virus, an emergent deadly paramyxovirus. Nature 436(7049):401–405
Nel LH (2001) Mokola virus: a brief review of the status quo. In Sixth SEARG meeting, Lilongwe/Malawi:80–86
O'Shea TJ, Cryan PM, Cunningham AA et al (2014) Bat flight and zoonotic viruses. Emerg Infect Dis 20(5):741
O'Sullivan J, Allworth A, Paterson D et al (1997) Fatal encephalitis due to novel paramyxovirus transmitted from horses. Lancet 349:93–95
Olival KJ, Islam A, Yu M et al (2013) Ebola virus antibodies in fruit bats, Bangladesh. Emerg Infect Dis 19:270–273
Olival KJ, Hayman DTS (2014) Filoviruses in bats: current knowledge and future directions. Viruses 6:1759–1788
Parashar UD, Sunn LM, Ong F et al (2000) Case-control study of risk factors for human infection with a new zoonotic paramyxovirus, Nipah virus, during a 1998–1999 outbreak of severe encephalitis in Malaysia. J Infect Dis 181:1755–1759
Paton NI, Leo YS, Zaki SR et al (1999) Outbreak of Nipah-virus infection among abattoir workers in Singapore. Lancet 354:1253–1256
Pawan J (1959) Rabies in the vampire bat of Trinidad with special reference to the clinical course and the latency of infection. Carib Med J 21:137–156
Paweska JT, Blumberg LH, Liebenberg C et al (2006) Fatal human infection with rabies-related Duvenhage virus, South Africa. Emerg Infect Dis 12:1965–1967
Peel AJ, Feldman K, Hayman DTS et al (2010) Widespread seroprevalence to Lagos bat virus and henipavirus in straw coloured fruit bats (*Eidolon helvum*) across East Africa. In: Proceedings of the 2nd International Berlin bat meeting: bat biology and infectious diseases, pp 15–16

Peel AJ, Sargan DR, Baker KS et al (2013) Continent-wide panmixia of an African fruit bat facilitates transmission of potentially zoonotic viruses. Nat Commun 4:2770
Pernet O, Schneider BS, Beaty SM et al (2014) Evidence for henipavirus spillover into human populations in Africa. Nat Commun 5:5342
Pfefferle S, Oppong S, Drexler JF et al (2009) Distant relatives of severe acute respiratory syndrome coronavirus and close relatives of human coronavirus 229E in bats, Ghana. Emerg Infect Dis 15:1377–1384
Philbey AW, Kirkland PD, Ross AD et al (1998) An apparently new virus (family Paramyxoviridae) infectious for pigs, humans, and fruit bats. Emerg Infect Dis 4:269–271
Pigott DM, Golding N, Mylne A et al (2014) Mapping the zoonotic niche of Ebola virus disease in Africa. Elife 3:e04395
Plowright RK, Foley P, Field HE et al (2011) Urban habituation, ecological connectivity and epidemic dampening: the emergence of Hendra virus from flying foxes (*Pteropus* spp.). Proc R Soc B 278:3703–3712
Plowright RK, Eby P, Hudson PJ et al (2015) Ecological dynamics of emerging bat virus spillover. Proc R Soc B 282(1798):20142124
Pooley S, Fa JE, Nasi R (2015) No conservation silver lining to Ebola. Conserv Biol 29(3):965–967
Poon L, Chu D, Chan K et al (2005) Identification of a novel coronavirus in bats. J Virol 79:2001–2009
Pourrut X, Kumulungui B, Wittmann T et al (2005) The natural history of Ebola virus in Africa. Microbes Infect/Inst Pasteur 7:1005–1014
Pourrut X, Souris M, Towner JS et al (2009) Large serological survey showing cocirculation of Ebola and Marburg viruses in Gabonese bat populations, and a high seroprevalence of both viruses in *Rousettus aegyptiacus*. BMC Infect Dis 9:159
Puechmaille SJ, Verdeyroux P, Fuller H, Gouilh MA, Bekaert M, Teeling EC (2010) White-nose syndrome fungus (*Geomyces destructans*) in bat, France. Emerg Infect Dis 16(2):290
Quammen D (2013) Spillover: animal infections and the next human pandemic. WW Norton & Company
Quan P-L, Firth C, Street C et al (2010) Identification of a severe acute respiratory syndrome coronavirus-like virus in a leaf-nosed bat in Nigeria. MBio 1:e00208–e00210
Racey PA, Hutson AM, Lina PHC (2012) Bat rabies, public health and european bat conservation. Zoonoses Public Health (Special Issue—Bats). 60:58–68
Racey PA (2015) The uniqueness of bats. In: Lin-Fa W, Cowled C (eds) Bats and Viruses: A New Frontier of Emerging Infectious Diseases. Wiley Blackwell. p. 1–22
Rahman SA, Hassan SS, Olival KJ et al (2010) Characterization of Nipah virus from naturally infected *Pteropus vampyrus* bats, Malaysia. Emerg Infect Dis 16(12):1990
Rahman MA, Hossain MJ, Sultana S et al (2012) Date palm sap linked to Nipah virus outbreak in Bangladesh, 2008. Vector-Borne Zoonotic Dis 12:65–72
Reynes J-M, Counor D, Ong S et al (2005) Nipah virus in Lyle's flying foxes, Cambodia. Emerg Infect Dis 11:1042–1047
Rihtarič D, Hostnik P, Steyer A, Grom J, Toplak I (2010) Identification of SARS-like coronaviruses in horseshoe bats (*Rhinolophus hipposideros*) in Slovenia. Arch Virol 155:507–514
Roberts BJ, Catterall CP, Eby P, Kanowski J (2012) Long-distance and frequent movements of the flying-fox *Pteropus poliocephalus*: implications for management. PLoS ONE 7:e42532
Rogers RJ, Douglas IC, Baldock FC et al (1996) Investigation of a second focus of equine morbillivirus infection in coastal Queensland. Aust Vet J 74(3):243–244
Roine R, Hillbom M, Valle M et al (1988) Fatal encephalitis caused by a bat-borne rabies-related virus. Clinical findings. Brain: a journal of neurology 111:1505–1516
Rønsholt L, Sørensen K, Bruschke C et al (1998) Clinically silent rabies infection in (zoo) bats. Vet Rec 142:519–520
Rota PA, Oberste MS, Monroe SS et al (2003) Characterization of a novel coronavirus associated with severe acute respiratory syndrome. Science 300:1394–1399
Roué SY, Nemoz M (2004) Unusual mortality in Schreiber's long fingered bat, at several nurseries. Wild Dis Assoc Newsl 14:4

Ruiz M, Chávez CB (2010) Rabies in Latin America. Neurol Res 32:272–277
Rupprecht CE, Gibbons RV (2004) Prophylaxis against rabies. New Engl J Med 351:2626–2635
Saéz AM, Weiss S, Nowak K, Lapeyre V, Zimmermann F et al (2015) Investigating the zoonotic origin of the West African Ebola epidemic. EMBO Mol Med 7(1):17–23
Salmón-Mulanovich G, Vásquez A, Albújar C, Guevara C (2009) Human rabies and rabies in vampire and nonvampire bat species, Southeastern Peru, 2007. Emerg Inf Dis 15:1308–1310
Samaratunga H, Searle J, Hudson N (1998) Non-rabies Lyssavirus human encephalitis from fruit bats: Australian bat Lyssavirus (pteropid Lyssavirus) infection. Neuropathol Appl Neurobiol 24:331–335
Sasaki M, Setiyono A, Handharyani E et al (2012) Molecular detection of a novel paramyxovirus in fruit bats from Indonesia. Virol J 9:240
Schatz J, Fooks AR, McElhinney L et al (2013) Bat rabies surveillance in Europe. Zoonoses Public Health 60(1):22–34
Schneeberger K, Czirják GÁ, Voigt CC (2013) Measures of the constitutive immune system are linked to diet and roosting habits of neotropical bats. PLoS ONE 8:e54023
Schneider L, Barnard B, Schneider H et al (1985) Application of monoclonal antibodies for epidemiological investigations and oral vaccination studies. In: Rabies in the tropics. Springer, Verlag
Schneider MC, Romijn PC, Uieda W et al (2009) Rabies transmitted by vampire bats to humans: an emerging zoonotic disease in Latin America? Rev panam de salud pública 25:260–269
Schnell MJ, McGettigan JP, Wirblich C, Papaneri A (2009) The cell biology of rabies virus: using stealth to reach the brain. Nat Rev Microbiol 8:51–61
Selimov MA, Tatarov AG, Botvinkin AD et al (1989) Rabies-related Yuli virus; identification with a panel of monoclonal antibodies. Acta Virol 33(6):542–546
Serra-Cobo J, Amengual B, Abellán C, Bourhy H (2002) European bat lyssavirus infection in Spanish bat populations. Emerg Inf Dis 8:413–420
Sétien A, Brochier B, Tordo N et al (1998) Experimental rabies infection and oral vaccination in vampire bats (*Desmodus rotundus*). Vaccine 16:1122–1126
Smith I, Wang LF (2013) Bats and their virome: an important source of emerging viruses capable of infecting humans. Curr Opin Virol 3(1):84–91
Smith C, Skelly C, Kung N, Roberts B, Field H (2014) Flying-fox species density-a spatial risk factor for Hendra virus infection in horses in Eastern Australia. PLoS ONE 9(6):e99965
Speare R, Skerratt L, Foster R et al (1997) Australian bat lyssavirus infection in three fruit bats from north Queensland. Commun Dis Intell 21:117–119
Steece R, Altenbach JS (1989) Prevalence of rabies specific antibodies in the Mexican free-tailed bat (*Tadarida brasiliensis mexicana*) at Lava Cave, New Mexico. J Wildl Dis 25:490–496
Streicker DG, Turmelle AS, Vonhof MJ et al (2010) Host phylogeny constrains cross-species emergence and establishment of rabies virus in bats. Science 329(5992):676–679
Streicker DG, Recuenco S, Valderrama W et al (2012) Ecological and anthropogenic drivers of rabies exposure in vampire bats: implications for transmission and control. Proc R Soc B 279:3384–3392
Swanepoel R, Leman PA, Burt FJ et al (1996) Experimental inoculation of plants and animals with Ebola virus. Emerg Inf Dis 2:321–325
Tang X, Zhang J, Zhang S et al (2006) Prevalence and genetic diversity of coronaviruses in bats from China. J Virol 80:7481–7490
Taniguchi S, Watanabe S, Masangkay JS et al (2011) Reston Ebolavirus antibodies in bats, the Philippines. Emerg Infect Dis 17:1559–1560
Thiriet D (2011) Conservation shouldn't be a popularity contest. Conversation:1–4
Tjørnehøj K, Fooks AR, Agerholm JS, Rønsholt L (2006) Natural and experimental infection of sheep with European bat lyssavirus type-1 of Danish bat origin. J Comp Pathol 134(2):190–201
Tong S, Conrardy C, Ruone S et al (2009) Detection of novel SARS-like and other coronaviruses in bats from Kenya. Emerg Inf Dis 15:482–485

Tong S, Zhu X, Li Y et al (2013) New world bats harbor diverse influenza A viruses. PLoS Pathog 9(10):e1003657
Towner JS, Pourrut X, Albariño CG et al (2007) Marburg virus infection detected in a common African bat. PLoS ONE 2:e764
Towner JS, Amman BR, Sealy TK et al (2009) Isolation of genetically diverse Marburg viruses from Egyptian fruit bats. PLoS Pathog 5:e1000536
Trimarchi C, Debbie J (1977) Naturally occurring rabies virus and neutralizing antibody in two species of insectivorous bats of New York State. J Wildl Dis 13:366–369
Turmelle A, Jackson F, Green D, McCracken G, Rupprecht C (2010) Host immunity to repeated rabies virus infection in big brown bats. J Gen Virol 91:2360–2366
van Thiel P, Van den Hoek J, Eftimov F et al (2008) Fatal case of human rabies (Duvenhage virus) from a bat in Kenya: The Netherlands, December 2007. Euro Surveill 13:118
van Thiel P-PA, de Bie RM, Eftimov F et al (2009) Fatal human rabies due to Duvenhage virus from a bat in Kenya: failure of treatment with coma-induction, ketamine, and antiviral drugs. PLoS Negl Trop Dis 3:e428
Vidgen ME, de Jong C, Rose K et al (2015) Novel paramyxoviruses in Australian flying-fox populations support host-virus coevolution. J Gen Virol 96(7):1619–1625
Virgin HW, Wherry EJ, Ahmed R (2009) Redefining chronic viral infection. Cell 138(1):30–50
Vogel G (2014) Are bats spreading ebola across sub-saharan Africa? Science 344:140
Voigt, CC, Phelps KL, Aguirre L, Schoeman MC, Vanitharani J, Zubaid A (2016) Bats and buildings: the conservation of synanthropic bats. In: Voigt, CC, Kingston, T (eds) Bats in the Anthropocene: conservation of bats in a changing world. Springer International AG, Cham, pp. 427–453
Wacharapluesadee S, Lumlertdacha B, Boongird K et al (2005) Bat Nipah virus, Thailand. Emerg Inf Dis 11:1949–1951
Warrilow D (2005) Australian bat lyssavirus: a recently discovered new rhabdovirus. In: Fu ZF (ed) The world of Rhabdoviruses. Springer, Berlin
Watanabe S, Masangkay JS, Nagata NK et al (2010) Bat coronaviruses and experimental infection of bats, the Philippines. Emerg Inf Dis 16:1217–1223
Wilkinson GS, South JM (2002) Life history, ecology and longevity in bats. Aging Cell 1(2):124–131
Woodroffe R (1999) Managing disease threats to wild mammals. Anim Conserv 2(03):185–193
World Health Organisation (2003) Summary of probable SARS cases with onset of illness from 1 November 2002 to 31 July 2003 (http://www.who.int/csr/sars/country/table2004_04_21)
World Health Organization (2014) Ebola virus disease, Fact Sheet. (http://www.who.int/mediacentre/factsheets/fs103/en)
Wibbelt G, Moore MS, Schountz T, Voigt CC (2010) Emerging diseases in Chiroptera: why bats? Biol Lett 6:438–440
Woo PC, Lau SK, Li KS et al (2006) Molecular diversity of coronaviruses in bats. Virology 351:180–187
Yaiw KC, Crameri G, Wang L et al (2007) Serological evidence of possible human infection with Tioman virus, a newly described paramyxovirus of bat origin. J Infect Dis 196:884–886
Yaiw KC, Bingham J, Crameri G et al (2008) Tioman virus, a paramyxovirus of bat origin, causes mild disease in pigs and has a predilection for lymphoid tissues. J Virol 82:565–568
Yang Y, Du L, Liu C et al (2014) Receptor usage and cell entry of bat coronavirus HKU4 provide insight into bat-to-human transmission of MERS coronavirus. Proc Nat Acad Sci USA 111:12516–12521
Yob JM, Field H, Rashdi AM et al (2001) Nipah virus infection in bats (order Chiroptera) in peninsular Malaysia. Emerg Inf Dis 7:439–441
Young PL, Halpin K, Selleck PW et al (1996) Serologic evidence for the presence in Pteropus bats of a paramyxovirus related to equine morbillivirus. Emerg Inf Dis 2:239–240
Zaki AM, Van Boheemen S, Bestebroer TM et al (2012) Isolation of a novel coronavirus from a man with pneumonia in Saudi Arabia. New Engl J Med 367:1814–1820
Zhang G, Cowled C, Shi Z et al (2013) Comparative analysis of bat genomes provides insight into the evolution of flight and immunity. Science 339(6118):456–460

Part III
Human-Bat Conflicts

Chapter 11
Impacts of Wind Energy Development on Bats: A Global Perspective

Edward B. Arnett, Erin F. Baerwald, Fiona Mathews, Luisa Rodrigues, Armando Rodríguez-Durán, Jens Rydell, Rafael Villegas-Patraca and Christian C. Voigt

Abstract Wind energy continues to be one of the fastest growing renewable energy sources under development, and while representing a clean energy source, it is not environmentally neutral. Large numbers of bats are being killed at utility-scale wind energy facilities worldwide, raising concern about cumulative impacts of wind energy development on bat populations. We discuss our current state of knowledge on patterns of bat fatalities at wind facilities, estimates of fatalities, mitigation efforts, and policy and conservation implications. Given the magnitude and extent of fatalities of bats worldwide, the conservation implications of understanding and mitigating bat fatalities at wind energy facilities are critically important and should be proactive and based on science rather than being reactive and arbitrary.

E.B. Arnett (✉)
Department of Natural Resource Management, Texas Tech University, Lubbock, TX, USA
e-mail: earnett@trcp.org

E.F. Baerwald
Department of Biological Sciences, University of Calgary, Calgary, Canada

F. Mathews
College of Life and Environmental Sciences, University of Exeter, Exeter, UK

L. Rodrigues
Instituto da Conservação da Natureza e das Florestas, Lisbon, Portugal

A. Rodríguez-Durán
Universidad Interamericana, Bayamón, Puerto Rico

J. Rydell
Biology Department, Lund University, 223 62 Lund, Sweden

C.C. Voigt and T. Kingston (eds.), *Bats in the Anthropocene: Conservation of Bats in a Changing World*, DOI 10.1007/978-3-319-25220-9_11

11.1 Introduction

Developing renewable energy alternatives has become a global priority, owing to long-term environmental impacts from the use of fossil fuels, coupled with a changing climate (Schlesinger and Mitchell 1987; McLeish 2002; Inkley et al. 2004) and because of growing concerns about negative effects from the use of nuclear power (Voigt et al. 2015a). Wind power is one of the fastest growing renewable energy sources worldwide (Fig. 11.1), in part due to recent cost-competitiveness with conventional energy sources, technological advances, and tax incentives (Bernstein et al. 2006). Although presently wind power contributes only about 4 % of the global electricity demand, some countries provide greater than 20 % of their demand from wind (e.g., Denmark [34 %] and Spain and Portugal [21 %]; World Wind Energy Association, www.wwindea.org). By the end of 2013, the Global Wind Energy Council reported that 318,105 MW of wind power capacity was installed worldwide (http://www.gwec.net/wp-content/uploads/2014/04/5_17-1_global-installed-wind-power-capacity_regional-distribution.jpg). The World Wind Energy Association (http://www.wwindea.org) projects that by 2020, more than 700,000 MW could be installed globally.

Wind energy development is not environmentally neutral, and impacts to wildlife and their habitats have been documented and are of increasing concern. Wind energy development affects wildlife through direct mortality and indirectly through impacts on habitat structure and function (Arnett et al. 2007; Arnett 2012; NRC 2007; Strickland et al. 2011). Bats are killed by blunt force trauma or barotrauma and may also suffer from inner ear damage and other injuries not readily noticed by examining carcasses in the field (Baerwald et al. 2008; Grodsky et al. 2011; Rollins et al. 2012; Fig. 11.2). Kunz et al (2007a) proposed several hypotheses that may explain why bats are killed and some of these ideas have subsequently been discussed by others (e.g., Cryan and Barclay 2009; Rydell et al 2010a). Collisions at turbines do not appear to be chance events, and bats probably are attracted to turbines either directly, as turbines may resemble roosts (Cryan 2008), or indirectly, because turbines attract insects on which the bats feed (Rydell et al. 2010b). Horn et al. (2008) and Cryan et al. (2014) provide video evidence of possible attraction of bats to wind turbines.

Regardless of causal mechanisms, bat fatalities raise serious concerns about population-level impacts because bats are long-lived and have exceptionally low reproductive rates, and their population growth is relatively slow, which

R. Villegas-Patraca
Unidad de Servicios Profesionales Altamente Especializados,
Instituto de Ecología, Coatepec, Veracruz, Mexico

C.C. Voigt
Department of Evolutionary Ecology, Leibniz Institute for Zoo and Wildlife Research,
Berlin, Germany

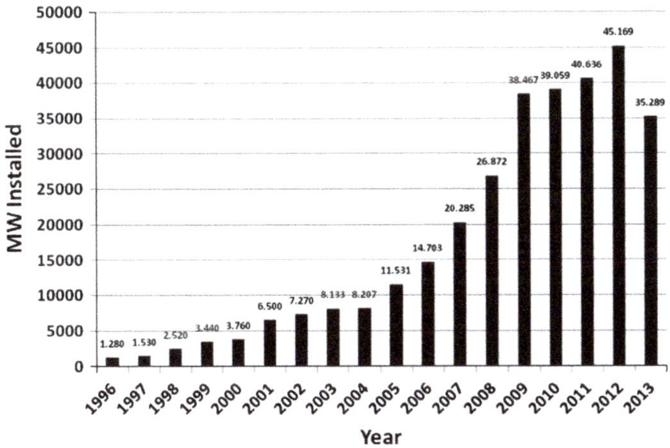

Fig. 11.1 Annual installed global wind energy capacity (MW) from 1996–2013 (modified from the Global Wind Energy Council, http://www.gwec.net/global-figures/graphs/)

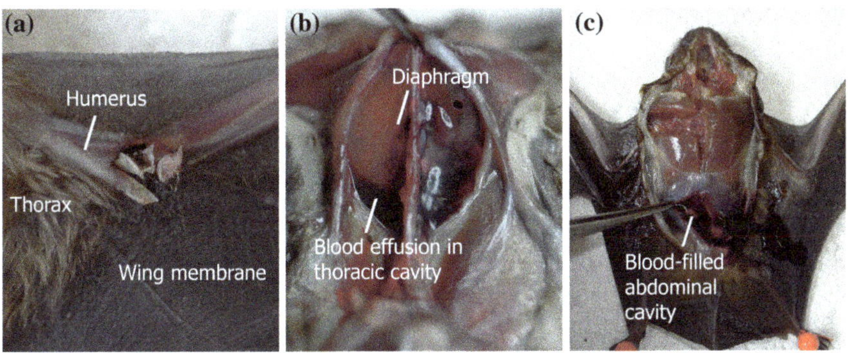

Fig. 11.2 Blunt force trauma (**a**) and barotrauma (**b**, **c**) in three noctule bats (*Nyctalus noctula*) killed at wind turbine in Germany. **a** Ventral view of an open fracture of the left humerus at the height of the elbow joint. **b** Ventral view of the opened abdominal cavity with blood effusion in the thoracic cavity visible behind the diaphragm (hemothorax). **c** Ventral view of opened carcass without bone fractures, but severe bleeding in the abdominal cavity (hemoabdomen) (picture courtesy: Gudrun Wibbelt, IZW)

limits their ability to recover from declines and maintain sustainable populations (Barclay and Harder 2003). Additionally, other sources of mortality cumulatively threaten many populations. For example, white-nosed syndrome causes devastating declines in bat populations in the USA and Canada (e.g., Frick et al. 2010), and national programs for improving insulation of buildings, particularly in Northern Europe, cause losses of roosting opportunities for bats such as the common pipistrelle (*Pipistrellus pipistrellus*; Voigt et al. 2016). Thus, high wind turbine mortality poses a serious threat to bats unless solutions are developed and implemented (Arnett and Baerwald 2013). In this chapter, we build on previous reviews of existing information (e.g., Arnett et al. 2008; Rydell et al. 2010a; Arnett and Baerwald 2013; EUROBATS 2014), synthesize information on bat fatalities at wind energy facilities worldwide, discuss unifying themes and policy and conservation implications, and offer insights for future directions of research and mitigation of bat fatalities at wind facilities.

11.2 Composition and Estimates of Bat Fatalities

We present information on estimates of bat fatalities as reported in published literature or publically available reports, but caution that studies had varying levels of effort, used different estimators (e.g., Huso 2011; Korner-Nievergelt et al. 2013) and different methods to quantify bias (Arnett et al. 2008; Strickland et al. 2011), thus biasing estimates. Also, most estimators fail to adequately account for unsearched area near turbines (Huso and Dalthorp 2013), which further biases estimates. Some studies report fatalities/turbine and others fatalities/MW of installed capacity. As such, data presented here offer a general and relative sense of fatalities within and among continents and do not represent quantitative comparisons.

11.2.1 North America

From 2000 to 2011 in the USA and Canada, annual bat fatality rates were highest at facilities located in the Northeastern Deciduous Forest (6.1–10.5 bats/MW; Fig. 11.3) and Midwestern Deciduous Forest-Agricultural (4.9–11.0 bats/MW) regions defined by Arnett and Baerwald (2013: 438). Average fatality rate in the

Fig. 11.3 Wind energy facilities on forested ridges in the eastern USA have consistently documented high fatality rates of bats (photograph by E.B. Arnett)

Great Plains region was moderately high (6 bats/MW, 95 % CI: 4.0–8.1 bats/MW), while the Great Basin/Southwest Desert region (1.0–1.8 bats/MW) consistently reports the least variable and lowest fatality rates for bats (Arnett et al. 2008; Arnett and Baerwald 2013; Johnson 2005). Wind energy facilities in this region occur in habitats generally offering few roosting resources, possibly (but untested) poor foraging opportunities, and may not be in migratory pathways, thus rendering these sites less risky to bats (Arnett and Baerwald 2013). However, facilities in other regions report high fatality rates of bats where there are large expanses of prairie and agricultural lands with few roosting resources, foraging opportunities, and likely migratory routes (e.g., Summer view Alberta, Canada, 8–14.6 bats/MW; Baerwald et al. 2008). Thus, current patterns in the Great Basin/Southwest region reported by Arnett and Baerwald (2013) may simply reflect biased reporting and an absence of evidence as opposed to evidence of absence (Huso and Dalthorp 2013).

Twenty-one of the 47 species of bats known to occur in the USA and Canada have been reported killed at wind energy facilities, and fatalities are skewed toward migratory species often referred to as "tree bats" that include hoary bats (*Lasiurus cinereus*; 38 %), eastern red bats (*Lasiurus borealis*; 22 %), and silver-haired bats (*Lasionycteris noctivagans*; 18.4 %) that comprise a total of 78.4 % of the recovered bat turbine fatalities in the USA and Canada (Arnett and Baerwald 2013). However, other species also are affected, sometimes seriously. Fatalities of the cave-living Brazilian free-tailed bats (*Tadarida brasiliensis*) are quite frequent in the southern USA during the maternity period in summer (Miller 2008; Piorkowski and O'Connell 2010). In the USA, two species listed as threatened or endangered under the Endangered Species Act also have been killed by turbines, the Indiana bat (*Myotis sodalis*) and Hawaiian hoary bat (*Lasiurus cinereus semotus*; Arnett and Baerwald 2013).

In the Oaxacan Isthmus region of Mexico, 32 of the 42 species of bats known to occupy this region (García-Grajales and Silva 2012; Briones-Salas et al. 2013) were found killed (Villegas-Patraca et al. 2012). These bats belonged to five different families (Mormoopidae, Molossidae, Vespertilionidae, Phyllostomidae, and Emballonuridae), although 52 % of the fatalities belonged to just two species, Davy's naked-backed bat (*Pteronotus davyi*; 40.2 %) and the ghost-faced bat (*Mormoops megalophylla*; 11.9 %), both of the family Mormoopidae. These two species are particularly abundant in the area studied and form colonies with thousands of individuals in caves (García-Grajales and Silva 2012). Both are aerial-hawking and relatively fast-flying bats (Bateman and Vaughan 1974; Adams 1989). Also, unlike those species killed most frequently in Holarctic regions of North America, these species do not tend to roost in trees. Ninety-seven percent of bat fatalities found at wind turbines are resident species. This differs considerably from the USA, Canada, and parts of northern Europe, suggesting that wind turbines are equally dangerous to resident cave bats assumed to be non-migratory as to migratory tree-roosting species. The common theme is rather that the most frequently killed species are adapted to flight and echolocation in the open air (e.g., bats that have a relatively high wing loading).

11.2.2 Europe

Rydell et al. (2010a) synthesized data from 41 sites in 5 countries in northwestern Europe and found that the Black Forest region in Germany ($n = 10$) had the highest annual fatality rates, averaging 10.5 bats killed/MW. Some regions in Germany had relatively low estimated annual fatality rates, averaging around 1.1–1.2 bats/MW (Rydell et al. 2010a), yet some of these studies did not control for carcass removal and searcher efficiency. The single comprehensive study that covered most parts of Germany did take the aforementioned field biases into account when estimating annual fatality rates of 10–12 bats per wind turbines, translating to 6–8 bats per MW produced (Korner-Nievergelt et al. 2013). Studies from mostly agricultural areas of Austria ($n = 3$), Switzerland ($n = 3$), and England ($n = 1$) yielded mean annual fatalities rates of 2.5, 5.3, and 0.6 bats killed/MW, respectively (Rydell et al. 2010a). In France, some particularly dangerous sites are located near water along the river Rhone in the east (Dubourg-Savage et al. 2011) and on the Atlantic coast in the west (Rydell et al. 2010a). In Spain, bat fatalities from 56 wind facilities ranged from 0.00 to 0.80 bats/MW per year (Camina 2012), but most studies did not correct for scavenger removal and searcher biases and therefore underestimate fatalities. In Portugal, annual fatality rates at 28 facilities ranged from 0.07 to 11.0/MW (L. Rodrigues, Instituto da Conservação da Natureza e das Florestas, unpublished data). Generally, data from Europe are inconsistently collected, rendering comparisons and generalizations across countries difficult. Nevertheless, it is clear that bats are frequently killed at wind turbines throughout the continent, with some facilities experiencing considerably higher fatality rates relative to others.

Members of EUROBATS recently synthesized data from several countries and reported 6429 documented bat kills of 27 species collected at wind facilities in Europe from 2003 to 2014 (EUROBATS 2014), but some studies used to derive estimates of fatality rates did not incorporate field bias or area corrections. The species of bats found most frequently at wind facilities across northern Europe were the common pipistrelle, common noctule (*Nyctalus noctula*), Nathusius' pipistrelle (*Pipistrellus nathusii*), and Leisler's bat (*Nyctalus leisleri*). In Germany, nearly 70 % of recorded deaths represent the latter three species and the particolored bat (*Vespertilio murinus*), all of which are long-distance migrants (Hutterer et al. 2005). Owing to its central geographical location on the European continent, Germany appears to provide ecological stepping stones for many long-distance bat migrants from northeastern populations (Steffens et al. 2004; Voigt et al. 2012). However, resident species or short-distance migrants, including common pipistrelle and northern bats (*Eptesicus nilssonii*), also are frequently killed in northern Europe (Rydell et al. 2010a). The majority (>90 %) of bats killed at wind turbines in southern Europe belong to the various pipistrelle and noctule species: common pipistrelle, Nathusius' pipistrelle, soprano pipistrelle (*Pipistrellus pygmaeus*), Kuhl's pipistrelle (*Pipistrellus kuhlii*), and Savi's pipistrelles (*Hypsugo savii*) and the common noctule, giant noctule (*Nyctalus lasiopterus*) and Leisler's bat (*Nyctalus leisleri*). Some of these are long-distance migrants (e.g., Nathusius' pipistrelle and common noctule) that often roost in tree holes, while others are resident and usually house-living species that do not migrate long distances (e.g., Kuhl's pipistrelle and Savi's pipistrelle). Rare species, such as the barbastelle (*Barbastella barbastellus*) and the *Myotis* and *Plecotus* spp., also are killed occasionally, but in smaller numbers. Thus, bats killed at wind turbines in southern Europe generally belong to the same genera as those in northern Europe (*Pipistrellus* and *Nyctalus* spp.), but include several non-migratory species such as Kuhl's and Savi's pipistrelles.

11.2.3 Africa

Little work has been done on wind energy facilities in Africa, and prior to 2012, no studies had been published from the continent. During a pilot study at a single turbine located in the Eastern Cape of South Africa, Doty and Martin (2012) found 18 carcasses of 2 species of bats—the Cape serotine (*Neoromicia capensis*) and Egyptian free-tailed bat (*Tadarida aegyptiaca*). No estimates of fatality rates were provided, likely because of small sample size of recovered carcasses and no bat carcasses were used during field bias trials. In the Western Cape of South Africa, Aronson et al. (2013) reported only one carcass of a Cape serotine. These studies confirm at least some species of bats are vulnerable to wind turbine mortality in South Africa, which could have implications for ecosystem function and conservation of bats in this region.

11.2.4 New Zealand and Australia

In Australia, Hall and Richards (1972) were the first to report bat fatalities at a wind facility in the world, and 22 white-striped free-tailed bats (*Tadarida australis*) were found over a 4-year period. Little work had been done in the region since this pioneering discovery, until Hull and Cawthen (2012) surveyed two wind facilities in Tasmania, where they recorded 54 bat fatalities from two species, Gould's wattled bats (*Chalinolobus gouldii*) and an unknown *Vespadelus* sp. More recently, Bennett (2012) found white-striped free-tailed bats at two turbines located in Victoria. While no estimates of fatality rates were provided for these studies, they indicate that some species of bats are at risk of wind turbine mortality in this part of the world.

11.2.5 South America, Central America, and the Caribbean

Few studies have been done in Latin American regions on bat fatalities caused by wind turbines. Puerto Rico hosts 13 species of bats of five families. Five of these 13 species belong to the family Phyllostomidae, which feed on fruits and nectar and forage in the understory and canopy (Gannon et al. 2005). It was originally speculated that these species would be at low risk for mortality caused by wind turbines based on their life histories and foraging patterns. Species in the family Molossidae also occur in Puerto Rico, and conversely, these species have been considered to be at higher risk to turbine collisions because they fly high in open spaces. Species from both families of bats have been detected during preconstruction surveys in areas where wind facilities were proposed. Twenty months of ongoing post-construction surveys in Puerto Rico revealed 30 carcasses from 11 of the 13 species, for a corrected mortality rate of about 10 bats/turbine /year (Rodríguez-Durán, Universidad Interamericana, unpublished data). Aside from the expected mortality of species in the family Molossidae, it was surprising that fruit and nectar feeding species of phyllostomids were followed in number of fatalities given their flight and foraging patterns. One important hazard for bats in this region relates to their use of hot caves as roosts (Rodríguez-Durán 2009; Ladle et al. 2012). Although little studied, these systems may be ubiquitous throughout parts of México, Panamá, Colombia, Venezuela, Brazil, and the Greater Antilles. Phyllostomids and mormoopids (family Mormoopidae) form large aggregations in hot caves and commute to foraging areas flying long distances at high altitude. This reliance on hot caves may place them at risk from wind facilities located near their feeding sites or along their commuting routes.

11.2.6 Asia

On the island of Taiwan off the Chinese mainland, wind facilities have been established along the western coastline, predominantly in former mangrove wetlands. Bat fatalities have been recorded at three of these facilities (C.H. Chou, Endemic Species Research Institute, unpublished data). Carcass searches and acoustic monitoring indicated regular feeding activity of bats near turbines in summer, and 51 dead bats were found. However, the study is ongoing and no field bias correction experiments have been conducted yet, so corrected fatality estimates are not available. The Japanese pipistrelle (*Pipistrellus abramus*), which is a non-migratory open-air foraging bat, was killed most frequently ($n = 39$). Six other species have also been found killed, although in smaller numbers (1–4 individuals for each species), namely Horikawa's brown bat (*Eptesicus serotinus horikawai*), common house bat (*Scotophilus kuhlii*), Chinese noctule (*Nyctalus plancyi velutinus*), Taiwanese golden bat (*Myotis formosus flavus*), a recently described mouse-eared bat (*Myotis secundus*), and Japanese long-fingered bat (*Miniopterus fuliginosus*).

Three other species have been observed foraging around the turbines, but have not yet been found during carcass searches. These species are the yellow-necked sprite (*Arielulus torquatus*), Taiwanese tube-nosed bat (*Murina puta*), and East Asian free-tailed bat (*Tadarida insignis*). Several of these species (e.g., yellow-necked sprite, Taiwanese golden bat, Taiwanese tube-nosed bat, Chinese noctule, Horikawa's brown bat, and *M. secundus*) are all island endemics, some of which occur in sparse and probably small and vulnerable populations. Nevertheless, the pattern conforms to that of most regions around the world, since the mortality predominantly (but not exclusively) affects species that feed in the open air (C.H. Chou, Endemic Species Research Institute unpublished data).

11.2.7 Conclusions

Bats are killed at wind turbines worldwide, and those fatalities are not restricted to migratory species at high latitudes, as previously suggested (e.g., Kunz et al. 2007a; Arnett et al. 2008). Hence, the bias toward tree-roosting migrants observed in North America and to some extent also in northern Europe is not consistent elsewhere. An emerging hypothesis is that bats that regularly move and feed in less cluttered and more open air-space are most vulnerable to collisions with wind turbines, regardless of continent, habitat, migratory patterns, and roost preferences. The species most often killed at wind turbines throughout Europe belong to aerial-hawking and relatively fast-flying, open-air species, and this is consistent with the pattern found in North America and Mexico. However, other species, including gleaning insectivores and even fruit feeders, also are killed occasionally. The vulnerability of tropical bat faunas is a potentially serious problem that must be addressed immediately and preferably before extensive wind facilities are planned and constructed.

While fatalities of endangered species like the Indiana bat are important from a legal perspective, they currently appear to be biologically irrelevant in comparison with those for hoary and eastern red bats, for example. However, fatalities of listed species worldwide may become increasingly important as wind energy development expands.

The paucity of studies in most regions of the world is alarming, particularly in Mexico, Central and South America, the Caribbean, Africa, New Zealand, and Australia. Notably, we could not find information on bat fatalities at wind facilities from mainland Asia, but the data from Taiwan indicate that the bat fauna of eastern Asia may be highly vulnerable at wind turbines. Turbine fatalities may be a serious threat to bats in, for example, China where wind energy development is substantial (Global Wind Energy Council, http://www.gwec.net/global-figures/graphs/#). This situation is further complicated by the fact that in most countries information gathered is sequestered either by wind energy companies or government agencies and not made readily available. The importance of having access to this information cannot be overstated for all regions of the world.

11.3 Patterns of Bat Fatality

11.3.1 Temporal Patterns

In the temperate Northern Hemisphere, most bat fatalities occur during late summer and early autumn. In the USA, fatalities peak in mid-July through early September in most parts of the country (Johnson 2005; Arnett et al. 2008; Baerwald and Barclay 2011; Arnett and Baerwald 2013). Studies from Europe demonstrate a similar pattern (e.g., in Germany, where most (about 90 %) bat fatalities at wind turbines occur between mid-July and the end of September; Brinkmann et al. 2011; Lehnert et al. 2014). Some studies from northern Europe and North America demonstrate smaller peaks of fatalities during spring (Arnett et al. 2008; Rydell et al. 2010a). In Greece and on the Iberian Peninsula of Spain and Portugal, the pattern is similar, with most (>90 %) fatalities in late summer (Georgiakakis et al. 2012; Camina 2012; Amorim et al. 2012), but in some places, particularly at high elevation sites, fatalities occur from May to October and without any obvious concentration in the late summer period (Dubourg-Savage et al. 2011; Camina 2012). Such consistent temporal patterns of fatality are helpful when predicting high-risk periods and applying some mitigation measures such as raising turbine cut-in speed (Arnett et al. 2011, Baerwald et al. 2009). Hull and Cawthen (2012) noted that fatalities predominantly occurred in autumn in Tasmania, where the climate is temperate. However, in the tropical Isthmus of Tehuantepec in Mexico, while 46 % of bat fatalities were found in the summer rainy season, no clear pattern in bat deaths associated with any season emerged.

In summary, while there are clear temporal patterns and a distinct late summer fatality peak in high-latitude temperate regions (north Europe and North America), the pattern becomes less obvious in warmer climates at lower temperate latitudes (south Europe) and temporal patterns may dissipate entirely in tropical regions (e.g., southern Mexico).

11.3.2 Spatial Patterns

Arnett and Baerwald (2013) noted that the spatial context of bat kills, both among turbines within a facility and among different facilities, could be useful for developing mitigation strategies. They hypothesized that if, for example, kills were concentrated at specific turbines, then curtailment, removal, or relocating that turbine may reduce bat deaths. However, if fatalities are broadly distributed, then facility-wide mitigation strategies would be necessary (Arnett et al. 2008). Thus far, studies worldwide have failed to detect specific turbines responsible for most fatalities at any given facility.

Other patterns at scales beyond individual turbines have been reported that may assist with assessing risk. Baerwald and Barclay (2011) found no differences in fatalities on the east vs. west side of a facility in southern Alberta, but the fatality rate was higher at the north end. Baerwald and Barclay (2011) hypothesized that because fall migrations are from north to south, higher fatality rates could be expected at the more northerly turbines first encountered by migrating bats. At a landscape scale, Baerwald and Barclay (2009) found both higher activity and fatality rates of bats at wind facilities near the foothills of the Rocky Mountains as compared to eastward prairie grasslands. They speculated that turbine proximity to stopover and roost sites in foothills habitat significantly increased fatality rates assuming that geographical landmarks are used for navigating migration routes and that bats judge nightly travel distances between suitable diurnal roosting sites.

11.3.3 Habitat Relationships

Relationships between bat fatalities and habitat or topographic characteristics may be useful for developing mitigation strategies (e.g., to avoid placing turbines near places where many bats move or forage, such as near open water sources, wetlands, or known roosts; Arnett et al. 2008; Arnett and Baerwald 2013; Rydell et al. 2010a). Johnson et al. (2004) did not find a significant relationship between the number of bat fatalities and any of the 10 cover types within 100 m of turbines at facilities in Minnesota or any relationship between fatalities and distance to nearest wetland or woodlot. In assessing the type of vegetation present in areas where the fatalities were found in wind facilities in the Isthmus of Tehuantepec, 79.6 % occurred in agricultural areas. In Oklahoma in 2004, Piorkowski and O'Connell (2010) found that turbines in eroded ravine topography accounted for higher fatality rates than those in areas of low topographic relief and reported some evidence that turbines in mixed cedar/pasture habitats killed more bats than those in cropland and prairie habitats. However, these patterns were not repeated in 2005 or for both years of the study when combined, and Piorkowski and O'Connell (2010) speculated that bats may have exhibited different habitat use patterns in different years or they did not measure factors better explaining annual differences they observed. Interestingly, Grodsky (2010) found that bat fatalities were actually lower near the Horicon Marsh in Wisconsin. Hull and Cawthen (2012) found no relationships between bat fatalities and proximity of turbines to the coast or vegetation. Hence, correlating high-risk locations with particular habitat types or topographic patterns has proven difficult and inconsistent.

Analyses of fatalities reported from Spain and Portugal, where most wind facilities are located on top of hills and mountains, suggest that the most significant environmental predictor of fatality rate is proximity to steep slopes with bare rock and no vegetation. Bare rock is warmed by the sun and radiates heat during the night, which likely facilities insect activity over the rocks (Ancilotto et al. 2014), possibly explaining higher fatality at sites near steep, rocky slopes. Alternatively, rocks on tops of hills and mountains might provide suitable roosts.

Piorkowski and O'Connell (2010) documented the first evidence of fatality of Mexican free-tailed bats at a North American wind facility that could be attributed to the site's proximity (~15 km) to a large maternity colony. In Wisconsin, Grodsky (2010) found no relationship between distances of turbines from a large hibernaculum (Neda Mine), but in this case, hibernating bats did not belong to the species most vulnerable to wind turbine mortality (see above). Georgiakakis et al. (2012) reported that the most frequently killed species at wind facilities in Greece exhibited different spatial patterns of fatality, speculating that this resulted from some turbines being located closer to roosts and/or commuting corridors. It may not be enough to consider the proximity of a facility to a maternity or hibernation site, but rather where it is located relative to feeding grounds or movement corridors (Arnett and Baerwald 2013). We are not aware of other studies demonstrating similar relationships or patterns with large maternity or winter roosts.

11.3.4 Climate and Weather Variables

Arnett (2005) was first to employ daily carcass searches and relate them to weather variables, discovering that most bats were killed on low-wind nights when power production appeared insubstantial. Based on this approach, Arnett et al. (2008) estimated that 82–85 % of bat fatalities at two facilities in the eastern USA occurred on nights with median nightly wind speeds of <6 m/s. Since this pivotal discovery, studies worldwide document that most bat fatalities occur during low-wind periods. In the USA, for example, Jain et al. (2011) found that maximum wind speeds when bat collisions likely occurred ranged from 2.4 to 5.3 m/s. Korner-Nievergelt et al. (2013) found that maximum collision rates of bats occurred at wind speeds between 3.5 and 5.7 m/s. Several other studies from Europe demonstrate a similar pattern (e.g., Amorim et al. 2012). Indeed, this consistency suggesting bat fatality is highest during lower wind speeds greatly assists predicting high-risk periods during which to apply operational mitigation.

Fatalities appear to increase as ambient temperature rises, a relationship observed in North America (e.g., Grodsky 2010; Young et al. 2011) and Europe (e.g., in Portugal; Amorim et al. 2012). Amorim et al. (2012) also found that bat fatalities increased with decreasing relative humidity. The effect of high temperature on fatality rate seems to apply both on the broader regional and climatic scales and according to daily changes in the weather (Dubourg-Savage et al. 2011 and unpublished data). Hence, at least in southern Europe, high fatality rates at wind turbines are most likely in warm and dry geographic areas (Mediterranean and low elevation) and also in warm weather (most common in late summer). In the end, this suggests that fatalities may be correlated with periods of high insect activity, which generally is most likely to occur under warm and dry conditions (Heinrich 1993).

Bat fatalities also have been correlated with other climatic factors that could assist with predicting high-risk periods. Baerwald and Barclay (2011) reported that species–specific fatalities were affected by greater moon illumination. They also observed that falling barometric pressure and the number of deaths were correlated and that whereas fatalities of silver-haired bats increased with increased activity of this species, moon illumination, and south-easterly winds, hoary bat mortality increased most significantly with falling barometric pressure. Interestingly, neither hoary bat activity nor fatality was influenced by any measured variables other than falling barometric pressure (Baerwald and Barclay 2011). Again, this could result from decreasing barometric pressure that triggers insect flight activity and therefore may motivate foraging efforts among bats by indicating a potential increase in food availability (Wellington 2011).

11.4 Offshore Wind Facilities

Potential impacts of offshore wind-energy development on bats are poorly understood, although observations in Europe and anecdotal accounts of bats occurring offshore suggest that impacts may occur. Bats are known to regularly migrate across the Baltic and North Seas and visit offshore facilities (Hutterer et al. 2005; Boshamer and Bekker 2008; Ahlén et al. 2009; Poerink et al. 2013; Rydell et al. 2014). Ahlén et al. (2009) recorded 11 species of bats flying and feeding over the sea up to 14 km from the shore. In spring and late summer, migrating bats are found along coastlines of the Baltic Sea and southeastern North Sea in northern Europe, including all offshore islands where observations have been made (Rydell et al. 2014). This suggests bats, including Nathusius' pipistrelles, soprano pipistrelles, and common noctules, migrate on a broad front across the Baltic Sea and along its coasts, using small islands for stopovers. Researchers in North America also have reported activity of bats in both near and offshore habitats, suggesting impacts are highly probable at facilities located in such places. Cryan and Brown (2007) discovered longitudinal movement by hoary bats from inland summer ranges to coastal regions during autumn and winter and suggested that coastal regions with non-freezing temperatures may be important wintering areas for hoary bats. Off the coast of Maryland, Johnson et al. (2011) recorded five species of bats, including eastern red bats, big brown bats (*Eptesicus fuscus*), hoary bats, tri-colored bats (*Perimyotis subflavus*), and silver-haired bats, on a barrier island and concluded these species used this island during migration, which could have implications for wind energy development near and offshore.

It seems likely that near and offshore wind facilities also will kill bats, but it is difficult or impossible to find bat fatalities at sea and no attempts to assess offshore turbine bat fatality have been made to date. Arnett and Baerwald (2013) suggested that impacts of the first several offshore wind-energy facilities proposed and built in North America, including those on inland waters such as the Great Lakes, be evaluated extensively both for fatalities and displacement effects.

They also suggested that a method for predicting fatalities at existing and planned wind facilities offshore will be required to understand impacts and develop mitigation strategies, because finding and retrieving dead birds and bats from water bodies will be a considerable challenge (Arnett et al. 2007; Arnett 2012).

11.5 Estimating Risk

Kunz et al. (2007b) found a positive correlation between post-construction bat activity and fatality from carcass searches conducted simultaneously. However, Kunz et al. (2007b) warned of several limitations of their analysis and noted that it was unclear whether pre-construction call rates could predict risk and level of post-construction fatality rates. When comparing 5 sites with fatality and activity data, and tall turbines (towers 65 m), Baerwald and Barclay (2009) found a significant positive relationship between post-construction activity and fatality at 5 wind facilities in Alberta. Amorim et al. (2012) and Korner-Nievergelt et al. (2013) also found increasing number of bat fatalities with increasing acoustic bat activity at facilities in Portugal and Germany, respectively. These studies correlating post-construction bat activity with fatality suggest that it may be possible to use indices of pre-construction bat activity to predict future fatality and, thus, risk and need for mitigation. However, while numerous studies have documented pre-construction activity of bats with hopes of inferring risk of collision mortality, these studies have yet to link with post-construction fatality data gathered from carcass searches. Hein et al. (2013) were the first to correlate pre-construction acoustic activity with post-construction fatalities from 12 paired study sites in the USA and found that no statistically significant relationship existed between bat fatalities/MW and bat passes/detector night and only a small portion of the variation in fatalities was explained by activity. Thus, Hein et al. (2013) concluded that prediction of risk prior to construction of a wind facility is highly variable and imprecise and acoustic data may not necessarily predict bat fatality in any reliable way. One explanation as to why correlations between pre-construction measurements of bat activity with similar measurements made post-construction or fatality estimates are weak could be that bats are attracted to the turbines once they are built and sites are used differently by at least some species (open-air bats) afterward (Horn et al. 2008; Kunz et al. 2007b; Arnett et al. 2008; Cryan et al. 2014).

Theoretical estimations of exposure risk of bats to collisions with turbines based on models may also improve our understanding of factors influencing fatality and the context of fatalities. Species distribution models developed in Italy suggest that 41 % of the region offers suitable foraging habitat for 2 species of bats vulnerable to wind turbines, Leisler's bat and the common pipistrelle, and these same areas encompass over 50 % of existing or planned wind farms (Roscioni et al. 2013). The authors believe fatality risk for these species is increased by the common proximity to forest edges, but this contradicts other findings from southern Europe, suggesting the opposite relationship (Dubourg-Savage et al. 2011).

Roscioni et al. (2014) further investigated habitat connectivity as a surrogate for assessing risks of wind facilities to bat migration and commuting in Italy. Using species distribution models, they found that most corridors used by bats were concentrated in an area where existing (54 %) and planned (72 %) wind facilities would interfere with important corridors connecting the western and the eastern parts of the region. In Portugal, mortality risk models indicated wind farms located in humid areas with mild temperatures and within 600 m of steep slopes had higher probabilities of mortality (Santos et al. 2013). They also demonstrated that high mortality risk areas overlapped greatly with the potential distribution of Leisler's bat in Portugal, suggesting that populations of this species may be at high risk to turbine fatalities (Santos et al. 2013). They also found that a large extent of the area predicted to be high risk for mortality overlapped with sites highly suitable for wind farm construction.

11.6 Cumulative Impacts

Estimates of fatalities, and thus any estimate of cumulative fatalities, are conditioned by field methodology for each study (e.g., search interval) and how each study did or did not account for sources of field sampling bias when calculating fatality rate estimates. Arnett and Baerwald (2013) synthesized information from 122 post-construction fatality studies (2000–2011) from 73 regional facilities in the USA and Canada and developed a regional weighted mean estimate of cumulative bat fatalities for the USA and Canada. Assuming fatality rates were (1) representative of all regional sites and (2) consistent from year to year without behavioral modification or mitigation, Arnett and Baerwald (2013) estimated cumulative bat fatalities in the USA and Canada ranged from 0.8 to 1.7 million over a 12-year period from 2000 to 2011. This estimate was projected to increase by 0.2–0.4 million bats in 2012 based on the assumptions and installed wind power capacity. Smallwood (2013) estimated 888,000 bats killed/year at wind facilities in the USA, while Hayes (2013) concluded that over 600,000 bats may have been killed by wind turbines in 2012 alone. However, neither of these estimates used all data available at the time they were published, nor did they weight their estimates by regionally collected data and installed wind energy capacity as Arnett and Baerwald (2013) did; the latter approach likely provides a more conservative and accurate estimate based on the studies and installed capacity from each region.

When controlling for field biases, an estimated 10–12 bats are killed annually at each wind turbine in Germany, if no mitigation measures have been implemented (Brinkmann et al. 2011). Assuming these numbers are representative of all types of wind turbines for all of Germany, it has been suggested that more than 200,000 bats were killed at onshore wind turbines in Germany, assuming no behavioral modification or mitigation measures were practiced (Voigt et al. 2015a). Over the past ten years of wind energy development, it is estimated that more than two million bats may have been killed by wind turbines in Germany, based on the reported large-scale development of wind turbines in that country (Berkhout et al. 2013; Voigt et al. 2015a).

Importantly, the context of wind turbine fatalities remains poorly understood, in part because little population data exist for most species of bats (O'Shea et al. 2003) and this hinders understanding population-level impacts, as well as effectiveness of mitigation measures. Population estimates for most species of bats around the world are lacking, and some bat populations are suspected or known to be in decline (e.g., Frick et al. 2010; Hutson et al. 2001; Ingersoll et al. 2013). Other populations, such as hibernating species in Europe, appear to be increasing (9 of 16 species examined by Van der Meij et al. (2014) increased at their hibernation sites from 1993 to 2011), but these species are not largely affected by wind turbines. In addition to natural and other forms of anthropogenic-induced mortality, wind turbine mortality further compounds population declines for many species of bats and warrants mitigation.

11.7 Mitigating Bat Mortality

As reported previously, most bat fatalities occur during relatively low-wind conditions over a relatively short period of time in late summer (Arnett et al. 2008) and operational adjustments under these conditions and during this time could reduce impacts on bats (Arnett 2005; Arnett et al. 2008; Kunz et al. 2007a). Behr and von Helversen (2006) were the first to examine operational mitigation in Germany, documenting around 50 % fewer bats killed at turbines having their cut-in speed (wind speed at which turbines begin producing electricity into the power grid) raised above the set manufacture's cut-in speed of 4.0 m/s. In the synthesis of operational mitigation studies in the USA and Canada, Arnett et al. (2013a) reported that most studies documented at least a 50 % reduction in bat fatalities when turbine cut-in speed was increased by 1.5 m/s above the manufacturer's cut-in speed, with up to a 93 % reduction in bat fatalities in one study (Arnett et al. 2011). Baerwald et al. (2009) demonstrated beneficial reductions (~60 %) with a low-speed idling approach. Young et al. (2011) discovered that feathering turbine blades (pitched 90° and parallel to the wind) at or below the manufacturer's cut-in speed resulted in up to 72 % fewer bats killed when turbines produced no electricity into the power grid. Arnett et al. (2013a) noted that studies failing to demonstrate statistically significant effects could be explained by lack of treatments being implemented during the study (i.e., winds were either too low or high to enable comparison of treatments). In Portugal, a mitigation study found that estimated mortality at turbine with raised cut-in speed was 0.3 bats/turbine compared to 1.6 bats/turbine at turbines operating normally, which resulted in a 78.5 % reduction in bat fatalities assuming all turbines at the facilities had raised cut-in speed (LEA 2010).

More recently, situation-dependent operation protocols, so-called algorithms, were developed for the operation of wind turbines. These algorithms consider a number of parameters such as ambient temperature, wind speed, season, and time of day as well as recorded bat activities for defining a set of operation rules for wind turbines (Korner-Nievergelt et al. 2013). However, these algorithms have been formulated for a single type of turbine and for a limited number of sites. Thus, the suggested algorithms may be unsuitable for other places with varied geographical and topographic characteristics, bat communities, and turbine types (Voigt et al. 2015a).

Few studies have disclosed actual power loss and economic costs of operational mitigation, but those that have suggest that <1 % of total annual output would be lost if operational mitigation was employed during high-risk periods for bat fatalities. While costs of lost power due to mitigation can be factored into the economics, financing, and power purchase agreements of new projects, altering turbine operations even on a limited-term basis potentially poses difficulties on existing projects. Although curtailment is relatively straightforward to implement on large modern turbines, for older models and for small to medium energy-generating turbines, there often is no way to remotely control or change cut-in speed; some turbines would require a technician to physically change turbine operating systems (which is not feasible). However, raising cut-in speed or altering blade angles to reduce rotor speed (termed "low-speed idling" by Baerwald et al. 2009) where blades are near motionless in low wind speeds remain the only proven solutions to mitigating bat kills at wind facilities. The fact that it may be difficult to apply these mitigation techniques to some old turbines should not compromise its use on contemporary turbines.

Other approaches to mitigating bat fatalities have been suggested, including projecting electromagnetic signals from small, portable radar units (Nicholls and Racey 2009) and ultrasonic broadcasts (Arnett et al. 2013b). However, the former approach has not been tested at large, utility-scale facilities, and none are yet being implemented broadly at wind energy facilities. Future studies of any mitigation approach must demonstrate greater or equal effectiveness to operational adjustments and also be cost-competitive with different operational strategies for mitigation.

11.8 Conservation Policy

In this section, we discuss a few selected issues regarding policy and regulation of wind facilities as they relate to wildlife impacts and successful integration of science, policy, and management to improve siting that minimizes risk to wildlife, including bats. This discussion is by no means exhaustive or comprehensive, but rather offers examples of policy issues from different regions of the world.

In the USA, the federal government's role in regulating wind power development is limited to projects occurring on federal lands, impacting federal trust species, or projects that have some form of federal involvement (e.g., interconnect with a federal transmission line) or require federal permits. The primary federal regulatory framework for protecting wildlife from impacts from wind power includes three laws—the US Migratory Bird Treaty Act, the Bald and Golden Eagle Protection Act, and the Endangered Species Act (ESA; GAO 2005; NRC 2007). Because wind-energy development has primarily occurred on non-federal land, regulating such facilities is largely the responsibility of state and local governments (GAO 2005). The primary permitting jurisdiction for wind-energy facilities in many instances is a local planning commission, zoning board, city council, or county board of supervisors or commissioners, and typically, these local jurisdictional entities regulate wind projects under zoning ordinances and building codes (GAO 2005), often without the basic knowledge needed to make informed decisions. Additionally, each state may enforce its laws regarding wind energy and wildlife impacts or establish cooperative efforts to address impacts. The US Fish and Wildlife Service has voluntary guidelines designed to help wind energy project developers avoid and minimize impacts of land-based wind projects on wildlife and their habitats (U.S. Fish and Wildlife Service 2012). In the USA, most species of bats, including migratory tree-roosting species killed most frequently by turbines, are not protected under federal, state, or provincial laws (Arnett 2012; Cryan 2011). Documented presence or fatality of species listed as threatened or endangered under the ESA (e.g., Indiana bat) does not necessarily mandate monitoring or mitigation as one might expect; rather, all efforts are voluntary even in cases involving a listed species, although threat of prosecution under the ESA increases when operators fail to collaborate or develop a conservation and mitigation plan.

Until recently, an Environmental Impact Assessment (EIA) in Canada was required under the Canadian Environmental Assessment Act (CEAA) when a Federal Authority initiated a wind project, granted any form of financial assistance or land for the project, and/or performed a regulatory duty in relation to the project, such as issuing a permit or license. Given that the Canadian Federal Government provided financial incentives for wind energy from 2002 to 2011, EIAs of wind energy projects were generally mandatory. However, wind energy projects no longer require federal environmental assessments (Canadian Environmental Assessment Act 2012), but projects may still require an environmental assessment if requested by the province or territory. Bats fall under the jurisdiction of the individual provinces (ten) and territories (three). As such, there are no Canada-wide bat and wind-energy policies or regulations; each province or territory sets their own policy and/or regulation regarding bats and wind energy projects (e.g., Ontario Ministry of Natural Resources 2011; Government of Alberta 2013). The Ontario Ministry of Natural Resources set a mortality threshold of 10 bats/turbine/year, which if exceeded triggers operational mitigation across the wind facility from 15 July to 30 September for the duration of the project (Ontario Ministry of Natural Resources 2011). This mortality threshold was based on fatality rates of bats at wind energy projects in Ontario and across North America. Ontario's guidelines do not explicitly consider cumulative effects (i.e., operational mitigation is only triggered by project-specific fatality rates). In Alberta, the integration of data, including acoustic data, collected during both the pre- and post-construction monitoring, helps guide the mitigation framework (Government of

Alberta 2013). For example, if less than five migratory-bat passes/detector night are recorded during pre-construction acoustic monitoring, then the project is considered to be a potentially acceptable risk, but if greater than ten migratory-bat passes/detector night were recorded, the project is considered to have a potentially high risk of bat fatalities and will likely require operational mitigation (Government of Alberta 2013). Unlike Ontario's guidelines, Alberta's mitigation framework explicitly considers cumulative effects (i.e., the proximity and risk at wind energy projects in the area are considered when determining the need for operational mitigation). Given the wide-ranging movement patterns of migratory tree bats and the tendency for wind energy projects to be clustered, from a conservation perspective, a policy which considers cumulative effects is superior to one that does not (Arnett et al 2013c).

Development of wind facilities in Mexico is regulated by laws and norms that have been enacted to achieve sustainable development. The General Law of Ecological Balance and Environmental Protection (GLEBEP) and its regulations are the main legal instruments the Mexican government has to protect ecosystems (www.semarnat.gob.mx). According to the GLEBEP, for a wind facility to be built, it is necessary to develop an EIA to determine the environmental feasibility of the project. The environmental authority has developed methodological guidelines for productive sectors to perform studies that meet the minimum information necessary for evaluation. In this case, the wind facilities are included in the energy sector, which includes, among others, hydroelectric, thermal, combined cycle plants, transmission lines, dams, and electrical substations. The first wind facility EIA (2000–2004) was completed with evaluations similar to those used for any other infrastructure (e.g., hydroelectric, thermoelectric, etc.) and was therefore not focused on impacts associated with wind facilities; collisions of birds and bats are not considered in the EIA. Thus, mitigation to reduce these impacts was not required by any regulatory authority at the first wind facilities in Mexico. Recently, the Mexican government has begun considering negative impacts on birds and bats and has incorporated measures including an annual monitoring program in these taxa during the entire cycle of wind energy projects. While there is no regulatory framework specifically for protection or conservation of bats in Mexico, there is an official standard that includes listings of flora and fauna found in risk categories similar to the Red List of the International Union for Conservation of Nature. The NOM-059-SEMARNAT-2010 (SEMARNAT 2010) includes the three categories of risk in order from most to least critical: in danger of extinction, threatened, and under special protection. The inclusion of species in each of these three categories is in accordance with technical and scientific criteria (SEMARNAT 2010). The NOM-059-SEMARNAT-2010 includes 38 species of bats, 19 of which warranted special protection. To date, however, only four species included in the NOM-059 have been found killed at wind facilities, and none of the species killed most frequently by turbines in Mexico are included in the NOM-59 given their abundance and wide distribution (Ceballos and Arroyo-Cabrales 2012; Ceballos et al. 2005). Apart from these legal instruments, there are no other legal mechanisms in Mexico to protect Mexican bats per se.

In the European Union, all species of bats, regardless of numbers, are strictly protected by law and it is illegal to deliberately kill or harm bats irrespective of any population effects. "Deliberately" means in this case that the actor is aware that activity may have an effect but still carries out the activity. The EIA Directive 85/337/EEC (amended to Council Directive 97/11/EC in 2011) legally requires an assessment to be carried out for all wind facilities with 5 or more turbines, or which are over 5 MW capacity. In addition, member states must restore or maintain their bat populations in favorable conservation status (Council Directive 92/43/EEC). All members of the EU have translated these directives into their domestic legislation which in theory should provide high levels of protection and a consistent way of handling the issue, based on the EUROBATS guidelines (Rodrigues et al. 2015). However, while bat issues are taken very seriously in some countries, this is not true in others. In the UK and Republic of Ireland, it is an offense to deliberately or recklessly kill or injure a bat or to deliberately disturb bats in a way that would significantly affect their local distribution or abundance, and detailed guidance is in place about the requirements for EIA.

The EUROBATS guidance (Rodrigues et al. 2015) proposes that turbines should not be placed closer than 200 m to woodland, whereas the current recommendations for the UK are that the blade tips should be at least 50 m from woodland or hedgerows. It is argued that a smaller buffer size is acceptable because the activity of bats found in the UK tends to decline rapidly with increasing distance from linear landscape features and woodlands (Natural England 2014). It is officially acknowledged that risk assessments for bats in the UK and Republic of Ireland are hampered by a lack of evidence in crucial areas (Bat Conservation Ireland 2012; Natural England 2014). Not only are collision rates unknown, but population estimates, and therefore inferences about the impact of turbine collisions on population viability, are uncertain. Standardized post-construction monitoring, including acoustic and carcass surveys, is recommended for sites identified as "high risk" (Bat Conservation Ireland 2012; Natural England 2014).

In reality, little post-construction monitoring occurs. This is at least partly because responsibility for requiring and enforcing survey conditions lies with local planning authorities, which are reluctant to impose conditions which may be open to legal challenge. A particular difficulty is that while generic guidance on survey designs is available, there is no standardized methodology and so it is extremely difficult to judge whether a particular level of bat activity would place a site as being in a "high risk" category: This point has already been raised at a legal review. Further, there has been no assessment of the relationship between pre-construction and post-construction acoustic surveys (or collision risk). It is therefore unclear how data collected pre-construction can be used to predict post-construction risk, particularly given evidence from the USA demonstrating a poor relationship between pre-construction activity surveys and bat fatalities (Hein et al. 2013). A final difficulty for Local Authorities is that the legal basis of bat protection relates to the conservation status of local populations (except in Scotland, where recklessly killing a bat is also an offense). Given that local population sizes are very poorly characterized, it is unclear how mitigation (such as raising turbine cut-in speed) could be enforced on the basis of reduced bat activity or bat fatalities.

In Sweden, there were no national guidelines until 2011 (Rydell et al. 2012), and more than ten years after the first turbines were built. Hence, many wind turbines in Sweden were constructed in poor locations with respect to bats, such as along the coast, and with operation permitted without any mitigation measures. After 2011, however, the Swedish Environmental Protection Agency concentrated resources to achieve national acceptance for the guidelines among decision makers, the industry and NGOs, and to implement them on all new project. In 2014, all proposed wind facilities are subject to a pre-construction survey and an evaluation regarding the risk to bats. In the Netherlands, all bats are species of annex II and/or IV of the European Habitats Directive EIA, and risk assessment and evaluation under Flora and Fauna law (research on what species are present) and possibly Nature Conservation law (when EHD annex II species are present) are obligatory. When risk species are present and fatalities cannot be excluded, a permit for construction is needed under FF law and NC law, and information on fatalities needs to be established using protocols and a curtailment may be required (Boonman et al. 2013; Limpens 2013).

Mitigation studies have shown that bat fatalities can be reduced substantially (e.g., Baerwald et al. 2009; Arnett et al. 2011). Although curtailing turbines holds great promise, the problem is that developing thresholds—those values that trigger some action—to mitigate bat kills is difficult, especially when supporting data are limited or imperfect (Arnett et al. 2013c). In Germany, recent models accounting for multiple environmental variables that predict and reduce collision rate and further minimize loss of energy production offer promise for mitigating bat fatalities (Korner-Nievergelt et al. 2013). However, these algorithms and most other operation protocols still tolerate an arbitrary number of bat fatalities (currently two bat fatalities per wind turbine per year where these algorithms are used in Germany) (Voigt et al. 2015a); any such fatality trigger in Europe is seemingly in opposition to current law, given that European legislation does not allow deliberate killing of any bat, regardless of population effects. Additionally, with increasing numbers of wind turbines, fixed annual "per capita" (i.e., per wind turbine) mortality rates may not be acceptable in light of limited bat population sizes, and the acceptance of a reduced fatality rate may not necessarily be consistent with national and EU legislation (Voigt et al 2015a). Regulatory authorities in the US state of Pennsylvania and the Canadian province of Ontario set thresholds for initiating curtailment based on the annual mean number of bats killed per turbine (28 and 10 bat fatalities/turbine, respectively Arnett et al. 2013c). Arnett et al. (2013c) argued that this approach sets a dangerous precedent and has several flaws, none the least of which is the assumption that bat populations are currently stable and remain so. This approach also ignores expanding development of wind turbines that will likely yield increasing bat fatalities per population or region. Policy and management efforts to mitigate bat fatalities and conserve bat populations affected by wind turbines should be proactive and based on science rather than being reactive and arbitrary.

11.9 Future Directions

Population data are generally lacking, and this not only impedes our understanding of actual impacts of wind turbines, but also impedes knowing the effectiveness of mitigation efforts. For example, we do not know whether raising turbine cut-in speed that might result in 50 % fewer bat fatalities will mitigate population-level impacts or simply delay inevitable losses (Arnett and Baerwald 2013). The lack of population data also makes it difficult to set triggers for mitigation (i.e., number of bats killed per turbine or MW that requires mitigation; but see Arnett et al. 2013c). However, population data are not likely to be available for most bat species in the near future, and thus, wind operators should practice the precautionary principle and implement operational mitigation at sites where bat fatalities are high, or are predicted to be high, even in the absence of population data.

Several knowledge gaps remain that must be filled in the immediate future. Most notably, many regions of the world lack any publicly available monitoring information on the impacts of wind energy facilities on bats and other wildlife (e.g., China). We strongly encourage wind energy developers and governments to end this trend and gather needed data to inform siting and operating wind facilities around the world. In addition to population studies and basic monitoring data described above, some key research priorities germane to all regions of the world include:

1. Evaluating effectiveness of pre-construction bat activity surveys in predicting future fatalities at wind facilities.
2. Determining whether approaches such as temperature profiles in relation to weather types and seasons (and different regions) or habitat suitability modeling for bats can effectively predict high-risk sites and be used by planning authorities and industry to help situate wind turbines in areas where the potential for conflict with bats is minimized.
3. Evaluate methods for assessing the risk and minimizing and mitigating impacts posed by offshore wind turbines to bats (including approaches such as radar and collision sensors).
4. Investigate the extent of migratory activity worldwide, particularly offshore, by international collaboration using a range of techniques [stable isotopes (Voigt et al. 2012; Lehnert et al. 2014; Baerwald et al. 2014), population genetics, potentially GPS tracking, etc.]. Although difficult, it may be possible to gather and pool acoustic data of activity patterns established through automated real-time recorders in numerous wind energy facilities to evaluate spatial and temporal patterns.
5. Future operational mitigation experiments should be designed to determine which factors (e.g., habitat, insect occurrence, temperature, wind, humidity, moon illumination) or combination of factors (Weller and Baldwin 2012) will best improve predictability of bat fatalities, while minimizing economic costs. Alternative mitigation approaches to operational adjustments should be proven equally or more effective at reducing bat fatalities at operating wind facilities before being accepted as viable mitigation approaches.

6. Detailed meta-analyses of existing data on cumulative fatality impacts and factors influencing fatalities are needed. It should be noted that data disclosure from many sites by some companies hinders such analyses.

We strongly encourage developers to follow guidelines (e.g., Kunz et al. 2007b; Rodrigues et al. 2015; Strickland et al. 2011; Bat Conservation Ireland 2012) consistently when implementing pre- and post-construction monitoring. Data should be placed into the public domain or preferably published in refereed journals. There are a number of policies, regulatory, and communication challenges we face in protecting bats while developing wind energy responsibly across the globe (Arnett 2012). Unless there is a federal, state, or provincial nexus, most research, siting, and mitigation efforts by wind energy developers and operating companies will be voluntary, likely without regard for cumulative effects. Sites that do trigger a regulatory nexus will usually be driven by endangered species issues (e.g., Indiana bats in the USA). It is apparent most local jurisdictional entities, regardless of country, lack experience in wildlife science, and unless they coordinate with their wildlife or natural resource agency specialists, concerns about bat fatalities may never be addressed in decision making for wind energy development. Another key issue is consistent application of regulations. The authors have encountered many situations where different individuals had varied interpretations of the same law or guidance policy, and this creates untenable situations, considerable uncertainty, consternation, and lack of trust among stakeholders that seems completely unnecessary and easily remedied (Arnett 2012).

We encourage continuing cooperation among all stakeholders, gathering needed information, avoiding construction in high-risk sites, considering cumulative effects, and implementing mitigation where needed even when no regulatory process is triggered (Arnett 2012; Arnett and Baerwald 2013). Wind energy developers should, however, be treated fairly and consistently to ensure proactive measures are implemented. Arnett (2012) noted that when some companies choose to cooperate, while others may not, unnecessary angst is generated and deters resolving wildlife impacts and other issues. Decision making must be based on the best available science. Also, consistent policy, accountability, effective siting and mitigation strategies, and a “level-playing field” for the industry (i.e., consistent requirements and incentives for all companies) are fundamental if we are to successfully develop wind energy that protects bats and other wildlife.

Acknowledgements We appreciate the invitation from C. Voigt and T. Kingston to prepare this chapter. Portuguese data were collected by AgriPro Ambiente, Bio3, Biota, Colmus, Ecomind, Ecosativa, Ecosfera, EDP, ENEOP2, EolFlor, LEA, Naturibérica, NOCTULA, Plecotus, PROCESL, Profico Ambiente, ProSistemas, STRIX, and Tecneira. We acknowledge C.-H. Chou at the Endemic Species Research Institute in Taiwan for letting us cite his important but yet unpublished data. This chapter was greatly enhanced by reviews from C. Hein, T. Kingston, H. Limpens, and 2 anonymous reviewers. We wish to dedicate this chapter to our friend Dr. Thomas Kunz for all of his wisdom, knowledge, and efforts in regard to wind energy impacts on bats.

References

Adams JK (1989) Pteronotus davyii. Mamm Species 346:1–5

Ahlén I, Baagøe HJ, Bach L (2009) Behavior of Scandinavian bats and foraging at sea. J Mammal 90:1318–1323

Amorim F, Rebelo H, Rodrigues L (2012) Factors and mortality at a wind farm in the Mediterranean region. Acta Chirop 14:439–457

Ancilotto L, Rydell J, Nardone V, Russo D (2014) Coastal cliffs on islands as foraging habitat for bats. Acta Chirop 16:103–108

Arnett EB (2005) Relationships between bats in Pennsylvania and West Virginia: an assessment of bat fatality search protocols, patterns of fatality, and behavioral interactions with wind turbines. Bat conservation international, Austin, Texas, USA http://www.batsandwind.org/pdf/postconpatbatfatal.pdf. Accessed 22 Dec 2014

Arnett EB (2012) Impacts of wind energy development on wildlife: challenges and opportunities for integrating science, management. In: Sands JP, DeMaso SJ, Brennan LA, Schnupp MJ (eds) Wildlife science: with management. Taylor and Francis, New York, pp 213–237

Arnett EB, Baerwald EF (2013) Impacts of wind energy development on bats: implications for conservation. In: Adams RA, Peterson SC (eds) Bat evolution, ecology, and conservation. Springer, New York, pp 435–456

Arnett EB, Inkley DB, Larkin RP, Manes S, Manville AM, Mason JR, Morrison ML, Strickland MD, Thresher R (2007) Impacts of wind energy facilities on wildlife and wildlife habitat. Wildl Soc Tech Rev 07-2 (The Wildlife Society, Bethesda, Maryland)

Arnett EB, Brown K, Erickson WP, Fiedler J, Henry TH, Johnson GD, Kerns J, Kolford RR, Nicholson CP, O'Connell T, Piorkowski M, Tankersley R Jr (2008) Patterns of fatality of bats at wind energy facilities in North America. J Wildl Manage 72:61–78

Arnett EB, Huso MMP, Schirmacher MR, Hayes JP (2011) Changing wind turbine reduces at wind facilities. Front Ecol Environ 9:209–214. doi:10.1890/100103

Arnett EB, Johnson GD, Erickson WP, Hein CD (2013a) A synthesis of studies to reduce at wind energy facilities in North America. Bat conservation international, Austin, Texas http://www.batsandwind.org/pdf/Operational%20Mitigation%20Synthesis%20FINAL%20REPORT%20UPDATED.pdf. Accessed 22 Dec 2014

Arnett EB, Hein CD, Schirmacher MR, Huso MMP, Szewczak JM (2013b) Evaluating the effectiveness of an ultrasonic acoustic deterrent for reducing. PLoS ONE 8(6):e65794. doi:10.1371/journal.pone.0065794

Arnett EB, Barclay RMR, Hein CD (2013c) Thresholds for bats killed. Front Ecol Environ 11:171. doi:10.1890/1540-9295-11.4.171

AronsonJB, Thomas AJ, Jordaan SL (2013) Bat fatality at a wind energy facility in the western Cape. Afr Bat Conserv News 31:9–12 (ISSN 1812-1268)

Baerwald EF, Barclay RMR (2009) Geographic variation in activity and fatality bats at wind energy facilities. J Mammal 90:1341–1349

Baerwald EF, Barclay RMR (2011) Patterns of activity and fatality bats at a wind energy facility in Alberta. J Wildl Manage 75:1103–1114

Baerwald EF, D'Amours GH, Klug BJ, Barclay RM (2008) Barotrauma is a significant cause of bat fatalities at wind turbines. Curr Biol 18(16):R695–R696

Baerwald EF, Edworthy J, Holder M, Barclay RMR (2009) A large-scale mitigtion experiment to reduce at wind energy facilities. J Wildl Manage 73:1077–1081

Baerwald EF, Patterson WP, Barclay RMR (2014) Origins patterns of bats killed in southern Alberta: evidence from stable isotopes. Ecosphere 5(article 118):1–17

Barclay RMR, Harder LM (2003) Life histories of bats: life in the slow lane. In: Kunz TH, Fenton MB (eds) Bat ecology. University of Chicago Press, Chicago, pp 209–259

Bat Conservation Ireland (2012) Wind turbine/wind development bat survey guidelines—version 2.8. Bat conservation Ireland http://www.batconservationireland.org/pubs/reports/BCIreland%20Wind%20Farm%20Turbine%20Survey%20Guidelines%20Version%202%208.pdf. Accessed 20 January 2015

Bateman GC, Vaughan TA (1974) Nightly activities of mormoopid bats. J Mammal 55:45–65

Behr O, von helversen O (2006) Gutachten zur Beein -trächtigung im freien Luftraum jagender und ziehender Fle -dermäuse durch bestehende Windkraftanlagen. Wirkungs -kontrolle zum Windpark "Rosskopf" (Freiburg i Br.) im Jahr 2005. Report to Regiowind GmbH & Co., Freiburg

Bennett E (2012) Hepburn wind farm bird and bat mortality survey interim report 11th July 2011–9th Jan 2012. Elmoby ecology http://hepburnwind.com.au/downloads/Hepburn%20Wind%20interim%20report%20Bird%20&%20Bat%20Monitoring.pdf. Accessed 25 Aug 2014

Berkhout V, Faulstich S, Görg P, Kühn P, Linke K, et al. (2013) Wind energy 2012. Fraunhofer-Institut für Windenergie und Energiesystemtechnik. IWES, Kassel

Bernstein MA, Griffin J, Lempert R (2006) Impacts on energy expenditures of use. Technical report prepared for the energy future coalition, RAND Corporation, Santa Monica, CL

Boonman M, Limpens HJGA, La Haye MJJ, van der Valk M, Hartman JC (2013) Protocollen leermuisonderzoek bij windturbines. Rapport 2013.28, Zoogdiervereniging & Bureau Waardenburg. 29 pp+ 1 bijlage

Boshamer JPC, Bekker JP (2008) Nathusius' pipistrelles (*Pipistrellus nathusii*) and other species of bats platforms in the Dutch sector of the North Sea. Lutra 51:17–36

Brinkmann R, Behr O, Niermann I, Reich M (2011) Entwicklung von Methoden zur Untersuchung und Reduktion des Kollisionsrisikos von Fledermausen an onshore-Windenergieanlagen. Schriftenreihe Institut fur Umweltplanung. Cuvillier Verlag Gottingen, p 457

Briones-Salas M, Peralta-Pérez My, García-Luis M (2013) Acoustic characterization of new species of bats for the state of Oaxaca. Theyra 4:15–32

Camina A (2012) At wind farms in—lessons to be learned. Acta Chirop 14:205–212

Canadian Environmental Assessment Act (2012) An ACT respecting the environmental assessment of certain activities and the prevention of significant adverse environmental effects. Assented to 29 June 2012, in force 06 July 2012. (Online). Government : Justice Laws http://laws-lois.justice.gc.ca/eng/acts/C-15.21/page-1.html. Accessed 6 Jan 2015

Ceballos G, Arroyo-Cabrales J (2012) Lista actualizada de los mamíferos de México 2012. Rev Mex Mastozoología 1:27–80

Ceballos G, Oliva G, coords. (2005) Los mamíferos silvestres de México. Comisión Nacional para el Conocimiento y Uso de la Biodiversidad y Fondo de Cultura Económica, México, DF

Cryan PM (2008) Mating behavior as a possible cause of bat fatalities at wind turbines. J Wildl Manage 72:845–849

Cryan PM (2011) Wind turbines as landscape impediments to connectivity of bats. Environ Law 41:355–370

Cryan PM, Barclay RMR (2009) Causes of bat fatalities at wind turbines: hypotheses and predictions. J Mammal 90:1330–1340

Cryan PM, Brown AC (2007) Migration of bats past a remote island offers clues towards the problem of bat fatalities at wind turbines. Biol Conserv 139(1–2):1–11

Doty AC, Martin AP (2012) Assessment of bat and avian mortality at a pilot wind turbine at Coega, Port Elizabeth, Eastern Cape, South Africa. New Zeal J Zool. doi:10.1080/03014223.2012.741068

Dubourg-Savage M-J, Rodrigues L, Santos H, Georgiakakis P, Papadotou E, Bach L, Rydell J (2011) Patterns of comparing north and south. Poster and abstract. In: Conference on wind energy and environmental impacts, Trondheim, Norway, May 2011

EUROBATS (2014) Report of the intersessional working group on wind turbines and bat populations. http://www.eurobats.org/sites/default/files/documents/pdf/Advisory_Committee/Doc.AC_.20.5.ReportIWGWindTurbines_0.pdf. Accessed 1 Aug 2015
Frick WF, Pollock JF, Hicks AC, Langwig KE, Reynolds DS, Turner GG, Butchkoski CM, Kunz TH (2010) An emerging disease causes regional population collapse of a common North American bat species. Science 329:679–682
Gannon MR, Kurta A, Rodríguez-Durán A, Willig MR (2005) Bats of Puerto Rico: an island focus and a Caribbean perspective. Texas Tech University Press, Lubbock
García-Grajales J, Silva AB (2012) Revisión al conocimiento de los murciélagos del estado de Oaxaca. Therya 3:277–293
Georgiakakis P, Kret E, Cárcamo B, Doutau B, Kafkaletou-Diez A, Vasilakis D, Papadatou E (2012) Bat fatalities at wind farms in north-eastern Greece. Acta Chirop 14:459–468
Cryan PM, Gorresen, PM, Hein CD, Schirmacher MR, Diehl RH, Huso MM, Hayman DTS, Fricker PD, Bonaccorso FJ, Johnson DH, Heist K, Dalton DC (2014) Behavior of bats at wind turbines. Proc Nat Acad Sci. doi:10.1073/pnas.1406672111
Government Accountability Office (GAO) (2005) Wind Power: impacts on wildlife and government responsibilities for regulating development and protecting wildlife. Report to Congressional Requesters, GAO-05-906, U.S. Government Accountability Office, Washington
Government of Alberta (2013) Bat mitigation framework for wind power development. In: Environment and sustainable resource development. Alberta, Canada
Grodsky SM (2010) Aspects of bird and bat mortality at a wind energy facility in southeastern Wisconsin: impacts, relationships, and cause of death. Thesis, University of Wisconsin, Madison
Grodsky SM, Behr MJ, Gendler A, Drake D, Dieterle BD, Rudd RJ, Walrath NL (2011) Investigating the causes of death for wind turbine-associated bat fatalities. J Mammal 92:917–925
Hall LS, Richards GC (1972) Notes on *Tadarida australis* (Chiroptera:Molossidae). Aust Mammal 1:46–47
Hayes MA (2013) Bats killed in large numbers at United States wind energy facilities. Bioscience 63:975–979
Hein CD, Gruver J, Arnett EB (2013) Relating pre-construction bat activity and post-construction bat fatality to predict risk at wind energy facilities: a synthesis. Bat conservation international, Austin, Texas http://www.batsandwind.org/pdf/Pre-%20Post-construction%20Synthesis_FINAL%20REPORT.pdf. Accessed 23 Dec 2014
Heinrich B (1993) The hot-blooded insects. Springer, Berlin
Horn J, Arnett EB, Kunz TH (2008) Behavioral responses of bats to operating wind turbines. J Wildl Manage 72:123–132
Hull CL, Cawthen L (2012) Bat fatalities at two wind farms in Tasmania, Australia: bat characteristics, and spatial and temporal patterns. New Zeal J Zool. doi:10.1080/03014223.2012.731006
Huso MMP (2011) An estimator of wildlife fatality from observed carcasses. Environmetrics 22:318–329. doi:10.1002/env.1052
Huso MMP, Dalthorp D (2013) Accounting for unsearched areas in estimating wind turbine-caused fatality. J Wildl Manage. doi:10.1002/jwmg.663
Hutson AM, Mickleburgh SP, Racey PA (2001) Global status and conservation action plan: microchiropteran bats. IUCN/SSC Chiroptera Specialist Group, Cambridge
Hutterer R, Ivanova T, Meyer-Cords C, Rodrigues L (2005) Bat migrations in Europe: a review of banding data and literature. Naturschutz und Biologische Vielfalt, Bonn 28:1–176
Ingersoll TE, Sewall BJ, Amelon SK (2013) Improved analysis of long-term monitoring data demonstrates marked regional declines of bat populations in the Eastern United States. PLoS ONE 8(6):e65907. doi:10.1371/journal.pone.0065907

Inkley DB, Anderson MG, Blaustein AR, Burkett VR, Felzer B, Griffith B, Price J, Root TL (2004) Global climate change and wildlife in North America. Wildl Soc Tech Rev 04-1 (The Wildlife Society, Bethesda, Maryland)

Jain AA, Koford RR, Hancock AW, Zenner GG (2011) Bat mortality and activity at a northern Iowa wind resource area. Am Mid Nat 165:185–200

Johnson GD (2005) A review of bat mortality at wind-energy developments in the United States. Bat Res News 46:45–49

Johnson GD, Perlik MK, Erickson WP, Strickland MD (2004) Bat activity, composition, and collision mortality at a large wind plant in Minnesota. Wildl Soc Bull 32:1278–1288

Johnson JB, Gates JE, Zegre NP (2011) Monitoring seasonal bat activi ty on a coastal barrier island in Maryland, USA. Environ Monit Assess 173:685–699

Korner-Nievergelt F, Brinkmann R, Niermann I, Behr O (2013) Estimating bat and bird mortality occurring at wind energy turbines from covariates and carcass searches using mixture models. PLoS ONE 8(7):e67997. doi:10.1371/journal.pone.0067997

Kunz TH, Arnett EB, Erickson WP, Hoar AR, Johnson GD, Larkin RP, Strickland MD, Thresher RW, Tuttle MD (2007a) Ecological impacts of wind energy development on bats: questions, research needs, and hypotheses. Front Ecol Environ 5:315–324

Kunz TH, Arnett EB, Cooper BM, Erickson WP, Larkin RP, Mabee T, Morrison ML, Strickland MD, Szewczak JM (2007b) Methods and metrics for studying impacts of wind energy development on nocturnal birds and bats. J Wildl Manage 71:2449–2486

Ladle RJ, Firmino JVL, Malhado ACM, Rodríguez-Durán A (2012) Unexplored diversity and conservation potential of Neotropical hot caves. Conserv Biol 26:978–982

LEA (2010) Monitorização dos efeitos da Medida de Minimização de Mortalidade do Parque Eólico do Outeiro Relatório final. Laboratório de Ecologia Aplicada da Universidade de Trás-os-Montes e Alto Douro. Vila Real, 78 pp

Lehnert LS, Kramer-Schadt S, Schonborn S, Lindecke O, Niermann I, Voigt CC (2014) Wind farm facilities in Germany kill noctule bats from near and far. PLoS ONE 9(8):e103106. doi:10.1371/journal.pone.0103106

Limpens HJGA, Boonman M, Korner-Nievergelt F, Jansen EA, van der Valk M, La Haye MJJ, Dirksen S, Vreugdenhil SJ (2013) Wind turbines and bats in the Netherlands—measuring and predicting. Report 2013.12, Zoogdiervereniging & Bureau Waardenburg. 77 pp+ 2 appendices

McLeish T (2002) Wind power. Nat N Eng 11:60–65

Miller A (2008) Patterns of avian and bat mortality at a utility-scalewind farm on the southern high plains. Thesis, Texas Tech University, Lubbock

National Research Council [NRC] (2007) Ecological impacts of wind-energy projects. National Academies Press, Washington

Natural England (2014) Natural England technical information note TIN051. Bats and onshore wind turbines, 3rd edn, March 2014 http://publications.naturalengland.org.uk/search?q=Bats+and+onshore+wind+turbines&num=100. Accessed 4 Jan 2015

Nicholls B, Racey PA (2009) The aversive effect of electromagnetic radiation on foraging bats—a possible means of discouraging bats from approaching wind turbines. PLoS ONE 7:e6246. doi:10.1371/journal.pone.0006246

O'Shea TJ, Bogan MA, Ellison LE (2003) Monitoring trends in bat populations of the United States and Territories: status of the science and recommendations for the future. Wild Soc Bull 31:16–29

Ontario Ministry of Natural Resources (2011) Bats and bat habitats: guidelines for wind power projects. Ontario, Canada

Piorkowski MD, O'Connell TJ (2010) Spatial pattern of summer bat mortality from collisions with wind turbines in mixed-grass prairie. Am Mid Nat 164:260–269

Poerink BJ, Lagerveld S, Verdaat H (2013) Bat activity in the Dutch offshore wind farm OWEZ and PAWP. IMARES Report no. C026/13 / tFC report no. 20120402. The Fieldwork Company, Wageningen, Holland http://edepot.wur.nl/248979. Accessed 29 Nov 2014

Rodrigues L, Bach L, Dubourg-Savage MJ, Karapandza B, Kovac D, Kervyn T, Dekker J, Kepel A, Bach P, Collins J, Harbusch C, Park K, Micevski B, Minderman J (2015) Guidelines for consideration of bats in wind farm projects – revision 2014. EUROBATS Publication Series no. 6 (English version). UNEP/EUROBATS Secretariat, Bonn, Germany, UNEP/EUROBATS Secretariat. http://www.eurobats.org/sites/default/files/documents/publications/publication_series/pubseries_no6_english.pdf. Accessed 1 Aug 2015

Rodríguez-Durán A (2009) Bat assemblages in the West Indies: the role of caves. In: Fleming TH, Racey P (eds) Island bats: evolution, ecology, and conservation. University of Chicago Press, Chicago, pp 265–280

Rollins KE, Meyerholz DK, Johnson GD, Capparella AP, Loew SS (2012) A forensic investigation into the etiology of bat mortality at a wind farm: barotrauma or traumatic injury? Vet Pathol 49:362–371

Roscioni F, Russo D, Di Febbraro M, Frate L, Carranza ML, Loy A (2013) Regional-scale modeling of the cumulative impact of wind farms on bats. Biodivers Conserv. doi:10.1007/s10531-013-0515-3

Roscioni F, Rebelo H, Russo D, Carranza ML Di Febbraro M, Loy A (2014) A modeling approach to infer the effects of wind farms on landscape connectivity for bats. Landscape Ecol. doi 10.1007/s10980-014-0030-2

Rydell J, Bach L, Dubourg-Savage M, Green M, Rodrigues L, Hedenstrom A (2010a) Bat mortality at wind turbines in northwestern Europe. Acta Chirop 12:261–274

Rydell J, Bach L, Dubourg-Savage MJ, Green M, Rodrigues L, Hedenström A (2010b) Mortality of bats at wind turbines links to nocturnal insect migration? Euro J Wildl Res 56:823–827

Rydell J, Engström H, Hedenström A, Larsen JK, Pettersson J, Green M (2012) The effect of wind power on birds and bats: a synthesis report. Report 6511, Swedish Environmental Protection Agency, Stockholm, Sweden

Rydell J, Bach L, Bach P, Guia Diaz L, Furmankiewicz J, Hagner-Wahlsten N, Kyheröinen EM, Lilley T, Masing M, Meyer MM, Petersons G, Suba J, Vasko V, Vintulis V, Hedenström A (2014) Phenology of migratory bat activity across the Baltic Sea and the south-eastern North Sea. Acta Chirop 16:139–147

Santos H, Rodrigues L, Jones G, Rebelo H (2013) Using species distribution modelling to predict bat fatality risk at wind farms. Biol Conserv 157:178–186

Schlesinger ME, Mitchell JFB (1987) Climate model simulations of the equilibrium climate response to increased carbon dioxide. Rev Geophys 25:760–798

Smallwood KS (2013) Comparing bird and bat fatality-rate estimates among North American wind-energy projects. Wildl Soc Bull 37:19–33

Steffens R, Zophel U, Brockmann D (2004) 40th anniversary bat marking Centre Dresden—evaluation of methods and overview of results. Sachsisches Landesamt für Umwelt und Geologie, p 132

Strickland MD, Arnett EB, Erickson WP, Johnson DH, Johnson GD, Morrison ML, Shaffer JA, Warren-Hicks W (2011) Comprehensive guide to studying wind energy/wildlife interactions. National Wind Coordinating Collaborative, Washington http://www.batcon.org/pdfs/wind/National%20Wind%20Coordinating%20Collaborative%202011_Comprehensive%20Guide%20to%20Studying%20Wind%20Energy%20and%20Wildlife%20Interactions.pdf. Accessed 4 Jan 2015

U.S. Fish and Wildlife Service (2012) Land-based wind energy guidelines. U.S. fish and wildlife service, Arlington, VA, USA http://www.fws.gov/windenergy/docs/WEG_final.pdf. Accessed 1 February 2015

Van der Meij T, Van Strien AJ, Haysom KA, Dekker J, Russ J, Biala K, Bihar Z, et al. (2014) Return of the bats? A prototype indicator of trends in European bat populations in underground hibernacula. Mammal Biol. doi:10.1016/j.mambio.2014.09.004

Villegas-Patraca R, Macías-Sánchez S, MacGregor-Fors I, Muñoz-Robles C (2012) Scavenger removal: bird and bat carcass persistence in a tropical wind farm. Acta Oecol 43:121–125

Voigt CC, Popa-Lisseanu AG, Niermann I, Kramer-Schadt S (2012) The catchment area of wind farms for European bats: a plea for international regulations. Biol Conserv 153:80–86

Voigt CC, Lehnert LS, Petersons G, Adorf F, Bach L (2015a) Wildlife and renewable energy: German politics cross migratory bats. Eur J Wildl Res (in press)

Voigt CC, Aguirre L, Phelps K, Schoeman C, Vanitharani J, Zubaid A (2016) Bats and buildings: the conservation of synanthropic bats. In: Voight CC, Kingston T (eds) Bats in the anthropocene: conservation of bats in a changing world. Springer International AG, Cham, pp. 427–453

Weller TJ, Baldwin JA (2012) Using echolocation monitoring to model bat occupancy and inform mitigations at wind energy facilities. J Wildl Manage 76:619–631

Wellington WG (2011) The effects of variations in atmospheric pressure upon insects. Can J Res 1946, 24d(2):51–70

Young DP, Bay K, Nomani S, Tidhar WL (2011) Nedpower mount storm wind energy facility post-construction avian and bat monitoring: July–October 2010. Western EcoSystems Technology, Inc, Cheyenne, Wyoming http://www.batsandwind.org/pdf/WV%20-%20Young%20et%20al.%202011%20-%20Mount%20Storm%20Fall%202010%20Report(%202-10-11).pdf. Accessed 4 Jan 2015

Chapter 12
Exploitation of Bats for Bushmeat and Medicine

Tammy Mildenstein, Iroro Tanshi and Paul A. Racey

Abstract Bat hunting for consumption as bushmeat and medicine is widespread and affects at least 167 species of bats (or c. 13 % of the world's bat species), in Africa, Asia, across the islands of Oceania, and to a lesser extent in Central and South America. Hunting is particularly prevalent among the large-bodied fruit bats of the Old World tropics, where half (50 %, 92/183) the extant species in the family Pteropodidae are hunted. Pteropodids that are hunted are six times more likely to be Red Listed as threatened: 66 % of species in IUCN threatened categories (CR, EN, VU, NT), compared to 11 % of species in the 'Least Concern' (LC) category. However, there still appears to be an information gap at the international level. One third of the hunted species on the Red List are not considered threatened by that hunting, and nearly a quarter of the bat species included in this review are not listed as hunted in IUCN Red List species accounts. This review has resulted in a comprehensive list of hunted bats that doubles the number of species known from either the IUCN Red List species accounts or a questionnaire circulated in 2004. More research is needed on the impacts of unregulated hunting, as well as on the sustainability of regulated hunting programs. In the absence of population size and growth data, legislators and managers should be precautionary in their attitude towards

T. Mildenstein (✉)
Department of Biology, Cornell College, Mt. Vernon, IA, USA
e-mail: TMildenstein@cornellcollege.edu

I. Tanshi
Department of Biological Sciences, Texas Tech University, Lubbock, TX, USA

I. Tanshi
Department of Animal and Environmental Biology, University of Benin, Benin City, Nigeria

P.A. Racey
Centre for Ecology and Conservation, College of Life and Environmental Sciences,
University of Exeter, Penryn Campus, Penryn TR10 9FE, UK

C.C. Voigt and T. Kingston (eds.), *Bats in the Anthropocene: Conservation of Bats in a Changing World*, DOI 10.1007/978-3-319-25220-9_12

hunting. Roost site protection should be a priority as it is both logistically simpler than patrolling bat foraging grounds and reduces the comparatively larger scale mortality and stress that hunting at the roost can cause. Education and awareness campaigns within local communities should demonstrate how bats are a limited resource and emphasize characteristics (nocturnal, slow reproducing and colonial) that make them particularly vulnerable to hunting pressure.

12.1 Introduction

Most of the chapters in this book (Voigt and Kingston 2016) consider negative consequences for bats from indirect effects of anthropogenic perturbations. In contrast, this chapter explores the direct exploitation of bats by humans for bushmeat and medicine.

A global review of bats as bushmeat was published in 2009 providing an overview of bat hunting based on published literature and a questionnaire widely distributed among bat biologists in 2004 (Mickleburgh et al. 2009). Here, we summarize what is currently known about the exploitation of bats for consumption and medicinal use, synthesizing the 2009 review with what has been published since and unpublished information the authors have gathered from colleagues. The result is a comprehensive list of hunted bats species that contains nearly twice as many species as known from either the IUCN Red List species accounts or the 2009 review. It is unclear whether the increased concern about hunting is the result of greater actual hunting pressure, or just represents our increased understanding of hunting impacts. What is clear is that reviews now explicitly attribute species declines and extinction risk to hunting pressure.

Most of the chapter discusses conservation needs in light of what is known about bat hunting. We summarize regional patterns in bat hunting and protection efforts and highlight areas of conservation concern. We provide details of current research aimed at learning more about hunting impacts, and we give examples of management activities to strengthen protection efforts against population-level effects of hunting. We end with recommendations for research to better understand the effects of hunting on bat populations as well as strategies for effectively managing hunting to support bat conservation. All common and scientific species names follow Simmons (2005).

12.2 Background

12.2.1 Overview of Bat Hunting

Bat hunting is widespread and affects at least 167 species of bats (or c. 13 % of the world's 1331+ bat species, Bat Conservation International 2015), occurring in Africa, Asia, across the islands of Oceania, and in some parts of Central and

South America (compiled from IUCN 2014; Mickleburgh et al. 2009, and personal communications by the authors; Appendix). Hunting is particularly prevalent on the large-bodied fruit bats (family Pteropodidae) in the Old World tropics, where half (50 %, 92/183) of all extant species experience hunting pressure (Mickleburgh et al. 2009; IUCN 2014). A much smaller proportion of insectivorous (<8 %, 75/962 species) are hunted, particularly members of the Emballonuridae, Hipposideridae and Molossidae in Asia and Southeast Asia, Vespertilionidae in North Africa and West and Central Asia, and Phyllostomidae in Brazil, Peru, and Bolivia as well as Rhinolophidae in sub-Saharan Africa, (Mickleburgh et al. 2009; Lizarro et al. 2010; IUCN 2014) (Appendix).

Bats are hunted for a variety of reasons, from their perceived medicinal properties e.g. Nicobar flying fox, *Pteropus faunulus*, Kingston et al. (2008); 'small bats' in Nepal (Tuladhar-Douglas 2008); fat from pteropodid species in Pakistan (Roberts 1997) to their use in ornate decoration such as the teeth of the Makira flying fox (*Pteropus cognatus*) used for necklaces (James et al. 2008). Bats are also hunted for sport by urban residents seeking country pursuits (e.g. large fruit bat hunting at Subic Bay, Philippines, S. Stier, pers. comm.) and tourists seeking exotic eating experiences (e.g. Pacific flying fox, *Pteropus tonganus*, hunting is offered as a recreation option at hotels in Vanuatu; A. Brooke pers. comm. in Hamilton and Helgen 2008). However, the most widespread reason for bat hunting, by far, is for consumption; all 167 species that are hunted are, at least in part, wanted for their meat as a source of protein. Bat meat ranges in value from a highly sought-after delicacy served at special ceremonies and traditional celebrations (e.g. *Pteropus mariannus* in the Mariana Islands) to "finger food" consumed in social drinking settings (e.g. many bat species in Southeast Asia, Mildenstein 2012; and in West Africa, M. Abedi-Lartey pers. comm.). Elsewhere, it provides an alternative source of protein for local people for whom meat is an expensive commodity (Jenkins and Racey 2008) and in extreme cases, bats are consumed as starvation food (Goodman 2006).

The intensity and frequency of bat harvesting varies from year round to periodical depending on the seasonality of the species, hunters' lifestyles, and/or local legislation. On the Islands of São Tomé and Príncipe, bats are hunted opportunistically for food all year round (Carvalho et al. 2014). In Southeast Asia, regular harvest of bats occurs in Indonesia and the Philippines (T. Mildenstein, unpublished data). A migratory species, the African straw-colored fruit bat (*Eidolon helvum*) is hunted whenever it is present in Accra (Fig. 12.1) and Kumasi, Ghana, between November and March (Kamins et al. 2011). Reduced hunting intensity in March is likely due to the northward migration of forest resident bats and/or the shift in hunters' occupation to farming. Similarly, in Madagascar, local legislation specifies two hunting seasons—one for fruit bats, and another for Commerson's leaf-nosed bat (*Hipposideros commersoni*) (Jenkins and Racey 2008), though actual hunting intensity is driven more by local availability (e.g. the lychee season for fruit bats). In some localities in east and western Nigeria, year round harvest of the Egyptian rousette (*Rousettus aegyptiacus*) is known (Fig. 12.1) (I. Tanshi pers. obs.), and *E. helvum* was documented as hunted during peak population periods in the southwest (Funmilayo 1978; Halstead 1977).

Fig. 12.1 Collection and sales of bats in Africa **a** *R. aegyptiacus* collected by a hunter with sticks from a limestone cave in Etapkini near Calabar, Nigeria (*credit* I. Tanshi), **b** Fruit bat kebab on sale in Kumasi, Ghana (*credit* M. Abedi-Lartey), **c** *E. helvum* and *H. monstrosus* on sale in a small market by the River Congo in Kisangani, DRC (*credit* Guy-C. Gembu)

Bushmeat is preferred to domestic livestock in many places because of the taste and perceived higher nutritional alue (Mbete et al. 2011, T. Mildenstein unpublished data). In locations where domestic meats and fish are generally preferred, such as Madagascar (Randrianandrianina et al. 2010), bushmeat becomes more important in periods of food shortage (Jenkins and Racey 2008). Similarly, on the island of Yap (Micronesia), hunting is socio-economically based, and bats are less desirable than seafood. Only people of lower social ranks with no access to the coast hunt fruit bats (Falanruw 1988). Consumption of bushmeat varies indirectly with the availability of other protein sources (e.g. in west Africa: Brashares et al. (2004)). In areas where bats are eaten, they are rarely the only available source of protein. The exception to this is in times of food insecurity, when people turn to bats as a food source, especially following natural disasters (e.g. typhoons: Aldabra flying fox, *Pteropus aldabrensis*, Mickleburgh et al. 2008a; Vanuatu flying fox, *Pteropus anetianus*, Helgen and Hamilton 2008a; Ontong Java flying fox, *Pteropus howensis*, Helgen and Allison 2008; Rodrigues flying fox, *Pteropus rodricensis*, Mickleburgh et al. 2008b; Samoan Flying Fox, *Pteropus samoensis* and *P. tonganus*, Brooke 2001, and *P. mariannus*, Esselstyn et al. 2006,

USFWS 2009) and during civil unrest (e.g., Bougainville monkey-faced fruit bat, *Pteralopex anceps antrata*, S. Hamilton, pers. comm.). Similarly, species found in low-lying areas (e.g. *P. aldabrensis* and *P. howensis*) may become increasingly important food to local communities as rising sea-levels destroy other food sources (Mickleburgh et al. 2008a; Helgen and Allison 2008).

Twenty years ago marked the end of a long period of international trade in the Pacific with many pteropodids being imported into Guam and the Northern Mariana Islands. Once local bat populations were depleted, bats were imported from other island groups and mainland Southeast Asia (e.g. Wiles and Payne 1986; Wiles 1992; Stinson et al. 1992). Protracted international effort eventually led in 1987–1989 to the addition of pteropodid species to the Appendices of the Convention on International Trade of Endangered Species (CITES), which has stopped legal trade of bats between nation states, although a black market still occurs (e.g. into Europe, Samuel 2013).

Currently, hunting of bats for trade tends to be locally-based, and not international, but varies widely in intensity. An extensive commercial chain of bat trade exists outside markets in Ghana (Kamins et al. 2011). Other high levels of trade, include that of the large flying fox (*Pteropus vampyrus*) in Kalimantan, Indonesia (Harrison et al. 2011) and of the Malagasy flying fox (*Pteropus rufus*) in Madagascar (Jenkins et al. 2007; Oleksy et al. 2015b). More commonly, bats are traded locally and on a lesser scale, with relatively few individuals sold in markets (e.g. *P. vampyrus* in the Philippines, Sheffers et al. 2012; and in Southeast Asia, Mickleburgh et al. 2009). Prices per bat range from <1 USD in Southeast Asia (Indonesia: Heinrichs 2004; the Philippines: T. Mildenstein unpublished data) to more than 130 USD when acquired through black market trading (e.g. *P. mariannus* on Guam and the Northern Mariana Islands, USFWS 2009).

12.2.2 Hunting Overview by Region

12.2.2.1 Africa

In total, 55 species of bats are hunted in Africa, including mainly abundant large-bodied fruit bats (Mickleburgh et al. 2009) such as *E. helvum*, Franquet's epauletted fruit bat (*Epomops franqueti*), Gambian epauletted fruit bat (*Epomophorus gambianus*), hammer-headed fruit bat (*Hypsignathus monstrosus*), *R. aegyptiacus* and medium-sized species like Angolan soft-furred fruit bat (*Myoncyteris angolensis*) (formerly *Lissonycteris*), Peter's lesser epauletted fruit bat (*Micropteropus pusillus*) and to a lesser degree insectivorous bats such as the large slit-faced bat (*Nycteris grandis*), Maclaud's horseshoe bat (*Rhinolophus maclaudi*), Ruwenzori horseshoe bat (*Rhinolophus ruwenzori*) and *Hipposideros* species. Although insectivorous bats are considered to be less palatable in many regions and may appear to be under low hunting pressure, (Kamins et al. 2011; Dougnon et al. 2012) this is not necessarily the case. Goodman (2006) showed that in addition to fruit bats, mainly the Malagasy straw-colored fruit bat (*Eidolon dupreanum*), *P. rufus,* and the Malagasy rousette (*Rousettus madagascariensis*), insectivorous *H. commersoni* is frequently hunted, especially during periods of food shortages. In addition, while fruit bats are probably the most commonly hunted group, 64 % of the 55 bat species hunted in Africa are animalivores (Appendix).

Hunting bats for food is common in West and Central African states where it can be a major threat to their populations (Funmilayo 1978; Mickleburgh et al. 2009; Kamins et al. 2011). Frequent bat hunting is recorded from Benin Republic, Ghana, Guinea, Liberia and Nigeria (Funmilayo 1978; Anstey 1991; Kamins et al. 2011; Dougnon et al. 2012) (Fig. 12.1), as well as in Cameroon, Congo Republic, Democratic Republic of Congo (DRC), Equatorial Guinea and Gabon. High levels of hunting have also been reported in the past from islands off Africa—the Comoros, Madagascar, Mauritius and Rodrigues and São Tomé and Príncipe as well as Pemba Island, Tanzania (Jenkins and Racey 2008; Carvalho et al. 2014), although conservation efforts have reduced this pressure in some of these islands (Trewhella et al. 2005).

While occasional bat hunting occurs in Mali and Zambia, there is almost no hunting in East Africa, except eastern Uganda, and bat hunting is rare in South Africa (Mickleburgh et al. 2009). Bats are also persecuted because of negative perceptions in Ethiopia (Mickleburgh et al. 2009) but that is not the focus of this chapter.

Although, Halstead (1977) reported the potential for sustainable harvesting of bats at the Ile Ife campus, over-exploitation of *E. helvum* in southwestern Nigeria was also evident (Funmilayo 1978). People in rural areas in southern Nigeria admit to eating bats occasionally, whereas in parts of eastern Nigeria, *R. aegyptiacus* is hunted intensively (Fig. 12.1). Over 3000 individuals of this species have been collected in one night from a cave in Buanchor village by several hunters who hunt more than once a month (I. Tanshi, unpublished).

North Africa and West and Central Asia. Bat hunting is less prevalent in North Africa and West and Central Asia. Of the 98 bat species that occur in this region, five are known to be hunted and these are for medicinal purposes, of the 98 bat species that occur in this region, five (all Vespertilionidae) are known to be hunted: long-fingered Myotis (*Myotis capaccinii*), Geoffroy's myotis (*Myotis emarginatus*), whiskered myotis (*Myotis mystacinus*), Natterer's myotis (*Myotis nattereri*), Maghrebian myotis (*Myotis punicus*) (Table 12.1, Appendix).

12.2.2.2 Asia

In Asia, hunting is known to affect 64 species, which represents the largest absolute number of hunted bat species in a region.

Southeast Asia. The hunting pressure on bats is greatest in Southeast Asia, where 56, or 17 % of the region's bat species are hunted (Table 12.1, Appendix). Bat hunting is widespread in 10 out of the 11 countries (Brunei, Cambodia, Indonesia, East Timor, Laos, Malaysia, Myanmar, Philippines, Thailand, and Vietnam). Only in Singapore are bats not thought to be hunted heavily (Mildenstein 2012; IUCN 2014).

Table 12.1 Proportion of bats hunted by region (Calculated by total number of extant bats species hunted divided by the total number of bat species in the region)

Taxon	Region	Total#	On Red List	Not on list	Total hunted	%hunted
Chiroptera		1146	97	70	167	14.6
	Caribbean islands	106	0	0	0	0.0
	East Asia	130	3	4	7	5.4
	Europe	42	0	0	0	0.0
	Meso America	177	0	0	0	0.0
	North Africa	41	3	1	4	9.8
	North America	49	0	0	0	0.0
	North Asia	43	0	0	0	0.0
	Oceania	173	25	15	40	23.1
	South America	249	0	8	8	3.2
	South and Southeast Asia	365	43	20	63	17.3
	SE	333	36	20	56	16.8
	South	114	8	5	13	11.4
	Sub-Saharan Africa	249	25	26	51	20.5
	West and Central Asia	94	1	0	1	11

High levels of hunting occur in Indonesia, where there is a long history of bat consumption (Fujita 1988) and large numbers of individuals are still sold in markets (e.g. *P. vampyrus*, Harrison et al. 2011; Sulawesi fruit bat, *Acerodon celebensis*, gray flying fox, *Pteropus griseus,* black flying fox, *Pteropus alecto*, Heinrichs 2004). Hunting pressure is also high in the Philippines, with a third (24/75) of its species known to be hunted. Although Philippine bats are protected from hunting by the Philippine Wildlife Act and the Philippine Cave Management Act, these laws are not well enforced, and hunting for personal consumption and local trade is widespread.

In Malaysia, hunting of some species is regulated, which may curb some of the hunting pressure but has not reduced hunting rates to sustainable levels (Epstein et al. 2009). The laws and levels of enforcement are different for the different regions of Malaysia. All bats are legally protected in Sarawak, but this is not the case in Sabah and peninsular Malaysia. Illegal hunting still occurs in orchards and by sport hunters in Sarawak at places where enforcement is lacking. Legal protection for Old World frugivorous bats is reviewed by Abdul-Aziz et al. (2015).

In Buddhist countries (Cambodia, Myanmar, Thailand and Vietnam), most roost sites of large fruit bats are found in the gardens around temples and monasteries because of the protection the monks provide (e.g. Ravon et al. 2014; T. Mildenstein unpublished data). Whether this degree of protection is sufficient to maintain stable populations of these species has yet to be investigated (Table 12.1, Appendix).

South Asia. In Bangladesh, large fruit bats are hunted for food by members of tribal groups (Mickleburgh et al. 2009). In India and Pakistan, bats are classified as vermin and are persecuted, although they are consumed infrequently, and more often killed for medicinal purposes (Noureen 2014). The exception is the Indian flying fox (*Pteropus giganteus*), which is eaten by indigenous forest-dwelling people (Mickleburgh et al. 2009). On the Andaman and Nicobar islands, black-eared flying fox (*Pteropus melanotus*) and *P. faunulus* are hunted and eaten on special occasions (Mickleburgh et al. 2009) (Table 12.1, Appendix).

North Asia. Bats are not specifically protected in China and many species are eaten, especially in southern China, where bats are found regularly in markets (Mickleburgh et al. 2009) (Table 12.1, Appendix). Requests from international agencies following the SARS outbreak, (which resulted in several hundred human deaths) that wildlife legislation be introduced in China prohibiting *inter alia* hunting and sale of bats have been ignored.

12.2.2.3 Pacific (Oceania)

Bats are often the only native mammals on remote Pacific Islands, and there is a long history of bat species being hunted in many of these areas. Bats are eaten on American Samoa, the Cook Islands and Niue, the Federated States of Micronesia, Fiji, Guam and the Northern Mariana Islands, New Caledonia, Palau, the Solomon Islands, and Vanuatu (Chambers and Esrom 1991; Mickleburgh et al. 2009). In total, 40 bat species are affected, 23 % of Oceania's bats, making this the region with the highest proportion of hunted bat species on the planet. The value of bat meat is highly variable in Oceania. It is a sought-after delicacy on Guam and the Mariana Islands, where the bats are strictly protected by the United States' Endangered Species Act (USFWS 2009). In contrast, in the nearby Federated States of Micronesia, the same bat species are rarely eaten (Mickleburgh et al. 2009). In American Samoa, (another United States territory), bats were consumed regularly in the past (Brooke 2001) but are now highly protected. Bat meat is also a delicacy in the Cook Islands, Niue, and Raratonga (Brooke and Tshapka 2002) and is a popular food on Fiji, New Caledonia, and Vanuatu (Mickleburgh et al. 2009).

12.2.2.4 South America

Bat hunting is much less common in South America, occurring in highly localized areas and affecting eight species in the families Phyllostomidae (7 spp.) and Vespertilionidae (1 sp.) (Table 12.1, Appendix).

12.3 Why Bat Hunting is a Conservation Problem

12.3.1 Negative Impacts on Bat Populations and Ecosystems

Bats are particularly vulnerable to the effects of hunting for a number of reasons. They are long-lived for their body size (five species live >30 years, Racey 2015) and reproduce slowly, with generally one young per year. They have a slow rate of fetal growth and long gestation periods (Racey and Entwistle 2000). Females and young bats are thus sensitive to hunting disturbance during a large portion of the year.

Bats are nocturnal, making them susceptible to hunting at their roost sites by day when humans can easily find them. This is especially a concern for the highly sought-after fruit bats in the Old World, which tend to roost conspicuously, aggregating in large numbers in the forest canopy (e.g. Mildenstein et al. 2008). Whether roosting colonies are in caves, cliffs, or trees, hunting at the roost site is likely to affect the entire colony. Hunting disturbance at the roost site causes injury to many bats from the spread of shot gun pellets, large-scale infant mortality when pups fall from fleeing mothers (Mildenstein and Stier unpublished data; R. Ulloa, pers. comm.), and higher stress levels as resting bats are startled and forced to flee from hunters (Van der Aa et al. 2006). In their survey of *P. rufus* in Madagascar, Mackinnon et al. (2003) recorded a high incidence of abandonment of historical roosts, which they attributed, at least in part, to high hunting pressure.

Finally, bat colonies are characterized by high roost site fidelity (e.g. Banack 1996; Brooke et al. 2000; Gumal 2004; Stier and Mildenstein 2005). Hence, bats may be reluctant to leave when hunting starts and may find it difficult to find alternative roost sites after fleeing hunters. Because bats are likely to eventually return to the preferred roost site, they are predictable prey for hunters. The overall effect of hunting at roost sites is reduction of bat population densities to a fraction of local carrying capacity (e.g. Mildenstein 2012).

These population-level impacts may also have negative ecological consequences. Some bat species play prominent roles in insect population control, pollination, seed germination and dispersal, and in many areas, bat species are essential to forest regeneration (e.g. large fruit bats are primary seed dispersers for hemiepiphytic figs, Shanahan 2001; Oleksy et al. 2015a). On isolated islands, where there is little ecological redundancy, bats are often recognized as keystone species due to their unique roles in seed dispersal (Shilton and Whittaker 2009). Mortality due to hunting may, therefore, have cascading effects on ecological communities (e.g. Mildenstein 2012) and ecosystem function (e.g. McConkey and Drake 2006).

12.3.2 Negative Impacts on Humans

The negative impacts of bat hunting extend beyond natural ecosystems to human communities. Bats in their natural ecological roles perform valuable ecosystem services beneficial to humans (e.g. insect suppression: Cleveland et al. 2006, pollination: Bumrungsri et al. 2008b, 2009, seed dispersal maintaining local watersheds: Banack 1998; Stier and Mildenstein 2005), all of which are reduced when bats are hunted. Bat colonies have also proved valuable as eco-tourism attractions supporting local economies (e.g., in Costa Rica, Lao People's Democratic Republic, Madagascar, the Philippines, and North America; examples in Pennisi et al. 2004). However, hunted bats that are wary of human presence often do not maintain colonies in locations easily viewed by people. Hence, reductions in bat populations as a result of hunting could have expensive ramifications on local communities' water supplies, agriculture, and eco-tourism industries.

Finally, the hunting of bats may also expose human communities to potentially zoonotic pathogens (Leroy et al. 2005). In the past decade, considerable attention has been paid to bats as natural reservoirs of emerging infectious diseases (Calisher et al. 2006). Studies that link infectious disease outbreaks to bats demonstrate the spillover potential through contact with bats or exposure to faeces and urine in bat habitats (reviewed by Plowright et al. 2015). Most notable are the Ebola virus outbreaks, which have attracted international attention. Leroy et al. (2009) suggest that the 2007 emergence of Ebola virus in the Occidental Kasai province of DRC could be attributable to the consumption of freshly killed bats. The authors trace the virus spread from a first patient with bat bushmeat contact to an outbreak of the disease in 260 persons resulting in 186 deaths in 2007. The re-emergence of the disease in 2014 may also have arisen from contact with bats (Saéz et al. 2015) and has proven far more deadly.

12.4 Overhunting as a Growing Concern for Conservation

Human communities have long exploited bat populations for consumption. Current hunting pressure, however, is likely to be much greater than historical pressure with increases in human population density, greater accessibility to natural areas, technological advances in bat capture methods and transport options, and relaxed adherence to cultural taboos (Brooke and Tschapka 2002; Millner-Gulland and Bennett 2003). Hence, bat hunting is likely to be unsustainable (Bradshaw et al. 2009), especially when coupled with other anthropogenic stressors (such as those described throughout this book).

Overhunting (commonly also "unregulated" hunting, although not all unregulated hunting is unsustainable, nor regulated hunting sustainable) is a globally-recognized threat to many wild species of animals (Robinson and Bennett 2000; Milner-Gulland and Bennett 2003). For bats, overhunting has been a conservation concern for over three decades (Lemke 1986; Mickleburgh et al. 1992, 2002, 2009; IUCN 2014). However, there has been a substantial lag time in our identification of which species are affected and assessment of the impact of hunting.

Twenty years ago, the conservation status of nearly half (78/160) of the Old World fruit bats was unknown due to lack of data (compiled from Mickleburgh et al. 1992). Today only 11 % (21/183) of the extant Old World fruit bat species on the Red List are considered data deficient (IUCN 2014). (These two reviews may differ slightly in their definitions of data deficient species.) This general increase in knowledge about bats includes a better understanding of the extent of hunting pressure. In the first conservation review, 49 (31 % of the total 160 known) Old World fruit bat species were recognized as hunted (Mickleburgh et al. 1992). Two decades later, nearly twice as many species ($N = 92$) are known to be hunted, representing over half of the 183 recognized species of Old World fruit bats (IUCN 2014) (Table 12.1).

Although there has been relatively little research explicitly focused on quantifying hunting impacts, the general level of concern about hunting effects on bat conservation has increased. Using Old World fruit bats as an example, in the first review, hunting was not considered a threat for most (60 %) of the hunted species (Mickleburgh et al. 1992). Now, all but five of these hunted species (25/30, 83 %) have been moved up to a higher threat status because of perceived pressures that hunting causes (IUCN 2014). Overhunting is a recognized factor in the loss of three (and probably also the little known fourth) of the now extinct fruit bat species (IUCN 2014) and a cause behind local extirpations within species' historic distributions (e.g. Polynesian sheath-tailed bat, *Emballonura semicaudata*, from Vanuatu, Helgen and Flannery 2002). Similarly, the declines of seven of the ten fruit bat species listed as critically endangered are attributed directly to hunting; the remaining three species are still virtually unknown (IUCN 2014) (Table 12.2).

The increased concern about bat hunting may be due to greater hunting pressure, or may just represent our increased awareness of hunting impacts. What is clear is that bat conservation biologists now explicitly attribute species declines and increased extinction risk to hunting. Seven hunted bat species previously assumed to be unaffected by hunting (Mickleburgh et al. 1992) now have hunting listed as a major threat (IUCN 2014). Most (68 %) of the species that are hunted, are listed as threatened by that hunting, while only 15 % of the hunted species are expected not to be affected. However, it should be pointed out that for the remaining 38 % of hunted species, reviews remain ambivalent about whether hunting is a problem. Similarly, in the review of bats as bushmeat carried out in 2004 (Mickleburgh et al. 2009), 59 % of questionnaire respondents said bat hunting

Table 12.2 Comparison of the conservation status of old world fruit bat species across two decades from two sets of species accounts: 1992 IUCN Action Plan (Mickleburgh et al. 1992); 2014 IUCN Red List; and for comparison, the 2009 Bats as Bushmeat review (Mickleburgh et al. 2009)

	IUCN Action Plan 1992	IUCN Red List 2014	Bushmeat Review 2009
# species considered	160	183	138 reports
# species hunted (% total)	49 (31 %)	92 (50 %)	82 (59 %)
# species perceived as threatened by hunting (% total hunted)	20 (40 %)	63 (68 %)	44 (54 %)
# data deficient species (% total species)	78 (49 %)	21 (11 %)	
# hunted species listed as LC (% total hunted)	29 (59 %)	10 (5.3 %)	
# hunted species not listed as hunted on IUCN list (% total hunted)	Unknown	18 (20 %)	

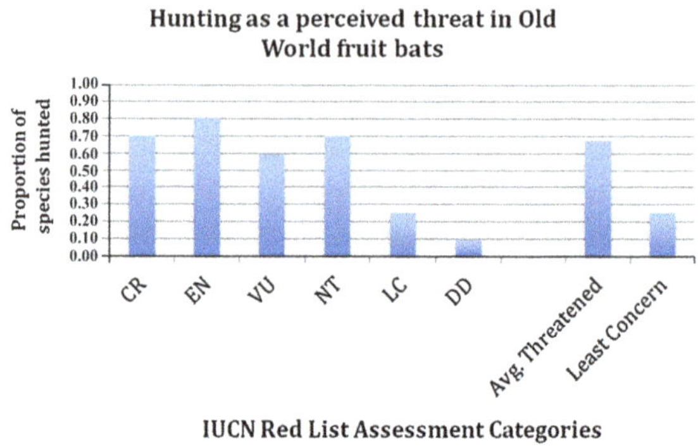

Fig. 12.2 Proportion of pteropodid species in IUCN categories listed as hunted. 69 % of species in threatened categories (*CR*, *EN*, *VU*, *NT*) list hunting as a threat, compared to 25 % of species in the "least concern" (*LC*) category

occurred in their region, and over half (54 %) of those species hunted were perceived to be negatively affected (Table 12.2).

The general consensus among biologists and managers is that hunting is a major conservation threat. Despite lacking measures of hunting impacts, there are many examples of population declines and extirpations of bats that are hunted (e.g. loss of historical bat roosts and reduced population sizes in Madagascar, Mackinnon et al. 2003; and in the Philippines, Heideman and Heaney 1992). Biologists studying Old World fruit bats currently rank hunting as the top conservation concern for this taxon (Mildenstein 2012). The IUCN Red List also reflects this concern. Fruit bat species that are known by the IUCN to be hunted are almost three times more likely to be listed as threatened ($N = 58$ spp. in IUCN categories: CR, EN, VU, NT compared to 21 species in the "LC" category) (Fig. 12.2). However, there still appears to be an information gap at the international level. Nearly 42 % (70/167) of the hunted species listed in this review are not listed by the IUCN Red List as threatened by hunting. Half of these (35 spp.) may be for lack of awareness, as they are not known to be hunted at all by the IUCN. The other half are described as hunted by the IUCN but not considered to be threatened by that hunting. In other words, these 35 species are described as hunted in their Red List species accounts, but then hunting is not included in the list of threats (compiled from IUCN 2014).

12.5 How Hunting Affects Bats

The least known area of bat biology is population dynamics, so it is difficult to extrapolate from hunting mortality rates to a quantitative assessment of hunting impacts on bat populations. Hence, one of the main conservation recommendations for protecting hunted species is the direct study of the population–level impacts of hunting (IUCN 2014).

Hunting does not necessarily lead to population declines in wild species. There are some examples of hunted bats that appear to have stable population sizes or where the effects of hunting are minimal. In these cases, hunting pressure is small relative to the bat population size due to effective law enforcement (e.g. *P. mariannus* on Rota, Mariana Islands, Mildenstein and Mills 2013), due to cultural/religious taboos (e.g. related to Muslim beliefs: *R. obliviosus* and *P. seychellensis* in the Comoros Islands, Sewall et al. 2003, 2007; *P. vampyrus* in the southern Philippines, Mildenstein 2012), and/or for the reasons given for the 35 species on the Red List that are hunted but not considered threatened by that hunting (see Appendix).

To evaluate the impacts of hunting on a bat population, research must compare the direct and indirect mortality rates of hunting with that population's capacity for growth. Falling short of these data-intensive lines of evidence, biologists have found other ways to provide inferences of hunting impacts, e.g., expert opinion, models of hunting and population growth, indices to measure population growth and/or hunting mortality, and by comparing hunted to non-hunted populations. Below, we describe the research that has contributed to knowledge of the impacts of hunting on bat populations to date.

12.5.1 Expert Opinion

Expert opinion surveys can be an efficient means of gathering information on conservation priorities when research is lacking. Because of the paucity of data on hunting impacts, much of the current concern about bat hunting is based largely on expert opinion derived from anecdotal evidence and observations of bat hunting impacts on local scales. Red List risk assessments for lesser known bats are often the result of consensus among biologists who have worked on the species. Conservation recommendations for most bat species that are hunted are based on perceived relationships between apparent bat population declines and levels of hunting that appear to be unsustainable (e.g. *Pteropus flanneryi*, Helgen et al. 2008a).

Surveys of bat biologists have been used to provide overviews of bats that are hunted and where. Most recently, Mickleburgh et al. (2009) conducted a literature review and global survey of bat biologists in 2004 to collate what is known about bat hunting. From 109 questionnaire respondents, there were 138 reports of bat consumption from which the authors provided a synthesis of bat hunting, identifying West Africa and Asia as the principal regions of conservation concern.

Expert opinion surveys have also shed light on hunting as the main threat and priority for conservation management to address. Mildenstein (2012) conducted surveys through questionnaires at two Southeast Asia regional bat conferences to learn about threats to fruit bat species. According to the 78 participants representing all Southeast Asian countries except East Timor, hunting is the main direct threat to fruit bats across this region.

Caveats. While expert opinion is a readily available source of information to identify conservation priorities in lieu of data, it does not replace systematically-acquired knowledge. There are many examples of subsequent research leading to recommendations that differ from expert opinion, especially when species-specific ecological distinctions are concerned (e.g. how to conserve co-occurring specialist and generalist species, Mildenstein 2012). It is incumbent upon conservation biologists to conduct research to verify priorities identified through expert opinion to focus conservation resources and efforts on the most urgent issues.

12.5.2 Determining Hunting Impacts on Bat Populations

To directly study the impacts of hunting, research must measure and compare hunting mortality rates to a bat populations' size and capacity for growth. To date, there are only a few studies that have tried to evaluate the sustainability of bat hunting. The first was Halstead's (1977) on the Ile-Ife campus of the University in Nigeria, which was unfortunately cut short by his return to the UK. Brooke and Tschapka (2002) modeled what would be "sustainable take" on Niue, based on the current bat population size and estimated reproductive rates. Comparing their modeled sustainable take to the numbers of bats hunted on Niue, they determined current hunting rates were unsustainable. Epstein et al. (2009) estimated potential harvest rates of *P. vampyrus* in Malaysia as a function of the number of hunting licenses issued. Incorporating these hunting mortality rates into their estimated bat population growth matrix, they projected declines in the Malaysian bat population using even the most conservative measures of hunting pressure. On Madagascar, Goodman (2006) extrapolated total hunting pressure on hipposiderids from a single hunter he interviewed. Comparing this estimated mortality rate to the local bat population surveyed, he then inferred hunting levels were detrimental, because take exceeded the breeding potential of the local bat populations.

Caveats. Rarely will information be available on harvest levels, population sizes, and reproductive rates for the same bat species. The studies described here work around missing information by using indices of bat harvest levels (e.g. licenses), estimates of reproduction rates from better known congenerics, and models of sustainable take based on rules of thumb from other harvested species (e.g. "RR" production method, named after its authors, Robinson and Redford 1991).

When indices, estimates, and models based on other species are used, there are caveats to consider. Researchers should be sure that the relationship between the index and the measure of interest is known and does not vary. Estimates based on similar species may differ from the species of interest. With population growth rates in particular, ignoring density-dependent factors could lead to inaccurate estimates of reproduction and population growth capability. Finally, models for determining sustainable hunting rates, such as the RR method used by Brooke and Tschapka (2002), predict a sustainable take rate of 40 % of the annual growth for species with life spans the length of fruit bats. However, this rate is based on similar species with potentially different life histories and may not take into account the other stressors that bats face today.

12.5.3 Measuring Hunting Mortality Rates

Surveys from local markets, hunters, and consumers can be used to gain insights into hunting rates and trends. Studies use a variety of methods to estimate hunting pressure on bats. In the Mariana Islands, Esselstyn et al. (2006) interviewed hunters of *P. mariannus* and found a 34 % increase in hunting pressure after a recent typhoon. In Madagascar, Goodman (2006) used his interview of a single bat hunter as a sample to extrapolate local hunting pressure on hipposiderids. In Niue, Brooke and Tschapka (2002) used government permit records to identify households with guns potentially used for hunting, and then interviewed a third of these potential hunters about their bi-monthly harvest rates. The authors used their results from this sample to extrapolate to hunting pressure over all hunters during the two month hunting season on the island. Kamins et al. (2011) interviewed a total of 551 Ghanians including hunters, vendors and consumers, demonstrating a high off-take of >128,000 *E. helvum* per year. A recent study on the same population used annulus markers in teeth to age bats and develop a static life table to determine age structured survival rates (Hayman et al. 2012). Markets were surveyed in Sulawesi to determine bat consumption rates of local people and number of bats exported to neighboring provinces for trade (Sheherazade and Tsang 2015). A questionnaire survey in 13 villages in Madagascar resulted in an estimate of 6500 bats taken each year (Razakarivony 2003) and staff at a roadside restaurant in western Madagascar reported serving about 30 *P. rufus* each day, which extrapolates to 10,000 a year (Racey et al. 2009). The largest roost counted during surveys at that time consisted of 5000 individuals (Mackinnon et al. 2003) which cast doubt on the reported rate of bat consumption. Nevertheless about 30 live *P. rufus* were present in panniers in the food storage area of the restaurant during a casual visit (Racey et al. 2009). Also in Madagascar, Oleksy et al. 2015b) interviewed hunters to learn about bat numbers taken as well as the location, time of night, and season in which the hunting occurred, to measure harvest rates.

Caveats. When using surveys of people's knowledge and opinions to collect information for conservation, it is important to remember the limitations of this source of information. Hunters can provide insight into numbers of bats killed, but not all of these are sold. So, when the study question involves bat trade, researchers must extend surveys to the end consumers of bats. To address this problem, the surveys by both Kamins et al. (2011) and Harrison et al. (2011) employed questionnaire approaches where all actors at different stages of the commodity chain were interviewed instead of restricting data collection only to market surveys.

There is a potential for market surveys to underrepresent the extent of fruit bat hunting, especially when many bats are not sold in regular or bushmeat markets (Kamins et al. 2011). Mbete et al. (2011) interviewed householders in Brazzaville and assembled a long list of bushmeat species consumed together with details of the markets from which they were purchased. Bats were conspicuous by their absence and enquiries (by PA Racey) revealed that they were 'marketed differently'.

In addition, hunter interviews have received conflicting reports on the accuracy of information gathered. For example, some studies report that hunters overestimate what they take, as has been shown for geese in the United States where goose hunting is legal (Andersen et al. 1996). Similarly, hunters who are being interviewed for their unique traditional knowledge may want to please or impress their interviewers, which could lead to overestimated harvest rates (e.g. indigenous Aeta people in the Philippines, T. Mildenstein and S. Stier, unpublished reports). However, in places where the species hunted is of perceived conservation concern, hunters may underestimate their take levels to avoid scrutiny, especially if hunting is illegal (e.g. for Mariana fruit bats, *P. mariannus*, in the Mariana Islands, T. Mildenstein unpublished reports).

To address the potential for inaccurate reports by hunters, some studies provide methods for hunters to report take anonymously. The study of typhoon impacts on hunting levels in the Mariana Islands, used a local hunter to collect data from other hunters (Esselstyn et al. 2006). In Madagascar, Oleksy et al. (2015b) provided a subpopulation of hunters with notebooks and monetary incentives to keep track of hunting off-take over time and return the information anonymously at the end of the study.

12.5.4 Estimating Hunting Impact from Population Declines

In cases where hunting mortality rates are unknown and/or difficult to measure, study of population trends in hunted areas can provide an indication of hunting impacts. Hunting is often assumed to be the cause behind measured population declines. For example, biologists interviewed local people at more than 30 bat roosting sites in Pakistan where bats were hunted, and the consensus of local knowledge suggested there were large-scale declines in bat populations in areas where hunting was common (Venkatesan 2007). The inference power of this type of study is stronger, however, if a comparison can be made to areas free from hunting. In comparisons of areas with and without hunting, bat population densities were 5–10 times larger when roost sites were protected (in the Philippines, Mildenstein 2012) and up 100 times larger when their entire habitat was protected (in the Mariana Islands, Mildenstein and Mills 2013).

Other studies have used indices of bat population size that are directly linked to hunted bats for measuring trends. Harrison et al. (2011) used questionnaires to survey hunters and market vendors across 12 key population centers in Kalimantan, Indonesia to gain insight into hunting intensity, seasonality, and market dynamics. They used capture rates by hunters and sales rates in markets as indices of the population size of bats in the wild. From reductions in the numbers of bats captured and the decreasing number of bats brought to market for sale, they inferred that the bat population in the wild was declining.

Caveats. Again, caution is warranted when inferences are based on surveys of people's opinions and care should be taken in planning surveys and interviews especially when researchers lack sociological training (St. John et al. 2013; see also Nuno and St. John 2014 for a review of survey techniques). Similarly, it is important to remember that population reductions in areas with hunting are the net result of many stressors, natural and human-caused (e.g. chapters of this book), and may not correlate directly with hunting pressure.

12.6 Conservation Management to Mitigate Hunting Impacts

12.6.1 Enforcement of Hunting Prohibition

For threatened populations, the elimination of hunting as a threat can produce population-level results within decades (see also roost site protection). For areas that have been not hunted, bat population sizes can be large, with densities (individuals/hectare of habitat) that are hundreds of times that of the same species in similar areas with hunting pressure (e.g. Mariana Islands, Mildenstein and Mills 2013; Tacio 2015). Similarly, eliminating hunting allows bat populations to recover. In American Samoa, typhoons and overhunting sent populations of *P. samoensis* and *P. tonganus* into a sharp decline, eventually triggering a hunting ban in 1992 (Craig et al. 1994a, b). Two decades later, the population sizes of these bats is much larger and considered stable (Brooke and Wiles 2008, A. Miles pers. comm.).

12.6.2 Regulated Hunting

Theoretically, hunting can be sustainable if regulated tightly to ensure population declines due to harvest do not exceed what bat populations can naturally replace given the range of environmental variability to which they are subjected (methods in Mills 2012). Although several countries permit hunting, bat hunting has generally proved difficult to regulate for sustainability for a number of reasons. In some places, hunting laws are hard to understand and therefore not followed by hunters. In Madagascar for example, bats can be hunted legally between May and August or February to May, depending on the species (Jenkins and Racey 2008). In addition, according to the regulations, hunting is allowed only during the day, and game species cannot be hunted at their roost. Hence bat hunting is technically impossible, but hunters harvest bats despite the rules (Racey et al. 2009).

In other countries, hunting regulations are poorly designed from a conservation perspective. In peninsular Malaysia, bat hunting is legal and numbers harvested are said to be regulated. However, while limits on the number killed are issued by the provincial government where the hunter resides, the license owners are allowed to hunt in multiple provinces which creates a potential for seasonal take that far exceeds the bat populations' capacities for regeneration (Epstein et al.

2009). In Niue, bat hunting is allowed for two months per year. Hunting levels are not regulated because of the assumption that the bat supply is unlimited. However, bats roost in sacred forest grounds that are taboo for humans to enter, making it impossible to obtain a count of the population size. Based on models of similar species, harvest rates exceed what is expected to be sustainable (Brooke and Tschapka 2002).

In other countries, harvest laws are not enforced or followed. For example, the main threat to the ornate flying fox (*Pteropus ornatus*) in New Caledonia is local hunting for food (Brescia 2007). There is a short legal fruit bat hunting season, which includes only the weekends of April with a quota of five bats/hunter. However, based on reports, there is widespread and substantial illegal hunting, including the commercial harvesting of these fruit bats (Brescia 2007).

Finally, there are a few countries where hunting is legal and unregulated. In Pakistan and India, bats are considered "vermin" or "pests" and hunting is actually encouraged by the government without concern for long-term sustainability or conservation (Noureen 2014).

Whenever hunting regulation is being considered, managers should be aware that sustainable off-take will have to be much lower than projected recruitment for a number of reasons. Current bat population sizes, distribution, and number of populations are a fraction of historical numbers. For example, mixed colonies of *P. vampyrus* and golden-crowned flying fox (*Acerodon jubatus*) in the Philippines are thought to once have been present on every major island in populations numbering in the hundreds of thousands (Heideman and Heaney 1992). Now, these bats have been extirpated from most islands. Fewer than 15 mixed colonies remain, often with less than 2000 individuals and dangerously small numbers of the endangered *A. jubatus*. (Mildenstein 2012). In addition to already being at historically low population sizes, other human-caused stressors (persecution, habitat fragmentation, global climate change and other perturbations described in this book) continue to act on bat populations, exacerbating detrimental effects of harvest. Hence, it is prudent for managers to be conservative when establishing regulated harvest limits, and to carefully monitor populations and adapt regulatory management as needed to meet sustainable goals.

12.6.3 Control of Guns, Ammunition, and Other Bat Hunting Tools

Gun control is expected to have a positive effect on bats. In those countries where private gun ownership is not allowed, *Pteropus* often benefits. After a coup d'état in the Seychelles in 1977, all guns were confiscated and the numbers of Seychelles flying fox (*Pteropus seychellensis*) rose having previously been of some conservation concern (Racey 1979; Nicoll and Racey 1981). A similar story occurred in Palau following the death of the President by gunshot wounds (A. Brooke, pers. comm.).

In Myanmar, private gun ownership is a crime with reportedly severe penalties. Perhaps as a result, bat populations are large and easily approached. Bats are still harvested. For example, *Pteropus* are catapulted to provide medicine for asthma, *Rousettus* is often netted at cave entrances and sold in a market close to Mandalay, and insectivorous bats are also caught at cave entrances, fried and sold as beer snacks (U Khin Maung Gyi, pers. comm.). However, the harvest rates and overall disturbance to bats in Myanmar are thought to be much lower without guns.

Bat hunting may also be regulated through control of capturing equipment. For example, in Sarawak, as in many countries, it is illegal to sell or buy mist nets without a permit. This method of protecting bats, however, is only effective if hunters use commercially-manufactured nets. Many bat hunters avoid the high cost and regulation of mist nets and make their own nets or hook and line traps from monofilament line and other inexpensive fishing materials (e.g. in the Philippines, Mildenstein 2012).

12.6.4 Roost Site Protection

Bats are most vulnerable at their day roosting sites. So, it is not surprising that bat populations settle in areas where they are most protected. In Buddhist countries like Cambodia, Myanmar, Thailand, and Vietnam, large fruit bats are commonly found in the gardens surrounding temples and monasteries (T. Mildenstein, unpublished data). The presence of monks and religious activities turn these areas into de facto sanctuaries for bats that would otherwise experience hunting pressure. In nearby non-Buddhist countries, such as Indonesia and the Philippines, the same fruit bat species colonize other "safe" spots such as privately protected lands and parks, especially in the forest interior, using topographical features that afford protection from people (e.g. along rivers, within mangrove islands, and on cliff edges, Mildenstein 2012). Active protection of roost sites alone (i.e. hunting still occurs away from the roost) has been shown to result in as much as ten times the number of roosting fruit bats for the same amount of forest habitat, and is especially important for sensitive species such as ecological specialists (Mildenstein 2012). Because of this, and the fact that roost sites are geographically predictable, conservation management by local government units and non-government organizations often target roost site protection.

Case Studies. Conservation efforts for the Pemba flying fox (*Pteropus voeltzkowi*) included roost protection through the setting up of community conservation clubs (Robinson et al. 2010). The recovery following these conservation programs led to the downgrading of the species' Red List threat assessment from Critically Endangered to Vulnerable.

Until recently, permanent nets were a regular method of hunting in the roost at Analalava, Madagascar by the people of Ambatondrazaka. The national NGO, Madagasikara Voakajy, initiated community-based protection of the fruit bat roost by incorporating payment for local rangers in a local peanut cooperative it funded.

Currently, hunting at the roost has ceased and the colony has increased from 200 to nearly 2000 individuals (Razafimanahaka 2013).

In Malaysia, the Wildlife Conservation Society has worked with local communities and the government in Sarawak to establish protected roosting areas of *P. vampyrus*. (M. Gumal, pers. comm. 2015). Four out of the five maternity roosting sites identified by Gumal (2004) are now protected for *P. vampyrus,* including: Loagan Bunut National Park, Sedilu National Park, Limbang Mangroves National Park, and Bruit National Park. A fifth maternity roost site at Bukit Sarang is in the preliminary proclamation stage for a National Park (M. Gumal pers. comm.).

In the Philippines, the Filipinos for Flying Foxes project is building on the successes of Bat Count-Philippines by developing bat roost site sanctuaries with local governments. The collaborating organizations (Philippine Biodiversity Conservation Foundation and Mabuwaya Foundation) are establishing community-managed roost site sanctuaries across the distribution of the endemic and endangered *A. jubatus* and studying bat population size increases and roost site fidelity in these newly protected roost sites (SOS 2012).

12.6.5 Education and Awareness Raising

One of the first steps toward conservation management of hunted bats is educating local communities. Bats are important to human communities in a number of ways, particularly for the valuable ecosystem services they provide, but local communities are often unaware of these. Because hunting, like other human-caused stressors, is tied directly to population declines in bats, it is important that human communities are aware of the trade-offs between temporary gains from bat exploitation and the risk of losing bats entirely from the region. Following a knowledge/attitude/behavior approach to understanding responsible environmental behavior (Hines et al. 1987), communities may come to appreciate bats and support bat conservation only after understanding their role in the environment (see Kingston 2016).

12.6.5.1 Knowledge

People are generally aware of bats present near their local communities. Bats are not cryptic animals, especially fruit bats that aggregate in large numbers by day using conspicuous roosting sites, and they often forage at night in fruiting and flowering trees on farms and in residential areas. Hence, local people's knowledge of bats often surpasses that of outside biologists, especially with respect to bat roosting locations, foraging habits, seasonal behaviors, and even threats (e.g. local community members' awareness of subtle seasonal changes in fruit bat diet of *P. mariannus,* Mildenstein and Mills 2013). It is, therefore, surprising how little is known about bat conservation status in these same areas. Population size and growth trends tend to be unknown by biologists and managers, much less by the non-scientific members of the local community. So, even though local people are aware of the disturbance they may be causing, they often have no idea of the severity of population-level consequences. Because bats appear to be numerous, popular belief is that humans can have only minimal impact on their populations. For example, the greater short-nosed fruit bat (*Cynopterus sphinx*) is believed by experts to be threatened by hunting in parts of its range, but in other parts, <1 % of local people surveyed believe the species could be threatened by their hunting (Johnson et al. 2003). Similarly, throughout the Philippines, bats are eaten regularly with little understanding of the impacts that harvest is causing. Hunters who join biologists on bat population counts commonly overestimate the population size by three orders of magnitude prior to the count and then are shocked when the counted population is in the hundreds or low thousands (Mildenstein et al. 2007; Mildenstein 2012).

Education and awareness programs. One of the most hunted bats in sub-Saharan Africa, *E. helvum* is the focus of members of the Eidolon Monitoring Network (EMN) who conduct education activities in areas near bat colonies (J. Fahr, pers. comm.). In Kenya and Nigeria, scientists and volunteers of the EMN carry out education programs in schools (Fig. 12.3) and among the general public (Tanshi et al. 2013). Education on islands around Africa has proven effective in drawing local attention to bat protection. Examples include the recovery program

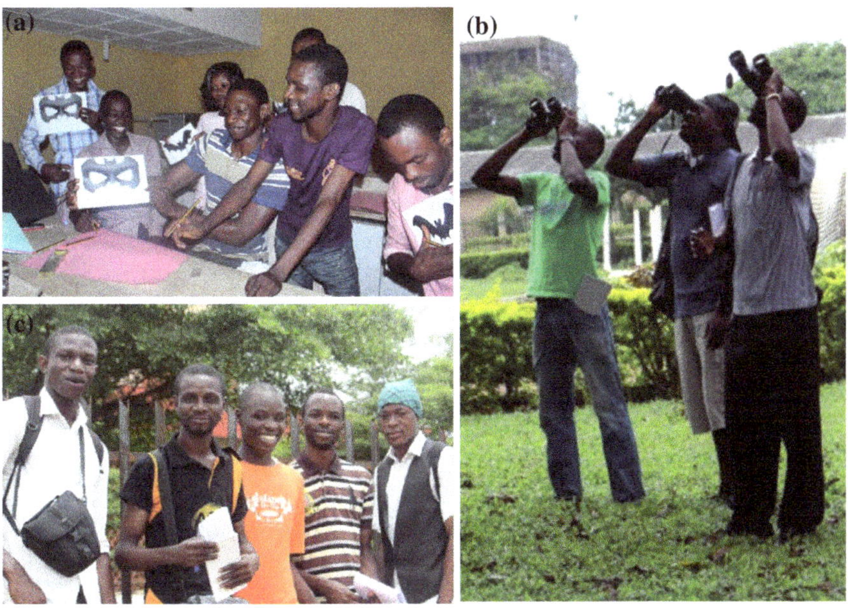

Fig. 12.3 Conservation education and bat population monitoring by volunteers in Eidolon Monitoring Network in Benin City, Nigeria, school students engage in conservation outreach event, **a** volunteers prepare conservation outreach materials, **b** volunteers counting straw-coloured fruit bats *Eidolon helvum* at King square, Ring Road, Benin City, **c** undergraduate student volunteers Eidolon Population Monitoring team from University of Benin, Benin City, Nigeria

for the *P. voeltzkowi* in Pemba Tanzania, for *P. rufus* in Madagascar, *P. rodricensis* in Rodrigues and the Comoro flying fox (*Pteropus livingstonii*) in the Comoros (Wilson and Graham 1992; Trewhella et al. 2005; O'Connor et al. 2006; Robinson et al. 2010; H. Doulton, pers. comm.).

12.6.5.2 Behavior—Local Commitment to Conservation of Bats and Bat Habitat

Finally, once communities that value bats become aware of the threats bats face, the may start to change their behaviors to support bat conservation (but see Kingston 2016). A multi-faceted education and awareness program in the Comoros Islands is a good example of how outreach can lead to changes in attitude and behavior that support conservation. Local citizens became involved in monitoring bat populations and directing conservation management (Trewhella et al. 2005).

12.6.5.3 Capacity Building of Local Rangers/PA Managers

Many programs include training and capacity building in their bat conservation efforts. Bat Count—Philippines held a national workshop in 2004 to train protected area managers in bat identification and monitoring techniques (Mildenstein et al. 2007; Mildenstein 2011). The project, Filipinos for Flying Foxes, is now working with local communities to establish sustainable management practices for bats (SOS 2012). The project trains local rangers and management staff to monitor their bat populations and encourages them to self-regulate their hunting pressure. In Dalaguete, Cebu, rangers have continued forest protection despite the inconsistencies in availability of their modest stipends provided by the local government (SOS 2012). In Divilacan, Northern Sierra Madre, Luzon, rangers receiving bat conservation training have elevated bat roost protection to the top priority in their regular monitoring activities (SOS 2012).

12.6.6 Stakeholder Engagement and Citizen Science

Collaborative conservation is more likely to be sustainable. In community-based conservation management, stakeholders from a varicty of factions within the community are required to work together to implement effective conservation practices. This often creates unlikely partnerships that bridge normal political, socioeconomic and religious divides. For example, former rebels work with local government officials to monitor bats in southern Mindanao, Philippines, a region known for often violent stand-offs between the Philippine government and Islamic separatists (LM Paguntalan pers. comm., SOS 2012). Uniting stakeholders toward the common goal of bat conservation, however, creates collaborative programs that prove to be robust to the changes that commonly lead to the demise of wildlife conservation programs (e.g. change in political administrations).

Validity of data. A frequent concern when working with citizen scientists, is that data gathered by untrained biologists may be less accurate and obscure the signal that is being studied (reviewed in Johnson 2008). However, community-based bat counts are perhaps a best case scenario for the use of citizen science. The data gathered are the number of bats observed, requiring just the ability to count and no other special training or equipment. Because bats, especially fruit bats, tend to aggregate, the population being counted is all in one place, by-passing many sources of error arising from sampling approaches to abundance assessment. Finally, the goal of community-based counts is detection of population trends across time, so that local communities can track impacts that hunting may be causing. Studies of count error in untrained observers show that while training and experience has a positive effect on count accuracy, counts made by untrained observers are as likely to detect population trend direction as those made by experienced biologists (Mildenstein 2012; Mildenstein and Mills 2013; Barlow et al. 2015).

Case studies. Population monitoring and roost protection for *P. rufus* in Madagascar provides a good example of the effectiveness of citizen science and participatory conservation efforts. Following the decline in populations of *P. rufus* in Madagascar from overhunting and habitat loss, the NGO Madagasikara Voakajy engaged local communities at four roost sites for the protection of the species. With the help of the local government, roost sites were designated for protection and firebreaks with bare ground areas constructed around roost sites. Local volunteers where trained to monitor the roosts of *P. rufus* using binoculars and hand tally counters and have continued to do so. In addition, the engagement of local people led to an interesting partnership where habitat restoration through tree planting is ongoing, while local farmers receive support through a crop seed loan system. Similarly, the local community is enforcing sustainable land use within the protected roost areas. The project organizers ascribe the success of the project to environmental education and outreach efforts, highlighting the benefits of local community engagement through citizen science and partnerships that improve local economies (Mahefatiana Ralisata pers. comm.).

In Asia, Filipinos for Flying Foxes also trains local bat stakeholders as citizen scientists. By providing these community members with the skills and experience to monitor their bat populations, the project is encouraging local stakeholders to conduct regular counts and to self-regulate their hunting pressure. So far, the project has visited more than 35 communities near to fruit bat roosts, and trained more than 200 local stakeholders in surveying and monitoring techniques. It is encouraging that after training, monitoring has continued by the local communities. Twelve communities have counted bats subsequent to training, and five of these have regularly conducted annual counts for 10 years after their training (Mildenstein 2011).

On Guam in the Mariana Islands, *P. mariannus* is a threatened species that must be monitored regularly by the US government under the US Endangered Species Act. Guam's last colony of *P. mariannus* has declined precipitously since the establishment of the invasive brown treesnake (*Boigia irregularis*, USFWS 2009). In the past 10 years the bats have no longer been aggregating in the historical colony location but rather are seemingly scattered in the forest, making population abundance assessments using traditional roost counting methods impossible. Given limited human resources and adherence to historical practices, biologists contracting with the US government have conducted fruit bat surveys sequentially using one or two observers from single observation sites in the forest on a survey morning. These surveys yield occasional bat sightings and location information but provide no basis on which to estimate the population size of the bats, which is essential to generate funding and motivate protective management of this formally-recognized USA national endangered species. In 2014, a different approach to surveying was initiated. Using 85 trained citizen scientists placed at observation stations throughout the forest, simultaneous observation permitted a survey of about 10 % of the forest habitat on Andersen Air Force Base. This collaborative project between the University of Guam and the U.S. Navy resulted in the first population size estimate for the threatened *P. mariannus* since the early 2000s. The survey also brought together local stakeholders representing 25 government and non-government organizations (including schools, environmental clubs, hunters, and local media) toward the common goal of supporting the conservation of a local endangered species (Fig. 12.4; Mildenstein et al. 2014).

Fig. 12.4 Citizen science support enabled the first population count of Mariana fruit bats on Andersen Air Force, Guam in nearly a decade. (Survey participants are showing the number of bats they counted on their raised fingers) (*credit* SSgt. M. White)

12.7 Recommendations for Conservation of Hunted Bats

12.7.1 More Research is Needed to Understand Hunting Impacts

One of the major challenges to bat conservation is lack of knowledge of how hunting affects bats, their population size, and distribution. Collecting these data requires trained biologists, sociologists, statisticians, and well-planned survey techniques and questionnaires that can address sensitive questions. For many regions and species, there is little or no population information on bats, so that population trends are unknown. Managers are therefore encouraged to start monitoring programs by which hunting impacts on population size can be tracked over time. Some bat conservation initiatives provide useful models for population size assessment and monitoring (e.g. Southeast Asian Bat Conservation Research Unit, Filipinos for Flying Foxes, WCS Malaysia, United States Department of Defense in the Mariana Islands, FFI Cambodia). Using the population size estimation and monitoring described, more studies are also needed that investigate the direct and indirect impacts of hunting on bat populations. For example, long term monitoring projects of both people and bats, could show trends in the correlations between number of hunted bats and bat population responses.

Finally, more needs to be learned about the people hunting bats and the conditions that lead to increased hunting (Cawthorn and Hoff 2015). If it is understood why people hunt (e.g. for protein? for livelihoods? to vary their diet? for

Fig. 12.5 Members of the women's peanut cooperative in Madagascar, which grows peanuts to supplement local protein supplies and uses a portion of the proceeds to pay rangers to protect fruit bat roosts (Razafimanahaka 2013)

tradition?) and what are the patterns in hunting intensity (e.g. seasonal, food insecurity), conservation managers can find creative solutions for mitigating hunting impacts (e.g. Razafimanahaka 2013; Fig. 12.5).

12.7.2 Research to Understand How to Protect Bats

There is a need to evaluate methods employed in bat protection programs. For example, roost site protection has been correlated with greater bat density (Mildenstein 2012). It is important to follow this up with research that demonstrates the effectiveness of this strategy and advises managers how to proceed. Important questions are: what is the bats' fidelity to protected roost sites? How quickly do bat populations increase with roost site protection? How quickly do bats become habituated to human presence?

Protection of habitat outside the roost area is also important, although much more difficult to enforce. Studies of the relationship between foraging habitat and bat population sizes would guide managers in their habitat-based conservation strategies.

Equally important is to understand where protection efforts are failing. There are many examples of regulated bat hunting leading to population declines. It is important to understand where laws and implementation are falling short and why.

12.7.3 Education/Outreach

Education and outreach in local communities is essential to successful hunting management campaigns. Many hunters do not perceive bats as a limited resource and are unaware of the effect they may be having on bat species' extinction risks. An obvious first step to bat conservation in hunted areas is therefore the dissemination of information on the bat population size, basic biology, and monitoring techniques, so that hunters can assess the impacts they are having.

Local communities should understand the benefits of bats and the valuable ecological services they provide, including their contribution to forest regeneration through seed dispersal. Communities that appreciate bats may be more likely to pursue conservation management. If people understand the array of risks of ignoring declining populations, they will be more inclined to exert effort to protect bats.

Local communities should also understand the human health risks of bat hunting and consumption. Bat handling, trade, preparation, and consumption by humans create a direct transmission route for disease spillover into human communities and warrants consideration in bat protection programs. Education about these risks is needed, as only a small percentage of participants in bat hunting and trade are aware of their risk of exposure to disease in Asia (Harrison et al. 2011) and Africa (Subramanian 2012). Finding a balance between the needs for bat conservation, sustainable harvesting and public health management is an important approach in the regions where regulated bat hunting is a goal (Halstead 1977).

Change in perceptions and attitudes towards conservation and wildlife in general require an effective outreach approach. Thus, education and public awareness projects should be designed to engage the audience as has been demonstrated to be effective in Madagascar (Racey 2013) and Latin America (Navarro 2013). The involvement of all stakeholders and policy makers in conservation outreach projects is crucial to the effectiveness of education programs by facilitating the enactment and enforcement of protective legislation (Robinson et al. 2010).

12.7.4 Protect Colony Locations at the Roost

Bat conservation through roost protection by local communities has been demonstrated to be effective for the recovery of previously declining populations (Mildenstein 2012; Fig. 12.6). The adoption of such roost protection programs in other countries could hold the key to sustaining populations. This is especially true for areas where fruit bat hunting is intense. If successful roost site protection programs could be demonstrated and published, these could be used as models for other areas (e.g. *P. rufus* populations in Madagascar—M. Ralisata pers. comm.; *P. vampyrus* in Malaysia, M. Gumal, pers. comm.; *P. vampyrus* and *A. jubatus* populations in the Philippines, SOS 2012).

Fig. 12.6 Tourists viewing formerly hunted fruit bats at their protected roost site in Mambukal Resort, Negros Occidental, Philippines (*credit* LM Paguntalan)

12.7.5 Regulated Hunting

In many areas where hunted bats are threatened, hunters do not want to extirpate bat populations, but they also do not want to lose the ability to hunt bats (Mildenstein 2012; Cawthorn and Hoffman 2015). In fact, some roost site protection campaigns are successful, because hunting outside the roost site is not discussed or prohibited, making it easier for hunters to respect roost site sanctuaries (T. Mildenstein pers. obs.; SOS 2012). Once communities understand that human disturbance has population-level impacts and that conservation management must balance negative impacts with the bats' innate ability to add to their population, community-level planning of a sustainable hunting program can ensue. Targets must be established for minimum population sizes and numbers of viable populations before hunting can be allowed. After thresholds are reached, sustainable harvest levels must be determined using adaptations of the well-developed harvest management practices for other species.

Finally, an effective enforcement and harvest regulation program must be designed that starts out conservatively, carefully tracking impacts of hunting on bat populations and making adjustments to hunting allowances as needed. Halstead (1977) described how regulated hunting of *E. helvum* at the University of Ile Ife in south western Nigeria can be mutually beneficial to the bat population, local community livelihoods, and managers of property where roosts are present. In places where hunting laws are in place but not respected or enforced, education and outreach are instrumental in garnering public support (as Madagasikara Voakajy has done for roosts of *P. rufus* in Madagascar).

12.7.6 Encourage Local Researchers and NGO's

A key to effective and sustainable conservation is to develop the capacity of local people, including local researchers and the establishment of local NGOs (Racey 2013). Few detailed studies report reliable estimates of bat hunting impacts on bat populations. Some studies may indeed have been conducted but remain as Masters or PhD theses or published as grey literature or in local journals, thereby limiting the distribution of such information. Because valuable results are not often published or accessible, current efforts to revise species account entries in the IUCN Red List have had to rely heavily on experts gathering unpublished information to determine conservation priorities for hunted species (T. Mildenstein, unpublished data). It is important that biologists are encouraged to publish their findings, even in lesser developed countries where there are few if any personal incentives for doing so (Milner-Gulland et al. 2010).

Finally, the establishment of local non-profit organizations creates a network for stakeholders and a bridge between local interests and conservation management. Such organizations play a critical role in ensuring the sustainability of conservation projects across political administration changes by engaging the local stakeholders and coordinating conservation activities in harmony with local needs (e.g. Figs. 12.5 and 12.6).

12.8 Conclusion

Conservation biologists' understanding of the role hunting plays in bat population declines has changed over the last three decades. Conservation concerns were originally focused on large scale hunting operations and especially international commercial trade (e.g. *Pteropus* spp. in the Pacific and Southeast Asia). After international trade was largely shut down in the late 1980s, conservation managers turned their attention to hunting within countries, still focusing on commercially hunted species as a highest priority (e.g. Mickleburgh et al. 1992).

Although commercially harvested species are still a high priority today, conservationists' concerns are no longer limited to species found in markets. With more research and experience, conservation managers have become increasingly aware of the negative impacts caused by hunting even on small scales, i.e. for personal use and/or local trade. Especially detrimental is hunting at roost sites, which can lead to a tenfold increase in population declines (Mildenstein 2012). This is probably because hunting disturbance at roosts also affects non-target individuals, including mothers and babies which are especially sensitive to disturbance. For hunted bat taxa (e.g. Old World fruit bats), hunting now ranks as a top threat among bat conservation biologists.

Research that quantifies the relationship between hunting rates and bat population declines is still lacking, but general awareness about the breadth of bat hunting effects has increased. The number of bat species known to be hunted is larger than in earlier reviews. Similarly, biologists now recognize that hunting is usually a threat to bats; the number of species documented as threatened by hunting is much larger now and includes many species that are not commercially hunted. However, for 28 % of the species known to be hunted according to IUCN Red List species accounts, hunting was either not considered to be a threat or not evaluated at all. Finally, information about a quarter (38/167) of the species listed as hunted by this review has come from sources other than the IUCN Red List, where there is no mention of hunting for those species. More research on population sizes and trends, hunting impacts, and effective management tools will provide very important information for bat conservation.

Research should also address the role of bat meat in local people's diets. Studies that focus on seasonal patterns in bat consumption and the dependency on bats as a source of protein would provide managers with information that would guide policy and conservation actions complementary to the dietary needs of local communities. Similarly, research on the use of bats for medicinal purposes should investigate patterns of bat use and ailments that bat consumption is said to cure. Study of the effects of bat use on ailing consumers in situ as well as in randomized trials could play a significant role in conservation by helping tease out whether the medicinal properties of bats have a real effect or are a myth. These investigations should begin with asthma and other chest complaints since anecdotes about the curative effects of bats for such conditions are widespread across the Old World tropics.

Along with conservation-focused research, public education and capacity building of local managers must also be encouraged to counter what is clearly a major threat to bat populations in the Old World tropics. Greater awareness about bats' reproductive characteristics of one young a year together with the ecological services bats provide will strengthen local communities' commitment to supporting conservation management. Hunters, in particular, begin to cooperate, even tracking their harvest rates as a community, when they realize that bats are a limited resource and that populations may be extirpated altogether if hunting pressure is not halted or highly regulated (Mildenstein 2011). Local managers can be empowered to track bat conservation and hunting with training in the simple and inexpensive field techniques needed to monitor bat population abundance, and these local stakeholders are key to creating sustainable monitoring programs.

Hunting has already led to the loss of four bat species in the last few decades. Without research, public education and awareness, and bolstering local managers' capacity to protect bats, unregulated hunting may well claim more many species.

Acknowledgment We thank Anne Brooke and three anonymous reviewers for helpful comments that improved the manuscript.

Appendix. List of Hunted Bat Species Showing Primary Use (Food or Medicine), Summarized by Region and Country. We Followed IUCN Regional Classification

IUCN region	Country in which hunted	Species	Status	Food	Medicine	Source
East Asia	China	*Cynopterus sphinx*	*LC*		x	Bates et al. (2008d), Mickleburgh et al. (2009)
	China	*Eonycteris spelaea*	*LC*	x		Francis et al. (2008c), Stebbings (1987)
	China	*Hipposideros pomona*	*LC*	x	x	Bates et al. (2008a)
	China	*Pteropus giganteus*	*LC*	x	x	Molur et al. (2008a), Mickleburgh et al. (2009)
	China	*Rousettus leschenaultii*	*LC*	x	x	Molur et al. (2002), Bates and Helgen (2008), Mickleburgh et al. (2009)
	Japan	*Pteropus pselaphon*	*CR*	x		Ishii and Maeda (2008)
	Taiwan	*Pteropus dasymallus*	*NT*	x		Heaney et al. (2008d)

IUCN region	Country in which hunted	Species	Status	Food	Medicine	Source
North Africa	Algeria, Morocco, Tunisia, Libya	*Myotis punicus*	*NT*		x	Aulagnier et al. (2008)
	Algeria, North Africa	*Myotis emarginatus*	*LC*		x	Hutson et al. (2008h)
	North Africa	*Myotis capaccinii*	*VU*		x	Hutson et al. (2008g)
	North Africa	*Myotis nattereri*	*LC*		x	Hutson et al. (2008a)
Oceania	American Samoa, Cook Islands and Niue, Fiji, New Caledonia, Vanuatu	*Pteropus tonganus*	*VU*	x		Hamilton and Helgen (2008),
	American Samoa, Fiji, Samoa	*Pteropus samoensis*	*NT*	x		Brooke and Wiles (2008)
	Fiji	*Mirimiri acro-donta/ Pteralopex acrodonta*	*CR*			Flannery (1995b)
	Fiji, Vanuatu	*Chaerephon bregullae/ Tadarida bregullae*	*EN*	x		Flannery (1995b), Palmeirim (2014)
	Fiji, Vanuatu	*Notopteris macdonaldi*	*VU*	x		Flannery (1995b), Palmeirim et al. (2007), Palmeirim (2008)
		Pteropus ualanus	*VU*	x		Wiles et al. (2008)
	Indonesia, Papua New Guinea	*Dobsonia moluccensis*	*LC*	x		Hutson et al. (2008j)
	Micronesia, Federated States of	*Pteropus yapensis*	*VU*	x		Wiles et al. (2008b)
	Micronesia, Federated States of; Caroline Islands	*Pteropus insu-laris/ Pteropus phaeocephalus*	*CR*	x		Helgen and Wiles (2010)
	Micronesia	*Pteropus molossinus*	*VU*	x		Buden et al. (2008)
	Micronesia, Guam and Commonweatlh of Northern Mariana Islands	*Pteropus mariannus*	*EN*	x		Falanruw and Manmaw (1992), Allison et al. (2008), Lemke (1992)
	New Caledonia	*Notopteris neocaledonica*	*VU*	x		Brescia (2008a), Boissenin and Brescia (2007)
	New Caledonia	*Pteropus ornatus*	*VU*	x		Brescia (2008b)
	New Caledonia	*Pteropus vetulus*	*VU*	x	x?	Brescia (2008c), Flannery (1995b)
	Palau	*Pteropus pelewensis*	*NT*	x		Wiles (2008), Wiles et al. (1997)

IUCN region	Country in which hunted	Species	Status	Food	Medicine	Source
	Papau New Guinea	*Pteropus hypomelanus*	*LC*	x		Francis et al. (2008a), Fujita and Tuttle (1991), Fujita (1988)
	Papua New Guinea	*Aproteles bulmerae*	*CR*	x		Flannery (1995b), Hutson et al. (2008m)
	Papua New Guinea	*Miniopterus magnater*	*LC*	x		Bonaccorso and Reardon (2008b), Cuthbert (2003a, b)
	Papua New Guinea	*Nyctimene aello*	*LC*	x		Bonaccorso and Helgen (2008), Cuthbert (2003a, b)
	Papua New Guinea	*Nyctimene cyclotis*	*DD*	x		Cuthbert (2003a)
	Papua New Guinea	*Pteralopex flanneryi*	*CR*	x		Helgen et al. (2008a)
	Papua New Guinea	*Pteropus neohibernicus*	*LC*	x		Salas et al. (2008)
	Papua New Guinea	*Rousettus amplexicaudatus*	*LC*	x		Csorba et al. (2008g), Utzurrum (1992)
	Papua New Guinea	*Syconycteris australis*	*LC*	x		Cuthbert (2003a)
	Papua New Guinea, possibly Cambodia and Vietnam	*Miniopterus pusillus*	*LC*	x		Cuthbert (2003a, b)
	Papua New Guinea, Solomon Islands	*Pteralopex anceps*	*EN*	x		Helgen et al. (2008c)
	Papua New Guinea, Vanuatu	*Miniopterus macrocneme*	*DD*	x		Bonaccorso and Reardon (2008a)
	Solomon Islands	*Pteralopex atrata*	*EN*	x		Helgen and Hamilton (2008b)
	Solomon Islands	*Pteralopex taki*	*EN*	x		Hamilton et al. (2008a)
	Solomon Islands	*Pteropus cognatus*	*EN*	x		James et al. (2008)
	Solomon Islands	*Pteropus nitendiensis*	*EN*	x		Leary et al. (2008a)
	Solomon Islands	*Pteropus rennelli*	*VU*	x		Hamilton et al. (2008b)
	Solomon Islands	*Pteropus tuberculatus*	*CR*	x		Leary et al. (2008b)
	Solomon Islands, Papau New Guinea	*Pteropus rayneri*	*NT*	x		Hamilton and Leary (2008), Bowen et al. (1997)
	Vanuatu	*Emballonura semicaudata*	*EN*	x		Bonaccorso et al. (2008), Chambers and Esrom (1991)
	Vanuatu	*Miniopterus tristis*	*LC*	x		Chambers and Esrom (1991)

IUCN region	Country in which hunted	Species	Status	Food	Medicine	Source
	Vanuatu	*Pteropus anetianus*	*VU*	x		Mickleburgh et al. (1992), Helgen and Hamilton (2008a), Chambers and Esrom (1991)
	Vanuatu	*Pteropus fundatus*	*EN*	x		Helgen and Hamilton (2008c), Chambers and Esrom (1991)
	Vanuatu, others?	*Aselliscus tricuspidatus*	*LC*	x		Bonaccorso et al. (2008), Chambers and Esrom (1991)
	Vanuatu, Papua New Guinea	*Miniopterus australis*	*LC*	x		Chambers and Esrom (1991)
South America	Bolivia	*Artibeus sp.*	*LC*			Lizarro et al. (2010)
	Bolivia	*Carollia perspicillata*	*LC*	x		Lizarro et al. (2010)
	Bolivia	*Desmodus rotundus*	*LC*			Lizarro et al. (2010)
	Bolivia	*Myotis sp.*	?			Lizarro et al. (2010)
	Brazil	*Glossophaga sp.*	*LC or DD*	x		Mickleburgh et al. (2009)
	Brazil	*sp. 1*		x		Lévi-Strauss (1979), Setz and Sazima (1987), Setz (1991)
	Brazil	*sp. 2*		x		Lévi-Strauss (1979), Setz and Sazima (1987), Setz (1991)
	Brazil	*sp. 3*		x		Lévi-Strauss (1979), Setz and Sazima (1987), Setz (1991)
South Asia	Bangladesh	*Pteropus giganteus*	*LC*	x	x	Mickleburgh et al. (2009), Molur et al. (2008a)
	India	*Hipposideros speoris*	*LC*		x	Molur et al. (2008b)
	India	*Latidens salimalii*	*EN*	x		Molur and Vanitharani (2008)
	India	*Megaderma lyra*	*LC*	x	x	Csorba et al. (2008a)
	India	*Megaderma spasma*	*LC*	x	x	Csorba (2008e)
	India	*Nyctalus montanus*	*LC*		x	Molur and Srinivasulu (2008)
	India	*Pteropus faunulus*	*VU*		x	Kingston et al. (2008); Singaravelan et al. (2009)
	India	*Pteropus melanotus*	*VU*	x		Hutson et al. (2008d), Mickleburgh et al. (2009)
	India	*Taphozous melanopogon*	*LC*	x		Csorba et al. (2008f), Molur et al. (2002)
	India	*Taphozous theobaldi*	*LC*	x	x	Bates et al. (2008e), Molur et al. (2002)
	India, Sri Lanka	*Hipposideros lankadiva*	*LC*	x	x	Molur et al. (2008c)

IUCN region	Country in which hunted	Species	Status	Food	Medicine	Source
	South Asia wide	*Rousettus leschenaultii*	*LC*		x	Molur et al. (2002), Bates and Helgen (2008), Mickleburgh et al. (2009)
	South Asia wide	*Cynopterus sphinx*	*LC*	x		Bates et al. (2008d), Molur et al. (2002)
South East Asia	Brunei (Borneo), Cambodia, Lao PDR, Malaysia, Thailand, Vietnam	*Chaerephon plicatus/ Tadarida plicata*	*LC*	x		Csorba et al. (2014)
	Brunei, Indonesia, Malaysia, Philippines, Thailand	*Pteropus vampyrus*	*NT*	x		Bates et al. (2008f), Clayton and Milner-Gulland (2000)
	Cambodia, Indonesia, Lao PDR, Philippines, Thailand, Vietnam	*Cynopterus brachyotis*	*LC*	x		Lacerna and Widmann (1999)
	Cambodia, Indonesia, Lao PDR, Thailand, Vietnam	*Cynopterus sphinx*	*LC*	x	x	Bates et al. (2008d), Johnson et al. (2003)
	Cambodia, Philippines, Thailand, Vietnam	*Eonycteris spelaea*	*LC*	x		Mickleburgh et al. (2009)
	Cambodia, Thailand	*Pteropus lylei*	*VU*	x		Bumrungsri et al. (2008b); Mickleburgh et al. (2009)
	Indonesia	*Acerodon celebensis*	*LC*	x		Hutson et al. (2008c), Clayton and Milner-Gulland (2000)
	Indonesia	*Acerodon humilis*	*EN*	x		Hutson et al. (2008b), Clayton and Milner-Gulland (2000)
	Indonesia	*Acerodon mackloti*	*VU*	x		Hutson et al. (2008i)
	Indonesia	*Cheiromeles parvidens*	*LC*	x		Csorba et al. (2008b)
	Indonesia	*Harpyionycteris celebensis*	*VU*	x		Hutson et al. (2008l)
	Indonesia	*Neopteryx frosti*	*EN*	x		Hutson et al. (2008k)
	Indonesia	*Pteropus alecto*	*LC*	x		Bergmans and Rozendaal (1988), Hutson et al. (2008n)
	Indonesia	*Pteropus caniceps*	*NT*	x		Hutson and Helgen (2008a)
	Indonesia	*Pteropus chrysoproctus*	*NT*	x		Hutson and Helgen (2008b)
	Indonesia	*Pteropus griseus*	*DD*	x		Heinrichs and Zahnke (1997)

IUCN region	Country in which hunted	Species	Status	Food	Medicine	Source
	Indonesia	*Pteropus lombocensis*	*DD*	x		Clayton and Milner-Gulland (2000), Helgen and Salas (2008a)
	Indonesia	*Pteropus melanopogon*	*EN*	x		Helgen and Salas (2008b)
	Indonesia	*Pteropus ocularis*	*VU*	x		Helgen and Salas (2008c)
	Indonesia	*Pteropus pohlei*	*EN*	x		Helgen and Bonaccorso (2008a)
	Indonesia	*Pteropus temminckii*	*VU*	x		Helgen and Bonaccorso (2008b)
	Indonesia	*Rousettus bidens*	*VU*	x		Helgen et al. (2008d)
	Indonesia	*Rousettus celebensis*	*LC*	x		Ruedas et al. (2008b)
	Indonesia	*Styloctenium wallacei*	*NT*	x		Ruedas et al. (2010), Mickleburgh et al. (2009)
	Indonesia	*Thoopterus nigrescens*	*LC*	x		Ruedas et al. (2008a), Mickleburgh et al. (2009)
	Indonesia, Thailand	*Rousettus leschenaultii*	*LC*	x	x	Mickleburgh et al. (2009)
	Lao PDR	*Hipposideros scutinares*	*VU*	x		Francis and Bates (2008)
	Lao PDR	*Tadarida latouchei*	*DD*	x		Francis and Maeda (2008)
	Lao PDR, Philippines, Thailand	*Taphozous melanopogon*	*LC*	x		Csorba (2008f), Magnus (2001)
	Lao PDR, Thailand, Vietnam	*Hipposideros armiger*	*LC*	x		Bates et al. (2008b)
	Malaysia	*Cheiromeles torquatus*	*LC*	x		Csorba et al. (2008c)
	Malaysia, Philippines, Thailand	*Pteropus hypomelanus*	*LC*	x		Francis et al. (2008a), Fujita and Tuttle (1991), Fujita (1988)
	Myanmar	*Hipposideros pomona*	*LC*	x	x	Bates et al. (2008a)
	Myanmar	*Rhinolophus marshalli*	*LC*	x		Bates (2003)
	Myanmar, Thailand	*Craseonycteris thonglongyai*	*VU*	x		Bates (2003)
	Philippines	*Acerodon jubatus*	*EN*	x		Mildenstein et al. (2008), Heaney and Heideman (1987)
	Philippines	*Acerodon leucotis*	*VU*	x		Ong et al. (2008b)
	Philippines	*Dobsonia chapmani*	*CR*	x		Heaney et al. (2008b), Heaney and Heideman (1987)
	Philippines	*Eonycteris robusta*	*NT*	x		Ong et al. (2008e)
	Philippines	*Harpyionycteris whiteheadi*	*LC*	x		Ong et al. (2008d), Mickleburgh et al. (2009)

IUCN region	Country in which hunted	Species	Status	Food	Medicine	Source
	Philippines	*Hipposideros coronatus*	*DD*	x		Gomez et al. (2008)
	Philippines	*Hipposideros pygmaeus*	*LC*	x		Heaney et al. (2008a)
	Philippines	*Macroglossus minimus*	*LC*	x		Mickleburgh et al. 2009, Francis et al. (2008b)
	Philippines	*Nyctimene rabori*	*EN*	x		Mickleburgh et al. (2009)
	Philippines	*Ptenochirus jagori*	*LC*	x		Heaney and Heideman (1987)
	Philippines	*Pteropus dasymallus*	*NT*	x		Heaney et al. (2008d)
	Philippines	*Pteropus leucopterus/ Desmalopex leucopterus*	*LC*	x		Ong et al. (2008a)
	Philippines	*Pteropus pumilus*	*NT*	x		Heaney et al. (2008c), Mickleburgh et al. (2009)
	Philippines	*Pteropus speciosus*	*DD*	x		Rosell-Ambal et al. (2008)
	Philippines	*Rhinolophus rufus*	*NT*	x		Ong et al. (2008c)
	Philippines	*Styloctenium mindorensis*	*DD*	x		Esselstyn (2008)
	Philippines, Thailand	*Hipposideros lekaguli*	*NT*	x		Csorba (2008d)
	Philippines, Thailand	*Rousettus amplexicaudatus*	*LC*	x		Csorba (2008g), Utzurrum (1992)
	SE Asia	*Megaderma spasma*	*LC*	x	x	Csorba (2008e)
	Thailand	*Hipposideros halophyllus*	*EN*	x		Bates et al. (2008b)
	Vietnam, Cambodia	*Megaderma lyra*	*LC*	x	x	Csorba et al. (2008a)
Sub Saharan Africa	Unspecified	*Epomophorus labiatus*	*LC*	x		Mickleburgh et al. (2008b)
	Unspecified	*Hipposideros gigas*	*LC*	x		Mickleburgh et al. (2008p)
	Unspecified	*Hipposideros jonesi*	*NT*	x		Mickleburgh et al. (2008g)
	Unspecified	*Hipposideros marisae*	*VU*	x		Mickleburgh et al. (2008h)
	Unspecified	*Hipposideros ruber*	*LC*	x		Mickleburgh et al. (2008i)
	Unspecified	*Hipposideros vittatus*	*NT*	x		Mickleburgh et al. (2008f)
	Unspecified	*Mops midas/Tadarida midas*	*LC*	x		Jenkins et al. (2014)
	Unspecified	*Myotis morrisi*	*DD*	x		Jacobs et al. (2008a)
	Unspecified	*Rhinolophus alcyone*	*LC*	x		Jacobs et al. (2008b)

IUCN region	Country in which hunted	Species	Status	Food	Medicine	Source
	Unspecified	*Rhinolophus guineensis*	*VU*	x		Fahr (2008a)
	Unspecified	*Rhinolophus hillorum*	*NT*	x		Jacobs et al. (2010)
	Unspecified	*Rhinolophus silvestris*	*DD*	x		Cotterill (2008)
	Unspecified	*Rhinolophus ziama*	*EN*	x		Fahr (2008d)
	Unspecified	*Rousettus lanosus*	*LC*	x		Mickleburgh et al. (2008n)
	Unspecified	*Taphozous mauritianus*	*LC*			Hutson et al. (2008e)
	Benin	*Epomophorus gambianus*	*LC*	x		Mickleburgh et al. (2008r)
	Benin, Cameroon, Congo Rep., Cote d'Ivoire, DRC, Equatorial Guinea, Ghana, Liberia, Nigeria, Tanzania, Uganda, Zambia	*Eidolon helvum*	*NT*	x	x	Halstead (1977), Kamins et al. (2011), Mickleburgh et al. (2008l)
	Benin, The Democratic Republic of the Congo, Equatorial Guinea, Nigeria	*Epomops franqueti*	*LC*	x		Colyn et al. (1987), Fa et al. (1995), Juste et al. (1995), Bennett Hennessey (1995)
	Comoros Islands	*Pteropus livingstonii*	*EN*	x		Trewhella et al. (1995)
	Congo Republic	*Lissonycteris angolensis*	*LC*	x		Wilson and Wilson (1991), Mickleburgh et al. (2008c)
	Congo Republic	*Nycteris grandis*	*LC*	x		Mickleburgh et al. (2008k), Bennett Hennessey (1995)
	Côte d'Ivoire	*Chaerephon ansorgei/ Tadarida ansorgei*	*LC*	x		Mickleburgh et al. (2008e)
	Equatorial Guinea	*Micropteropus pusillus*	*LC*	x		Juste et al. (1995), Fa (2000)
	Equatorial Guinea, Nigeria	*Rousettus aegyptiacus*	*LC*	x		Fa et al. (1995), Fa (2000), Benda et al. (2008)
	Guinea	*Hipposideros lamottei*	*CR*	x		Mickleburgh et al. (2008q)
	Guinea	*Rhinolophus maclaudi*	*EN*	x		Fahr (2008b), Fahr et al. (2002), Fahr and Ebigbo (2003)
	Guinea	*Rhinolophus ruwenzorii*	*VU*	x		Fahr et al. (2002), Fahr and Ebigbo (2003), Fahr (2008c)
	Madagascar	*Chaerephon jobimena/ Tadarida jobimena*	*LC*	x		Andriafidison et al. (2014a)

IUCN region	Country in which hunted	Species	Status	Food	Medicine	Source
	Madagascar	*Eidolon dupreanum*	*VU*	x		Andriafidison et al. (2008a), Jenkins and Racey (2008)
	Madagascar	*Emballonura atrata*	*LC*	x		Jenkins et al. (2008b)
	Madagascar	*Hipposideros commersoni*	*NT*	x		Goodman (2006), Jenkins and Racey (2008)
	Madagascar	*Miniopterus gleni*	*LC*	x		Andriafidison et al. (2008b), Goodman (2006), Goodman et al. (2008)
	Madagascar	*Miniopterus majori*	*LC*	x		Jenkins and Rakotoarivelo (2008)
	Madagascar	*Miniopterus manavi*	*LC*	x		Andriafidison et al. (2008c), Golden (2005)
	Madagascar	*Mops leucostigma*	*LC*	x		Andriafidison et al. (2014b)
	Madagascar	*Mormopterus jugularis*	*LC*	x		Andriafidison et al. (2008d)
	Madagascar	*Myzopoda aurita*	*LC*	x		Jenkins et al. (2008a)
	Madagascar	*Pteropus rufus*	*VU*	x		Andriafidison et al. (2008e), Jenkins and Racey (2008)
	Madagascar	*Rousettus madagascariensis*	*NT*	x		Andriafidison et al. (2008f), Jenkins and Racey (2008)
	Madagascar	*Scotophilus robustus*	*LC*	x		Andriafidison et al. (2008g)
	Madagascar	*Triaenops furculus*	*LC*	x		J. Razafimanahaka pers. comm
	Madagascar	*Triaenops rufus*	*LC*	x		Goodman (2006), Andriafidison et al. (2008h)
	Mauritius	*Pteropus niger*	*VU*	x		Hutson and Racey (2013), Mickleburgh et al. (2009)
	Mauritius	*Pteropus rodricensis*	*CR*	x		Mickleburgh et al. (2008d), Mickleburgh et al. (2009)
	Rwanda	*Rhinolophus hilli*	*CR*	x		Fahr (2010)
	São Tomé and Príncipe	*Chaerephon tomensis*	*EN*	x		Carvalho et al. (2014)
	São Tomé and Príncipe	*Miniopterus newtoni*	*DD*	x		Carvalho et al. (2014)
	São Tomé and Príncipe	*Myonycteris brachycephala*	*EN*	x		Carvalho et al. (2014)
	Seychelles	*Pteropus seychellensis*	*LC*	x		Carvalho et al. (2014)
	Tanzania (Pemba)	*Pteropus voeltzkowi*	*VU*	x		Mickleburgh et al. (2008m)
	The Democratic Republic of the Congo, Nigeria	*Hypsignathus monstrosus*	*LC*	x		Mickleburgh et al. (2008j), Mickleburgh et al. (2009)
	Unspecified	*Myotis mystacinus*	*LC*		x	Hutson et al. (2008f)

References

Abdul Aziz S, Olival KJ, Bumrungsri S, Richards GC, Racey PA (2016) The conflict between fruit bats and fruit growers: species, legislation and mitigation. In: Voigt CC, Kingston T (eds), Bats in the Anthropocene: conservation of bats in a changing world. Springer International AG, Cham, pp. 377–420

Allison A, Bonaccorso F, Helgen K, James R (2008) *Pteropus mariannus*. The IUCN Red List of threatened species. Version 2014.3. www.iucnredlist.org. Downloaded on 18 Mar 2015

Andersen WL, Thornberg DD, Whitton RM (1996) Estimating goods harvest in southern Illinois quota zones. Wildl Soc Bull 24:233–237

Andriafidison D, Cardiff SG, Goodman SM et al () *Eidolon dupreanum*. The IUCN Red List of threatened species. Version 2014.3. www.iucnredlist.org. Downloaded on 17 Mar 2015

Andriafidison D, Cardiff SG, Goodman SM et al (2008b) *Miniopterus gleni*. The IUCN Red List of threatened species. Version 2014.3. www.iucnredlist.org. Downloaded on 15 Mar 2015

Andriafidison D, Cardiff SG, Goodman SM et al (2008c) *Miniopterus manavi*. The IUCN Red List of threatened species. Version 2014.3. www.iucnredlist.org. Downloaded on 15 Mar 2015

Andriafidison D, Cardiff SG, Goodman SM et al (2008d) *Mormopterus jugularis*. The IUCN Red List of threatened species. Version 2014.3. www.iucnredlist.org. Downloaded on 17 Mar 2015

Andriafidison D, Cardiff SG, Goodman SM et al (2008e) *Pteropus rufus*. The IUCN Red List of threatened species. Version 2014.3. www.iucnredlist.org. Downloaded on 18 Mar 2015

Andriafidison D, Cardiff SG, Goodman SM et al (2008f) *Rousettus madagascariensis*. The IUCN Red List of threatened species. Version 2014.3. www.iucnredlist.org. Downloaded on 18 Mar 2015

Andriafidison D, Cardiff SG, Goodman SM et al (2008g) *Scotophilus robustus*. The IUCN Red List of threatened species. Version 2014.3. www.iucnredlist.org. Downloaded on 18 Mar 2015

Andriafidison D, Cardiff SG, Goodman SM et al (2008h) *Triaenops rufus*. The IUCN Red List of threatened species. Version 2015.2. www.iucnredlist.org. Downloaded on 18 Mar 2015

Andriafidison D, Cardiff SG, Goodman SM et al (2014a) *Chaerephon jobimena*. The IUCN Red List of threatened species. Version 2014.3. www.iucnredlist.org. Downloaded on 15 Mar 2015

Andriafidison D, Cardiff SG, Goodman SM et al (2014b) *Mops leucostigma*. The IUCN Red List of threatened species. Version 2014.3. www.iucnredlist.org. Downloaded on 17 Mar 2015

Anstey S (1991) Wildlife utilisation in liberia. The findings of a national survey 1989–1990. Report for WWF-UK, Godalming, UK

Aulagnier S, Juste J, Karataş A et al. (2008) *Myotis punicus*. The IUCN Red List of threatened species. Version 2014.3. www.iucnredlist.org. Downloaded on 18 Mar 2015

Banack SA (1996) Flying foxes, genus *Pteropus*, in the Samoan Islands: interactions with forest communities. Dissertation. University of California, California

Banack SA (1998) Diet selection and resource use by flying foxes (genus Pteropus). Ecol 79:1949–1967

Barlow KE, Briggs PA, Haysom KA et al (2015) Citizen science reveals trends in bat populations: the national bat monitoring programme in Great Britain. Biol Conserv 182:14–26

Bat Conservation International (2015) 1331 and counting. Bats magazine. Winter 2015. P 16

Bates PJJ (2003) Conservation Status of the Bat and Bird Fauna of the Limestone Karst Areas of Mon and Kayin States, Myanmar (Burma). Unpublished Report for Fauna and Flora International 100 % Fund, Fauna and Flora International, Cambridge, UK

Bates P, Helgen K (2008) *Rousettus leschenaultii*. The IUCN Red List of threatened species. Version 2014.3. www.iucnredlist.org. Downloaded on 18 Mar 2015

Bates P, Bumrungsri S, Csorba G et al (2008a) *Hipposideros pomona*. The IUCN Red List of threatened species. Version 2014.3. www.iucnredlist.org. Downloaded on 21 Mar 2015

Bates P, Bumrungsri S, Francis C, Csorba G (2008b) *Hipposideros armiger*. The IUCN Red List of threatened species. Version 2015.2. www.iucnredlist.org. Downloaded on 27 July 2015
Bates P, Bumrungsri S, Francis C et al (2008c) *Hipposideros halophyllus*. The IUCN Red List of threatened species. Version 2015.2. www.iucnredlist.org. Downloaded on 27 July 2015
Bates P, Bumrungsri S, Molur S, Srinivasulu C (2008d) *Cynopterus sphinx*. The IUCN Red List of threatened species. Version 2014.3. www.iucnredlist.org. Downloaded on 17 Mar 2015
Bates P, Bumrungsri S, Walston J et al. (2008e) *Taphozous theobaldi*. The IUCN Red List of threatened species. Version 2015.2. www.iucnredlist.org. Downloaded on 27 July 2015
Bates P, Francis C, Gumal M et al. (2008f) *Pteropus vampyrus*. The IUCN Red List of threatened species. Version 2014.3. www.iucnredlist.org. Downloaded on 15 Mar 2015
Benda P, Aulagnier S, Hutson AM et al (2008) *Rousettus aegyptiacus*. The IUCN Red List of threatened species. Version 2014.3. www.iucnredlist.org. Downloaded on 18 Mar 2015
Bennett Hennessey A (1995) A study of the meat trade in Ouesso, Republic of Congo. Wildlife Conservation Society, New York
Bergmans W, Rozendaal FG (1988) Notes on collections of fruit bats from Sulawesi and some off-lying islands (Mammalia, Megachiroptera). Verh Leiden 248:5–74
Boissenin M, Brescia F (2007) Inventaire des colonies de mégachiroptère en Province nord de la Nouvelle-Calédonie. Institut Agronomique néo-Calédonien (IAC), Unpublished report
Bonaccorso F, Allison A (2008) *Emballonura semicaudata*. The IUCN Red List of Threatened Species 2008: Version 2014.3. http://www.iucnredlist.org. Downloaded on 18 May 2015
Bonaccorso F, Helgen K (2008) *Nyctimene aello*. The IUCN Red List of threatened species. Version 2014.3. www.iucnredlist.org. Downloaded on 18 Mar 2015
Bonaccorso F, Reardon T (2008a) *Miniopterus macrocneme*. The IUCN Red List of threatened species. Version 2014.3. www.iucnredlist.org. Downloaded on 15 Mar 2015
Bonaccorso F, Reardon T (2008b) *Miniopterus magnater*. The IUCN Red List of threatened species. Version 2014.3. www.iucnredlist.org. Downloaded on 18 Mar 2015
Bonaccorso F, Leary T, Hamilton (2008) *Aselliscus tricuspidatus*. The IUCN Red List of threatened species. Version 2014.3. www.iucnredlist.org. Downloaded on 15 Mar 2015
Bowen-Jones E, Abrutut D, Markham B, Bowe S (1997) Flying foxes on Choiseul (Solomon Islands)-the need for conservation action. Oryx 31:209–217
Bradshaw CJA, Sodhi NS, Brook BW (2009) Tropical turmoil: a biodiversity tragedy in progress. Front Ecol Environ 7:79–87
Brashares JS, Arcese P, Sam MK et al (2004) Bushmeat hunting, wildlife declines, and fish supply in west Africa. Science 306:1180–1183
Brescia F (2007) Amélioration des connaissances et recommandations pour la sauvegarde des populations de mégachiroptères (roussettes) en Province Nord de la Nouvelle-Calédonie. Institut Agronomique néo-Calédonien, Unpublished report
Brescia F () *Notopteris neocaledonica*. The IUCN Red List of threatened species. Version 2014.3. www.iucnredlist.org. Downloaded on 15 Mar 2015
Brescia F (2008b) *Pteropus ornatus*. The IUCN Red List of threatened species. Version 2014.3. www.iucnredlist.org. Downloaded on 18 Mar 2015
Brescia F (2008c) *Pteropus vetulus*. The IUCN Red List of threatened species. Version 2014.3. www.iucnredlist.org. Downloaded on 18 Mar 2015
Brooke AP (2001) Population status and behaviours of the Samoa flying fox, *Pteropus samoensis*, on Tutuila Island, American Samoa. J Zool 254:309–319
Brooke AP, Tschapka M (2002) Threats to overhunting of the flying fox *Pteropus tonganus* (Chiroptera: Pteropodidae) on Niue Island, South Pacific Ocean. Biol Conserv 103:343–348
Brooke A, Wiles G (2008) *Pteropus samoensis*. The IUCN Red List of threatened species. Version 2014.3. www.iucnredlist.org. Downloaded on 18 Mar 2015
Brooke AP, Solek C, Tualaulelei A (2000) Roosting behavior of colonial and solitary flying foxes in American Samoa (Chiroptera: Pteropodidae). Biotropica 32(2):338–350
Buden D, Wiles G, Helgen K et al (2008) *Pteropus molossinus*. The IUCN Red List of threatened species. Version 2014.3. www.iucnredlist.org. Downloaded on 18 Mar 2015

Bumrungsri S, Harbit A, Benzie C, Carmouche K, Sridith K, Racey PA () The pollination ecology of two species of *Parkia* (Mimosaceae) in southern Thailand. J. Tropl Ecol 24:467–475

Bumrungsri S, Suyanto A, Francis C (2008b) *Pteropus lylei*. The IUCN Red List of threatened species. Version 2014.3. www.iucnredlist.org. Downloaded on 18 Mar 2015

Bumrungsri S, Spraoiaya E, Chongsiri I, Sridith K, Racey PA (2009) Does no bat mean no durian—the pollination ecology of semi-wild durian (*Durio zibethinus* L.) in Southern Thailand. J Trop Ecol 25:85–92

Calisher CH, Childs JE, Field H E, Holmes KV, Schountz T (2006). Bats: important reservoir hosts of emerging viruses. Clinical microbiology reviews, 19(3):531–545

Carvalho M, Palmeirim JM, Rego FC et al (2014) What motivates hunters to target exotic or endemic species on the island of São Tomé, Gulf of Guinea? Oryx 1–9

Cawthorn DM, Hoffman LC (2015) The bushmeat and food security nexus: a global account of the contributions, conundrums and ethical collisions. In press at Food Security International, 20 pp

Chambers MR, Esrom D (1991) The fruit bats of Vanuatu. Bat News 20:4–5

Clayton L, Milner-Gulland EJ (2000) The trade in wildlife in north Sulawesi, Indonesia. In: Robinson JG, Bennett E (eds) Hunting for sustainability in tropical forests. Columbia University Press, New York, pp 473–496

Cleveland CJ, Betke M, Federico P et al (2006) Economic value of the pest control service provided by Brazilian free-tailed bats in south-central Texas. Front Ecol Environ 4:238–243

Colyn M, Dudu A, Mbaelele M (1987) Exploitation du petit et moyen gibier des forests ombrophiles du Zaire. Nat et Faune 3:22–39

Cotterill FPD (2008) *Rhinolophus silvestris*. The IUCN Red List of threatened species. Version 2014.3. www.iucnredlist.org. Downloaded on 18 Mar 2015

Craig P, Morrell TE, So'Oto K (1994a) Subsistence harvest of birds, fruit bats and other game in American Samoa, 1990–1991. Pac Sci 48:344–352

Craig P, Trail P, Morrell TE (1994b) The decline of fruit bats in American Samoa due to hurricanes and overhunting. Biol Conserv 69:261–266

Craig P, Syron W (1992) Fruit bats in American Samoa: their status and future. In: Wilson DE, Graham GL (eds) Pacific Island Flying Foxes: Proceedings of an International Conservation Conference pp. 145–149

Csorba G, Bates P, Molur S, Srinivasulu C (2008a) *Megaderma lyra*. The IUCN Red List of threatened species. Version 2014.3. www.iucnredlist.org. Downloaded on 15 Mar 2015

Csorba G, Bumrungsri S, Francis C et al (2008b) *Cheiromeles parvidens*. The IUCN Red List of threatened species. Version 2014.3. www.iucnredlist.org. Downloaded on 15 Mar 2015

Csorba G, Bumrungsri S, Francis C et al (2008c) *Cheiromeles torquatus*. The IUCN Red List of threatened species. Version 2014.3. www.iucnredlist.org. Downloaded on 17 Mar 2015

Csorba G, Bumrungsri S, Francis C et al (2008d) *Hipposideros lekaguli*. The IUCN Red List of threatened species. Version 2014.3. www.iucnredlist.org. Downloaded on 18 Mar 2015

Csorba G, Bumrungsri S, Francis C et al (2014). *Chaerephon plicatus*. The IUCN Red List of threatened species. Version 2014.3. www.iucnredlist.org. Downloaded on 15 Mar 2015

Csorba G, Bumrungsri S, Helgen K et al (2008e) *Megaderma spasma*. The IUCN Red List of threatened species. Version 2014.3. www.iucnredlist.org. Downloaded on 15 Mar 2015

Csorba G, Bumrungsri S, Helgen K et al (2008f) *Taphozous melanopogon*. The IUCN Red List of Threatened Species. Version 2014.3. www.iucnredlist.org. Downloaded on 21 Mar 2015

Csorba G, Rosell-Ambal G, Ingle N (2008g) *Rousettus amplexicaudatus*. The IUCN Red List of threatened species. Version 2014.3. www.iucnredlist.org. Downloaded on 18 Mar 2015

Cuthbert R (2003a) The impact of hunting on wildlife populations in Papua New Guinea: report 2. Unpublished Report for Fauna and Flora International 100 % Fund, Fauna and Flora International, Cambridge, UK

Cuthbert R (2003b) The sustainability of hunting in Papua New Guinea: a preliminary analysis. Unpublished Report for Fauna and Flora International 100 % Fund, Fauna and Flora International Cambridge, UK

Dougnon TJ, Djossa BA, Youssao I et al (2012) Bats as bushmeat in Benin: yield in carcass and meat quality of the fruit bats *Eidolon helvum* (Kerr, 1792) and *Epomophorus gambianus* (Ogilby, 1835). Int J Sci Adv Technol 2:81–90

Epstein JH, Olival KJ, Pulliam JRC et al (2009) *Pteropus vampyrus*, a hunted migratory species with a multinational home-range and a need for regional management. J App Ecol 46:991–1002

Esselstyn J (2008) *Styloctenium mindorensis*. The IUCN Red List of threatened species. Version 2015.2. www.iucnredlist.org. Downloaded on 27 July 2015

Esselstyn JA, Amar A, Janeke D (2006) Impact of post-typhoon hunting on Mariana fruit bats (*Pteropus mariannus*). Pac Sci 60:531–539

Fa JE (2000) Hunted animals in Bioko Island, West Africa: sustainability and future. In: Robinson JG, Bennett E (eds) Hunting for sustainability in tropical forests. Columbia University Press, New York, pp 168–198

Fa JE, Juste J, Perez De Val J, Castroviejo J (1995) Impact of market hunting on mammal species in Equatorial Guinea. Conserv Biol 9:1107–1115

Fahr J () *Rhinolophus guineensis*. The IUCN Red List of threatened species. Version 2014.3. www.iucnredlist.org. Downloaded on 18 Mar 2015

Fahr J (2008b) *Rhinolophus maclaudi*. The IUCN Red List of threatened species. Version 2014.3. www.iucnredlist.org. Downloaded on 18 Mar 2015

Fahr J (2008c) *Rhinolophus ruwenzorii*. The IUCN Red List of threatened species. Version 2014.3. www.iucnredlist.org. Downloaded on 18 Mar 2015

Fahr J (2008d) *Rhinolophus ziama*. The IUCN Red List of threatened species. Version 2014.3. www.iucnredlist.org. Downloaded on 18 Mar 2015

Fahr J (2010) *Rhinolophus hilli*. The IUCN Red List of threatened species. Version 2014.3. www.iucnredlist.org. Downloaded on 18 Mar 2015

Fahr J, Ebigbo NM (2003) A conservation assessment of the bats of the Simandou Range, Guinea, with the first record of Myotis welwitschii (Gray 1866) from West Africa. Acta Chiropterol 5:125–141

Fahr J, Vierhaus H, Hutterer R, Kock D (2002) A revision of the *Rhinolophus maclaudi* species group with the description of a new species from West Africa (Chiroptera: Rhinolophidae). Myotis 40:95–126

Falanruw M (1988) On the status, reproductive biology, and management of fruit bats in Yap. Micronesica 21:39–51

Falanruw MC, Manmaw CJ (1992) Protection of flying foxes on Yap Islands. In: DE Wilson, GL Graham (eds), Proceedings of an international conservation conference on pacific island flying foxes, US Fish and Wildlife Service Biological Report 90(23). US Department of the Interior, Fish and Wildlife Service, Washington, DC, pp 150–154

Flannery T (1995a) Mammals of New Guinea (Revised and Updated Edition). Reed Books, Chatswood,

Flannery TF (1995b) Mammals of the South-West Pacific, Moluccan Islands. Reed Books, Chatswood, Australia and Cornell University Press, Ithaca, USA

Francis C, Bates P (2008) *Hipposideros scutinares*. The IUCN Red List of threatened species. Version 2014.3. www.iucnredlist.org. Downloaded on 15 Mar 2015

Francis C, Maeda K (2008) *Tadarida latouchei*. The IUCN Red List of threatened species. Version 2014.3. www.iucnredlist.org. Downloaded on 17 Mar 2015

Francis C, Rosell-Ambal G, Bonaccorso F et al. (2008a) *Pteropus hypomelanus*. The IUCN Red List of threatened species. Version 2014.3. www.iucnredlist.org. Downloaded on 18 Mar 2015

Francis C, Rosell-Ambal G, Sedlock J et al. (2008b) *Macroglossus minimus*. The IUCN Red List of threatened species. Version 2015.2. www.iucnredlist.org. Downloaded on 27 July 2015

Francis C, Rosell-Ambal G, Tabaranza B et al. (2008c) *Eonycteris spelaea*. The IUCN Red List of threatened species. Version 2015.2. www.iucnredlist.org. Downloaded on 27 July 2015

Fujita M (1988) Flying foxes and economics. BATS 6:4–9

Fujita MS, Tuttle MD (1991) Flying foxes (Chiroptera: Pteropodidae): threatened animals of key ecological and economic importance. Conserv Biol 5:455–463

Funmilayo O (1978) Fruit bats for meat: are too many taken? Oryx 14:377–378

Golden CD (2005) Eaten to endangerment: mammal hunting and the bushmeat trade in Madagascar's Makira Forest. Undergraduate Thesis, Harvard University

Gomez R, Ong P, Rosell-Ambal G et al (2008) *Hipposideros coronatus*. The IUCN Red List of threatened species. Version 2015.2. www.iucnredlist.org. Downloaded on 27 July 2015

Goodman SM (2006) Hunting of microchiroptera in southwestern Madagascar. Oryx 40:225–228

Goodman SG, Ratrimomanarivo FH, Ranivo J, Cardiff S (2008) The hunting of microchiropteran bats in different portions of Madagascar. Afri Bat Conserv News 16:4–7

Gumal MT (2004) Diurnal home range and roosting trees of a maternity colony of *Pteropus vampyrus natunae* (Chiroptera: Pteropodidae) in Sedilu, Sarawak. J Trop Ecol 20:247–258

Halstead LB (1977) Fruit bats—an example of wildlife management. Nigerian Field 42:50–56

Hamilton S, Helgen K (2008) *Pteropus tonganus*. The IUCN Red List of threatened species. Version 2014.3. www.iucnredlist.org. Downloaded on 15 Mar 2015

Hamilton S, Leary T (2008) *Pteropus rayneri*. The IUCN Red List of threatened species. Version 2014.3. www.iucnredlist.org. Downloaded on 18 Mar 2015

Hamilton S, Helgen K, James R et al. (2008a) *Pteralopex taki*. The IUCN Red List of threatened species. Version 2014.3. www.iucnredlist.org. Downloaded on 18 Mar 2015

Hamilton S, Helgen K, Pilgrim J (2008b) *Pteropus rennelli*. The IUCN Red List of threatened species. Version 2014.3. www.iucnredlist.org. Downloaded on 18 Mar 2015

Harrison ME, Cheyne SM, Darma F et al (2011) Hunting of flying foxes and perception of disease risk in Indonesian Borneo. Biol Conserv 144:2441–2449

Hayman DTS, McCrea R, Restif O et al (2012) Demography of straw-colored fruit bats in Ghana. J Mammal 93(5):1393–1404. doi:10.1644/11-MAMM-A-270.1

Heaney LR, Heideman PD (1987) Philippine fruit bats: endangered and extinct. BATS 5:3–5

Heaney L, Balete D, Ong P et al (2008a) *Hipposideros pygmaeus*. The IUCN Red List of threatened species. Version 2015.2. www.iucnredlist.org. Downloaded on 27 July 2015

Heaney L, Ong P, Tabaranza B et al. (2008b) *Dobsonia chapmani*. The IUCN Red List of threatened species. Version 2014.3. www.iucnredlist.org. Downloaded on 17 Mar 2015

Heaney L, Rosell-Ambal G, Tabaranza B et al (2008c) *Pteropus pumilus*. The IUCN Red List of threatened species. Version 2014.3. www.iucnredlist.org. Downloaded on 18 Mar 2015

Heaney L, Rosell-Ambal G, Tabaranza B, Izawa M (2008d) *Pteropus dasymallus*. The IUCN Red List of threatened species. Version 2014.3. www.iucnredlist.org. Downloaded on 18 Mar 2015

Heideman PD, Heaney LR (1992) Acerodon jubatus species account. In: Mickleburgh SP, Hutson AM, Racey PA (eds) Old world fruit bats: an action plan for their conservation. IUCN/SSC Chiroptera Specialist Group, Switzerland, pp 37–38

Heinrichs S (2004) Sulawesi fruit bat conservation and education campaign. Unpublished report for the Fauna and Flora International 100 % Fund, Fauna and Flora International, Cambridge, UK

Heinrichs S, Zahnke K (1997) The fruit bats of Sulawesi. BATS 15:11–15

Helgen K, Allison A (2008) *Pteropus howensis*. The IUCN Red List of threatened species. Version 2014.3. www.iucnredlist.org. Downloaded on 15 Nov 2014

Helgen K, Bonaccorso F (2008a) *Pteropus pohlei*. The IUCN Red List of threatened species. Version 2014.3. www.iucnredlist.org. Downloaded on 18 Mar 2015

Helgen K, Bonaccorso F (2008b) *Pteropus temminckii*. The IUCN Red List of threatened species. Version 2014.3. www.iucnredlist.org. Downloaded on 18 Mar 2015

Helgen KM, Flannery TF (2002) Distribution of the endangered Pacific sheathtail bat *Emballonura semicaudata*. Aust Mammal 24:209–212

Helgen K, Hamilton S (2008a) *Pteropus anetianus*. The IUCN Red List of threatened species. Version 2014.3. www.iucnredlist.org. Downloaded on 18 Mar 2015

Helgen K, Hamilton S (2008b) *Pteralopex atrata*. The IUCN Red List of threatened species. Version 2014.3. www.iucnredlist.org. Downloaded on 18 Mar 2015

Helgen K, Hamilton S (2008c) *Pteropus fundatus*. The IUCN Red List of threatened species. Version 2014.3. www.iucnredlist.org. Downloaded on 15 Mar 2015
Helgen K, Salas L (2008a) *Pteropus lombocensis*. The IUCN Red List of threatened species. Version 2015.2. www.iucnredlist.org. Downloaded on 27 July 2015
Helgen K, Salas L (2008b) *Pteropus melanopogon*. The IUCN Red List of threatened species. Version 2014.3. www.iucnredlist.org. Downloaded on 18 Mar 2015
Helgen K, Salas L (2008c) *Pteropus ocularis*. The IUCN Red List of threatened species. Version 2014.3. www.iucnredlist.org. Downloaded on 18 Mar 2015
Helgen K, Wiles G (2010) *Pteropus insularis*. The IUCN Red List of threatened species. Version 2015.2. www.iucnredlist.org. Downloaded on 27 July 2015
Helgen K, Hamilton S, Leary T (2008a) *Pteralopex flanneryi*. The IUCN Red List of threatened species. Version 2014.3. www.iucnredlist.org. Downloaded on 18 Mar 2015
Helgen K, Hamilton S, Leary T, Bonaccorso F (2008b) *Pteralopex anceps*. The IUCN Red List of threatened species. Version 2014.3. www.iucnredlist.org. Downloaded on 18 Mar 2015
Helgen K, Kingston T, Ruedas L, Maharadatunkamsi D (2008c) *Rousettus bidens*. The IUCN Red List of threatened species. Version 2014.3. www.iucnredlist.org. Downloaded on 18 Mar 2015
Hines JM, Hungerford HR, Tomera AN (1987) Analysis and synthesis of research on responsible environmental behavior: a meta-analysis. J Environ Educ 18:1–8
Hutson AM, Helgen K (2008a) *Pteropus caniceps*. The IUCN Red List of threatened species. Version 2014.3. www.iucnredlist.org. Downloaded on 18 Mar 2015
Hutson AM, Helgen K (2008b) *Pteropus chrysoproctus*. The IUCN Red List of threatened species. Version 2014.3. www.iucnredlist.org. Downloaded on 18 Mar 2015
Hutson AM, Racey PA (2013) *Pteropus niger*. The IUCN Red List of threatened species. Version 2014.3. www.iucnredlist.org. Downloaded on 18 Mar 2015
Hutson AM, Aulagnier S, Spitzenberger F (2008a) *Myotis nattereri*. The IUCN Red List of threatened species. Version 2014.3. www.iucnredlist.org. Downloaded on 18 March 2015
Hutson AM, Kingston T, Helgen K (2008b) *Acerodon humilis*. The IUCN Red List of threatened species. Version 2014.3. www.iucnredlist.org. Downloaded on 17 Mar 2015
Hutson AM, Kingston T, Helgen K, Sinaga U (2008c) *Acerodon celebensis*. The IUCN Red List of threatened species. Version 2014.3. www.iucnredlist.org. Downloaded on 17 Mar 2015
Hutson AM, Kingston T, James D et al (2008d) *Pteropus melanotus*. The IUCN Red List of threatened species. Version 2014.3. www.iucnredlist.org. Downloaded on 18 Mar 2015
Hutson AM, Racey P, Ravino J et al (2008e) *Taphozous mauritianus*. The IUCN Red List of threatened species. Version 2014.3. www.iucnredlist.org. Downloaded on 18 Mar 2015
Hutson AM, Spitzenberger F, Aulagnier S, Coroiu I (2008f) *Myotis mystacinus*. The IUCN Red List of threatened species. Version 2014.3. www.iucnredlist.org. Downloaded on 18 Mar 2015
Hutson AM, Spitzenberger F, Aulagnier S et al (2008g) *Myotis capaccinii*. The IUCN Red List of threatened species. Version 2014.3. www.iucnredlist.org. Downloaded on 18 Mar 2015
Hutson AM, Spitzenberger F, Aulagnier S, Nagy Z (2008h) *Myotis emarginatus*. The IUCN Red List of threatened species. Version 2014.3. www.iucnredlist.org. Downloaded on 18 Mar 2015
Hutson AM, Suyanto A, Helgen K (2008i) *Acerodon mackloti*. The IUCN Red List of threatened species. Version 2014.3. www.iucnredlist.org. Downloaded on 17 Mar 2015
Hutson AM, Suyanto A, Helgen K et al (2008j) *Dobsonia moluccensis*. The IUCN Red List of threatened species. Version 2014.3. www.iucnredlist.org. Downloaded on 17 Mar 2015
Hutson AM, Suyanto A, Helgen K (2008k) *Neopteryx frosti*. The IUCN Red List of threatened species. Version 2014.3. www.iucnredlist.org. Downloaded on 18 Mar 2015
Hutson AM, Suyanto A, Kingston T, Helgen K (2008l) *Harpyionycteris celebensis*. The IUCN Red List of threatened species. Version 2014.3. www.iucnredlist.org. Downloaded on 18 Mar 2015
Hutson T, Helgen K, Flannery T, Wright D (2008m) *Aproteles bulmerae*. The IUCN Red List of threatened species. Version 2014.3. www.iucnredlist.org. Downloaded on 17 Mar 2015

Hutson T, Suyanto A, Helgen K et al (2008n) *Pteropus alecto*. The IUCN Red List of threatened species. Version 2014.3. www.iucnredlist.org. Downloaded on 18 March 2015
Ishii N, Maeda K (2008) *Pteropus pselaphon*. The IUCN Red List of threatened species. Version 2014.3. www.iucnredlist.org. Downloaded on 18 Mar 2015
IUCN (2014) The IUCN Red List of threatened species. Version 2014.3. www.iucnredlist.org. Downloaded on 24 Apr 2015
Jacobs D, Cotterill FPD, Taylor PJ (2008a) *Myotis morrisi*. The IUCN Red List of threatened species. Version 2014.3. www.iucnredlist.org. Downloaded on 21 Mar 2015
Jacobs D, Cotterill FPD, Taylor PJ (2008b) *Rhinolophus alcyone*. The IUCN Red List of threatened species. Version 2014.3. www.iucnredlist.org. Downloaded on 18 Mar 2015
Jacobs D, Cotterill FPD, Taylor PJ (2010) *Rhinolophus hillorum*. The IUCN Red List of threatened species. Version 2014.3. www.iucnredlist.org. Downloaded on 18 Mar 2015
James R, Hamilton S, Helgen K (2008) *Pteropus cognatus*. The IUCN Red List of threatened species. Version 2014.3. www.iucnredlist.org. Downloaded on 18 Mar 2015
Jenkins RK, Racey PA (2008) Bats as bushmeat in Madagascar. Madagascar Conserv Dev 3(1):22–30
Jenkins RKB, Rakotoarivelo AR (2008) *Miniopterus majori*. The IUCN Red List of threatened species. Version 2014.3. www.iucnredlist.org. Downloaded on 15 Mar 2015
Jenkins RK, Racey PA, Andriafidison D, Razafindrakoto N, Razafimahatratra E, Rabearivelo A, Razafimanahaka HJ (2007) Not rare, but threatened: the endemic Madagascar flying fox *Pteropus rufus* in a fragmented landscape. Oryx 41(2):263–271
Jenkins RKB, Rakotoarivelo AR, Ratrimomanarivo FH, Cardiff SG (2008a) *Myzopoda aurita*. The IUCN Red List of threatened species. Version 2014.3. www.iucnredlist.org. Downloaded on 17 Mar 2015
Jenkins RKB, Rakotoarivelo AR, Ratrimomanarivo FH, Cardiff SG (2008b) *Emballonura atrata*. The IUCN Red List of threatened species. Version 2015.2. www.iucnredlist.org. Downloaded on 27 July 2015
Jenkins RKB, Racey PA, Ranivo J et al (2014) *Mops midas*. The IUCN Red List of threatened species. Version 2014.3. www.iucnredlist.org. Downloaded on 17 Mar 2015
Johnson DH (2008) In defense of indices: the case of bird surveys. J Wildl Manag 72:857–868
Johnson A, Singh S, Dongdala M, Vongsa O (2003) Wildlife hunting and use in the nam ha protected area: implications for rural livelihoods, biodiversity conservation. Wildlife Conservation Society, Vientiane, Lao PDR
Juste J, Fa JE, Perezdel Val J, Castroviejo J (1995) Market dynamics of bushmeat species in Equatorial Guinea. J Appl Ecol 32:454–467
Kamins AO, Restif O, Ntiamoa-Baidu Y et al (2011) Uncovering the fruit bat bushmeat commodity chain and the true extent of fruit bat hunting in Ghana, West Africa. Biol Conserv 144:3000–3008
Kingston T (2016) Cute, creepy or crispy—how values, attitudes and norms shape human behavior toward bats. In: Voigt CC, Kingston T (eds) Bats in the anthropocene: conservation of bats in a changing world. Springer International AG, Cham, pp. 571–588
Kingston T, Molur S, Srinivasulu C (2008) *Pteropus faunulus*. The IUCN Red List of threatened species. Version 2014.3. www.iucnredlist.org. Downloaded on 18 Mar 2015
Lacerna ID, Widmann P (1999) Biodiversity utilisation in a Tagbanua community, southern Palawan, Philippines. In: Goeltenboth F, Milan P, Asio VB (eds), Procedures of the international conference on applied tropical ecology, 8–10 Sept 1998. Visayas State College of Agriculture, BayBay, Leyte, Philippines, pp 52–64
Leary T, Hamilton S, James R (2008a) *Pteropus nitendiensis*. The IUCN Red List of threatened species. Version 2014.3. www.iucnredlist.org. Downloaded on 15 Mar 2015
Leary T, Hamilton S, James R (2008b) *Pteropus tuberculatus*. The IUCN Red List of threatened species. Version 2014.3. www.iucnredlist.org. Downloaded on 18 Mar 2015
Lemke TO (1986) Marianas fruit bats near extinction. BATS 3:1

Lemke TO (1992) Status of the Marianas fruit bat (*Pteropus mariannus*) in the Northern Mariana Islands north of Saipan. In: Proceedings of an international conservation conference on pacific island flying foxes. US Fish and Wildlife Service Biological Report. 90(23)
Leroy EM, Kumulungui B, Pourrut X, Rouquet P, Hassanin A, Yaba P, Swanepoel R (2005) Fruit bats as reservoirs of Ebola virus. Nature 438(7068):575–576
Leroy EM, Epelboin A, Mondonge V, Pourrut X, Gonzalez JP, Muyembe-Tamfum J, Formenty P (2009) Human Ebola outbreak resulting from direct exposure to fruit bats in Luebo, Democratic Republic of Congo, 2007. Vector-Borne Zoonotic Dis 9(6):723–728
Lévi-Strauss C (1979) Tristes Tropicoes. Edicoes 90, Lisbon, Portugal
Lizarro D, Galarza MI, Aguirre LF (2010) Traffic and trade of Bolivian bats. Rev Bol Ecol y Cons Amb 27:63–75
Mackinnon JL, Hawkins CE, Racey PA (2003) Pteropodidae, in Goodman SP, Benstead JP (eds) The natural history of Madagascar. University of Chicago Press, Chicago, pp 1299–1302
Magnus N (2001) Bushmeat utilisation in Phu Kheio Wildlife Sanctuary, Northeast Thailand. MSc thesis, University of Alberta, Edmonton, Canada
Mbete RA, Banga-Mboko H, Racey PA et al (2011) Household bushmeat consumption in Brazzaville. Trop Conserv Sci 4:187–202
McConkey KR, Drake DR (2006) Flying foxes cease to function as seed dispersers long before they become rare. Ecol 87:271–276
Mickleburgh SP, Hutson AM, Racey PA (1992) Old world fruit-bats—an action plan for their conservation. IUCN, Gland
Mickleburgh SP, Hutson AM, Racey PA (2002) A review of the global conservation status of bats. Oryx 36:18–34
Mickleburgh SP, Hutson AM, Bergmans W et al. (2008a) *Pteropus aldabrensis*. The IUCN Red List of threatened species. Version 2014.3. www.iucnredlist.org. Downloaded on 15 Nov 2014
Mickleburgh S, Hutson AM, Bergmans W (2008b) *Epomophorus labiatus*. The IUCN Red List of threatened species. Version 2014.3. www.iucnredlist.org. Downloaded on 18 Mar 2015
Mickleburgh S, Hutson AM, Bergmans W (2008c) *Lissonycteris angolensis*. The IUCN Red List of threatened species. Version 2014.3. www.iucnredlist.org. Downloaded on 18 Mar 2015
Mickleburgh S, Hutson AM, Bergmans W (2008d) *Pteropus rodricensis*. The IUCN Red List of threatened species. Version 2014.3. www.iucnredlist.org. Downloaded on 18 Mar 2015
Mickleburgh S, Hutson AM, Bergmans W, Cotterill FPD (2008e) *Chaerephon ansorgei*. The IUCN Red List of threatened species. Version 2014.3. www.iucnredlist.org. Downloaded on 17 Mar 2015
Mickleburgh S, Hutson AM, Bergmans W, Cotterill FPD (2008f) *Hipposideros vittatus*. The IUCN Red List of threatened species. Version 2014.3. www.iucnredlist.org. Downloaded on 15 Mar 2015
Mickleburgh S, Hutson AM, Bergmans W, Fahr J (2008g) *Hipposideros jonesi*. The IUCN Red List of threatened species. Version 2014.3. www.iucnredlist.org. Downloaded on 21 Mar 2015
Mickleburgh S, Hutson AM, Bergmans W, Fahr J (2008h) *Hipposideros marisae*. The IUCN Red List of threatened species. Version 2014.3. www.iucnredlist.org. Downloaded on 21 Mar 2015
Mickleburgh S, Hutson AM, Bergmans W, Fahr J (2008i) *Hipposideros ruber*. The IUCN Red List of threatened species. Version 2014.3. www.iucnredlist.org. Downloaded on 15 Mar 2015
Mickleburgh S, Hutson AM, Bergmans W, Fahr J (2008j) *Hypsignathus monstrosus*. The IUCN Red List of threatened species. Version 2014.3. www.iucnredlist.org. Downloaded on 18 Mar 2015
Mickleburgh S, Hutson AM, Bergmans W, Fahr J (2008k) *Nycteris grandis*. The IUCN Red List of threatened species. Version 2014.3. www.iucnredlist.org. Downloaded on 17 Mar 2015

Mickleburgh S, Hutson AM, Bergmans W, Fahr J, Racey PA (2008l) *Eidolon helvum*. The IUCN Red List of threatened species. Version 2014.3. www.iucnredlist.org. Downloaded on 17 Mar 2015
Mickleburgh S, Hutson AM, Bergmans W, Howell K (2008m) *Pteropus voeltzkowi*. The IUCN Red List of threatened species. Version 2014.3. www.iucnredlist.org. Downloaded on 18 Mar 2015
Mickleburgh S, Hutson AM, Bergmans W, Howell K (2008n) *Rousettus lanosus*. The IUCN Red List of threatened species. Version 2014.3. www.iucnredlist.org. Downloaded on 18 Mar 2015
Mickleburgh S, Hutson AM, Bergmans W et al (2008o) *Pteropus seychellensis*. The IUCN Red List of threatened species. Version 2014.3. www.iucnredlist.org. Downloaded on 18 Mar 2015
Mickleburgh S, Hutson AM, Bergmans W (2008p) *Hipposideros gigas*. The IUCN Red List of threatened species. Version 2015.2. www.iucnredlist.org. Downloaded on 27 July 2015
Mickleburgh S, Hutson AM, Bergmans W, Fahr J (2008q) *Hipposideros lamottei*. The IUCN Red List of threatened species. Version 2015.2. www.iucnredlist.org. Downloaded on 27 July 2015
Mickelburgh S, Hutson AM, Bergmans W (2008r) *Epomophorus gambianus*. The IUCN Red List of threatened species. Version 2014.3. www.iucnredlist.org. Downloaded on 18 Mar 2015
Mickleburgh SP, Waylen K, Racey PA (2009) Bats as bushmeat: a global review. Oryx 43:217–234
Mildenstein TL (2011) Bat count Philippines, final report to the conservation leadership program. Available on-line at http://www.conservationleadershipprogramme.org/media/2014/11/001706F_Philippines_FinalReport_Bat-Count-Philippines_21stJune2011.pdf
Mildenstein TL (2012) Conservation of endangered flying foxes in the Philippines: effects of anthropogenic disturbance and research methods for community-based conservation. Ph.D. thesis, University of Montana, United States
Mildenstein TL, Mills LS (2013) Mariana fruit bat conservation through research and local capacity building. Final Report for Cooperative Agreement Number: N40192-11-2-8005, prepared for NAVFAC Marianas
Mildenstein TL, Cariño AB, Stier SC (2007) Bat count 2003, final report to conservation leadership programme. Available on-line http://www.conservationleadershipprogramme.org/media/2014/11/001003_2003_Philippines_FR_-Bat-Count-2003.pdf
Mildenstein T, Cariño A, Paul S et al (2008) *Acerodon jubatus*. The IUCN Red List of threatened species. Version 2014.3. www.iucnredlist.org. Downloaded on 17 Mar 2015
Mildenstein TL, Alexander IR, Miller R (2014) Mariana fruit bat monitoring on Andersen Air Force Base. Final Report for Cooperative Agreement Number: N40192-13-2-8007, prepared for NAVFAC Marianas and U.S. Air Force Base
Mills LS (2012) Conservation of wildlife populations: demography, genetics, and management, 2nd edn. Blackwell Publishing, Oxford
Milner-Gulland EJ, Bennett EL (2003) Wild meat: the bigger picture. Tren Ecol Evol 18:351–357
Milner-Gulland EJ, Fisher M, Browne S, Redford KH, Spencer M, Sutherland WJ et al (2010) Do we need to develop a more relevant conservation literature? Oryx 44:1–2
Molur S, Srinivasulu C (2008) *Nyctalus montanus*. The IUCN Red List of threatened species. Version 2014.3. www.iucnredlist.org. Downloaded on 18 Mar 2015
Molur S, Vanitharani J (2008) *Latidens salimalii*. The IUCN Red List of threatened species. Version 2014.3. www.iucnredlist.org. Downloaded on 18 Mar 2015
Molur S, Marimuthu G, Srinivasulu C et al. (2002) Status of South Asian Chiroptera: Conservation Assessment and Management Plan (C.A.M.P.) Workshop Report. Zoo Outreach Organization/CBSG-South Asia, India
Molur S, Srinivasulu C, Bates P, Francis C (2008a) *Pteropus giganteus*. The IUCN Red List of threatened species. Version 2014.3. www.iucnredlist.org. Downloaded on 18 Mar 2015

Molur S, Yapa W, Srinivasulu C (2008b) *Hipposideros speoris*. The IUCN Red List of threatened species. Version 2014.3. www.iucnredlist.org. Downloaded on 15 Mar 2015
Molur S, Yapa W, Srinivasulu C (2008c) *Hipposideros lankadiva*. The IUCN Red List of threatened species. Version 2015.2. www.iucnredlist.org. Downloaded on 27 July 2015
Navarro L (2013) Communication and education about bats: a powerful tool for changing misconceptions. Under the 43rd international bat research conference. San José, Costa Rica
Nicoll ME, Racey PA (1981) The Seychelles fruit bat *Pteropus seychellensis*. Afr J Ecol 19:361–364
Noureen S (2014) Ecology, roosting habits and reproduction in the Megachiroptera bats of pothwar region, Pakistan. Ph.D. Dissertation, Arid Agriculture University Rawalpindi, 240 pp
Nuno A, St. John FAV (2014) How to ask sensitive questions in conservation: a review of specialized questioning techniques. Biol Conserv 89:5–15
O'Connor T, Riger P, Jenkins RKB (2006) Promoting fruit bat conservation through education in Madagascar. J Int Zoo Educ Assess 42:26–33
Oleksy R, Racey PA, Jones G (2015a) High-resolution GPS tracking reveals habitat selection and the potential for long-distance seed dispersal by Madagascan flying foxes *Pteropus rufus*. Glob Ecol Cons 3:678–692
Oleksy R, Randrianandrianina F, Jenkins RKB (2015b) Commercial hunting of foraging fruit bats in Western Madagascar. Afr Bat Conserv News 37:3–5
Ong P, Rosell-Ambal G, Tabaranza B et al (2008a) *Pteropus leucopterus*. The IUCN Red List of threatened species. Version 2014.3. www.iucnredlist.org. Downloaded on 18 Mar 2015
Ong P, Rosell-Ambal G, Tabaranza B et al (2008b) *Acerodon leucotis*. The IUCN Red List of threatened species. Version 2014.3. www.iucnredlist.org. Downloaded on 17 Mar 2015
Ong P, Rosell-Ambal G, Tabaranza B, Heaney L (2008c) *Rhinolophus rufus*. The IUCN Red List of threatened species. Version 2014.3. www.iucnredlist.org. Downloaded on 18 Mar 2015
Ong P, Rosell-Ambal G, Tabaranza B et al (2008d) *Harpyionycteris whiteheadi*. The IUCN Red List of threatened species. Version 2014.3. www.iucnredlist.org. Downloaded on 18 Mar 2015
Ong P, Rosell-Ambal G, Tabaranza B et al (2008e) *Eonycteris robusta*. The IUCN Red List of threatened species. Version 2014.3. www.iucnredlist.org. Downloaded on 18 Mar 2015
Palmeirim J (2008) *Notopteris macdonaldi*. The IUCN Red List of threatened species. Version 2014.3. www.iucnredlist.org. Downloaded on 15 March 2015
Palmeirim J (2014) *Chaerephon bregullae*. The IUCN Red List of threatened species. Version 2014.3. www.iucnredlist.org. Downloaded on 15 Mar 2015
Palmeirim JM, Champion A, Naikatini A et al (2007) Distribution, status and conservation of the bats of the Fiji Islands. Oryx 41(4):509–519
Pennisi LA, Holland SM, Stein TV (2004) Achieving bat conservation through tourism. J Ecot 3:195–207
Plowright RK, Eby P, Hudson PJ et al (2015) Ecological dynamics of emerging bat virus spillover. Proc Biol Sci 282:2014–2124
Racey PA (1979) Two bats in the Seychelles. Oryx 15:148–152
Racey PA (2013) Bat conservation: past, present and future. In: Adams RA, Pedersen SC (eds) Bat evolution, ecology, and conservation. Springer, Berlin, pp 517–532
Racey PA (2015) The uniqueness of bats. In: Wang L, Cowled C (eds) Bats and viruses. Wiley, New York
Racey PA, Entwistle AC (2000) Life history and reproductive strategies in bats. In: Crighton E, Krutzsch PH (eds) Reproductive biology of bats. Academic Press, NY, pp 363–414
Racey PA, Goodman SM, Jenkins RKB (2009) The ecology and conservation of malagasy bats. In: Fleming TH, Racey PA (eds) Island bats—ecology, evolution and conservation. Chicago University Press, p 369–404
Randrianandrianina FH, Racey PA, Jenkins RKB (2010) Hunting and consumption of mammals and birds by people in urban areas of western Madagascar. Oryx 44:411–415
Ravon S, Furey NM, Hul V, Capelle J (2014) A rapid assessment of flying fox (*Pteropus* spp.) colonies in Cambodia. Camb J Nat Hist 2014:14–18

Razafimanahaka J (2013) When peanuts help to conserve fruit bats. Madagasikara Voakajy Newsletter on-line. http://www.madagasikara-voakajy.org/index.php?option=com_content&view=article&id=161:when-peanuts-help-to-conserve-fruit-bats&catid=19:livelihoods&Itemid=4
Razakarivony HV (2003) Etude d'impact de la prédation humaine sur la rousette *Pteropus rufus* (Tiedemann, 1808) dans la région de Morondava. Memoire DEA, Départment de Biologie Animale, Université d'Antananarivo
Roberts TJ (1997) The mammals of Pakistan, 2nd edn. Oxford University Press, Pakistan
Robinson JG, Bennett EL (2000) Carrying capacity limits to sustainable hunting in tropical forests. In: Robinson JG, Bennett EL (eds) Hunting for sustainability in tropical forests. Columbia University Press, New York, pp 13–30
Robinson JG, Redford KH (1991) Sustainable harvest of neotropical forest mammals. In: Robinson JG, Redford KH (eds) Neotropical wildlife use and conservation. University of Chicago Press, Chicago, pp 415–429
Robinson JE, Bell DJ, Saleh FM et al (2010) Recovery of the Vulnerable Pemba flying fox *Pteropus voeltzkowi*: population and conservation status. Oryx 44:416–423
Rosell-Ambal G, Heaney L, Helgen K (2008) *Pteropus speciosus*. The IUCN Red List of threatened species. Version 2014.3. www.iucnredlist.org. Downloaded on 18 Mar 2015
Ruedas L, Kingston T, Helgen K, Sinaga J (2008a) *Thoopterus nigrescens*. The IUCN Red List of threatened species. Version 2014.3. www.iucnredlist.org. Downloaded on 18 Mar 2015
Ruedas L, Kingston T, Sinanga U, Maharadatunamsi D (2008b) *Rousettus celebensis*. The IUCN Red List of threatened species. Version 2014.3. www.iucnredlist.org. Downloaded on 18 Mar 2015
Ruedas L, Kingston T, Sinanga U (2010) *Styloctenium wallacei*. The IUCN Red List of Threatened Species. Version 2014.3. www.iucnredlist.org. Downloaded on 18 Mar 2015
Saéz AM, Weiss S, Nowak K et al (2015) Investigating the zoonotic origin of the West African Ebola epidemic. EMBO Mol Med 7:17–23
Salas L, Helgen K, Hamilton S (2008) *Pteropus neohibernicus*. The IUCN Red List of threatened species. Version 2014.3. www.iucnredlist.org. Downloaded on 18 Mar 2015
Samuel H (2013) Frozen porcupines and bats confiscated in Paris exotic food raid. The Telegraph, World News, 6 Dec 2013. Accessed on-line http://www.telegraph.co.uk/news/worldnews/europe/france/10500982/Frozen-porcupines-and-bats-confiscated-in-Paris-exotic-food-raid.html
Setz EZF (1991) Animals in the Nambiquara diet: methods of collection and processing. J Ethnobiol 11:1–22
Setz EZ, Sazima I (1987) Bats eaten by Nambiquara Indians in Western Brazil. Biotropica 19:190
Sewall BJ, Granek EF, Trewhella WJ (2003) The endemic Comoros Islands fruit bat *Rousettus obliviosus*: ecology, conservation, and red list status. Oryx 37(3):344–352
Sewall BJ, Granek EF, Moutui MFE, Trewhella WJ, Reason PF, Rodrìguez-Clark KM, Säid I, Caroll JB, Masefield W, Toilibou N, Vély M, Faissoili BM, Feistner TC, Wells S (2007) Conservation action plan for livingstone's flying fox: a strategy for an endangered species, a diverse forest and the Comorian people, 52 pp
Shanahan M, So S, Gompton SG, Gorlett R (2001) Fig-eating by vertebrate frugivores: a global review. Biol Rev 76: 529–572
Sheffers BR, Corlett RT, Diesmos A, Laurance WF (2012) Local demand drives a bushmeat industry in a Philippine forest preserve. Trop Conserv Sci 5(2):133–141
Sheherazade J, Tsang SM (2015) Quantifying the bat bushmeat trade in North Sulawesi, Indonesia, with suggestions for conservation action. Glob Ecol Cons 3:324–330
Shilton LA, Whittaker RJ (2009) The role of pteropodid bats in reestablishing tropical forests on Krakatau. In: Fleming TH, Racey PA (eds) Island bats. Chicago University Press, Chicago, pp 176–215

Simmons N (2005) Order Chiroptera. In: Wilson DE, Reeder DM (eds) Mammal species of the world: a taxonomic and geographic guide. John Hopkins University Press, Baltimore, pp 312–529

Sinavaiana C, Enright J (1992) The cultural significance of the flying foxes in Samoa: a legendary view. In: Wilson DE, Graham GL (eds) Proceeding of an international conservation conference on Pacific Island flying foxes. US Fish and Wildlife Service Biological Report, vol 90, pp 36–38

Singaravelan N, Marimuthu G, Racey PA (2009) Do fruit bats deserve to be listed as vermin in the Indian Wildlife (Protection), Amended Acts? A critical review. Oryx 43(04):608–613

SOS (2012) Filipinos for flying foxes project page. International Union for Conservation of Nature (IUCN) Save Our Species Fund. http://www.sospecies.org/sos_projects/mammals/filipinos_for_flying_foxes/

Stebbings B (1987) Bats, tequila and the King of Siam. BBC Wildlife, Jan 1987, 4–8

Stier SC, Mildenstein TL (2005) Dietary habits of the world's largest bats. J Mammal 86:719–728

Stinson DW, Glass PO, Taisacan EM (1992) Declines and trade in fruit bats on Saipan, Tinian, Aguijan, and Rota. In: Wilson DE, Graham GL (eds) Pacific island flying foxes: proceedings of an international conservation conference, vol 90. US Fish and Wildlife Service Biological Report No 23, pp 61–67

St. John FAV, Keane AM, Milner-Gulland EJ (2013) Effective conservation depends upon understanding human behaviour. In: MacDonald DW, Willis KJ (eds) Key topics in conservation biology. Wiley, New York

Subramanian (2012) Zoonotic disease risk and the bushmeat trade: assessing awareness among hunters and traders in Sierra Leone. Ecohealth 4:471–482

Tacio HD (2015) To the batcave: a visit to Samal Island's world record-holding bat colony. GMA News Online, Phillipines. http://www.gmanetwork.com/news/story/269418/lifestyle/travel/to-the-batcave-a-visit-to-samal-island-s-world-record-holding-bat-colony

Tanshi I, Labiran AJ, Chukwuyem SF et al (2013) Bat conservation in Nigeria: volunteer contribution to bat population survey and public awareness. In: Olajuyigbe SE, Coker OM, Olaleru F (eds) Proceedings of the 4th annual biodiversity conference. Lagos, Nigeria

Trewhella WJ, Reason PF, Bullock RJ et al (1995) Conservation of *Pteropus livingstonii*: catching fruit bats in the Comores. Myotis 32–33:297–305

Trewhella WJ, Rodriguez-Clark KM, Corp N et al (2005) Environmental education as a component of multidisciplinary conservation programmes: lessons from conservation initiatives for critically endangered fruit bats in the Western Indian Ocean. Conserv Biol 19:75–85

Tuladhar-Douglas W (2008) The use of bats as medicine among the Newars. J Ethno 28:69–91

USFWS (2009) Draft revised recovery plan for the Mariana fruit bat or fanihi (*Pteropus mariannus mariannus*), pp 1–83

Van der Aa PJH, Lorica MRP, Komdeur J (2006) The hormonal and behavioral response to the presence and activities of humans in three co-roosting flying fox species (*Acerodon jubatus, Pteropus vampyrus* and *P. hypomelanus*) in Boracay and Mambukal in the Philippines. Acta Zool Sini 52:827–837

Venkatesan A (2007) Status of Indian flying fox *Pteropus giganteus* in Bengaluru. Bat Net News 8:13–15

Voigt CC, Kingston T (2016) Bats in the Anthropocene: conservation of bats in a changing world. Springer International AG, Cham

Wiles GJ (1992) Recent trends in the fruit bat trade on Guam. In: Wilson DE, Graham GL (eds) Pacific island flying foxes: proceedings of an international conservation conference, vol 90. U.S. Fish and Wildlife Service Biological Report No 23, pp 52–60

Wiles G (2008) *Pteropus pelewensis*. The IUCN Red List of threatened species. Version 2014.3. www.iucnredlist.org. Downloaded on 18 Mar 2015

Wiles GJ, Payne NH (1986) The trade in fruit bats *Pteropus* spp. on Guam and other Pacific islands. Biol Conserv 38:143–161

Wiles GJ, Engbring J, Otobed D (1997) Abundance, biology and human exploitation in the Palau Islands. J of Zool Lond 241:203–227

Wiles G, Helgen K, Allison A (2008a) *Pteropus ualanus*. The IUCN Red List of threatened species. Version 2014.3. www.iucnredlist.org. Downloaded on 18 Mar 2015

Wiles G, Helgen K, Allison A (2008b) *Pteropus yapensis*. The IUCN Red List of threatened species. Version 2014.3. www.iucnredlist.org. Downloaded on 18 Mar 2015

Wilson DE, Graham GL (eds) (1992) Pacific Island flying foxes: proceedings of an international conservation conference, vol 90. US Fish and Wildlife Biological Report 23. U.S. Department of the Interior, United States

Wilson VJ, Wilson BLP (1991) La chasse traditionnelle et commerciale dans le sud-ouest du Congo. Tauraco Res Rep 4:279–289

Chapter 13
The Conflict Between Pteropodid Bats and Fruit Growers: Species, Legislation and Mitigation

Sheema Abdul Aziz, Kevin J. Olival, Sara Bumrungsri, Greg C. Richards and Paul A. Racey

Abstract Pteropodid bats damage a wide range of fruit crops, exacerbated by continuing loss of their natural food as forests are cleared. In some countries where such damage occurs, bats are not legally protected. In others, as a result of pressure from fruit growers, legal protection is either not implemented or overridden by legislation specifically allowing the killing of bats. Lethal control is generally ineffective and often carried out with shotguns making it an animal welfare issue, as many more animals are injured or orphaned than are killed. Here, we review the literature and current state of the conflict between fruit growers and pteropodids and describe a wide range of potential mitigation techniques. We compile an extensive list of bats and the fruit crops on which they feed where this has resulted in conflicts, or could lead to conflict, with fruit growers. We also discuss the legal status of bats in some countries where such conflicts occur. We found the most effective means of preventing bat damage to crops is the use of fixed nets (that generally prevent entanglement) covering a whole orchard. Netting individual trees, or fruit panicles, using small net bags, is also effective. Management methods that assist netting include pruning to maintain low stature of trees. These

S.A. Aziz (✉)
Rimba, Muséum National d'Histoire Naturelle, Paris, France
e-mail: sheema.abdulaziz@gmail.com

K.J. Olival
EcoHealth Alliance, New York, USA
e-mail: olival@ecohealthalliance.org

S. Bumrungsri
Department of Biology, Prince of Songkla University, Songkhla, Thailand
e-mail: sarabumrungsri@gmail.com

G.C. Richards
Bat Fauna Consultant, Canberra, Australia
e-mail: batman3812@bigpond.com

C.C. Voigt and T. Kingston (eds.), *Bats in the Anthropocene: Conservation of Bats in a Changing World*, DOI 10.1007/978-3-319-25220-9_13

exclusion techniques are the best management options considering both conservation and public health issues. Although lights, sonic and ultrasonic noises, noxious smells and tastes have been used to deter bats from eating fruit, there have been no large-scale systematic trials of their effectiveness. Nevertheless, broadcasting the sound of discharging shotguns followed by the sound of wounded bats has proved effective in Australia. The use of decoy fruit trees is the least investigated method of mitigation and requires detailed knowledge of the natural diet of the bat species involved. The few studies of dietary preferences undertaken to date suggest that bats prefer non-commercial fruit when it is available, and we highlight this as an area for future research.

13.1 Introduction

The Old World bat family Pteropodidae is distributed throughout the tropics and subtropics of Australasia, Africa and Oceania (Marshall 1983; Mickleburgh et al. 1992). It comprises 196 species (Simmons 2005) that feed primarily on fruit, flowers (nectar, pollen, petals and bracts) and leaves of at least 188 plant genera from 64 families (Lobova et al. 2009; Fleming and Kress 2011), although some species have also been recorded eating insects (e.g. Clulow and Blundell 2011; Scanlon et al. 2013). Fujita (1988) and Fujita and Tuttle (1991) used the term 'flying foxes' to refer to all bats in the family Pteropodidae, but we restrict this term to the 70 species in the genera *Pteropus* and *Acerodon* (IUCN 2014), following the definition outlined by Kingston (2010), and use the term 'fruit bats' for the remainder.

Although bat–plant interactions were first recorded in 1772, it is now known that coevolution has shaped these complex interrelationships over millennia, producing bat-flower and bat-fruit syndromes (Marshall 1983; Fleming et al. 2009). This phytophagous diet results in valuable ecosystem services of pollination and seed dispersal (Kunz et al. 2011). Pteropodid bats are responsible for propagating at least 289 species of plants, of which 186 provide economically important resources and products including fruits, drinks, foods, ornamental plants, timbers, fibres, tannins, dyes, medicines, and animal fodder (Fujita and Tuttle 1991; Lobova et al. 2009). In addition, large populations of flying foxes are necessary to maintain the health of Old World tropical forests (Fujita and Tuttle 1991; Nyhagen et al. 2005; McConkey and Drake 2006). Such healthy functioning ecosystems ultimately provide humans with additional benefits such as climate regulation, nutrient cycling, water filtration, and erosion control (Kunz et al. 2011).

P.A. Racey
Centre for Ecology and Conservation, University of Exeter, Cornwall Campus, Penryn, Cornwall, UK
e-mail: p.a.racey@exeter.ac.uk

S.A. Aziz
University of Nottingham Malaysia Campus, Semenyih, Malaysia

Dependence on plant materials has also led to a long history of interactions between these bats and humans, particularly at shared food resources. Flying foxes with striped faces are depicted in aboriginal cave paintings in Kimberley, Australia, attributed to the Bradshaw people, between 17,000 and 60,000 years ago. Whether they brought such stripe-faced bats to the area or idolised the bat pollinator of a favoured tree, the baobab, in their drawings, is unknown. Genetic studies showed that baobab seeds were brought on their journey from Ethiopia to Australia, as an important provider of food, nutrients and building materials. This ancient rock art may be the first human recognition of the ecosystem services of pteropodids (Richards et al. 2012).

Despite the documented benefits of bats, negative attitudes towards them persist among the general public (Marshall 1983; Fujita and Tuttle 1991; Kunz et al. 2011). Pteropodid bats, in particular flying foxes, are frequently shot, persecuted and even legally culled as agricultural pests (Bumrungsri et al. 2009; Epstein et al. 2009). In Thailand, for example, farmers of durian (*Durio zibethinus*) set nets to catch the dawn bat (*Eonycteris spelaea*) visiting their trees when in flower and leave the bats to die, because they see that flowers fall after bat visits and conclude that bats have destroyed them (S. Bumrungsri, unpublished). In fact, the flowers fall naturally after the bats have pollinated them, but unhelpful misconceptions such as this exacerbate the conflict between bats and humans. In addition, pteropodid bats are hunted intensively for food and medicinal uses (including commercial trade), leading to severe declines throughout their range (Epstein et al. 2009; Mickleburgh et al. 2009; Harrison et al. 2011). Estimates based on current deforestation rates in Southeast Asia project that many pteropodids may become globally extinct by the end of this century (Lane et al. 2006), with flying foxes being of particular concern in Southeast Asia due to intense hunting pressure (Struebig et al. 2007; Meyer et al. in press). The Old World Fruit Bat Action Plan (Mickleburgh et al. 1992) helped stimulate research on pteropodids but is now out of date and is being revised. The conservation status of these bats has worsened since the plan was published, and a quarter of all species are now endangered (IUCN 2014).

Here, we review the current state of knowledge regarding human–pteropodid interactions by geographic region, legal policies affecting pteropodid bats, and methods of mitigating the damage they cause to fruit crops.

13.2 The Extent of Feeding by Bats on Fruit Crops and Its Implications

The fruit crops which bats have been reported to damage are listed in Table 13.1.

13.2.1 The Mediterranean

Madkour (1977), writing about the Egyptian fruit bat (*Rousettus aegyptiacus*) in Egypt, stated that it was ‘a highly dangerous fruit pest’ and that ‘its control is

Table 13.1 Bat species recorded foraging on fruit crops, by country. Fruit species, common name, bat species and countries where conflict/potential conflict has been recorded, and sources of information

Fruit species	Common name	Bat species	Countries	References
Agave sislana	Sisal (flowers only)	*Pteropus rufus*	Madagascar	Oleksy et al. (2015)
Anacardium occidentale	Cashew	*Pteropus hypomelanus*	Malaysia	S.A. Aziz, unpublished
		Pteropus lylei	Thailand	Weber et al. (2015)
		Pteropus voeltzkowi	Tanzania (Pemba)	Entwistle and Corp (1997)[b]
		Unknown	Guinea	Van Mele et al. (2009)
Ananas comosus	Pineapple	*Cynopterus sphinx*	Sri Lanka	Yapa et al. (1999)
		Pteropus dasymallus	Japan (Ryukyu)	Vincenot et al. (2015)
		Pteropus giganteus	Sri Lanka	Yapa et al. (1999)
		Rousettus leschenaultii		
Annona sp.	Custard Apple/Soursop	*Rousettus aegyptiacus*	Egypt	Madkour (1977)
Areca catechu	Areca Palm	*Cynopterus sphinx*	India	Chakravarthy and Girish (2003)
		Pteropus giganteus	Myanmar	Win and Mya (2015)
Arecaceae spp.	Palms	*Cynopterus sphinx*	Sri Lanka	Phillips (1980)
		Pteropus giganteus		
		Rousettus leschenaultii		
Artocarpus altilis	Breadfruit	*Pteropus voeltzkowi*	Tanzania (Pemba)	Entwistle and Corp (1997)[b]
Artocarpus heterophyllus	Jackfruit	*Pteropus giganteus*	Myanmar	Win and Mya (2015)
Artocarpus lacucha	Monkey Jack	*Pteropus lylei*	Thailand	Weber et al. (2015)[b]
Azadirachta indica	Neem	*Pteropus giganteus*	Myanmar	Win and Mya (2015)
Cajanus cajan	Pigeon Pea	*Pteropus giganteus*	Myanmar	Win and Mya (2015)

(continued)

Table 13.1 (continued)

Fruit species	Common name	Bat species	Countries	References
Carica papaya	Papaya	*Cynopterus sphinx*	Sri Lanka	Yapa et al. (1999)
		Pteropus giganteus	Myanmar	Win and Mya (2015)
		Pteropus lylei	Thailand	Weber et al. (2015)[b]
		Rousettus leschenaultii	Sri Lanka	Yapa et al. (1999)
		Pteropus voeltzkowi	Tanzania (Pemba)	Entwistle and Corp (1997)[b]
		Pteropus spp.	Australia	Ratcliffe (1931)
Ceiba pentandra	Kapok	*Pteropus giganteus*	Myanmar	Win and Mya (2015)
Ceratonia siliqua	Carob	*Rousettus aegyptiacus*	Cyprus	Del Vaglio et al. (2011)[b]
			Israel	Galil et al. (1976)
				Moran and Keider (1993)
				Izhaki et al. (1995)
			Lebanon	Lewis and Harrison (1962)[b]
Citrus paradisi	Grapefruit	*Rousettus aegyptiacus*	Israel	Galil et al. (1976)
				Moran and Keider (1993)
		Pteropus alecto	Australia	Ratcliffe (1931)[a]
		Pteropus conspicillatus		Waples (2002)
		Pteropus poliocephalus		
		Rousettus aegyptiacus	Israel	Galil et al. (1976)
				Moran and Keider (1993)
		Pteropus alecto	Australia	Ratcliffe (1931)[a]
		Pteropus conspicillatus		
Citrus reticulata	Mandarin	*Rousettus aegyptiacus*	Egypt	Madkour (1977)
		Pteropus poliocephalus	Australia	Rogers (2002)
Citrus sinensis	Orange	*Rousettus aegyptiacus*	Egypt	Madkour (1977)
Citrus sp.	Citrus	*Rousettus aegyptiacus*	Cyprus	Del Vaglio et al. (2011)[a]

(continued)

Table 13.1 (continued)

Fruit species	Common name	Bat species	Countries	References
Citrus spp.	Citrus fruits	*Rousettus aegyptiacus*	Turkey	Albayrak et al. (2008)
		Pteropus dasymallus	Japan (Ryukyu)	Vincenot et al. (2015)
		Pteropus spp.	Australia	Ratcliffe (1931)[a]
Coccinia grandis	Ivy Gourd	*Pteropus lylei*	Thailand	Weber et al. (2015)
Coffea spp.	Coffee	*Cynopterus* spp.	Indonesia	Huang et al. (2014)
		Pteropus poliocephalus	Australia	Waples (2002)
Dimocarpus longan	Longan	*Pteropus niger*	Mauritius	V. Dooblad, pers. comm.
Diospyros kaki	Persimmon	*Pteropus poliocephalus*	Australia	Rogers (2002)
		Rousettus aegyptiacus	Israel	Korine et al. (1999)
			Turkey	Albayrak et al. (2008)
Diospyros malabarica	Gaub tree	*Pteropus lylei*	Thailand	Weber et al. (2015)[b]
Diospyros rhodocalyx	Tako Naa	*Pteropus lylei*	Thailand	Weber et al. (2015)[b]
Durio zibethinus	Durian (flowers only)	*Eonycteris spelaea*	Thailand	S. Bumrungsri, unpublished
		Pteropus hypomelanus	Malaysia	S.A. Aziz, unpublished
Eriobotrya japonica	Loquat	*Pteropus dasymallus*	Japan (Ryukyu)	Vincenot et al. (2015)
		Rousettus aegyptiacus	Cyprus	Del Vaglio et al. (2011)
			Israel	Izhaki et al. (1995)
				Korine et al. (1999)
			Turkey	Albayrak et al. (2008)
Eucalyptus deglupta	Eucalyptus	*Pteropus vampyrus*	Malaysia	Fujita and Tuttle (1991)
Eugenia aromatica	Clove	*Pteropus voeltzkowi*	Tanzania (Pemba)	Entwistle and Corp (1997)
Euphoria malaiense	Mata Kucing	*Pteropus hypomelanus*	Malaysia	S.A. Aziz, unpublished
Ficus carica	Common Fig	*Rousettus aegyptiacus*	Israel	Korine et al. (1999)
			Turkey	Albayrak et al. (2008)

(continued)

Table 13.1 (continued)

Fruit species	Common name	Bat species	Countries	References
Ficus rumphii	Rumpf's Fig	*Pteropus giganteus*	Myanmar	Win and Mya (2015)
		Pteropus lylei	Thailand	Weber et al. (2015)[b]
Ficus sp.	Fig	*Rousettus aegyptiacus*	Cyprus	Del Vaglio et al. (2011)[a]
Ficus spp.	Figs	*Rousettus aegyptiacus*	Egypt	Madkour (1977)
			Lebanon	Lewis and Harrison (1962)
		Pteropus spp.	Australia	Ratcliffe (1931)
Fragaria sp.	Strawberry	*Rousettus aegyptiacus*	Egypt	Madkour (1977)
Lansium parasiticum	Langsat	*Cynopterus brachyotis*	Malaysia	Fujita (1988)
		Pteropus spp.		
		Pteropus hypomelanus		S.A. Aziz, unpublished
Litchi chinensis	Lychee/Litchi	*Pteropus niger*	Mauritius	V. Dooblad, pers. comm.
		Pteropus poliocephalus	Australia	Rogers (2002)
				Waples (2002)
		Pteropus rodricensis	Mauritius (Rodrigues)	Price (2013)
		Rousettus aegyptiacus	Israel	Galil et al. (1976)
				Moran and Keider (1993)
Malphigia emarginata	Acerola	*Pteropus lylei*	Thailand	Weber et al. (2015)[b]
Malus domestica	Apple	*Rousettus aegyptiacus*	Egypt	Madkour (1977)
			Israel	Galil et al. (1976)
				Moran and Keider (1993)
Malus sp.	Apple	*Rousettus aegyptiacus*	Turkey	Albayrak et al. (2008)

(continued)

Table 13.1 (continued)

Fruit species	Common name	Bat species	Countries	References
Mangifera indica	Mango	*Cynopterus sphinx*	Guinea	Van Mele et al. (2009)
			Sri Lanka	Phillips (1980)
				Yapa et al. (1999)
		Pteropus giganteus	Maldives	Dolbeer et al. (1988)
			Myanmar	Win and Mya (2015)
			Sri Lanka	Phillips (1980)
				Yapa et al. (1999)
		Pteropus hypomelanus	Malaysia	S.A. Aziz, unpublished
		Pteropus lylei	Thailand	Weber et al. (2015)
				S. Bumrungsri, unpublished
		Pteropus niger	Mauritius	Dooblad, unpublished
		Pteropus poliocephalus	Australia	Waples (2002)
		Pteropus spp.		Ratcliffe (1931)[a]
		Pteropus rodricensis	Mauritius (Rodrigues)	Price (2013)
		Pteropus voeltzkowi	Tanzania (Pemba)	Entwistle and Corp (1997)[b]
		Rousettus aegyptiacus	Egypt	Madkour (1977)
			Israel	Izhaki et al. (1995)
		Rousettus leschenaultii	Sri Lanka	Phillips (1980)
				Yapa et al. (1999)
		Unknown	India	Singaravelan (2002)
Manilkara zapota	Sapota/Sapodilla	*Pteropus giganteus*	India	Chakravarthy and Girish (2003)
		Pteropus lylei	Thailand	Weber et al. (2015)[b]
Morinda angustifolia	Ye-yo	*Pteropus giganteus*	Myanmar	Win and Mya (2015)
Morus alba	White Mulberry	*Rousettus aegyptiacus*	Egypt	Madkour (1977)

(continued)

Table 13.1 (continued)

Fruit species	Common name	Bat species	Countries	References
Morus nigra	Black Mulberry/ Blackberry	*Rousettus aegyptiacus*	Egypt	Madkour (1977)
Morus sp.	Mulberry	*Rousettus aegyptiacus*	Cyprus	Del Vaglio et al. (2011)
			Turkey	Albayrak et al. (2008)
Musa paradisiaca	Banana	*Cynopterus sphinx*	Sri Lanka	Yapa et al. (1999)
		Pteropus giganteus		
		Rousettus leschenaultii		
Musa sapientum	Banana	*Pteropus voeltzkowi*	Tanzania (Pemba)	Entwistle and Corp (1997)[b]
Musa sp.	Banana	*Pteropus lylei*	Thailand	Weber et al. (2015)[b]
Musa spp.	Bananas	*Pteropus dasymallus*	Japan (Ryukyu)	Vincenot et al. (2015)
		Pteropus lylei	Thailand	Weber et al. (2015)
				S. Bumrungsri, unpublished
		Pteropus poliocephalus	Australia	Rogers (2002)
		Pteropus spp.		Waples (2002)
				Ratcliffe (1931)[a]
		Rousettus aegyptiacus	Egypt	Madkour (1977)
			Israel	Izhaki et al. (1995)
Nephelium lappaceum	Rambutan	*Cynopterus brachyotis*	Malaysia	Fujita (1988)
		Cynopterus sphinx	Sri Lanka	Yapa et al. (1999)
		Pteropus giganteus		
		Pteropus hypomelanus	Malaysia	S.A. Aziz, unpublished
		Pteropus vampyrus		Fujita (1988)
				K.J. Olival, unpublished
		Rousettus leschenaultii	Sri Lanka	Yapa et al. (1999)
Oroxylum indicum	Midnight Horror	*Pteropus giganteus*	Myanmar	Win and Mya (2015)

(continued)

Table 13.1 (continued)

Fruit species	Common name	Bat species	Countries	References
Persea americana	Avocado	*Pteropus rufus*	Madagascar	Oleksy et al. (2015)
Phoenix dactylifera	Date Palm	*Rousettus aegyptiacus*	Israel	Izhaki et al. (1995)
				Korine et al. (1999)
			Turkey	Albayrak et al. (2008)
Phoenix sylvestris	Sugar/Silver Date Palm	*Cynopterus sphinx*	Bangladesh	Khan et al. (2011)
		Pteropus giganteus		
		Rousettus leschenaultii		
Phoenix spp.	Dates	*Rousettus aegyptiacus*	Egypt	Madkour (1977)
			Lebanon	Lewis and Harrison (1962)
Prunus dulcis	Almond	*Pteropus giganteus*	Maldives	Dolbeer et al. (1988)
Prunus persica	Nectarine	*Pteropus poliocephalus*	Australia	Comensoli (2002)
	Peach	*Rousettus aegyptiacus*	Egypt	Madkour (1977)
			Turkey	Albayrak et al. (2008)
Prunus sp.	Apricot	*Rousettus aegyptiacus*	Egypt	Madkour (1977)
			Turkey	Albayrak et al. (2008)
Prunus sp.	Cherry	*Rousettus aegyptiacus*	Turkey	Albayrak et al. (2008)
Prunus sp.	Plum	*Rousettus aegyptiacus*	Egypt	Madkour (1977)
			Turkey	Albayrak et al. (2008)
Prunus spp.	Stone fruit: a fruit with flesh or pulp enclosing a stone, such as a peach, nectarine, plum, or cherry.	*Pteropus poliocephalus*	Australia	Ratcliffe (1931)
				Rogers (2002)
				Waples (2002)
		Pteropus scapulatus		Ratcliffe (1931)

(continued)

Table 13.1 (continued)

Fruit species	Common name	Bat species	Countries	References
Psidium guajava	Guava	*Cynopterus sphinx*	India	Singaravelan (2002)
			Sri Lanka	Yapa et al. (1999)
		Rousettus leschenaultii	Sri Lanka	Yapa et al. (1999)
		Pteropus dasymallus	Japan (Ryukyu)	Vincenot et al. (2015)
		Pteropus giganteus	India	Chakravarthy and Girish (2003)
			Maldives	Dolbeer et al. (1988)
			Myanmar	Win and Mya (2015)
			Sri Lanka	Yapa et al. (1999)
		Pteropus lylei	Thailand	Weber et al. (2015)[b]
		Pteropus poliocephalus	Australia	Waples (2002)
		Pteropus voeltzkowi	Pemba (Tanzania)	Entwistle and Corp (1997)[b]
Punica granatum	Pomegranate	*Rousettus aegyptiacus*	Egypt	Madkour (1977)
			Israel	Galil et al. (1976)
				Moran and Keider (1993)
			Turkey	Albayrak et al. (2008)
		Pteropus giganteus	Myanmar	Win and Mya (2015)
Pyrus communis	Pear	*Rousettus aegyptiacus*	Israel	Galil et al. (1976)
				Moran and Keider (1993)
Pyrus sp.	Pear	*Rousettus aegyptiacus*	Cyprus	Del Vaglio et al. (2011)[a]
			Egypt	Madkour (1977)
Rosaceae	Pome fruit: a fleshy fruit, such as an apple or pear with several seed chambers and an outer fleshy part	*Pteropus poliocephalus*	Australia	Ratcliffe (1931)
				Waples (2002)
		Pteropus scapulatus		Waples (2002)

(continued)

Table 13.1 (continued)

Fruit species	Common name	Bat species	Countries	References
Saccharum officinarum	Sugar Cane	*Pteropus dasymallus*	Japan (Ryukyu)	Vincenot et al. (2015)
Sandoricum kotejape	Santol	*Pteropus lylei*	Thailand	Weber et al. (2015)[b]
				S. Bumrungsri, unpublished
Sesbania grandiflora	Agate/Hummingbird Tree (flowers only)	*Pteropus* spp.	Thailand	S. Bumrungsri, unpublished[b]
Syzygium aqueum	Water Apple	*Cynopterus brachyotis*	Malaysia	Fujita (1988)
		Pteropus spp.		
Syzygium cumini	Jambul/Jamun	*Pteropus giganteus*	Myanmar	Win and Mya (2015)
			Pakistan	Roberts (1997)
		Pteropus lylei	Thailand	Weber et al. (2015)[b]
		Pteropus voeltzkowi (flowers only)	Tanzania (Pemba)	Entwistle and Corp (1997)[b]
Syzygium jambos	Rose Apple	*Pteropus giganteus*	Myanmar	Win and Mya (2015)
		Pteropus lylei	Thailand	Weber et al. (2015)[b]
Syzygium javanicum	Java Apple	*Pteropus lylei*	Thailand	S. Bumrungsri, unpublished
Syzygium samarangense	Java Apple	*Pteropus lylei*	Thailand	Weber et al. (2015)[b]
Syzygium spp.	Jambu	*Pteropus hypomelanus*	Malaysia	S.A. Aziz, unpublished
Tamarindus indica	Tamarind	*Pteropus giganteus*	Myanmar	Win and Mya (2015)
		Pteropus lylei	Thailand	Weber et al. (2015)
		Pteropus rufus	Madagascar	Oleksy et al. (2015)

(continued)

Table 13.1 (continued)

Fruit species	Common name	Bat species	Countries	References
Theobroma cacao	Cocoa	Unknown	Papua New Guinea	Hicks (1967)
Vitis vinifera	Grape	*Cynopterus sphinx*	India	Verghese (1998)
				Srinivasulu and Srinivasulu (2002)
		Pteropus spp.	Australia	Ratcliffe (1931)[a]
		Rousettus aegyptiacus	Turkey	Albayrak et al. (2008)
Ziziphus jujuba	Jujube	*Pteropus giganteus*	Myanmar	Win and Mya (2015)

[a]Feeding impact was reported to be small enough to be negligible
[b]Although the fruit species may not have been grown commercially, there may still be conflict with bats eating garden fruit of local people

of great economic importance'. He reported that there were records of the species attacking cultivated fruit trees, and during the course of his study, they were recorded feeding on apple (*Malus domestica*), apricot (*Prunus* sp.), banana (*Musa* sp.), custard apple (*Annona* sp.), date (*Phoenix* sp.), mandarin (*Citrus reticulata*), mango (*Mangifera indica*), mulberry (black *Morus nigra* and white *M. alba*), orange (*Citrus sinensis*), peach (*Prunus persica*), pear (*Pyrus* sp.), plum (*Prunus* sp.), pomegranate (*Punica granatum*) and strawberry (*Fragaria* sp.). However, there was no mention of whether this was recorded from in situ observations or feeding trials in captivity.

Korine et al. (1999) showed that *R. aegyptiacus* in Israel ate mainly non-commercial fruits and also to a lesser extent leaves and pollen, challenging the assumption that this species is a major agricultural pest. Out of 14 identified plant species comprising its diet, only four are grown commercially in Israel: persimmon (*Diospyros kaki*), loquat (*Eriobotrya japonica*), fig (*Ficus carica*) and date (*Phoenix dactylifera*), with the largest component consisting of figs (*Ficus* spp.). The perception of *R. aegyptiacus* as a pest (Harrison 1964; Moran and Keidar 1993) led to conflict with farmers, resulting in extermination programs that reduced its population in the country (Korine et al. 1999; Hadjisterkotis 2006). As these control measures involved widespread fumigation of caves by the authorities, using the chlorinated hydrocarbon lindane, many populations of cave-dwelling insectivorous bats were also drastically reduced (Makin and Mendelssohn 1987). Other reports from Israel detail bats consuming commercial fruits such as apples, bananas, carobs (*Ceratonia siliqua*), dates, grapefruits (*Citrus paradisi*), lychees (litchi; *Litchi sinensis*), mandarins, pears and pomegranates (Galil et al. 1976; Moran and Keider 1993; Izhaki et al. 1995). However, the overall extent of actual damage to fruit crops is unknown and requires further detailed investigation.

In Lebanon, *R. aegyptiacus* was observed feeding on carobs, dates and figs. Its preference for dates and figs in particular, which are also cultivated for human consumption, caused it to be the only bat species considered to be of economic importance there. Farmers used shotguns to kill bats, and even though fruit such as dates could be protected by cloth bags or nets before ripening, this was seldom done. Some farmers were even known to starve populations of bats in caves by placing nets over the roost entrance (Lewis and Harrison 1962).

Qumsiyeh (1980) initially stated that the population of *R. aegyptiacus* in Jordan was increasing. However, more than a decade later, Qumsiyeh et al. (1992) concluded that the species was already under threat due to destruction of its roost sites, even though the issue of fruit crop damage had yet to be investigated in the country.

Albayrak et al. (2008) reported that in the Mediterranean region of Turkey, *R. aegyptiacus* fed on both wild and commercially grown fruits. Their study identified 15 different species, of which 13 were marketed: plum, loquat, apple (*Malus* sp.), fig (*F. carica*), pomegranate, grape (*Vitis vinifera*), persimmon, date, mulberry (*Morus* sp.), cherry (*Prunus* sp.), peach, apricot and citrus (*Citrus* sp.). Wild fruits were consumed only during the winter. They concluded that fruit bats could have a considerable impact on fruit crops, with farmers in Hatay Province claiming that bats consumed 10–15 % of their loquat harvest. Fruit bats were thus considered to be serious pests and were subsequently killed regularly. According to Spitzenberger (1979), in the past, this involved fumigating roost caves or walling up their entrances. Harrison and Bates (1991) reported that farmers caged their commercial fruit trees in order to protect them from *R. aegyptiacus*. However, a more recent study in 2012 found that local people in Turkey did not consider bats to be as much of a problem as birds (E. Coraman, pers. comm.).

In Cyprus, only anecdotal reports were previously available for the diet of *R. aegyptiacus*. A preliminary assessment of its diet was carried out by Del Vaglio et al. (2011) from droppings, in order to determine the bats' real impact on crops. The diet consisted mostly of fruits, several species of which were the same as those reported by Korine et al. (1999) for Israel and Albayrak et al. (2008) for Turkey. The species is an opportunistic forager, with non-native plants forming an important component of its diet, yet Del Vaglio et al. (2011) concluded that its damage to economically important plants in Cyprus is negligible. Their study found that the bat fed mainly on wild fruits and escaped ornamental plants and that only five out of the 11 plant species it consumed—citrus, fig, loquat, mulberry and plum—were grown as commercial fruit crops in Cyprus.

13.2.2 Africa and the Indian Ocean

In Guinea, mango and cashew (*Anacardium occidentale*) farmers listed fruit bats among the mammals (together with monkeys, squirrels and other rodents) that cause damage to their harvest. The bats target ripening mangoes and cause significant damage, identified by a visible seed protruding from the eaten lower part of the fruit. However, only 4 % of farmers identified fruit bats as pests, compared to 92 % who identified fruit flies as the most significant pest. Overall, according to the farmers, five species of insects, followed by squirrels, caused greater damage than bats. The majority of farmers did not carry out any pest management (Van Mele et al. 2009).

Entwistle and Corp (1997) examined the diet of *Pteropus voeltzkowi*), which is endemic to the island of Pemba, off the coast of Tanzania. They found that it consisted of a high proportion of cultivated fruit grown on 'shamba' plots, in particular mangoes which formed a key component of the diet during the duration of the study. Breadfruit (*Artocarpus altilis*) was also consumed. In addition, interview surveys with villagers and students yielded additional cultivated fruit species among the food plants of this bat species (Table 13.1).

In the Indian Ocean, Dolbeer et al. (1988) described the Indian flying fox (*Pteropus giganteus*) as a major cause of damage to almonds (*Prunus dulcis*), guavas (*Psidium guajava*) and mangoes in the Maldives, although losses were not quantified. In Mauritius, the Agricultural Research and Extension Unit of the Food and Agricultural Research Council estimated an overall average of 10 % of orchard lychee fruit was damaged by the Mauritian flying fox (*Pteropus niger*). Ten trees were studied in each of three orchards and damage to individual fruit averaged 2, 7, and 17 %. In contrast, a smaller study of four longan trees (*Dimocarpus longan*) recorded damage to all fruit panicles. Mangoes were also damaged at a rate of 10–30 % (V. Dooblad, pers. comm.). In contrast, a recent study by Ramlugun (2013) in a lychee orchard found that high winds and introduced birds, but not bats, resulted in fruit losses of 30 and 9.5 %, respectively. On Rodrigues (an autonomous island of the Republic of Mauritius), losses to backyard mango and lychee production were estimated at about 36 %, much of which was perceived to be to the Rodrigues flying fox (*Pteropus rodricensis*) (Price 2013).

Oleksy et al. (2015) carried out GPS tracking of the Madagascan flying fox (*Pteropus rufus*) to determine its foraging movements and habitat selection in south-eastern Madagascar. The study revealed that this species has a strong preference for feeding on the nectar and pollen of sisal (*Agave sislana*) in overgrown plantations. These bats also feed on the fruits of avocado (*Persea americana*), lychee, mango and tamarind (*Tamarindus indica*). However, it was not mentioned whether this causes any conflict issues.

13.2.3 Indian Subcontinent

Bats are causing increasing economic loss in the Indian grape industry due to a decline in wild fruits and flowers, coupled with the increase of grape-growing areas. Verghese (1998) first drew attention to the damage caused by the greater short-nosed fruit bat (*Cynopterus sphinx*) to grapes in Bangalore where the vines are trained to grow on overhead trellises. They entered the vineyard through canopy gaps in the bower, not from the sides, and consumed only the juice of the fruit, while the pulp, seed and skin were discarded. Signs of bat damage included these remnants littered at the base of the grape vine, as well as grape bunches with bare stalks. The damage was greater in parts of the vineyard adjacent to open spaces, suggesting that growing non-commercial trees around the entire vineyard might deter bat foraging. Damage was lower in vines situated closer to an adjacent mango orchard.

Similarly in Andhra Pradesh State, Srinivasulu and Srinivasulu (2002) showed that the magnitude of damage caused by *C. sphinx* to grapes varied with the distance of the vines from the periphery of the vineyard. Damage was extensive (90 %) at the periphery, but none was recorded in the centre, where the higher density of the vines made approach flights difficult. In contrast, in Tamil Nadu State, *C. sphinx* was not known to damage grape crops; rather, eight bird species were the primary crop pests. There were also no records of this bat species damaging sapota (*Manilkara zapota*) in that state; however, considerable damage was reported to mango and guava crops (Singaravelan 2002).

In the State of Karnataka, Chakravarthy and Girish (2003) recorded losses of 18 % of areca nuts (*Areca catechu*) caused by a population of 3500–4000 *P.*

giganteus and 2–28 *C. sphinx*. Bats also damaged 13–22 % of sapota fruits, although higher levels of damage were inflicted by birds. Up to 28 % of guava fruits were also damaged by bats. *C. sphinx* has also been recorded as damaging mango and guava in Tamil Nadu State but did not damage sapota (Singaravelan 2002).

In Bangladesh, *P. giganteus* feeds on date palm sap (*Phoenix sylvestris*), widely harvested in the country as a beverage in the winter months (December–March) (Luby et al. 2006). Infrared camera traps have recorded *P. giganteus* and other pteropodid species (*Cynopterus* spp. and *Rousettus leschenaultii*) drinking from clay pots used to collect the sap at night (Khan et al. 2011), although the magnitude of the loss has not been investigated. This bat–plant sap interaction is a route for the transmission of zoonotic disease (see 13.3).

In Pakistan, *P. giganteus* is also labelled as vermin due to a perception that it raids fruit crops in orchards (Mahmood-Ul-Hassan et al. 2010). Apart from areca nuts, sapota and guava, it is also blamed for heavy economic losses of crops of mango and jamun (*Syzygium cumini*) (Roberts 1997; Chakravarthy and Girish 2003). However, a dietary study conducted by Mahmood-Ul-Hassan et al. (2010) in Lahore found that *P. giganteus* feeds primarily on wild figs rather than commercial crops. The study concluded that the perception of *P. giganteus* as a pest is a misconception, and its economic value as a pollinator is far greater for the fruit industry.

In Sri Lanka, according to Yapa et al. (1999), fruit farmers claimed that pteropodid bats (*C. sphinx, R. leschenaultii* and *P. giganteus*) damage fruit trees in plantations and home gardens. Fruits that were specifically identified as suffering heavy damage by *C. sphinx* were mango and rambutan (*Nephelium lappaceum),* with mango thought to be particularly vulnerable in monoculture plantations. Bananas (*Musa paradisiaca*), papayas (paw-paws; *Carica papaya*) and even pineapples (*Ananas comosus*) were apparently also targeted. Their study concluded that *C. sphinx* was ‘capable of causing heavy damage’ and could thus potentially be a ‘major fruit pest’. Earlier, casual records collected by Phillips (1980) also reported guava, mango, soursop and several palm species being consumed by pteropodids in Sri Lanka.

13.2.4 Southeast Asia

Although there are 95 species of pteropodids, including 31 flying foxes (IUCN 2014) in Southeast Asia, there is little published information on fruit crop damage caused by bats. Perception of damage is however widespread and has implications for conservation. For example, it may explain the Malaysian government’s reluctance to provide full protection for the nation’s flying foxes (large flying fox *Pteropus vampyrus* and island flying fox *P. hypomelanus*) by halting licensing which has led to unsustainable hunting (Epstein et al. 2009). So far, little attempt has been made to investigate the issue of conflict or quantify the economic loss.

Fujita (1988) reported that pteropodid bats, specifically flying foxes and the lesser dog-faced fruit bat (*Cynopterus brachyotis*), are considered pests by orchard owners interviewed in Malaysia and Indonesia and are therefore shot when they visit these orchards. Fruit growers considered bats to be particularly problematic for rambutan, langsat (*Lansium parasiticum*) and water apple (*Syzygium aqueum*), which are all important market fruits. The owner of one of the largest langsat orchards in Peninsular Malaysia revealed that if measures were not taken to protect his fruit crop several days prior to harvest, 20 % of the crop would be lost to bats. However, he also considered that simple protective measures could be undertaken such as shining bright lamps, lighting fires under the trees, or shooting to scare the bats away, in which case the damage would be negligible. This same orchard owner also appeared to display an understanding of the importance of pteropodids as seed dispersers—he considered that almost all of the langsat trees in his village resulted from seeds dropped by bats. His langsat fruit was typically harvested for sale in the local market.

Fujita and Tuttle (1991) conducted some preliminary investigations into bat pest control in Malaysia and Indonesia, interviewing six plantation/orchard owners and six professional hunters. Owners employed bounty hunters to eradicate bats during flowering and fruiting seasons who could earn up to USD 3 per bat, shooting as many as 100 in one night from a single plantation. A group of three to five hunters regularly patrolled an orchard, using bright lights to locate the bats. According to one hunter, up to seven bats could be hit with a single shot (Fujita 1988). These bats were killed in disproportionately large numbers despite plantation/orchard owners reporting that more significant damage was caused by other animals such as giant squirrels (*Ratufa* spp.), pig-tailed macaques (*Macaca nemestrina*), binturong (*Arctictis binturong*), Timor deer (*Cervus timorensis)* and bearded pigs (*Sus barbatus).* A professional hunter employed by a pulp and paper plantation in Sabah (Malaysian Borneo) reported that in 1983 alone he purchased 2000 rounds of ammunition for sport shooting of flying foxes that were attracted to the eucalyptus flowers. He also reported that bats were killed in the thousands annually during 1983 and 1984, but that their numbers had been drastically reduced by 1985 (Fujita and Tuttle 1991). Using population models based on roost census data and numbers of hunting permits issued in Peninsular Malaysia, Epstein et al. (2009) found that rates of hunting were unsustainable and would lead to local extinction of *P. vampyrus*.

Gumal et al. (1998) acknowledged that in Sarawak (Malaysian Borneo), an increase in commercial fruit crops, coupled with the loss of habitats such as beach forests, mangroves and peat swamps, has resulted in flying foxes foraging in orchards and farms. This encroachment has led to them being labelled as pests, and it is reasonable to assume that a similar situation occurs in Sabah and Peninsular Malaysia.

On Tioman Island (Peninsular Malaysia), *P. hypomelanus* was reported by local people to feed on a wide range of cultivated fruit trees in their villages, where the bats also roost. This happens despite the fact that wild food resources are still widely available in nearby largely intact forest and has resulted in conflict with villagers despite the fruit being cultivated for personal consumption rather than a source of livelihood. Seeds of mango, cashew and rambutan have been found beneath day roosts, and people also frequently reported that the bats feed on langsat, mata kucing (*Euphoria malaiense*) and various types of *Syzygium* fruits. Durian (*D. zibethinus*) pollen has been found in flying fox faeces, and camera-trapping in durian trees has confirmed that *P. hypomelanus* feeds on durian flowers. Preliminary observations of feeding behaviour suggest that only the nectar is sought, leaving the flowers intact on the branch, and as such, these bats probably perform an important pollination service. Yet some villagers believe that the bats damage or remove the flowers, thereby affecting fruit set (S.A. Aziz, unpublished).

Farmers in Peninsular Malaysia use large, treble fishing hooks and monofilament line set in fruit orchard trees to capture flying foxes. This inhumane method is often lethal, and its efficacy in protecting crops has not been tested. One male *P. vampyrus* used in a satellite telemetry study was captured in a rambutan orchard in Johor, Malaysia, using this method and released after sustaining minor injuries (Epstein et al. 2009; K.J. Olival, unpublished). Gumal et al. (1998) concluded that there is a need to investigate non-lethal methods for protecting orchards and fruit gardens against bats.

In 2005, a newspaper article highlighted the overall decline of *Pteropus* in Malaysia, attributing it to logging and hunting (Teoh 2005). Interestingly, it cautioned that this would negatively affect cash crops such as durian (*D. zibethinus*), petai (*Parkia speciosa*), rambutan and langsat, highlighting the flying fox's role as a pollinator for these trees. However, some confusion may have arisen between flying foxes (*Pteropus* spp., *Acerodon* spp.) and smaller fruit bats such as *E. spelaea*, since Fujita (1988) and Fujita and Tuttle (1991) use the term to refer to all bats of the family Pteropodidae.

In southeast Thailand, fruit farmers stated that Lyle's flying fox (*Pteropus lylei*) damages less than 10 % of harvestable mangoes, and far fewer bananas, water apples (*Syzygium javanicum*) and santol (*Sandoricum koetjape*). Damage is reduced when fruit trees are mixed compared to monoculture systems. Farmers with mango monocultures treated flying foxes as pests, but most farmers with mixed fruit orchards did not regard them as such (S. Bumrungsri, unpublished). According to these farmers, these flying foxes feed mainly on several fig species, especially *F. religiosa* which is regarded as a sacred tree in Buddhist Thailand. These figs are common in the landscape, particularly in temples. Flying foxes also feed on flowers of the agate or hummingbird tree (*Sesbania grandiflora*), commonly found across South and Southeast Asia and in villages in Thailand, where the flowers and young pods are consumed by people. Farmers also mentioned that flying foxes forage in groups of 10–15 individuals and keep returning to the same feeding area on consecutive nights (S. Bumrungsri, unpublished).

More recently, Weber et al. (2015) conducted GPS tracking of *P. lylei* in central Thailand. Tracked bats were found to forage mostly in farmland, plantations and gardens. All 34 recorded food plant species were noted to also be useful to local people, though not necessarily as fruits for sale or consumption. Thirty-one species were identified as fruit resources, and an unspecified 42 % of these were cash crops (therefore, the only species listed in Table 13.1 are ones that the authors know are cultivated by people in Southeast Asia for either fruits or flowers). Only mango, cashew, banana and tamarind were mentioned specifically as having high economic value or as being cultivated crops. Mangoes were also the most frequently eaten fruit, followed by bananas and tamarind. Such competition for resources between bats and humans was acknowledged as a potential source of conflict. Local farmers confirmed that flying foxes are hunted as an orchard pest in this area.

In Indonesia, Huang et al. (2014) have studied *Cynopterus* feeding in coffee (*Coffea* spp.) plantations in Sumatra. Most growers (93 % of 16 interviewed) reported that bats visit their plantations. Coffee berries are taken to feeding perches and the beans discarded after the pericarp is eaten. This study is now investigating the potential of marketing bat-discarded coffee beans as a premium wildlife product.

A recent dietary study on *P. giganteus* in the Mandalay region of central Myanmar (Win and Mya 2015) also interviewed local villagers to determine the extent of conflict between flying foxes and fruit tree owners. The bats were found to feed on 24 fruit species, 13 of which were also eaten by people. Of these, only three—guava, mango and tamarind—were of commercial importance. *Morinda angustifolia* and *Azadirachta excelsa* are used for medicinal purposes, while *Ceiba pentandra* is still used for stuffing pillows (a practice that is dying out in other Southeast Asian countries). Despite this, local people view the bats positively, and no conflict was reported. The authors of the study concede that a superabundance of mangoes is one reason why people are still willing to tolerate a certain amount of fruit loss.

13.2.5 Australia and Papua New Guinea

Australia has the oldest and most comprehensive records documenting the issue of flying foxes and fruit crop damage. Ratcliffe (1931) provided detailed reports on 'depredations' by flying foxes on both commercial orchards and garden trees in New South Wales (NSW) and Queensland. Although flying foxes were known to feed on bananas, citrus fruits, mangoes and grapes, losses were not significant enough for the bats to be considered 'economic pests'. Slight losses were reported for papayas, some losses for pome fruit (fruits of the family Rosaceae having several seed chambers and an outer fleshy part, such as an apple or pear) and stone fruit (fruits of the genus *Prunus* with flesh or pulp enclosing a stone, such as a peach, nectarine, plum, or cherry), and heavy losses for figs (*Ficus* spp.). For some fruits such as bananas, mangoes and papayas, the regular practice of picking them before they ripen was often sufficient to avoid heavy losses to flying foxes.

Despite these known losses, the extent of flying fox damage to commercial fruit has seldom been quantified in Australia, even in more recent reports. Eby (1995) refers to 'substantial financial loss to growers' and lists a relatively large number of commercial exotic fruits on which *Pteropus* spp. feed, although damage was of particular concern to growers of stone fruits and banana. Stacey (1990) refers to heavy stone fruit losses during the prolonged drought conditions of 1986, with bats eating immature green fruit. Waples (2002) reported that most requests for licences to shoot flying foxes in NSW came from growers of stone fruits and lychee, but that damage was also reported to guava, mango, banana, pome fruits and coffee. Signs of damage include broken branches, clawed fruit and fruit remnants under trees (Comensoli 2002). Ullio (2002) reported that from 1995 to 2000, fruit growers in NSW suffered an annual gross market value loss of AUD 10.4 million due to consumption by flying foxes. When taking into account the resulting loss to affiliated industries such as packaging, employment, transport and marketing, the overall financial loss was estimated to be around AUD 26 million annually. Prior to 1998, on the north coast of NSW, only stone fruit, lychee and persimmon were consistently eaten by flying foxes in significant quantities (Rogers 2002). Losses increased dramatically from 1998, particularly in orchards without netting. The stone fruit industry suffered a loss of AUD 4–6 million (not including preharvest costs, which usually exceed AUD 20,000). Sixty per cent of orchards without netting suffered losses of 50–100 %, around AUD 45,000 per grower. The mandarin industry reports losing at least 40 % of its annual crop, while in 2001, an individual lychee grower reported a loss of more than AUD 500,000 in the unnetted section of her orchard (Rogers 2002). Comensoli (2002) measured the damage caused by flying foxes to his nectarine (*P. persica*) orchard, estimating that 20 % of ripe fruit was damaged over a period of 19 days, reducing the annual profit from his entire crop by 16 %. In Queensland, orchardists also suffered particularly high crop damage in the summer of 1998. The estimated total loss for that season was approximately AUD 10 million (Teagle 2002), with some growers having lost up to 90 % of their crop (Dewhurst 1998). It should be noted, however, that the above estimates of orchard losses have not been verified and originate primarily from growers.

Due to this perceived high economic loss, the Australian fruit industry considers species of *Pteropus* to be its main vertebrate crop pests (Ullio 2002). Yet it has been acknowledged, even among fruit growers, that increasing feeding by flying fox on commercial crops is due to the loss of natural food resources as the rainforests, heathland and *Melaleuca* swamps of Australia's eastern seaboard have been increasingly cleared for urban development (e.g. Bicknell 2002; Biel 2002; Gough 2002; Rogers 2002). As a result, Biel (2002) and Rogers (2002) proposed that fruit growers should be financially compensated for economic loss and that the wider community should also bear the cost of mitigation and biodiversity conservation.

Details of bat–grower conflict in Papua New Guinea are scant, but a report by Hicks (1967) stated that bats and birds together caused the loss of 8.7 % of cocoa (*Theobroma cacao*) pods from an orchard from 1962 to 1965.

13.2.6 The Pacific

Luskin (2010) studied the foraging behaviour of the Pacific flying fox (*Pteropus tonganus*) in a landscape mosaic in Fiji. He found that mean foraging density was four times higher in farmland compared to native dry forest, with high foraging competition taking place almost completely in farmland alone. Severe deforestation has resulted in a large bat population that has shifted away from feeding on flowers in forests to feeding more on fruits in farms. However, no observations were made on which type of fruits suffered predation. Farmland resources, with their higher fecundity, now appear to be the staple of *P. tonganus*' diet. Daily, crepuscular mass migration from forests to farmlands has reduced feeding density in forests, thus reducing the aggressive feeding interactions needed to catalyse effective seed dispersal necessary for forest regeneration (McConkey and Drake 2006). The loss of this ecological role could be disastrous for Pacific tropical dry forest, which is a critically endangered habitat (Myers et al. 2000). Also, while the abundance of farmlands has buffered the flying fox population from the effects of extensive deforestation, further research is needed to determine what damage or effects this may have on fruit crops, as well as flying fox nutrition.

In Japan, previous studies on the Ryukyu flying fox (*Pteropus dasymallus*) focused on diet and did not report any conflict with humans (e.g. Funakoshi et al. 1993; Nakamoto et al. 2007, 2009; Lee et al. 2009). In the Ryukyu Archipelago, Nakamoto et al. (2007) reported that Orii's flying fox (*P. dasymallus inopinatus*) on Iriomote-jima Island is a generalist forager, with almost 50 % of its diet consisting of cultivated or naturalised plants. The majority (67.9 %) of its diet throughout the year is composed of fruits. Although its main food resource is *Ficus microcarpa*, the subspecies appeared to adopt a varied diet through intense use of abundant planted trees, as a response to unstable food conditions in an urban environment. Some of these plants are from gardens, parklands and walkways, but others are agricultural plants from plantations. On Iriomotejima Island, Lee et al. (2009) found that the Yaeyama subspecies (*P.d. yayeyamae*) had a comparably less diverse diet and was more abundant in forest compared to cultivated areas, with figs again dominating its diet. Yet bats were still observed in larger groups frequenting villages containing fruit trees. Neither study mentioned predation of economically important fruit as being an issue of concern, and conflict with humans had not been previously identified by anyone as a threat for this particular species.

However, a more recent study by Vincenot et al. (2015) has revealed for the first time that farmers do indeed kill *P.d. yayeyamae*, illegally, because it feeds on crops. Face-to-face interviews and direct observations have shown that flying foxes are frequently killed either through netting, poison or physical beatings, to stop them from feeding in plantations of banana, citrus, guava, loquat, pineapple and sugar cane (*Saccharum officinarum*). This persecution has clearly contributed to continuing declines that were noticeable to interview respondents, and which contradicts the IUCN's decision in 2008 to downgrade the Red List status of *P. dasymallus* from endangered (EN) to near threatened (NT).

13.3 Food-Borne Zoonotic Disease Risk from Pteropodid Bats

An additional concern to crop damage caused by pteropodid bats is the potential for zoonotic disease transmission via fruit contaminated with bat excreta (i.e. saliva, urine, faeces). Old World fruit bats are natural reservoirs to a number of such diseases, including several emerging viruses that have limited or no pathogenicity in their bat hosts but high fatality rates in people. These include Ebola viruses (Leroy et al. 2005), Marburg virus (Towner et al. 2009), Nipah virus (Rahman et al. 2013), Hendra virus (Halpin et al. 2000), and lyssaviruses in Australia (Mackenzie et al. 2003) and Thailand (Lumlertdacha et al. 2005). While the transmission pathway for each virus is not always known, there is compelling evidence, in a small number of cases, that points to a food-borne route, most notably multiple spillover events of Nipah virus from *Pteropus giganteus* to people in Bangladesh (see below). Filoviruses (Ebola and Marburg) are also of great consequence to human health, as evident from the large west Africa outbreak of Zaire Ebola virus that began in early 2014. Much remains unknown about the natural hosts and ecology of filoviruses in bats (Olival and Hayman 2014), but Ebola virus may be transmitted from bats to humans through faeces (Swanepoel et al. 1996), but most likely through direct contact with blood (i.e. preparing hunted bats) (Leroy et al. 2009) or via contact with dead-end host carcasses (e.g. gorillas) (Leroy et al. 2004). Recent experimental studies have shown that Marburg virus can be excreted in bat saliva, answering important questions about its potential zoonotic spread via the oral route (Amman et al. 2014a). It has been postulated that bats and gorillas may share Ebola virus through contact at shared fruit resources, but this has not been verified and additional research is needed to better understand the ecological connections between bats and other mammal hosts in the transmission of these diseases (Groseth et al. 2007; Olival and Hayman 2014).

Henipaviruses (Hendra and Nipah viruses) are recently emerged paramyxoviruses that originate primarily from *Pteropus* spp. as their natural reservoir. Transmission of Hendra virus in Australia and Nipah virus in Malaysia from bats to intermediate or amplifying domestic animal hosts (horses and pigs, respectively) likely occurred though consumption of partially chewed fruit contaminated with bat saliva or ingestion of bat urine under bat foraging sites (Field et al. 2001; Chua et al. 2002). Henipaviruses have been shown experimentally to remain viable on the surface of mango and in other tropical fruit juices (lychee and papaya) from 2 h to 2 days depending on temperature and pH (Fogarty et al. 2008). Similarly, Chua et al. (2002) successfully isolated Nipah virus from a fruit in the wild that was partially eaten by *P. hypomelanus*. Thus, the risk of oral transmission of henipaviruses to humans via consumption of partially chewed fruit exists, although it is likely to be low. However, direct transmission of Nipah virus from bats to people occurs in Bangladesh nearly every year through the consumption of date palm sap, presumably contaminated with urine, saliva or faeces from infected *P. giganteus* (Luby et al. 2006; Rahman et al. 2012). Preventive measures are being used to block bats' access to date palm sap collection pots and reduce the risk of Nipah virus transmission (Nahar et al. 2010). Other mitigation measures that reduce the overall damage of crops by pteropodid bats will further mitigate any risk, however small, of zoonotic disease transmission via this route. Culling bat populations as a form of disease control is rarely effective and often has the opposite effect of increasing transmission and risk. This was recently demonstrated during an attempt to eradicate a population of *R. aegyptiacus* as a form of Marburg virus control, where prevalence of the virus significantly increased after the cull (Amman et al. 2014b). Additional approaches to reducing bat–human contact at potential disease interfaces should be developed, and disease mitigation should be carried out in a way that reduces risk without impacting bat populations.

13.4 Legislative Approach to Reducing Pteropodid Damage to Crops

13.4.1 Australia

Australia has 13 species of pteropodids, seven of which are flying foxes. Some are listed under the federal government's Environmental Protection and Biodiversity Conservation (EPBC) Act 1999 and several state wildlife protection laws.

Flying foxes became protected species in the State of New South Wales (NSW) in 1986 under the National Parks and Wildlife Act 1974. Since then, farmers and fruit growers have been required to obtain licences from the NSW National Parks and Wildlife Service (NPWS) in order to shoot flying foxes to protect their crops (Waples 2002). Licences are granted only when a NPWS representative has visited the orchard to inspect and assess whether the damage is severe enough to warrant culling (Comensoli 2002). Each licence allows a maximum of 50 flying foxes to be shot, and no more than two licences can be granted per landowner per season. Licence holders are required to submit reports on actual numbers of flying foxes killed (Waples 2002). However, in practice, this licensing system is far from perfect, as compliance monitoring and enforcement are neither practical nor feasible, and therefore, records can be unreliable (McLachlan 2002; Waples 2002; Thiriet 2010).

In 2001, the NSW government changed the listing of the grey-headed flying fox (*Pteropus poliocephalus*) from Protected to Vulnerable under the NSW Threatened Species Conservation Act 1995 (Eby and Lunney 2002). This resulted in negative reactions from the commercial fruit industry (e.g. Biel 2002; Bicknell 2002; Bower 2002; Comensoli 2002; Thiriet 2010), as it meant that even if shooting of the threatened species were still permitted for crop protection, it would be subject to a tighter licensing system, resulting in socio-economic repercussions, particularly for small growers (Bower 2002; Comensoli 2002; Ullio 2002; Waples 2002). The state government subsequently continued to allow shooting of the species for crop protection (Thiriet 2010). However, at the time of writing, the NSW government has now banned shooting of flying foxes as an orchard control method (G. Richards, unpublished).

In July 2011, in order to eliminate the need to issue shooting licences and to mitigate flying fox damage to crops, the NSW government introduced a AUD 5 million scheme to subsidise the cost of installing netting for commercial orchardists in the Sydney Basin and Central Coast regions, where impacts occur every year. Once a netting subsidy has been received, the orchardist is no longer eligible for a shooting licence for the netted area of the property. Subsidies are intended to meet half the cost of installing netting and are capped at AUD 20,000 per hectare. Orchardists are responsible for all ongoing maintenance and replacement costs. Not only are flying foxes (and parrots) excluded from the fruit crops, but hail damage is also reduced. This often means that the cost of netting is recovered in the season following its installation. Because netting in now subsidised, from July 2015, licences to shoot flying foxes as a crop protection measure will only be issued where damage to orchards is the result of special circumstances (e.g. the orchard is on terrain too steep to net). The issuing of such licences will eventually be phased out.

P. poliocephalus and the spectacled flying fox (*P. conspicillatus*) were listed as Vulnerable under the EPBC Act 1999, in 2001 and 2002, respectively. One year after its federal listing, *P. poliocephalus* was also listed as Vulnerable in the State of Victoria. Neither the little red flying fox (*P. scapulatus*) nor the black flying fox (*P. alecto*) is listed as threatened under any Australian legislation, and the State of Queensland has yet to list any flying fox species as threatened (Thiriet 2010).

In 2002, the State of Queensland banned the use of electric shocks for crop protection, though this was on grounds of animal cruelty rather than conservation. Prior to this, orchardists could receive a damage mitigation permit for electrocuting flying foxes on overhead grids. The use of such electric grids to kill a keystone species was later found to be in breach of the EPBC Act 1999 (which had led to the listing of *P. conspicillatus*), although this was construed as a negative impact on the world heritage values of a nearby Wet Tropics World Heritage Area rather than a biodiversity conservation issue. Shooting of *P. poliocephalus* and *P. conspicillatus* was still allowed for the purpose of crop protection, with an annual limit of up to 1.5 % of the lowest agreed national population estimate for the species. A quota of 30 animals per orchardist per month was implemented. However, in 2008, the state banned all shooting of flying foxes, again due to concerns over animal cruelty (Thiriet 2010).

In 2012, Queensland reintroduced shooting of flying foxes causing damage to commercial fruit, including *P. poliocephalus* and *P. conspicillatus*. However, shooting quotas for these two species are less than for the little red and black flying foxes, *P. scapulatus* and *P. alecto*. Fruit growers require permits to shoot, which are granted only if they can prove that non-lethal methods of control have failed. Such permits allow the use of shotguns and heavy shot on stationary but not on flying bats. Clear X-ray evidence in Australia (Richards et al. 2012; Divljan et al. 2009) and palpation of lead shot in live and dead bats in Madagascar, the Seychelles (P.A. Racey, unpublished) and Mauritius (V. Tatayah, pers. comm.) reveal that the use of shotguns results in wounding and is inhumane, because death is not instantaneous. Also, Thiriet (2010) pointed out that some bats that are shot may be lactating, and their young left behind in the colony will eventually starve to death. Shotguns were however banned in the Seychelles in the 1970s. The toxic effects of lead shot have been well documented for birds (Mateo 2009), and it is likely to have similar effects in bats.

In both Queensland and NSW, there has been very little (if any) monitoring by relevant authorities of numbers of bats shot in orchards. The only known scientific study was conducted near Sydney in 2007 (Divljan et al. 2009). Over a 140-day period, a total of 164 dead or injured flying foxes were collected and data were compiled from 136 carcasses. Eighty or so bats per week exceeded the number allowed by permits. The sex ratio was strongly skewed towards females (1:1.73), of which 54 (65 %) were lactating at the time. Thirteen of these were shot while carrying their dependent young, while 41 pups would have been left behind in the roost to die. Hence, the total estimate of flying foxes that died due to shooting in the orchard over the two-week period was 205. Collected bats suffered from various injuries, and at least 30 % (44 % including the pups left in the camp) were alive and unattended more than 8.5 h after shooting (Richards et al. 2012). This is in contravention of the definition of 'humane killing' and the Prevention of Cruelty to Animals Act 1979.

13.4.2 Cyprus

In Cyprus, *R. aegyptiacus* was officially declared a pest by the Department of Agriculture in the early 1900s. Destruction campaigns and programs to eradicate the species began in the late 1920s. As in Israel, fumigation of caves also depleted populations of insectivorous bats. In addition, bats were shot, with the government offering free cartridges and payment to participating hunters as well as payment for dead bats. These control campaigns finally ended in 1990 after there were very few bats left (Hadjisterkotis 2006). The species became legally protected after Cyprus law No. 24 of 1988 ratified the Convention on the Conservation of European Wildlife and Habitats. This was made possible when Cyprus became a candidate for European Union membership. As the Convention previously only protected insectivorous bats, in 1993 Cyprus added *R. aegyptiacus* to the EU list of protected bats in Annexes II and IV of the council directive 92/42/EEC on the conservation of natural habitats and of wild flora and fauna (Hadjisterkotis 2006).

13.4.3 Israel

In Israel, two laws protect animals outside nature reserves or national parks. 'The law for the protection of wild animals' concerns hunting and is considered to be stronger legislation than `The law for the protection of natural values'. The former aims mainly to regulate hunting (what, how and where?) and lists all protected mammals, including some non-local species. The second law aims to protect aniChironax melanocephalus are listed asmals, plants, fossils and speleothems.

R. aegyptiacus is protected by neither law and is considered a pest. Although it is legal to kill fruit bats, cruel killing is forbidden by the 'Animal welfare act'.

Fruit bat colonies are protected in national parks and nature reserves, but if the bats' foraging sites are outside protected areas, then they may be legally killed.

Israeli conservationists have had protracted negotiations with the Ministry of Agriculture regarding Israel joining the EUROBATS agreement. Although that is likely to happen in the near future, a derogation will be sought to maintain the pest status of *R. aegyptiacus*, at least for the immediate future (A. Streit, pers. comm.).

13.4.4 Japan

Pteropus dasymallus is one of the only two pteropodid species found in Japan, and as such, it is protected at both national and prefectural levels. Both the Daito (*P. dasymallus daitoensis*) and Erabu (*P.d. dasymallus*) subspecies are listed as critically endangered (CR) on the IUCN Red List, but *P.d. inopinatus* and *P.d. yayeyamae* are not even listed, and the latter two subspecies are only considered as NT in prefectural assessments (Vincenot et al. 2015).

Despite a severe lack of data on the population and conservation status of this species, the IUCN identified its threats only as habitat destruction, electrocution on power cables and occasional accidental entanglement in nets (Heaney et al. 2008). Yet Vincenot et al. (2015) have uncovered evidence of conflict between *P.d. yayeyamae* and humans on all fruit production islands in the Yaeyama archipelago that they visited. The only island without conflict, Kuroshima, focuses on cattle production instead. This conflict has led to severe declines in flying fox populations throughout the archipelago. It is likely that a similar situation occurs on Okinawa Island, where *P.d. inopinatus* occurs, as there is higher urbanisation and more agricultural fields there, and crop destruction by flying foxes was reported there in a 2013 Japanese-language news article. In the light of this new evidence, the conservation status of this species needs to be carefully reassessed, with population monitoring and conservation programmes being clearly necessary requirements.

13.4.5 Malaysia

In Malaysia, wildlife is governed under three distinct legislative systems according to the three main geopolitical regions: Peninsular Malaysia, Sabah and Sarawak. Protection of the country's two species of flying fox (*P. hypomelanus* and *P. vampyrus*) varies within and across each of the main geopolitical regions. In Peninsular Malaysia, the Department of Wildlife and National Parks (also known as PERHILITAN) regulates wildlife policy and hunting. For nearly 40 years, wildlife conservation policy was determined by the Protection of WildLife Act of 1972, which listed both flying fox species under Schedule II, or Protected Wild (Game) Animals. Hunting of both species is allowed with a permit, and there are no seasons or limits to the numbers of permits that may be issued by a state in Peninsular Malaysia. In 1990, under the Protection of Wild Life Amendment Order, a bag limit was set that allowed 50 bats to be shot under a single permit and the time of hunting was limited to 0700–1200 h and 0500–0700 h each day. Each licence costs MYR 25 (USD 8) (Teoh 2005). No other bats are listed. As in Australia, such a licensed hunting system is difficult to monitor and regulate.

A study by Epstein et al. (2009) evaluated the abundance and roost distribution of *P. vampyrus* in Peninsular Malaysia, finding that the number of hunting licences issued had doubled since 1996, and concluded that current levels were unsustainable and likely to cause local extinction within 6–81 years. Further, these estimates of hunting pressure from licence data were likely to be an underestimate as they did not include illegal hunting, and there was also a provision in the 1972 Act that allowed killing, shooting or removal of an unspecified number of any wild animal that is 'causing damage or there is reason to believe that it is about to cause serious damage to crops, vegetables, fruit, growing timber…if reasonable efforts to frighten away the wild animal have failed'.

The study by Epstein et al. (2009) was highlighted in the media (Burns 2009; Kandasamy 2009) and prompted a response from PERHILITAN that they would consider implementing a hunting ban as part of the then current review of the act. However, when the act was repealed in 2010 by the new Wildlife Conservation Act 2010 (Act 716), flying foxes had still not been moved from the 'Protected' list to the 'Totally Protected' list, meaning that licensed hunting is still permitted, and the provision for protecting crops (Part VI, sec 54) is also still permitted in the new legislation.

In February 2012, following lobbying and recommendations from conservation research group Rimba, the Terengganu state government implemented a state-wide moratorium on hunting of flying foxes (Rimba 2012). Prior to this, the State of Johor had banned hunting of all wildlife when its Sultan issued a royal decree to this effect (Charles and Benjamin 2010). However, to date, no nation-wide hunting ban has been announced by PERHILITAN at the federal level, and other bat species remain unprotected.

As in Peninsular Malaysia, in Sabah, the two flying fox species are currently listed under Schedule 3, sections 2, 25(2) as 'Protected species of animals for which hunting licence is required' under the Wildlife Conservation Enactment of 1997. However, there is no clear provision for shooting animals to protect crops without a licence, and no other bat species are legally protected.

In Sarawak, research by Gumal et al. (1998) successfully resulted in all bat species in the state being listed as 'Protected' in May 1998, under Part II of the Sarawak Wildlife Ordinance 1998 (with the exception of *Cheiromeles torquatus* that is listed in Part I, as 'Totally Protected'). The Sarawak Forest Department does not allow legal bat hunting and has implemented some of the strictest policies in Malaysia to regulate guns and ammunition and decrease the extent of wildlife poaching.

13.4.6 Mauritius and Madagascar

In 2006, the government of Mauritius proposed changes in legislation to allow culling of the only pteropodid on the island, the endemic *P. niger,* as a result of losses of marketable fruit, principally lychees. Any effect of this change was confounded, however, by existing legislation that prohibited the discharge of firearms after dark or with the aid of lights, and in the event, in one year, only six bats were officially killed. The proposal to cull an endemic species (albeit only in fruit orchards) on an island where two species (small Mauritian flying fox *P. subniger* and Rodrigues flying fox *P. rodricensis*) had already become extinct as a result of cyclones, habitat loss and overhunting was a major factor in the upgrading of the Red List status of *P. niger* in 2008 from Vulnerable to Endangered. Pressure on the government from growers of commercial fruit, particularly lychees, but also longans and mangoes, resulted in surveys of bat numbers by the National Parks and Conservation Service (NPCS). In November/December 2010, 49–56,000 bats were counted at 47 roost sites. This was broadly consistent with the results of an earlier count by Robyn (2007) of 12–16,000 bats at 24 of 57 known roosts. As a result, the Red List status of *P. niger* was downgraded from Endangered to Vulnerable in 2013.

Despite assurances from NPCS that there were no plans to cull bats, the National Terrestrial Diversity and National Parks bill was being considered by parliament in May 2012 and has been the subject of a public consultation. It allows for the culling of species that have attained high numbers and pest status. Irrespective of this, the Mauritian Wildlife Foundation, the main conservation NGO on the island, reports that up to 2000 bats are shot annually by hunters and fruit growers.

In Madagascar, bats are ‘animaux gibiers,’ i.e. game animals, and can be hunted legally although there is a close season coinciding with pregnancy and lactation. Officially, licences are required by hunters, but in practice, this is not usually observed, as enforcement is challenging to implement. Some hunters observe the close season (P.A. Racey, unpublished).

13.4.7 South Asia

In India, all pteropodid species with the exception of the Critically Endangered *Latidens salimalii* are categorised as vermin and included as such in Schedule V of the Indian Wildlife (Protection) Act 1972 and Amended Acts. However, only three of the thirteen species—*P. giganteus*, *R. leschenaultii* and *C. sphinx*—feed extensively on commercial fruit, and the remaining ten species forage mainly in forest where they play an important role in pollination and seed dispersal, and there is no evidence that they visit commercial orchards. The Indian government has ignored successive attempts by conservationists to have forest bats delisted (Singaravelan et al. 2009).

In Bangladesh, the newly revised Wildlife Preservation and Security Act 2012 protects all species of bats. Hunting is prohibited without government permission and a licence, and offenders can face imprisonment and/or a fine (Act translated from Bengali by A. Islam, pers. comm.).

In Pakistan, *P. giganteus* is listed in the fourth schedule of the Punjab Wildlife (Protection, Preservation, Conservation and Management) Act 1974, which specifically includes animals that have no legal protection and can be hunted.

In Sri Lanka, the Fauna and Flora Protection Ordinance 1937 (amended 2009) provides protection for all bat species in the country, and hunting is strictly prohibited. Bat roosts such as caves are not currently protected, but the Department of Wildlife Conservation is currently in discussion to protect such sites as refuges by law (W. Yapa, pers. comm.).

13.4.8 Thailand

In Thailand, all species of *Pteropus* (*P. hypomelanus*, *P. intermedius*, *P. lylei*, *P. vampyrus*), nectarivorous bats (*E. spelaea*, *Macroglossus minimus* and *M. sobrinus*) and *Chironax melanocephalus* are listed as 'protected animals' under the Wildlife Protection and Reservation Act 1992. Another 13 bat species found in Thailand, including all *Cynopterus* and *Rousettus,* are not protected. However, all animals are protected within designated areas, which include national parks, wildlife sanctuaries, and religious establishments (temples, mosques).

Out of a population of 38,000 bats forming 16 colonies of *P. lylei* in central Thailand, 90 % (13 colonies) are found in temples (Boonkird and Wanghongsa 2004), and thus their roosting colonies are well protected. In contrast, most known colonies of *P. vampyrus* and *P. hypomelanus* are found outside protected areas and therefore suffer from hunting and roost disturbance, except for colonies on oceanic islands. Generally, due to cheaper prices and greater abundance of fruit crops in Thailand, along with smaller population sizes of flying foxes, *Pteropus* spp. are less likely to be regarded as crop pests. However, smaller fruit bats such as *Cynopterus* spp. and *Rousettus* spp. are common and are still regarded as pests. Hunting and selling of flying foxes is widely known to be illegal. Attempts should be made to protect roosting sites outside designated protected areas.

13.5 Non-lethal Methods of Mitigation

13.5.1 Netting and Associated Tree Management

The only demonstrably effective method of preventing loss of fruit to bats and birds is full exclusion netting. The country in which this has been deployed to the greatest extent and most successfully is Australia where some large fruit orchards are enclosed in nets supported by cables, frames or posts (Minifie and Willis 1990; Campbell and Greer 1994; Gough 1992; Stacey 1992; Hall and Willis 1992). The netting has a mesh size of about 48 mm, is erected well above the height of the trees and is also attached to the ground at the edges (Fig. 13.1). Such orchards extend to 90 ha in area (G.C. Richards, unpublished), and the nets protect the crops from bats, other mammals (including possums), birds and hail. Estimates of the cost per hectare of netting vary widely from AUD 6,000 (from a conservationist) to AUD 60,000 (from a fruit grower) (Don't Shoot Bats 2013). Several

Fig. 13.1 Nets supported by frames in Australia over apples and stone fruit (*Photograph* Greg Richards)

state governments now subsidise the erection of netting for orchardists, and most of those interviewed considered that the structures pay for themselves at the first harvest.

But while netting may work for some, issues remain with its implementation in Australia, resulting in poor uptake among some growers (Gough 2002; Ullio 2002). Exclusion netting is costly and thus may not be economically feasible, particularly for smallholdings when flying fox damage is inconsistent and unpredictable from year to year (Slack 1990; Tidemann et al. 1997; Bower 2002; Gough 2002; McLachlan 2002; Rogers 2002; Ullio 2002). Many growers are reluctant to take on this added financial burden and are unlikely to net their crops (Ullio 2002). Bicknell (2002) pointed out that the financial cost of maintaining netting is too great, and it brings an added risk as a fire hazard. Comensoli (2002) further stated that the annual cost of leasing finance for netting outweighed the actual cost of flying fox damage to his crops. He and Ullio (2002) also pointed out that netting creates a microclimate within the orchard that results in poor fruit yield and inferior fruit quality—an experience echoed by other growers, with at least one case of netting in a lychee orchard resulting in a fungal disease (Bicknell 2002).

However, many orchards can be netted as long as they qualify for a state government subsidy, which is often 50 % of the cost. For example, the netted orchard shown in Fig. 13.1 was able to recover its costs at least by the second season, and with 18 ha (40 acres) now protected not just from bats, but also birds and hailstones, it produces top-quality fruit at high prices and with an environmentally-friendly tag. It used to be thought that because flying fox damage was sporadic and netting might have detrimental effects on crop growth, permanently netting an orchard was not viable for some orchards (Comensoli 2002; Rogers 2002; Ullio 2002). However, now that the industry has settled on a mesh size of 48 mm, so insect pollinators can freely access fruit trees, detrimental effects are no longer reported.

Netting is still not suitable for the banana industry, where plantations are often located on steep slopes that are impossible to net (Bower 2002; Rogers 2002; McLachlan 2002). In some cases, farmers who cannot afford to net have been forced out of business by heavy losses to flying foxes and other frugivores (Rogers 2002). This industry should investigate specific options and provide research funding for trials of innovative ideas, such as solar-powered ripening bags. To ensure that bananas ripen evenly, in Australia each bunch is covered with a plastic bag so that the ethylene by-product is evenly distributed. Once flying foxes smell the ripening aroma, they home in on bunches that they know to be palatable. A solar-powered bag with a low-voltage electric barrier would humanely deter flying foxes, and they would eventually learn not to tear bags open to feed.

A decade after the earlier reports, it is becoming increasingly accepted that netting of orchards is the only method of ensuring their full protection. In Australia, consumer expectations of high-quality fruit are acknowledged by major supermarket chains, so all fruit must be unmarked. Netting that excludes flying foxes, parrots and hail is now considered an industry standard by large producers, so it is only small family orchards that usually do not install nets. Because netting entire orchards is expensive (Reilly and Slack 1990), it can only be undertaken when large-scale farming of cash crops makes it cost-effective and justifies the investment.

Commercial crops are also protected by netting in Israel (Korine et al. 1999) and Thailand (Fig. 13.2) (S. Bumrungsri, unpublished), where fixed nets that cover the trees are most effective, although some growers also use mist nets despite their untested efficacy. They are usually lethal to bats, which are not always removed from the nets (C. Korine, pers. comm.). In Thailand, some fruit farmers also erect mist nets in their orchards, leaving tens or hundreds of nectarivorous bats (e.g. *E. spelaea*) to die (S. Bumrungsri, unpublished).

In some countries, such as Mauritius, entire lychee trees are netted and the government encourages this by subsidising 75 % of the cost of 10 nets per grower. However, individual growers may have 200–300 trees, and the method is applicable only to relatively low-growing orchard trees and not to the much older and larger 'backyard' trees which produce a significant proportion of the national lychee crop.

In Thailand, entire longan trees or groups of trees are covered by either plastic sheets or netting and the former also accelerate ripening (Fig. 13.3) (S. Bumrungsri, unpublished). Farmers actively prune these trees after harvesting, in order to maintain their low stature so that the trees are easily covered with netting during the next fruiting season. A cheaper method of mitigation used in

Fig. 13.2 Netted longan orchard in Thailand (*Photograph* Sara Bumrungsri)

Fig. 13.3 Longan trees covered by nets in Thailand (*Photograph* Sara Bumrungsri)

Mauritius and Cambodia involves the use of panicle nets consisting of small net bags commonly used to package vegetables, which are of an appropriate size to fit over an immature lychee panicle and can be reused from year to year (Fig. 13.4).

In Mauritius, trees are also pruned to make it easier to cover them with nets, but this involves some loss of productivity until they grow new fruit-bearing branches. When new orchards are planted, dwarf varieties are recommended (as bats prefer feeding on taller trees) and trees are now more widely spaced and are kept pruned to a height that facilitates the deployment of panicle or whole-tree nets. This lowers fruit production for the first three fruiting seasons, but production increases after that. When nets are supported by frames or poles, trees must also be pruned so that they do not grow into the net. This active pruning technique can also be applied to rambutan and lychee. Some nets are removed at the end of the fruiting season and replaced at the beginning of the next. However, most growers leave the nets in place for several seasons (V. Tatayah, pers. comm.).

Much fruit is picked before it has fully ripened and becomes attractive for flying foxes. Mango farmers in north Queensland stated in interviews (G.C. Richards, unpublished) that they harvested their crop just at the onset of ripening, well before it became soft enough to be highly attractive to bats. As well as being too hard to bruise during transport to markets as a high-quality product, there was negligible loss to the growers. By the time flying fox raids began to increase, the remaining crop was high on the trees and difficult to harvest and was left for wildlife.

Fig. 13.4 Net bags enclosing lychee panicles in Mauritius (*Photograph* Vikash Tatayah, Mauritian Wildlife Foundation)

In Bangladesh, a simple cost-effective method is used to prevent bats (including *P. giganteus* and smaller fruit bats, *Cynopterus* and *Rousettus* spp.) and other pests (e.g. birds) from accessing date palm sap during collection. This involves the use of bamboo skirts that cover the top of the collection pot and the shaved part of the palm tree (Nahar et al. 2010) and has been enthusiastically adopted by palm sap collectors (gachhis) (Fig. 13.5). Without it, date palm sap contaminated with bat faeces and urine is of lower quality and value, and, importantly, the risk of Nipah virus transmission is also reduced by using bamboo skirts over collection pots (Nahar et al. 2010).

13.5.2 Decoy Crops

A decoy crop produces less valuable or non-commercial fruit which is more attractive to bats than the crop to be harvested. Before selecting a plant species as a decoy crop, the feeding habits and preferences of the bats should be established. There have been many relevant studies. For example, in the Indian Ocean, Racey and Nicoll (1984) listed the food plants of the Seychelles flying fox (*Pteropus seychellensis*), while Nyhagen et al. (2005) did so for *P. niger* on Mauritius. Bollen and van Elsacker (2002) and Long and Racey (2007) studied the diet of the Madagascan flying fox (*Pteropus rufus)* in Madagascar and showed that bats feeding within 100 km of one another shared few food plants. The diet of another

Fig. 13.5 Bamboo skirt to prevent bats from accessing palm sap in Bangladesh. *Photograph* JH Epstein/EcoHealth Alliance

Malagasy endemic *Eidolon dupreanum* was described by Picot et al. (2007). Stier and Mildenstein (2005) studied the dietary habits of *P. vampyrus* and *Acerodon jubatus* in the Philippines. Parry-Jones and Augee (2001) and Williams et al. (2006) investigated food resources and the effect of food availability on the occupation of urban areas by *P. poliocephalus* in Australia, where Richards (1990) also described the diet of *P. conspicillatus*. Bumrungsri et al. (2007) reported on the diet of two species of *Cynopterus* in Thailand, and Hodgkison et al. (2003, 2004) studied nine fruit bat species in Peninsular Malaysia.

However, only a few studies have sought rigorously to establish feeding preferences: Korine et al. (1998) for *R. aegyptiacus*, Yapa et al. (1999) for *C. sphinx*, Nelson et al. (2005) for the Pacific flying fox (*Pteropus tonganus*), and Andrianaivoarivelo et al. (2012) for the Madagascan rousette (*Rousettus madagascariensis*). Bats were briefly taken into captivity to assess their fruit preferences. The first study compared fruits preferred by bats with those eaten by birds, and found that while bats ate 100 % of the introduced fruit species they were offered, only 14 % of native fruit species offered to them were actually consumed—suggesting that *R. aegyptiacus* only became common in the eastern Mediterranean with the introduction of new cultivated plants (Korine et al. 1998). The second study found that out of three different types of fruit offered, guava was the most preferred, followed by sea almond (*Terminalia catappa*) and mango, with fully ripe fruits being preferred over semi-ripe fruits. It concluded that this provided some support for farmers' claims that bats caused damage to their crops (Yapa et al. 1999). The third study tested fruit choice in relation to nutritional requirements. Flying foxes were found to prefer low-calcium, high-sugar fruits such as papayas, but although sugar was the primary basis for fruit selection, pregnant and lactating females required greater amounts of calcium. However, the flying foxes in this study consistently avoided figs, which are excellent sources of calcium (Nelson et al. 2005). In the last study, bats were found to prefer native and commercially unimportant figs (*F. polita*), rose apple (*Syzygium jambos*) and mountain apple (*S. malaccense*) to the cash crops of lychees and persimmon (Andrianaivoarivelo et al. 2012). These important results provide a perspective on the dietary preferences of pteropodids and should be repeated with other species.

There is convincing evidence that planting *Muntingia calabura*, which is very attractive to *C. sphinx*, can lessen the impact of these bats on commercial fruit. Singaravelan and Marimuthu (2006) showed that *C. sphinx* visited *Muntingia* more than any other wild or commercial fruit and recommended that it is planted around fruit orchards. Verghese's (1998) study on grapes in India found that less bat damage occurred closer to a mango orchard and suggested that presence of these trees deters the bats from feeding on grapes. However, it may be that the fruit bats simply show a stronger preference for feeding on mangoes (e.g. Ayensu 1974; Mahmood-Ul-Hassan et al. 2010). It would thus be useful to compare the results of Verghese's (1998) study with a similar study in the adjacent mango orchard.

Law et al. (2002) recommended planting trees which fruit in spring in Australia to relieve the flying fox damage suffered by orchardists at that time of year. Although the effectiveness of these decoy crops is yet untested, there is evidence that *P. poliocephalus* will cease consumption of commercial fruit if alternative native foods become available (Eby 1990). However, in order to be effective, the selection of plant species must be based on their high productivity and attractiveness to bats as well as producing fruit at the same time as the commercial crop. Local site conditions must also match the specific needs of the plant in order to ensure optimum growth. Most importantly, these food trees should not be planted in the immediate vicinity of orchards but located away from commercial fruit-growing areas in order to attract the bats away from orchards (Law et al. 2002). The authors also suggest that planting *Syzygium* around commercial fruit trees may reduce the feeding of bats on the latter. As these planting schemes still need to be tested for effectiveness, Law et al. (2002) suggested monitoring results through regular mapping and identifying dietary changes in the bats.

13.5.3 Deterrents/Aversion Agents

There is some evidence that strong smells such as rotting fish may deter bats from approaching ripening fruit and trials to investigate this are currently under way in Thailand (S. Bumrungsri, unpublished). Bicknell (2002) suggested that smoke could be used as an aversion agent, since it is known among Australian orchardists that it is disliked by flying foxes. On Tioman Island in Malaysia, anecdotal information from local communities relates that people build fires under roost trees in order to smoke out flying foxes, although the efficacy of this method is only temporary as it does not deter them from returning (S.A. Aziz, unpublished).

Over the last 30 years in Australia, deterrents used by fruit growers have included flashing and rotating lights, electronic distress sounds, gas-operated bird scare guns, electric shocks, and smell and taste deterrents. However, most of these are used in isolation and their effectiveness has not been systematically assessed, with results being mixed and most evidence anecdotal (Ullio 2002). A project to trial smell and taste deterrents was carried out by the Queensland Parks and Wildlife Service (QPWS) and the Queensland Flying-fox Consultative Committee (QFCC) in 2000. This involved three different commercial products for repelling animals, but none provided complete protection, and the results were ultimately inconclusive due to the small scale of the testing. A plant secondary compound was also tested, with more promising results, and further trials were planned (Teagle 2002), although the outcome is unknown. Bicknell (2002) considered that shooting to frighten, rather than shooting to kill, could also be an effective method.

A noise deterrent was developed in Australia in the late 1990s that reduced orchard crop losses caused by *P. conspicillatus* and *P. poliocephalus*, which was an adaptation of a bird deterrent known as the 'Phoenix Wailer' (Phoenix Agritech Canada Ltd). In essence, it was a sound system with four stereo channels. Each channel had a speaker in the centre of the crop and another at a corner. Sounds were randomly played on each channel, with the sound appearing to come from the centre of the stereo pair. Pellet scars on wing membranes of a large proportion of flying foxes captured in Australia indicate that they had been targeted using shotguns, and therefore, the deterrent system also reproduced a shooting scenario. Sounds of humans (motorbikes, dogs barking) came from one channel, then randomly from another channel came sounds of shotguns, and then from another the screams of a wounded flying fox. Trials in several fruit-growing areas were successful, but the results were not accepted by the industry, which instead called for government trials although these were not implemented. The fruit-growing industry itself did not support independent trials, so this novel approach to mitigation has not been adopted (G.C. Richards, unpublished).

An ultrasonic repeller (Ultrason-X; Bird-X Inc, Chicago) was ineffective at preventing damage to longan panicles by *P. niger* in Mauritius. A similar device (Sonixgate, Tikod Trade Ltd. Tel-Aviv www.batman.co.il) is used in Israel in lychee orchards where it is popular with users, although its effectiveness has not been independently established (C. Korine, pers. comm.). Bomford and O'Brien (1990) reviewed the effectiveness of several sonic deterrent devices in animal damage control, although most tests did not involve bats. They pointed out that the efficacy of ultrasonic deterrents for bats was controversial, and there was no evidence that such devices had practical value. They concluded that broadcasting distress or alarm calls was probably the most promising noise deterrent method.

13.5.4 Combined Methods of Mitigation

In India, partially covering vulnerable sections of the canopy of fruit trees, illumination and scaring with noises saved 4.5, 6 and 11 % of the fruits of sapota, respectively (Chakravarthy and Girish 2003). However, the effectiveness of these methods was temporary, and for longer term protection, three methods were recommended: planting non-commercial species of figs attractive to the bats; dividing orchards into smaller plots so that trees may be covered with sprigs of foliage, thatch or nylon net; and covering bunches of grapes with dry sprigs of foliage, netting, use of firecrackers or electric fencing. Also in India, Verghese (1998) found that grapes in vineyards could be protected from bat damage if nylon netting is erected around the trellis-grown bower up to bower height, combined with using twigs and briers to cover canopy gaps in the bower.

A combination of lights, noises and plastic flags is widely used in Mauritius (V. Tatayah, pers. comm.).

13.5.5 Biological Control Agent—Weaver Ants Oecophylla longinoda

During an interview survey in Guinea, west Africa, almost half of farmers reported that bats fear the weaver ant *Oecophylla longinoda*, and more than half appreciate that orchards with abundant weaver ants experience less fruit damage due to the ants' protective role, possibly because bats are repelled by the smell of the ants. However, 40 % of farmers also felt that the weaver ant itself is also a form of pest, as it rolls up leaves and is a nuisance during harvest (Van Mele et al. 2009). Yet this species is considered by entomologists and ecologists to be a potential biological control agent (Van Mele 2008). Lokkers (1990) has also suggested the potential of using weaver ants to reduce fruit damage by bats in Australia. However, this proposed method would require a native weaver ant species and requires further research and field trials.

13.6 Recommendations and Issues for Future Consideration

According to opinions from both conservationists and some orchardists, shooting is not an effective means of mitigating flying fox damage to fruit crops, particularly when animal numbers are high (Hall and Richards 1987a, b; McLachlan 2002; Ullio 2002; Thiriet 2010). Most fruit damage occurs when the bats' native food supply is drastically lowered due to droughts or nectar washout (i.e. when heavy rain washes the nectar off the flowers), and killing does not prevent damage in orchards under high flying fox pressure. Because of the bats' mobility, shooting will not stop bats from foraging, as a continuous stream of animals will move into the site from further afield (Martin and McIlwee 2002). Shooting flying foxes has thus become an unnecessary persecution. Hundreds of thousands have been killed in Australia's east coast, even though for 80 years it has been known to be ineffective.

The most effective method to date for reducing crop loss not just to bats, but also to birds and hailstones, is full exclusion netting. If growers' estimates of orchard losses to animals have some credence, then their exclusion is the most appropriate management option. The implementation of such an effective mitigation measure should thus be explored in other countries that experience problems of bat damage to fruit crops. This requires full cooperation between the fruit industry, relevant managing authorities from the government, scientists and conservationists.

In addition to this, below we summarise some main issues that warrant more detailed attention and action in order to resolve the conflict between pteropodid bats and fruit growers.

13.6.1 Better Knowledge of Pteropodid Diet and Foraging Preferences

Studies from Cyprus, Israel, Madagascar and Pakistan have shown that fruit bats prefer native wild fruits compared to commercial fruit crops (Korine et al. 1999; Mahmood-Ul-Hassan et al. 2010; Del Vaglio et al. 2011; Andrianaivoarivelo et al. 2012). These findings can be used as a compelling argument in mitigating conflict with orchardists (Del Vaglio et al. 2011) and to prevent deliberate killing of bats for crop protection. However, examples from Australia, Fiji, India, Japan and Malaysia show that depletion of food resources due to habitat loss can drive flying foxes to feed in fruit orchards (Gumal et al. 1998; Verghese 1998; Tidemann 1999; Nakamoto et al. 2007; Luskin 2010). Any mitigation efforts therefore must ensure that wild food sources continue to be maintained in the long term, and where these have been depleted, tree-planting must be carried out to replenish the loss. This is especially important because Nelson et al. (2000) showed that there are negative nutritional consequences for flying foxes which change their diet from native to agricultural fruits.

However, Biel (2002) reported that even when much native blossom was available nearby, *P. poliocephalus* still preferred to feed on fruit crops, and Bower (2002) stated that it appears to prefer lychees 'over all naturally available foods'. Rogers (2002) reported that flying foxes on the North Coast of NSW were initially only a problem for the stone fruit, lychee and persimmon industries, but that once more of these orchards started adopting netting, the bats began moving on to bananas, coffee, mandarins and mangoes. This was exacerbated in 1999 and 2000 by a decrease in native food, with McLachlan (2002) reporting a similar issue for the 2000/2001 season. Yet there is some evidence that flying foxes will cease to feed on commercial fruit crops if their native food sources again become available in the wild (Eby 1990; Andrianaivoarivelo et al. 2012). The solution may thus consist of a careful selection of preferred tree species planted in appropriate locations away from fruit orchards (Law et al. 2002). Such methods are as yet unproven and require further trials and research.

13.6.2 Funding Interventions and Research to Mitigate the Pteropodid–Grower Conflict

Bicknell (2002) advocated an urgent need for funding research into non-lethal aversion agents to mitigate flying fox damage. Such research funds have not yet been made available in Australia. Individual government authorities have been reluctant to take ownership of the problem, while industry organisations do not view it as an industry-wide issue, as the majority of fruit growers in some parts of the country are not affected. Apart from research into specific mitigation methods, there is also a need to study netted orchards in order to determine the effects of netting—not just on the environment created under the net and on the ripening fruit, but also the implications of excluding other potential pollinators such as birds and insects. Ultimately, however, aversion agents and cheaper methods would be a preferred method for many orchardists in Australia compared to netting or even culling (Ullio 2002), and funds should be provided to develop and test such methods (Bicknell 2002; Bower 2002; Thiriet 2010). Thiriet (2010) also suggested that the dearth of such funding is caused by negative community attitudes and political considerations, which may influence the inaccurate conservation status of some species of flying foxes, such as Least Concern as in Queensland. The unpopularity of these species must thus be overcome in order to attract appropriate research funding.

Australian orchardists maintain that it is the government's responsibility, not theirs, to fund the research (Bicknell 2002) because they believe it was not orchards which caused the habitat loss driving this problem (Tidemann 1999). Bicknell (2002) pointed out that orchards provide flying foxes with food when wild resources are scarce. He also highlighted how government departments are responsible for releasing large areas of flying fox habitat for logging and agriculture and that therefore, the financial burden of protecting crops from flying foxes should be borne by the authorities. Biel (2002) echoed this concept of 'public good conservation', stating that flying fox conservation benefits the wider community, and cited examples of other projects that utilised the community benefit approach. He pointed out that the loss of native flying fox food in Australia was caused by ‘the people who lived in the cities’, since most fruit orchards were established on land that had already been previously cleared for cattle grazing. Fruit growers could thus be said to have revegetated the land, and therefore, it is unfair that they alone should bear the cost of protecting flying foxes. Martin and McIlwee (2002) agree with this and recommend that the cost of netting should be subsidised by federal funding.

13.6.3 Education of Growers and the Public

Apart from research into damage mitigation methods, there is also an urgent need to educate fruit growers and increase their awareness on the ecological and economic importance of pteropodid bats. Huang et al. (2014) found that in Sumatra, less than 20 % of coffee growers interviewed ($n = 16$) were aware that bats were pollinators, and none were aware that bats dispersed seeds. As demonstrated by Bumrungsri et al. (2008, 2009), pteropodids are major pollinators for some commercially important fruit trees, even when the bats may not be the most frequent visitors. In northern Queensland, *P. conspicillatus* plays a vital role in rainforest reproduction through pollination, and although these bats may affect the fruit industry, there has never been a full study of the economic value of flying foxes as pollinators of eucalypt hardwood forests in Australia. The majority of timber trees harvested on the east coast produce nectar and pollen only at night (P. Birt, unpublished). This means that as flying fox populations decrease, fewer timber trees will join the logging cycle.

More importantly, pteropodid bats play a major role in seed dispersal. The further that a fruit is dispersed from the parent tree, the greater the chance of any resulting seedlings surviving to maturity. Fighting over feeding territories (the squabbling heard at night) leads to the loser departing with a fruit in its mouth, and consuming it at a distance. This has been termed the ‘raiders versus residents’ seed dispersal model (Richards 1990), tested by McConkey and Drake (2006) in Tonga, where they showed that once numbers of flying foxes declined below a threshold where there was no conflict over feeding territories, then seed dispersal away from the parent tree ceased.

Therefore, while economic estimates of fruit loss are an important first step in quantifying the problem of crop damage, a cost–benefit analysis that takes into account the positive economic impact of ecosystem services provided by the same bat species (e.g. pollination and seed dispersal to maintain healthy forests) is needed. The resulting data must be shared and communicated to growers and orchardists. They will form a crucial component in mitigating conflict, as at least one example from an Australian orchardist shows that ignorance can lead to opinions that the ecological role of flying foxes is negligible. The orchardist claimed to have seen no evidence that flying foxes are essential to forests (Bicknell 2002). This shows that in some cases, feeding on fruit crops can create a negative bias among fruit growers against the beneficial aspects of bats. Therefore, bat conservation must also extend to educating and raising awareness of the farmers and fruit growers who bear the brunt of bat damage to their crops.

In Australia, NSW Agriculture suggested that in order to address and overcome the conflict between fruit growers and flying foxes, a NSW Flying-fox Consultative Committee should be formed along the lines of the Queensland Flying-fox Consultative Committee (QFFCC). This would include a dedicated Flying-fox Management Unit that would be responsible for population monitoring, community liaison, research and compensation to growers (Bower 2002). The QFFCC's role included providing a forum for multi-stakeholder consultation, providing advice on policy development, developing strategies to address crop damage, and disseminating information to stakeholders (Teagle 2002). This model of collaborative approach involving all interest groups should be adopted in other countries where pteropodid feeding is a serious concern for their respective fruit industries. However, support for this type of consultation is entirely dependent upon the policies of the government of the time. No consultative committees now exist in New South Wales or Queensland, and the government of the latter state actively supports the destruction of flying foxes in orchards.

In the Mascarene Islands, public education programmes about *P. rodricensis* on Rodrigues since 1998 have led to a sense of pride in this species among the inhabitants, despite the fact that the increasing bat population (>20,000 individuals on an island area of 109 km^2) damages the fruit on backyard trees and causes some discontent. While Rodriguans complain about loss of fruit, they are still tolerant of the bats, as a result of positive messages in schools and communities, and are less inclined to call for culling as a solution to the crop-raiding issue (V. Tatayah, pers. comm.).

13.7 Conclusions

Pteropodid bats can reduce the harvestable yield of a wide range of fruit crops, resulting in economic losses that can be severe. However, this problem appears to be caused, and exacerbated, by continuing loss of the bats' natural food, which happens when humans clear natural forests. Lethal methods to reduce fruit crop damage are ineffective and problematic, and thus, the best solution is to implement non-lethal mitigation such as fixed nets, deterrents and decoy trees. In some instances, a combination of some or all of these non-lethal methods may be required. However, further research and trials are required for some of these methods, and these would be aided by ecological research focusing specifically on feeding behaviour and dietary preferences of those pteropodid species implicated in crop damage. In addition, there is an urgent need to educate fruit growers, authorities and the general public about the important benefits and ecosystem services provided by pteropodid bats. Such information may work best when presented in economic terms and measurements, such as cost–benefit analyses, to make it immediately relevant to economies and livelihoods.

Acknowledgements We are grateful to C. Korine, N. Singaravelan and V. Tatayah for their comments on an earlier draft; to T. Kingston for helpful editing; and to N. Islam, A. Tsoar and W. Yapa for their assistance.

References

Albayrak I, Asan N, Yorulmaz T (2008) The natural history of the Egyptian fruit bat, *Rousettus aegyptiacus,* in Turkey (Mammalia: Chiroptera). Turk J Zool 32:11–18

Amman BR, Jones MEB, Sealy TK et al (2014a) Oral shedding of Marburg virus in experimentally infected Egyptian fruit bats (*Rousettus aegyptiacus*). J Wildl Dis (ahead of print)

Amman BR, Nyakarahuka L, McElroy AK (2014b) Marburgvirus resurgence in Kitaka mine bat population after extermination attempts, Uganda. Emerg Infec Dis 20(10):1761–1764

Andrianaivoarivelo RA, Jenkins RKB, Petit EJ et al (2012) *Rousettus madagascariensis* (Chiroptera: Pteropodidae) shows a preference for native and commercially unimportant fruits. Endanger Species Res 19:19–27

Ayensu EA (1974) Plant and bat interactions in West Africa. Ann Mo Bot Gard 61:702–727

Bicknell JR (2002) The need for aversion agents for managing flying-foxes on crops and the difficulties in attracting research funds. In: Eby P, Lunney D (eds) Managing the grey-headed flying-fox as a threatened species in NSW. Royal Zoological Society of New South Wales, Mosman, NSW, pp 63–69

Biel E (2002) The cost to orchardists in the management of the grey-headed flying-fox. Who pays? A community benefit approach. In: Eby P, Lunney D (eds) Managing the grey-headed flying-fox as a threatened species in NSW. Royal Zoological Society of New South Wales, Mosman, NSW, pp 47–52

Bollen A, Van Elsacker L (2002) Feeding ecology of *Pteropus rufus* (Pteropodidae) in the littoral forest of Sainte Luce, SE Madagascar. Acta Chiropterol 4:33–47

Bomford M, O'Brien PH (1990) Sonic deterrents in animal damage control: a review of device tests and effectiveness. Wildl Soc Bull 18:411–422

Bower C (2002) Management issues in minimisation of damage by flying-foxes to horticultural crops. In: Eby P, Lunney D (eds) Managing the grey-headed glying-fox as a threatened species in NSW. Royal Zoological Society of New South Wales, Mosman, NSW, pp 77–79

Boonkird K, Wanghongsa S (2004) On the population number and distribution of flying foxes (*Pteropus lylei*) in central plain. In: 2003 Annual Report of Wildlife Research Division. Department of National Parks, Wildlife and Plant Conservation, Bangkok, pp 89–100 (In Thai)

Bumrungsri S, Leelapaibul W, Racey PA (2007) Resource partitioning in sympatric *Cynopterus* bats in lowland tropical rainforest, Thailand. Biotropica 39:241–248

Bumrungsri S, Harbit A, Benzie C et al (2008) The pollination ecology of two species of *Parkia* in southern Thailand. J Trop Ecol 24:467–475

Bumrungsri S, Sripaoraya E, Chongsiri T et al (2009) The pollination ecology of durian (*Durio zibethinus*) in southern Thailand. J Trop Ecol 25:85–92

Burns J (2009) Extinction threat to flying fox. BBC Online, 25 August, http://news.bbc.co.uk/2/hi/science/nature/8221132.stm. Accessed 5 Nov 2013

Campbell J, Greer N (1994) Controlling pest losses—why, how, how much? In: Bird and bat control for horticulture and aquaculture, Nambour, Queensland, 18 May 1994. Seminar proceedings, Department of Environment and Heritage, Queensland, pp 38–41

Chakravarthy AK, Girish AC (2003) Crop protection and conservation of frugivorous bats in orchards of hill and coastal regions of Karnataka. Zoos' Print J 18:1169–1171
Charles L, Benjamin N (2010) Ruler bans hunting of wildlife. The Star, 5 March
Chua KB, Koh CL, Hooi PS et al (2002) Isolation of Nipah virus from Malaysian island flying-foxes. Microb Infect 4:145–151
Clulow S, Blundell AT (2011) Deliberate insectivory by the fruit bat *Pteropus poliocephalus* by aerial hunting. Acta Chiropterol 13:201–205
Comensoli P (2002) The impact upon fruit growers of a decision to list the grey-headed flying-fox as a Vulnerable species under the NSW Threatened Species Conservation Act. In: Eby P, Lunney D (eds) Managing the grey-headed flying-fox as a threatened species in NSW. Royal Zoological Society of New South Wales, Mosman, NSW, pp 53–55
Del Vaglio MA, Nicolau H, Bosso L et al (2011) Feeding habits of the Egyptian fruit bat *Rousettus aegyptiacus* on Cyprus island: a first assessment. Hystrix Ital J Mamm 22:281–289
Dewhurst (1998) Correspondence from Queensland Fruit and Vegetable Growers to DPI. Quoted by Teagle S 2002 op cit
Divljan A, Parry-Jones K, Eby P (2009) Report on deaths and injuries to grey-headed flying-foxes, *Pteropus poliocephalus* shot in an orchard near Sydney, NSW. Aust Zool 35:698–710
Dolbeer RR, Fiedler LR, Rasheed H (1988) Management of fruit bat and rat populations in the Maldive Islands, Indian Ocean. In: Crabb AC, Marsh RE (eds) Proceedings of the vertebrate pest conference, vol 13, pp 112–118. University of California, Davis
Don't Shoot Bats 2013. Lethal crop protection is not effective. http://www.dontshootbats.com/crop-protection.html. Accessed 18 November 2013
Eby P (1990) Seed dispersal and seasonal movements by grey-headed flying foxes and the implications for management. In: Slack J (ed) Flying-fox workshop proceedings. Wollongbar Agricultural Institute, Wollongbar, pp 28–32
Eby P (1995) The biology and management of flying foxes in NSW. NSW National Parks and Wildlife Service, Hurstville, NSW
Eby P, Lunney D (2002) Managing the grey-headed flying-fox *Pteropus poliocephalus* as a threatened species: a context for the debate. In: Eby P, Lunney D (eds) Managing the grey-headed flying-fox as a threatened species in NSW. Royal Zoological Society of New South Wales, Mosman, NSW, pp 1–15
Entwistle AC, Corp N (1997) The diet of *Pteropus voeltzkowi*, an endangered fruit bat endemic to Pemba Island, Tanzania. Afr J Ecol 35:351–360
Epstein JH, Olival KJ, Pulliam JRC et al (2009) *Pteropus vampyrus*, a hunted migratory species with a multinational home-range and a need for regional management. J Appl Ecol 46:991–1002
Field H, Young P, Yob JM et al (2001) The natural history of Hendra and Nipah viruses. Microb Infect 3:307–314
Fleming TH, Kress WJ (2011) A brief history of fruit and frugivores. Acta Oecol 37:521–530
Fleming TH, Geiselman C, Kress WJ (2009) The evolution of bat pollination—a phylogenetic perspective. Ann Bot 104:1017–1043
Fogarty R, Halpin K, Hyatt AD et al (2008) Henipavirus susceptibility to environmental variables. Virus Res 132:140–144
Fujita F (1988) Flying foxes and economics. BATS Magazine 6(1). http://www.batcon.org/index.php/media-and-info/bats-archives.html?task=viewArticle&magArticleID=318. Accessed 3 March 2013
Fujita MS, Tuttle MD (1991) Flying foxes (Chiroptera, Pteropodidae)—threatened animals of key ecological and economic importance. Cons Biol 5:455–463
Funakoshi K, Watanabe H, Kunisaki T (1993) Feeding ecology of the northern Ryukyu fruit bat, *Pteropus dasymallus dasymallus*, in a warm-temperate region. J Zool Lond 230:221–230
Galil J, Stein M, Horoviz A (1976) On the origin of the sycamore fig (*Ficus sycomorus* L.) in the Middle East. Gard Bull XXIX:191–205

Gough JD (1992) Drift nets of the northern rivers. In: Blade K (ed) Fruit crop protection seminar. NSW National Parks and Wildlife Service, Australia, pp 14–17

Gough J (2002) The increasing need for netting fruit orchards against bat and bird damage and the increasing problems in affording netting. In: Eby P, Lunney D (eds) Managing the grey-headed flying-fox as a threatened species in NSW. Royal Zoological Society of New South Wales, Mosman, NSW, pp 56–57

Groseth A, Feldmann H, Strong JE (2007) The ecology of Ebola virus. Trends Microbiol 15:408–416

Gumal MT, Irwan M, Brandah CJ et al (1998) The ecology and role of the large flying fox (*Pteropus vampyrus*) in Sarawakian rain forests—*Keluang embet embet*. Hornbill 2:82–95

Hadjisterkotis E (2006) The destruction and conservation of the Egyptian fruit bat *Rousettus aegyptiacus* in Cyprus: a historic review. Eur J Wildl Res 52:282–287

Hall LS, Richards GC (1987a) Crop protection and management of flying-foxes (Chiroptera: Pteropodidae). Aust Mammal 10:137–139

Hall LS, Richards GC (1987b) The flying-fox problem in eastern Australia. In: Proceedings of the 1987 Australian vertebrate pest control conference, Coolangatta, pp 279–283

Hall J, Willis B (1992) Netting orchards against flying foxes, birds and hail. In: Blade K (ed) Fruit crop protection seminar. NSW National Parks and Wildlife Service, Australia, pp 24–28

Halpin K, Young PL, Field HE, Mackenzie JS (2000) Isolation of Hendra virus from pteropid bats: a natural reservoir of Hendra virus. J Gen Virol 81:1927–1932

Harrison DL (1964) The Mammals of Arabia. Insectivora, Chiroptera, Primates. Ernest Bern Limited, London

Harrison DL, Bates PJJ (1991) The Mammals of Arabia. Harrison Zoological Museum, 2nd edn. Lakeside Printing, London

Harrison ME, Cheyne SM, Darma F et al (2011) Hunting of flying foxes and perceptions of disease risk in Indonesian Borneo. Biol Cons 144:2441–2449

Heaney L, Rosell-Ambel G, Tabaranza B et al (2008) Pteropus dasymallus. IUCN Red List of Threatened Species. Version 2015.3. http://www.iucnredlist.org. Accessed 25 July 2015

Hicks PG (1967) Observations on the diseases and conditions of cocoa pods in Papua and New Guinea—Pod losses 1962–1965. P N G Agr J 19:5–9

Hodgkison R, Balding ST, Zubaid A et al (2003) Fruit Bats (Chiroptera: Pteropodidae) as seed dispersers and pollinators in a lowland Malaysian rain forest. Biotropica 35:491–502

Hodgkison R, Balding ST, Zubaid A et al (2004) Temporal variation in the relative abundance of fruit bats (Megachiroptera: Pteropodidae) in relation to the availability of food in a lowland Malaysian rain forest. Biotropica 36:522–533

Huang JC-C, Jazdzyk E, Nusalawo M, et al (2014) Is Bat coffee a potential "wing-wing" tool for biodiversity conservation in southwestern Sumatra? Paper presented at the 16th international bat research conference, Hotel Herradura, San José, 11–15 Aug 2013, Bat Res News (in press)

IUCN (2014) IUCN Red List of Threatened Species. Version 2014.2. http://www.iucnredlist.org. Accessed 10 Nov 2014

Izhaki I, Korine C, Arad Z (1995) The effect of bat (*Rousettus aegyptiacus*) dispersal on seed germination in eastern Mediterranean habitats. Oecologia 101:335–342

Kandasamy D (2009) Flying foxes facing extinction. The Malay Mail, 2 Sept

Khan MSU, Hossain J, Gurley ES et al (2011) Use of infrared camera to understand bats' access to date palm sap: implications for preventing Nipah virus transmission. EcoHealth 7:517–525

Kingston T (2010) Research priorities for bat conservation in Southeast Asia: a consensus approach. Biodivers Conserv 19:471–484

Korine C, Izhaki I, Arad Z (1998) Comparison of fruit syndromes between the Egyptian fruit-bat (*Rousettus aegyptiacus*) and birds in East Mediterranean habitats. Acta Oecol 19:147–153

Korine C, Izhaki I, Arad Z (1999) Is the Egyptian fruit-bat *Rousettus aegyptiacus* a pest in Israel? An analysis of the bat's diet and implications for its conservation. Biol Cons 88:301–306

Kunz TH, de Torrez EB, Bauer D et al (2011) Ecosystem services provided by bats. Ann NY Acad Sci 1223:1–38

Lane DJW, Kingston T, Lee BPY-H (2006) Dramatic decline in bat species richness in Singapore, with implications for Southeast Asia. Biol Cons 131:584–593

Law B, Eby P, Somerville D (2002) Tree-planting to conserve flying-foxes and reduce orchard damage. In: Eby P, Lunney D (eds) Managing the grey-headed flying-fox as a threatened species in NSW. Royal Zoological Society of New South Wales, Mosman, NSW, pp 84–90

Lee Y-F, Takaso T, Chiang T-Y et al (2009) Variation in the nocturnal foraging distribution of and resource use by endangered Ryukyu flying foxes (*Pteropus dasymallus*) on Iriomotejima Island, Japan. Contr Zool 78:51–64

Leroy EM, Rouquet P, Formenty P et al (2004) Multiple Ebola virus transmission events and rapid decline of central African wildlife. Science 303:387–390

Leroy EM, Kumulungui B, Pourrut X et al (2005) Fruit bats as reservoirs of Ebola virus. Nature 438:575–576

Leroy EM, Epelboin A, Mondonge V et al (2009) Human Ebola outbreak resulting from direct exposure to fruit bats in Luebo, Democratic Republic of Congo, 2007. Vector-Borne Zoonotic Dis 9:723–728

Lewis RE, Harrison DL (1962) Notes on bats from the Republic of Lebanon. Proc Zool Soc Lond 138:473–486

Lobova TA, Geiselman CK, Mori SA (2009) Seed dispersal by bats in the neotropics. New York Botanical Garden, New York (Mem New York Bot Gard 101)

Lokkers C (1990) Colony dynamics of the green tree ant (*Oecophylla smaragdina* Fab.) in a seasonal tropical climate. PhD dissertation, James Cook University

Long E, Racey PA (2007) An exotic plantation crop as a keystone resource for an endemic megachiropteran, *Pteropus rufus* in Madagascar. J Trop Ecol 23:1–11

Luby SP, Rahman M, Hossain MJ et al (2006) Foodborne transmission of Nipah virus, Bangladesh. Emerg Infect Dis 12:1888–1894

Lumlertdacha B, Boongird K, Wanghongsa S et al (2005) Survey for bat lyssaviruses, Thailand. Emerg Infect Dis 11:232–236

Luskin MS (2010) Flying foxes prefer to forage in farmland in a tropical dry forest landscape mosaic in Fiji. Biotropica 42:246–250

Mackenzie JS, Field HE, Guyatt KJ (2003) Managing emerging diseases borne by fruit bats (flying foxes), with particular reference to henipaviruses and Australian bat lyssavirus. J Appl Microbiol 94:59S–69S

Madkour G (1977) *Rousettus aegyptiacus* (Megachiroptera) as a fruit eating bat in A.R. Egypt. Agric Res Rev 55:167–172

Mahmood-Ul-Hassan M, Gulraiz TL, Rana SA et al (2010) The diet of Indian flying-foxes (*Pteropus giganteus*) in urban habitats of Pakistan. Acta Chiropterol 12:341–347

Makin D, Mendelssohn H (1987) Israel wipes out its bats. Ecologist 16:1–2

Marshall AG (1983) Bats, flowers and fruit: evolutionary relationships in the Old World. Biol J Linn Soc 20:115–135

Martin L, McIlwee AP (2002) The reproductive biology and intrinsic capacity for increase of the grey-headed flying-fox *Pteropus poliocephalus* (Megachiroptera), and the implications of culling. In: Eby P, Lunney D (eds) Managing the grey-headed flying-fox as a threatened species in NSW. Royal Zoological Society of New South Wales, Mosman, NSW, pp 91–108

Mateo R (2009) Lead poisoning in birds in Europe and the regulations adopted by different countries. In: Watson RT, Fuller M, Pokras M et al (eds) Ingestion of lead from spent ammunition: implications for wildlife and humans. The Peregrine Fund, Boise, Idaho, pp 71–98

Meyer C, Struebig M, Willig M (in press) Responses of tropical bats to habitat fragmentation, logging and deforestation. In: Kingston T, Voigt CC (eds) Bats in the anthropocene. Springer, New York, pp 63–94

McConkey KR, Drake DR (2006) Flying foxes cease to function as seed dispersers long before they become rare. Ecology 87:271–276

McLachlan B (2002) NPWS operational management of commercial crop damage by flying-foxes—licensing in practice, a far north coast perspective. In: Eby P, Lunney D (eds) Managing the grey-headed flying-fox as a threatened species in NSW. Royal Zoological Society of New South Wales, Mosman, NSW, pp 80–83

Mickleburgh S, Hutson AM, Racey PA (1992) Old World Fruit Bats—an action plan for their conservation. IUCN, Gland
Mickleburgh S, Waylen K, Racey PA (2009) Bats as bushmeat—a global review. Oryx 43:217–234
Minifie G, Willi B (1990) Duranet cloth and supporting systems. In: Slack JM (ed) Flying fox workshop proceedings, Wollongbar Agricultural Institute, Wollongbar. NSW Agriculture and Fisheries, Australia, pp 47–55
Moran S, Keidar H (1993) Checklist of vertebrate damage to agriculture in Israel. Crop Prot 12:171–182
Myers N, Mittermeier RA, Mittermeier CG et al (2000) Biodiversity hotspots for conservation priorities. Nature 403:853–858
Nahar N, Sultana R, Gurley ES et al (2010) Date palm sap collection: exploring opportunities to prevent Nipah transmission. EcoHealth 7:196–203
Nakamoto A, Kinjo K, Izawa M (2007) Food habits of Orii's flying-fox, *Pteropus dasymallus inopinatus*, in relation to food availability in an urban area of Okinawa-jima Island, the Ryukyu Archipelago, Japan. Acta Chiropterol 9:237–249
Nakamoto A, Kinjo K, Izawa M (2009) The role of Orii's flying-fox (*Pteropus dasymallus inopinatus*) as a pollinator and seed disperser on Okinawa-jima Island, the Ryukyu Archipelago, Japan. Ecol Res 24:405–414
Nelson SL, Miller MA, Heske EJ et al (2000) Nutritional consequences of a change in diet from native to agricultural fruits for the Samoan fruit bat. Ecography 23:393–401
Nelson SL, Masters DV, Humphrey SR et al (2005) Fruit choice and calcium block use by Tongan fruit bats in American Samoa. J Mammal 86:1205–1209
Nyhagen DF, Turnbull SD, Olesen JM et al (2005) An investigation into the role of the Mauritian flying fox, *Pteropus niger*, in forest regeneration. Biol Cons 122:491–497
Oleksy R, Racey PA, Jones G (2015) High-resolution GPS tracking reveals habitat selection and the potential for long-distance seed dispersal by Madagascan flying foxes *Pteropus rufus*. Glob Ecol Conserv 3:678–692
Olival KJ, Hayman DTS (2014) Filoviruses in Bats: current knowledge and future directions. Viruses 6:1759–1788
Parry-Jones KA, Augee ML (2001) Factors affecting the occupation of a colony site in Sydney, New South Wales by the grey-headed flying-fox *Pteropus poliocephalus* (Pteropodidae). Aust Ecol 26:47–55
Phillips WWA (1980) Manual of Mammals of Sri Lanka (Part 1). Wild Life and Nature Protection Society of Sri Lanka
Picot M, Jenkins RBK, Ramilijoana O et al (2007) The feeding ecology of *Eidolon dupreanum* (Pteropodidae) in eastern Madagascar. Afr J Ecol 45:645–650
Price V (2013) Trouble in paradise: mapping human-wildlife conflict in the western Indian Ocean. Unpublished MSc thesis, Imperial College, London, UK
Qumsiyeh MB (1980) New records of Bats from Jordan. Säugetierkd Mitt 1:36–39
Qumsiyeh MB, Disi AM, Amr ZS (1992) Systematics and distribution of the bats (Mammalia: Chiroptera) of Jordan. Dirasat 19B:101–118
Racey PA, Nicoll ME (1984) Mammals. In: Stoddart DR (ed) Biogeography and ecology of the Seychelles Islands. Dr W Junk, The Hague, pp 607–626
Rahman MA, Hossain MJ, Sultana S et al (2012) Date palm sap linked to Nipah virus outbreak in Bangladesh, 2008. Vector-Borne Zoonotic Dis 12:65–72
Rahman SA, Hassan L, Epstein JH et al (2013) Risk factors for Nipah virus infection among pteropid bats, Peninsular Malaysia. Emerg Infect Dis 19:51–60
Ramlugun TMRRS (2013) An assessment of fruit bat damage caused by the Mauritian fruit bat (*Pteropus niger*) in a litchi (*Litchi chinensis*) orchard. BSc thesis, University of Mauritius
Ratcliffe FN (1931) The flying fox (*Pteropus*) in Australia. Report for co-operative work conducted on behalf of the Council for Scientific Research, the New South Wales Department of Agriculture, and the Queensland Home Secretary's Department, Melbourne

Reilly T, Slack J (1990) Cost and returns of netting low-chill stonefruit orchards. In: Slack JM (ed) Flying fox workshop proceedings. Wollongbar Agricultural Institute, Wollongbar, 22 August 1990. NSW Agriculture and Fisheries, Australia, pp 61–65

Richards GC (1990) The Spectacled flying fox, *Pteropus conspicillatus*, in north Queensland. 2. Diet, feeding ecology and seed dispersal. Aust Mammal 13:25–31

Richards GC, Hall LS, Parish S (2012) A natural history of Australian Bats: working the night shift. CSIRO Publishing, Melbourne

Roberts TJ (1997) The Mammals of Pakistan. Oxford University Press, Karachi

Robyn S (2007) Suivi et essai d'estimation de la population de chauve-souris frugivore (*Pteropus niger*) a l'Ile Maurice (monitoring and population estimate of a fruit bat [*Pteropus niger*] in Mauritius). Université Paul Cézanne, Memoir de Stage

Rogers J (2002) The economic and social implications of flying-fox predation on the north coast of NSW. In: Eby P, Lunney D (eds) Managing the grey-headed flying-fox as a threatened species in NSW. Royal Zoological Society of New South Wales, Mosman, NSW, pp 58–62

Rimba (2012) Special update: Terengganu protects flying foxes! http://myrimba.org/2012/02/16/special-update-terengganu-protects-flying-foxes/. Accessed 19 May 2013

Scanlon AT, Petit S, Sternberg LDS (2013) Insectivory in Fijian flying foxes (Pteropodidae). Aust J Zool 61:342–349

Singaravelan N (2002) Foraging behaviour of fruit bats in orchards. PhD thesis, Madurai Kamaraj University

Singaravelan N, Marimuthu G (2006) *Muntingia calabura*—an attractive food plant of *Cynopterus sphinx*—deserves planting to lessen orchard damage. Acta Chiropterol 8:239–245

Singaravelan N, Marimuthu G, Racey PA (2009) Do all fruit bats deserve to be listed as vermin in the Indian Wildlife (Protection) & Amended Acts—a critical review. Oryx 43:608–613

Simmons N (2005) Chiroptera. In: Wilson DE, Reeder DAM (eds) Mammals of the World—a taxonomic and geographic reference. Johns Hopkins University Press, Baltimore, pp 312–529

Slack J (1990) Flying-fox damage in low-chill stone fruit orchards. In: Slack JM (ed) Flying-fox workshop proceedings. NSW Agriculture and Fisheries, Wollongbar, NSW, pp 55–60

Spitzenberger F (1979) Die Säugetierfauna Zyperns. Teil II: Chiroptera, Lagomorpha, Carnivora und Artiodactyla. Ann Nat Hist Mus 82:439–465

Srinivasulu C, Srinivasulu B (2002) Greater short-nosed fruit bat (*Cynopterus sphinx*) foraging and damage in vineyards in India. Acta Chiropterol 4:167–171

Stacey P (1990) Fruit industry focus-exotic fruit industry, NSW North Coast. In: Slack JM (ed) Flying fox workshop proceedings, Wollongbar Agricultural Institute. NSW Agriculture and Fisheries, Australia, pp 13–16

Stacey P (1992) To net or not to net. In: Blade K (ed) Fruit crop protection seminar. NSW National Parks and Wildlife Service, Australia, pp 20–23

Stier SC, Mildenstein TL (2005) Dietary habits of the world's largest bats: the Philippine flying foxes, *Acerodon jubatus* and *Pteropus vampyrus lanensis*. J Mammal 86:719–728

Struebig MJ, Harrison ME, Cheyne SM et al (2007) Intensive hunting of large flying foxes *Pteropus vampyrus natunae* in Central Kalimantan, Indonesian Borneo. Oryx 41:390–393

Swanepoel R, Leman PA, Burt FJ et al (1996) Experimental inoculation of plants and animals with Ebola virus. Emerg Infect Dis 2:321–325

Teagle S (2002) Queensland Flying-fox Consultative Committee—formation, outcomes and future strategies. In: Eby P, Lunney D (eds) Managing the grey-headed flying-fox as a threatened species in NSW. Royal Zoological Society of New South Wales, Mosman, NSW, pp 109–116

Teoh TH (2005) Logging and hunting killing bats. The Star, 5 June

Thiriet D (2010) Flying fox conservation laws, policies and practices in Australia—a case study in conserving unpopular species. Australas J Nat Resour Law Policy 13:161–194

Tidemann CR (1999) Biology and management of the grey-headed flying-fox, *Pteropus poliocephalus*. Acta Chiropterol 1:151–164

Tidemann R, Kelson SI, Jamieson G (1997) Flying-fox damage to orchard fruit in Australia—incidence, extent and economic impact. Aust Biol 10:179–186

Towner JS, Amman BR, Sealy TK et al (2009) Isolation of genetically diverse Marburg viruses from Egyptian fruit bats. PLoS Path 5:e1000536

Ullio L (2002) To net or not to net that is the question! But is it the answer? In: Eby P, Lunney D (eds) Managing the grey-headed flying-fox as a threatened species in NSW. R Zool Soc New South Wales, Mosman, NSW, pp 70–76

Van Mele P (2008) The importance of ecological and socio-technological literacy in R&D priority setting: the case of a fruit innovation system in Guinea, West Africa. Int J Agric Sust 6:183–194

Van Mele P, Camara K, Vaysierres JF (2009) Thieves, bats and fruit flies: local ecological knowledge on the weaver ant *Oecophylla longinoda* in relation to three 'invisible' intruders in orchards in Guinea. Int J Pest Manage 55:57–61

Verghese A (1998) Non-destructive control of the bat, *Cynopterus sphinx* Vahl (Chiroptera: Pteropodidae) in grapes (*Vitis vinifera* Linnaeus) in India. Int J Pest Manage 44:81–85

Vincenot CE, Koyama L, Russo D (2015) Near threatened? First report of unsuspected human-driven decline factors in the Ryukyu flying fox (*Pteropus dasymallus*). Mammal Biol 80:273–277

Waples K (2002) Review of the NPWS policy on the mitigation of commercial crop damage by flying-foxes. In: Eby P, Lunney D (eds) Managing the grey-headed flying-fox as a threatened species in NSW. R Zool Soc New South Wales, Mosman, NSW, pp 39–46

Weber N, Duengkae P, Fahr J et al (2015) High-resolution GPS tracking of Lyle's flying fox between temples and orchards in central Thailand. J Wildl Manage 79:957–968

Williams NSG, McDonnell MJ, Phelan GK et al (2006) Range expansion due to urbanization: increased food resources attract grey-headed flying-foxes (*Pteropus poliocephalus*) to Melbourne. Aust Ecol 31:190–198

Win SS, Mya KM (2015) The diet of the Indian Flying Fox *Pteropus giganteus* (Brünnich. 1782) (Chiroptera: Pteropodidae) in Myanmar—conflicts with local people? J Threat Taxa 7:7568–7572

Yapa WB, Kumarasinghe J, Digana PMCB et al (1999) Food preferences (three food types) of Sri Lankan short-nosed fruit bat *Cynopterus sphinx* (Chiroptera) in a semi-natural condition. Vidyodaya J Sci 8:109–116

Chapter 14
Bats and Buildings: The Conservation of Synanthropic Bats

Christian C. Voigt, Kendra L. Phelps, Luis F. Aguirre, M. Corrie Schoeman, Juliet Vanitharani and Akbar Zubaid

Abstract Humans have shared buildings with bats for thousands of years, probably as early as first humans built primitive huts. Indeed, many bat species can be defined as synanthropic, i.e., they have a strong ecological association with humans. Bats have been observed using buildings as roosting and foraging sites, temporary shelters, for reproduction and hibernation. A synanthropic lifestyle may result in direct fitness benefits owing to energetic advantages in warmer roosts, which may ultimately lead to more rapid gestation and faster development of juveniles, or by being less exposed to natural predators in urban environments. All

---

Luis F. Aguirre, M. Corrie Schoeman, Juliet Vanitharani and Akbar Zubaid: Contributed equally, listed alphabetically.

C.C. Voigt (✉)
Department of Evolutionary Ecology, Leibniz Institute for Zoo and Wildlife Research, Berlin, Germany
e-mail: ccvoigt@googlemail.com

K.L. Phelps
Department of Biological Sciences, Texas Tech University, Lubbock, USA

L.F. Aguirre
Centro de Biodiversida y Genetica, UniversidadMajor de San Simon, Cochabamba, Bolivia

M. Corrie Schoeman
School of Life Sciences, University of KwaZulu-Natal, Durban, South Africa

J. Vanitharani
Department Advanced Zoology and Biotechnology and Research Centre, Sarah Tucker College, Palayamkottai, Tirunelveli, Tamil Nadu, India

A. Zubaid
School of Environmental and Natural Resource Sciences, Faculty of Science and Technology, Universiti Kebangsaan Malaysia, 43600 Bangi, Malaysia

C.C. Voigt and T. Kingston (eds.), *Bats in the Anthropocene: Conservation of Bats in a Changing World*, DOI 10.1007/978-3-319-25220-9_14

these benefits may allow bats to use buildings as stepping stones to exploit habitats otherwise devoid of roosting structures and may even lead to the expansion of geographic ranges. Yet, the coexistence with humans also comes with some risks. Bats may be exposed to chemical pollutants, particularly preservation chemicals used on lumber or during pest control measures. Bats may also be at risk of direct persecution or they may die accidently if trapped within buildings. In general, eviction of bats from buildings should follow the general rule of avoidance–mitigation–compensation. When considering conservation measures for synanthropic bats, it is most important to assess the role of the building for different life stages of bats. Construction work at buildings should be conducted in a manner that minimizes disturbance of bats. Artificial roosts can replace lost roosts, yet bats will often not accept alternative roosts. Demographic changes in human populations may lead to the abandonment of buildings, for example, in rural areas and to increased conflicts in urban areas when old buildings are replaced by new buildings or when previously unoccupied space in buildings is renovated. We advocate maintenance and enhancement of roosts for synanthropic bats, in addition to outreach and education campaigns, to improve the tolerance of humans for synanthropic bats.

14.1 Introduction

14.1.1 What Is the Purpose of This Review?

Bats are nocturnal mammals that spend the daytime in dark places (Kunz 1982; Kunz and Lumsden 2003). Usually, they depend on natural roosting structures such as caves, crevices, foliage, branches, tree trunks, and hollows among many others. Bats most likely used buildings as roosts when humans started to build primitive huts thousands of years ago. Indeed, some bat species, such as the hairy split-faced bat, *Nycteris macrotis*, inhabit thatched huts in Africa that are likely similar to the earliest buildings of humans (Poché 1975).

In this chapter, we focus on bats that use man-made buildings that are coinhabited by humans. We refer to these bats as synanthropic species, or species that are "ecologically associated with humans (Merriam-Webster's dictionary). We do not use synanthropic species in the context of bats living in anthropogenically shaped landscapes; rather, this topic is covered in Chap. 2 (Reichel-Jung and Threlfall 2015). Nonetheless, we address certain aspects of bats living in other man-made structures unoccupied or abandoned by humans.

Synanthropic bat species have benefited from the expansion of human populations, and some species have likely expanded their geographic distribution as humans moved into new habitats worldwide. Yet this close association has disadvantages when synanthropic bats are faced with threats from humans. Currently, about a quarter of all bat species are considered threatened or near threatened, and one of the most prominent threats is loss of roost sites and disturbance at roosts (Mickleburgh et al. 2002). Therefore, this chapter is timely, and we hope that it will contribute to the conservation of synanthropic bats.

14.1.2 Relevant Natural History Features of Synanthropic Bats

Along the fast–slow continuum of life, bats are considered to be in the slow lane, even though most other mammals of similar size are in the fast lane (Barclay and Harder 2003; Bielby et al. 2007). Bats have low reproductive rates that are associated with exceptionally long life spans, a feature most obvious in insectivorous bats from temperate zones (Wilkinson and South 2002; Munshi-South and Wilkinson 2010). Long life spans may predispose bats to inhabit relatively permanent structures, such as in buildings, since some bat species are loyal to their roost over many years and form long-term social relationships with other colony members (Kerth et al. 2011).

Similar to other small mammals, bats exhibit relatively high mass-specific metabolic rates (McNab 2002). Many bats are also heterothermic, reducing their body temperature and consequently metabolic rate, during periods of adverse conditions, such as low resource abundance (insects, fruits, or nectar), low ambient temperature, or high rainfall (Geiser 2004). Most notably, temperate zone bats employ extended torpor when they hibernate in winter. Apart from hibernation, almost all bats use torpor on a daily basis as an energy-saving strategy (Speakman and Thomas 2003; Willis et al. 2006). During daytime torpor, bats may use passive rewarming when ambient temperatures peak during the warmest part of the day (Turbill et al. 2003). The use of radiant heat created by the exposure of building exteriors to sunshine likely saves synanthropic bats significant amounts of energy since they do not depend on endogenous heat production in brown adipose tissue (Geiser and Drury 2003). This could be a selective advantage for bats using sun-exposed buildings instead of dark caves as daytime roosts or hibernacula (Lausen and Barclay 2006; Halsall et al. 2012).

14.1.3 Which Bat Species Use Buildings?

The order Chiroptera comprises 19 living families, with at least one species in each family known to roost in buildings (Figs. 14.1 and 14.2), with the notable exceptions of Furipteridae, Mystacinidae, Myzopodidae, Natalidae, and Thyropteridae. Quite often, only local residents are aware of the occurrence of synanthropic bat species. The chapter on bats and urbanization (Reichel-Jung and Threlfall 2015) provides a meta-analytic perspective on bats living in urban landscapes. Many of the species included in their analysis also roost in buildings; thus, the general patterns derived from their study may also hold true for aspects of roost choice in synanthropic bats.

Fig. 14.1 Example of synanthropic bats that use both natural roosts and buildings. The greater sac-winged bat, *Saccopteryx bilineata*, shown here in Costa Rica, forms colonies in the cavities formed by large buttress roots of canopy trees. In the absence of such trees, this species will roost on the exterior walls of buildings (or inside if the building is abandoned as shown in the *right picture*; © *left picture* Knörnschild M, right picture Voigt CC)

Fig. 14.2 Colony of *Megaderma lyra* under a tin roof of a building in India

14.1.4 Human–Bat Conflict in Buildings and the Legal Protection of Synanthropic Bats

Buildings constructed specifically as human dwellings are usually well maintained and protected against opportunistic invasions by unwanted animals. Unfortunately, synanthropic bats are unwanted by most humans, which generate conflicts (Gareca et al. 2007). Accordingly, synanthropic bats are persecuted virtually worldwide, even if the legal framework may define this action as criminal. Documented cases of humans removing bats from buildings are apparent across the entire geographic range of synanthropic bats (e.g., Merzlikin 2002), but most cases remain unnoticed by law enforcement agencies even where bats are legally protected. Indeed, bats are legally protected in only a few countries. For example, bats are protected in countries of the European Union according to the Habitats Directive (Council Directive 92/43/EEC). Also, migratory bats are specifically protected in countries that have signed the United Nations convention on the "Conservation of Migratory Species of Wild Animals" (Lyster 1989). In some countries, conservationists have established action plans for threatened bat species, including suggestions for protecting synanthropic bats (Aguirre et al. 2010). However, these recommendations have not yet been converted into some form of legal framework. In African and Asian countries, bats are not protected under specific legislation. In summary, the level of protection of synanthropic bats by national or international legislation is highly variable and clearly deficient.

14.2 How Do Bats Find and Use Buildings?

Since most bat species are not capable of constructing their own roosts (Kunz 1982; Kunz and Lumsden 2003), they depend largely on preexisting roosting structures, either of natural or of artificial origin. Therefore, roost sites are likely a limited resource for bats (Kunz 1982; Kunz and Lumsden 2003), such that buildings may constitute an important substitute for natural roosts (Lisón et al. 2013). Buildings may resemble rocks or cavelike structures, which may attract bats into crevices or attics. Once one or a few bats establish a roost in a building, other bats may recognize the newly established roosts by olfactory or acoustic cues. The importance of nonsocial information such as visual and temperature-related cues and social sensory cues, e.g., conspecific echolocation calls, has recently been confirmed as important information for the common noctule bat, *Nyctalus noctula*, to initiate roost exploration (Ruczyński et al. 2007). Presumably, noctule bats use the same set of cues for exploring buildings as temporary shelters or hibernacula (Bihari 2004; Kozhurina and Gorbunova 2004; Szodoray-Parádi et al. 2004; Cel'uch and Kaňuch 2005; Cel'uch et al. 2006; Bačkor et al. 2007).

14.2.1 Buildings as Foraging Sites

Buildings are rarely used by bats as foraging sites, although abandoned buildings may develop into small urban ecosystems. For example, Aspetsberger et al. (2003) found that cockroaches (Blattodea: Blaberidae), sharing the space under the metal roof of a building with little free-tailed bats, *Chaerephon pumilus*, comprised more than 60 % of the diet of the bats. Yet, most observations of foraging at buildings are bats hunting insects around illuminated buildings. Artificial lighting is known to attract insects, and consequently, bats may chase insects close to illuminated buildings (Rydell 1991, 1992; Rydell and Racey 1995; Pavey 1999; Rowse et al. 2015).

14.2.2 Buildings as Shelters During Foraging Bouts

Buildings provide structures that can be used by bats as a temporary shelter. For example, buildings are often used by bats as a shelter to digest food items gathered during their most recent foraging bout (Ormsbee et al. 2007). This behavior has been observed in many species, including tropical carnivorous species such as the greater false vampire bat, *Megaderma lyra*, in India (Subbaraj and Balasingh 1996), and the greater slit-faced bat, *Nycteris grandis*, in southern Africa (Fenton et al. 1990) as well as temperate insectivorous bats such as Leisler's bat, *Nyctalus leisleri*, in Europe (Shiel et al. 1999), and the pallid bat, *Antrozous pallidus*, in the USA (Lewis 1994). In general, the temporary use of buildings by foraging bats may be the first step toward a more permanent occupation of buildings.

14.2.3 Buildings as Maternity Roosts

Females of many synanthropic bats use buildings as maternity roosts. Sometimes adult males share the same roost, but often the sexes are segregated. According to our literature survey, at least 35 bat species form maternity colonies in buildings. Energetic advantages and reduced predation risk may be benefits for female bats that give birth and raise their young in buildings. Harbusch and Racey (2006) reported that the serotine bat, *Eptesicus serotinus*, selected old buildings with slate roofing for maternity roosts, largely because such buildings tend to have small holes and fissures allowing easy access. Also, such buildings offered suitable temperatures of about 22 °C during gestation and lactation periods, a critical parameter for the survival of offspring (Harbusch and Racey 2006). Further, many species that form maternity colonies in buildings show high levels of site fidelity and natal philopatry, with female young returning to the same roosts to reproduce when they mature (Harbusch and Racey 2006). This could initiate a tradition of using buildings instead of natural roosts in local bat populations.

14.2.4 Buildings as Swarming Sites

Several European bats, such as common pipistrelle bats, *Pipistrellus pipistrellus*, and parti-colored bats, *Vespertilio murinus*, swarm at large buildings during autumn (Kanuch et al. 2010; Šuba et al. 2010). Usually, swarming occurs after juveniles have fledged and as they start to disperse from their natal roost. In Marburg, Germany, common pipistrelles swarm between mid-August and late September not only at tall buildings, such as historic towers, castles, and churches, but also at large multistory buildings. Interestingly, bat researchers recorded almost exclusively juvenile bats during swarming events (Kanuch et al. 2010; Šuba et al. 2010), and therefore, it was argued that swarming was related to roost exploration (Smit-Viergutz and Simon 2000). Yet, a social function of swarming behavior has also been suggested, for example, for *Vespertio murinus* (Kanuch et al. 2010; Šuba et al. 2010). To the best of our knowledge, swarming of bats at buildings has not been observed in countries outside of Europe.

14.2.5 Buildings as Hibernacula

Many bat species are known to hibernate in buildings, presumably because building interiors rarely reach freezing temperatures, turning them into ideal hibernation sites for bats. For many of these species, natural hibernacula include not only caves, rock crevices, and rock screes, but also tree hollows. Michaelsen et al. (2013) reported that in Norway, hibernating bats prefer anthropogenic structures rather than natural subground hibernacula, but the reason for this preference was unknown. Bats, such as big brown bats, *Eptesicus fuscus*, hibernating in walls of heated buildings expose themselves to ambient temperatures of 2–5 °C which are created by the balance between warm interior temperatures from heated rooms and cold ambient temperatures from the outside (Whitaker and Gummer 1992). In addition, bats hibernating in buildings may also benefit from occasional passive rewarming, when being exposed to mild exterior temperatures. *Nyctalus noctula* usually forms maternity colonies in tree roosts, yet in Central and northern Europe, they frequently use prefabricated buildings, i.e., multistory buildings consisting of prefabricated concrete walls that are assembled at the construction site. Throughout continental Europe, large numbers of noctule bats hibernate in such buildings in crevices at about 5–10 m aboveground, sometimes forming winter aggregations of a few thousand individuals (Zahn et al. 2000; Kozhurina and Gorbunova 2004; Cel'uch and Kaňuch 2005; Cel'uch et al. 2006). Bats in subtropical and tropical zones may also use buildings during adverse conditions and employ torpor, yet their biology is largely unknown and therefore in need of further studies.

14.3 Benefits of a Synanthropic Lifestyle in Bats

14.3.1 Increased Fitness of Bats Using Buildings

Bats would not use buildings as roosts without a proximate (ecological or physiological) or ultimate (evolutionary) benefit. In the following, we will discuss three potential benefits for bats using buildings, which seem to be linked to increased fitness over the short or long term.

Reduced predation risk In general, bats face only a few predators compared to non-volant mammals of similar size (Sibly and Brown 2007). Yet some birds, mammals, and even invertebrates hunt bats on a regular basis (Gillette and Kimbrough 1970; Speakman 1991; Altringham 1996; Nyffeker and Knörnschild 2013). Roosts in buildings could reduce the exposure of bats to predators if predators avoid anthropogenic environments. For example, snakes and giant centipedes hunt neotropical bats at the entrance of caves, and many of these species are less abundant or even absent in an urban environment (Molinari et al. 2005; Esbérard and Vrcibradic 2007). In North America, big brown bats, *E. fuscus*, seem to be less exposed to predators when roosting in buildings than in natural roosts (Lausen and Barclay 2006). However, clustered emergence of bats from roosts in buildings may point to antipredatory behavior in synanthropic bats in urban environments (Speakman et al. 1995; Duvergé et al. 2008; but see Irwin and Speakman 2003).

Energetic benefits Bats may survive periods of adverse weather conditions, such as heavy rain or low ambient temperatures, by roosting in a warm and dry building. The energetic benefits for bats roosting in buildings may manifest particularly during critical life history stages, such as reproduction and hibernation.

Buildings may provide conditions that are beneficial for reproducing female bats. For example, elevated ambient temperatures in attics seem to be ideal for pregnant and lactating bats. Angolan free-tailed bats, *Mops condylurus*, inhabit maternity roosts under corrugated steel roofs of houses that often exceed 40 °C during the day (Maloney et al. 1999), enabling them to maintain ideal growth conditions throughout the reproductive period without expending a lot of energy (Vivier and van der Merwe 2007). Their use of hot roosts may even be linked to increased reproductive rates (Bronner et al. 1999). Higher roost temperatures in attics seem to be also favorable for the development of juveniles of European greater mouse-eared bats, *Myotis myotis*. This species forms large clusters of individuals in natural cave roosts, presumably to benefit from huddling and sharing of body heat (Dietz et al. 2009). In buildings, however, greater mouse-eared bats usually form smaller colonies, and these smaller clusters may be energetically feasible only because *Myotis myotis* may benefit from exogenous instead of endogenous heat when roosting in warm attics (Zahn 1999). The use of different locations depending on reproductive state has been confirmed for other species as well, including Rafinesque's big-eared bats, *Corynorhinus rafinesquii* (Roby et al. 2011). Similar to attic-roosting *Myotis myotis*, thermal benefits have also been suggested for *Eptesicus fuscus*. Pregnant big brown bats rarely entered torpor when roosting at favorable ambient conditions in buildings (Lausen and Barclay 2006). The avoidance of torpor may be advantageous for fetal development. For example, big brown bats gave birth earlier when roosting in buildings than when roosting in natural roosts. Furthermore, juveniles from buildings fledged one to two weeks earlier than conspecifics born in natural roosts (Lausen and Barclay 2006). Similarly, building-dwelling bats gave birth earlier than their conspecifics roosting in foliage or trees (Kurta 2010). These temporal differences could translate to important advantages for building-roosting juveniles; for example, they have more time to explore new roosts and foraging sites and to prepare for the onset of hibernation.

Hibernating bats may also benefit from thermal advantages in buildings. For example, thermoregulation of *E. fuscus* hibernating in buildings was more similar to that of tree-dwelling species than to that of cave-hibernating conspecifics (Halsall et al. 2012). The authors argued that bats hibernating in buildings may benefit to a larger extent from passive rewarming (Halsall et al. 2012), which may lead to massive savings of crucial fat depots (Turbill et al. 2003; Geiser and Drury 2003). This notion is also supported by the observation that some bats, such as *Nyctalus noctula*, choose crevices behind sun-exposed walls when hibernating in buildings (Bihari and Bakos 2001).

In addition to thermal benefits, synanthropic bats may also benefit by reducing their travel distance and thus time to foraging sites, resulting in substantial energetic savings from reduced commuting distances (Knight and Jones 2009).

Presence of social or mating partners If buildings are selected as roost sites by a single bat, conspecifics may follow to benefit from social advantages (Kerth 2008). These secondary social benefits for synanthropic bats are identical to those of conspecific roosting in natural roosts. Briefly, bats that form large colonies in buildings may be less exposed to predators because of the dilution effect. They may as well benefit from information transfer and by cooperation among members of the same social unit. Clustered emergence of bats from a roost may constitute an antipredator behavior (Speakman et al. 1995), yet clustered emergence may be disrupted in large colonies due to bottleneck effects (Speakman et al. 1999). Gillam et al. (2011) found non-random patterns when pit-tagged *Eptesicus fuscus* emerged from buildings, indicating that these bats may form social bonds that likely influence their foraging. Information transfer might also be involved during swarming at buildings as observed in some temperate zone bats (Kanuch et al. 2010; Šuba et al. 2010). Finally, bats may explore buildings in search of mating partners. For example, buildings are known to be used as mating roosts in a number of species, such as greater sac-winged bats, *Saccopteryx bilineata* (Bradbury and Emmons 1974; Bradbury and Vehrencamp 1976), greater mouse-eared bats, *Myotis myotis* (Dietz et al. 2009), spear-nosed bat, *Phyllostomus hastatus* (Santos et al. 2003), and free-tailed bats, such as *Tadarida brasiliensis* and *Mops condylurus* (Vivier and van der Merwe 2001).

14.3.2 Enhanced Access to Habitats by Using Buildings as Ecological Stepping Stones

Extending the aforementioned argument that bats may benefit from using buildings as shelters by shortening travel distances to foraging habitats, one could argue that bats may even be able to explore and exploit new habitats by using buildings as ecological stepping stones. For example, some uniform and homogenous agricultural habitats, such as the former prairies of the Midwestern USA, are nearly void of roosting structures. Therefore, it is almost impossible for aerial-hawking insectivorous bats to use these habitats, unless artificial roosting structures are available. Here, buildings may present pivotal resources for bats to survive in an otherwise hostile environment. Farm buildings, villages, and cities may create structurally complex islands used by bat colonies (Coleman and Barclay 2012a), and this could possibly lead to an increase in local species richness. Some synanthropic bats, such as *Mops condylurus*, are capable of using exceedingly hot roosts (40 °C) which allow them to colonize habitats that other bats with a lower tolerance toward high roost temperatures are not able to exploit (Maloney et al. 1999), suggesting that heat tolerance might be favorable for bats with a synanthropic lifestyle.

In forested areas, buildings may provide roosting structures for cave-roosting bats, i.e., for bats that do not use tree hollows or crevices. By using buildings as roosts, these bats may gain access to other habitats. For example, in a forest habitat in Central Europe, bats that typically do not occupy tree cavities, such as *Eptesicus serotinus* and *Vespertilio murinus*, will instead inhabit buildings. By doing this, they gain access to insect-rich forest habitats (Mazurska and Ruczyński 2008).

Buildings can also provide roosting sites for cave-roosting bats in urban areas. For example, *Otomops martiensseni* exploits buildings only in the city of Durban, South Africa, while elsewhere in its range it uses caves as roosts. Despite the reduced availability of food and intensive large-scale agricultural land use in the surrounding landscape, the species is quite common in Durban (Fenton et al. 2002). Similarly, Moutou's free-tailed bat, *Mormopterus francoismoutoui*, uses a variety of human structures (e.g., roof slats, window shutters) across the island of La Réunion, Mauritius, yet it was thought to be restricted to roosts in lava tubes and crevices along cliff faces before the colonization of the island by European settlers (~AD 1500; Goodman et al. 2008a). Seemingly, this species has profited from the large-scale changes that occurred on this island over the past centuries. In summary, buildings may present an important resource for synanthropic bats that could increase foraging ranges of individual bats as well as the diversity of local bat assemblages.

14.3.3 Expansion of Geographic Ranges

The use of buildings as roosts may also lead to the expansion of a species' geographic range (Kunz and Reynolds 2003). Some temperate bat species such as greater mouse-eared bats, *Myotis myotis*, and lesser and greater horseshoe bats, *Rhinolophus hipposideros* and *Rhinolophus ferrumequinum*, respectively, predominantly form maternity roosts in caves in southern Europe but occupy mostly attics of large buildings (e.g., churches and castles) in more northern regions of their geographic ranges where cave temperatures are too cold to host cave-roosting maternity colonies (Dietz et al. 2009). The notch-eared bat, *Myotis emarginatus*, also uses buildings as maternity roosts in the northernmost part of their range in Europe (Dekker et al. 2013). Frafjord (2007) observed a small nursery colony of the northern bat, *Eptesicus nilssonii*, in the attic of a cabin at the northern extent of the species range. The roost was only occupied when human inhabitants heated the house for their own use, giving support to the idea that bats benefited from the warmer roost temperatures. The use of buildings as a driving force to reach more northern limits of their geographic ranges (in the Northern hemisphere) has been suggested for *E. nilssonii* and soprano pipistrelles, *Pipistrellus pygmaeus*, in Norway (Michaelsen et al. 2004), *Pipistrellus pipistrellus*, in Sweden (Ahlen et al. 2004), and *Eptesicus fuscus*, in North America (McAlpine et al. 2002). Bats may also benefit from favorable thermal conditions in buildings at higher elevations and may thus go beyond their normal elevational range. For example, a maternity colony of the rare eastern small-footed myotis, *Myotis leibii*, was found roosting in a high-elevation cabin above the previously known elevational limits for this species (O'Keefe and LaVoie 2011).

A similar argument can be made for hibernating bats in buildings. Strelkov (2002) made the point that the ability of some European bat species, such as *Nyctalus noctula*, to hibernate in buildings may have enabled them to overwinter in more northern regions than when using exclusively natural roosts. By doing so, *Nyctalus noctula* are closer to their breeding ranges when arousing from hibernation in spring, which gives them an advantage in terms of time and energy in relation to conspecifics that migrate to more southern areas. This could lead to the expansion of this species' geographic range northward.

14.4 Negative Consequences of a Synanthropic Lifestyle in Bats

14.4.1 Decreased Fitness Owing to Direct Threats

Humans The foremost direct threat for synanthropic bats are humans. The co-occupancy of buildings by bats and humans gives rise to various conflicts. Interestingly, many early papers that discuss bats in buildings deal largely with the eradication or control of bats roosting in buildings (e.g., Silver 1935; Daver 1953; Kunz et al. 1977; Barclay 1980). These papers were gradually replaced by descriptive papers about the biology of synanthropic bats and eventually by those focusing on conservation topics. Nonetheless, eradication of bats from buildings is an eminent, yet mostly undocumented, problem, in all regions of the world. Unfortunately, there are no data available on the number of bats killed each year by closing entrances to daytime roosts in buildings, by destroying roosts, or by fumigating or poisoning bats. In many countries, bats are considered pests or vermin. Therefore, eradication of whole colonies is commonly practiced. In some African countries, synanthropic molossids are consumed by humans as a delicacy (Goodman et al. 2008b), and the bats' distinct odor is regarded favorably from a culinary perspective (Allen et al. 1917), yet synanthropic bats as a form of bushmeat is rather the exception (Mildenstein and Tanshi 2015).

Buildings as traps Buildings may act as traps for bats, when bats that enter a building through open windows or structural gaps are unable to find the exit (Gaisler 1998). In Europe, *Pipistrellus pipistrellus* is most often trapped during autumn swarming (Pfalzer and Weber 2007; Kanuch et al. 2010). Bats may also be killed if they become trapped in wire mesh that is used to protect buildings from feral pigeons (König and Neumann 1996).

Predators Bats may also be killed by synanthropic predators. Some birds of prey, e.g., European kestrels and tawny and barn owls, specialize on bats that use buildings as roosts (Kovats et al. 2008; Lesiński et al. 2013; Mikula et al. 2013). Indeed, the high density of kestrels in Rome, Italy, is thought to be related to the abundance of feral pigeons, swifts, and bats (Salvati et al. 1999). In the Neotropics, great kiskadees, *Pitangus sulphuratus*, have been observed hunting *Myotis nigricans* and *Myotis albescens* when bats emerged from a building (Fischer et al. 2010). African goshawks, *Accipiter tachiro*, have attacked molossid bats, *Mops condylurus* and *Chaerephon pumilus*, near their roosts in buildings (Fenton et al. 1994). Synanthropic bats that fall to the ground or fly close to the ground may be captured and killed by domestic cats (Bruijn 1990; Ancillotto et al. 2013). Snakes and invertebrates have also been observed hunting bats in or at buildings (Esbérard and Vrcibradic 2007; Nyffeler and Knörnschild 2013).

In some cases, natural predators may have devastating effects on bats, in particular when they specialize on hunting emerging bats at the entrance of roosts. Synanthropic owls are especially efficient predators of bats in or around buildings. An effective protective measure is to install a small water hose above the entrance of a colony that is triggered by the presence of a perching predator (pers. communication K. Kugelschafter, Fig. 14.3). Since owls are puzzled by the sudden stream of water, they immediately leave the entrance without any harm done to the predator (pers. commun. K. Kugelschafter).

Pathogens Mühldorfer et al. (2012) reported that one-third of bat deaths in Germany were due to bacterial infections. According to this study, viral infections were less important as a natural cause of death, even though rabies infections are documented in some populations of synanthropic bats (O'Shea et al. 2012; Racey et al. 2013). Yet it is unknown to what extent bats suffer from rabies and whether disease dynamics are exacerbated in synanthropic species because of their specific choice of buildings.

14.4.2 Decreased Fitness Owing to Indirect Threats

Roosting in buildings, particularly in urban environments, may provide fitness benefits for bats. However, Coleman and Barclay (2012b) concluded that urban bats did not perform better in terms of body condition, reproductive rate, and number of weaned juveniles compared to rural bats. Indeed, bats seemed to perform best in the transition zone between urban and rural sites, and thus, the authors summarized that the process of urbanization may be universally detrimental to bats (Coleman and Barclay 2012b).

Fig. 14.3 Method to repel owls from the entrance of bat colonies in Germany. Water flows from the overhead water hose when the perching owl (*Strix aluco*) interrupts a light beam at the entrance to a colony of greater mouse-eared bats (*Myotis myotis*) (copyright Kugelschafter K)

Chemical pollutants Indirect threats for synanthropic bats are numerous, yet because of their subtle nature, it is more difficult to pinpoint indirect mortality risks. Chemical pollutants are likely an indirect threat for bats roosting in buildings (Mitchell-Jones et al. 1989; Bayat et al. 2014). Wood used in buildings is usually treated by chemicals such as chlorinated hydrocarbons to prevent rotting caused by fungi or insects. Bats roosting in buildings are in close contact with wooden structures and thus may be exposed to chemical preservatives such as lindane, pentachlorophenol, and pyrethroids (Racey and Swift 1986; Boyd and Myhill 1988; Mitchell-Jones et al. 1989; Shore et al. 1990, 1991; Bennet and Thies 2007). Most often, bats do not die immediately after contact with treated wood but instead suffer sublethal effects that cause, for example, immune suppression (Corrao et al. 1985; Clark and Shore 2001). Sometimes, bats are eradicated from buildings using poisons, and these poisons continue to persist so that bats may be exposed to toxic residues for extended periods after application. Poisoning is usually a gradual process that is exacerbated in temperate bats via the accumulation of toxic compounds in fat tissue and seasonal mobilization of these compounds during migration and hibernation (Mitchell-Jones et al. 1989; Bayat et al. 2014). Since the introduction of alternative bat-friendly chemical treatments of wood in buildings, mortality caused by chlorinated hydrocarbons has decreased markedly (Bayat et al. 2014), yet from a global perspective, the problem of slow poisoning of synanthropic bats in buildings remains an issue.

Parasites Patterson et al. (2007) found that bats inhabiting relatively permanent roosts, such as caves and buildings, carry more ectoparasites than bats that roost in temporal structures, such as leaves or tree hollows. Buildings may provide ectoparasites, e.g., streblid flies, bed bugs (*Cimex* spp.), or reduviid bugs (Triatominae; Reduviidae), an ideal substrate for egg laying and larval development. As female bats may be immunosuppressed during pregnancy, they may suffer from heavy parasite infestation during reproduction (Christe et al. 2000; Pearce and O'Shea 2007). The combined effect of inflammation and immune challenge may then increase oxidative stress and consequently reduce longevity in house-dwelling bats (Schneeberger et al. 2013; Lilley et al. 2014). Endoparasites are poorly studied in synanthropic bats. *Leishmania braziliensis* occurs in Brazilian house-dwelling bats, yet it is unknown whether roost choice and colony dynamics are different from those of conspecifics roosting in natural roosts and whether building roosts may impose a higher risk of contracting these parasites (Shapiro et al. 2013).

14.5 Consequences for Humans Sharing Buildings with Bats

14.5.1 Benefits of Sharing a Building with Bats

There are several direct benefits for humans when sharing buildings with synanthropic bats. Bats provide essential ecological services (e.g., pest suppression, pollination, seed dispersal) near houses, villages, and cities (Jones et al. 2009; Kunz et al. 2011; Ghanem and Voigt 2012). For example, synanthropic bats, such as molossids, feed on large quantities of insects that are vectors of human diseases, such as dengue, yellow fever, and chikungunya fever (Andrianaivoarivelo et al. 2006; Goodman et al. 2008b). In tropical and subtropical regions, bats are important seed and pollen dispersers. Orchards in house gardens may largely benefit from the cost-free ecosystem services provided by pollinating bats. Insectivorous bats have the ability to reduce insect herbivory in temperate forests (Böhm et al. 2011), tropical forests (Kalka et al. 2008), and tropical agricultural fields (Williams-Guillén et al. 2008; Maas et al. 2013). Thus, the presence of synanthropic bats comes with large, yet mostly unacknowledged, benefits to humans. Lastly, bats are an integral component of our natural heritage, and thus, they have intrinsic value (Soulé 1985).

14.5.2 Pathogen and Parasite Exposure

Viruses Bats inhabiting buildings may be reservoir hosts of viruses. For example, North American *Eptesicus fuscus* and Eurasian *Eptesicus serotinus* are both synanthropic species roosting in buildings, and they are known for their relatively high prevalence of rabies (Zorya 2002; O'Shea et al. 2012; Racey et al. 2013). In Dutch populations of *Eptesicus serotinus*, bats exhibited a 21 % seroprevalence for lyssavirus (Van der Poel et al. 2005), yet is unknown how many of these positive cases were infectious. In another Dutch study, 30 % of sampled bats that bit humans tested positive for European bat lyssaviruses (Takumi et al. 2009). Other synanthropic bat species may carry lyssaviruses, such as the molossid bats *Tadarida brasiliensis* or *Nyctinomops macrotis* in North and South America, or vespertilionid bats such as *Eptesicus furinalis* in South America (Clark et al. 1996; Uieda 1998; Passos et al. 1998; de Almeida et al. 2011; Favi et al. 2012) or nycterid bats such as *Nycteris thebaica* in Zimbabwe (Foggin 1988). In Kenya, SARS-like coronaviruses (CoVs) were identified in a *Chaerephon* spp. (Tong et al. 2009), and in South Africa, bat-derived CoVs that are closely related to the MERS-CoV were found in *Neoromicia capensis* (Corman et al. 2014). Frequent roost switching of synanthropic bats may increase the transmission risk of the rabies virus to humans (Ellison et al. 2007), particularly when humans try to evict bats from houses (Streicker et al. 2013). In general, precautionary measures should be taken when handling synanthropic bats: (1) Do not touch or handle bats without gloved hands, and (2) in case of a bat bite, immediately proceed to the appropriate facility for post-exposure prophylactics. A more detailed treatment of bat-related diseases is provided in Chap. 10 (Schneeberger and Voigt 2016).

Bacteria Bacterial infections are one of the primary causes of natural death in temperate bats (Mühldorfer et al. 2012), and many of the documented bacterial strains are relevant to human health. For example, bats may act as a reservoir for *Bartonella/Burkholderia* bacteria, which can be transmitted to humans via bed bugs (Saenz et al. 2013). Bat ticks, specifically *Argas vespertilionis*, collected from a human-inhabited building were documented to carry *Borrelia*, *Rickettisa*, and *Ehrlichia* species (Socolovschi et al. 2012). *Staphylococcus nepalensis* was detected in guano samples from mixed *M. myotis* and *M. blythii* summer roosts, and guano in or near buildings may pose a significant threat to human health (Vandzurova et al. 2013). To our knowledge, no direct infection of humans with bat-related bacterial strains has been described. Overall, synanthropic bats have the potential to transmit zoonotic diseases, yet as outlined by Mühldorfer et al. (2011), there is no evidence, at least for temperate zone bats, that they pose a greater health risk to humans than other wildlife species.

Parasites Besides bat-specific ectoparasites, bats may also carry generalist ectoparasites that could infect humans as well. For example, bed bugs (*Cimex* spp.) could possibly switch between bat roosts and rooms inhabited by humans (Pearce and O'Shea 2007). Bat ticks have been suggested to cause inflammatory responses in humans living in a building with bats in the attic (Labruna et al. 2014). Ticks associated with bats, and known to bite humans, may also be carriers of bacteria or viruses that can cause disease in humans. For example, *Carios kelleyi* collected from residential and community buildings in Jackson County, Iowa, tested positive for *Rickettsia* (Loftis et al. 2005). In addition, some endoparasites are threats to human health, yet many depend on an invertebrate host as a vector for transmission to humans. For example, in Brazil, *Leishmania braziliensis* occur in some synanthropic bat species that serve as a reservoir host for leishmaniasis but require sand flies as a vector (Shapiro et al. 2013).

Fungus Environments soiled with large accumulations of guano may harbor *Histoplasma capsulatum*, a fungal pathogen that causes histoplasmosis. When roosts in attics, roofs, and other rooms are not cleaned on a regular basis, guano accumulates creating a greater risk to humans (Bartlett et al. 1982; Martins et al. 2000). Humans can develop histoplasmosis after inhaling the microscopic spores of *H. capsulatum*, often while participating in activities that disturb a heavily contaminated environment. While histoplasmosis is rarely fatal, infections in individuals with weakened immune systems can become severe (Martins et al. 2000), yet it is questionable that infections by *H. capsulatum* can be traced back to bats.

14.5.3 Noise, Odor, Dust, and Activity

Although echolocation calls emitted by most bats in open space are not audible to humans, many social vocalizations of bats are noticeable because they are typically below the 20 kHz auditory threshold of humans. These vocalizations may be particularly evident at times of the year when pups use contact calls to attract their returning mothers. Such vocalizations combined with noises caused by terrestrial locomotion of bat inhabitants, e.g., molossid bats moving through small crevices below tin roofs, can be a nuisance for human inhabitants. In addition, humans sometimes complain about bat-related odors and dust (Razafindrakoto et al. 2011).

14.5.4 Harmful Bats

Bat feces is suggested to have antigenic properties, causing skin rashes in susceptible humans (Alonso et al. 1998), yet detailed studies are lacking. To our knowledge, there is only one bat species worldwide that could be directly harmful to humans. The common vampire bat, *Desmodus rotundus*, consumes mammalian blood but is restricted to Latin America. Although this species feeds primarily on livestock animals, e.g., cattle (Delpietro et al. 1992; Voigt and Kelm 2006), vampire bats may feed on sleeping humans not protected inside buildings (Schneider et al. 2001; Carvalho-Costa et al. 2012). Though vampire bats are not known to inhabit occupied buildings, in some areas of South America, these bats inhabit abandoned buildings next to occupied houses (Mialhe 2013). Besides the potential of contracting rabies via a bite, humans can suffer from inflammation, secondary infections, and blood loss. Overall, humans are not a regular victim for vampire bats.

14.5.5 Destruction of Buildings Caused by Bat Excreta

Bats may inhabit buildings over many years, or even centuries, and accumulated feces and urine may cause severe damage to buildings. For example, bat guano was the cause of damage to some buildings of the UNESCO World Heritage Centre—Angkor monuments in Cambodia. Apparently, salts in excretions of bats are eroding the sandstone of some ancient buildings (Hosono et al. 2006). In most cases, structural damage can be prevented by removing accumulations of guano. Plastic sheets can be placed over exposed structures to protect them and facilitate the removal of urine and guano; in addition, wooden boards placed directly under roosts may also be helpful in collecting bat excreta from roosts inside buildings.

14.6 Conservation of Bats in Buildings: Avoidance, Mitigation, and Compensation

The protection of synanthropic bats and their roosts should occur in a tri-level hierarchical pattern. First, it should be determined whether bat roosts can be preserved, e.g., left untouched, even when construction work is carried out near the roost. Second, if construction work affects the roost, developers and architects should mitigate the impact on the bat colony (mitigation). Lastly, if bat roosts are going to be lost, when, for example, barns are converted into apartments (Briggs 2004), appropriate compensation measures should be practiced in order to offer bats an alternative roost. Although this general approach may not be applicable in all countries, particularly when the legal framework is lacking, we will elaborate on it in the remainder of the chapter.

Conservation guidelines for bats in buildings have been formulated in various countries, including those from the European Union (Table 14.1; Marnell and Prsetnik 2010). Conservation networks (Kingston et al. 2016, Chap. 16) could use these and our recommendations to develop further region-specific guidelines for the protection of local synanthropic bats.

14.6.1 General Considerations for the Conservation of Bats in Buildings

Monitoring of colonies Monitoring of bat colonies, particularly maternity colonies, in buildings needs to be conducted with appropriate care (Kunz and Reynolds 2003). In some countries, it is legally forbidden to disturb bats in their roosts, particularly during the maternity period. Kunz and Reynolds (2003) suggested conducting evening emergence counts at roost exits to monitor maternity colonies without disturbing bats.

Table 14.1 List of Web-based resources pertaining to the conservation of synanthropic bats (sorted alphabetically according to continent or country)

Country	Web address
EU	http://www.eurobats.org/sites/default/files/documents/publications/publication_series/pubseries_no4_english_2nd_edition.pdf
France	http://www.sfepm.org/chiropteres.htm
Australia	http://ausbats.org.au/#/bats-in-your-house/4569171536
Latin America	http://www.relcomlatinoamerica.net/images/PDFs/PROTOCOLO.pdf
Germany	http://www.nabu.de/tiereundpflanzen/saeugetiere/fledermaeuse/aktivwerden/01506.html
Ireland	http://www.batconservationireland.org
Italy	http://biocenosi.dipbsf.uninsubria.it/chiroptera/
Netherlands	http://www.vzz.nl
Russia	http://zmmu.msu.ru/bats/popular/v_dome.htm
UK	http://www.bats.org.uk/pages/bats_and_buildings.html
UK	http://www.bedsbatgroup.org.uk/wordpress/?page_id=3429
UK	http://jncc.defra.gov.uk/page-2861
UK	http://www.naturalengland.org.uk/ourwork/regulation/wildlife/species/bats.aspx
USA	http://www.conservewildlifenj.org/protecting/projects/bat/buildings/
USA	http://www.nature.nps.gov/biology/wns/assets/docs/2012BatsInBuildingsWebinarOdegard.pptx

Life stages of bats For effective protection of synanthropic bats, it is crucial to understand the purpose of the buildings being used as roosts by bats. We have outlined several possibilities for why bats use buildings. Since bats may be particularly vulnerable during their reproductive period and during hibernation, roosts that are used by bats during these life stages are of prime concern for conservation efforts. The central recommendation for such roosts is to leave them untouched, unless gradual deterioration of the building may destroy the roost.

Human occupancy Usually, disturbance of synanthropic bats by humans is detrimental to colonies. For example, de Boer et al. (2013) showed for the Netherlands that hibernacula in buildings were more suitable for bats when disturbance by humans was low. However, it should be noted that some studies report that synanthropic bats tend to leave roosts when humans no longer use buildings, possibly because buildings are no longer heated (Frafjord 2007). In Poland, Sachanowicz and Wower (2013) found evidence that the gradual deterioration of buildings caused an impoverishment of species in the local assemblages of attic-dwelling bats. Therefore, human occupancy of buildings may be a benefit in some circumstances and a disadvantage in others, depending on the species involved and the specific life stages.

Interior of roosts The size and spatial structure of building interiors affects the occupancy by synanthropic bats. For example, the availability of sufficient space and optimal microclimatic conditions seem to be beneficial for attic-dwelling bats, such as the endangered Townsend's big-eared bat *Corynorhinus townsendii*

(Betts 2010) and Rafinesque's big-eared bat, *Corynorhinus rafinesquii* (Loeb and Zarnoch 2011). In addition to roost compartments, relatively higher ambient temperatures in roost interiors are also relevant for bats inhabiting buildings (Entwistle et al. 1997). *Eptesicus fuscus* prefer old buildings with galvanized (tin) roofs that are also taller than surrounding buildings, most likely because of higher temperatures and wider temperature gradients in these buildings (Williams and Brittingham 1997). For some hibernating bats, the size and number of hiding places may contribute to the quality of hibernacula in buildings.

Exterior of roosts Synanthropic bats not only depend on suitable roosting interiors, but also depend on the quality of the surrounding environment, e.g., for foraging or drinking. Suitable roost entrances are critical for some bats, particularly for fast-flying species with a low ability to maneuver (Neubaum et al. 2007). For example, *Nyctalus noctula* roosting in buildings preferred roosts that were located at the top floors (Bihari 2004; Cel'uch and Kaňuch 2005). Molossids, e.g., *Chaerephon ansorgei*, and vespertilionids, e.g., *Neoromicia capensis*, that inhabit crevices or narrow spaces under roofs are capable of landing and crawling through narrow roost entrances, whereas horseshoe bats, e.g., *Rhinolophus clivosus*, and slit-faced bats, e.g., *Nycteris thebaica*, require an opening large enough to fly through since they usually do not crawl (Monadjem et al. 2010). Other species, such as *Pipistrellus pipistrellus*, are generalists with respect to their roost preference, i.e., they do not prefer specific structural attributes of buildings (Jenkins et al. 1998). It is also noteworthy that some species may require several roosts in separate buildings to establish a stable colony, e.g., greater horseshoe bats, *Rhinolophus ferrumequinum* (Maltagliati et al. 2013), eastern pipistrelles, *Pipistrellus subflavus* (Whitaker 1998), and *Eptesicus fuscus* (Ellison et al. 2007; O'Shea et al. 2012).

Additional landscape elements, such as vegetation and water sources, have been suggested to promote bat populations in cities (Neubaum et al. 2007). Trees in the vicinity of roosts were beneficial for pipistrelle bats, *Pipistrellus pipistrellus*, not only as foraging grounds but also as a protection against aerial predators, thus enabling bats to increase their nocturnal foraging activity substantially by emerging earlier from their roost (Jenkins et al. 1998). Brown long-eared bats, *Plecotus auritus*, preferred buildings situated close to woodland and water (Entwistle et al. 1997; Moussy 2011).

Illumination of buildings at night by streetlamps reduces the quality of roosts for some bats. For example, European *Rhinolophus ferrumequinum*, *Myotis emarginatus, and Myotis oxygnathus* emerged later at sunset from roosts when buildings were illuminated. Also, body mass and forearm length were smaller in juveniles from illuminated buildings than in those from not illuminated. In the worst case, roosts are abandoned after direct lighting of the buildings in which the roost is located (Boldogh et al. 2007).

Eviction of bats from roost Eviction of bats from houses is practiced worldwide, yet it is against the law in some countries. The corresponding authorities may grant concessions if there is no alternative to the exclusion of bats from roosts. Yet, in many countries, it is a legal requisite that appropriate measures are practiced to compensate for the loss of a roost. The permanent closure of roost exits or the destruction of a roost should only be considered during times when bats are not using the roost, e.g., outside the reproduction or hibernation period. Otherwise, bats may be trapped and killed, which is against animal welfare. If roosts are destroyed or closed, bats may switch to alternative roosts (Neilson and Fenton 1994). After eviction of *Eptesicus fuscus* from buildings, females produced fewer offspring at alternative sites, even though foraging behavior remained constant (Brigham and Fenton 1986). Relocation of bats to nearby habitats usually fails because bats will return to their original roost in most cases. Lastly, the permanent eviction of bats from roosts may increase the frequency of roost switching. In the case of species with a high prevalence of rabies infections, it is predicted that the rabies transmission risk may increase due to more, and possibly undirected, movements of evicted bats around buildings (e.g., *Eptesicus fuscus*, Streicker et al. 2013). Therefore, roost closures might have unforeseen and unwanted side effects for public health.

14.6.2 Avoiding or Mitigating Roost Losses in Buildings

Roosts are key resources for bats since many species are limited by roost availability (Kunz 1982; Kunz and Lumsden 2003). Therefore, a prime conservation effort should be the protection of existing roosts and possibly the enhancement of their quality. If private or commercial development of buildings is an inevitable conflict with synanthropic bats, appropriate measures should be practiced, particularly when the species is endangered and/or protected.

Reduction of human disturbance Disturbance of bat roosts in buildings can lead to a variety of outcomes, ranging from direct effects when people disturb building roosts to indirect effects of noise and light pollution. Bats seem to adjust quickly to noise, yet as Rowse et al. (2015) point out, some species may be quite sensitive to artificial light. For example, *Pipistrellus pipistrellus* are quite tolerant to artificial light during foraging, but altered their emergence behavior when exposed to different light intensities at their daytime roost (Down et al. 2003). Directing artificial light at roost entrances may have a negative impact on bats roosting in buildings (Boldogh et al. 2007). Adjusting the regime of artificial light near a colony and reducing the light spill from neighboring buildings or streetlamps should be considered to improve the quality of roosts in buildings.

It is important to recognize that human visits to hibernacula of bats in buildings might cause bats to arouse from hibernation, a process that is energetically costly and causes bats to deplete their fat depots which increases the risk of starvation (Speakman and Thomas 2003). Therefore, it is necessary to cease visitations to known hibernacula to minimize impacts on hibernating bats.

The impact of disturbance caused by structural work in buildings, e.g., renovation of roof structures or attics, can have severe consequences for synanthropic bats. Indeed, colonies will abandon roosts because of this disturbance. To minimize these negative impacts, construction work should only take place during the annual period when bats are not using the roost. For example, renovation of attics used by bats as maternity roosts should only occur after juveniles fledge or when colony members leave the roost to hibernate in another location. Minimizing disturbance is also vital for protecting hibernacula, and construction work at these sites should not be conducted during the hibernation period.

Conservation of smaller-sized roosts Reduction of the size of bat roosts inside buildings might be acceptable if the only alternative is the complete loss of a roost. For example, attics or barns are sometimes converted to apartments or houses, respectively. If bats are roosting in an attic or barn, a small part of it could be separated from the space used by humans and this smaller space could be designated for the exclusive use by bats. However, it should be noted that many bat species roosting in attics or barns prefer large and complex structures with some variation in microclimate conditions. A decrease in size and structural complexity of the roost space may lead to the gradual decline in colony size and possibly complete loss. Therefore, a reduction in roost size may best be accompanied by the provision of new artificial roosts that are suitable for the specific bat species (Figs. 14.4 and 14.5; Kunz and Reynolds 2003).

Fig. 14.4 Artificial bat roost on the exterior of the Leibniz Institute for Zoo and Wildlife Research building in Berlin, Germany. *Nyctalus noctula* use the roost during autumn

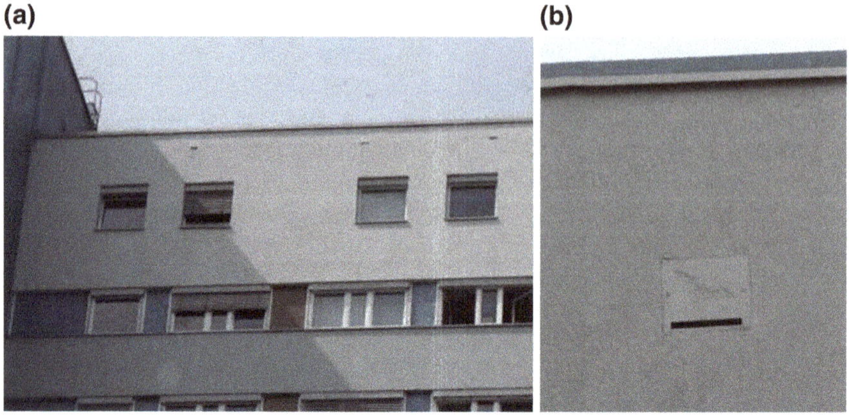

Fig. 14.5 Artificial bat roosts embedded into the external insulation layer of a renovated public building in Berlin, Germany: **a** row of artificial roosts within the top floor of a seven-story building; **b** detail of a single artificial roost (the horizontal exit is at the base). Such roosts are suitable as hibernation sites and stopover sites during migration for noctule bats, *Noctula noctula*, in Europe, yet they may not host as many individuals as buildings before renovation

14.6.3 Compensating for Lost Roosts

Sometimes it is inevitable that roosts in buildings are lost. The addition of artificial bat boxes near previously occupied buildings can successfully compensate in some instances. For example, colonies of *Pipistrellus pygmaeus* and *Plecotus auritus* and various other species throughout Europe benefited from artificial roosts when the original roost was destroyed (Anonymous 2006; Beck and Schelbert 1999). Artificial bat roosts were also provided for and accepted by South American *Molossus molossus* when roosts in buildings were destroyed (Alberico et al. 2004). In North America, *Eptesicus fuscus*, and *Myotis lucifugus*, will occupy artificial bat boxes installed at buildings that formerly housed colonies (Brittingham and Williams 2000). For example, the Bat House Research Project in the Kruger National Park, South Africa, has recently provided new accommodation for bats in the Letaba Rest Camp in an effort to help identify the most effective way to remove bats from buildings within the park (http://www.krugerpark.co.za/krugerpark-times-2-11-bat-accommodation-19864.html). Similar attempts to provide alternative roosting structures for synanthropic bats have been successful in the USA; for example, artificial roosts have been built on the campus of the University of Florida to host populations of *Tadarida brasiliensis* and other native bats (https://www.flmnh.ufl.edu/index.php/bats/home/).

These success stories should not imply that roosts in building are replaceable by artificial structures and that bats will readily occupy artificial roosts. Sometimes, for unknown reasons, bats avoid artificial roosts in buildings completely. Therefore, protection of existing roosts should be considered prior to attempting the use of artificial roosts.

14.6.4 Loss of Roosts Due to Demographic Changes in the Human Population

Demographic changes in human populations of many countries are turning rural areas into areas nearly devoid of humans. As a result, buildings are abandoned and, due to a lack of maintenance, deteriorate over time. Shortly after abandonment, many synanthropic bat species benefit, likely due to the reduced disturbance by humans. Deserted buildings may provide new roosting structures for bats, e.g., for *Hipposideros nicobarulae* in Myanmar (Douangboupha et al. 2012). Yet in the long run, synanthropic bats may vanish from these sites when buildings deteriorate (Sachanowicz and Wower 2013). Another effect of demographic changes involves movement and thus concentration of people in urban areas. Following this, previously unused buildings, even in industrial areas, or unoccupied space under the roof of buildings are converted into houses or apartments to host the influx of people in cities. This may cause losses of roosting opportunities for synanthropic bats. In China, like in many Asian countries, a vast number of old buildings are demolished during the process of modernization and this reduces the density of roosts significantly for synanthropic bats (Zhang et al. 2009).

14.7 Examples of Good Practice

14.7.1 Example 1: The Outreach Program for the "Bat-Friendly House"

To conserve synanthropic bat species, education appears to be the prime method to protect bat roosts in buildings. Kingston (2016, Chaps. 17 and 18) address various outreach approaches. Here, we focus on a specific German-based conservation program called "bat-friendly house." Directed by a consortium of nonprofit organizations (spearheaded by the "Naturschutzbund" Germany) and federal and local authorities and bat conservationists, the program has created a "Bat-Friendly House" award for owners who protect bat colonies in their buildings. The major goal of this program is to support populations of synanthropic bats by maintaining or even enhancing their roosts and to involve local people in the protection of bats. Several hundred houses have been deemed bat-friendly in the federal states of Hessen, Schleswig-Holstein and Northrine-Westfalia and others in Germany. The award ceremony is usually accompanied by a press campaign to raise awareness about the conservation of bats that use building as their roosts. Similar programs have been initiated in other EU countries.

14.7.2 Example 2: Renovated Buildings Designated for Bat Conservation Purposes

There are many examples of buildings that were renovated successfully to mitigate the human–bat conflict or to protect endangered bats. The details of the vast majority of these cases have not been documented or published. Yet, it is encouraging to read about some of the examples on Web pages or in the gray literature of nongovernmental organizations (Table 14.1). The EUROBATS publication, available at www.eurobats.org, provides examples of successful projects throughout Europe. Many of these examples underline that the details of specific conservation efforts depend largely on the biology of the target bat species and on local circumstances, ranging from the building in question, the overall legal framework, and the funding agencies and the authorities and persons involved. We have summarized some general features in the next section that might be relevant for many synanthropic bats, but we cannot provide a comprehensive overview of all projects. We have also refrained from repeating case studies that have already been described in detail at other places. Instead, we focus on a single example that we consider successful because it combines efficiently the practical aspects of protecting a building for an endangered bat species, preservation and enhancement of suitable habitats, and a community-based outreach program to facilitate the acceptance and thus continuation of the project beyond the funding period.

Protection of the last maternity colony of greater horseshoe bats, *Rhinolophus ferrumequinum*, in Germany According to surveys over the past decades, populations of greater horseshoe bats, *Rhinolophus ferrumequinum*, are on the decline throughout Europe (Ransome and Hutson 2000; Dietz et al. 2009; Spitzenberger et al. 2010). Although some parts of southern Germany were inhabited previously by this species, today they are virtually absent from Germany except for a maternity colony found in 1992 in Hohenburg, a small village located in northeastern Bavaria adjacent to a large military training area. Because of its rarity, this species is categorized as "Threatened by Extinction" in the national red list of mammals for Germany.

The colony occupied a house and adjacent farm buildings that were built in the sixteenth century. Since the 1980s, the buildings have not been inhabited or used by humans (Fig. 14.6). Thus, the complex deteriorated and was nearly to the point of collapse when the colony was discovered. In 1992, there were 21 adults, yet it is unknown how large the colony had been before its discovery.

After initial monitoring of the bats in the colony and their feeding habits, it was decided in 2011 to apply for a grant from the European Union which supports biodiversity projects. Since the funding scheme required complementary funding sources, the applicants, namely the "Landesbund für Vogelschutz e.V.," a German NGO devoted to protecting national biodiversity, contacted additional partners, such as "Bayerischer Naturschutzfonds," "Bundesanstalt für Immobilienaufgaben," and "Naturpark Hirschwald," to reach the critical financial needs for achieving the conservation plan.

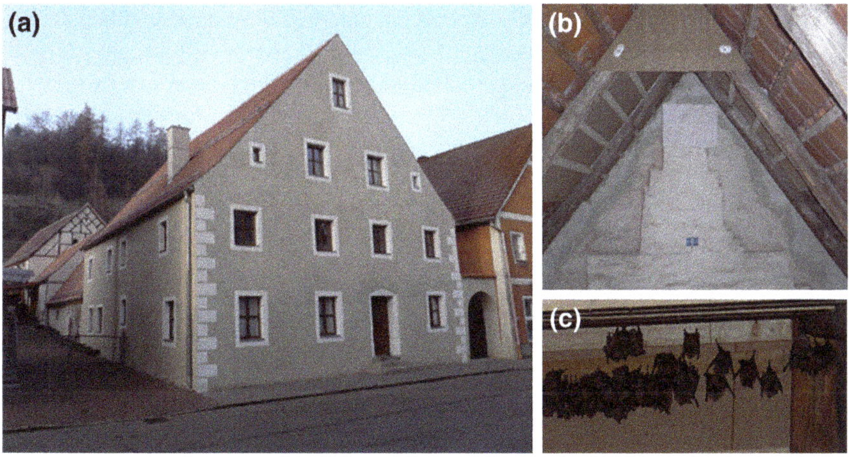

Fig. 14.6 Building complex that hosts the last maternity roost of the greater horseshoe bat, *Rhinolophus ferrumequinum*, in Germany (**a**). Bats most often use the attic of the largest backyard building (**b**). The attic ceiling functions as a heat trap where warm, upward moving air is trapped; this is the preferred roosting area for the colony (**c**)

Based on an initial investment made by the German government in support of small and intermediate companies during the 2011 bank crisis, it was first decided to renovate the complex of buildings after bats left for hibernation in nearby caves. Developers were faced with the difficult task of renovating a building complex that was protected by law, while at the same time keeping the roosting requirements of greater horseshoe bats in mind. The majority of space inside the building complex was designated for the exclusive use by bats. The ground floor level of the main buildings was transformed into an education center and some office space for the project coordinator. The fact that several attics and rooms with variable microclimatic conditions were available to the colony likely contributed to the success of the project. This is consistent with observations of roost use by greater horseshoe bats elsewhere. For example, Maltagliati et al. (2013) pointed out that the largest nursery colony of *Rhinolophus ferrumequinum* in Italy uses several buildings. The Hohenburg house was carefully modified to include some further beneficial structures for bats. For example, workers built a so-called heat dome inside the attic where warm, upward moving air is trapped in a structure that is used by bats as a roosting site (Fig. 14.6b, c). Furthermore, they created a 1-m^2 pool of water (3 cm depth) at which bats may drink. Finally, roost exits were constructed in a way that prevents predators, e.g., stone martens and domestic cats, from entering the building.

Second, it was understood that horseshoe bats would not survive if adjacent habitat structures degenerate by forest succession. Therefore, they designed a strategy to protect and indeed improve habitat structures for *Rhinolophus ferrumequinum*, a strategy that has proven successful for other synanthropic bats as well

(Murphy et al. 2012). Accordingly, a strategic plan was developed to protect foraging habitats and enhance the quality of landscape elements. Efforts are currently underway to convert forests into so-called Hutewald, which is an extensive forest used by livestock as pasture that resembles a landscape park. Furthermore, nearby hibernacula in natural caves were protected by fences to minimize disturbance of roosting bats.

Finally, the project includes an outreach program in which local people are informed about the progress of the project and engaged in fulfilling the working plan. The education center is equipped with monitors to provide real-time views into the bat colony. Project workers explain the goals of the project and show visitors the emergence of colony members at dusk. By using bat detectors and by direction observation, visitors learn firsthand about the biology of this fascinating species. Lastly, interested people might also visit the Web page of the project and observe bats using the Webcam (http://www.lbv.de/unsere-arbeit/life-natur-projekte/life-projekt-hufeisennase.html). Local hotels and restaurants in Hohenburg and adjacent villages have benefited from tourists and bat enthusiasts who come to this area for the single purpose of learning more about the Hohenburg colony of *Rhinolophus ferrumequinum*.

Since its discovery, the size of the colony has increased fourfold, numbering 94 adult *Rhinolophus ferrumequinum* and 37 juveniles in 2013 (pers. comm. Rudolf Leitl). Currently, efforts are underway to provide appropriate roosting structures in buildings and protected hibernacula in the nearby area to offer a suite of habitats for the expanding Hohenburg colony with the ultimate goal to establish further colonies in the larger region.

14.8 Synthesis and Outlook

Synanthropic bats are, by definition, in close contact with humans. Although this contact bears some risks to both humans and bats, it also provides opportunities to promote bat conservation. Practical aspects regarding the conservation of synanthropic bats in buildings, such as how to construct a new roost or enhancement of an existing building roost, should be one part of conservation efforts. From our point of view, it is equally important to engage in outreach programs and communicate with building owners about the conservation value of synanthropic bats (see also Kingston et al. 2016). With respect to research directions, we identify the following questions that need to be addressed:

1. What sensory cues do bats use to explore buildings as potential night or day roosts?
2. What are the differences in microclimate between natural and building roost sites, particularly in tropical and subtropical regions?
3. Is use of building roosts a learned behavior? Do local populations establish a tradition of inhabiting buildings?

4. Is swarming behavior unique to European bats?
5. Are there differences in the way bats use buildings between areas or continents where buildings have been in place for many centuries compared with areas where humans have only built houses recently.
6. Do tropical and subtropical bats also use buildings for extended periods of torpor, similar to hibernation of temperate zone bats?
7. What is the selective benefit for synanthropic bats inhabiting roosts in buildings compared with conspecifics inhabiting natural roosts? Why do some species commonly hibernate in buildings and others do not (see also Rintoul and Brigham 2014)?
8. Do tropical and subtropical bats exhibit similar expansions of geographic ranges when thermal benefits of using buildings as roosts are not the predominant driving benefit?
9. Is it possible to estimate the monetary value of ecosystem services provided by synanthropic bats?
10. To what extent have the geographic ranges of synanthropic bats changed in response to the coinhabitation of buildings?

Apart from these basic research questions, we need to engage in larger conservation efforts to protect synanthropic bats in developing countries, taking into account their ecological and economic value. Synanthropic bats face an uncertain future in many temperate countries due to political measures and specific programs to improve building standards, e.g., building modernization in the European Union that involves increased insulation of exterior walls has led to the large-scale eviction of synanthropic bats from buildings. We also see a strong incentive to coordinate conservation efforts to protect populations of synanthropic bats. Bats that live in the same buildings as humans could be ambassadors for the conservation of bats if other successful outcomes are replicated and publicized to a general audience. We conclude that synanthropic bats coinhabiting buildings with humans may provide good opportunities to teach humans in both urban and rural environments about wildlife species, particularly bats.

Acknowledgments We thank Drs. Mark Brigham, Scott Reynolds, and Allen Kurta for constructive comments that helped improve the chapter.

References

Aguirre LF, Moya MI, Arteaga BLL, Galarza MMI, Vargas EA, Barboza Marquez K, Peñaranda DA, Pérez-Zubieta JC, Terán VMF and Tarifa T (2010) Plan de acción para la conservación de los murciélagos amenazados de Bolivia. BIOTA-PCMB, MMAA-VBCC-DGB, UICN-SSC-BSG, CBG-UMSS. Cochabamba, Bolivia

Ahlen I, Bach L, Johansson T (2004) Första kolonin av pipistrell funnen i Sverige: a common pipistrelle *Pipistrellus pipistrellus* colony found in Sweden. Fauna och Flora 99:16–17
Alberico M, Saavedra-R CA, Paredes HG (2004) Criterios para el diseno e istalacion de casas para murcielagos: Proyecto CPM (Cali, Valle del Cauca, Colombia). Actualidades Biologicas 26:5–11
Allen JA, Lang H, Chapin JP (1917) The American Museum Congo expedition collection of bats. Am Mus Nat Hist 37:405–478
Alonso A, Irañeta SG, Rodríguez SM, Scavini LM, Rodríguez SR (1998) Bat feces as an indoor allergen. J Investig Allergol Clin Immunol 8:365–369
Altringham JD (1996) Bats: biology and behavior. Oxford University Press, United Kingdom
Ancillotto L, Serangeli MT, Russo D (2013) Curiosity killed the bat: domestic cats as bat predators. Mamm Biol 78:369–373
Andrianaivoarivelo AR, Ranaivoson N, Racey PA, Jenkins RKB (2006) The diet of three synanthropic bats (Chiroptera: Molossidae) from eastern Madagascar. Acta Chiropterol 8:439–444
Anonymous (2006) A review of the success of bat boxes in houses. Scottish Natural Heritage commissioned report 160:1–41
Aspetsberger F, Brandsen D, Jacobs DS (2003) Geographic variation in the morphology, echolocation and diet of the little free-tailed bat, *Chaerephon pumilus* (Molossidae). Afr Zool 38:245–254
Barclay RMR (1980) Comparison of methods used for controlling bats in buildings. J Wildl Manage 44:502–506
Barclay RMR, Harder LD (2003) Life histories of bat: life in the slow lane. In: Kunz TH, Fenton MB (eds) Bat ecology. Chicago University Press, pp 209–25
Bartlett PC, Vonbehren LA, Tewari RP, Martin RJ, Eagleton L, Isaac MJ, Kulkarni PS (1982) Bats in the belfry: an outbreak of histoplasmosis. Am J Pubic Health 72:1369–1372
Bayat S, Geiser F, Kristiansen P, Wilson SC (2014) Organic contaminants in bats: trends and new issues. Env Intern 63:40–52
Beck A, Schelbert B (1999) Fledermauskästen als Ersatz für zerstörte Quartiere an Bauten. Mitt der Aargauischen Naturforschenden Ges 35:115–127
Bennet BS, Thies ML (2007) Organochlorine pesticide residues in guano of Brazilian free-tailed bats, *Tadarida brasiliensis* Saint-Hilaire, from east Texas. B Environ Contam Tox 78:191–194
Betts BJ (2010) Thermoregulatory mechanisms used in a maternity colony of Townsend's big-eared bats in northeastern Oregon. Northwest Nat 91:288–298
Bielby J, Mace GM, Binina-Emonds OR, Cardillo M, Gittleman JL, Jones KE, Orme CD, Purvis A (2007) The fast-slow continuum in mammalian life history: an empirical reevaluation. Am Nat 169:748–757
Bihari Z (2004) The roost preference of *Nyctalus noctula* (Chiroptera, Vespertilionidae) in summer and the ecological background of their urbanization. Mammalia 68:329–336
Bihari Z, Bakos J (2001) Roost selection of *Nyctalus noctula* (Chiroptera, Vespertilionidae) in urban habitat. In: Woloszyn BW (ed) Proceedings of the European Bat Research Symposium, Poland, pp 29–39
Boehm SM, Wells K, Kalko EKV (2011) Top-down control of herbivory by birds and bats in the canopy of temperate broad-leaved oaks (*Quercus robur*). PLoS ONE 6(4):e17857
Boldogh S, Dobrosi D, Samu P (2007) The effects of the illumination of buildings on house-dwelling bats and its conservation consequences. Acta Chiropterol 9:527–534
Boyd IL, Myhill DG (1988) Uptake of gamma-HCH (lindane) by pipistrelle bats and its effect on survival. Environ Pollut 51:95–111
Bradbury JW, Emmons L (1974) Social organization of some Trinidad bats. I Emballonuridae Z Tierpsychol 36:137–183
Bradbury JW, Vehrencamp SL (1976) Social organization and foraging in Emballonurid bats. I. Field studies. Behav Ecol Sociobiol 1:337–381
Briggs P (2004) Effect of barn conversion on bat roost sites in Hertfordshire, England. Mammalia 68:353–364

Brigham RM, Fenton MB (1986) The influence of roost closure on the roosting and foraging behaviour of *Eptesicus fuscus* (Chiroptera: Vespertilionidae). Can J Zool 64:1128–1133

Brittingham MC, Williams LM (2000) Bat boxes as alternative roosts for displaced bat maternity colonies. Wildlife Soc B 28:197–207

Bronner G, Maloney SK, Buffenstein R (1999) Survival tactics within thermally-challenging roosts: heat tolerance and cold sensitivity in the Angolan free-tailed bat, *Mops condylurus*. S Afr J Zool 34:1–10

Bruijn Z (1990) Domestic cat *Felis catus* as a predator of bats. Lutra 33:30–34

Carvalho-Costa FA, Tedesqui VL, Monteiro MDN, Bóia MN (2012) Outbreaks of attacks by hematophagous bats in isolated riverine communities in the Brazilian Amazon: a challenge to rabies control. Zoonoses Public Health 59:272–277

Cel'uch M, Kaňuch P (2005) Winter activity and roosts of the noctule (*Nyctalus noctula*) in an urban area (central Slovakia). Lynx 36:39–45

Cel'uch M, Danko Š, Kaňuch P (2006) On urbanisaton of and *Pipistrellus pygmaeus* in Slovakia. Vespertilio 9–10:219–221

Christe P, Arlettaz R, Vogel P (2000) Variation in intensity of a parasitic mite (*Spinturnix myoti*) in relation to the reproductive cycle and immunocompetence of its bat host (*Myotis myotis*). Ecol Lett 3:207–212

Clark DR, Shore RF (2001) Chiroptera. In: Shore RF, Rattner BA (eds) Ecotoxicology of wild mammals. Wiley, pp 159–214

Clark DR, Lollar A, Cowman DF (1996) Dead and dying Brazilian free-tailed bats (*Tadarida brasiliensis*) from Texas: rabies and pesticide exposure. Southwest Nat 41:275–278

Coleman JL, Barclay RMR (2012a) Urbanization and the abundance and diversity of prairie bats. Urban Ecosyst 15:87–102

Coleman JL, Barclay RMR (2012b) Influence of urbanization on demography of little brown bats (*Myotis lucifugus*) in the prairies of North America. PLoS ONE 6(5):e20483

Corman V, Ithete NL, Richards LE, Schoeman MC, Preisler W, Drosten C, Drexler JF (2014) Rooting the phylogenetic tree of MERS-Coronavirus by characterization of an ancestral virus from an African Bat. J Virol. doi:10.1128/JVI.01498-14

Corrao A, Catalano E, Zava B (1985) Destructive effects of chlorinated pesticides on a bats colony (Chiroptera). Mammalia 49:125–130

Daver SR (1953) Eradicating bats from bungalows. J Bombay Nat Hist Soc 51:734–735

de Almeida MF, Favoretto SR, Martorelli LFA, Trezza-Netto J, Campos ACD, Ozahata CH, Sodre MM, Kataoka APAG, Sacramento DRV, Durigon EL (2011) Characterization of rabies virus isolated from a colony of *Eptesicus furinalis* bats in Brazil. Rev Inst Med Trop Sao Paulo 53:31–37

de Boer WF, van de Koppel S, de Knegt HJ, Dekker JJA (2013) Hibernation site requirements of bats in man-made hibernacula in a spatial context. Ecol Appl 23:502–514

Dekker JJA, Regelink JR, Jansen EA, Bringkmann R, Limpens HJGA (2013) Habitat use by female Geoffroy's bats (*Myotis emarginatus*) at its two northernmost maternity roosts and the implications for their conservation. Lutra 56:111–120

Delpietro HA, Marchevsky N, Simonetti E (1992) Relative population densities and predation of the common vampire bat (*Desmodus rotundus*) in natural and cattle-raising areas in northeast Argentina. Prev Vet Med 14:13–20

Dietz C, von Helversen O, Nill D (2009) Bats of Britain, Europe and northwest Africa. A & C Black, London

Douangboupha B, Bumrungsri S, Satasook C, Soisook P, Bu SSH, Aul B, Harrison DL, Pearch MJ, Thomas NM, Bates PJJ (2012) A new species of small Hipposideros (Chiroptera: Hipposideridae) from Myanmar and a reevaluation of the taxon *H. nicobarulae* Miller, 1902 from the Nicobar islands. Acta Chiropterol 13:61–78

Down NC, Beaton V, Guest J, Polanski J, Robinson SL, Racey PA (2003) The effects of illuminating the roost entrance on emergence behavior of *Pipistells pygmaeus*. Biol Conserv 111:247–252

Duvergé PL, Jones G, Rydell R, Ransome RD (2008) Functional significance of emergence timing in bats. Ecography 23:32–40
Ellison LE, O'Shea TJ, Neubaum DJ, Bowen RA (2007) Factors influencing movement probabilities of big brown bats () in buildings. Ecol Appl 17:620–627
Entwistle AC, Racey PA, Speakman JR (1997) Roost selection by the brown long-eared bat *Plecotus auritus*. J Appl Ecol 34:399–408
Esbérard CEL, Vrcibradic D (2007) Snakes preying on bats: new records from Brazil and a review of recorded cases in the Neotropical region. Rev Bras Zool 24:848–853
Favi M, Bassaletti A, Lopez J, Rodriguez L, Yung V (2012) Epidemiological description of rabies reservoir in bats in the metropolitan region: Chile 2000–2009. Rev Chil de Infectologia 28:223–228
Fenton MB, Swanepoel CM, Brigham RM, Cebek J, Hickey MBC (1990) Foraging behaviour and prey selection by large slit-faced bats (*Nycteris grandis*; Chiroptera: Nycteridae). Biotropica 22:2–8
Fenton MB, Rautenbach IL, Smith SE, Swanepoel CM, Grosell J, van Jaarsveld J (1994) Raptors and bats: threats and opportunities. Anim Behav 48:9–18
Fenton MN, Taylor PJ, Jacobs DS, Richardson EJ, Bernard E, Bouchard S, Debaeremaeker KR, ter Hofstede H, Hollis L, Lausen CL, Lister JS, Rambaldini D, Ratcliffe JM, Reddy E (2002) Researching little-known species: the African bat *Otomops martiensseni* (Chiroptera: Molossidae). Biodivs Conserv 11:1583–1606
Fischer E, Munin RL, Longo JM, Fischer W, De Souza PR (2010) Predation on bats by great kiskadees. J Field Ornithol 81:17–20
Foggin CM (1988) Rabies and rabies-related viruses in Zimbabwe: historical, virological and ecological aspects. Dissertation, University of Zimbabwe
Frafjord K (2007) Foretrekker nordflaggermus *Eptesicus nilssonii* oppvarma hus? Do northern bats *Eptesicus nilssonii* prefer heated building as roosts? Fauna 60:239–245
Gaisler J (1998) Bat thanatocenosis with *Eptesicus serotinus*, *Vespertilio murinus* and *Nyctalus leisleri* in a building. Folia Zool 47:313–315
Gareca E, Rey-Ortiz G, Aguirre LF (2007) Relación entre el conocimiento acerca de los murciélagos y las actitudes de cinco grupos sociales de Cochabamba. In: Aguirre LF (ed) Historia natural, distribución y conservación de los murciélagos de Bolivia. Fundación Simón I, Patiño, Santa Cruz, pp 99–103
Geiser F (2004) Metabolic rate and body temperature reduction during hibernation and daily torpor. Annu Rev Physiol 66:239–274
Geiser F, Drury RL (2003) Radiant heat affects thermoregulation and energy expenditure during from torpor. J Comp Physiol 173:55–60
Ghanem SJ, Voigt CC (2012) Increasing awareness of ecosystem services provided by bats. Adv Stud Behav 44:279–302
Gillam EH, O'Shea TH, Brigham RM (2011) Nonrandom patterns of roosts emergence in big brown bats, *Eptesicus fuscus*. J Mammal 92:1253–1260
Gillette DD, Kimbrough JD (1970) Chiropteran morality. In: Slaughter BH, Walton DW (eds) About bats. Southern Methodist University Press, pp 262–283
Goodman SM, Jansen van Vuuren B, Ratrimomanarivo F, Probst J-M, Bowie RCK (2008a) Specific status of populations in the Mascarene Islands referred to *Mormopterus acetabulosus* (Chiroptera: Molossidae), with description of a new species. J Mammal 89:1316–1327
Goodman SM, Ratrimomanarivo FH, Ranivo J, Cardiff SG (2008b) The hunting of microchiropteran bats in different portions of Madagascar. Afr Bat Con News 16:4–7
Halsall AL, Boyles JG, Whitaker JO (2012) Body temperature patterns of big brown bats during winter in a building hibernaculum. J Mammal 93:497–503
Harbusch C, Racey PA (2006) The sessile serotine: the influence of roost temperature on philopatry and reproduction phenology of *Eptesicus serotinus* (Schreber, 1774)(Mammalia: Chiroptera). Acta Chiropterol 8:213–229
Hosono T, Uchida E, Suda C, Ueno A, Nagagawa T (2006) Salt weathering of sandstone at the Angkor monuments, Cambodia: identification of the origins of salts using sulfur and strontium isotopes. J Archeol Sci 33:1541–1551

Irwin NR, Speakman JR (2003) Azorean bats *Nyctalus azoreum*, cluster as they emerge from roosts, despite the lack of avian predators. Acta Chriopterol 5:185–192
Jenkins EV, Laine T, Morgan SE, Cole KR, Speakman JR (1998) Roost selection in the pipistrelle bat, *Pipistrellus pipistrellus* (Chiroptera: Vespertilionidae), in northeast Scotland. Anim Behav 56:909–917
Jones G, Jacobs DS, Kunz TH, Willig MR, Racey PA (2009) Carpe noctem: the importance of bats as bioindicators. Endangered Species Res 8:93–115
Kalka MB, Smith AR, Kalko EKV (2008) Bats limit arthropods and herbivory in a tropical forest. Science 320:71
Kanuch P, Fornuskova A, Bartonicka T, Bryja J, Rehak Z (2010) Do two cryptic pipistrelle bat species differ in their autumn and winter roosting strategies within the range of sympatry? Folia Zool 59:102–107
Kerth G (2008) Causes and consequences of sociality in bats. Bioscience 58:737–755
Kerth G, Perony N, Schweitzer F (2011) Bats are able to maintain long-term social relationships despite the high fission-fusion dynamics of groups. P Zool Soc Lond. doi:10.1098/rspb.2010.2718
Kingston (2016) Cute, Creepy, or crsipy- how values, attitudes and norms shape human behavior toward bats. In: Voigt CC, Kingston T (eds) Bats in the anthropocene: conservation of bats in a changing world. Springer International AG, Cham, pp 571–588
Kingston T, Aguirre L, Armstrong K, Mies R, Racey P, Rodríguez-Herrera, Waldien D (2016) Networking networks for global bat conservation. In: Voigt CC, Kingston T (eds) Bats in the anthropocene: conservation of bats in a changing world. Springer International AG, Cham, pp 539–566
Knight T, Jones G (2009) Importance of night roosts for bat conservation: roosting behaviour of the lesser horseshoe bat *Rhinolophus hipposideros*. Endangered Species Res 8:79–86
König H, Neumann F (1996) Drahtgitter zur Taubenabwehr als Fledermausfallen. Wire mesh as unintentional bat traps. Fauna und Flora (Rheinland-Pfalz) 21:141–143
Kovats D, Habarics B, Urban H (2008) Épületlakó denevérfajok populációdinamikai vizsgálata gyöngybaglyok jelenlétében, a Szatmár-Beregi Tájvédelmi Körzet területén. The changes of population of house-dwelling bat colonies in the presence of barn owls (*Tyto alba*) in the Szatmar-Bereg landscape protection area (Hungary). Deneverkutatas 4:59–73
Kozhurina EI, Gorbunova YA (2004) Hibernation of bats in the Volga Delta. Plecotus 7:104–105
Kunz TH (1982) Roosting ecology of bats. In: Kunz TH (ed) Ecology of bats. Plenum Press, New York, pp 1–56
Kunz TH, Lumsden LF (2003) Ecology of cavity and foliage roosting bats. In: Kunz TH, Fenton MB (eds) Bat ecology. University of Chicago Press, Chicago, pp 3–90
Kunz TH, Reynolds DS (2003) Bat colonies in buildings. In: O'Shea TJ, Bogan MA (eds) Monitoring trends in bat populations of the United States and territories: problems and prospects. US Geological Survey, Biological resources discipline, information and technology report, pp 91–102
Kunz TH, Anthony ELP, Rumage WT III (1977) Mortality of little brown bats following multipiple pesticide applications. J Wildl Manag 41:476–483
Kunz TH, Braun de Torrez E, Bauer DM, Lobova TA, Fleming TH (2011) Ecosystem services provided by bats. Ann Ny Acad Sci 1223:1–38
Kurta A (2010) Reproductive timing, distribution, and sex ratios of tree bats in Lower Michigan. J Mammal 91:586–592
Labruna MB, Marcilia A, Ogrzewalska M, Barros-Attesti DM, Dantas-Torres F, Fernandes AA, Leite RC, Venzal JM (2014) New records and human parasitism by *Ornithodoros mimon* (Acari: Argasidae) in Brazil. J Med Entomol 51:283–287
Lausen CL, Barclay RMR (2006) Benefits of living in a building: big brown bats (*Eptesicus fuscus*) in rocks versus buildings. J Mammal 87:362–370
Lesiński G, Kasprzyk K, Gryz J (2013) Bats taken by the tawny owl in relation to its roosting sites. North-west J Zool 8:247–251
Lewis SE (1994) Night roosting ecology of pallid bats (*Antrozous pallidus*) in Oregon. Am Midl Nat 132:219–226

Lilley TM, Stauffer J, Kanerva M, Eeva T (2014) Interspecific variation in redox status regulation and immune defense in five bat species: the role of ectoparasites. Oecologia 175:811–823

Lisón F, Palazón JA, Calvo JF (2013) Effectiveness of the natura 2000 network for the conservation of cave-dwelling bats in a Mediterranean region. Anim Conserv 16:528–537

Loeb SC, Zarnoch SJ (2011) Seasonal and multiannual roost use by Rafinesque's big-eared bats in the coastal plain of South Carolina. Conserv Manage East Big-Eared Bats—US Forest Serv Gen Techn Rep SRS 145:111–121

Loftis AD, Gill JS, Schriefer ME, Levin ML, Eremeeva ME, Gilchrist MJR, Dasch GA (2005) Detection of *Rickettsia*, *Borrelia*, and *Bartonella* in *Carios kelleyi* (Acari: Argasidae). J Med Entomol 43:473–480

Lyster S (1989) The convention on the conservation of migratory species of wild animals (the "Bonn Convention"). Nat Resour J 29:979–1000

Maas B, Cough Y, Tscharntke T (2013) Bats and birds increase crop yield in tropical agroforestry landscapes. Ecol Lett 16:1480–1487

Maloney SK, Bronner G, Buffenstein R (1999) Thermoregulation in the Angolan free-tailed bat *Mops condylurus*: a small mammal that uses hot roosts. Physiol Zool 72:385–396

Maltagliati G, Agenlli P, Cannicci S (2013) Where and at what time? Multiple roost use and emergence time in greater horseshow bats (*Rhinolophus ferrumequinum*). Acta Chiropterol 15:113–120

Marnell F, Prsetnik P (2010) Protection of overground roosts for bats. Eurobats Publication Series 4, pp 1–57

Martins P, Neves C, Lopes AA, Santos Q, Araújo NN, Pereira M (2000) Histoplasmosis presenting as acute respiratory distress syndrome after exposure to bat feces in a home. Braz J Infect Dis 4:103–106

Mazurska K, Ruczynski I (2008) Bats select buildings in clearings in Bialowieza Primeval forest. Acta Chiropterol 10:331–338

McAlpine DF, Muldoon F, Forbes GJ, Wandeler AI, Makepeace S, Broders HG, Goltz JP (2002) Over-wintering and reproduction by the big brown bat, *Eptesicus fuscus*, in New Brunswick. Can Field Nat 116:645–647

McNab BK (2002) The physiological ecology of vertebrates: a view from energetics. Cornell University Press, New York

Merzlikin IR (2002) On some causes of bat mortality in Sumy region (Northeastern Ukraine). Plecotus 113–115

Mialhe PJ (2013) Characterization of *Desmodus rotundus* (E. Goeffroy, 1810) (Chiroptera, Phyllostomidae) shelters in the municipality of Sao Pedro—SP. Braz J Biol 73:521–526

Michaelsen TC, Tore C, Grimstad KJ, Anonby JE (2004) Noen interessant funn av dagoppholdssted for flaggermus. Some interesting discoveries of day roosts for bats in Norway. Fauna (Oslo) 57:54–61

Michaelsen TC, Olsen O, Grimstad KJ (2013) Roosts used by bats in late autumn and winter at northern latitudes in Norway. Folia Zool 62:297–303

Mickleburgh SP, Hutson AM, Racey PA (2002) A review of the global conservation status of bats. Oryx 36:18–34

Mikula P, Hromada M, Tryjanowski P (2013) Bats and swifts a food of the European kestrel (*Falco tinnunculus*) in a small town in Slovakia. Ornis Fennica 90:178–185

Mildenstein T, Tanshi I (2016) Direct exploitation of bats. In: Voigt CC, Kingston T (eds) Bats in the anthropocene: conservation of bats in a changing world. Springer International AG, Cham, pp 325–363

Mitchell-Jones AJ, Cooke AS, Boyd IL, Stebbings RE (1989) Bats and remedial timber treatment chemicals. Mammal Rev 18:93–110

Molinari J, Gutiérrez EE, Asecencão AA, Nassar JM, Arends A, Márquez RJ (2005) Predation by giant centipedes, *Scolopendra gigantea*, on three species of bats in a Venezuelan cave. Carribean J Sci 41:340–346

Monadjem A, Taylor PJ, Cotterill FPD, Schoeman MC (2010) Bats of central and southern Africa: a biogeographic and taxonomic synthesis. Wits University Press, Johannesburg

Moussy C (2011) Selection of old stone buildings as summer day roost by the brown long-eared bat *Plecotus auritus*. Acta Chiropterol 13:101–111
Mühldorfer K, Speck S, Wibbelt G (2011) Diseases in free-ranging bats from Germany. BMC Vet Res 7:61
Mühldorfer K, Speck S, Kurth A, Lesnik R, Freuling C, Müller T, Kramer-Schadt S, Wibbelt G (2012) Diseases and causes of death in European bats: dynamics in disease susceptibility and infection rates. PLoS ONE 6:e29773
Munshi-South J, Wilkinson GS (2010) Bats and birds: exceptional longevity despite high metabolic rates. Ageing Res Rev 9:12–19
Murphy SE, Greenaway F, Hill DA (2012) Patterns of habitat use by female brown long-eared bats presage negative impacts of woodland conservation management. J Zool (Lond) 288:177–183
Neilson AL, Fenton MB (1994) Responses of little brown myotis to exclusion and to bat houses. Wildl Soc Bull 22:8–14
Neubaum DJ, Wilson JR, O'Shea TJ (2007) Urban maternity roost selection by big brown bats in Colorado. J Wildl Manage 71:728–736
Nyffeler M, Knörnschild M (2013) Bat predation by spiders. PLoS ONE 8:e58120
O'Keefe JM, LaVoie M (2011) Maternity colony of eastern small-footed myotis (*Myotis leibii*) in a historic building. Southeast Nat 10:381–383
O'Shea TJ, Neubaum DJ, Neubaum MA, Cryan PM, Ellison LE, Stanley TR, Rupprecht CE, Pape WH, Bowen RA (2012) Bat ecology and public health surveillance for rabies in an urbanizing region of Colorado. Urban Ecosyst 14:665–697
Ormsbee PC, Kiser J, Perlmeter SI (2007) The importance of night roosts to the ecology of forest bats. In: Lacki MJ, Hayes JP, Kurta A (eds) Bats in forests: conservation and management. John Hopkins University Press, Baltimore, pp 129–152
Passos EC, Carrieri ML, Dainovskas E, Camara M, Silva MMS (1998) Isolamento do virus rabico em morcego insectivoro, *Nyctinomops macrotis*, no municipo de Diadema, SP (Brasil)—Isolation of rabies virus in an insectivorous bat *Nyctinomops macrotis*, in southeastern Brazil. Rev Saude Publica 32:74–76
Patterson BD, Dick CW, Dittmar K (2007) Roosting habits of bats affect their parasitism by bat flies (Diptera: Streblidae). J Trop Ecol 23:177–189
Pavey CR (1999) Foraging ecology of the two taxa of large-eared horseshoe bat, *Rhinolophus philippinensis*, on Cape York Peninsula. Australian Mammal 21:135–138
Pearce RD, O'Shea TJ (2007) Ectoparasites in an urban population of big brown bats (*Eptesicus fuscus*) in Colorado. J Parasitol 83:518–530
Pfalzer G, Weber C (2007) Invasionen der Zwergfledermaus (*Pipistrellus pipistrellus*) im Stadtgebiet von Kaiserslautern (BRD, Rheinland-Pfalz). Mass invasions of the common pipistrelle bat (*Pipistrellus pipistrellus*) in the city of Kaiserslautern (Germany, Rhineland-Palatine). Fauna und Flora in Rheinland-Pfalz Beiheft 35:212–219
Poché RM (1975) The bats of National Park W, Niger, Africa. Mammalia 39:39–50
Racey P, Swift SM (1986) The residual effects of remedial timber treatments on bats. Biol Conserv 35:205–214
Racey PA, Hutson AM, Lina PHC (2013) Bat rabies, public health and European bat conservation. Zoonoses Public Health 60:58
Ransome R, Hutson AM (2000) Action plan for the conservation of the greater horseshoe bat in Europe (*Rhinolophus ferrumequinum*). Nature and environment 109, Council of Europe Publishing, Strasbourg
Razafindrakoto N, Harwll A, Jenkins R (2011) Bats roosting in public buildings: a preliminary assessment from Moramang, Eastern Madagascar. Madagascar Conserv Dev 5:85–88
Reichel-Jung K, Threlfall CG (2016) Urbanization and its effects on bats—a global meta-analysis approach. In: Voigt CC, Kingston T (eds) Bats in the anthropocene: conservation of bats in a changing world. Springer International AG, Cham, pp 13–28
Rintoul JLP, Brigham RM (2014) The influence of reproductive condition and concurrent environmental factors on torpor and foraging patterns in female big brown bats (*Eptesicus fuscus*). J Comp Physiol B 184:777–787

Roby PL, Gumbert MW, Sewell PL, Brewer SW (2011) Characteristics of roosts used by Rafinesque's big-eared bat (*Corynorhinus rafinesquii*) on camp Mackall, North Carolina. Conserv Manage East Big-Eared Bats—US For Serv Gen Tech Rep SRS 145:101–110

Rowse EG, Lewanzik D, Stone EL, Harris S, Jones G (2016) Dark matters: the effects of artificial lighting on bats. In: Voigt CC, Kingston T (eds) Bats in the anthropocene: conservation of bats in a changing world. Springer International AG, Cham, pp 187–207

Ruczyński I, Kalko EKV, Siemers BM (2007) The sensory basis of roost finding in a forest bat, *Nyctalus noctula*. J Exp Biol 217:3607–3615

Rydell J (1991) Seasonal use of illuminated areas by foraging northern bats *Eptesicus nilssoni*. Ecography 14:203–207

Rydell J (1992) Exploitation of insects around streetlamps by bats in Sweden. Funct Ecol 6:744–750

Rydell J, Racey PA (1995) Street lamps and the feeding ecology of insectivorous bats. Sym Zool S 67:291–307

Sachanowicz K, Wower A (2013) Assemblage structure and use of anthropogenic roosts by bats in the eastern Carpathians: case study in the Bieszczady National Park (SE Poland). Ital J Zool 80:139–148

Saenz VL, Maggi RG, Breitschwerdt EB, Kim J, Vargo EL, Schal C (2013) Survey of *Bartonella* spp. in US bed bugs detects *Burholderia multivorans* but not *Bartonella*. PLoS ONE 8:e73661

Salvati L, Manganaro A, Fattorini S, Piattella E (1999) Density, nest spacing, breeding success and diet of a kestrel *Falco tinnunculus* urban population. Alauda 67:47–52

Santos M, Aguirre LF, Vázquez LB, Ortega J (2003) Phyllostomus hastatus. Mamm Species Acc 722:1–6

Schneeberger K, Voigt CC (2016) Zoonotic viruses and conservation of bats. In: Voigt CC, Kingston T (eds) Bats in the anthropocene: conservation of bats in a changing world. Springer International AG, Cham, pp 263–282

Schneeberger K, Czirják GÁ, Voigt CC (2013) Inflammatory challenge increases measures of oxidative stress in a free-ranging long-lived mammal. J Exp Biol 216:414–4519

Schneider MC, Aron J, Santos-Burgoa C, Uieda W, Ruiz-Velazco S (2001) Common vampire bat attacks on humans in a village of the Amazon region of Brazil. Pan Am J Pub Hlth 25:260–268

Shapiro JT, de Costa Lima Junior MS, Dorval ME, de Oliveira França A, Cepa Matos Mde F, Bordignon MO (2013) First record of *Leishmania braziliensis* presence detected in bats, Mato Gross do Sul, southwest Brazil. Acta Tropica 128:171–174

Shiel CB, Shiel RE, Fairley JS (1999) Seasonal changes in the foraging behaviour of Leisler's bats (*Nyctalus leisleri*) in Ireland as revealed by radio-telemetry. J Zool 249:347–358

Shore RF, Boyd IL, Leach DV, Stebbings RE, Myhill DG (1990) Organochlorine residues in roof timber treatments and possible implications for bats. Environ Pollut 64:179–188

Shore RF, Myhill DG, French MC, Leach DV, Stebbings RE (1991) Toxicity and tissue distribution of pentachlorophenol and permethrin in pipistrelle bats experimentally exposed to treated timber. Envioron Pollut 73:101–118

Sibly RM, Brown JH (2007) Effets of body size and lifestyle on evolution of mammal life histories. Proc Nat Acad Sci 104:17707–17712

Silver J (1935) Eliminating bats from buildings, vol 109. United States Department of Agricultural leaflet, pp 1–5

Smit-Viergutz K, Simon M (2000) Eine vergleichende Analyse des sommerlichen Schwärmverhaltens der Zwergfledermaus (45 kHz Ruftyp, *Pipistrellus pipistrellus* Schreber, 1774) an den Invasionsorten und am Winterquartier. A comparative analysis of the summertime swarming behavior of the common pipistrelle (*Pipistrellus pipistrellus* Schreber, 1774, 45 kHz call type) at the sites of the invasion and in the hibernaculum. Myotis 38:69–89

Socolovschi C, Kernif T, Raoult D, Parola P (2012) Borrelia, Rickettsia, and Ehrlichia species in bat ticks, France, 2010. Emer Infect Dis 18:1966–1975

Soulé ME (1985) What is conservation biology? Bioscience 35:727–734
Speakman JR (1991) The impact of predation by birds on bat populations in the British Isles. Mammal Rev 21:123–142
Speakman JR, Thomas DW (2003) Physiological ecology and energetics of bats: In: Kunz TH, Fenton MB (eds) Bat ecology. Chicago University Press, pp 430–409
Speakman JR, Stone RE, Kerslake JL (1995) Temporal patterns in the emergence behaviour of pipistrelle bats, *Pipistrellus pipistrellus*, from maternity colonies are consistent with an anti-predator response. Anim Behav 50:1147–1156
Speakman JR, Irwin N, Tallach N, Stone R (1999) Effect of roost size on the emergence behaviour of pipistrelle bats. Anim Behav 58:787–795
Spitzenberger F, Weiss E, Sackl P (2010) Massive population decline of the critically endangered greater horseshoe bat, *Rhinolophus ferrumequinum* (Schreber, 1774) in Styria, south-eastern Austria, between the mid-1990s and 2009 (Mammalia, Chiroptera). Joannea Zool 11:5–17
Streicker DG, Franka R, Jackson FR, Rupprecht CE (2013) Anthropogenic roost switching and rabies virus dynamics in house-roosting big brown bats. Vector-borne Zoonot 13:498–504
Strelkov PP (2002) Material on wintering of migratory bat species (Chiroptera) on the territory of the former USSR and adjacent regions, Part 2. *Nyctalus noctula*. Plecotus 5:35–56
Šuba J, Vietniece D, Petērsons G (2010) The parti-coloured bat *Vespertilio murinus* in Riga (Latvia) during autumn and winter. Environ Exp Biol 8:93–96
Subbaraj R, Balasingh J (1996) Night roosting and lunar phobia in Indian false vampire bat *Megaderma lyra*. J Bombay Nat Hist Soc 93:1–7
Szodoray-Parádi F, Szodoray-Parádi A, Sike T (2004) Building dwelling bats survey in Satu Mare County, Romania. Sau Mare, Studii şi Comunicăre, Seria Ştiinţele Naturale 4–5:196–200
Takumi K, Lina PHC, van der Poel WHM, Kramps JA, Van Der Giessen JWB (2009) Public health risk analysis of European bat lyssavirus infection in the Netherlands. Epidemiol Infect 137:803–809
Tong S, Conrardy C, Ruone S, Kuzmin IV, Guo X, Tao Y, Niezgoda M, Haynes L, Agwanda B, Breiman RF, Anderson LJ, Rupprecht CE (2009) Detection of novel SARS-like and other coronaviruses in bats from Kenya. Emerg Infect Dis 15:482–485
Turbill C, Körtner G, Geiser F (2003) Natural use of heterothermy by a small, tree-roosting bat during summer. Physiol Biochem Zool 76:868–876
Uieda W (1998) Rabies in the insectivorous bat *Tadarida brasiliensis* in southeastern Brazil. Rev Saude Publ 23:484–485
Van der Poel WH, Van der Heide R, Verstraten ERAM, Takumi K, Lina PHC, Kramps JA (2005) European bat lyssaviruses, the Netherlands. Emerg Infect Dis 11:1854–1859
Vandzurova A, Backor P, Jarvorsky P, Pristas P (2013) *Staphylococcus nepalensis* in the guano of bats (Mammalia). Vet Microbiol 164:116–121
Vivier L, van der Merwe M (2001) Aspects of the histology of reproduction in the female Angolan free-tailed bat *Mops condylurus* (Microchiroptera: Molossidae) in Mpumalanga, South Africa. J Zool 254:495–504
Vivier L, van der Merwe M (2007) The incidence of torpor in winter and summer in the Angolan free-tailed bat, *Mops condylurus* (Microchiroptera: Molossidae), in a subtropical environment, Mpumulanga, South Africa. Afr Zool 42:50–58
Voigt CC, Kelm DH (2006) Host preference of the common vampire bat (*Desmodus rotundus*; Chiroptera) assessed by stable isotopes. J Mammal 87:1–6
Whitaker JO Jr (1998) Life history and roost switching in six summer colonies of eastern pipistrelles in buildings. J Mammal 79:651–659
Whitaker JO Jr, Gummer SL (1992) Hibernation of the big brown bat, *Eptesicus fuscus*, in buildings. J Mammal 73:312–316
Wilkinson GS, South JM (2002) Life history, ecology and longevity in bats. Aging Cell 1:124–131

Williams LM, Brittingham MC (1997) Selection of maternity roosts by big brown bats. J Wildl Manage 61:359–368
Williams-Guillén K, Perfecto I, Vandermeer J (2008) Bats limit insects in a neotropical agroforestry system. Science 320:70
Willis CKR, Brigham RM, Geiser F (2006) Deep, prolonged torpor by pregnant, free-ranging bats. Naturwissenschaften 93:80–83
Zahn A (1999) Reproductive success, colony size and roost temperature in attic-dwelling bat *Myotis myotis*. J Zool 247:275–280
Zahn A, Christoph C, Christoph L, Kredler M, Reitmeier A, Reitmeier F, Schachenmeier C, Schott T (2000) Die Nutzung von Spaltenquartieren an Gebaeuden durch Abendsegler (*Nyctalus noctula*) in Suedostbayern. The use of crevices in buildings as roosts by the noctule (*Nyctalus noctula*) in southeast Bavaria. Myotis 37:61–76
Zhang LB, Zhu GH, Jones G, Zhang SY (2009) Conservation of bats in China: problems and recommendations. Oryx 43:179–182
Zorya AV (2002) Contribution of bats to spreading rabies on the territory of Kharkov region and prevention of conflict sutations "man-bat". Plecotus et al. Supplement: pars specialis 118–120

Chapter 15
Conservation Ecology of Cave Bats

Neil M. Furey and Paul A. Racey

Abstract Caves and other subterranean sites such as mines are critical to the survival of hundreds of bat species worldwide, since they often provide shelter for most of a nation's bat fauna. In the temperate zone, caves provide roosts for hibernation and for some species, breeding in summer, whereas in warmer regions, they support high species richness year round and enormous colonies that maintain substantial ecosystem services. Due to the solubility of the substrate, the highest densities of caves occur in karst landscapes. Given their importance for bats, relatively few studies have investigated factors involved in cave selection, although current evidence suggests that the density and size of caves are the best predictors of species diversity and population sizes. Thermal preferences have been established for some cave-dwelling species as well as their vulnerability to disturbance, particularly during hibernation and reproduction. Growth in limestone quarrying and cave tourism industries worldwide severely threatens cave-dwelling bats, in addition to loss of foraging habitat, hunting for bushmeat, incidental disturbance and disruptive guano harvesting. Apparent declines of cave bats in Europe and North America also pose serious concerns, as do global climate change predictions. The main conservation response to threats to cave bats in these continents

N.M. Furey (✉)
Fauna & Flora International (Cambodia Programme), No. 19, Street 360,
Boeng Keng Kang 1, PO Box 1380, Phnom Penh 12000, Cambodia
e-mail: neil.furey@fauna-flora.org

P.A. Racey
Centre for Ecology and Conservation, College of Life and Environmental Sciences,
University of Exeter, Cornwall Campus, Treliever Road, Penryn TR10 9EZ, UK
e-mail: p.a.racey@exeter.ac.uk

C.C. Voigt and T. Kingston (eds.), *Bats in the Anthropocene: Conservation of Bats in a Changing World*, DOI 10.1007/978-3-319-25220-9_15

has been gating, but this remains relatively untested as a means of protecting colonies in other regions. Research on sustainable harvesting of bats as bushmeat and their responses to different types of human disturbance at caves and loss of surrounding foraging habitats is required. More caves of outstanding importance for bats at national and international levels also require protection.

15.1 Introduction

Bats fly mainly at night and spend the day in roosts which provide shelter from extremes of temperature, other climatic variables and predators. The most widely used day roosts occur in caves and because of the global abundance of surface carbonate rock (Fig. 15.1), in karstic caves. However, caves in other rock formations, as well as mines, wartime fortifications and other underground situations, are also used by roosting bats, because all provide a relatively cool and constant environment compared to that outside. Although the term 'cave-dependent' is often applied to bats, and will be used in this review, it is recognized that while their need for day roosts is incontrovertible, dependency is difficult to establish. The threats to such roosts and the bats they shelter also have much in common and for that reason this chapter will consider all such roost types, which will often be referred to collectively as caves. We aim to review the importance of such sites for the maintenance of bat biodiversity. We consider those factors which make caves important for bats and whether bats select caves with particular features. Our main aim however is to highlight the threats to bats in caves and the ways in which these may be mitigated.

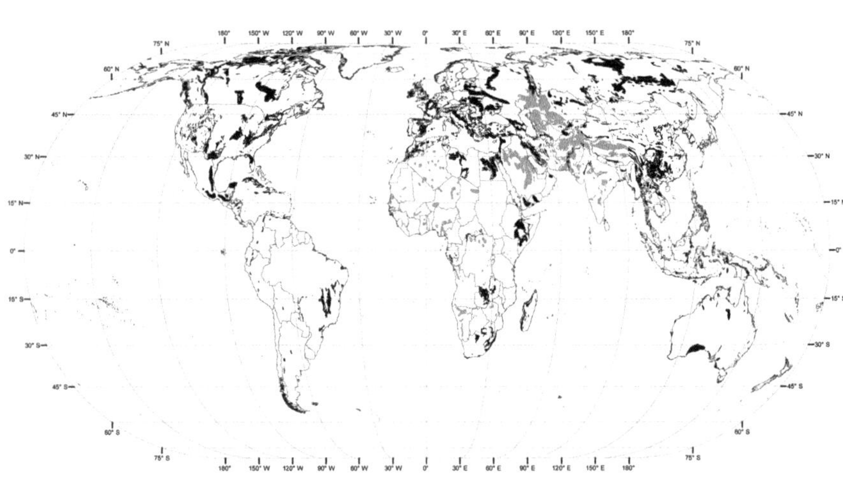

Fig. 15.1 Global distribution of carbonate rocks (© Paul Williams, University of Auckland, NZ)

15.2 Why Do Cave Bats Matter?

The largest aggregations of living vertebrates are found in caves, and in the 1950s and early 1960s, midsummer colonies of adult Mexican free-tailed bats (*Tadarida brasiliensis*) in 17 caves in the south-western USA were estimated to total 150 million individuals (McCracken 2003; Russell and McCracken 2006) (Fig. 15.2). In contrast, the largest number of tree-roosting bats in any location is currently estimated at 8 million for the straw-coloured fruit bat (*Eidolon helvum*) in a small area of swamp forest in Kasanka National Park, Zambia (Racey 2004). Large aggregations are characteristic of molossid bats in caves in both Old and New Worlds and despite repeated efforts to harness modern technology such as Doppler radar (Horn and Kunz 2008) and thermal infrared video (Betke et al. 2008), accurate counting of the numbers involved has proved elusive. Not surprisingly however, the evening emergence of such colonies attracts significant numbers of tourists around the world every year. For example, an amphitheatre at the entrance to Carlsbad caverns, New Mexico has allowed visitors to observe the dusk departure (and dawn return) of a large colony of *T. brasiliensis* over several decades, although the US National Parks Service have banned the use of flash photography in recent times because of concerns that it disturbs the bats (Altringham 2011).

The survival of many bat species worldwide depends upon natural caves and other underground sites such as mines (Mickleburgh et al. 2002). For instance, of

Fig. 15.2 Evening emergence of *T. brasiliensis* from Frio cave in Texas, USA (© Merlin D. Tuttle, Bat Conservation International, www.batcon.org)

39 bat species in temperate America (north of Mexico), 18 rely substantially on caves (46 %), including 13 species that dwell in them all year round, while the remaining five depend on caves for hibernation sites (McCracken 1989). Of the 40 European bat species for which information is available, 28 are found in caves during hibernation and a few all year round (Dietz et al. 2009). Arita (1993a) documented similarly high levels of occupancy in subtropical Mexico, where 60 of the 134 bat species known (45 %) regularly use caves. Even higher occupancy has been found in China, where 77 % of the known bat fauna (101 of 131 species) roosts in caves and other subterranean habitats (Luo et al. 2013) and similar figures exist for Puerto Rico and North Vietnam (Rodriguez-Durán 2009; Furey et al. 2010). Because cave-roosting bats spend at least half their lives inside caves (Kunz 1982), protection of these sites is central to their conservation. Due to the solubility of calcium carbonate, caves are found in particularly high density in karstic areas and research in Southeast Asia suggests they may serve as population reservoirs subsidizing bat species diversity in fragmented landscapes that could otherwise decline over time (Struebig et al. 2009).

The ecological services provided by cave bats have been documented in recent years (Boyles et al. 2011; Kunz et al. 2011). In Texas, *T. brasiliensis* fly up to 900 meters before dispersing to forage over crops, and include in their diet important pests such as cotton bollworm moth (*Helicoverpa zea*). The proportion of such pests in their faeces allows the economic value of such predation to be estimated, which includes a reduction in the number of costly pesticide applications required (Cleveland et al. 2006). In Thailand, the wrinkle-lipped free-tailed bat (*Chaerephon plicatus*) consumes economically significant amounts of white-backed planthoppers (*Sogatella furcifera*) which are major pests of rice crops (Leelapaibul et al. 2005; Wanger et al. 2014). The dawn bat (*Eonycteris spelaea*) which forms colonies of up to 20,000 individuals in SE Asian caves (Medway 1958) is the primary pollinator of durian (*Durio zibethinus*), a high value fruit (Bumrungsri et al. 2009) and a commonly eaten tree bean (*Parkia speciosa*) (Bumrungsri et al. 2008), alongside other economically important plant species (Bumrungsri et al. 2013).

Mining the guano of cave-dwelling bats is a worldwide phenomenon as the undigested remains of insects are rich in nitrogen and phosphates (Gillieson 1996). This is particularly true in Asia, where bat guano is a major source of fertilizer whose sale and use features prominently in many local economies (Leh and Hall 1996; Leelapaibul et al. 2005; Aye 2006). This has resulted in overharvesting and disturbance of cave roosting bats (Bumrungsri et al. 2013), exacerbated by cave modifications made to assist guano extraction (Elliot 1994). Allied to this, the guano produced by bats constitutes a primary source of energy in cave ecosystems and survival of a considerable proportion of the terrestrial invertebrate fauna in tropical caves is dependent upon its continued deposition. These communities include a suite of highly-adapted and narrowly-endemic arthropods (often referred to as guanophiles or guanobionts) which complete their entire life cycle in or around guano piles (Deharveng and Bedos 2012).

15.3 Life in Caves

Caves confer important advantages in the form of permanent, thermally stable and humid environments which protect bats against inclement weather and reduce loss of body water (Kunz 1982; Gunn 2003; Avila-Flores and Medellin 2004). Added to this are potential benefits in reduced predation risk and thermoregulatory advantages derived from aggregating in large numbers. Disbenefits may include increased commuting costs in foraging, higher incidence of parasites and disease transmission, and possibly greater intraspecific foraging competition (Kunz 1982). Although caves that permit human access are necessarily the ones about which most is known, bats also roost extensively in smaller inaccessible rock cavities. For example, four overwintering colonies of noctule bats (*Nyctalus noctula*), comprising about a thousand individuals, were reported in vertical crevices 1–2 m in both height and depth and 5–7 cm in breadth in calcareous lakeside cliffs in Romania (Barbu and Sim 1968). In Madagascar, Malagasy straw-coloured fruit bats (*Eidolon dupreanum*) often roost in such crevices high on cliffs, where they are less accessible to hunters (Mackinnon et al. 2003).

While caves have the disadvantage of being uncommon in many areas and may be located far from suitable foraging sites (Bradbury 1977), roost fidelity is greatest among bat species that use caves and buildings (Lewis 1995). This is thought to stem in part from their permanency, although many caves are unsuitable as roosts, particularly those that are too cold or warm to promote efficient thermoregulation (Kunz 1982). Caves can be viewed as largely azonal habitats because they share a similar environment across all latitudes and all macroclimates. While some bats with a very restricted distribution such as Kitti's hog-nosed bat (*Craseonycteris thonglongyai*) are found only in karstic caves, others, such as many European species, are found equally in caves, disused mines, railway tunnels, wartime fortifications, churches and domestic roof spaces. Beyond local variations, temperature is the most basic physical factor distinguishing abiotic environments in tropical versus temperate caves (Deharveng and Bedos 2012). In temperate regions, caves provide roosting sites for hibernation and for some species, breeding in summer, and in tropical regions, where bats do not hibernate, they support very large colonies and high bat diversity (Rodriguez-Durán and Lewis 1987; Monadjem et al. 2010; Furey et al. 2011).

The majority of temperate zone bat species hibernate in caves, and a few taxa continue to roost there throughout the year including the summer period of reproduction (Dietz et al. 2009; Nagy and Postawa 2010). In early autumn, thousands of bats swarm at the entrances of caves each night, flying in and out, although most leave before dawn. The sex ratio of swarming bats is heavily skewed towards males and it is thought that mating occurs, and since the swarming bats come from many different colonies, that this provides an opportunity for gene flow and also for mothers to show their young where to hibernate (Parsons and Jones 2003; Parsons et al. 2003a, b). Bats are also attracted to potential hibernacula by the echolocation calls of conspecifics (Avery et al. 1984). In temperate regions, the mating that began at swarming continues in some species until spring. During hibernation, males often arouse and copulate with torpid females (Thomas et al. 1979). Analysis of the ambient temperature records at which bats were found torpid in the wild reveal that these range from −10 to 21 °C, with a mode of 6 °C for vespertilionid bats (n = 29 species) and 11 °C for rhinolophids (n = 5 species) (Webb et al. 1995).

In Europe, bats make extensive use of subterranean fortifications, such as those of the Maginot line constructed between France and Germany before World War 1 and the 30 km of underground tunnels built at Nietoperek in eastern Poland during World War 2 on the strategic route from Warsaw to Berlin. The integral drainage system of the latter was subsequently vandalized so parts of the system are now flooded and there is a range of humidities and airflows. Annual hibernation counts are carried out and 37,000 bats of eight species were recorded in 2013, making this the most important hibernaculum in Central Europe, with comparable numbers of bats to many natural European caves (Kokurewicz et al. 2013). Use of wartime structures by bats is also common in the Netherlands and de Boer et al. (2013) found that internal size-related variables had the greatest positive effect on hibernation site suitability. Of the 45 bat species in North America, 28 roost in old mines, which are the only known roosts of the Curacaoan long-nosed bat (*Leptonycteris curasoae*) in the USA (Pierson 1998). With over 300,000 abandoned mines in the state of Nevada alone, guidelines were required for their evaluation as a conservation resource and to resolve potential conflicts (Riddle 1995), and these have been adapted for wider use by Bat Conservation International (Tuttle and Taylor 1994).

In contrast to temperate regions, cave environments in the tropics are typically more stable and uniformly inhabited (Brosset 1966). Though species diversity and population sizes in a cave can fluctuate between different seasons, many tropical bat species roost in caves throughout the year so that reproduction occurs there (McWilliam 1982; McDonald et al. 1990; Siles et al. 2007; Rodriguez-Durán 2009; Monadjem et al. 2010; Furey et al. 2011). Studies of their social organization have revealed that males defend groups of females in erosion domes in the ceilings of karstic caves in several species including the greater spear-nosed bat (*Phyllostomus hastatus*) in the Neotropics (McCracken and Bradbury 1981) and Hildegarde's tomb bat (*Taphozous hildegardeae*) in coastal limestone caves in Kenya (McWilliam 1988). Recent evidence also suggests that *E. spelaea* may adopt a similar harem social organisation in Thailand (Bumrungsri et al. 2013). The abundance of crevices and cavities in caves is believed to facilitate population substructuring and the defense of these roosts by harem males, with clear benefits for both sexes since males achieve most copulations in a cluster they protect, and females gain protection for themselves and their offspring (Bradbury 1977; McCracken and Bradbury 1981). McCracken (1993) has shown how lactating female *T. brasiliensis* in huge maternity colonies locate their own young on cave walls by spatial memory, together with the sound and scent of their young.

A defining characteristic of karst areas—the abundance of calcium as the cation of calcium carbonate—has been suggested as a driver of the use of karstic caves by insectivorous bats. Insects are a poor source of calcium and several studies have shown that calcium levels in the bones of female bats are lowest during lactation, as calcium is mobilized and transferred to the young in milk (Kwiecinski et al. 1987; Booher and Hood 2010). This led Barclay (1994, 1995) to suggest that shortage of calcium may be a greater constraint on reproduction than meeting its energy demands and that one of the reasons that bats roost in karst caves is that they acquire calcium by licking the walls, which is a common observation (Codd et al. 1999). There has been only one study to test this hypothesis, which was not supported by the evidence, since bats were distributed across all underground sites in a wide range of geological formations and were not concentrated in karst landscapes (Bernard and Davison 1996). Nevertheless, the fact that Adams et al. (2003) captured more female and juvenile bats over water holes with harder water (indicating higher calcium levels) suggests that environmental calcium is important, particularly for reproducing females and their young.

15.3.1 Cave Selection

The numbers and diversity of bats found in caves are influenced by their dimensions, structural complexity and microclimate, the availability of food in the surrounding landscape, parasite and predation pressure, human disturbance, historical use by bats, their maneuverabilty in flight and interactions between species. Considering how important caves are for global and local bat biodiversity, there have been relatively few studies of these factors. For instance, half of the bat species known from a 155 km^2 karst reserve in North Vietnam (21/42) used a single large cave over a 23 month period (Furey et al. 2011), whereas in Malaysia, Struebig et al. (2009) found that a single area of karst caves had a dominant influence on bat assemblage composition at non-karst sites up to 11 km away through the presence of two cave-dwelling species.

Brunet and Medellin (2001) revealed a positive relationship between species richness and cave surface area in central Mexico. Roost site diversity as indicated by spatial variation in relative humidity and the presence of erosion domes in cave ceilings (Fig. 15.3) was associated with this species-area relationship. Consistent with this, Arita (1996) found that the largest caves in the northern Yucatan Peninsula of Mexico harbored the most diverse assemblages and largest populations, including several species of conservation concern. At a national level however, Arita (1993b) found that few of the vulnerable species of Mexican bats roost in caves with high species richness or large populations, suggesting that conservation plans based solely on diversity would not adequately protect the country's cave bat fauna. Non-random associations are also common among bats roosting in the hot caves of Puerto Rico and Rodriguez-Durán (1998) speculated that interspecific variation in peak emergence times associated with temporal differences in foraging patterns might allow these caves to support more bats than would be possible in a monospecific colony or random assemblage of species.

Fig. 15.3 Cave roost of *Taphozous melanopogon* in an erosion dome in Thailand (single bat to left of the main group is *Eonycteris spelaea* (© Pipat Soisook)

In a study of the cave complex in Ankarana National Park in the limestone massif of northern Madagascar, Cardiff (2006) found that longer caves, more complex caves, those with larger entrances or with entrances at lower elevation and those with less temporal variation in ambient temperature all had significantly higher bat species richness. In a similar study in the karstic Bemaraha National Park in western Madagascar, Kofoky et al. (2007) found that species richness and abundance was low in all but one of 16 caves—Anjohikinakina, which contained five species and over 9000 individuals of one. This cave was difficult to access and, unlike some of the others in the national park, was seldom visited by tourists.

These findings are broadly reflected in East Asia. In a study of 255 subterranean sites in central and eastern China, Luo et al. (2013) found that bat species richness was positively correlated with cave size and negatively correlated with human disturbance. The incidence of nationally threatened and endemic species was also positively correlated with species richness, which was greater in caves formerly used for tourism than in abandoned mines. In a study of 25 subterranean sites in Funiu Mountain (eastern China), Niu et al. (2007) similarly found that bat species distributions were highly dependent on the type and size of roost, with large caves supporting unusually high species richness and abundances. Over 80 % of the bats recorded were located on the southern side of the mountain which was attributed to climatic differences (higher annual rainfall and average temperatures) and the higher incidence of large caves there.

Nagy and Postawa (2010) further explored the relationship between cave variables and bat occupancy during the hibernation and breeding seasons in 79 caves in mountainous areas of Romania. Maternity colonies were divided between species that select either high or low temperatures, whereas winter aggregations were divided across three groups: (i) species that prefer high temperatures and hibernate at low altitudes, (ii) species preferring mid- to high elevations and low temperatures, and, (iii) species that hibernate in large, cold cave systems with permanent water flow. Piksa et al. (2013) also found that the species richness and assemblage structure of hibernating bats varied altitudinally across 70 caves in the nearby Carpathian mountains of southern Poland, such that stepped changes occurred in assemblage structure that reflected zones observed in vegetation. Geographical location and temperature were found to be the most important factors influencing overall species occurrence by Nagy and Postawa (2010) and their results support Brunet and Medellin's (2001) conclusion that high cave densities provide suitable conditions for large populations of different bat species.

The influence of external environment or "ecological context" on cave selection by bats appears little studied, particularly in terms of access to factors such as food and water. Nevertheless, there seems little doubt that, as in foliage-roosting species, persistent degradation and loss of foraging habitats is likely to threaten the viability of cave-dwelling populations as a result of increased nightly commuting costs and poorer foraging conditions reducing individual fitness (Kingston 2013). For instance, in a comparative study of pristine and modified forests in Vietnamese karst, Furey et al. (2010) found that although species richness was only slightly reduced, the abundance of cave-dwelling rhinolophids and hipposiderids in disturbed and degraded forests was less than a third of that in primary forest, despite comparable sampling effort and availability of caves. In addition, as cave-dwelling species in Asia differ considerably in their wing morphology and thus vagility (Furey 2009), it would appear likely that progressive isolation of cave roosts in anthropogenic landscapes will differentially affect species with weaker dispersal abilities (Fig. 15.4). However, these potential population and species losses may be mitigated to some extent by increases in the abundance of species that use human-made habitats (Mendenhall et al. 2014).

15.3.2 Influence of Cave Microclimate

There are several microclimatic factors which may determine the selection of caves and the location of roost sites within them—temperature, relative humidity and airflow, which are interrelated, and, light intensity. There have been several studies investigating the importance of cave temperature, but the most instructive, extending over 15 years, took place in the Guelhemergroeve mines in South Limburg, Netherlands, where limestone has been mined since the Middle Ages (Daan and Wichers 1968). Nine species of vespertilionid and rhinolophid bats are found there. Two, which approach the northern border of their distribution in South Limburg (Geoffroy's myotis *Myotis emarginatus* and lesser horseshoe bat *Rhinolophus hipposideros*) arrive early and hibernate in the warmer distal end of

Fig. 15.4 Forested karst hills surrounded by wet rice cultivation in North Vietnam (© Neil Furey)

the tunnel system all winter, until late spring (Fig. 15.5a). Three species (the barbastelle *Barbastella barbastellus*, Natterer's bat *M. nattereri* and long-eared bats *Plecotus auritus*) are found in small numbers in protected positions in the mine entrance and stay for the shortest periods, with frequent arrivals and departures (Fig. 15.5b). The four remaining species (the pond bat *M. dasycneme*, the mouse-eared bat *M. myotis*, Daubenton's bat *M. daubentonii* and the whiskered bat *M. mystacinus*) which comprise 80 % of the bats occupying the mines in winter and whose distribution extends further north in Europe, arrive late, in November and December. They hibernate initially in the warmer distal end of the tunnel system where they hang in exposed situations, but as winter progresses, they arouse and move progressively closer to the cold entrance, where they hibernate in crevices, presumably to avoid air currents (Fig. 15.5c). This movement, which has been termed 'internal migration', reflects a preference for lower temperature as winter progresses. So why don't these bats hibernate at the entrance at the beginning of winter? This may reflect selection for higher relative humidity which maintains the condition of the bats' delicate wing membranes but which decreases as temperature increases. At the beginning of the winter, the entrance is relatively warm and humidity is lower than at the distal end of the cave. As the winter gets colder, bats move to take advantage of the rising humidity at the entrance. The arousals and movements which characterize internal migration would also appear to reflect the fact that the fat is metabolized at a faster rate in the warmer distal end of the tunnel system, and that the same amount of fat will last longer if the metabolic rate of the bats is lower at the colder entrance (Daan and Wichers 1968). In the UK, Ransome (1968, 1971) has also shown that the greater horseshoe bat (*Rhinolophus ferrumequinum*) requires a series of hibernacula providing a range of airflow patterns and temperature regimes.

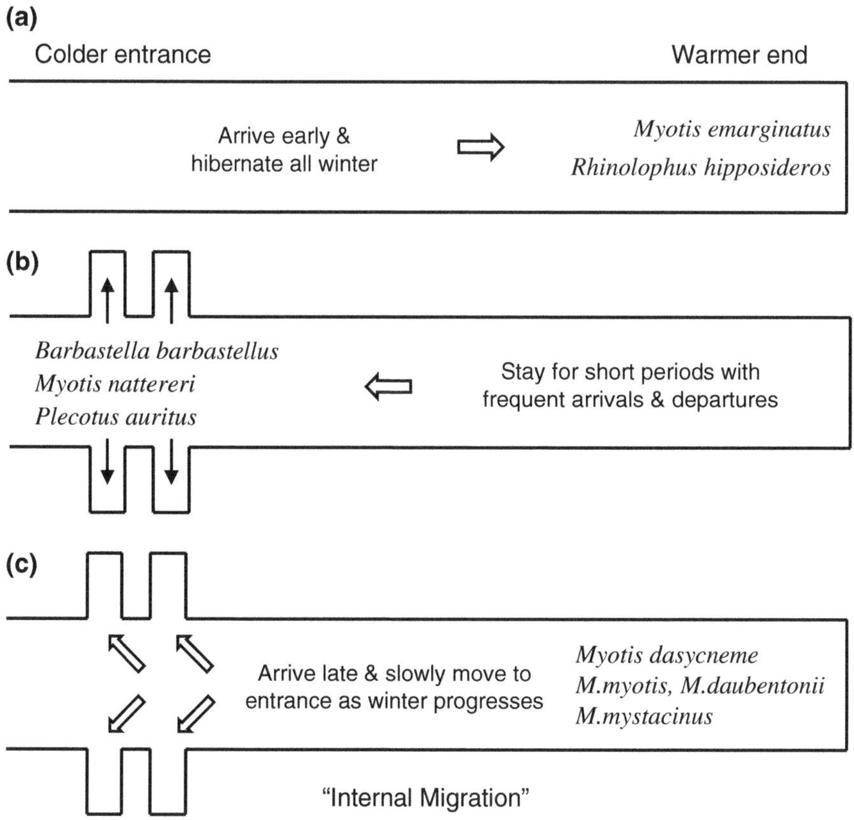

Fig. 15.5 a–c Differences in use of South Limburg limestone mines among nine bat species, four of which exhibit 'internal migration'. (after Daan and Wichers 1968)

In isolated mountain ranges in California, the Californian leaf-nosed bat (*Macrotus californicus*) uses geothermally heated winter roost sites in abandoned mines, with stable year round temperatures of about 29 °C, which minimizes energy expenditures. They also have an energetically frugal pattern of foraging which relies on visual prey detection. These two factors have allowed this most northerly representative of the Phyllostomidae to invade the temperate zone (Bell et al. 1986).

Although in temperate regions bats use caves mainly for hibernation, some species continue to occupy them throughout the year and young are born there. Slight differences in summer temperature between caves are important and young of the same bat species in caves with higher temperature grow faster and reach adult dimensions sooner. Growth rates of known-age young of the gray myotis (*M. grisescens*), a nationally endangered species endemic to several eastern states in the USA, were compared between a colony of 600 in a cave at 13.9 °C and a colony of 2200 at 16.4 °C. Significantly increased growth rates in the latter resulted in mean attainment of first flight at 24 days of age compared with 33 days in the former. The young reared at the higher cave temperature have an extra week to increase their foraging efficiency and their body mass before hibernation begins which may be crucial to their overwintering survival (Tuttle 1976).

These temperatures are however cool compared with the hot caves of the tropics which fall into two categories so far as bats are concerned. The first are heated by convection, with hot air rising from the plain below and entering a vertical cave at higher elevation. The endemic Australian ghost bat (*Macroderma gigas*) roosted in such caves on Mount Etna, Queensland during pregnancy and lactation, and because females experience thermoneutral conditions and do not have to expend energy to maintain a high constant body temperature, they can divert more energy to growing a foetus and producing milk (P. Racey unpublished). The second type of hot cave is heated by the bats themselves. These are characterized by a small entrance, at floor level, opening onto a series of chambers, along which a temperature gradient is established. Species with low basal metabolic rates (BMR), as measured in the laboratory (the Antillean ghost-faced bat *Mormoops blainvillei*, the sooty mustached bat *Pteronotus quadridens* and Leach's single-leaf bat *Monophyllus redmani*) selected the distal hotter end of the temperature gradient, which, at 35 °C, approached thermoneutrality. Large numbers of bats are needed to maintain such a high temperature, and in Cucaracha cave, Puerto Rico, half a million bats roost in the distal chamber (Rodriguez-Durán and Lewis 1987; Rodriguez-Durán and Soto-Centeno 2003; Rodriguez-Durán 2009; Ladle et al. 2012).

In Mexico, Avila-Flores and Medellin (2004) found that heterothermic species in the family Vespertilionidae used colder caves with the widest temperature range (1.6–29.8 °C) whereas homeothermic species in the four exclusively tropical families Emballonuridae, Mormoopidae, Phyllostomidae and Natalidae occupied warmer caves (14.5–37.5 °C). Within these caves, precise homeotherms, with a narrow range of body temperatures, occupied cooler roosts than more labile homeotherms. Body size and temperature were negatively correlated. The smallest homeothermic insectivorous species, weighing less than 10 g, consistently occupied roosts with temperatures greater than 20 °C, often 25 °C, whereas only the largest homeothermic insectivores were found in caves with temperatures as low as 16 °C. Frugivorous, nectarivorous and sanguivorous bats were found in a wide range of temperatures but often less than 20 °C. No trends could be detected so far as relative humidity was concerned, and, overall, temperature was the most important physical variable influencing roost selection.

The last physical variable is light intensity, which is also the least studied because until recently, recording it accurately has not been possible. Some insectivorous bat species are more light tolerant than others, such as the Seychelles sheath-tailed bat (*Coleura seychellensis*) which often roosts in open boulder caves (Bambini et al. 2006), although light levels in these caves are not thought to be a major factor in roost selection. In contrast, Old World fruit bats (Pteropodidae) rely on vision and those species which roost in caves, such as *E. dupreanum* in Madagascar, do so within sight of the entrance (Cardiff et al. 2009). Within this bat family, the genus *Rousettus* has evolved a form of echolocation involving clicking with its tongue and is capable of roosting deep in caves (MacKinnon et al. 2003; Waters and Vollrath 2003). Gould (1988) raised the possibility that the wing-clapping of *E. spelaea* may aid their navigation in the dark caves where they roost. Similar wing-clapping is reported in the bare-backed bat (*Dobsonia moluccensis*) which also roosts in dark caves (Churchill 2008).

15.3.3 Importance of Bats for Cave Ecosytems

Due to the absence of primary production and general scarcity of food underground, most life in caves is invertebrate and largely dependent on energy sources from the surface such as penetrating tree roots and organic debris washed in by percolating waters or floods (Gillieson 1996). While bat guano appears to be less significant for cave-restricted invertebrates (often referred to as troglobites or troglobionts) inhabiting temperate caves, a considerable proportion of the terrestrial fauna in tropical caves depends upon its continued deposition (Deharveng and Bedos 2012). The significance of this lies in the fact that subterranean invertebrates are globally diverse and caves are thought to rank among the hottest of biodiversity hotspots (*sensu* Myers et al. 2000) worldwide in terms of their levels of species endemism and threat (Gilbert and Deharveng 2002; Whitten 2009).

It has long been assumed that guano accumulations support less invertebrate diversity and few narrowly-endemic species compared to low-energy cave habitats. However, this view is challenged by the recent discovery of a huge radiation of typically guanobiotic Cambalopsid millipedes across Southeast Asia, whereby each karst area harbors one or two site-endemic species (Golovatch et al. 2011). Further, as most tropical karsts have yet to be investigated and cave-restricted species new to science continue to be discovered in virtually every survey (both troglobionts and guanobionts), the era of tropical cave biodiversity exploration has clearly only just begun. Notwithstanding this, due to the major contribution guanobionts make to overall cave diversity, disturbance to bats is increasingly regarded as one of the most serious threats to tropical cave invertebrates. Paradoxically, this concern is probably more relevant to common and widespread bat species (e.g. *C. plicatus* in Asia) than rarer or non-colonial species however, since the former produce the most guano in cave ecosystems (Deharveng and Bedos 2012).

15.4 Conservation Threats

Due to their low annual reproductive rates, bat populations take a relatively long time to recover from population losses associated with human activities (Racey and Entwistle 2000). Slow population growth rates thus exacerbate existing threats to bat populations. This poses a particular problem for cave-dwelling bats, particularly species which are gregarious and colonial, as any intrusion into the relatively small and confined spaces that caves provide tends to affect the entire aggregation (McCracken 1989). The fact that large numbers of individuals are often concentrated into only a few specific roost sites results in high potential for disturbance (Sheffield et al. 1992). It also increases the potential for Allee effects—recently redefined as a positive relationship between any component of individual fitness and either numbers or density of conspecifics (Stephens et al. 1999).

Caves have a long history of human use, with the earliest direct evidence of occupation dating back to at least 700,000 BP (from the Peking person site near Beijing, China: Gillieson 1996). Originally providing havens for prehistoric hunter-gatherers, caves across the world have since served a remarkable range of purposes. These include military fortifications and wartime refuges, horticultural uses, sanatoria for patients with respiratory and other ailments, sites for religious worship and burial, storage and dumping facilities, sources of water and fertilizer, and finally, destinations for opportunistic recreation and commercial tourism. More generally, because karst is highly porous, the integrity of caves in karst areas depends on complex interactions between hydrology, biology and geomorphology within their catchments. As the health of broader subterranean communities is strongly influenced by their surrounding environment, activities impacting cave-dwelling life consequently include those affecting the surface environment (Watson et al. 1997; Vermeulen and Whitten 1999).

15.4.1 Seasonality and Climate Change

Cave-dwelling bats are especially vulnerable to disturbance during periods of torpor and hibernation. Although hibernating bats periodically arouse, such arousals are energetically expensive and can account for 75 % of winter energy expenditure (Thomas et al. 1990). Disturbance can thus cause premature arousal which can deplete critical energy reserves to the extent that the bat is unable to survive the winter (Thomas 1995). As a result, human disturbance is widely regarded as a significant cause of over-winter mortality in temperate zone bats (Sheffield et al. 1992; Mitchell-Jones et al. 2007).

The lethal effects of repeated arousals during hibernation have been dramatically illustrated by the death of many millions of bats hibernating in caves in North America following infection with white-nose syndrome (*Geomyces destructans*—now renamed *Pseudogymnoascus destructans*: Minnis and Lindner 2013). This fungus invades the skin and irritates the bats, causing them to arouse repeatedly, until their stored body fat is exhausted and they starve to death (Reeder and Moore 2013). The situation regarding hibernation in the seasonal tropics is less well documented, although as many regions at higher tropical latitudes periodically experience cold winters (e.g. North Vietnam 18–23°N, lowest temperatures varying from −3.4 to 6.0 °C: Van et al. 2000), insectivorous bats inhabiting such areas are likely to undergo bouts of torpor during the coldest periods when they rely on stored body fat. Storage of spermatozoa in the reproductive tract of overwintering bats is a key reproductive adaptation of those living at temperate latitudes (Racey 1979) and the elevation of body temperature associated with frequent arousals from hibernation is thought to compromise the viability of stored sperm by encouraging their phagocytosis by leucocytes (Guthrie 1933; Racey 1975).

Disturbance during pregnancy, lactation and weaning is widely recognized as highly detrimental to recruitment in bat populations (McCracken 1989; Sheffield et al. 1992; Jubertie 2000; Mitchell-Jones et al. 2007). Protection during these periods is consequently also central to cave bat conservation. As reproduction is energetically expensive (Racey and Speakman 1987), many bat species time the event so that lactation, the most costly stage (Kurta et al. 1989), coincides with peak food availability (Racey and Entwistle 2000). This peak may also occur during weaning for many species (Bernard and Cumming 1997). In temperate zone bats, parturition and lactation occur in summer, whereas in the seasonal tropics, growing evidence suggests reproductive activity for many insectivorous, frugivorous and nectarivorous bats is associated with rainfall, with lactation occurring during the peak rainy season (Racey and Entwistle 2000).

The likelihood that reproductive cycles will be affected by global climate change warrants attention as such effects are predicted to be significant in temperate zone bats (Jones and Rebelo 2013). As with other taxa, climate change is predicted to alter the distribution of bat species (Scheel et al. 1996; La Val 2004; Rebelo et al. 2010; Hughes et al. 2012). Altered distribution patterns are also anticipated for hibernating species due to changes in energetic demands (Humphries et al. 2002). Range shifts have already been observed in the case of Kuhl's pipstrelle (*Pipistrellus kuhlii*), which has moved northwards from Mediterranean regions into Central and Eastern Europe over 15 years (Sachanowicz et al. 2006). In projecting the effects of several climate change scenarios on 171 bat species in Southeast Asia, Hughes et al. (2012) found only 1–13 % (1–22 spp.) showed no reductions in their current ranges. Though range expansions were projected for some species, it was perceived that this might challenge those with poor dispersal abilities. This could pose a particular problem for cave-dwelling bats in Asian karst areas, since widespread isolation of karst outcrops has already occurred (Struebig et al. 2009; Furey et al. 2010) (Fig. 15.6). Even species capable of shifting their ranges in response to the rapid rate of current climate change may be hampered by the limited availability of suitable caves and potential time required for suitable foraging habitats to develop (Rebelo et al. 2010).

15.4.2 Incidental Disturbance

Although intentional disturbance of cave-dwelling bats as a result of vandalism and other causes is well documented and widespread, unintentional disturbance

Fig. 15.6 Land use changes leading to isolation of the Gunung Kanthan karst outcrop in Ipoh, Malaysia (created by Kendra Phelps © Google Earth)

can pose an even greater threat due to the many other reasons that humans use caves (McCracken 1989) such as opportunistic recreation, camping, caving excursions, dumping refuse and use as storage facilities. For example, the importance of the Nietoperek fortifications in Western Poland as a bat hibernaculum was first brought to the attention of bat biologists outside the Iron Curtain by a Russian plan to dump radioactive waste there. The plan was shelved as a result of a successful campaign by conservationists. Throughout Poland, groups known as "bunkermen" meet socially in underground fortifications where they may disturb the bats.

Thomas (1995) showed that non-tactile disturbance from seemingly innocent cave visits during hibernation periods can cause bats to arouse and maintain significantly greater flight activity for up to eight hours afterwards. Such arousals are highly detrimental to their over-winter survival and non-tactile disturbance during other critical periods such as reproduction may lead to: (1) death of young that lose their roost-hold and fall to the cave floor, (2) females abandoning the roost for less ideal sites where prospects for reproductive success may be reduced, (3) greater energy expenditure among females and less efficient energy transfer to young (translating into slower growth of young and increased foraging demands on females), (4) reductions in the thermoregulatory benefits of a roost as a result of decreased numbers of bats frequenting the site (McCracken 1989; Sheffield et al. 1992).

As a result, uncontrolled human disturbance often leads to decreases in numbers of bats roosting in caves and mines (Tuttle 2013). For instance, disturbance in caves in West Virginia, USA, occupied by the Indiana myotis (*M. sodalis*) and Townsend's big-eared bat (*Corynorhinus townsendii*) resulted in a decline from 1137 bats to 286 in one cave and from 560 to 168 in another (Stihler and Hall 1993). Conversely, when ten caves were protected by grilling and fencing, *M. sodalis* populations increased, from 1615 to 6297 bats (290 %) and *P. townsendii* from 3455 to 7491 (117 %). Because fencing is more easily vandalized, gating is considered by many as more successful at preventing disturbance, although some bat species do not tolerate gates and it is important to establish the bat-preferred design.

15.4.3 Extractive Industries

Limestone quarrying for cement and construction materials presents a severe threat to cave-dwelling bats in karst areas as it can result in the total loss of outcrops (Fig. 15.7), leaving few options for remediation. Global demand for cement alone was projected to increase by 4.1 % per annum to 3.5 billion tons in 2013 despite the western financial crisis (Sutherland et al. 2012) (Fig. 15.8). This is believed to pose perhaps the greatest threat to cave bats in Southeast Asia, as the region has the highest annual quarrying rates in the tropics and these appear to be increasing faster than in other regions, at 5.7 % per year (Clements et al. 2006; Kingston 2010). In contrast, the impact of smaller artisanal mining operations

Fig. 15.7 Quarrying of limestone hills in southern Cambodia (© Neil Furey/Fauna & Flora International)

Fig. 15.8 Quarrying of limestone in the Petersburg mines of South Limburg, Netherlands (© Joep Orbons)

appears largely unevaluated so far, though such operations are widespread and commonly target cave sediments in countries such as Vietnam (N. Furey unpublished, Tordoff et al. 2004).

Groundwater abstraction and land cover changes in the catchments of caves can affect their environments in several ways. Though empirical data on the effects on bats appear to be few, abiotic changes include altered hydrological cycles (particularly where natural land cover is replaced with impervious surfaces such as roads) and altered cave microclimates as a result of sedimentation blocking voids for percolating water in overlying rocks (Watson et al. 1997). Removal of vegetation at cave entrances may also alter airflows and temperatures within a cave to such an extent that its habitable portions are reduced or eliminated (Sheffield et al. 1992). Conversely, alien plants may overgrow cave entrances and prevent their use as roosts (Gerlach and Taylor 2006) and other invasive species such as feral cats have been identified as predators of cave-dwelling bats (Rodriguez-Durán et al. 2010; Tuttle 2013). More dramatically, large water projects can flood caves through reservoir creation and groundwater recharge efforts. For instance, recharge efforts led to violent flooding of the Valdina Farms sinkhole in Texas in 1987, with the loss of a colony of four million *T. brasiliensis* and a rare colony of Peter's ghost-faced bat (*Mormoops megalophylla*) (Elliot 2004).

Cave-dwelling bats are especially vulnerable to harvesting for consumption and trade due to their aggregation into these confined spaces. Reviews of global patterns in bat hunting for bushmeat indicate that this is common in the Old World tropics, but with possible exceptions, does not appear to be widespread or having a significant effect in other regions (Mickleburgh et al. 2009; Mildenstein et al. 2016). Though a problem in Madagascar and in many African countries, the threat to cave-dwelling bats appears to be most widespread and acute in the Asian tropics, where bat harvesters target many species. For instance, Hall et al. (2002) reported dramatic declines in *E. spelaea* and the greater naked bat (*Cheiromeles torquatus*) during their successive surveys of Niah caves, Sarawak compared with numbers recorded in earlier surveys by Medway (1958) and attribute this to hunting for human consumption. While Asian hunters often target species that are large, colonial and/or abundant (e.g. *E. spelaea*, *Rousettus* spp., *C. torquatus*, *C. plicatus* and bent-winged bats *Miniopterus* spp.), smaller bats are also taken and harvesting activities are highly likely to negatively affect other species sharing the same caves (Hutson et al. 2001; Mickleburgh et al. 2009). Given the scale and severity of bushmeat hunting on bats, there is a pressing need for research on sustainable harvesting.

Despite its widespread occurrence, accounts of the impact of guano harvesting upon cave-dwelling bats appear to remain largely anecdotal. This may stem in part from the difficulty of accurately monitoring large bat colonies, although rates of guano accumulation and harvesting records reflect their size (Fig. 15.9).

Fig. 15.9 Entrance to Tarum Cave in western Cambodia (*main picture*) where 200–400 sacks of bat guano (*inset picture*) produced by the largest colony of *Chaerephon plicatus* in the country have been harvested every month since 1995 (© Neil Furey/Fauna & Flora International)

Fig. 15.10 Sale of decorative cave speleothems near the Vietnam-China border (© Neil Furey/ Fauna & Flora International)

It is generally acknowledged that insensitive harvesting operations can be highly detrimental to cave bat populations (Hutson et al. 2001), particularly where cave modifications are undertaken to facilitate guano extraction (Elliot 1994). Similar concerns apply to the harvesting of cave swiftlet (*Aerodramus* spp. and *Collocalia* spp.) nests in Southeast Asia (Suyanto and Struebig 2007) since trade in these has expanded greatly in recent decades, causing significant disturbance to bats sharing the same caves (Wiles and Brooke 2013). In both instances, the perceived benefits of continued harvests can encourage local communities to protect the producers (Leh and Hall 1996; Bates 2003), although research to identify and validate sustainable harvesting practices is clearly needed. Lastly, harvesting of speleothems for decorative purposes represents another widespread practice in Southeast Asia whose impacts on cave bats appear to remain largely unevaluated (Fig. 15.10).

15.4.4 Cave Tourism

Cave tourism, which began in the late nineteenth century, has dramatically increased threats to all life in caves. Around 20 million people worldwide were estimated to visit caves recreationally each year in the mid-1990s and the industry has burgeoned in East Asia more recently (Gillieson 1996; Zhang et al. 2009; Furey et al. 2011; Luo et al. 2013) (Fig. 15.11). Development of caves for tourism typically involves the introduction of artificial lighting and physical modifications to cave substrates in the form of entrance structures, stairs, walkways, and car parks. Alongside disturbance caused by their presence, cave visitors create marked fluctuations in temperature, relative humidity and carbon dioxide concentrations, all of which can lead to roost abandonment. For instance, commercialization of Fourth Chute Cave in Quebec, Canada resulted in abandonment of the largest hibernacula of eastern small-footed myotis (*M. leibii*) known at the time in eastern North America (Mohr 1972).

Mann et al. (2002) explored behavioural responses of a maternity colony of 1000 cave myotis (*M. velifer*) by experimentally exposing the colony to cave tours. High light intensity had the most detrimental effect with bat activity levels and flight increasing with proximity to tour routes and when tour groups talked. All of these behavioural responses increased as the maternity season progressed. Consistent with this, in a review of 225 subterranean sites in China, Luo et al. (2013) showed that recreational activities had pronounced detrimental effects on the numbers of bat species and presence of species of special conservation concern. Almost 90 % of the sites were found to be disturbed and only 15 % of natural caves were unaffected by disturbance. Concerns about the impact of cave tourism on Chinese bats have also been raised by Niu et al. (2007) and Zhang et al. (2009)

Fig. 15.11 Cave visitation during the annual Tet holiday in North Vietnam (© Neil Furey)

and Olson et al. (2011) also found that numbers of hibernating bats significantly increased in Cadomin cave, Canada after restrictions on visitors were enforced. In Madagascar, associations between tourism and hunting of cave-dwelling pteropodids were noted by Cardiff et al. (2009) who speculated these might be due to improved access facilitating hunting activities when tourists are absent. Cardiff et al. (2012) also analyzed the effects of tourism on the Malagasy rousette (*Rousettus madagascariensis*) and found that maintaining a minimum distance of 12 m between tourists and roosting bats and avoiding their illumination caused the least disturbance.

One of the few detailed investigations of the effects of tourism on bat numbers has been carried out in the Dupnisa cave system in the Thrace region of Turkey, one of the largest aggregations of bats in southeast Europe with mean numbers for 15 species of 25,000 in winter and 4000 in summer (Paksuz and Özkan 2012). The maximum number of bats recorded during a single survey of the three connected caves is 56,000. The total length of the system exceeds 2.5 km and tourists are admitted to about 400 m of two hibernation caves during summer but excluded from the cave containing maternity colonies. The assertion by Paksuz and Özkan (2012) that mean bat numbers using the cave have increased, significantly so in the maternity cave, since it was opened to tourism in July 2003 has been challenged by Furman et al. (2012) who concluded that there has been a 20 and 60 % reduction in the two hibernation caves in February and March 2003–08 respectively compared with their own surveys in March 2001 and a 90 % reduction in the maternity cave in April and May 2002–07, compared with their survey in April 2001 (Furman and Özgül 2004). Furman et al. (2012) point out that the comparison made by Paksuz and Özkan (2012) is ambiguous as it contrasts the construction period (including the early days of tourism) with the later period after construction was finished and the system was opened to tourists, and they provide no data for the period before any construction work started. The only significant increase in bat numbers reported by Paksuz and Özkan (2012) was in the cave closed to tourists and bats in caves accessible to tourists may have moved to the less disturbed cave. This inconsistency is significant as the development of the Dupnisa system for tourism may be followed in other cave complexes.

15.4.5 Insights from Long Term Studies

The most distinctive feature of several European studies of bats in karst is their duration. Bats were counted in 32 limestone mines in South Limburg from 1943, yearly in half of them, to 1987 (Weinreich and Voshaar 1992). Two factors affected the caves directly over this period—the erection of grills to prevent unauthorized entry, and the cultivation of mushrooms. Intensive mushroom culture reduced the number of bats by 90 % after three years. However, extensive culture, involving much less disturbance, and the installation of grills, had no significant effect on bat numbers. Population trends for eight species for which the most complete data sets are available reveal that from 1943 until 1987, four species, *R. hipposideros*, *M. myotis*, *M. emarginatus* and *M. nattereri* showed steep declines; three *M. mystacinus/brandti*, *M. dasycneme* and *P. auritus* remained fairly stable and one, *M. daubentonii*, showed a dramatic increase. A possible explanation of this increase is the intensification of agriculture following the second world war which led to the eutrophication of fresh waters. This in turn resulted in an increase in chironomid flies on which *M. daubentonii* feeds.

The total number of bats hibernating in these 32 limestone mines decreased overall by two-thirds between 1943 and 1957. Numbers then stabilized and from 1977 returned to their former level. Although the study revealed the negative effects of intensive mushroom cultivation, other factors are implicated, in particular the cessation of banding bats during hibernation after 1957. In addition, the first decade after the second world war coincided with unlimited use of agricultural pesticides, which took its toll on many species of wildlife (Carson 1962). Restrictions were placed on the use of the most toxic and persistent of these pesticides from 1968 to 1973 and that coincides with the beginning of recovery of bat numbers (Weinreich and Voshaar 1992).

The second long running study involving both summer and winter bat populations is located in the Czech and Slovak republics. This began in 1948 and 89,000 bats of 23 species were banded in the following 52 years, approximately a third of which are found in karst caves (Gaisler et al. 2003). A regular winter census has taken place in one of the tourist caves, Sloupsko, in the Moravian karst (Zukal et al. 2003). Bats were originally banded in both summer and winter roosts but once the practice of arousing bats during hibernation to band them was abandoned, the numbers using the cave increased, as in the Dutch study. The recovery rate is remarkably high for a banding study—27 % for *M. myotis* and 18 % for *R. hipposideros*, as is the revealed longevity—37 years for the 35 g *M. myotis*.

Fifteen of the 23 bat species found in the Czech and Slovak republics are found in the Moravian karst, but the bat community in winter is dominated by *M. myotis* which accounts for about half of the bats visible during the census (Zukal et al. 2003). The numbers of *M. myotis* and *R. hipposideros*, species of conservation concern throughout their European distribution, increased dramatically in the 1990s and this is attributed to good management. Apart from the winter census, no underground activity is permitted during hibernation. Vehicular traffic on the access roads to the caves is limited throughout the year and cars and lorries are prohibited in the main valleys.

15.4.6 Declines in Cave Bats

Although the difficulties of accurate counting have confounded assessments of trends in numbers of bats using caves, most available information points to declines. Dumitresco and Orghidan (1963) reported 'more than a hundred thousand' common pipistrelles (*Pipistrellus pipistrellus*) in the Sura Mare cave in

Romania. Half a century later, only 34,000 individuals were recorded there, and, a total of only 150,000 individuals of 22 species in 79 caves throughout the country (Nagy and Postawa 2010). Dramatic declines have also occurred in numbers of *T. brasiliensis* in caves in the south-western USA (McCracken 2003) (Table 15.1).

The declines at Carlsbad cavern have been attributed to the use of the organochlorine insecticides DDT and dieldrin (Geluso et al. 1976, 1981), which were subsequently banned. There is no evidence however that the declines at Eagle Creek shared the same cause (McCracken, 1986). Other factors may also have contributed to these declines, such as the boring of a shaft through the main bat roosting area in Carlsbad to facilitate guano mining, which altered temperature, relative humidity and airflow within the roost (McCracken 1986). A major guano mining operation, involving the installation of electric lights, may have led to the complete abandonment of U-Bar cave, New Mexico by bats (McCracken 1986).

Equally dramatic declines have been recorded in Mexico as a result of attempts by cattle ranchers to control vampire bats by burning car tyres and dynamiting, with equally lethal effects on non-target bat species, and also mining (S. Walker pers. comm. in Hutson et al. 2001) (Table 15.2). In the intervening years however, better-targeted vampire control and other conservation initiatives in Mexico have halted or reversed these trends (Medellin 2003).

Table 15.1 Declines in cave bats in USA

Colony	Year	Estimated size
Carlsbad cavern, New Mexico	1936	8.7×10^6
	1957	4.0×10^6
	1973	218,000
Eagle Creek cave, Arizona	1963	$25–50 \times 10^6$
	1969	30,000

Table 15.2 Declines in cave bats in northern Mexico

Cave	State	Historical population	1991 population
La Ojuela	Durango	184,000	0
Tio Bartola	Nuevo Leon	4×10^6	30,000
La Boca	Nuevo Leon	Millions	100,000
Del Marviri	Sinaloa	940,000	250,000
El Omo	Tamaulipas	Millions	0
Del Guano	Tamaulipas	440,000	125,000
Quintero	Tamaulipas	567,000	30,000
La Mula	Tamaulipas	303,000	100,000

15.5 Conservation Responses

Not surprisingly, conservation responses to threats facing cave-dwelling bats are strongly linked—though by no means confined—to the growth of organizations across the world dedicated to conserving all bat species. This subject was reviewed by Racey (2013) who suggests that while bat conservation has made much progress in Europe and North America and is growing in strength in Central and South America and parts of Asia and Australasia, half of the world remains a "conservation void" so far as bats are concerned. This conservation void includes most of Africa, all of the Middle East, much of the Russian Federation and all of the former Russian republics, together with most of Asia, including China, Mongolia and Tibet.

15.5.1 National and International Initiatives for the Protection of Cave Bats

The Council of Europe reviewed underground habitats and proposed selection criteria for their protection (Jubertie 1992). This was followed by IUCN's Guidelines for Cave and Karst Protection prepared by the World Commission on Protected Areas Working Group (Watson et al. 1997), although it was realized at the time by one of the authors that more detailed treatment of biodiversity issues was required (Hamilton-Smith 2001). This was begun, but never completed. Among the many broader cave-related publications that have appeared (e.g. Gunn 2003; White and Culver 2012), the treatment of Vermeulen and Whitten (1999) for East Asia is notable in explicitly addressing the threat to cave biodiversity from tourism and exploitation of limestone for industrial purposes by providing options for impact assessment, site selection, mitigation and national management of karst areas.

National academic societies have also produced guidelines for the protection of bat roosts, particularly those in caves (e.g. Sheffield et al. 1992) and among the bat conservation NGOs, the UK's Bat Conservation Trust was among the first to produce a conservation code for cave visitors (Hutson et al. 1988). Since then, a variety of organizations have produced materials to raise public awareness of bats at karst caves around the world. In Madagascar for instance, Madagasikara Voakajy have produced guidelines in three languages (Malagasy, English and French) for tourists visiting caves in the Bemaraha karst, with clear instructions about minimizing disturbance to bats. The international speleological community has also, in general, been sensitive to the potential effects of their activities on bats and other cave fauna and codes of ethics have been published by national caving societies in several regions. An important advance within the caving community has also been the replacement of carbide lamps, the combustion products of which are toxic, with electric torches.

The development of a network of protected areas including many sites of outstanding importance for cave-dwelling bats across the European Union (known as Natura 2000) has been viewed as an important step change in European bat conservation, although its effectiveness in protecting the foraging habitats of cave bats in the region has been questioned (Lison et al. 2013). Allied to this, the Advisory Committee of Eurobats (an intergovernmental agreement for the protection of European populations of bats) has a working group on underground sites, in addition to other groups dealing with related subjects such as surveillance and monitoring and wind farms. These have resulted in well illustrated guidelines for the protection and management of subterranean sites and lists of important sites across Europe, which are freely available from the Eurobats website (Mitchell-Jones et al. 2007).

The United States Endangered Species Act provides strong protection for several cave-dwelling species, and individual states maintain their own lists of endangered and threatened species and species of special conservation concern, as do agencies such as the Forest Service and Bureau of Land Management. Alongside four sub-national bat groups which address bat conservation issues in the western, mid-western, northeast and southeast states respectively, Bat Conservation International has mounted successful programs for bat conservation in American caves and mines, in addition to initiatives aimed at building capacity and protecting cave bats in many other countries such as the Philippines (Racey 2013).

In Central America, concerns about the plight of predominantly or wholly cave-dwelling species such as *T. brasiliensis*, *L. curasoae* and Mexican long-nosed bat (*Leptonycteris nivalis*) that migrate between Mexico and the southern United States led to the establishment in 1994 of the Program for the Conservation of Mexican Bats (PCMM: Programma para la Conservacion de los Murcielagos de Mexico). In 2007, this in turn led to the launch of the Latin American Network for Bat Conservation (RELCOM: Red Latinoamericana para la conservacion de los Murcielagos), an alliance of organizations and individuals in 22 countries (including the Caribbean) concerned with bat conservation. In South and Southeast Asia, the respective regional equivalents are Chiropteran Conservation Information Network for South Asia (CCINSA) and Southeast Asian Bat Conservation Research Unit (SEABCRU), whereas in Africa, a new network Bat Conservation Africa was formed by bat conservationists in 2013 which encompasses 19 African countries and the West Indian Ocean islands (Kingston et al. 2016).

15.5.2 Development of Gating

The large number of abandoned mines in the USA, the need to maintain public safety and to conserve resident bats has led to extensive gating of both mines and caves (Dalton and Dalton 1995; Vories et al. 2004). Gating has long been a controversial subject within the speleological community (Kennedy 2006), and similarly for bat conservationists, not least because gates installed at cave entrances from the 1950s through the early 1970s virtually always led to roost abandonment (Tuttle 1977). Much has been learnt through trial and error however, so that studies of more recently installed cave gates provide grounds for optimism (e.g. Pierson et al. 1991, Stihler and Hall 1993; Decher and Choate 1995; Fant et al. 2009). For instance, Martin et al. (2003) recorded a statistically significant increase in the numbers of *M. grisescens* from 60,130 to 70,640 bats between 1981 and 2001 in 25 gated caves in Oklahoma, USA. However, recent studies before and after gating hibernacula of *M. sodalis* reported decreased rates of growth for increasing populations and the reverse for declining populations (Crimmins et al. 2014). Berthinussen et al. (2014) summarise the mixed results of ten gating studies on three continents.

Because fencing is more easily vandalized, gating is recognized by many as more successful at preventing disturbance. However, species such as *T. brasiliensis* cannot tolerate gates due to their flight geometry and large colony sizes, and for bat species that can, it is critical to establish their preferred design. Improperly designed gates can alter cave environments by restricting air circulation, causing population declines (Richter et al. 1993). For example, *M. grisescens* requires an open flyway above gates, whereas *P. townsendii* will tolerate full gates with horizontal bars (Tuttle 1977). In the UK, Pugh and Altringham (2005) examined the effect of different sizes of horizontal gate spacings on numbers of Natterer's bats (*Myotis nattereri*) entering swarming sites in autumn and provided clear recommendations for future gate design. While a detailed treatment of the subject is beyond the scope of the present chapter, the proceedings of a multidisciplinary meeting to develop gate design provides a wide variety of well-illustrated examples of gated caves and mines (Vories et al. 2004) and similarly useful advice is given in Hildreth-Werker and Werker (2006), Mitchell-Jones et al. (2007) and Fant et al. (2009).

Nevertheless, a great deal remains to be learnt about the reactions of bats to gates, even in the USA and Canada, where most of the voluminous information available is anecdotal with few systematic studies conducted to date (Sherwin and Altenbach 2004; Spanjer and Fenton 2005). For instance, Vories et al. (2004) recommended studies of the effects of gates on cave microclimate, wind tunnel assessments of the airflow characteristics of different gate designs, the acoustic signatures of the gates and their possible interference with echolocation calls. Since gates have the potential of protecting cave bats from disturbance but little or no information exists on the responses to gating of hundreds of species across the world, clearly much research lies ahead.

15.5.3 Artificial Hibernacula and Maternity Roosts

In the UK, members of local conservation groups have constructed many artificial caves, often from a series of concrete sewer pipes to which roosting cavities are added. Unfortunately, the rate of occupancy, even over a 25 year period of monitoring, has been so low that such an approach cannot be recommended as a means of mitigating threats to cave-dwelling bats. Furthermore, Berthinussen et al. (2014) found no published evidence for the effects of providing artificial hibernacula for bats to replace sites lost to development. However, in Brittany, France and County Clare, Ireland, houses were constructed to serve as maternity roosts and hibernacula for *R. ferrumequinum* and *R. hipposideros* respectively. The former was used by over 100 individuals in summer and winter and the latter by 220 hibernating bats (Eurobats 2014).

The Combe Down stone mine complex in the UK is one of the twenty most important hibernacula in Europe. Because engineering work was required to stabilise the complex in order to protect the parts of Combe Down village that were above it, and both UK species of *Rhinolophus* roost there during summer and winter, extensive mitigation was devised by Ransome (2010). This included the construction of three incubation chambers, each inside a different mine. Each chamber was a small underground room partly maintained at ca 27 °C with a roof lined with mesh-covered plywood from which the bats could hang. Both rhinolophid species used the chambers, and subsequently gave birth and reared young there. In 2014, there were two underground breeding colonies of *R. hipposideros*, each of some 40–70 adults and 30 young and a colony of *R. ferrumequinum* with about 120 adults and a dozen pups (R. Ransome pers. comm.). This initiative stemmed from Ransome's earlier success in the improving the survivorship of young of *R. ferrumequinum* by installing electric tubular heaters in the maternity roost at his study colony in the roof of a mansion (Ransome 1998).

Some success has also been obtained in the UK by protecting sites already known to be used by bats, often by gating (S. Thompson pers. comm.; Hutson 1993). Examples are small chalk caves entered by a vertical shaft (known as dene holes in the South of England) and ice houses (brick-lined domed structures) constructed mainly underground, often on a hillside close to water. The success of concentrating conservation efforts on known roosts was exemplified recently by the massive enlargement of a small bat cave on a hillside above a much larger cave occupied by 20,000 bats which was to be inundated by the Balikesir Havran irrigation dam in Turkey. Floor material containing invertebrates and guano was transferred from the old to the new cave and up to 13,000 bats took up residence there (Irfan Albayrak and Eda Türkyiğit pers. comm.).

More generally, the survival of bats that overwinter in caves has been helped in summer by the widespread use of bat boxes of various sizes and designs. Berthinussen et al. (2014) summarise the results of 22 studies of artificial roost structures across the world. In the USA, substantial progress has been achieved and some artificial roosts now provide the only hope of recovery for local populations. For instance, two large bat houses built in 1991 and 2010 in Gainesville, Florida contained 300,000 bats in 2012 (mostly *T. brasiliensis*), more than all the natural roosts in the state combined (Tuttle 2013).

15.5.4 Recent Initiatives

Due to concerns that unsustainable guano harvesting practices are increasing and threatening millions of bats worldwide, guidelines have been prepared to help address the clear need for harvesting standards that minimize negative impacts on these and other cave-dwelling organisms. The guidelines cover generic aspects such as baseline assessments, guano extraction methods and policies for site management and monitoring, and have been adopted by the IUCN (IUCN SSC 2014).

The establishment of a new Cave Invertebrate Specialist Group within the IUCN Species Survival Commission in 2013 is also promising. Unlike cave-dwelling bats, many of the invertebrates that live permanently in caves are highly range-restricted endemics incapable of dispersing to other sites (Vermeulen and Whitten 1999). As such animals are highly susceptible to environmental change and hence extinction, many will undoubtedly qualify on distributional grounds alone as Critically Endangered or Endangered in the Red List assessments that the specialist group intends to undertake as a priority.

Given the importance of bat guano to cave biodiversity in the tropics (Deharveng and Bedos 2012), additional justifications for protection of sites inhabited by cave-dwelling bats are likely to emerge. These should in turn lead to conservation outcomes either as a result of the influence the IUCN Red List exerts on national legislation for protected areas development, wildlife protection and/or EIA processes, or by alternatively triggering the environmental safeguards of multilateral institutions such as the World Bank that commonly engage in development projects in tropical karst areas (Vermeulen and Whitten 1999). Greater cooperation between bat conservationists and invertebrate biologists is therefore clearly in the interests of conserving cave life as a whole.[1]

15.6 Future Directions

Caves and other subterranean sites are critical to the survival of hundreds of bat species worldwide. Karst caves in particular are experiencing unprecedented disturbance due to their increasingly realised potential for the construction and tourism industries. As these threats are heavily exacerbated by loss of foraging habitats, bushmeat hunting, incidental disturbance and disruptive guano harvesting, research and allied conservation actions are urgently needed to reduce the impacts of these activities. Extermination attempts due to disease fears, such as the recent depopulation of bats at Kitaka mine in Uganda are also a concern (Amman et al. 2014). Since disturbance during critical periods such as reproduction are particularly detrimental to population recruitment, studies to determine when these

[1]The Cave Invertebrate Specialist Group can be contacted at CISG@fauna-flora.org.

occur in poorly documented regions have clear conservation relevance. Similarly, the use of gates to protect cave bat colonies is relatively untested outside Europe and North America and could do much to mitigate threats in other regions.

In recent years, some international conservation NGOs have worked with major cement manufacturers such as Holcim and Lafarge in some karstic areas to identify the most important caves so far as bat and other biodiversity is concerned and to avoid disturbing them. In other areas, multinational corporations destroy caves with impunity. There is a need therefore for international protocols that protect cave biodiversity while satisfying the demand for cement and construction materials (Whitten 2012; BirdLife et al. 2014) and it would be appropriate for the IUCN to take the lead in this respect. The priority for bat biologists is to collate and develop the existing fragmented information on caves in both the Old and New World tropics and identify sites of outstanding importance at national and international levels. These can then be protected from exploitation.

Acknowledgements We thank Tony Hutson and Shirley Thompson for providing unpublished information, Tony Whitten, Louis Deharveng and Dave Waldien for their advice, Armando Rodriguez-Durán, Tony Mitchell-Jones, Gabor Csorba and an anonymous referee for helpful comments, and Kendra Phelps for bringing the Gunung Kanthan karst to our attention. We are also grateful to Paul Williams, Pipat Soisook, Merlin Tuttle and Pter Lina for providing images and to the editors, T. Kingston and C. Voigt, for their invitation to contribute to this volume.

References

Adams RA, Pedersen SC, Thibault KM et al (2003) Calcium as a limiting resource to insectivorous bats: can water holes provide a supplemental mineral source? J Zool (London) 260:184–189

Altringham JD (2011) Bats: from evolution to conservation. Oxford University Press, Oxford

Amman BR, Nyakarahuka L, McElroy AK et al (2014) Marburgivirus resurgence in Kitaka mine bat population after extermination attempts, Uganda. Emerg Infect Dis 20:1761–1764

Arita HT (1993a) Rarity in neotropical bats: correlations with phylogeny, diet and body mass. Ecol Appl 3:506–517

Arita HT (1993b) Conservation biology of the cave bats of Mexico. J Mamm 74:693–702

Arita HT (1996) The conservation of cave-roosting bats in Yucatan, Mexico. Biol Conserv 76:177–185

Avery MI, Racey PA, Fenton MB (1984) Short distance location of hibernaculum by little brown bats (*Myotis lucifugus*). J Zool (London) 204:588–590

Avila-Flores R, Medellin RA (2004) Ecological, taxonomical and physiological correlates of cave use by Mexican bats. J Mammal 85:675–687

Aye NN (2006) Ecology and economic importance of *Tadarida plicata* (Buchannan, 1800), free-tailed bat in some areas of Myanmar. Dissertation, University of Yangon

Bambini L, Blyth A, Bradford T et al (2006) Another Seychelles endemic close to extinction: the emballonurid bat *Coleura seychellensis*. Oryx 40:310–318
Barbu P, Sin G (1968) Observatii asupra hibernarii speciei *Nyctalus noctula* (Schreiber 1774) in faleza lacului Razelm - Capel Dolosman - Dobrogea. St Si Cerc Biol Seria Zooloogie T 20 NR. 3 P 291–297. (in Romanian - English summary)
Barclay RMR (1994) Constraint on reproduction by flying vertebrates: energy and calcium. Am Nat 144:1021–1031
Barclay RMR (1995) Does energy or calcium availability constrain reproduction in bats? In: Racey PA, Swift SM (eds) Ecology, evolution, and behaviour of bats. Clarendon Press, Oxford, pp 245–258
Bates PJJ (2003) Conservation status of the bat and bird fauna of the limestone karst areas of Mon and Kayin States, Myanmar (Burma). Unpublished report for Fauna & Flora International 100 % fund, Fauna & Flora International, Cambridge
Bell GP, Bartholomew GA, Nagy KA (1986) The role of energetics, water economy, foraging behavior and geothermal refugia in the distribution of the bat *Macrotus californicus*. J Comp Physiol B 156:441–450
Bernard RTF, Cumming GS (1997) African bats: evolution of reproductive patterns and delays. Q Rev Biol 72:253–274
Bernard RTF, Davison A (1996) Does calcium constrain reproductive activity in insectivorous bats? Some empirical evidence for Schreiber's long fingered bat (*Miniopterus schreibersii*). S Afr J Zool 31:218–220
Berthinussen A, Richardson OC, Altringham J (2014) Bat conservation: global evidence for the effects of interventions. Synyopses of conservation evidence, Pelagic Publishing Ltd
Betke M, Hirsh DE, Makris NC et al (2008) Thermal imaging reveals significantly smaller Brazilian free-tailed bat colonies than previously estimated. J Mammal 89:18–24
BirdLife, FFI, IUCN et al (2014) Joint briefing paper on extraction and biodiversity in limestone areas. Cambridge
Booher CM, Hood WR (2010) Calcium utilisation during reproduction in big brown bats (*Eptesicus fuscus*). J Mammal 91:952–959
Boyles JG, Cryan PM, McCracken GF et al (2011) Economic importance of bats in agriculture. Science 332:41–42
Bradbury JW (1977) Social organization and communication. In: Wimsatt WA (ed) Biology of bats, vol 3. Academic Press, New York, pp 1–72
Brosset A (1966) La biologie des Chiropteres. Masson, Paris
Brunet AK, Medellin RA (2001) The species-area relationship in bat assemblages of tropical caves. J Mammal 82:1114–1122
Bumrungsri S, Harbit A, Benzie C et al (2008) The pollination ecology of two species of *Parkia* in southern Thailand. J Trop Ecol 24:467–475
Bumrungsri S, Sripaoraya E, Chongsiri T et al (2009) The pollination ecology of durian (*Durio zibethinus*, Bombacaceae) in southern Thailand. J Trop Ecol 25:85–92
Bumrungsri S, Lang D, Harrower C et al (2013) The dawn bat, *Eonycteris spelaea* Dobson (Chiroptera: Pteropodidae) feeds mainly on pollen of economically important food plants in Thailand. Acta Chiropt 15:95–104
Cardiff SG (2006) Bat cave selection and conservation in Ankarana, northern Madagascar. Dissertation, Columbia University, Columbia
Cardiff SG, Ratrimomanarivo FH, Rembert G et al (2009) Hunting, disturbance and roost persistence of bats in caves at Ankarana, northern Madagascar. Afr J Ecol 47:640–649
Cardiff SG, Ratrimomanarivo FH, Goodman SG (2012) The effect of tourist visits on the behaviour of *Rousettus madagascariensis* (Chiroptera: Pteropodidae) in the caves of Ankarana, northern Madagascar. Acta Chiropt 14:479–490
Carson R (1962) Silent spring. Houghton Mifflin, Boston
Churchill SK (2008) Australian bats. Allen and Unwin, Sydney
Clements R, Sodhi NS, Schilthuizen M et al (2006) Limestone karsts of Southeast Asia: imperiled arks of biodiversity. Bioscience 56:733–742

Cleveland CJ, Betke M, Federico P et al (2006) Economic value of the pest control service provided by Brazilian free-tailed bats in south-central Texas. Front Ecol Environ 4:238–243

Codd J, Clark B, Sanderson K (1999) Drinking by the common bent-winged bat *Miniopterus schreibersii* and calcium in cave water. Bat Res News 40:9–10

Crimmins SM, McKann PC, Szymanski JA et al (2014) Effects of cave gating on population trends at individual hibernacula of the Indiana bat (*Myotis sodalis*). Acta Chiropt 16:129–137

Daan S, Wichers HJ (1968) Habitat selection of bats hibernating in a limestone cave. Sonderdruck aus Z f Säugetierkunde Bd 33:262–287

Dalton DC, Dalton VM (1995) Mine closure methods including a recommended gate design. In: Riddle BR (ed) Inactive mines as bat habitat: guidelines for research, survey, monitoring and mine management in Nevada. Biological Resources Research Centre, University of Nevada, Reno, pp 130–135

de Boer WF, van de Koppel S, de Knegt HJ et al (2013) Hibernation site requirements of bats in man-made hibernacula in a spatial context. Ecol Appl 23:502–514

Decher J, Choate JR (1995) *Myotis grisescens*. Mamm Species 510:1–7

Deharveng L, Bedos A (2012) Diversity patterns in the tropics. In: White WB, Culver DC (eds) Encyclopedia of caves. Academic Press, Chennai, pp 238–250

Dietz C, von Helversen O, Nill D (2009) Bats of Britain. Europe and Northwest Africa. A & C Black, London

Dumitresco M, Orghidan T (1963) Contribution a la connaissnace de la biologie de *Pipistrellus pipistrellus* Schreiber. Annales de Spéléol. 18:511–517

Elliot WR (1994) Conservation of Texas caves and karst. In: Elliot WR, Veni G (eds) The caves and karst of Texas. Convention Guidebook, National Speleological Society, Alabama, pp 85–97

Elliot WR (2004) Protecting caves and cave life. In: Culver DC, White WB (eds) Encyclopedia of caves. Elsevier, Amsterdam, pp 458–467

Eurobats (2014) Report of the inter-sessional working group on man-made purpose-built bat roosts. Doc.EUROBATS.StC9-AC19.34

Fant J, Kennedy J, Powers R et al (2009) Agency guide to cave and mine gates. American Cave Conservation Association, Bat Conservation International and Missouri Department of Conservation, USA

Furey NM (2009) Bat assemblages in Vietnamese karst: diversity, reproduction, echolocation and ecomorphology. Dissertation, Aberdeen University

Furey NM, Mackie IJ, Racey PA (2010) Bat diversity in Vietnamese limestone karst areas and the implications of forest degradation. Biodivers Conserv 19:1821–1838

Furey NM, Mackie IJ, Racey PA (2011) Reproductive phenology of bat assemblages in Vietnamese karst and its conservation implications. Acta Chiropt 13:341–354

Furman A, Özgül A (2004) The distribution of cave-dwelling bats and conservation status of underground habitats in Northwestern Turkey. Biol Cons 120:243–248

Furman A, Coraman E, Bilgin R (2012) Bats and tourism: a response to Paksuz and Ozkan. Oryx 46:330

Gaisler J, Hanak V, Hanzal V et al (2003) Results of bat banding in the Czech and Slovak Republics (1948–2000). (in Czech with English summary) Vespertilio 7:3–61

Geluso KN, Altenbach JS, Wilson DE (1976) Bat mortality: pesticide poisoning and migratory stress. Science 194:184–186

Geluso KN, Altenbach JS, Wilson DE (1981) Organochlorine residues in young Mexican free-tailed bats from several roosts. Amer Midl Nat 105:249–257

Gerlach J, Taylor M (2006) Habitat use, roost characteristics and diet of the Seychelles sheath-tailed bat *Coleura seychellensis*. Acta Chiropt 8:129–139

Gilbert J, Deharveng L (2002) Subterranean ecosystems: a truncated functional biodiversity. Bioscience 52:473–481

Gillieson D (1996) Caves: processes, development and management. Blackwell publishers, UK

Golovatch SI, Geoffrey JJ, Mauries JP et al (2011) New species of the millipede genus *Glyphiulus* Gervais, 1847 from the *granulatus*-group (Diplopoda: Spirostreptida: Cambalopsidae. Arthropoda Selecta 20:65–114
Gould E (1988) Wing-clapping sounds of *Eonycteris spelaea* (Pteropodidae) in Malaysia. J Mammal 69:378–379
Gunn J (2003) Climate of caves. In: Gunn J (ed) Encyclopedia of caves and karst science. Fitzroy Dearborn, Routledge, New York, pp 228–230
Guthrie MJ (1933) Notes on the seasonal movements and habits of some cave bats. J Mammal 14:1–19
Hall LS, Richards G, Abdullah MT (2002) The bats of Niah national park, Sarawak. Sarawak Museum Journal 57:28–255
Hamilton-Smith E (2001) Current initiatives in the protection of karst biodiversity. Natura Croatica 10:229–242
Hildreth-Werker V, Werker JC (2006) Cave conservation and restoration. National Speleological Society, Alabama
Horn JW, Kunz TH (2008) Analyzing NEXRAD doppler radar images to assess nightly dispersal patterns and population trends in Brazilian free-tailed bats (*Tadarida brasiliensis*). Integr Comp Biol 48:24–39
Hughes AC, Satasook C, Bates PJJ et al (2012) The projected effects of climatic and vegetation changes on the distribution and diversity of Southeast Asian bats. Glob Change Biol 18:1854–1865
Humphries MM, Thomas DW, Speakman JR (2002) Climate-mediated energetic constraints on the distribution of hibernating animals. Nature 418:313–316
Hutson AM (1993) Action plan for the conservation of bats in the United Kingdom. The Bat Conservation Trust, London
Hutson AM, Mickleburgh SP, Racey PA (comp.) (2001) Microchiropteran bats: global status survey and conservation action plan. IUCN/SSC Chiroptera Specialist Group. IUCN, Gland, Switzerland, and Cambridge, UK
Hutson AM, Mickleburgh SP, Mitchell-Jones AJ (1988) Bats underground, a conservation code. Fauna and Flora Preservation Society, Nature Conservancy Council and the Vincent Wildlife Trust, UK
IUCN SSC (2014) IUCN SSC guidelines for minimizing the negative impact to bats and other cave organisms from guano harvesting. Ver. 1.0. IUCN, Gland
Jones G, Rebelo H (2013) Responses of bats to climate change: learning from the past and predicting the future. In: Adams RA, Pedersen SC (eds) Bat evolution, ecology, and conservation. Springer, New York, pp 457–478
Jubertie C (1992) Underground habitats and their protection. Convention on the conservation of European wildlife and natural habitats, standing committee, Council of Europe
Jubertie C (2000) Conservation of subterranean habitats and species. In: Wilkens H, Culver DC, Humphreys WF (eds) Ecosystems of the World: subterranean ecosystems. Elsevier, Amsterdam, pp 691–700
Kennedy J (2006) On cave gates. In: Hildreth V, Werker JC (eds) Cave conservation and restoration. National Speleological Society, Alabama, pp 147–165
Kingston T (2010) Research priorities for bat conservation in Southeast Asia: a consensus approach. Biodivers Conserv 19:471–484
Kingston T (2013) Response of bat diversity to forest disturbance in Southeast Asia: insights from long-term research in Malaysia. In: Adams RA, Pedersen SC (eds) Bat evolution, ecology, and conservation. Springer, New York, pp 169–185
Kingston T, Aguirre L, Armstrong K et al (2016) Networking networks for global bat conservation. In: Voigt CC, Kingston T (eds) Bats in the anthropocene: conservation of bats in a changing world. Springer International AG, Cham, pp 539–566

Kofoky AF, Andriafidison D, Ratrimomanarivo F et al (2007) Habitat use, roost selection and conservation of bats in Tsingy de Bemaraha National Park, Madagascar. Biodivers Conserv 16:1039–1053

Kokurewicz T, Rusiński M, Weigle A (2013) 12th International winter bat census in Natura 2000 site "Nietoperek" (PLH080003) (Lubuskie County)—12th January 2013. Report from the census and the results. National Foundation for Environmental Protection, Warszawa, Wrocław 20th May 2013 (in Polish)

Kunz TH (1982) Roosting ecology. In: Kunz TH (ed) Ecology of bats. Plenum, New York, pp 1–55

Kunz TH, de Torrez EB, Bauer D et al (2011) Ecosystem services provided by bats. Ann NY Acad Sci 1223:1–38

Kurta A, Bell GP, Nagy KA et al (1989) Energetics of pregnancy and lactation in free-ranging little brown bats (*Myotis lucifugus*). Physiol Zool 62:804–818

Kwiecinski GG, Krook L, Wimsatt WA (1987) Annual skeletal changes in the little brown bat *Myotis lucifugus lucufugus* with particular reference to pregnancy and lactation. Am J Anat 178:410–420

La Val RK (2004) Impact of global warming and locally changing climate on tropical cloud forest bats. J Mammal 85:237–244

Ladle RJ, Firmino JVL, Malhado ACM et al (2012) Unexplored diversity and conservation potential of neotropical hot caves. Conserv Biol 26:978–982

Leelapaibul W, Bumrungsri S, Pattanawiboon A (2005) Diet of wrinkle-lipped free-tailed bat (*Tadarida plicata* Buchannan, 1800) in central Thailand: insectivorous bats potentially act as biological pest control agents. Acta Chiropt 7:111–119

Leh C, Hall LS (1996) Preliminary studies on the production of guano and the socioeconomics of guano collection in Niah Cave, Sarawak. Sarawak Mus J 71:25–38

Lewis SA (1995) Roost fidelity of bats: a review. J Mammal 76:481–496

Lison F, Palazon JA, Calvo JF (2013) Effectiveness of the Natura 2000 Network for the conservation of cave-dwelling bats in a Mediterranean region. Anim Conserv 16:528–537

Medway L (1958) 300,000 bats. Sarawak Mus J 8:667–679

Luo J, Jiang T, Lu G et al (2013) Bat conservation in China: should protection of subterranean habitats be a priority? Oryx 47:526–531

MacKinnon JL, Hawkins CE, Racey PA (2003) Pteropodidae. In: Goodman SM, Benstead JP (eds) The natural history of Madagascar. The University of Chicago Press, Chicago and London, pp 1299–1302

Mann SL, Steidl RJ, Dalton VM (2002) Effects of cave tours on breeding *Myotis velifer*. J Wildl Manage 66:618–624

Martin KW, Leslie DM, Payton ME et al (2003) Internal cave gating for protection of colonies of the endangered gray bat (*Myotis grisescens*). Acta Chiropt 5:143–150

McCracken GF (1986) Why are we losing our Mexican free-tailed bats? Bats 3:1–4

McCracken GF (1989) Cave conservation: special problems of bats. Am Nat Speleol Soc Bull 51:47–51

McCracken GF (1993) Locational memory and female-pup reunions in Mexican free-tailed bat maternity colonies. Anim Behav 45:811–813

McCracken GF (2003) Estimates of population sizes in summer colonies of Brazilian free-tailed bats (*Tadarida brasiliensis*). In: O'Shea TJ, Bogan MA (eds) Monitoring trends in bat populations in the United States and territories: problems and prospects. Geolological Survey, US, pp 21–30

McCracken GF, Bradbury JW (1981) Social organisation and kinship in the polygynous bat *Phyllostomus hastatus*. Behav Ecol Sociobiol 8:11–34

McDonald JT, Rautenbach IL, Nel JAJ (1990) Roosting requirements of five bat species at De Hoop Guano Cave, southern Cape Province of South Africa. S Afr J Wildl Res 20:157–161

McWilliam AN (1982) Adaptive responses to seasonality in four species of Microchiroptera in coastal Kenya. Dissertation, University of Aberdeen

McWilliam AN (1988) The reproductive cycle of male tomb bats, *Taphozous hildegardeae* (Chiroptera: Emballonuridae), in a seasonal environment of the African tropics. J Zool (London) 215:433–442
Medellin RA (2003) Diversity and conservation of bats in Mexico: research priorities, strategies and Actions. Wildl Soc B 31:87–97
Mendenhall CD, Karp DS, Meyer CFJ et al (2014) Predicting biodiversity change and averting collapse in agricultural landscapes. Nature 509:213–217
Mickleburgh SP, Hutson AM, Racey PA (2002) A review of the global conservation status of bats. Oryx 36:18–34
Mickleburgh SP, Waylen K, Racey PA (2009) Bats as bushmeat: a global review. Oryx 43:217–234
Mildenstein T, Tanshi I, Racey PA (2016) Exploitation of bats for bushmeat and medicine. In: Voigt CC, Kingston T (eds) Bats in the anthropocene: conservation of bats in a changing world. Springer International AG, Cham, pp 325–363
Minnis AM, Lindner DL (2013) Phylogenetic evaluation of *Geomyces* and allies reveals no close relatives of *Pseudogymnoascus destructans* comb.nov. in bat hibernacula of eastern North America. Fungal Biol. doi:10.1016/j.funbio.2013.07.001
Mitchell-Jones AJ, Bihari Z, Masing M et al. (2007) Protecting and managing underground sites for bats. EUROBATS Publication series No. 2. UNEP/EUROBATS Secretariat, Bonn, Germany
Mohr CE (1972) The status of threatened species of cave-dwelling bats. Bull Nat Speleol Soc 34:33–47
Monadjem A, Taylor PJ, Cotterill FPD et al (2010) Bats of Southern and Central Africa—a biogeographic and taxonomic synthesis. Wits University Press, Johannesburg
Myers N, Mittermeier RA, Mittermeier CG et al (2000) Biodiversity hotspots for conservation priorities. Nature 403:853–858
Nagy ZL, Postawa T (2010) Seasonal and geographical distribution of cave-dwelling bats in Romania: implications for conservation. Anim Conserv 14:74–86
Niu H, Wang N, Zhao L et al (2007) Distribution and underground habitats of cave-dwelling bats in China. Anim Conserv 10:470–477
Olson CR, Hobson DP, Pybus MJ (2011) Changes in population size of bats at a hibernaculum in Alberta, Canada, in relation to cave disturbance and access restrictions. Northwestern Naturalist 92:224–230
Paksuz S, Özkan B (2012) The protection of the cave bat community in the Dupnisa cave system, Turkey, following opening for tourism. Oryx 46:130–136
Parsons KN, Jones G (2003) Dispersion and habitat use by *Myotis daubentonii* and *Myotis nattereri* during the swarming season: implications for conservation. Anim Conserv 6:283–290
Parsons KN, Jones G, Greenaway F (2003a) Swarming activity of temperate zone microchiropteran bats: effects of season, time of night and weather conditions. J Zool (London) 261:257–264
Parsons KN, Jones G, Davidson-Watts I et al (2003b) Visitation of underground sites by bats during late summer and autumn: implications for conservation. Biol Conserv 111:63–70
Pierson ED (1998) Tall trees, deep holes and scarred landscapes—conservation biology of North American bats. In: Kunz TH, Racey PA (eds) Bat biology and conservation. Smithsonian Institution Press, Washington, pp 309–325
Pierson ED, Rainey WE, Koontz DM (1991) Bats and mines: experimental mitigation for Townsend's big-eared bat at the McLaughlin Mine in California. Thorne ecological institute proceedings V: issues and technology in the management of impacted wildlife. Thorne Ecological Institute, Boulder, pp 31–42
Piksa K, Nowak J, Zmihorski M et al (2013) Nonlinear distribution pattern of hibernating bats in caves along an elevational gradient in mountain (Carpathians, southern Poland). PLOSone 8:e68066

Pugh M, Altringham JD (2005) The effect of gates on cave entry by swarming bats. Acta Chiropt 7:293–299

Racey PA (1975) The prolonged survival of spermatozoa in bats. In: Duckett JC, Racey PA (eds) The biology of the male gamete. Biol J Linn Soc 7(Suppl 1):385–416

Racey PA (1979) The prolonged storage and survival of spermatozoa in Chiroptera. In: Comparative aspects of reproduction in chiroptera. J Reprod Fertil 56:391–402

Racey PA (2004) 8,000,000 fruit bats, Africa's best kept wildlife secret. BATS 1:1–5

Racey PA (2013) Bat conservation: past, present and future. In: Adams RA, Pedersen SC (eds) Bat evolution, ecology, and conservation. Springer, New York, pp 517–532

Racey PA, Entwistle AE (2000) Life history and reproductive strategies of bats. In: Crichton EG, Krutzsch PH (eds) Reproductive biology of bats. Academic Press, San Diego, pp 363–468

Racey PA, Speakman JR (1987) The energy costs of pregnancy and lactation in heterothermic bats. Symp Zool Soc Lond 57:107–125

Ransome RD (1968) The distribution of the greater horseshoe bat, *Rhinolophus ferrumequinum*, in relation to environmental factors. J Zool 154:77–112

Ransome RD (1971) The effect of ambient temperature on the arousal frequency of the hibernating greater horseshoe bat, *Rhinophus ferrumequinum*, in relation to site selection and the hibernation state. J Zool 164:357–371

Ransome RD (1998) The impact of maternity roost conditions on populations of greater horseshoe bats. Engl Nat Res Rep 292:1–80

Ransome RD (2010) Environment and Ecology. In: Lord F (ed) The Stabilisation of combe down stone mines: the saving of a village. Combe Down Stone Mines Project, Bath and North East Somerset Council, pp 73–77

Rebelo H, Tarroso P, Jones G et al (2010) Predicted impact of global climate change on European bats in relation to their biogeographic patterns. Glob Change Biol 16:561–576

Reeder DM, Moore MS (2013) White-Nose syndrome: a deadly emerging infectious disease of hibernating bats. In: Adams RA, Pedersen SC (eds) Bat evolution, ecology, and conservation. Springer, New York, pp 413–434

Richter AR, Humphrey SR, Cope JB et al (1993) Modified cave entrances: thermal effect on body mass and resulting decline of endangered Indiana bats (*Myotis sodalis*). Cons Biol 7:407–415

Riddle BR (ed) (1995) Inactive Mines as bat habitat: guidelines for research, survey, monitoring and mine management in Nevada. Biological Resources Research Center, University of Nevada, Reno

Rodriguez-Durán A (1998) Nonrandom aggregations and distribution of cave-dwelling bats in Puerto Rico. J Mamm 79:141–146

Rodriguez-Durán A (2009) Bat assemblages in the West Indies: the role of caves. In: Fleming TH, Racey PA (eds) Island bats: evolution, ecology, and conservation. University of Chicago Press, Chicago and London, pp 265–280

Rodriguez-Durán A, Lewis AR (1987) Patterns of population size, diet and activity time for a multispecies assemblage of bats at a cave in Puerto Rico. Carib J Sci 23:352–360

Rodriguez-Durán A, Soto-Centeno JA (2003) Temperature selection by tropical bats roosting in hot caves. J Therm Biol 28:465–468

Rodriguez-Durán A, Perez J, Montalban MA et al (2010) Predation by free-roaming cats on an insular population of bats. Acta Chiropt 12:359–362

Russell AL, McCracken GF (2006) Population genetic structure of very large populations: the Brazilian free-tailed bat *Tadarida brasiliensis*. In: Akbar Z, McCracken GF, Kunz TH (eds) Functional and evolutionary ecology of bats. Oxford University Press, Oxford, pp 225–247

Sachanowicz K, Wower A, Bashta AT (2006) Further range extension of *Pipistrellus kuhlii* (Kuhl, 1817) in central and eastern Europe. Acta Chiropt 8:543–548

Scheel D, Vincent TLS, Cameron GN (1996) Global warming and the species richness of bats in Texas. Conserv Biol 10:452–464

Sheffield SR, Shaw JH, Heidt GA et al (1992) Guidelines for the protection of bat roosts. J Mammal 73:707–710

Sherwin RE, Altenbach JS (2004) The responses of bats to gates. In: Vories KC, Throgmorton D, Harrington A (eds) Proceedings of bat gate design: a technical interactive forum held March 4–6, 2002 at the Red Lion Hotel, Austin Texas. Office of Surface Mining, Illinois, pp 333–338

Siles L, Munoz A, Aguirre LF (2007) Bat diversity in three caves in a montane forest of Bolivia. Ecotropica 13:67–74

Spanjer GR, Fenton MB (2005) Behavioural responses of bats to gates at caves and mines. Wildl Soc Bull 33:1101–1112

Stephens PA, Sutherland WJ, Freckleton RP (1999) What is the Allee effect? Oikos 87:185–190

Stihler CW, Hall JS (1993) Endangered bat populations in West Virginia caves gated or fenced to reduce human disturbance. Bat Res News 34:130

Struebig MJ, Kingston T, Zubaid A et al (2009) Conservation importance of limestone karst outcrops for Palaeotropical bats in a fragmented landscape. Biol Conserv 142:2089–2096

Sutherland WJ, Aveling R, Bennun L et al (2012) A horizon scan of global conservation issues for 2012. Trends Ecol Evol 27:12–18

Suyanto A, Struebig MJ (2007) Bats of the Sangkulirang limestone karst formations, East Kalimantan—a priority region for Bornean bat conservation. Acta Chiropt 9:67–95

Thomas DW (1995) Hibernating bats are sensitive to tactile disturbance. J Mammal 76:940–946

Thomas DW, Fenton MB, Barclay RMR (1979) Social behaviour of the little brown bat, *Myotis lucifugus* I. Mating behaviour. Behav Ecol Sociobiol 6:129–136

Thomas DW, Dorais M, Bergeron JM (1990) Winter energy budgets and cost of arousals for hibernating little brown bats, *Myotis lucifugus*. I. Mating behaviour. J Mammal 71:475–479

Tordoff AW, Bao TQ, Tu ND et al (eds) (2004) Sourcebook of existing and proposed protected areas in Vietnam, 2nd edn. BirdLife International in Indochina and Ministry of Agriculture and Rural Development, Hanoi

Tuttle MD (1976) Population ecology of the gray bat (*Myotis grisescens*): factors influencing growth and survival of newly volant young. Ecology 57:587–595

Tuttle MD (1977) Gating as a means of protecting cave-dwelling bats. In: Aley T, Hodes D (eds) National cave management symposium 1976. Speleobooks, Albuquerque, New Mexico, pp 77–82

Tuttle MD (2013) Threats to bats and educational challenges. In: Adams RA, Pedersen SC (eds) Bat evolution, ecology, and conservation. Springer, New York, pp 363–391

Tuttle MD, Taylor D (1994) Bats and mines. Resource Publication No 3. Bat Conservation International, Austin Texas

Van NK, Hien NT, Loc PK et al (2000) Bioclimatic diagrams of Vietnam. Vietnam National University Publishing House, Hanoi

Vermeulen JJ, Whitten AJ (1999) Biodiversity and cultural property in the management of limestone resources. The World Bank, Washington

Vories KC, Throgmorton D, Harrington A (2004) Proceedings of bat gate design: a technical interactive forum held March 4–6, 2002 at the Red Lion Hotel, Austin Texas. Office of Surface Mining, Alton, Illinois

Wanger TC, Darras K, Bumrungsri S et al (2014) Bat pest control contributes to food security in Thailand. Biol Cons 171:220–223

Waters DA, Vollrath C (2003) Echolocation performance and call structure in the megachiropteran fruit-bat *Rousettus aegyptiacus*. Acta Chiropt 2003:209–219

Watson J, Hamilton-Smith E, Gilleson D et al (eds) (1997) Guidelines for cave and karst protection. IUCN, Gland, Switzerland and Cambridge, UK

Webb PI, Speakman JR, Racey PA (1995) How hot is a hibernaculum? A review of the temperatures at which bats hibernate. Can J Zool 74:761–765

Weinreich JA, Voshaar JHO (1992) Population trends of bats hibernating in marl caves in the Netherlands (1943–1987). Myotis 30:75–84

Whitten T (2009) Applying ecology for cave management in China and neighbouring countries. J App Ecol 46:520–523

Whitten T (2012) Protecting biodiversity. International Cement Review, June 2012

Wiles GJ, Brooke AP (2013) Conservation threats to bats in the tropical pacific islands and insular Southeast Asia. In: Fleming TH, Racey PA (eds) Island bats: evolution, ecology, and conservation. University of Chicago Press, Chicago and London, pp 405–459

White WB, Culver DC (eds) (2012) Encyclopedia of caves. Academic Press, Chennai

Zhang L, Zhu G, Jones G (2009) Conservation of bats in China: problems and recommendations. Oryx 43:179–182

Zukal J, Rehak Z, Kovarik M (2003) Bats of the Sloupsko-sosuvske jeskyne cave (Moravian Karst, Central Moravia). (In Czech with English summary) Lynx (Praha) n.s. 34: 205–220

Part IV
Conservation Approaches, Educational and Outreach Programs

Chapter 16
The Roles of Taxonomy and Systematics in Bat Conservation

Susan M. Tsang, Andrea L. Cirranello, Paul J.J. Bates and Nancy B. Simmons

Abstract Taxonomy—the description, naming, and classification of organisms—and systematics—the study of the evolutionary relationships of organisms—are both crucial components in conservation, providing a necessary framework for any conservation initiative. With more than 200 new bat species identified or raised from synonymy in the past decade and additional taxa described monthly, the Age of Discovery is ongoing for bats. New taxonomic and systematic discoveries clarify the status of populations, and the recognition of distinct species and lineages allows appropriate conservation strategies to be crafted, increasing the likelihood of recovery. In addition to identifying species and specimens, taxonomists care for vouchers, provide species lists for localities, and communicate taxonomic ideas to non-experts, especially through descriptions, keys, and field guides. Taxonomists can also provide conservation planning tools such as inventory data, estimates of extinction risk and extinction rate, and information for defining protected areas. Despite the importance of taxonomy, a lack of financial and institutional support impedes the training and employment of taxonomists and such factors need to be

S.M. Tsang (✉)
Department of Biology, City College and the Graduate Center at the City University of New York, New York, NY 10031, USA
e-mail: stsang@amnh.org

S.M. Tsang · A.L. Cirranello · N.B. Simmons
Department of Mammalogy, Division of Vertebrate Zoology, American Museum of Natural History, New York, NY 10024, USA
e-mail: andreacirranello@gmail.com

N.B. Simmons
e-mail: simmons@amnh.org

P.J.J. Bates
Harrison Institute, Bowerwood House, St. Botolph's Road, Sevenoaks, Kent TN13 3AQ, UK
e-mail: pjjbates2@hotmail.com

C.C. Voigt and T. Kingston (eds.), *Bats in the Anthropocene: Conservation of Bats in a Changing World*, DOI 10.1007/978-3-319-25220-9_16

overcome. Taxonomic and systematic discoveries, especially those involving cryptic species and unrecognized diversity, are rapidly increasing with the advent of modern genetics. Researchers must be cautious to argue from multiple lines of evidence when naming new species and be clear about the species concept they employ, as these have wide ranging impacts beyond taxonomy. Creating new ties between taxonomists and non-experts will be crucial in conservation of a diverse range of organisms in increasingly fragile landscapes.

16.1 Introduction

Global biodiversity is being lost at an unprecedented rate as a result of environmental change and human activity. Like other organisms, bats are at risk and many populations and species are threatened. As of 2013, the International Union for Conservation of Nature and Natural Resources (IUCN) Red List recognized 81 bat species as Near Threatened, 95 as Vulnerable, 51 as Endangered, 26 as Critically Endangered, and 5 as Extinct (IUCN 2014). It is clear that decisions must be made now to combat ongoing loss of species and populations. However, appropriate management decisions cannot be made without a marriage among conservation biologists, taxonomists, and legislators. Before conservation strategies can be implemented, the species composition of a locality must be well understood; otherwise, the effectiveness of any conservation effort cannot be accurately quantified.

Clearly defining species boundaries—while often difficult—is crucial to basic research and conservation. Some level of agreement on the organisms and populations considered part of any species is necessary for studying and tracking the health of organisms and ecosystems. Taxonomy—the description, naming, and classification of organisms—provides this necessary framework. Taxonomy, along with classification, often is conflated with systematics (Schuh 2000), which is more properly defined as the study of the diversification and evolutionary relationships of organisms through time. Despite often being used interchangeably, they are distinctly different, though systematic research includes recognition of taxa (i.e., taxonomy) as a necessary ingredient to reconstructing the past. Phylogenies produced by systematists provide a crucial foundation for examining biological phenomena and hypotheses, such as adaptive radiation or biogeographic scenarios, some of which are important for informing conservation decisions. Phylogenies help predict where biodiversity hotspots may be located, inform how distinct populations may be from one another, and identify unique lineages that preserve critical genetic diversity. Without systematics, other aspects of natural history lose their historical framework; and without taxonomy, systematics loses its basic operational unit. This chapter will demonstrate the many ways in which taxonomy and systematics have contributed to past conservation efforts and how they will continue to enrich protection of bat species globally.

16.2 The Continuing Age of Discovery

Taxonomy is not a dead science; the Age of Discovery is ongoing, especially for bats (Fig. 16.1). The number of bats discovered in the last couple of decades is higher than expected when compared to other mammalian orders (Reeder et al. 2007). With each subsequent volume of *Mammal Species of the World* (Honacki et al. 1982; Wilson and Reeder 1993, 2005), the number of recognized bat species has increased dramatically, with new species described from every corner of the world. Between publication of the last edition in 2005 and the end of 2013, nearly 200 new bat species were described or resurrected from synonymy, including 120 species new to science (Table 16.1), putting the total number of bat species at just over 1300 at the time of writing of this chapter. The continuing high rate of discovery (or recognition) of new bats can be a potential impediment to conservation since it is difficult to assess the status of each newly discovered species within a short period of time, and because it is difficult to make management plans in the absence of abundance or natural history information (both of which are typically lacking for newly recognized taxa). However, new discoveries may clarify the status of isolated populations, and the recognition of these distinct species can allow appropriate conservation and management strategies to be crafted.

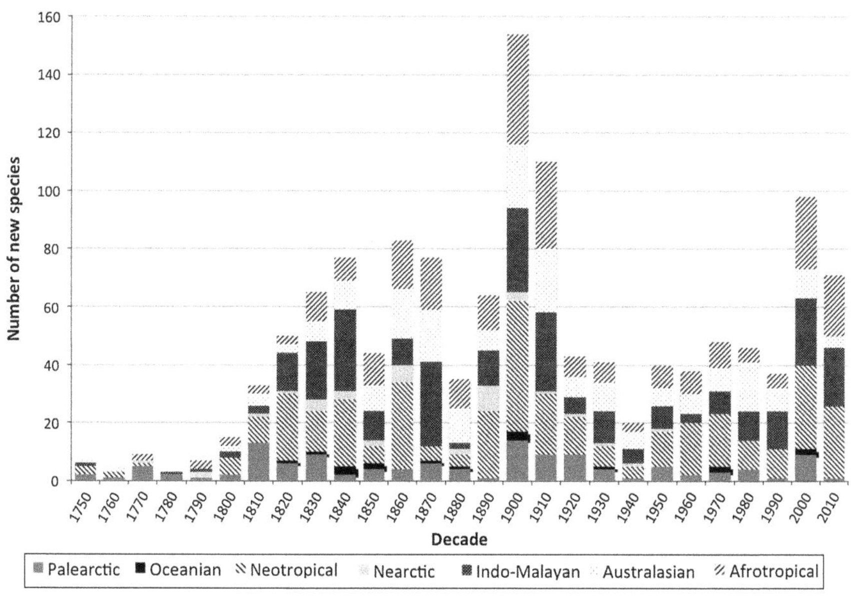

Fig. 16.1 Number of new bat species described per decade since 1750. Species were categorized to zoogeographic region (as defined by Newton 2003) of discovery according to type localities. Species since 2010 only reflect discoveries prior to the writing of this chapter (early 2014). New species are constantly being described from the tropics, with rates of discovery in the Afrotropics and Indo-Malayan regions catching up with the Neotropics

Table 16.1 New bat species by family since the previous edition of Mammal Species of the World (Wilson and Reeder 2005)

Family	Species	Year	Family	Species	Year
Phyllostomidae	*Anoura cadenai*	2006	*Vespertilionidae*	*Barbastella beijingensis*	2007
	Anoura carashina	2010		*Eptesicus lobatus*	2009
	Anoura fistulata	2005		*Eptesicus taddei*	2006
	Carollia benkeithei	2006		*Glischropus bucephalus*	2011
	Carollia manu	2004		*Harpiola isodon*	2006
	Carollia monohernandezi	2004		*Hypsugo lanzai*	2011
	Chiroderma vizottoi	2010		*Kerivoula kachinensis*	2004
	Dryadonycterix capixaba	2012		*Kerivoula titania*	2007
	Lonchophylla cadenai	2006		*Murina balaensis*	2013
	Lonchophylla chocoana	2004		*Murina beelzebub*	2011
	Lonchophylla fornicata	2007		*Murina bicolor*	2009
	Lonchophylla orcesi	2005		*Murina chrysochaetes*	2011
	Lonchophylla orienticollina	2009		*Murina eleryi*	2009
	Lonchophylla pattoni	2006		*Murina fionae*	2012
	Lonchophylla peracchii	2013		*Murina gracilis*	2009
	Lophostoma kalkoae	2012		*Murina guilleni*	2013
	Lophostoma yasuni	2004		*Murina harpioloides*	2008
	Micronycteris buriri	2011		*Murina harrisoni*	2005
	Micronycteris giovaniae	2007		*Murina jaintiana*	2012
	Micronycteris yatesi	2013		*Murina lorelieae*	2011
	Platyrrhinus albericoi	2005		*Murina pluvialis*	2012
	Platyrrhinus angustirostris	2010		*Murina recondita*	2009
	Platyrrhinus fusciventris	2010		*Murina shuipuensis*	2011
	Platyrrhinus guianensis	2014		*Murina tiensa*	2007
	Platyrrhinus ismaeli	2005		*Murina walstoni*	2011

(continued)

Table 16.1 (continued)

Family	Species	Year	Family	Species	Year
	Platyrrhinus masu	2005		*Myotis annatessae*	2013
	Platyrrhinus matapalensis	2005		*Myotis badius*	2011
	Platyrrhinus nitelinea	2009		*Myotis dieteri*	2005
	Sturnira bakeri	2014		*Myotis diminutus*	2011
	Sturnira burtonlimi	2014		*Myotis handleyi*	2013
	Sturnira koopmanhilli	2006		*Myotis indochinensis*	2013
	Sturnira perla	2011		*Myotis izecksohni*	2011
	Sturnira sorianoi	2005		*Myotis lavali*	2011
	Uroderma bakeri	2014		*Myotis midastactus*	2014
	Xeronycteris vieirai	2005		*Myotis phanluongi*	2008
Emballonuridae	*Coleura kibomalandy*	2012		*Neoromicia robertsi*	2012
	Paraemballonura tiavato	2006		*Neoromicia roseveari*	2013
	Peropteryx pallidoptera	2010		*Nyctophilus corbeni*	2009
Thyropteridae	*Thyroptera devivoi*	2006		*Nyctophilus shirleyae*	2009
	Thyroptera wynneae	2014		*Pipistrellus hanaki*	2004
Molossidae	*Chaerephon atsinanana*	2010		*Pipistrellus raceyi*	2006
	Chaerophon jobimena	2004		*Plecotus strelkovi*	2005
	Eumops wilsoni	2009		*Rhogeessa bickhami*	2012
	Molossus alvarezi	2011		*Rhogeessa menchuae*	2012
	Mops bakarii	2008		*Scotophilus andrewreborii*	2014
	Mormopterus eleryi	2008		*Scotophilus ejetai*	2014
	Mormopterus francoismoutoui	2008		*Scotophilus livingstonii*	2014
	Mormopterus halli	2014		*Scotophilus marovaza*	2006
	Mormopterus kitcheneri	2014		*Scotophilus tandrefana*	2005
	Mormopterus lumsdenae	2014		*Scotophilus trujilloi*	2014

(continued)

Table 16.1 (continued)

Family	Species	Year	Family	Species	Year
Pteropodidae	*Casinycteris campomaanensis*	2014		*Tylonycteris pygmaeus*	2008
	Desmalopex microleucopterus	2008	*Natalidae*	*Natalus lanatus*	2005
	Dyacopterus rickartii	2007	*Myzopodidae*	*Myzopoda schliemanni*	2006
	Epomorphorus anselli	2004	*Miniopteridae*	*Miniopterus aelleni*	2008
	Pteralopex flanneryi	2005		*Miniopterus brachytragos*	2009
	Pteropus allenorum	2009		*Miniopterus egeri*	2011
	Pteropus coxi	2009		*Miniopterus griffithsi*	2009
	Styloctenium mindorensis	2007		*Miniopterus mahafaliensis*	2009
	Thoopterus suhaniahae	2012		*Miniopterus maghrebensis*	2014
Rhinolophidae	*Rhinolophus chiewkweeae*	2005		*Miniopterus mossambicus*	2013
	Rhinolophus cohenae	2012		*Miniopterus petersoni*	2008
	Rhinolophus huananus	2008		*Miniopterus sororculus*	2007
	Rhinolophus indorouxi	2013	*Hipposideridae*	*Asellia arabica*	2011
	Rhinolophus kahuzi	2013		*Hipposideros boeadi*	2007
	Rhinolophus mabuensis	2012		*Hipposideros einnaythu*	2011
	Rhinolophus mossambicus	2012		*Hipposideros griffini*	2012
	Rhinolophus schnitzleri	2011		*Hipposideros khaokhouayensis*	2006
	Rhinolophus smithersi	2012		*Hipposideros khasiana*	2006
	Rhinolophus willardi	2013		*Paratriaenops pauliani*	2008
	Rhinolophus xinanzhongguoensis	2009		*Triaenops menamena*	2009
Rhinopomatidae	*Rhinopoma hadramauticum*	2009		*Triaenops parvus*	2009

Most species discoveries in the past decade have been made in Neotropical families, though more genetic and phonic data on Rhinolophidae and Hipposideridae are now uncovering greater species diversity in the Paleotropics

Species discoveries and recognition may bring attention to previously overlooked areas or act as a symbol of local pride. For example, in the Mekong Delta of Vietnam, a sixteenth-century Khmer pagoda in Soc Trang City called Wat Matahup, or Chua Doi—the Bat Pagoda—is home to a mixed colony of thousands of flying foxes (*Pteropus vampyrus* and *Pteropus lylei*), which are listed by the IUCN as Near Threatened and Vulnerable, respectively. The pagoda is a cultural and historic icon and the only pagoda in the region with a resident bat colony. The locals feel a sense of pride, as these rare bats roost only in the trees within the temple grounds. Monks actively protect the bats from increasing hunting pressure. This interest has resulted in the creation of bat and sustainability education campaigns by locals. These programs are aimed at educating young children on the importance of the bats to the ecosystem.

16.3 The Role of the Taxonomist in Conservation

The most basic contribution of the taxonomist to conservation is to identify and name the species being protected (Table 16.2). Being unable to differentiate among species makes it virtually impossible to manage wildlife, leads to poor decision-making, and causes unforeseen ecological consequences. Taxonomists are often the only people who can identify an animal—an underappreciated skill. For bats, this is of special importance as bats are an extremely diverse group, and many bat species are cryptic and therefore cannot be readily identified by amateurs and other biologists based on obvious external features. Taxonomists also form the backbone of any museum system. They are responsible for identification of voucher specimens that include whole organisms, skins, skeletons, skulls, and, increasingly, frozen tissues. Along with other museum personnel, they are responsible for ensuring that these specimens are preserved as a reference for future researchers. Natural history collections curate and maintain critical data associated with specimens including species identification, locality, sex, date of collection, collector, and other pertinent information. Much important taxonomic work takes place in these collections, with major taxonomic revisions of museum material often clarifying the status of particular species.

One of the most common requests to taxonomists from other researchers is for a species list for a particular locality. Without an easy way to identify species, non-taxonomists may not be able to accurately interpret collected data that are relevant to conservation, including information on habitat, geographic distribution, abundance, and basic features of ecology (e.g., roost sites for bats). Field researchers collect these ecological data; but many field researchers only observe animals and do not collect vouchers. Their observations—e.g., "bat species X and Y occur in caves all along the northwest coast"—form the basis of our understanding of fauna and species distributions alike. But, without vouchers, current and future research may not actually address the questions at hand. What happens when species Y is later recognized to be three species? What happens if species X has been

Table 16.2 The process of describing a new species can be broken down into two parts: the research necessary prior to description and the publication

Part 1: Research prior to description	
Collect data corroborating unique species identity	
Several lines of evidence	
Morphological, genetic, behavioral, ecological, phonic	
Literature review	
Is it a variant?	
Was it previously reduced to synonymy?	
Is it a new record in that area for a known species?	
Did it use the wrong name?	
Visit reference collections	
Compare to reference, voucher, or type (if possible) specimens of similar species	
Collaborate with systematist if necessary	
Part 2: Publication	
What kind of paper?	
Species description	Redescription
Revision	Synopsis
Review	Catalog
Monograph	Phylogeny
Checklist	Subspecies description
Description of Higher Taxon	
Create scientific binomial following rules set by ICZN	
Establish type specimen(s) and type locality	
Sections	
Diagnosis (distinguishing characters only)	
Description (all traits)	
Taxonomic characters	Color
Life history characters	Quantitative characters
Life stages	Behavioral/ecological characters
Discussion—significance?	
Ecology	
Distribution	
Material examined	

The above-mentioned table is derived from taxonomic procedures described in Winston (1999), a reference which is recommended by the International Commission on Zoological Nomenclature (ICZN)

misidentified? In such circumstances, how are we to know which bats are really present in the area? Effective gathering, consolidation, and analysis of data for conservation efforts require accurate species identifications as well as collection of voucher material, if possible.

Taxonomists must also communicate their work to non-experts, including other biologists. The taxonomic literature is notoriously inaccessible to non-specialists as it is often filled with obscure terminology and outdated names. Many historically important papers were published in journals that are not accessible to researchers in developing countries. Taxonomic revisions are not always readily available and widely circulated, allowing old names to persist in the literature and, more recently, Web-based faunal lists. This may complicate species delimitation and confuse consolidation of other ecologically important information necessary for effective conservation.

Best practices for species identification/documentation include the following: (1) use of a broad range of data to support species identifications, including morphological, genetic, and (if relevant) echolocation data; (2) a thorough review of the literature for names applied to the group(s) or specimens examined so that the oldest valid name is used; and (3) publication in an open-access journal for the broadest possible exposure. Examples of recent papers that use one or more of these best-practice approaches are as follows:

1. Larsen et al. (2010), who raised a previously recognized South American subspecies of *Artibeus*, *Artibeus jamaicensis aequatorialis*, to full species, *A. aequatorialis*, based on combined morphometric, mitochondrial, and AFLP (amplified fragment length polymorphism) data. The paper provided detailed context, including a review of the history of research on the species and a literature review of previous work on the genus. The study also provided a clear species account of *A. aequatorialis* and was published in the widely available journal *Zootaxa*.
2. Taylor et al. (2012), who recognized, on the basis of distinct echolocation calls, possible cryptic species within the *Rhinolophus hildebrandtii* complex of southern Africa. Subsequently, he described four new species supported by a combination of acoustic, morphometric, and molecular data.
3. Buden et al. (2013), who revised the Micronesian species *Pteropus insularis*, recognizing two subspecies, *P. pelagicus pelagicus* and *P. pelagicus insularis*. The authors examined a series of specimens and evaluated morphological features and conducted a thorough literature review of past names prior to revising the taxonomy of this species.
4. Velazco et al. (2014), who described the new species *Thyroptera wynneae* from South America. In this case, the morphological data unambiguously supported specific status for the collected voucher material, despite there being several other congeners found in sympatry.

The studies of *P. pelagicus* and *T. wynneae* were both published in the open-access journal, *ZooKeys* and *American Museum Novitates*, respectively, and are readily available to researchers from developing countries.

Products produced by taxonomists for use by experts and educated non-experts alike include keys and descriptions. Keys use mutually exclusive statements that help lead users to identifications of unknown organisms. Good keys use diagnostic features illustrated by line drawings or photographs to differentiate between species and include redundancy to ensure correct identifications at earlier steps. Incomplete keys often cause problems when they are the only means available to identify an animal. A good key enhances the work of land managers and other decision makers as well as researchers studying ecology, zoonotic diseases, and agriculture by allowing them to identify easily confused species and to access updated information on taxonomic nomenclature.

Taxonomists must understand the skills and facilities that are available where the key will be used. Keys that rely on external characters from a living animal must take precedence over features that can only be seen in museum preparations or with the use of a microscope (although some craniodental data may be needed to supplement external characters, especially in bats). Microscopes may not be available under field conditions, or at all at the locality under study. Extracting and cleaning skulls, or measuring morphological features requires training. Good examples of accessible keys are Barquez et al. (1993), which is available bilingually, and Taylor (2000), which includes acoustic profiles. Both of these keys use easily distinguishable external characteristics along with illustrations to assist in identification.

Taxonomists sometimes also produce field guides, drawing on knowledge of collection records, phylogenetic relationships, species distributions, and natural history to enlighten experts and non-experts alike. Field guides engage the scientifically literate public and can act as an illuminating form of outreach for bats. Top-notch field guides, such as those by Francis (2008) for the mammals of Southeast Asia and Reid (2009) for the mammals of Central America, are produced by experts and include detailed notes on species identification, natural history, distribution maps, and color illustrations or high-quality photographs. While not quite a field guide, Bat Conservation International freely provides species profiles on their Web site for all 47 species of North American bats. It is likely that Web-based field guides, or mobile device apps, will come to play a larger role in field identifications in the future, and these resources will benefit from attention by taxonomists during their development.

16.4 Taxonomy and International Agreements

The importance of taxonomy is recognized by the Convention on Biological Diversity's (CBD) Global Taxonomy Initiative program. Inadequate taxonomic information is recognized as an obstacle to making informed management decisions in conservation, sustainable use of resources, and availability of genetic resources (Secretariat of the Convention on Biological Diversity 2008). The legally binding CBD was signed by 193 governments in 1992–1993 at the UN Conference for Environment and Development. Article 7 (identification and monitoring), Article 12 (research and training), and Article 17 (public awareness and education) of the CBD directly address the need for taxonomic research to be conducted and used for conservation. Furthermore, the strategy plan for 2011 to 2020 specifically referenced the need to "improve the status of biodiversity and by safeguarding ecosystems, species, and genetic diversity" (Secretariat of the Convention on Biological Diversity 2012). The CBD indicates a willingness of governments to recognize the importance of taxonomy in resolving environmental challenges.

The importance of taxonomy in protecting species is most immediately visible under the Convention for International Trade of Endangered Species (CITES) agreement. With 179 Parties having now joined the Convention, to which they agree to voluntarily adhere, CITES provides a rank system with varying degrees of protection to more than 35,000 plant and animal species. Under CITES, all *Acerodon* and *Pteropus* species, or flying foxes, are listed as Appendix I or II. Appendix I species are deemed as threatened by extinction and all international trade is prohibited except for non-commercial purposes (e.g., scientific research). Appendix II affords protection to species that are not currently threatened, but may become threatened without controlled trade. Appendix II also protects similar-looking species in order to discourage illegal wildlife trafficking. All members of *Acerodon* and *Pteropus* are listed at both the genus and species level because many species have very restricted ranges and some are endangered, but species identification—especially by non-experts—is extremely difficult. The only non-pteropodid currently listed by CITES is the Uruguay population of the white-lined broad-nosed bat (*Platyrrhinus lineatus*), which is listed under Appendix III. Appendix III species are protected within a signatory country, but that signatory country has indicated it requires extended cooperation from other countries to prevent exploitation.

The importance of taxonomy in international agreements is also evident in the Convention on Migratory Species' (CMS) EUROBATS Agreement, which originally recognized 37 species, but now includes all 52 bat species (both migratory and non-migratory) in Europe, North Africa, and the Middle East. EUROBATS sets legal protection standards and develops and promotes management and conservation strategies across international borders, with 35 of 63 states within the targeted range as signatories. Revisions to the number of species listed, with an increase of 7 new species since 1995, are due to continuing taxonomic work in the region (CMS 2013).

16.5 Taxonomy as a Conservation Planning Tool

Taxonomy may be used as part of conservation either directly (e.g., generating species lists, defining hotspots in need of protection, inventories and monitoring, providing global perspective) or indirectly (e.g., estimating extinction risk, estimating rate of extinction). For example, the Southeast Asian Bat Conservation Research Unit (Kingston 2010) identified advancement in taxonomy and systematics research as a regional priority even though this consortium focuses on capacity building and conservation, not taxonomy. What follows is a summary of practices that conservation biologists currently employ, and also new perspectives and methods that taxonomy and systematics may bring to conservation management.

16.5.1 A Basic Question: What is a Species?

Effective species conservation requires defined taxonomic units that reflect biological reality and can be documented and tracked through space and time using survey and inventory methods appropriate to the organism and ecosystem. Defining and identifying such units is frequently much harder than it sounds. The most commonly used taxonomic unit in conservation biology is the species, though populations are occasionally considered unique enough to merit protection (Justice Department et al. 1996). Species are considered by both scientists and the public to be real, physical entities worthy of conservation. The fact that species have names makes it easier for non-experts to understand and protect them. However, species concepts in biology are far from simple (Cracraft 1989; de Queiroz 1998; Wheeler and Meier 2000; Baker and Bradley 2006; de Queiroz 2011) and applying a set of practical rules to standardize species units is helpful for making species lists in any given area. Taxonomic units for conservation recovery planning must acknowledge the ever-evolving nature of these units in natural systems. While methods of species definition and recognition are debated among researchers [e.g., reproductive isolation for the Biological Species Concept, monophyly for the Phylogenetic Species Concept (PSC), and genetic divergence for the Genetic Species Concept (GSC)], each recognizes that species are composed of populations and that by their nature they are dynamic, not static, units (de Queiroz 2005).

Compared to species of insects and birds, bats are relatively taxonomically stable, and issues related to new cryptic taxa are relatively minor in the sense that cryptic bat species are usually confined to within the boundaries of what was previously considered a single species (Jones et al. 2009). Cryptic species excepted, new information or the application of new species concepts has not tended to change species limits in most bat taxa, suggesting that species limits in bats (or at least those subject to revisionary studies within the last 25 years) are already defined to maximize stability (e.g., buffering against phylogenetic uncertainty) (Lee 2005). Despite hopes to the contrary, it seems unlikely that all taxonomists will ever agree on a single species concept, even for taxa within a relatively restricted group such as Chiroptera. A variety of factors influence the species concept employed in different studies: available data (e.g., morphology, molecules, echolocation calls, behavior), past history of work on the group, type(s) of training received by the researchers, sample sizes in the study, and available analytical tools may all play a role. In this context, it is important for taxonomists to be explicit about the species concept they employ in a study in order to make their data and conclusions transparent to other researchers.

16.5.2 Listing Species for Protection

The species lists that taxonomists assemble form the basic units used by international, national, and local authorities that provide protection to wildlife. Quantitative analysis has shown that the longer a species has been placed on a list of threatened or endangered species, the more likely it is to recover (Taylor et al. 2005). Many agencies have taxonomic standards that must be met prior to inclusion in a listing. For example, the IUCN requires that names be validly published in accordance with Codes (e.g., The International Code for Zoological Nomenclature or ICZN), and checklists, such as *Mammal Species of the World* (Wilson and Reeder 2005), should be employed where possible. The IUCN accepts the following taxa for listing: species, subspecies, varieties (only for plants), and geographically separate subpopulations. It may also allow undescribed species to be listed under extraordinary circumstances. International legislation includes multilateral environmental agreements (e.g., CITES and CMS) that directly support bat conservation, but other free-trade agreements can also uphold the goals of conservation by combating illegal wildlife trade and promoting species persistence. For example, the North American Free Trade Agreement created the Commission for Environmental Cooperation to identify and address reasons for the decline of widespread species such as the monarch butterfly (Commission for Environmental Cooperation 2010).

Protection on the national level may vary from country to country, but in most cases the species is the unit of concern. In addition to protecting species, many nations recognize the importance of protecting habitats as well; examples include both the Endangered Species Act (ESA) in the USA (US Fish and Wildlife Service 2013), the Species at Risk Act (SARA) in Canada (Species at Risk Act 2013), and the Habitats Directive of the European Union (European Commission 2014). These pieces of legislations all rely on a species list to provide protections with the listing process critical to successful conservation. Within the USA, there have been numerous critiques of the ESA from both scientific perspectives (e.g., Rohlf 1991; Pennock and Dimmick 1997; but see Waples 1998) and policy perspectives (Doremus 1997). Often species listed as threatened by IUCN are not similarly recognized as such by ESA. Taxa listed by the ESA include subspecies that are not listed by the IUCN; three of the eleven bat species on the ESA's threatened and endangered list are listed at the subspecific level (Table 16.3). Within the EU, Annex II of the Habitats Directive calls for the establishment of a Special Area of Conservation to protect recognized species, and Annex IV calls for a strict protection regime across the entire natural range of the species in the EU (Council Directive 92/43/EEC, European Commission 2014).

Differences in listing among countries and NGOs, such as IUCN, may reflect different definitions of "threatened" or "endangered," or reflect the varying ways that priority is afforded to a taxon during assessment. Monotypic genera are sometimes afforded greater priority in evaluation and listing than species, down to the level of population. The phylogenetic uniqueness of a species is an important factor in conservation assessments (IUCN 1980; McNeely et al. 1990; Tisdell 1990). Consequently, the taxonomic mindset of specialists on the group ("splitters" versus "lumpers") may play a very critical role in their decisions concerning when and if a taxon is afforded protection.

There are a handful of instances in which recognition of a new species has resulted in direct conservation action. In Thailand, the discovery of Kitti's

Table 16.3 Conservation status of bat species protected under the US's Endangered Species Act (ESA) compared to the global IUCN Red List

Species name according to ESA	US ESA	IUCN
Leptonycteris nivalis	E	E
Tadarida brasiliensis		LC
Macrotus californicus		LC
Myotis grisescens	E	NT
Diphylla ecaudata		LC
Lasiurus cinereus semotus	E	LC
Choeronycteris mexicana		NT
Myotis sodalis	E	E
Leptonycteris curasoae		V
Leptonycteris (curasoae) yerbabuenae	E	V
Pteropus tokudae	E	EX
Pteropus mariannus	T	E
Plecotus rafinesquii		LC
Corynorhinus (Plecotus) townsendii		LC
Corynorhinus (Plecotus) townsendii ingens	E	
Corynorhinus (Plecotus) townsendii virginia	E	
Eumops underwoodi		LC
Eumops floridanus	E	CR
Emballonura semicaudata rotensis	C	E
Non-American bats		
Craseonycteris thonglongyai	E	V
Aproteles bulmerae	E	CR
Pteropus rodricensis	E	CR
Hipposideros ridleyi	E	V
Emballonura semicaudata semicaudata	C	E

Blanks represent lack of listing (ESA) or lack of recognition of species or subspecies (IUCN). The ESA also lists and extends protection to some foreign bat species to discourage people under American jurisdiction from further contributing to species decline. Listing of foreign species may increase in situ conservation action and provide limited financial assistance and training
ESA abbreviations: *E* Endangered, *T* Threatened, *C* Candidate
IUCN abbreviations: *EX* Extinct, *CR* Critically Endangered, *E* Endangered, *V* Vulnerable, *NT* Near Threatened, *LC* Least Concern
Source US Fish and Wildlife Service Environmental Conservation Online System, Species Report, Listed Species; IUCN Red List

hog-nosed bat (*Craseonycteris thonglongyai*; Hill 1974) and the recognition of the distinctiveness of the taxon with the definition of a new family, led to the creation of the 500 km^2 Sai Yok National Park in 1980 under the Wildlife Animal Reservation and Protection Act, B.E. 2535. However, a population subsequently discovered outside the park in Myanmar is not protected, and relatively little is known from its status. The Myanmar population is genetically distinct from the Thai population but morphologically indistinguishable from it (e.g., cryptic), raising questions about whether or not it should be considered a distinct taxon or simply an isolated population (Bates et al. 2001; Pereira et al. 2007; Puechmaille et al. 2011). These discoveries have led to changes in the dynamic of conservation for *Craseonycteris*, since conservation priorities are often related to species range sizes. Similarly, ongoing discovery of cryptic species in Africa, such as Rosevear's serotine (*Neoromicia roseveari*), has led to calls for protection of the Upper Guinean forests, which are threatened by rampant human disturbance (Monadjem et al. 2013).

There are times when national recognition of a species as endangered comes too late, resulting in extinction. In some cases, this is in part due to taxonomic confusion—a circumstance that underlines the importance of taxonomy for conservation. The Christmas Island pipistrelle (*Pipistrellus murrayi*) is an unfortunate example from Australia. The only native insectivorous bat on Christmas Island, it was once widespread but underwent dramatic population declines by the mid-1990s (Beeton et al. 2010). The reasons for this decline remain unclear, but likely include introduction of non-native species (e.g., common wolf snake, feral cats, giant centipedes, and yellow crazy ants) that either disturbed roost sites or preyed on bats (Lumsden et al. 2007). It is also possible that control efforts focused on yellow crazy ants (*Anoplolepis gracilipes*) might have inadvertently poisoned the bats (Beeton et al. 2010). The muddled taxonomic history of the Christmas Island pipistrelle apparently contributed to poor management decisions. Koopman (1973, 1993) considered *P. murrayi* to be a synonym of *P. tenuis*, a common Southeast Asian species, apparently based on general morphological similarity. Hill and Harrison (1987) treated *P. murrayi* as a separate species based on the presence of a distinctive baculum, but this gained little attention at the time. Lack of a focused taxonomic treatment of the pipistrelle species complex resulted in lack of any real consensus about the status of the Christmas Island pipistrelle. The Australian government was slow to act upon findings from a long-term monitoring program, which recommended captive breeding programs for the Christmas Island pipistrelle in 2006 (Martin et al. 2012). It was only after genetic studies by Beeton et al. (2010) corroborated that *P. murrayi* was a distinct species that an emergency response was initiated in 2009 (Martin et al. 2012). However, these efforts came too late—the Christmas Island pipistrelle apparently became extinct in 2009 (Lumsden 2009).

Placing a species on international or national lists may be a prerequisite for local conservation actions such as habitat restoration or protection. The Indiana bat (*Myotis sodalis*) is listed as Endangered under the US's ESA. As such, the species is protected in the USA, meaning that commercial expansion must take into consideration the levels of disturbance to the population before development or operation may proceed in a given area. This has led to US Fish and Wildlife guidelines for businesses such as coal mining companies and wind farms (e.g., US Fish and Wildlife Service 2012) that describe development without harming local wildlife, such as Indiana bat populations. In a recent case against a wind energy company in West Virginia that failed to perform a due-diligence survey prior to development, the courts ordered an injunction against the company and required that it apply for incidental take permits before continuing operations. The wind turbines were allowed to be powered on only in the winter when the bats were hibernating (Woody 2009). In another case, a bat habitat restoration project has been proposed in Ozark National Forest, Arkansas, after ice storm-damaged acres of forest. The idea in this case is to ensure there will be enough healthy stands of trees for the Indiana bat (USDA 2012).

16.5.3 Downsides of Species Listing

Although well intentioned, adoption of global endangered species lists may in some cases be detrimental to more localized protection and conservation efforts. Many countries, and some subnational units, have simply adopted the IUCN Red List of species into their legislation. This practice can be inappropriate, as is recognized by IUCN itself. The criteria used in the IUCN list are specifically designed to identify the species that are most endangered at a global level, not within a region, nation, or specific locality. Consequently, the IUCN has issued "Guidelines for Application of IUCN Red List Criteria at Regional and National Levels" (IUCN 2012) to aid in the application of IUCN principles to more regional surveys. National governments that adopt IUCN listings in their entirety typically do not conduct their own taxonomic and systematic assessment of the species and population status of species that reproduce in or regularly visit the region within their borders. The IUCN advises using the globally derived Red List to set regional conservation priorities under only two conditions: (1) when there are a high number of endemics or threatened near endemics in the region, and (2) when there are little to no data concerning the species within a region. In all other situations, the IUCN advises following IUCN guidelines to assess extinction risk at the geographic scale of interest (local, national, and regional) and publishing Red Lists at this scale. Full compliance with the guidelines allows the country or region to state that their regional Red List follows the IUCN system.

Application of global lists at the local level may miss some species that need local protection. Alternatively, negative conservation outcomes may result if local values are compromised as a result of uncritical national protection of IUCN-listed species. For example, if the presence of a protected species impedes economic development, landowners in a region may destroy the species' habitat or deny the existence of that species to avoid local legal consequences stemming from its IUCN listing (Possingham et al. 2002). Planners and legislators need to appreciate that there are many dimensions to threat and protection and provide landowners and other stakeholders with incentives to protect endangered species.

16.5.4 Inventory and Monitoring Programs

Monitoring bat populations can be an important tool in efforts to understand the condition of an ecosystem, since bats have long been recognized as good indicator species (Fenton et al. 1992; Medellín et al. 2000; Jones et al. 2009). An indicator species is one whose presence, absence, and condition is suggestive of environmental health (Noss 1990). Since bats provide many ecosystem functions, such as pollination and seed dispersal, they are intrinsically linked to plant populations where they live (Fujita and Tuttle 1991; McConkey and Drake 2006). The predation of bats on insects may also reflect arthropod abundance and species diversity (Kalka et al. 2008). Bats can also be indicative of global climatic shifts. For instance, *Pteropus alecto* and *Pteropus poliocephalus* experienced increasingly frequent massive die-offs during extreme heat spikes in Australia (Welbergen et al. 2008). In early 2014, a record-breaking heat wave in central and eastern Australia resulted in one of the most catastrophic die-offs ever recorded—more than 45,000 flying foxes of the three native species (*P. alecto*, *P. poliocephalus*, and *Pteropus scapulatus*) died and more than 1000 juveniles were orphaned (Welbergen et al. 2014). These mass mortality events appear to coincide with the increasing frequency and intensity of extreme climate events that are predicted for Australia due to climate change (IPCC 2012).

Collection of voucher specimens, while sometimes controversial, is widely regarded by systematists and taxonomists as critical to inventory projects (Voss and Emmons 1996; Simmons and Voss 1998, 2009). Vouchers are necessary for any future work such as reassessments of the initial study or further extension of the initial work when new information or methods become available. Vouchers, including tissue samples, are especially necessary when species are cryptic or nearly so—some bat species can only be identified by minute morphological differences, (e.g., cranial characters, or in small vespertilionids, the baculum (penis bone) (Hill and Harrison 1987) or by molecular means (e.g., Clare et al. 2013). Vouchers are also necessary to provide type specimens (minimally a holotype but preferably also paratypes) if a new species is discovered (ICZN 2012).

In some regions of the world, taxonomists may be the only biologists with active research programs and therefore may be the only scientists positioned to collect the population and ecological data required for conservation assessments. They may also be the only biologists on hand to provide information about threats to species at particular localities. These taxonomists often have studied species throughout their ranges and are able to offer a more accurate assessment of conservation status by thinking globally instead of locally. For example, for the current revision of the Old World Fruit Bat Action Plan, the team leaders have reached out to a number of bat researchers, many of whom are taxonomists, to determine the most appropriate IUCN Red List status for each species. Most of the current specialist groups of the IUCN Red List include at least one taxonomic expert. This allows for the establishment of international versus national priorities and the creation of appropriate management strategies at the correct taxonomic level. For instance, in Britain, all bats and their roosts are protected by multiple domestic and international laws, even though a majority of these species are listed as Least Concern by IUCN (Bat Conservation Trust 2013). The UN's Global Biodiversity Outlook 3 also repeatedly references trends in population size and diversity of different taxa (Secretariat of the Convention on Biological Diversity 2010). These trends are based on species-specific data—data that are worthless without proper taxonomic identifications of the species in question.

To counter the lack of taxonomic experts during surveys, a technique called parataxonomic sorting was introduced in the late 1980s for entomological surveys in the Neotropics (Janzen 1991) and subsequently for plant surveys (Baraloto et al. 2007; Abadie et al. 2008). Parataxonomy focused on the use of "morphospecies" to sort collected specimens into Recognizable Taxonomic Units (RTUs) (Cranston and Hillman 1992; Oliver and Beattie 1993) as a preliminary method of assessment in the absence of enough taxonomic expertise. RTUs are not truly recognized biological species, and the sorting method is recognized as non-scientific, but the efficiency of the method quickly turned a preliminary sorting method into a source of data for biological surveys. However, results of parataxonomy are inconsistent and these methods have been criticized for the low quality of data and incorrectly grouped individuals (Krell 2004; Baraloto et al. 2007). Parataxonomy is largely uninformative when it comes to inventories, biogeographic studies, area selection for conservation, autecology, and habitat comparisons although it may still be useful in limited capacities for global comparisons of gross species richness or single-site descriptions of species richness of some taxa (Krell 2004). However, bats are particularly ill-suited to parataxonomic efforts because taxa are difficult to distinguish and the process of collecting specimens is time-intensive and requires specialized training and permits that are difficult to obtain even when one is an expert. Parataxonomy seems to hold little promise for chiropteran studies.

16.5.5 Defining Protected Areas

One commonly used method for defining protected areas is identification of "biodiversity hotspots" with "exceptional concentrations of endemic species…experiencing exceptional loss of habitat" (Myers et al. 2000). Generally, methods of prioritizing areas for conservation based on measuring endemicity, phylogenetic diversity, or taxon richness represent variations of the hotspot approach—they all measure some proxy for species representation and identify areas for conservation based on these variables. Such methods stand in contrast to area selection approaches that focus on threatened or degraded habitats. The hotspot approach to choosing protected areas has been criticized as susceptible to taxonomic instability (Isaac et al. 2004). Some authors have suggested that hotspots should use higher level taxonomy to identify areas that warrant protection and sidestep issues related to unstable taxonomy (Balmford et al. 2000; Amori and Gippoliti 2003). Genera and species were found to be highly correlated and may select for the same priority areas, whereas family and order are not very informative (Balmford et al. 2000; Amori and Gippoliti 2003). This approach may be inappropriate for bats, as young, rapid radiations may result in higher species diversity than would be predicted based on generic diversity. For example, in the Paleotropics, site-wide diversity is primarily driven by only a few genera (e.g., *Hipposideros*, *Rhinolophus*, *Kerivoula*) (Kingston et al. 2003). This pattern is also seen in the Neotropics, although to a lesser extent (e.g., *Artibeus*/*Demanura*,

Micronycteris, *Lophostoma/Tonatia*) (Voss and Emmons 1996; Simmons and Voss 1998). Selection of hotspots based on species richness would value a site where selection based on genera would not, potentially leading to missed conservation opportunities.

To combat issues related to taxonomic stability, conservation should implement approaches that emphasize the uniqueness of taxa or areas (Gippoliti and Groves 2012). Newer methods for conservation often emphasize evolutionary uniqueness in concert with extinction risk when choosing priority protection areas (Collen et al. 2011). However, regardless of taxonomic resolution, the hotspot approach may be unlikely to reduce extinction risk in areas such as the Andes, where high species richness is correlated with areas with low human disturbance (Fjeldså 2000). The hotspot approach in this case ignores species at greatest risk in areas with high levels of human contact and may result in directing more resources to areas that require little intervention. Complementarity takes into account human development, selecting sites that may not have high biodiversity, but would result in conservation of more species in the area.

The relative taxonomic stability of bats means that some conservation decisions may be easier to make. But it does not mean that hotspot approaches are always appropriate. Instead of focusing on overall species richness, some area selection approaches focus on an umbrella species, or a variation thereof: focal, keystone, flagship, or threatened species (Lambeck 1997; Roberge and Angelstam 2004), with the assumption that protection of their habitat will benefit other organisms in the area. This approach often focuses on "charismatic megafauna," such as tigers, elephants, and primates, that are large-bodied as these species tend to have larger area requirements (Roberge and Angelstam 2004) and overlooks species with specialized habitat requirements or niche habitats, such as limestone karsts, that are irrelevant to large animals. Such niche habitats may be crucial to the survival of rare and endemic taxa with small ranges and narrow niches, such as threatened bat species such as Kitti's hog-nosed bat (*C. thonglongyai*) and the Thailand leaf-nosed bat (*Hipposideros halophyllus*).

16.5.6 Estimating Extinction Risk and Extinction Rate: The Role of Phylogenetics

Much emphasis is placed on extinction risk by conservation biologists in relation to climate change, habitat fragmentation, and habitat loss, but we cannot determine current rates of extinction and compare them to past rates of extinction without accurate knowledge of global biodiversity and updated phylogenies. Phylogenies allow researchers to test hypotheses related to character trait evolution, including traits related to natural history and extinction risk (Jones et al. 2003). Shared ecological traits from any one clade are by definition non-independent since all the species in a clade are linked by common ancestry. Analysis of patterns requires the removal of the historical signal in the data through the phylogenetic comparative method, otherwise known as "correcting" for phylogeny (Felsenstein 1985). Taxonomic bias for risk of extinction and for susceptibility to invasion is a known issue for conservation biologists (Fisher and Owens 2004). It may not be possible to conduct detailed research on every at-risk species within a short time span, but the comparative method allows for a quick assessment of conservation priorities based on shared risk of extinction in vulnerable clades. This may also provide perspective on causes of species decline. All of these data may allow for conservation actions to be taken sooner rather than later, with early action being more cost-effective and more successful (Fisher and Owens 2004).

Jones et al. (2003) conducted a multivariate analysis of correlation between extinction risk in bats (represented by IUCN threat level) and various natural history and morphological traits known to correlate with extinction risk in other taxa (Purvis et al. 2000; Isaac et al. 2005; Forero-Medina et al. 2009). Jones et al. (2003) found extinction risk to be highly correlated with evolutionary history, meaning clades shared similar levels of threat. Correlation of extinction risk with evolutionary history indicates the necessity of accounting for the phylogenetic history of clades when making such determinations, opening opportunities for determining the critical factors for clades. For example, geographic range size was the most important predictor of extinction risk across Chiroptera, though it was found to be an order of magnitude higher in pteropodids, which have smaller mean geographic ranges, than other bats. Among non-pteropodid bats, larger body size, larger group size, and low wing aspect ratios were significantly correlated with higher extinction risk. In pteropodids, smaller litter size was significantly correlated with extinction risk. These findings explained approximately half of the variance in extinction risk, and more work remains to be done. In a recent study of vespertilionid bats, those in threatened categories were more likely to be dietary specialists than those listed as Least Concern (Boyles and Storm 2007). As robust phylogenies are assembled and more ecological data are collected, the comparative method will be of great use for identifying important contributors to extinction risk in bats.

High genetic variation is generally thought to be associated with lower extinction risk, as species with greater amounts of variation are more able to adapt to changing environmental conditions (Lacy 1997; Hermisson and Pennings 2005). Endangered species generally have reduced genetic variability and, even after their numbers have recovered, may not be able to recover genetic variability and thus still face high extinction risk (Frankham 2005). While recovery rates may vary depending on how long populations were bottlenecked, a slow recovery would be predicted for bats, as they generally have low reproductive rates. Since populations recover too slowly, there would be a greater loss of genetic diversity as well. Rapid and irreversible loss of genetic diversity further increases extinction risk of a species and underscores the need for preemptive conservation action. However, reduced genetic variability must be shown to be truly a recent bottleneck through anthropogenic disturbance, as in the case of sea otters being impacted by the fur trade (Larson et al. 2002). In other mammalian species, such as cheetahs (Menotti-Raymond and O'Brien 1993) and wolverines (Schwartz et al. 2007), low genetic variability is a result of previous historical demographic changes (e.g., bottlenecking from Pleistocene glacial cycles, founder effects at periphery of distribution). Phylogenetic research is necessary to approximate expected levels of genetic variation before statements about genetic health of an endangered species can be made. There are currently no examples in bats using these methods, but comparisons of extant populations to historical specimens in museum collections may help determine whether threatened populations are experiencing anthropogenic bottlenecks. This research is now made possible by new methods in high-throughput sequencing of ancient DNA from degraded material (Gilbert et al. 2007; Mason et al. 2011; Dabney et al. 2013) and modeling of heterochronous data (Ho et al. 2007; Navascués et al. 2010; Drummond et al. 2012).

Estimations of speciation and extinction rates may also be made from phylogenies (Ricklefs 2007; Fitzjohn et al. 2009; but see Rabosky 2010 about the need for inclusion of fossil data) using speciation–extinction models derived from birth–death models in population ecology. Greater availability of time-calibrated phylogenetic trees now makes this method viable for estimating the likelihood that a clade will go extinct during a particular time slice. However, these estimates of likelihood of speciation and extinction are tied to the completeness of the phylogeny, meaning more phylogenetic work must be completed if these estimates are to be used for making predictions about species diversity in that clade. These model-based methods allow researchers to investigate speciation or extinction rates as compared to random chance. Anthropogenic effects on extinction can thus be more accurately assessed. Such research may also be used as a second test of hypotheses of species loss in concert with current methods favored by conservation biologists, such as species–area relationship and endemics–area relationship (e.g., Lane et al. 2006).

From phylogenetic studies, researchers now know that some species may be the only remaining representative of an old lineage, while others are one of many in very diverse clades. These old lineages, known as relict species, have genes and traits that have survived from deep timescales and tell a tale of resilience (and luck) in the face of regime shifts and faunal turnover. These taxa may have survived previous major extinction events, and researchers can study them to understand how species may continue to survive in the face of the current extinction crisis (Habel and Assmann 2010). Relict species may also represent the only living relatives of fossil taxa, allowing systematists to place fossil taxa correctly in a tree. Representing both extant and extinct taxa is necessary for accurate estimates of extinction rates (Rabosky 2010). How accurate these estimates may be for bats is still unclear, as there are few dated phylogenies and the only study in non-volant mammalian extinction rates found that clade age was not correlated to higher extinction (Verde Arregoitia et al. 2013). *Mystacina tuberculata* is an example of a relict species. It is a New Zealand endemic and the sole extant representative of an entire family that was once more widespread. The fossil record of mystacinids includes the bat genus *Icarops* from the Oligocene and Miocene of Australia (Hand et al. 2001), but the family also includes *Mystacina robusta*, a species that went extinct in historic times (Daniel 1990). Even with molecular tools, researchers have had difficulties resolving the sister taxon of Mystacinidae, likely a result of deep, rapid radiations that created short internal branches with conflicting phylogenetic signal (Kennedy et al. 1999).

16.6 Impediments to Taxonomic Research

A decline in both amateur and professional taxonomists has been documented (e.g., Stuessy and Thomas 1981; Hopkins and Freckleton 2002), with reductions or elimination of jobs in museums and universities for those trained in taxonomy. There are few skilled and trained bat taxonomists, slowing fieldwork as well as the publication of comprehensive taxonomic revisions, species lists, field guides, and popular works on bats. In part, this appears due to what has been described as a "classic market failure" for taxonomy (Aylward et al. 1993; Hoagland 1996). Taxonomy is an "externalized" cost:

> Growing out of a tradition of reciprocity and collegiality, taxonomists frequently do not charge clients directly for their specialized services and products, such as identifications and biodiversity databases, even though the users of these services and products now extend far beyond their fellow taxonomists. These service activities are often ancillary to a taxonomist's basic monographic work, for which he or she receives grant funds, or subsidizes on his own or through his employers. The cost of doing taxonomy is not factored into most biodiversity or ecology projects. Research grants (even in taxonomy) and ecological monitoring activities rarely include funds for the curation and care of voucher specimens, or the establishment and maintenance of museums. (Hoagland 1996)

The result? A reluctance by employers to hire those who do not bring in funds and cause a perceived drain on the institution, and a reluctance by students to pursue taxonomy as a career in favor of fields offering more money and jobs. While there are a growing number of young bat taxonomists in the developing world (Anwarali Khan et al. 2010; Douangboubpha et al. 2012; Soisook et al. 2013) where educational institutions are newly committed to developing and protecting local biodiversity, the lack of funds for taxonomy still presents a substantial impediment (Aylward et al. 1993). The few taxonomic experts in developed countries that still remain are discouraged from pursuing taxonomy in regions of the world where both the biota and their ecosystems are most understudied due to a combination of stricter local specimen export laws and lack of funding. Additionally, the low impact factor of taxonomic journals is a major impediment for academics at non-museum institutions whose performance reviews for promotion hinge largely upon the impact factor of journals in which they publish (Venu and Sanjappa 2011).

In most scientific fields, including other disciplines of systematics, specialists have grouped themselves in associations that publish journals and act as lobbies to promote their discipline and defend their members. However, there exists no international or national scientific society specifically devoted to the promotion of taxonomy, the publication of general papers on the discipline, its theoretical background, its history, or its problems and its future. In part as a result, taxonomists are typically under-represented in official or unofficial bodies that play significant roles in shaping scientific policies, budgets, and definition of priorities. Yet, taxonomists are critically needed for research in understudied groups, such as bats, especially in developing countries. Without any formalized society, it becomes difficult to pass on the expertise and shared standards that are essential to all other fields in biology, including conservation.

The reduction in numbers of taxonomists in institutions in developed countries and the increase seen in developing countries is complicated by a great deal of historical baggage. Type specimens (the actual specimens to which scientific names are attached) and important taxonomic literature are still based in institutions in developed countries, and there is still an imperative need for repatriation of information as well as capacity building outside these centers. Capacity building can occur at three different levels: individual (build individual ability to contribute to taxonomy), institutional (modernize museum infrastructure and policies, increase the level of curatorial proficiency in staff), and societal (engage the public in understanding and learning about biodiversity and being held accountable for it). Lack of access to available information is then also a part of the taxonomic impediment to conservation, not just lack of research in the discipline.

Progress has been made recently to increase accessibility of resources housed in institutions in developed countries. Digitization of type specimens of bats by some of the larger museums (e.g., American Museum of Natural History), increased availability of literature through online sources, increased training in developing countries, and increased collaborations between Western taxonomists with young taxonomists from developing countries have begun to counter gaps in knowledge and training. Collections research fellowships are now available at some institutions to provide researchers with funds needed for visiting museums and inspecting specimens first-hand. Developing countries now see an increase in new bat taxa described in international, open-access journals by in-country scientists. New, well-maintained, and actively used natural history collections now exist in places like the University of Phnom Penh, Cambodia; Prince of Songkla University, Thailand; and the National University of Laos, thanks to local support and funding by NGOs such as the Darwin Initiative, the Systematics Association, and the MacArthur Foundation. Older collections in species-rich tropical countries, such as at the National Museum of the Philippines, the Museo de Zoología-Mamíferos, Pontifica Universidad Católica del Ecuador, and the Museu Nacional, Universidade Federal do Rio de Janeiro in Brazil, have refurbished outdated collections spaces and benefited from increased access to information and increased local capacity as talented local scientists have helped reignite interest in conservation and biodiversity initiatives.

Museum collections and historic taxonomic descriptions themselves may, counterintuitively, present impediments to taxonomic research. While today's taxonomists use morphological and genetic data (when available) to establish species limits, such modern methods have only come to the fore recently. Many older species names are attached to poorly preserved type specimens, sometimes dry skins, museum taxidermy mounts, or specimens that survived long sea voyages pickled in rum or other spirits. These specimens may be so damaged that viewing important features, or any features, from the published descriptions is impossible, leading to confusion regarding the recognition of the species in question. In some cases, the type specimens have been lost or destroyed and new type specimens (known as neotypes) must be designated, again introducing the possibility of confusion. Older names are often based on brief and sometimes inadequate descriptions that fail to provide sufficient detail to facilitate distinction from similar species. Even when faunas have been well surveyed, these issues of taxonomy frequently cause confusion about the number and identity of species inhabiting a particular region. Taxonomic confusion may contribute to the inability to properly attribute a name to organisms or integrate new data, barring species from protection that they may have been granted had they been accurately recognized and complicating conservation efforts.

16.7 Conservation in the Era of Molecular Phylogenetics

Molecular tools have given systematists new ways to resolve phylogenies and population networks and thus new ways to delimit species and other units of conservation concern. Genetics has created new ways of thinking about what a species is, and this has led to healthy debates about species delimitation. In some countries such as Germany, conservation legislation takes into account the genetics of organisms as well as their species limits. The Nationale Strategie zur Biologischen Vielfalt (National Strategy for Biological Diversity of Germany, BMU 2007) recognizes that the entire gene pool of a species must be protected. While this may not always be possible, the reason for this approach is based on the desire to protect distinct lineages.

Populations are often locally adapted and may be on different evolutionary trajectories even within what is recognized as a single species. The term Evolutionary Significant Unit or ESUs was originally coined to reflect the importance of these units in conservation decisions (Ryder 1986; Moritz 1994). ESUs may be at the species level or below and ESU definitions generally include the idea that the ESU is currently geographically isolated from other ESUs, that there is genetic differentiation at neutral markers, or that there is local phenotypic variation. The term ESU has since changed to reflect both evolutionary processes along with ecological exchangeability. The crosshair analysis advocated by Crandall et al. (2000) uses tests of null hypotheses in four categories (genetic, ecological, recent, and history) to determine whether populations should be considered ESUs or not. Species are not static, but evolving; if given enough time, ESUs may evolve into entities that require a different taxonomic status, e.g., a population may become a new species. ESUs may represent unique gene pools and may be of special conservation concern; proper conservation action can be taken only if they are recognized.

Molecular genetics has also allowed researchers to identify cryptic species, species that are morphologically indistinguishable (or nearly so) but exhibit significant genetic divisions that form species boundaries (Pfenninger and Schwenk 2007). These discoveries have helped systematists further understand the mechanisms that drive the speciation process, such as sympatric reproductive isolation without morphological differentiation, but they also have conservation implications (Bickford et al. 2007). Cryptic species represent a previously unrecognized part of the biota of a region and thus may be important to conservation biologists who are interested in identifying and understanding biodiversity hotspots. In bats, many previously unrecognized cryptic species are now being found through molecular assays even in very well-studied areas (Mayer et al. 2007). Early results from bar coding work in Southeast Asia suggest that the number of bat species may be twice that currently recognized (Francis et al. 2010). The level of discovery of new taxa in the last decade has generally corroborated this estimate (Table 16.1).

A classic example of a cryptic species hiding in plain sight is the European pipistrelle (*Pipistrellus pipistrellus*). One of the most common bats throughout its range, the European pipistrelle was not recognized as a cryptic complex until echolocation data suggested the presence of more than one species of pipistrelle occurring in sympatry throughout much of Europe (Jones and van Parijs 1993; Barlow 1997; Barlow and Jones 1997). Since the early echolocation studies, mitochondrial data (Hulva et al. 2004), microsatellite data (Hulva et al. 2010), information on foraging (Davidson-Watts and Jones 2005), and habitat selection data (Davidson-Watts et al. 2006) have further corroborated the split of the European pipistrelle into two distinct species (*P. pipistrellus* and *Pipistrellus pygmaeus*). Similar echolocation studies in Southeast Asia of hipposiderid bats (Kingston et al. 2001; Thabah et al. 2006) and African *Rhinolophus* (Taylor et al. 2012) have shown that these groups likely contain many cryptic species that can be distinguished by distinct phonic profiles, but not so easily by morphology. In many cases, molecular work remains to be conducted to clarify the numbers and limits of species in these complexes.

Molecular tools can now be used to characterize biodiversity in a more efficient manner than could be done in the past, particularly in poorly studied regions of the world. However, these tools must be used with caution, as not every new mitochondrial clade warrants recognition as a distinct species—some genes are known to be hypervariable and poor indicators of species limits (Engstrom et al. 2004; Lohse 2009; Galtier et al. 2009). The phylogenetic signal for hybridizing species may look very similar to incomplete lineage sorting (e.g., both phenomena would result in non-monophyletic trees) and therefore requires more genetic data and stricter quantitative assessments of genetic data to test different evolutionary scenarios (Maddison 1997; Yu et al. 2012). Many molecular studies of bats published in recent years have failed to review important elements such as the morphology or echolocation call structure of putative species, or have failed to include a sufficient number of genes or individuals. Mitochondrial clades may point to the need for more research into a potential species complex, but such clades cannot be readily assumed to represent a new species. The recent discovery of multiple allopatric mitochondrial lineages of *Pteronotus parnellii* (Clare et al. 2011) *Chrotopterus auritus*, *Glossophaga soricina*, and *Saccopteryx bilineata* (Clare 2011) indicates that deep divergences may exist within these species, but further study of genetic, morphological, or behavioral characters is needed as noted by these authors. Even in well-studied regions, such as Europe, cryptic species may have only been recently recognized as new phylogenetic methods and more nuclear data have become available, such as the Natterer's bat (*Myotis nattereri*) complex (Salicini et al. 2011). Mitochondrial divergence may also reflect sex-based differences in dispersal rather than new species. For example, Ozark big-eared bats (*Corynorhinus townsendii ingens*) have low levels of mitochondrial divergence between caves, but their microsatellite data indicate that there is likely male-mediated gene flow between populations (Weyandt et al. 2005).

Examples exist of cases where mitochondrial data have been misleading in bats. A cautionary tale is that of two subspecies of *Myotis lucifugus* (*M. lucifugus* and *M. carissima*), which exhibited enough mitochondrial divergence that they could have been recognized as separate species on the basis of molecular evidence alone (Dewey 2006). However, analyses of ten additional nuclear markers have shown that both these subspecies are experiencing high levels of gene flow, resulting in the absence of population structure even if these were historically separate populations. Additionally, no morphological characters diagnose the mitochondrial clades (Lausen et al. 2008). Consequently, there is no justification, despite the mitochondrial indicators, for recognizing these subspecies as separate species.

In contrast, Goodman et al. (2009) used a combined molecular and morphological dataset to resolve cryptic species in *Miniopterus manavi*. This study demonstrates a "best practices" approach to resolving widespread species complexes. Previous research using only mitochondrial data suggested that *M. manavi* in Madagascar and the Comoros represented unique lineages. However, sampling was limited and the relationships between clades were not fully resolved (Weyeneth et al. 2008). Using increased geographic sampling and morphological comparisons of type specimens, each of the clades was more clearly defined. *Miniopterus aelleni* was recognized as a new species, and its species diagnosis and description was accompanied by photographs of a live individual and skulls, and illustrations of dental characters (Goodman et al. 2009). Despite the relative rarity of *M. aelleni* to *M. manavi* on Madagascar, both species were found in several protected areas and the authors did not suggest further conservation action.

Extensive sampling throughout the geographic range of the relevant species is needed when attempting to resolve the relationships within a species complex. Simulation data suggests that more complete taxonomic sampling improves phylogenetic accuracy (Pollock et al. 2002). Too much missing data, either in the form of missing characters (e.g., missing genetic loci or using only mitochondrial data for some taxa) or missing taxa (e.g., incomplete geographic sampling) can lead to unresolved trees or incorrect inferences through phenomena such as long-branch attraction (Wiens 2003, 2006).

Currently, the field of molecular phylogenetics is undergoing a major shift away from locus-by-locus data collection to next-generation sequencing methods (also called high-throughput sequencing) that will allow for the collection of massive datasets in a relatively short period of time (Faircloth et al. 2012; Lemmon et al. 2012; Lemmon and Lemmon 2012). As prices fall and computational pipelines are developed to deal with the influx of data, taxonomically complex problems may be resolved by the increased availability of molecular character data. Genomic advances will also allow for detection of signs of natural selection in recent history (e.g., Pickrell et al. 2009 in humans; vonHoldt et al. 2010 in dogs), which could be used to determine how recent historical events such as climate change or human disturbance have affected natural populations. Having more data may not be the only solution to taxonomic problems, however—more powerful computational models means greater ability to analyze multilocus datasets that are already available. By taking cues from population genetics and phylogeography, historical models can now be incorporated into analyses to understand the effects of microevolutionary processes on species histories (Edwards and Beerli 2000). Establishing that a tip on a phylogenetic tree is truly representative of a species, and not just a genetic lineage, is fundamental to the goals of systematics and necessary prior to further analyses about speciation and diversification (Edwards 2009).

16.8 The Problem of "Taxonomic Inflation"

Taxonomic inflation caused by improper species delimitation can have profound effects on conservation, as biodiversity hotspots may be misidentified, or conservation priorities are selected based on poor evidence. With the advent of molecular phylogenies, imprudent application of the PSC or the GSC has been criticized for greatly inflating the number of recognized species in mammals, where many subspecies have been raised to full species rank. The examples cited by critics, such as Zachos et al. (2013) for Cetartiodactyla and Isaac et al. (2004) and Mace (2004) for Primates, however, are not due to application of the PSC or molecular phylogenetics; instead, they are generally due excessive splitting of inadequate datasets. For instance, critics cite splitting the mainland serow (*Capricornus sumatraensis*) into six species from one as evidence of taxonomic inflation. Yet the split of this species was based on pelage characteristics and was complicated by small sample sizes (Groves and Grubb 2011), and as such it has nothing to do with a new understanding of genetics. While the mainland serow may not have warranted such splitting, the critiques against taxonomic inflation ignore the fact that newly recognized species in these complexes may reflect biological reality (Gippoliti and Groves 2012; Gutiérrez and Helgen 2013). A more comprehensive set of data may be needed to confirm species boundaries, but new research should not be thrown out in favor of older taxonomy just because the latter is more convenient. Like other branches of science, our knowledge, and views of taxonomy change, other researchers also need to embrace this aspect of defining species.

Gippoliti and Groves (2012) responded to criticisms of taxonomic inflation by citing several examples of how integrative modern taxonomy (including multiple lines of evidence) has positively affected conservation. Critics of taxonomy are not wholly against the findings of modern taxonomy. For example, Zachos et al. (2013) recognized several legitimate cases of cryptic species in African elephants, giraffes, and European badgers. In each case, multiple lines of evidence corroborated species boundaries and warranted species-level recognition. Critics of taxonomic inflation seek the same comprehensive data collection that taxonomists do and generally make the same recommendations that we have outlined above. If uncertainty surrounding preliminary mitochondrial data exists, decision makers should determine if clades of interest correspond to any ESU or other management units (Miralles and Vences 2013), not throw out the new taxonomic information entirely.

It is important for taxonomists to state methods used to delimit species so that new candidate assessments can be easily made in the future. Explicit enumeration of methods, species concepts, and data makes taxonomic assessments more repeatable and testable by others. Clearly written species descriptions based on multiple lines of evidence help maintain the species identity over time, reducing confusion in the long run about the species and its associated name. A recent study in the Malagasy lizard genus *Madascincus* found that different species-delimitation protocols (e.g., Bayesian Assignment Test, HaploWeb, or Generalized Mixed Yule Coalescent Approach) result in wildly different recognized species, with the Bayesian Assignment test approach being in the most agreement with integrative taxonomy (Miralles and Vences 2013). Clearly stating methods can also reduce noise from new species concepts or new data, since it can be quickly determined if this new information will change how the species is viewed and understood. If species limits are known to be stable, that helps maintain the credibility of the lists that legislators and agencies so heavily rely upon for conservation.

16.9 Conclusion

The Age of Discovery is not over for chiropteran taxonomists, who play a critical role in efforts to ensure the documentation and protection of bat diversity by providing a necessary framework for conservation initiatives. Use of a broad range of data (morphological, molecular, behavioral, acoustic) has had a marked effect on the number of bat species identified in the past decade; molecular and acoustic data have indicated that there may be numerous cryptic bat species that cannot be successfully identified using morphology alone.

In addition to identifying species and caring for museum specimens, taxonomists create species lists for localities and communicate taxonomic ideas to non-experts, especially through species lists, descriptions, keys, and field guides. These activities lead to important opportunities for outreach via public exhibits at home institutions or in the field. Taxonomists also provide conservation planning tools such as inventory data, estimates of extinction risk, and information to help define protected areas. These activities allow researchers and government agencies to lower extinction risks and improve the likelihood of species recovery.

More training should be provided to non-taxonomic experts through short workshops focusing on specimen collection and identification techniques. When conducting research, taxonomists may provide the first close-up look at bats to local populations. Capitalizing on this opportunity to inform people about their local biota through leaflets, talks, and training, can advance local and regional conservation goals.

Impediments to the training of new taxonomists remain substantial, including a lack of funding for the identification and storage of voucher specimens, the absence of a taxonomy "lobby" and journal devoted to taxonomic practice, and the low status often accorded to taxonomic publications. However, accessibility to museum materials in developed countries—both voucher material and literature—is increasing through ongoing digitization efforts. Worldwide interest in local biodiversity is also increasing and new bat taxonomists, with new or growing collections, are now practicing around the globe. It is our hope that all taxonomists advocate for appropriate management strategies for bats on a global scale by reaching out to local populations, non-expert scientists, and legislators; effectively communicating complex scientific ideas and listening to local concerns; and continuing to provide a robust scientific basis for conservation as we work to prevent bat extinctions in the Anthropocene.

References

Abadie J-C, Andrade C, Machon N, Porcher E (2008) On the use of parataxonomy in biodiversity monitoring: a case study on wild flora. Biodivers Conserv 17:3485–3500

Amori G, Gippoliti S (2003) A higher–taxon approach to rodent conservation priorities for the 21st century. Anim Biodivers Cons 2:1–18

Anwarali Khan FA, Solari S, Swier VJ et al (2010) Systematics of Malaysian woolly bats (Vespertilionidae: *Kerivoula*) inferred from mitochondrial, nuclear, karyotypic, and morphological data. J Mamm 91:1058–1072

Aylward BA, Echeverria J, Fendt L, Barbier EB (1993) The economic value of species information and its role in biodiversity conservation: case studies of Costa Rica's National Biodiversity Institute and Pharmaceutical Prospecting. Report to the International Swedish Development Authority

Baker RJ, Bradley RD (2006) Speciation in mammals and the genetic species concept. J Mamm 87:643–662

Balmford A, Lyon AJE, Lang RM (2000) Testing the higher-taxon approach to conservation planning in a megadiverse group: the macrofungi. Biol Cons 93:209–217

Baraloto C, Ferreira E, Walthier F (2007) Limitations and applications of parataxonomy for community forest management in southwestern Amazonia. Ethnobot Res Appl 5:77–84

Barlow KE (1997) The diets of two phonic types of the bat *Pipistrellus pipistrellus* in Britain. J Zool 243:597–609
Barlow KE, Jones G (1997) Function of pipistrelle social calls: field data and a playback experiment. Anim Behav 53:991–999
Barquez RM, Giannini NP, Mares MA (1993) Guide to the bats of Argentina. Oklahoma Museum of Natural History, Norman
Bat Conservation Trust (2013) Bats and the law. Available via http://www.bats.org.uk/pages/bats_and_the_law.html. Accessed 30 Dec 2013
Bates PJJ, Nwe T, Swe KM, Bu SSH (2001) Further new records of bats from Myanmar (Burma), including *Craseonycteris thonglongyai* Hill 1974 (Chiroptera: Craseonycteridae). Acta Chirop 3:33–41
Beeton B, Burbidge A, Grigg G, et al. (2010) Final report of the christmas Island expert working group to the minister for environment protection, heritage and the arts. Available via http://www.environment.gov.au/resource/final-report-christmas-island-expert-working-group. Accessed 30 Dec 2013
Bickford D, Lohman DJ, Sodhi NS et al (2007) Cryptic species as a window on diversity and conservation. TREE 22:148–155
BMU (2007) Nationale strategie zur biologischen vielfalt (National Strategy on Biological Diversity), Bonn. Available via http://www.bmub.bund.de/themen/natur-arten/naturschutz-biologische-vielfalt/nationale-strategie. Accessed 30 Dec 2013
Boyles JG, Storm JJ (2007) The perils of picky eating: dietary breadth is related to extinction risk in insectivorous bats. PLoS ONE 2:e672
Buden DW, Helgen KM, Wiles GJ (2013) Taxonomy, distribution, and natural history of flying foxes (Chiroptera, Pteropodidae) in the Mortlock Islands and Chuuk State, Caroline Islands. ZooKeys 345:97
Clare EL (2011) Cryptic species? Patterns of maternal and paternal gene flow in eight neotropical bats. PLoS ONE 6:e21460
Clare EL, Adams AM, Maya-Simões AZ et al (2013) Diversification and reproductive isolation: cryptic species in the only new world high-duty cycle bat, *Pteronotus parnellii*. BMC Evol Biol 13:26
Clare EL, Lim BK, Fenton MB, Hebert PDN (2011) Neotropical bats: estimating species diversity with DNA barcodes. PLoS ONE 6:e22648
CMS (2013) UNEP/EUROBATS agreement on the conservation of populations of European bats. Available via http://www.eurobats.org. Accessed 30 Dec 2013
Collen B, Turvey ST, Waterman C et al (2011) Investing in evolutionary history: implementing a phylogenetic approach for mammal conservation. Phil Trans R Soc B 366:2611–2622
Commission for Environmental Cooperation (2010) Conserving the monarch butterfly and promoting sustainable livelihoods. Available via http://www.cec.org/monarch. Accessed 3 Jul 2014
Cracraft J (1989) Species as entities of biological theory. In: What the philosophy of biology is. Springer, Berlin, pp 31–52
Crandall KA, Bininda-Emonds ORP, Mace GM, Wayne RK (2000) Considering evolutionary processes in conservation biology. TREE 15:290–295
Cranston P, Hillman T (1992) Rapid assessment of biodiversity using "biological diversity technicians". Aust Biol 5:144–155
Dabney J, Knapp M, Glocke I et al (2013) Complete mitochondrial genome sequence of a Middle Pleistocene cave bear reconstructed from ultrashort DNA fragments. PNAS 110:15758–15763
Daniel MJ (1990) Greater short-tailed bat. In: King C (ed) The handbook of New Zealand Mammals. Oxford University Press, Auckland, pp 131–136
Davidson-Watts I, Jones G (2005) Differences in foraging behaviour between *Pipistrellus pipistrellus* (Schreber, 1774) and *Pipistrellus pygmaeus* (Leach, 1825). J Zool 268:55–62

Davidson-Watts I, Walls S, Jones G (2006) Differential habitat selection by *Pipistrellus pipistrellus* and *Pipistrellus pygmaeus* identifies distinct conservation needs for cryptic species of echolocating bats. Biol Cons 133:118–127
De Queiroz K (1998) The general lineage concept of species and the defining properties of the species category. Monism, Pluralism, Unity and Diversity, pp 49–89
De Queiroz K (1999) The general lineage concept of species and the defining properties of the species category. Species: New Interdisc Essays 49–89
De Queiroz K (2011) Branches in the lines of descent: Charles Darwin and the evolution of the species concept. Biol J Linn Soc 103:19–35
De Queiroz K (2005) Ernst Mayr and the modern concept of species. PNAS 102:6600–6607
Dewey TA (2006) Systematics and phylogeography of North American *Myotis* (Chiroptera: Vespertilionidae). PhD thesis, University of Michigan
Doremus H (1997) Listing decisions under the endangered species act: why better science isn't always better policy. Wash Univ Law Q 75:1029
Douangboubpha B, Sanamxay D, Xayaphet V et al (2012) First record of *Sphaerias blanfordi* (Chiroptera: Pteropodidae) from Lao PDR. Trop Nat Hist 12:117–122
Drummond AJ, Suchard MA, Xie D, Rambaut A (2012) Bayesian phylogenetics with BEAUti and the BEAST 1.7. Mol Biol Evol 29:1969–1973
Edwards SV (2009) Is a new and general theory of molecular systematics emerging? Evolution 63:1–19
Edwards SV, Beerli P (2000) Perspective: gene divergence, population divergence, and the variance in coalescence time in phylogeographic studies. Evolution 54:1839–1854
Engstrom TN, Shaffer HB, McCord WP (2004) Multiple data sets, high homoplasy, and the phylogeny of softshell turtles (Testudines: Trionychidae). Syst Biol 53:693–710
European Commission (2014) The habitats directive. Available via http://ec.europa.eu/environment/nature/legislation/habitatsdirective. Accessed 3 Jul 2014
Faircloth BC, McCormack JE, Crawford NG, et al. (2012) Ultraconserved elements anchor thousands of genetic markers spanning multiple evolutionary timescales. Syst Biol 1–35
Felsenstein J (1985) Phylogenies and the comparative method. Am Nat 125:1–15
Fenton MB, Acharya L, Audet D et al (1992) Phyllostomid bats (Chiroptera: Phyllostomidae) as indicators of habitat disruption in the Neotropics. Biotropica 24:440–446
Fisher DO, Owens IPF (2004) The comparative method in conservation biology. TREE 19:391–398
Fitzjohn RG, Maddison WP, Otto SP (2009) Estimating trait-dependent speciation and extinction rates from incompletely resolved phylogenies. Syst Biol 58:595–611
Fjeldså J (2000) The relevance of systematics in choosing priority areas for global conservation. Environ Cons 27:67–75
Forero-Medina G, Vieira MV, de Grelle CEV, Almeida PJ (2009) Body size and extinction risk in Brazilian carnivores. Biota Neotropica 9:45–50
Francis CM (2008) A guide to the Mammals of Southeast Asia. Princeton University Press, Princeton
Francis CM, Borisenko AV, Ivanova NV et al (2010) The role of DNA barcodes in understanding and conservation of mammal diversity in Southeast Asia. PLoS ONE 5:e12575
Frankham R (2005) Genetics and extinction. Biol Cons 126:131–140
Fujita MS, Tuttle MD (1991) Flying foxes (Chiroptera: Pteropodidae): threatened animals of key ecological and economic importance. Cons Biol 5:455–463
Galtier N, Nabholz B, Glémin S, Hurst GDD (2009) Mitochondrial DNA as a marker of molecular diversity: a reappraisal. Mol Ecol 18:4541–4550
Gilbert MTP, Tomsho LP, Rendulic S et al (2007) Whole-genome shotgun sequencing of mitochondria from ancient hair shafts. Science 317:1927–1930
Gippoliti S, Groves CP (2012) "Taxonomic inflation" in the historical context of mammalogy and conservation. Hystric It J Mamm 23:8–11

Goodman SM, Maminirina CP, Weyeneth N et al (2009) The use of molecular and morphological characters to resolve the taxonomic identity of cryptic species: the case of *Miniopterus manavi* (Chiroptera, Miniopteridae). Zool Scr 38:339–363
Groves C, Grubb P (2011) Ungulate taxonomy. Johns Hopkins University Press, Baltimore
Gutiérrez EE, Helgen KM (2013) Oudated taxonomy blocks conservation. Nature 495:314
Habel JC, Assmann T (2010) Relict species: phylogeography and conservation biology, Springer
Hand S, Archer M, Godthelp H (2001) New Miocene *Icarops* material (Microchiroptera: Mystacinidae) from Australia, with a revised diagnosis of the genus. Mem Australas Palaeontol 25:139–146
Hermisson J, Pennings PS (2005) Soft sweeps: molecular population genetics of adaptation from standing genetic variation. Genetics 169:2335–2352
Hill JE (1974) A new family, genus and species of bat (Mammalia: Chiroptera) from Thailand. Bull Br Mus Nat Hist 27:301–336
Hill JE, Harrison DL (1987) The baculum in the Vespertilioninae (Chiroptera: Vespertilionidae) with a systematic review, a synopsis of *Pipistrellus* and *Eptesicus* and the descriptions of a new genus and subgenus. Bull Br Mus Nat Hist 52:225–305
Ho SYW, Kolokotronis S-O, Allaby RG (2007) Elevated substitution rates estimated from ancient DNA sequences. Biol Lett 3:702–705
Hoagland KE (1996) The taxonomic impediment and the Convention of Biodiversity. Assoc Syst Coll Newsl 24(61–62):66–67
Honacki JH, Kinman KE, Koeppl JW (1982) Mammal species of the world: a taxonomic and geographic reference, 1st edn. Alan Press, Lawrence
Hopkins GW, Freckleton RP (2002) Declines in the numbers of amateur and professional taxonomists: Implications for conservation. Anim Cons 5:245–249
Hulva P, Fornůsková A, Chudárková A et al (2010) Mechanisms of radiation in a bat group from the genus *Pipistrellus* inferred by phylogeography, demography and population genetics. Mol Ecol 19:5417–5431
Hulva P, Horácek I, Strelkov PP, Benda P (2004) Molecular architecture of *Pipistrellus pipistrellus*/*Pipistrellus pygmaeus* complex (Chiroptera: Vespertilionidae): further cryptic species and mediterranean origin of the divergence. Mol Phylogenet Evol 32:1023–1035
ICZN (2012) International code of zoological nomenclature. Available via http://iczn.org/iczn/index.jsp. Accessed 3 Jul 2014
IPCC (2012) Summary for policymakers. In: Field CB, Barros V, Stocker TF et al (eds) Managing the risks of extreme events and disasters to advance climate change adaptation. Cambridge University Press, Cambridge, pp 1–9
Isaac NJB, Jones KE, Gittleman JL, Purvis A (2005) Correlates of species richness in mammals: body size, life history, and ecology. Am Nat 165:600–607
Isaac NJB, Mallet J, Mace GM (2004) Taxonomic inflation: its influence on macroecology and conservation. TREE 19:464–469
IUCN (2014) The IUCN red list of threatened species. Version 2014.1. Available via http://www.iucnredlist.org. Accessed 3 Jul 2014
IUCN (2012) Guidelines for application of IUCN red list criteria at regional and national levels 4.0. IUCN, Gland. Available via http://www.iucnredlist.org/technical-documents/categories-and-criteria. Accessed 3 Jul 2014
IUCN (1980) World conservation strategy: living resource conservation for sustainable development. IUCN-UNEP-WWF, Gland. Available via https://portals.iucn.org/library/efiles/documents/WCS-004.pdf. Accessed 30 Dec 2013
Janzen DH (1991) How to save tropical biodiversity: the national biodiversity institute of Costa Rica. Am Ent 37:159–171
Jones G, Jacobs DS, Kunz TH et al (2009) Carpe noctem: the importance of bats as bioindicators. End Spec Res 8:93–115
Jones G, van Parijs SM (1993) Bimodal echolocation in pipistrelle bats: are cryptic species present? Proc Roy Soc Lond B 251:119–125

Jones KE, Purvis A, Gittleman JL (2003) Biological correlates of extinction risk in bats. Am Nat 161:601–614

Justice Department, Fish and Wildlife Service, National Oceanic and Atmospheric Administration (1996) Policy regarding the recognition of distinct vertebrate population. Fed Reg 61:4721–4725. Available via https://www.federalregister.gov/articles/1996/02/07/96-2639/policy-regarding-the-recognition-of-distinct-vertebrate-population-segments-under-the-endangered. Accessed 30 Dec 2013

Kalka MB, Smith AR, Kalko EKV (2008) Bats limit arthopods and herbivory in a tropical forest. Science 320:2008

Kennedy M, Paterson AM, Morales JC et al (1999) The long and short of it: branch lengths and the problem of placing the New Zealand short-tailed bat, *Mystacina*. Mol Phylogenet Evol 13:405–416

Kingston T (2010) Research priorities for bat conservation in Southeast Asia: a consensus approach. Biodivers Conserv 19:471–484

Kingston T, Francis CM, Akbar Z, Kunz TH (2003) Species richness in an insectivorous bat assemblage from Malaysia. J Trop Ecol 19:67–79

Kingston T, Lara MC, Jones G et al (2001) Acoustic divergence in two cryptic *Hipposideros* species: a role for social selection? Proc Roy Soc Lond B 268:1381–1386

Koopman K (1973) Systematics of Indo-Australian *Pipistrellus* Period. Biol 75:113–116

Koopman K (1993) Order Chiroptera. In: WIlson DE, Reeder DM (eds) Mammal species of the world: a taxonomic and geographic reference, 2nd edn. Smithsonian Institution Press, Washington, DC, pp 137–241

Krell F-T (2004) Parataxonomy versus taxonomy in biodiversity studies—pitfalls and applicability of "morphospecies" sorting. Biodivers Conserv 21:795–812

Lacy RC (1997) Importance of genetic variation to the viability of mammalian populations. J Mamm 78:320–335

Lambeck RJ (1997) Focal species: a multi-species umbrella for nature conservation. Cons Biol 11:849–856

Lane DJW, Kingston T, Lee BP-H (2006) Dramatic decline in bat species richness in Singapore, with implications for Southeast Asia. Biol Cons 131:584–593

Larsen PA, Marchán-Rivadeneira MR, Baker RJ (2010) Taxonomic status of Andersen's fruit-eating bat (*Artibeus jamaicensis aequatorialis*) and revised classification of *Artibeus* (Chiroptera: Phyllostomidae). Zootaxa 2648:45–60

Larson S, Jameson R, Etnier M et al (2002) Loss of genetic diversity in sea otters (*Enhydra lutris*) associated with the fur trade of the 18th and 19th centuries. Mol Ecol 11:1899–1903

Lausen CL, Delisle I, Barclay RMR, Strobeck C (2008) Beyond mtDNA: nuclear gene flow suggests taxonomic oversplitting in the little brown bat (*Myotis lucifugus*). Can J Zool 86:700–713

Lee MSY (2005) Choosing reference taxa in phylogenetic nomenclature. Zool Scr 34:329–331

Lemmon AR, Emme S, Lemmon EM (2012) Anchored hybrid enrichment for massively high-throughput phylogenomics. Syst Biol 61:727–744

Lemmon AR, Lemmon EM (2012) High-throughput identification of informative nuclear loci for shallow-scale phylogenetics and phylogeography. Syst Biol 61:745–761

Lohse K et al (2009) Can mtDNA barcodes be used to delimit species? A response to Pons et al. (2006). Syst Biol 58:439–442

Lumsden L (2009) The extinction of the christmas Island pipistrelle. Aust Bat Soc Newsl 33:21–25

Lumsden L, Schulz M, Ashton R, Middleton D (2007) Investigation of threats to the christmas Island pipistrelle. A report to the Department of the Environment and Water Resources

Mace GM (2004) The role of taxonomy in species conservation. Proc Roy Soc Lond B 359:711–719

Maddison WP (1997) Gene trees in species tree. Syst Biol 46:523–536

Martin TG, Nally S, Burbidge AA et al (2012) Acting fast helps avoid extinction. Cons Lett 5:274–280

Mason VC, Li G, Helgen KM, Murphy WJ (2011) Efficient cross-species capture hybridization and next-generation sequencing of mitochondrial genomes from noninvasively sampled museum specimens. Genome Res 21:1695–1704
Mayer F, Dietz C, Kiefer A (2007) Molecular species identification boosts bat diversity. Frontiers in Zoology 4:4
McConkey KR, Drake DR (2006) Flying foxes cease to function as seed dispersers long before they become rare. Ecology 87:271–276
McNeely JA, Miller KR, Reid WV et al (1990) Conserving the world's biological diversity. International Union for Conservation of Nature and Natural Resources/World Resources Institute/Conservation International/World Wildlife Fund/U.S. World Bank, Gland
Medellín RA, Equihua M, Amin MA (2000) Bat diversity and abundance as indicators of disturbance in neotropical rainforests. Cons Biol 14:1666–1675
Menotti-Raymond M, O'Brien SJ (1993) Dating the genetic bottleneck of the African cheetah. PNAS 90:3172–3176
Miralles A, Vences M (2013) New metrics for comparison of taxonomies reveal striking discrepancies among species delimitation methods in *Madascincus* lizards. PLoS ONE 8:e68242
Monadjem A, Richards L, Taylor PJ, Stoffberg S (2013) High diversity of pipistrelloid bats (Vespertilionidae: *Hypsugo*, *Neoromicia*, and *Pipistrellus*) in a West African rainforest with the description of a new species. Zool J Linn Soc 167:191–207
Moritz C (1994) Defining "evolutionarily significant units". TREE 9:373–375
Myers N, Mittermeier RA, Mittermeier CG et al (2000) Biodiversity hotspots for conservation priorities. Nature 403:853–858
Navascués M, Depaulis F, Emerson BC (2010) Combining contemporary and ancient DNA in population genetic and phylogeographical studies. Mol Ecol 10:760–772
Newton I (2003) The speciation and biogeography of birds. Academic Press, Amsterdam
Noss RF (1990) Indicators for monitoring biodiversity: a hierarchical approach. Cons Biol 4:355–364
Oliver I, Beattie AJ (1993) A possible method for the rapid assessment of biodiversity. Cons Biol 7:562–568
Pennock DS, Dimmick WW (1997) Critique of the evolutionarily significant unit as a definition for "distinct population segments" under the U.S. Endangered Species Act. Cons Biol 11:611–619
Pereira MJR, Rebelo H, Teeling EC et al (2007) Status of the world's smallest mammal, the bumble-bee bat *Craseonycteris thonglongyai*, in Myanmar. Oryx 40:456
Pfenninger M, Schwenk K (2007) Cryptic animal species are homogeneously distributed among taxa and biogeographical regions. BMC Evol Biol 7:121
Pickrell JK, Coop G, Novembre J et al (2009) Signals of recent positive selection in a worldwide sample of human populations. Genome Res 19:826–837
Pollock DD, Zwickl DJ, McGuire JA, Hillis DM (2002) Increased taxon sampling is advantageous for phylogenetic inference. Syst Biol 51:664–671
Possingham HP, Andelman SJ, Burgman MA et al (2002) Limits to the use of threatened species list. TREE 17:503–507
Puechmaille SJ, Gouilh MA, Piyapan P et al (2011) The evolution of sensory divergence in the context of limited gene flow in the bumblebee bat. Nat Commun 2:573
Purvis A, Gittleman JL, Cowlishaw G, Mace GM (2000) Predicting extinction risk in declining species. Proc Roy Soc Lond B 267:1947–1952
Rabosky DL (2010) Extinction rates should not be estimated from molecular phylogenies. Evolution 64:1816–1824
Reeder DM, Helgen KM, Wilson DE (2007) Global trends and biases in new mammal species discoveries. Occ Pap Mus Texas Tech Univ 269:1–35
Reid F (2009) A field guide to the Mammals of central America and South Mexico, 2nd edn. Oxford University Press, Oxford

Ricklefs RE (2007) Estimating diversification rates from phylogenetic information. TREE 22:601–610

Roberge J-M, Angelstam P (2004) Usefulness of the umbrella species concept as a conservation tool. Cons Biol 18:76–85

Rohlf DJ (1991) Six biological reasons why the endangered species act doesn't work—and what to do about it. Cons Biol 5:273–282

Ryder OA (1986) Species conservation and systematics: the dilemma of subspecies. TREE 1:9–10

Salicini I, Ibáñez C, Juste J (2011) Multilocus phylogeny and species delimitation within the Natterer's bat species complex in the Western Palearctic. Mol Phylogenet Evol 61:888–898

Schuh RT (2000) Biological systematics: principles and applications. Cornell University Press, Ithaca

Schwartz MK, Aubry KB, Mckelvey KS et al (2007) Inferring geographic isolation of wolverines in California using historical DNA. J Wildl Manage 71:2170

Secretariat of the Convention on Biological Diversity (2008) Guide to the global taxonomy initiative

Secretariat of the Convention on Biological Diversity (2012) Strategic plan for biodiversity 2011–2020, including aichi biodiversity targets, rio. Available via http://www.cbd.int/sp. Accessed 30 Dec 2013

Secretariat of the Convention on Biological Diversity (2010) Global biodiversity outlook 3. Continuum (Minneapolis, Minn) 19. Available via http://www.cbd.int/gbo3. Accessed 30 Dec 2013

Simmons NB, Voss RS (1998) The mammals of Paracou, French guiana: a neotropical lowland rainforest fauna. Part 1, Bats. Bull Am Mus Nat Hist 237

Simmons NB, Voss RS (2009) Collection, preparation, and fixation of specimens and tissues. In: Kunz TH, Parsons S (eds) Ecological and behavioral methods for the study of bats, 2nd edn. Johns Hopkins University Press, Baltimore, pp 849–864

Soisook P, Karapan S, Satasook C, Bates PJJ (2013) A new species of *Murina* (Mammalia: Chiroptera: Vespertilionidae) from peninsular Thailand. Zootaxa 3746:567–579

Species at Risk Act (2013) Species at risk act, SC 2002, c 29. Available via http://canlii.ca/t/52010. Accessed 4 Jul 2014

Stuessy TS, Thomas KF (1981) Trends, priorities, and needs in systematic biology. Association of Systematics Collections, Lawrence

Taylor MFJ, Suckling KF, Rachlinski JJ (2005) The effectiveness of the endangered species act: a quantitative analysis. Bioscience 55:360

Taylor PJ (2000) Bats of Southern Africa. Identification and Conservation, University of Natal Press, Pietermaritzburg, Guide to their Biology

Taylor PJ, Stoffberg S, Monadjem A et al (2012) Four new bat species (*Rhinolophus hildebrandtii* complex) reflect plio-pleistocene divergence of dwarfs and giants across an afromontane archipelago. PLoS ONE 7:e41744

Thabah A, Rossiter SJ, Kingston T et al (2006) Genetic divergence and echolocation call frequency in cryptic species of *Hipposideros larvatus* s. l. (Chiroptera: Hipposideridae) from the Indo-Malayan region. Biol J Linn Soc 88:119–130

Tisdell C (1990) Economics and the debate about preservation of species, crop varieties and genetic diversity. Ecol Econ 2:77–90

U.S. Fish and Wildlife Service (2013) Endangered species act. Available via http://www.fws.gov/endangered/laws-policies. Accessed 4 Jul 2014

U.S. Fish and Wildlife Service (2012) U.S. Fish and wildlife service land-based wind energy guidelines. Arlington, Virginia. Available via http://www.fws.gov/windenergy/docs/weg_final.pdf. Accessed 30 Dec 2013

USDA (2012) Indiana bat habitat restoration project scoping notice. Available via http://www.fs.usda.gov/Internet/FSE_DOCUMENTS/stelprdb5357273.pdf. Accessed 30 Dec 2013

Velazco PM, Gregorin R, Voss RS, Simmons NB (2014) Extraordinary local diversity of disk-winged bats with the description of a new species and comments on roosting behavior. Am Mus Nov 1–28

Venu P, Sanjappa M (2011) The impact factor and taxonomy. Curr Sci 101:2011

Verde Arregoitia LD, Blomberg SP, Fisher DO (2013) Phylogenetic correlates of extinction risk in mammals: species in older lineages are not at greater risk. Proc Roy Soc Lond B 280:20131092

vonHoldt BM, Pollinger JP, Lohmueller KE et al (2010) Genome-wide SNP and haplotype analyses reveal a rich history underlying dog domestication. Nature 464:898–902

Voss RS, Emmons LH (1996) Mammalian diversity in neotropical lowland forests: a preliminary assessment. Bull Am Mus Nat Hist:230

Waples R (1998) Evolutionary significant units, distinct population segments, and the endangered species act: reply to Pennock and Dimmick. Publications, Agencies and Staff of the US Department of Commerce Paper 433

Welbergen J, Booth C, Martin J (2014) Killer climate: tens of thousands of flying foxes dead in a day. The Conversation. Available via http://theconversation.com/killer-climate-tens-of-thousands-of-flying-foxes-dead-in-a-day-23227. Accessed 3 Jul 2014

Welbergen JA, Klose SM, Markus N, Eby P (2008) Climate change and the effects of temperature extremes on Australian flying-foxes. Proc Roy Soc Lond B 275:419–425

Weyandt SE, van den Bussche RA, Hamilton MJ, Leslie DM Jr (2005) Unraveling the effects of sex and dispersal: ozark big-eared bat (*Corynorhinus townsendii ingens*) conservation genetics. J Mamm 86:1136–1143

Weyeneth N, Goodman SM, Stanley WT, Ruedi M (2008) The biogeography of *Miniopterus* bats (Chiroptera: Miniopteridae) from the Comoro Archipelago inferred from mitochondrial DNA. Mol Ecol 17:5205–5219

Wheeler QD, Meier R (2000) Species concepts and phylogenetic theory: a debate. Columbia University Press, New York

Wiens JJ (2006) Missing data and the design of phylogenetic analyses. J Biomed Inform 39:34–42

Wiens JJ (2003) Missing data, incomplete taxa, and phylogenetic accuracy. Syst Biol 52:528–538

Wilson DE, Reeder DM (1993) Mammal species of the world: a taxonomic and geographic reference, 2nd edn. Smithsonian Institution Press, Washington DC

Wilson DE, Reeder DM (2005) Mammal species of the world: a taxonomic and geographic reference, 3rd edn. Johns Hopkins University Press, Baltimore

Winston JE (1999) Describing species: practical taxonomic procedure for biologists. Columbia University Press, New York

Woody T (2009) Judge halts wind farm over bats. New York Times Green Blog. Available via http://green.blogs.nytimes.com/2009/12/10/judge-halts-wind-farm-over-bats. Accessed 30 Dec 2013

Yu Y, Degnan JH, Nakhleh L (2012) The probability of a gene tree topology within a phylogenetic network with applications to hybridization detection. PLoS Genet 8:e1002660

Zachos FE, Apollonio M, Bärmann EV et al (2013) Species inflation and taxonomic artefacts—a critical comment on recent trends in mammalian classification. Mamm Biol 78:1–6

Chapter 17
Networking Networks for Global Bat Conservation

Tigga Kingston, Luis Aguirre, Kyle Armstrong, Rob Mies, Paul Racey, Bernal Rodríguez-Herrera and Dave Waldien

Abstract Conservation networks link diverse actors, either individuals or groups, across space and time. Such networks build social capital, enhance coordination, and lead to effective conservation action. Bat conservation can benefit from network approaches because the taxonomic and ecological diversity of bats, coupled with the complexity of the threats they face, necessitates a wide range of expert knowledge to effect conservation. Moreover, many species and issues transcend political boundaries, so conservation frequently requires or benefits from international cooperation. In response, several regional bat conservation networks have arisen in recent years, and we suggest that, with the globalization of threats to

T. Kingston (✉)
Department of Biological Sciences, Texas Tech University, Lubbock, USA
e-mail: tigga.kingston@ttu.edu

L. Aguirre
Centro de Biodiversidad y Genética, Universidad Mayor de San Simón, Cochabamba, Bolivia
e-mail: laguirre@fcyt.umss.edu.bo

K. Armstrong
Australasian Bat Society, Inc., Lindfield, Australia
e-mail: kyle.n.armstrong@gmail.com

K. Armstrong
Australian Centre for Evolutionary Biology and Biodiversity, The University of Adelaide, Adelaide, SA, Australia

K. Armstrong
South Australian Museum, Adelaide, SA, Australia

R. Mies
Organization for Bat Conservation, Bloomfield Hills, USA
e-mail: rmies@batconservation.org

C.C. Voigt and T. Kingston (eds.), *Bats in the Anthropocene: Conservation of Bats in a Changing World*, DOI 10.1007/978-3-319-25220-9_17

bats, there is now a need for a global network to strengthen bat conservation and provide a unified voice for advocacy. To retain regional autonomy and identity, we advocate a global network of the regional networks and develop a roadmap toward such a meta-network using a social network framework. We first review the structure and function of existing networks and then suggest ways in which existing networks might be strengthened. We then discuss how regional gaps in global coverage might best be filled, before suggesting ways in which regional networks might be linked for global coverage.

17.1 Introduction

Individuals have formed groups to address conservation issues for decades, but with the application of network theory to social settings, we can now gain insights on the consequence of the structure of conservation-oriented groups for group function. Networks comprise nodes that are linked together by some form of interaction. In social networks, nodes (or actors) are typically individuals, but they may also be groups or entities in their own right, linked by relationships that typically reflect socially oriented values such as friendship, reputation, altruism, and reciprocity (Fig. 17.1).

Conservation networks link actors involved in conservation activities across space (Guerrero et al. 2013). A network may be specifically formed to address a management objective, or arise organically and informally through stakeholder interactions. Interest in network approaches to conservation and natural resource governance (e.g., Bodin and Prell 2011) has been precipitated by the growing realization that top-down centralized approaches often fail to engage stakeholders, are rarely adaptive to local conditions, and as a consequence often fail to achieve sustainable conservation outputs (Bodin and Crona 2009). Regardless of the specific issue, conservation networks have three implicit objectives: (i) The network builds *social capital* [information, resources, knowledge, connections held by the group (Putnam 2000) or individual actors (Portes 1998)] (Newman and Dale 2007); (ii) the network strengthens relationships among activities in a system such that their common effectiveness is enhanced (*coordination*—Hessels 2013); and (iii) that the increase in social capital and coordination will have *agency* (Newman and Dale

P. Racey
Centre for Ecology and Conservation, University of Exeter in Cornwall, Penryn, UK
c-mail: p.a.raccy@cxctcr.ac.uk

B. Rodríguez-Herrera
Escuela de Biología, Universidad de Costa Rica, San Pedro Montes de Oca, Costa Rica
e-mail: bernal.rodriguez@ucr.ac.cr

D. Waldien
Bat Conservation International, Austin, USA
e-mail: dwaldien@batcon.org

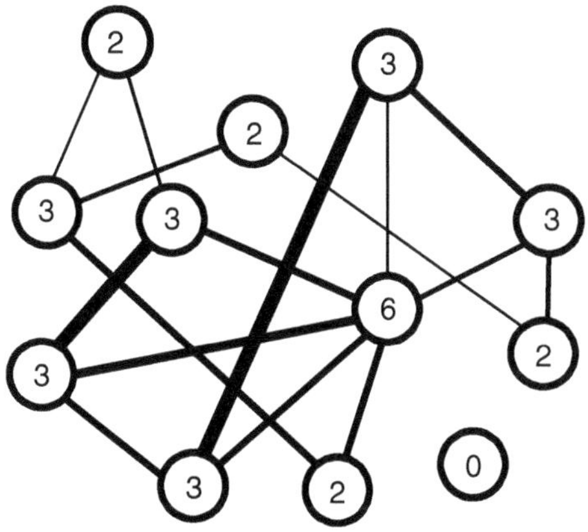

Fig. 17.1 A simple social network. *Circles* are nodes (or actors) connected to one another by links (*straight lines*), also called vectors. Links may be bi- or unidirectional and can be weighted by the strength of the connection between nodes, depicted here by link thickness. Bidirectional links may differ in strength (weight) with direction, for example, if a local coordinator in a bat conservation network commonly sends more information out than she receives, but this has been omitted for clarity. The number of links connected to a node is the degree centrality, shown here within each node. The mean degree for this network is 2.67, and the network density is 0.24 (16/66)

2007), i.e., ability of a group to turn social capital derived from the network into conservation action.

Bat conservation may be facilitated by network approaches for several reasons. First, conservation networks can be particularly effective in dealing with issues operating at multiple spatial and temporal scales and thereby preventing mismatches between the scale at which conservation actions are undertaken and that of the problem (Guerrero et al. 2013). Bat conservation is susceptible to scale mismatches in both space and time. From a geographical perspective, coordinated effort across political boundaries may be required to ensure species' protection across their entire range and to manage migratory species. The Agreement on the Conservation of Populations of European Bats (UNEP EUROBATS), which came into force in 1994, was set up under the Convention on the Conservation of Migratory Species of Wild Animals (CMS), precisely for these reasons. Thirty-five of the 63 range states have acceded to the Agreement, which aims to protect all 52 species of European bats. In the Paleotropics, larger Pteropodidae are known to move across borders [e.g., *Eidolon helvum* (Richter and Cumming 2008), *Pteropus* spp. (Epstein et al. 2009; Breed et al. 2010)], while the continuous north–south latitudinal orientation of the Americas has promoted seasonal migration across borders in several genera (Popa-Lisseanu and Voigt 2009). Stable taxonomy is essential for conservation (Tsang et al. 2015) and similarly may require international cooperation to resolve taxonomic conundrums and test systematic hypotheses of taxa distributed across multiple countries (e.g., Ith et al. 2011). Commercial trade in *Pteropus* spp. for human consumption and traditional medicine has imperiled many species, particularly in the Pacific Islands and western Indian Ocean Islands (Mickleburgh et al. 2009; Mildenstein et al. 2016). Although one *Acerodon* and 10 species of *Pteropus* are listed under Appendix I of Convention on International Trade in Endangered Species of Wild Fauna and Flora (CITES) and the remainder together with *Acerodon* spp. on Appendix II (June 2014), illegal trade will likely continue without coordinated international enforcement among parties.

From a temporal standpoint, because bats are long-lived (Wilkinson and South 2002) decades of observations/monitoring may be required to detect population numbers responding to disturbance or management (Meyer et al. 2010). Moreover, long-term efforts deploying standardized methods across funding cycles and staff turnover require substantial training and coordination. The UK's National Bat Monitoring Programme was established in 1996, but it took a further 15 years of work before statistically robust population trends could be estimated, and then for "only" 10/11 of the UK's 17 breeding species (Barlow et al. 2015). The enormous citizen science effort is spearheaded and coordinated by the Bat Conservation Trust (BCT), a network of 100+ local bat groups. In addition, long-term social or political change may be needed to address particular threats to bats, particularly if the threat is embedded in cultural practices or superstitious beliefs (Kingston 2016).

Second, the social capital and coordination brought by a network approach are important because bats are so diverse taxonomically and ecologically that few practitioners can hold knowledge of more than a handful of species; most researchers are taxonomically or geographically limited. Similarly, varied skill sets are required to garner the basic knowledge that underpins conservation efforts (e.g., taxonomy, ecology, acoustics, genetics and phylogenetics, population monitoring, disease ecology, outreach/engagement, policy), and many issues require an integrative approach to conservation action. Finally, bat research expertise is patchily distributed in many parts of the world, residing in particular institutes within countries, or absent entirely from some countries. Connecting experts through a network accelerates both knowledge transfer among them and the development of capacity in underrepresented areas.

Given the potential for networks to coordinate and strengthen bat conservation, it is not surprising that several bat networks have evolved over the last 25 years. The purpose of this chapter is to review the structure and function of existing bat conservation networks and to discuss the ways in which application of social network theory might strengthen existing networks, facilitate the establishment of new networks, and ultimately guide efforts to link regional networks into a global network of networks.

17.2 Existing Bat Conservation Networks

We focus our review on networks that have conservation as a primary mission and that encompass two or more countries, namely Agreement on the Conservation of Populations of European Bats (UNEP EUROBATS); the Australasian Bat

Society (ABS); Bat Conservation Africa (BCA); BatLife Europe; BCT; Chiroptera Conservation and Information Network of South Asia (CCINSA); North American Bat Conservation Alliance (NABCA); Red Latinoamericana para la Conservación de los Murciélagos (Latin American Bat Conservation Network) (RELCOM); and Southeast Asian Bat Conservation Research Unit (SEABCRU) (Table 17.1,

Table 17.1 Summary information for existing bat conservation networks

Name (acronym)	Agreement on the Conservation of Populations of European Bats (UNEP EUROBATS)
Web presence	Web site: http://www.eurobats.org
Founded	1994
Geographical scope	63 range states (countries) of Europe, North Africa, and the Middle East
Structure	An agreement to which range states (countries) accede and thereby becoming parties. Working group substructure
Membership	35 range states have acceded of a possible 63
Communication	Electronic newsletter, Web presence, annual Meeting of the Advisory Committee (AC), four-yearly Meeting of the Parties to the Agreement. Inter-sessional working groups report to AC, resulting documents published/available on Website
Leadership	EUROBATS is now part of the United Nations Environment Programme and is administered by an executive secretary, with a small administrative staff. An Advisory Committee (AC) comprising invited representatives from range state government departments, Statutory Nature Conservation Organizations (SNCOs), NGOs, and observers meets annually to prepare resolutions for adoption by parties to the Agreement (the signatory governments) who meet every four years
Funding	Member states pay an annual subscription. EUROBATS established the separately funded European Projects Initiative to provide grants of up to 10,000 Euros
Mission and objectives	(1) Exchange information and coordinate international research and monitoring initiatives; (2) arrange the Meetings of the Parties and the Advisory and Standing Committee Meetings; (3) stimulate proposals for improving the effectiveness of the Agreement and attract more countries to participate in and join the Agreement; (4) stimulate public awareness of the threats to European bat species and what can be done at all levels to prevent their numbers dwindling further
Primary activities	(1) The fifteen intersessional working groups produce authoritative reports which help to inform conservation practice. (2) The annual Meetings of the Advisory Committee, in addition to providing valuable opportunities for exchanging ideas about best practice in bat conservation, produce resolutions which are presented to and generally adopted by the four-yearly Meeting of the Parties. An example is the resolution on rabies, the full text of which appears on the Web site, which urged signatories to the Agreement which had not already done so, to introduce surveillance programs. That was successful and several more range states introduced such programs. (3) European Bat Night is an annual awareness-raising activity. (4) The Year of the Bat 2011–2012 was introduced initially as a European Initiative but quickly went global

(continued)

Table 17.1 (continued)

Name (acronym)	Agreement on the Conservation of Populations of European Bats (UNEP EUROBATS)
Major successes	The commitment of 35 European governments to conserve bat populations

Name (acronym)	Australasian Bat Society, Inc (ABS)
Web presence	Web site: http://ausbats.org.au/. Facebook: Australasian Bat Society, e-mail Discussion List, Twitter, YouTube uploads
Founded	1992 (origins 1964)
Geographical scope	Australasia: Australia, New Zealand, Melanesia
Structure	A conservation society with an elected executive team, plus various subcommittees and formalized positions that are created as required
Membership	Researchers, environmental consultants, wildlife rehabilitators, advocates, land managers, naturalists, and educators c. 350 members
Communication	Biennial conference, biannual newsletter, Web presence, quarterly executive meeting, and e-mails (online)
Leadership	Executive committee elected by membership for 2-years term. Comprise President and 2 VPs, Secretary, Treasurer, Editor, Membership Officer. Advisory "extended executive" of past office bearers and helpers. Informal positions—public officer, bat night coordinator, communications officer, social media officer, sponsorship officer, conservation officer, media spokesperson
Funding	Membership subscriptions, conference registrations and sponsorship, advertising in newsletter, account interest, donations, fundraising events
Mission and objectives	*Mission* "To promote the conservation of all populations of all species of bats in Australasia." *Objectives* Encourage membership, disseminate information and outreach materials, advocate for bat conservation and management by advising decision makers, encourage bat research, fund raising, organize biennial conference, build relations and work with other organizations, promote ethical and humane practices in study of bats, support carer and rehabilitation organizations, maintain a public fund for donations
Primary activities	Biennial research conference and workshops, liaising with Local and State Government on issues of bat management and conservation (e.g., flying fox dispersals, bats in mines and bridges, threatened species), produce fact sheets and position statements about bat–human conflict issues (e.g., shooting as control method for flying foxes), media statements on selected issues, survey standards, assist all levels of Government with their information and policy documents, community education events ("Bat Nights" talks and walks)
Major successes	Input to Government policy—Guidance Notes, Action Plans, Conservation Status listings, threatened species survey guidelines. 16 well-attended biennial conferences. 42 editions of newsletter since 1993, plus other similar periodicals since 1964, integration of wildlife carers, significant promotion of bats to the public

Name (acronym)	BatLife Europe
Web presence	Web site: http://batlife-europe.info. Facebook: BatLifeEurope
Founded	2011

(continued)

Table 17.1 (continued)

Name (acronym)	BatLife Europe
Geographical scope	Europe and North Africa
Structure	Country-based network comprising national conservation NGOs ("partner organizations"), usually 1 per country. 33 partners from 30 countries (2013)
Membership	NGOs involved in bat conservation, but not necessarily exclusively so. Membership to NGOs open
Communication	Newsletter, Web presence, triennial conference (European Bat Research Symposium). Trustees meet up to 6× per year online
Leadership	Board of 14 trustees nominated and elected by partner organizations every three years at a meeting of partners at the European Bat Research Symposium. The Board is run by the Chair, with support from the Vice Chair, Secretary, and Treasurer
Funding	Partner NGOs pay an annual subscription or are sponsored by another member. Small grants
Mission and objectives	To promote the conservation of all wild bat species and their habitats throughout Europe, for the benefit of the public. Objectives focus on the following: (1) communication and knowledge sharing; (2) identifying priorities for action; (3) developing projects; and (4) building capacity and international support
Primary activities	Member of the European Habitats Forum seeking to influence European environmental policies, active within the Eurobats Agreement. Disseminates knowledge and experience to build capacity across network (workshops planned). Working on development of a European biodiversity indicator based on bat hibernation surveillance data
Major successes	Bringing together 33 NGO's to form the network. Capacity building survey completed to guide development actions. Contributed to the Pan European Indicator and the European Union Bat Action Plan

Name (acronym)	Bat Conservation Africa (BCA)
Web presence	Web site: http://www.batconafrica.net. Facebook: Bat Conservation Africa Google Listserv: batconafrica@googlegroups.com
Founded	2013
Geographical scope	Africa and the island nations of the western Indian Ocean
Structure	Organized around six regions (southern, eastern, central, western, northern Africa, and western Indian Ocean Islands)
Membership	Individuals joining the list serve, c. 80 members from 25 countries
Communication	List serve and e-mail
Leadership	Steering Committee of representatives from each region, led by a Chair and Vice Chair selected by the Steering Committee. External Advisory Committee to be established
Funding	
Mission and objectives	*Vision* Bats and humans live in harmony in Africa. *Mission* To create a platform for the promotion of bat conservation in Africa. *Objectives* (1) Establish a platform for information sharing; (2) capacity building-skills transfer, education and training, leadership, resources; (3) identify and promote regional conservation priorities; and (4) identify and respond to knowledge gaps on African bats

(continued)

Table 17.1 (continued)

Name (acronym)	Bat Conservation Africa (BCA)
Primary activities	Current emphasis on establishing network operations and lines of communication. Future emphasis on meeting objectives with targeted activities
Major successes	

Name (acronym)	Bat Conservation Trust (BCT)
Web presence	Web site: www.bats.org.uk. Facebook: Bat Conservation Trust. Twitter @_BCT_, LinkedIn Forum
Founded	1990
Geographical scope	England, Wales, Scotland, Northern Ireland (UK)
Structure	Networks c. 100 local Bat Groups
Membership	5600 members including members of the public, volunteers, ecologists and environmental consultants, government workers, academics and teachers
Communication	Newsletters (adult and youth), monthly e-bulletins (general, bat workers, National Monitoring Programme), Web presence, annual national conferences, and separate annual conferences/forums in Scotland and Wales. Regional meetings biennially
Leadership	BCT is a fully constituted NGO and registered charity and must conform to the regulations of the Charity Commissioners in England and Wales and the Office of the Scottish Charity Regulator in Scotland. It is governed by a board of 12 trustees with elected officers. The board appoint the CEO. There are presently 30–35 staff
Funding	Donors, government conservation agencies, charitable trusts and foundations, Heritage Lottery Fund, contracts for service provision (e.g., National Bat Helpline), fees for conferences and training, membership fees, donations from public and major donors
Mission and objectives	*Vision* A world where bats and people thrive together in harmony. *Mission* To secure the future of bats in a changing world. Key objectives that lead work conducted—*Discover* To establish the capacity of the landscape to support viable populations of bats. *Act* To secure and enhance bat populations to the full capacity of the landscape. *Inspire* To win the level of support required to achieve and maintain these bat populations
Primary activities	Monitoring bats, conservation research, landscapes for bats, buildings, development and planning, biodiversity policy and lobbying, training and best practice for professionals, bat crime investigations, education and engagement
Major successes	Establishing and growing the National Bat Monitoring Programme (trends for 10 of UK's 17 breeding species). Lead on Biodiversity Action Plans for bats, which led to targeted advice for buildings industry and woodland managers, and establishment of bat crime investigations, and a training program for professionals whose work affects bats. Public education effectively changed people's attitudes to bats in UK

Name (acronym)	Chiroptera Conservation and Information Network of South Asia (CCINSA)
Web presence	Web page: www.zooreach.org/Networks/Chiroptera/Chiroptera.html

(continued)

Table 17.1 (continued)

Name (acronym)	Chiroptera Conservation and Information Network of South Asia (CCINSA)
Founded	1999
Geographical scope	South Asia (Bangladesh, Bhutan, India, Nepal, Maldives, Pakistan, Sri Lanka and Afghanistan)
Structure	None
Membership	Academic, government, NGO, teachers, volunteers c. 270 members
Communication	Newsletter
Leadership	Founded and run by Sally Walker with help from staff and President invited by her
Funding	Support for workshops from Zoo community, plus other small grants
Mission and objectives	*Mission* To encourage and promote the study of bats of South Asia, by organizing and running a network of bat specialists, and to provide them useful services. *Objectives* (1) To maintain a check list and database of bats; (2) implement a program of bat research training workshops; (3) develop and disseminate outreach materials; and (4) lobbying for the protection of bats
Primary activities	Organizing and conducting workshops on techniques for studying bats, lobbying for specific causes by contacting appropriate governmental departments
Major successes	Development of bat conservation community in S Asia, 9 workshops with 251 participants. Established Pterocount, a program using volunteers to monitor local populations of *Pteropus giganteus*. Successful public education program and dissemination of outreach materials. Successfully lobbied to get two threatened bats moved from Schedule V ("vermin") to Schedule I (absolute protection) of the Indian Wildlife Protection Act 1972

Name (acronym)	North American Bat Conservation Alliance (NABCA)
Web presence	Facebook: North American Bat Conservation Alliance
Founded	1997 as North American Bat Conservation Partnership, 2008 as Alliance, relaunched 2013
Geographical scope	Canada, USA, Mexico
Structure	A federation of working groups and organizations in North America
Membership	Working groups and organizations involved in bat conservation. Membership to working groups open. c. 500 individuals
Communication	Annual open meeting at varied national or international professional meetings (2014 onward), tied biennially to North American Society for Bat Research meeting. Monthly conference calls among organizing committee. List serves with quarterly summaries (planned)
Leadership	Organizing committee comprising representatives form member organizations and working groups. Leadership to rotate between USA, Canada, Mexico
Funding	
Mission and objectives	To promote the conservation of bats in North America by facilitating collaboration, coordinating priorities, and elevating awareness, for the benefits of bats, people, and their ecosystems

(continued)

Table 17.1 (continued)

Name (acronym)	North American Bat Conservation Alliance (NABCA)
Primary activities	Facilitating communication among bat working groups across North America, developing conservation priorities, and assisting the bat community in addressing important issues impacting the conservation of North American bats
Major successes	List of conservation priorities completed. Trilateral agreement to promote cooperation in the conservation of bat populations in North America. Letter of Intent signed by representatives of Environment Canada, secretariat of the Environment and Natural Resources for the United Mexican States, and the Fish and Wildlife Service of the USA, April 2015

Name (acronym)	Red Latinoamericana para la Conservación de los Murciélagos [Latin American Bat Conservation Network) (RELCOM)]
Web presence	Web site: http://www.relcomlatinoamerica.net/. Blog: http://reddemurcielagos.blogspot.com/. Facebook: Relcom Murciélagos. iNaturalist: (http://www.inaturalist.org/projects/murcielagos-de-latinoamerica-y-el-caribe): Groups: Yahoo RELCOM. Twitter: @Relcom
Founded	2007
Geographical scope	Latin America and the Caribbean
Structure	Country-based network constituted by local Programs for Bat Conservation (PCMs), one program per country. 5 countries at foundation, 22 countries currently
Membership	1 PCM per country, but PCM membership open to all interested in bat welfare, large academic membership. c. 800 people
Communication	Quarterly newsletter, Web presence, biennial conference (since 2014), subregional initiatives (e.g., Central and South America)
Leadership	Acting General Coordinator (AGC) elected by 51 % majority of voting members, one from each PCM, during General Assembly. Serves 3 years. AGC appoints a board of directors with individual responsibilities for research, conservation, and education. Board also includes Elected GC and Past GC. Governed by Bylaws approved by General Assembly
Funding	Donors support General Assembly. PCM's generate local funding, apply for national and international academic and conservation grants, sell merchandizing and have membership contributions
Mission and objectives	Guarantee the persistence of healthy bat species and viable populations in Latin America and that in all the countries their importance is acknowledged and recognized. *Research* Promote and stimulate the generation of scientific knowledge that contributes to the conservation of bats and their habitats. *Education and public outreach* Spread the knowledge about bats over the civil society and involve local people in their conservation. *Conservation* Promote the implementation of specific actions and policies aimed at preserving the species and bat populations in Latin America
Primary activities	Promotion and designation of Important Bat Conservation Areas/Sites. Conservation research projects. Task force for rapid response to problems associated with vampire bats and rabies. Public outreach supported by traveling education kit. Capacity building within and outside PCMs

(continued)

Table 17.1 (continued)

Name (acronym)	Red Latinoamericana para la Conservación de los Murciélagos [Latin American Bat Conservation Network) (RELCOM)]
Major successes	Creation and consolidation of Important Bat Conservation Areas/Sites. Publication of action plans for threatened species. Delisting of *Leptonycteris yerbabuenae* reflects the success of conservation action by one of RELCOM associates from Mexico (PCMM)
Further reading	Aguirre et al. (2014)

Name (acronym)	Southeast Asian Bat Conservation Research Unit (SEABCRU)
Web presence	Web site: http://www.seabcru.org. Facebook: Southeast Asian Bat Conservation Research Unit (SEABCRU)
Founded	2007
Geographical scope	SE Asia: Brunei, Cambodia, East Timor, Indonesia, Laos, Malaysia, Myanmar, Philippines, Thailand, Singapore, Vietnam
Structure	Organized around four conservation priorities—flying foxes, cave bats, forest bats, and taxonomy and systematics
Membership	Open to all interested in SE Asian bats, core membership comprises those with research background c. 400
Communication	Web site, Facebook, conferences, workshops
Leadership	Led by Principal Investigator while supported by NSF, with Steering Committee comprising experts in the priority research areas (2–3 per priority) from SE Asia, USA, UK. Steering Committee supported by student teams from USA and SE Asia (3–4 per priority)
Funding	Established with funds from BAT Biodiversity Partnership. 5-years grant from US's National Science Foundation (NSF) as a Research Coordination Network (2011–2016)
Mission and objectives	*Mission* To provide an organizational framework to coordinate and implement research, capacity building, and outreach to promote the conservation of Southeast Asia's diverse but threatened bat fauna. *Objectives* under NSF funding: (1) Effect a regional assessment of the distribution, abundance, and status of SE Asian bats through the implementation of research activities centered on the four priority areas. The SEABCRU network will develop standardized research protocols for each priority and train Southeast Asian bat researchers in the protocols through a series of workshops. (2) Recruit students and researchers to the SEABCRU, engage them in the research priorities, promote effective international communication, and stimulate collaboration
Primary activities	Conferences and expert workshops to develop protocols, training workshops to build capacity across the region. Establish a regional database for bat locality data. Online community of practice
Major successes	Protocols for research rolling out in 2015. 3 international conferences organized, international workshops in Thailand (2012), Cambodia (2013), Myanmar (2014), Vietnam (2014)
Further reading	Kingston (2010), Kingston et al. (2012)

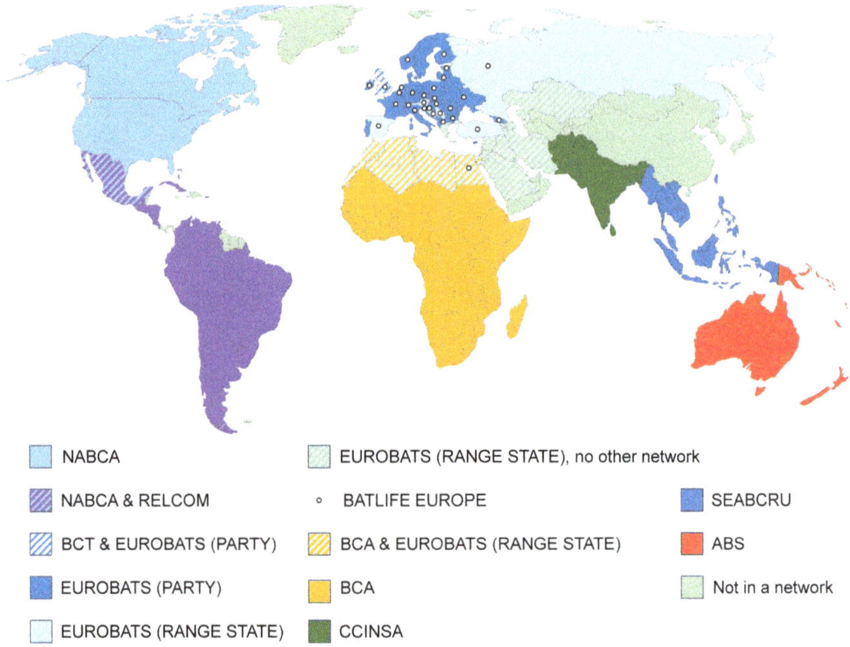

Fig. 17.2 Map of the world with coverage provided by existing bat conservation networks. Countries that are not within a network are filled with *light pink*. Note that some networks require active membership of nations, so countries may fall within the geographic scope of a network but not be members (RELCOM, EUROBATS, BatLife Europe). For networks based on individual membership, geographic scope is illustrated (*BCA*, *CCINSA*, *SEABCRU*, *ABS*). Network acronyms as in Table 17.1

Fig. 17.2). We recognize that there are a growing number of very active national networks (e.g., Asian Bat Research Institute, Bat Association of Taiwan, Bat Study and Conservation Group of Japan, and Indian Bat Conservation Research Unit), as well as NGOs such as Bat Conservation International (BCI) and the Lubee Bat Conservancy, discussed in Racey (2013). The IUCN Bat Specialist Group has a global network structure, but its primary role is to provide member expertise to the IUCN in support of Red List assessments and the development of Action Plans (e.g., Mickleburgh et al. 1992; Hutson et al. 2001). In addition, the North American Society for Bat Research (NASBR) is a large and active network, but the Society's mission is the promotion and development of the scientific study of bats, which it achieves by organizing an annual symposium. Although scientific study extends to conservation and public education, and the society puts forth resolutions on conservation issues, conservation is not the primary focus of the network, so is not included in this review. Together, our focal eight conservation networks unite bat researchers and conservation practitioners in over 130 countries, but major gaps persist and geographical coverage within networks is heterogeneous. Despite active national groups in Japan and Taiwan, as a region East Asia lacks coverage, as does Central Asia, the Middle East (although Israel, Jordan, Lebanon, Syria, and Saudi Arabia are included as range states within EUROBATS and BatLife Europe), and much of the Russian Federation.

17.2.1 Commonalities of Existing Networks

17.2.1.1 Origins and Activities

Most of the networks were founded as a response to the prevalence and intensity of threats to bat populations, lack of scientific knowledge about bats to support conservation action and changes to public policy, and to combat the contribution of public antipathy or ignorance to bat conservation issues. The common overarching goal in all cases is to halt declines and support sustainable populations. To achieve this goal, common foci or organizational themes are research, education/outreach, and conservation. In regions with few bat researchers, or high variance in expertise, research also encompasses building local academic and sometimes volunteer capacity to implement research, typically through workshops and development and sharing of guidance documents (e.g., CCINSA, RELCOM, SEABCRU, BatLife Europe, EUROBATS).

Most networks see themselves as providing a regional organizational framework, guiding or coordinating local activities, and facilitating transboundary communication and capacity building. They aim to realize broader-scale impacts and identify priorities for action at larger scales (NABCA, SEABCRU, RELCOM, BatLife Europe, EUROBATS). Several networks are also instigating, or already implementing, region-wide initiatives, with particular focus on surveying and monitoring populations (BCT, NABCA, RELCOM, SEABCRU, BatLife Europe), data collation and storage (SEABCRU, BatLife Europe, BCT), and evaluation and priority-setting of species, habitats, and threats (all).

Several networks play a direct role in policy development and implementation. In some cases, individuals or groups representing the network act as advisors to governments, in others the network directly lobbies decision makers. Because of its conspicuous foundation in published science and other scientific activities, the ABS has had a strong advisory role at all levels of Government in Australia, having major input into guidance notes (the information used to assess major development proposals by Government), producing action plans and associated recommendations for Conservation status listing, and survey guidelines for threatened listed species, and making submissions to parliamentary inquiries. As a member of the Wildlife and Countryside Link, BCT regularly contributes to joint responses on bat-relevant issues to government bodies, while EUROBATS is a network of parties to an agreement directly influencing conservation policy, as it pertains to bats, in member states. Networks may also take a more direct lobbying approach. CCINSA has been working for years to move India's fruit bats from Schedule V of the Wildlife Act of India 1972, which defines them as vermin that can be exterminated without legal penalty. Two threatened species were afforded protection (moved to Schedule I), but the influence of the agricultural lobby has kept the remaining 12 species on Schedule V (Singaravelan et al. 2009). RELCOM has been lobbying for the creation and acquisition of legal status of Areas and Sites of Importance for the Conservation of Bats across Latin America (see Sect. 17.4.1) and promoting the implementation of bat conservation action plans.

17.2.1.2 Structure and Membership

Most of the networks exhibit substructure. In many cases, independent subgroups hold membership to the network. These are national Bat Conservation Programs (PCMs) in RELCOM, national conservation NGOs in BatLife Europe, range states in EUROBATS, local bat groups in BCT, and regional working groups in NABCA. Thematic structure is seen in some networks. SEABCRU is organized around four conservation priorities; the ABS has subcommittees addressing flying fox issues, outreach and education, and a small-grants program; EUROBATS has intersessional working groups, reporting on key conservation issues (15 currently); and RELCOM is implementing key strategies organized by subregion (e.g., Central and South America). Individual membership is varied, whereas some networks formed around a core of bat researchers in academic settings (SEABCRU, RELCOM), others have greater representation of members from NGOs (BatLife Europe), Statutory Nature Conservation Organizations/Agencies (SNCOs) and government departments (NABCA, EUROBATS), volunteer members of the public (BCT), or a combination (ABS, BCA). As networks mature, membership tends to diversify. The ABS was founded by bat researchers as a scientific society in 1992 (with an informal origin associated with a research newsletter launched in 1964), but now includes members from universities, government, other conservation societies, and private industry.

17.2.1.3 Challenges to Network Sustainability

By far the greatest challenge to network scope and sustainability is funding. Outside Europe, the networks do not have a paid staff or executive (with the exception of a small staff in CCINSA) and are run by volunteers. While volunteer origins and membership often confer network strength (Bodin and Crona 2009), time constraints can slow or limit responses to new challenges. Moreover, although several networks have a core of conservation researchers that remains relatively stable, as network activities can to some extent be integrated with their research agenda, there may be high turnover of volunteers involved with local activities (outreach programs, surveys etc.). Maintaining or rebuilding capacity because of volunteer turnover is a challenge, e.g., for PCMs within RELCOM.

Generally, it is a lot easier to attract funding for specific projects and programs than for staff or volunteer compensation, but these projects may be short term and tied to specific areas. Conservation solutions that require long-term monitoring with standardized methodologies (mandatory for statistical inference of success or failure of interventions) often lack "innovation appeal" to referees and funding organizations. Access to core or unrestricted funding which can be used for key strategic work, or to maintain basic network administration, is hard to secure. BCT has managed to grow its unrestricted income through donations, membership, legacies, and community fundraising, with some success, but this takes time and investment, and can be hard to maintain during periods of economic downturn. Ironically, while lack of protective legislation hampers conservation progress for some networks, protective legislation can lead to negative attitudes toward bats in other areas, particularly during recessions when protection of species can be seen as a barrier to economic growth. In addition, perceived "exaggerated" bat protection efforts can lead to reluctance among citizens to admit to the occurrence of bats in their property at all, for fear of losing partial control over their property.

In a social network, links between actors are almost entirely based on forms of communication, so mechanisms for communication (from face-to-face to online contact) are critical for the success of a network, particularly when members are geographically dispersed. All the bat conservation networks have a Web presence for interaction and/or issue newsletters, and many have regular face-to-face meetings, but gaps in communication can cause network stress, particular when node diversity is high (i.e., members come from many different backgrounds and perspectives). Effective communication is critical if network members differ in their position on a key issue. For example, tensions between the core actors in BCT and supporters and volunteers in 2006 over BCT's stance on a government study of rabies in bats generated very strong concerns (Racey et al. 2013). This led to a review and new model of working with volunteers (partner and network agreements, regular meetings and communication) which proved very beneficial.

17.3 What We Can Learn from Theories of Network Structure and Function

17.3.1 Network Structure and Function

Network functioning describes the process by which certain network conditions lead to various network-level outcomes (Provan and Kenis 2008). Network structure influences individual and group agency, that is, the ability of a group to turn social capital derived from the network into conservation action at the network level. Network structure can be thought of as a map of the relationships (links) between the nodes (actors) in the network. Not all actors are connected to each other. Degree centrality measures the number of links an actor has, and betweenness centrality describes the extent to which an actor links actors that are otherwise disconnected (Burt 1992). The distribution of degree and betweenness centrality across the network is used to characterize network-level characteristics such as network density (number of existing ties divided by the number of possible ties—a measure of degree) and network centrality (variability in degree among network members) (Wasserman and Faust 1994). In general, a network with high density (one with many highly connected actors) (e.g., Fig. 17.3a) facilitates rapid transfer of knowledge and development of trust, is resilient to the loss of individual actors, and promotes collective action (Bodin and Crona 2009). High link density would therefore seem to be a desirable network characteristic. However, there can be trade-offs. Very high link density can lead to network homogenization and homophily. In a homogenized network, all nodes share similar knowledge and perspectives, which limits responses to novel problems, decreasing network resilience. Homophily describes the tendency for people to interact with individuals with characteristics similar to themselves, whether by preference or restricted opportunities (McPherson and Smith-Lovin 1987) and can lead to reluctance to interact with dissimilar others, promoting a "them versus us" environment (Newman and Dale 2005, 2007). Homophily can also restrict individual freedom (Portes 1998) and discourage dissenting opinions (Newman and Dale 2007). Homophily consequently hinders innovation by cutting off actors from needed information and imposing social norms that discourage innovation and inhibit links to dissimilar others (bridging ties).

More typically, the degree of individual actors varies quite widely. Centralized networks in which a few individuals are highly connected (Fig. 17.3b) similarly have benefits and costs. Central actors can prioritize and coordinate activities resulting in effective collective action (Sandström and Carlsson 2008), but this is most effective if problems are relatively simple and short-term. Long-term planning and more complex solutions require a more decentralized structure to access different knowledge and expertise more readily (Bodin and Crona 2009). Moreover, high network centrality can leave the network vulnerable to the removal or dysfunctionality of a few central actors, and to asymmetries of influence and power (Ernstson et al. 2008).

Betweenness (linking disconnected actors), also described as bridging (bridging links and bridging actors), is important in several regards. First, bridging links reduce the path lengths (shortest distance between actors) and network diameter (longest distance) and create "small world" networks (Watts 2003) that can lead to the rapid dissemination and penetration of ideas across the network. Second, bridging actors can connect disparate subgroups. The extent to which a network comprises cohesive subgroups is referred to as network cohesion or modularity (Bodin and Crona 2009) (Fig. 17.3d). Subgroups may hold different sets of knowledge and skills that can be vital to the resolution of a complex problem, but this expertise must be integrated across the network through bridging links. If subgroups are poorly connected (Fig. 17.3c), they can tend internally toward homophily and homogenization (Bodin and Crona 2009).

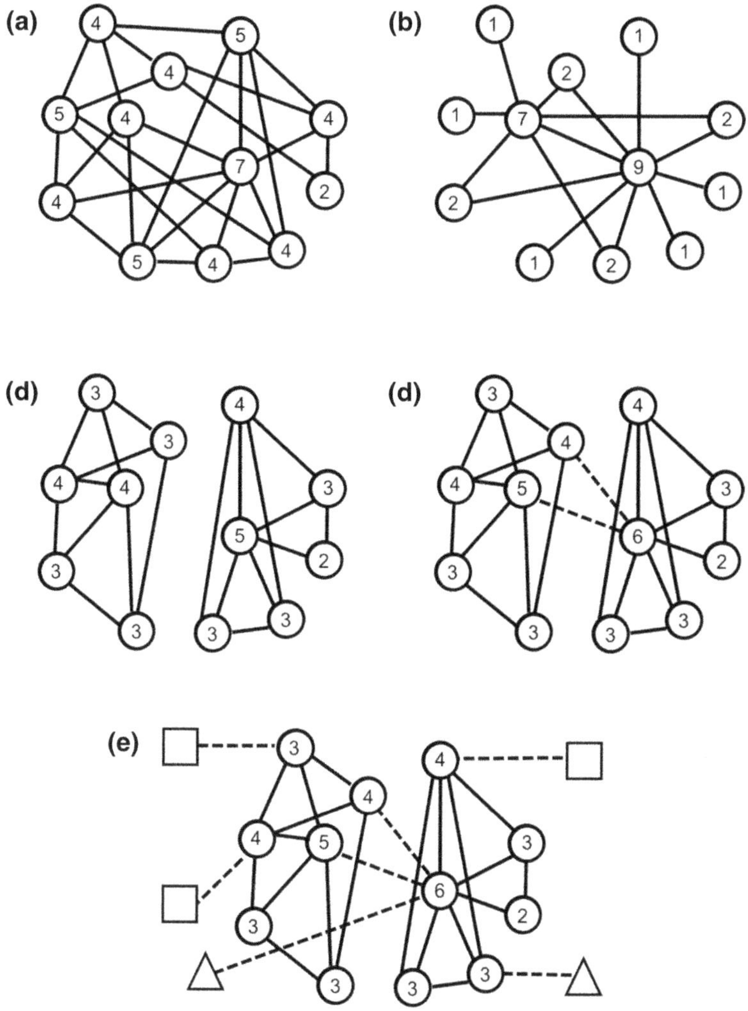

Fig. 17.3 Archetypal network configurations of the social network presented in Fig. 17.1. **a** A highly connected network, with no clear modularity (subgroups) (mean degree 4.33, network density 0.38). **b** A highly centralized network, in which two actors who are highly connected reducing mean degree (2.50) and network density (0.23). **c** Extreme modularity in which the network divided into two isolated subgroups. The subgroups are highly connected or cohesive (mean degree 3.33 and density 0.67). **d**. Network with high modularity with two distinguishable, cohesive subgroups, connected by bridging links (*dashed lines*). **e** Network with high modularity but connected subgroups (**d**) with peripheral ties to actors outside the network (*open squares* and *triangles*)

Just as the distribution of links between actors can vary across the network, the links themselves may vary both qualitatively (type of link) and quantitatively (strength). Links can be a form of communication, a collaboration, an agreement, knowledge, or data transfer. The strength of the link can be suggested by simple frequency counts (number of new joint conservation projects started), or more holistically as suggested by Granovetter (1973): "The strength of a tie is a (probably linear) combination of the amount of time, the emotional intensity, and intimacy (mutual confiding), and the reciprocal services which characterize the tie" (p. 1361). Actors linked by strong (or bonding) ties are more likely to influence one another, promoting mutual learning and sharing of resources but at the price of information redundancy and social "imprisonment" (Borgatti and Foster 2003). Weak or "bridging" ties promote the sharing of diverse information as they are usually between dissimilar others. On one hand, this promotes network resilience and adaptability to change, but on the other hand, these links may be broken more easily.

17.3.2 Structural Characteristics of Effective Conservation Networks: Within Subgroup Cohesion, Across Subgroup Collaboration, Bridging Actors, and Peripheral Actors

Given the trade-offs between network characteristics outlined above, is there such a thing as an "ideal" network structure for effective conservation? Recent reviews (Vance-Borland and Holley 2011; Mills et al. 2014) suggest that polycentric networks in which multiple, heterogeneous subgroups are linked by bridging ties maintain the greatest diversity of response options. Each subgroup has high within-group cohesion so is characterized by dense linkages (high degree centrality, strong or bonding ties) among people sharing specific knowledge that work together productively—enhancing knowledge development (Bodin et al. 2006; Bodin and Crona 2009). Within the network as a whole, there are multiple subgroups, which differ in the knowledge areas and expertise (subgroup diversity—Newman and Dale 2007), developing the diversity of knowledge held by the network as a whole (Bodin et al. 2006; Ernstson et al. 2008; Bodin and Crona 2009; Sandström and Rova 2010). Such functional diversity enhances network adaptability and resilience (Newman and Dale 2007; Mills et al. 2014), cultivates creativity (Aslan et al. 2014) and obviates internal turf battles in large networks (Reuf et al. 2003). Critical to network success are bridging relationships (actors with high betweenness centrality) among the diverse subgroups to promote sharing of expert knowledge and counter tendencies toward subgroup homophily. Network sustainability and adaptability are further enhanced if there are connections to actors outside the network (peripheral actors) who hold specialized knowledge, skills, or resources. Put simply, we can identify four network characteristics indicative of success—within subgroup cohesion, across subgroup collaboration, availability of bridging actors, and inclusion of peripheral actors (Fig. 17.3e).

Network structure tends to evolve through time naturally as the goals of actors change, or the success of actors leads to greater engagement and linking. Structure and transitions can and often should also be managed more actively. For example, while diverse, polycentric networks may be a valid end-goal structure, centralized networks with a few highly motivated actors already connected to many others are good for the initial phase of forming groups (Olsson et al. 2004; Crona and Bodin 2006), and several of the bat networks began with a handful of well-connected actors (ABS, BCT, SEABCRU, and RELCOM). Once the network is more established, managed transitions can increase modularity and long-term decentralization. Moreover, during periods of stability, actors should be provided with opportunities to develop new relational ties with others, which can then be drawn upon in times of change (Olsson et al. 2006). Ideally, rather than simply increasing connectivity among all network members, inspection of network maps and data can be used to implement "network weaving"—the strategic development of new relationships among actors for their mutual benefit and to enhance overall network agency or response to a specific challenge (e.g., a new threat to bats) (Vance-Borland and Holley 2011).

17.4 Toward a Global Network of Networks

17.4.1 Do We Need a Global Network?

A global network of networks can certainly build social capital among bat researchers and conservationists, and facilitate knowledge transfer and capacity building. Moreover, the existing networks are diverse, collectively holding knowledge and skills that range from taxonomy to advocacy. Connectivity among networks could rapidly increase functional diversity, resilience, and adaptability of both individual networks and a global network of networks. It could also provide a platform to develop bridging ties to peripheral actors with greater expertise and skills in key areas, notably lobbying and environmental education. Such a meta-network could also provide a venue for discussion of issues at the global level and for explicit requests for assistance with critical issues. This assistance could be in terms of technical or strategic advice, or collaborative projects that combine resources for the common goal. But is there a need for global agency? We suggest that there are several sets of circumstances in which a global network might facilitate conservation efforts.

First, some issues are genuinely of global concern or can benefit from prioritization efforts at the global scale. For example, habitat loss is a global issue, and the use of standardized, objective criteria to identify critical biodiversity areas worldwide can galvanize and support protection efforts, and provide a basis for monitoring. The Important Bird Areas (IBAs) Program, initiated by BirdLife

International over 30 years ago, now comprises a network of over 10,000 IBAs and has had a major impact on the development of protected areas worldwide to ensure sustainable bird populations (BirdLife International 2008). RELCOM recently launched a similar program for bats in Latin America—Areas and Sites of Importance for the Conservation of Bats (Areas or Sitios para la Conservación (AICOMs/SICOMs) (Aguirre and Barquez 2013) and to date have identified 60 Areas and Sites, including 17 binational AICOMs. A coordinated initiative by a global network to develop this program worldwide could reap similar benefits for bat diversity, particularly if the network develops mechanisms to support and monitor protection of the sites after designation. Similarly, global priority-setting at the species level requires coordinated effort. While this remains the remit of the IUCN, problems arise integrating national evaluations with the global effort. Although the IUCN provides guidelines for the application of Red List criteria at regional and national levels (IUCN 2012), the guidelines and criteria are arguably difficult to apply where data are sparse, as is the case for many bat species. This has led to a proliferation of different national methods, even within regions [e.g., Aguirre et al. 2009—Bolivia, Sánchez et al. 2007—Mexico, US Endangered Species Act (ESA 1973, as amended)], which are difficult to integrate within and across regions. A global network could discuss and develop common criteria to establish the conservation status of bats at local and national scales, and provide a clearer link or integration to the global IUCN Red List assessments.

Second, several conservation issues that originated in certain areas are now "going global"—knowledge gained by regional networks could be vital for rapid responses in other parts of the world. For example, the impact of wind energy installations on bat populations has hitherto been of most concern and best studied in North America and Europe (Arnett et al. 2015). However, 103 countries used wind power on a commercial basis in 2013, with the most dynamic markets with highest growth rates in Latin America, eastern Europe, and for the first time Africa (WWEA 2014), drawing many networks into the development of guidelines to minimize bat fatalities. A global network allows for the rapid synthesis and dissemination of expertise and advocacy materials (e.g., white papers/position statements/research summaries of mitigation approaches) to support efforts in areas lacking direct experience of an issue. Similar issues are being (or could be) realized across multiple regions or globally include the role of bats as reservoir hosts in zoonotic infectious diseases (Schneeberger and Voigt 2016), white-nose syndrome (Frick et al. 2015), and hunting of bats (Mildenstein et al. 2016).

Third, a global network secures the diversity of expertise to respond to future threats. It is noteworthy that some of the biggest threats facing bats today were unimagined less than 20 years ago, with no mention in edited volumes (e.g., Kunz and Racey 1998) or action plans (Mickleburgh et al. 1992; Hutson et al. 2001) of mortality at wind installations, white-nose syndrome, or the role of bats in emerging infectious diseases (EIDs) and the attendant consequences for public and government perceptions of bats. We do not know what new threats to bats might emerge in the coming decades, nor whence they might originate. A global network would facilitate coordinated responses and support for regional issues.

Finally, a global network would provide a means for current and emergent critical issues to become widely known and, critically, could act as a single voice to promote bat conservation through global positions on recurrent, widespread issues such as wind installations, habitat loss and the protection of critical sites, EIDs. A unified voice and global position could also be key in local or national issues where governments, resilient to the dogged efforts of the local group, might be swayed by unified international scrutiny or outrage. Many of the regional networks have faced such challenges. For example, in Australia, the ABS is in urgent need of support to keep up with the number and scale of political issues and administrative actions surrounding flying foxes, and it is conceivable that unified global advocacy might have prompted earlier, precautionary, action as the Christmas Island Pipistrelle (*Pipistrellus murrayi*) declined to (presumed) extinction. Some suggestion that international opinion can influence local decisions comes from Mauritius. In 2006, the prime minister of Mauritius was heavily lobbied by British conservationists to void a cull of *Pteropus niger*, planned to placate fruit farmers. The lobbyists' influence is uncertain as the cull went ahead, *but* its success was limited by existing, *observed*, legislation precluding the discharge of firearms after dark.

We believe a global network can play a key role in bat conservation in the coming decades. However, it must retain the personality of each regional network and promote local bat conservation. Based on the effectiveness of polycentric diverse networks outlined above (Sect. 17.3.2), we envisage a global network as a meta-network of regional networks (Table 17.1) linked by bridging ties among members to generate an emergent, but decentralized global network of networks. To reach this end requires that existing regional networks be supported and strengthened, the establishment of new networks in areas of the world currently not covered, and the development of bridging links across regional networks to provide global coverage.

17.4.2 Strengthening Existing Networks

From our review of characteristics of successful conservation networks (Sect. 17.3.2), existing networks might consider activities that increase the number and strength of links among its actors. This increases mean degree, with redundancy improving resilience to member loss (Folke et al. 2005), and greater connectivity facilitating knowledge transfer. Face-to-face events (conferences, workshops, etc.) as well as online social networks (e.g., Facebook) provide for bidirectional communication among actors and an increase in connectivity through establishment and strengthening of social bonds. Although online social ties are often weak (Burke et al. 2010), they may nevertheless cultivate and crystallize otherwise ephemeral relationships established face-to-face (Ellison et al. 2007; Lewis and West 2009).

While organizations may not be in the position to conduct a full social network analysis to guide explicit network weaving (as advocated by Prell et al. 2008, 2009), development can still be strategic. Identifying and connecting or developing "missing nodes" is an important aspect of network strengthening—are there individuals, themes, perspectives, knowledge, and countries missing from the network? Do actors exist but are not connected, or does the network need to encourage the development of new capacity?

Establishing connections to existing actors not currently in a network increases network diversity and hence adaptability, which in turn is central to maintaining social capital (Newman and Dale 2007). In Southeast Asia, Myanmar has had an active bat research community for at least a decade, but for political reasons it has been difficult to connect it to the rest of the SEABCRU, a situation that the SEABCRU has actively sought to rectify with a workshop in 2014, now that political landscape has changed. From a knowledge perspective, early in SEABCRU development it became clear that the network lacked expertise in disease ecology, despite the fact that Southeast Asia is an emerging disease hotspot (Jones et al. 2008), and actively recruited an actor from Ecohealth Alliance to fill that expertise gap. As a network grows, actors with specific management skills needed to run the network may need to be recruited. BCT actively headhunted to achieve a skill mix for the board of trustees that included strategy, organizational development, funding, marketing, legal, financial, HR, bat research, and conservation as well as volunteers perspectives.

In many cases, actors or nodes may not currently exist. Lack of expertise and capacity was one of the driving motivations behind the establishment of CCINSA, a network that has focused much of its efforts on training workshops. The role that this can play in establishing new nodes is illustrated by the growth of activities in Nepal, following a CCINSA workshop in 2007. Participants went on to establish two organizations involved in bat conservation—Small Mammal Conservation and Research Foundation (2009) and Natural Resources Research and Conservation Centre (2010). RELCOM began with representatives from five countries (Brazil, Bolivia, Costa Rica, Guatemala, and Mexico) and grew network membership by actively recruiting key bat conservationists and researchers from across Latin America. In countries lacking expertise (e.g., in Central America), senior leaders from RELCOM actively built capacity through courses and workshops and identified local members needed to fill the gaps in region-wide representation. This approach grew RELCOM from five to 22 countries in just five years, and most of the remaining gaps are being filled by organizations actively petitioning to join.

The SEABCRU five-year plan allocated year three for the identification and filling of gaps in the SEABCRU network. In accord with the SEABCRU's thematic approach, gaps were defined as areas lacking expertise in, but facing, one or more of the four major threats. Activities center on fostering capacity to fill these gaps. These include a flying fox workshop in Cambodia (2013) to train biodiversity researchers in monitoring protocols, dietary studies, bat–farmer conflict resolution, and disease ecology, and a similar workshop focused on cave bat conservation in southern Vietnam (2014).

Filling in network gaps that lack existing actors can be challenging, and several networks have encountered difficulties, despite having identified clear targets. Efforts have generally been hampered by lack of funds to support foundational events (e.g., workshops), lack of suitable liaisons in the target area that can anchor events, and political constraints. Political constraints may be current (countries restricting international relations because of war or ideology), or historical. As an example of the latter, the majority of countries in Central and South America are now members of RELCOM, but the Guianas of northeastern South America have greater, recent European affiliations (comprising French Guiana, an overseas department of France), Guyana (British Guiana until independence in 1966), and Surinam (part of Dutch Guiana until 1975). These countries support high bat diversity, face similar conservation challenges to the rest of the continent, and lack local research capacity, but colonial and immigration history have limited their integration with Latin America, and hence with RELCOM.

Established networks should also work to develop links to other conservation stakeholders (Mills et al. 2014—scale-crossing to peripheral actors; Fig. 17.3e). Obvious "peripheral actors" include those engaged in similar issues (e.g., raptor fatalities at wind installations) or habitats (e.g., RAMSAR wetland groups). Perhaps, the most intuitive and common peripheral actors for bat conservation networks are cave groups. Cave groups have contributed to bat surveys from the Philippines to the USA. The Australian Speleological Federation played a major role in gathering bat knowledge in Australia in the late 1950s, and the legacy of this interaction is embodied in the ABS constitution, which seeks "to establish and maintain links, and work cooperatively, with other organizations within and outside Australia which share similar aims and objectives to the Society." More recently, the ABS became part of the Places You Love alliance of more than 40 green groups in response to pressure to weaken Australian environmental laws and has increased interaction with other smaller bat conservation and wildlife rehabilitation groups in Australia. Similarly, BatLife Europe works with "collaborating organizations," such as local NGOs, museums, and companies, to exchange information and participate in activities.

Networks should be cognizant that, as discussed above, the most effective network structure may change through time. As the network becomes more established and grows, knowledge and responsiveness can be enhanced by transitioning from a centralized structure (Fig. 17.3b) to one with greater modularity (Fig. 17.3d). RELCOM is actively transitioning to a more modular structure through the establishment of subregional groups (Central and South America), while maintaining the strong bonds already established. This structure allows the network to respond more effectively to the issues in each subregion. For example, Central America is in need of greater capacity building, as local PCMs are comprised of very young researchers, whereas expertise is more established in South America. The network is further subdividing South America into the Andes, Amazon, Southern Cone, and Caribbean to reflect the dominant conservation issues: wind turbines and habitat fragmentation in the Andes; habitat destruction in Amazonia; wind turbines in the Southern Cone; and bat migration and roost loss associated with hurricanes in the Caribbean.

As described above (Sect. 17.2.1), most of the bat conservation networks are already modular, comprising subgroups defined geographically or thematically.

Geographical subgroups are likely to be more cohesive initially (as actors within them know each other), but may tend toward homophily over time. In some cases, there may not be sufficient actors to make up a geographic subunit, as was the case with the SEABCRU at its foundation. Thematic groups promote functional diversity of the network as a whole, but it may take time for trust and strong bonds to develop within them. Ultimately, a mix of both is desirable, with members from geographical groups sitting on different thematic teams. This "jigsaw" strategy (Aronson and Patnoe 2011) promotes cooperative learning as expert knowledge developed in thematic groups is returned to the geographical groups. Currently, EUROBATS includes elements of this strategy with intersessional working group members drawn from member states. This strategy also ensures a variety of weak (bridging) and strong (bonding) ties among more actors, and explicit network weaving (Prell et al. 2008, 2009).

Network centrality is further decreased if the leadership structure transitions to a rotational one with elected officers serving for specified terms, as several of the networks do (e.g., RELCOM, ABS, BCA). Rotational leadership also avoids cliques and encourages different viewpoints. Conversely, failure to decentralize leaves the network vulnerable to loss of central actors, homophily, and poor long-term recruitment. Networks should also maintain ongoing recruitment programs to replace people, who leave, and maintain network heterogeneity (Newman and Dale 2007).

17.4.3 Filling Regional Gaps—Establishing New Networks

Major regional gaps include East Asia (covering China, Japan, North Korea, South Korea, Mongolia), Central Asia (Kazakhstan, Kyrgyzstan, Tajikistan, Turkmenistan, Uzbekistan, Afghanistan), the Middle East (18 countries), and the Russian Federation (Fig. 17.2).

The first question, rather similar to that when filling in gaps in existing networks, is to determine whether expertise (possible actors/nodes) already exists and just needs connecting in these regions, or if the area is completely lacking expertise. In East Asia, there are several active national groups, namely the Asian Bat Research Institute, Bat Study and Conservation Group of Japan, and the Bat Association of Taiwan, as well as individual actors in Mongolia and China, which could be the kernels of a regional network. Similarly, the EUROBAT range state members Israel, Jordan, Lebanon, Syria, and Saudi Arabia could serve as nodes in establishing a Middle East network.

A limited number of actors (be they individuals or national groups) should not hinder the development of a network, provided of course the actors can commit to the venture. Rather, based on the general principle that founding networks are most likely to succeed if they are fairly centralized (Olsson et al. 2004; Crona and Bodin 2006), the best approach at foundation is to identify a few actors in the region that are well connected with others (high betweenness), which could be brought together to establish or strengthen links needed to form a network. If a handful of central actors are already connected this is ideal, otherwise it is *essential* to spend time building trust and fostering interpersonal relationships (and skills) before getting into issues (Newman and Dale 2007; Cheruvelil et al. 2014). Many of the existing networks (e.g., BCT, RELCOM, SEABCRU) started with a small group of people that were already connected with strong bonds (positive interactions going back many years). In several cases, the group already had the characteristics of a network (social capital, coordination) with agency directed at a specific task. In the UK, BCT evolved from the Mammal Society Bat Group. In Australia, the ABS was preceded by the Australian Bat Banding Scheme (1960), and a collective effort to produce the first bat identification guide. Core members of what was to become the SEABCRU first came together to organize the 1st SE Asian International Bat Conference (2007). Similarly, RELCOM was created by five existing Bat Conservation Programs during the 15th International Bat Research Conference in Mérida, México (2007). Because these actors also had high betweenness (lots of links to others), they were then able to pull in diverse people to build the network. Conversely, networks may struggle to persist beyond foundation if the founding actors do not have or develop strong ties to one another and/or have low betweenness (few links to others).

The diversity of actors involved during network formation should also be considered. High diversity of members can avoid structural homophily (Prell et al. 2008; Cheruvelil et al. 2014), but there must be sufficient commonality of perspectives and expectations among members to provide cohesive network objectives and to develop and strengthen links. Diversity of actors in terms of age, career stage, and nationality has generally proven productive, and although new networks might begin with a fairly centralized structure, thought can still be given to internal structure and subgroups with inclusion of actors with diverse expertise (e.g., SEABCRU steering committee included specialists in each of the four priority research areas) or from different nationalities (e.g., RELCOM). However, communication (and hence link strength) can falter during network formation when actors come from different institutional backgrounds and hence mandates (e.g., academic, non-governmental, governmental, consultancy). In essence, social capital builds more readily when actors are diverse, but not so diverse that agendas and modes of communication differ. As the network matures, it becomes easier to integrate and capitalize on different perspectives. Whereas several of the younger networks largely comprise members with similar backgrounds (e.g., SEABCRU, RELCOM—academic, NABCA working groups drawn from government agencies, NGOs), older networks, such as the ABS, have broader membership that include representation in universities, government, other conservation societies, and private industry.

Early development of a network's mission and objectives can help establish network identity and guide membership decisions and help actors clarify what it means to be part of the network versus an independent researcher, conservationist, or NGO. Moreover, actors that are expected to play a role in the network need to be included or consulted during the establishment process. Given that most actors in bat conservation networks are volunteers, networks will be more sustainable if actors are not only committed to the overall goals of the network but also see increases in personal social capital that lead to tangible benefits. Identifying objectives that contribute to the core network mission requires collective input, but benefit actors directly can be invaluable. Benefits may accrue to the subgroup (e.g., NGO, PCM), but also to the individual in the form of publications, research proposals, or databases that facilitate their own research or applied conservation objective. For example, the SEABCRU explicitly identified publications that met the network's objectives by synthesizing regional conservation knowledge (Abdul-Aziz et al. 2016; Mildenstein et al. 2016) or resolving multi-national taxonomic concerns (e.g., Ith et al. 2011), and is currently developing a regional echolocation call library for acoustic surveying and monitoring of bat diversity in anthropogenic landscapes. Social capital built through the network can also be mobilized to apply for conservation research funding for collaborative teams from within the network. RELCOM partnered with BCI to offer seed grants for its members, and several PCMs have joined together to conduct research, such as a project on the study of migratory patterns of *Leptonycteris curasoae* (IUCN Red List as Vulnerable), which involves participants from Venezuela, Colombia, Aruba, Bonaire, and Curacao. EUROBATS launched the European Projects Initiatives with maximum grants of 10,000 euros to address urgent site- or species-based conservation issues or to fund training workshops in range states. Priority is given to transboundary projects and those promoting international cooperation between the parties and range states to the Agreement.

Fostering the development of expertise in regions with none, essentially developing sufficient nodes to actually support a network, is a significant challenge. Nonetheless, basic network principles apply, and supporting a few actors who can develop (or have) strong bonds between them and are linked to many others will likely maximize success. Broad initiatives to identify enthusiastic, key actors might target vertebrate biodiversity specialists, as it is relatively easy to transfer bat research techniques and knowledge to bird and small mammal researchers. Interest in bat diversity and conservation in Bangladesh (Group for Conservation and Research on Bats) grew out of projects on bats and EID at veterinary institutes (Nurul Islam pers. comm.), providing another avenue for identifying key actors. Involving interested actors in the activities of existing networks and the global network can expose them to the value of network approaches and suggest organizational modes.

17.4.4 Networking Networks for Global Coverage

Our vision is of a global network resulting from bridging ties across regional networks. As such, it would be a largely decentralized entity, but overseen by a coordinating committee drawn from the member networks. To foster bridging ties and accelerate exchange of best practice, thematic subgroups could be identified (e.g., research, outreach, policy) and populated with members from each network. Working groups, similar to those of EUROBATS, to address specific issues of global or multi-regional concern would further weave the network together. Such a jigsaw approach would additionally disseminate expertise back to the regional networks.

Other approaches to develop and sustain bridging ties are offered by the network literature. "Board interlocks" (Borgatti and Foster 2003) develop ties among organizations through a member of one organization sitting on the governing body of another. With so many regional networks, this might be a little unwieldy, but initial efforts might focus on the thematic subgroups, with members attending events run by other networks. In some cases, members from one network may lead a training event of another. For example, SEABCRU steering committee member Neil Furey was the key resource person for a 2014 CCINSA workshop in Bangladesh.

Joint ventures (e.g., collaborative conservation projects and joint symposia) and inter-organizational alliances provide access to information and knowledge resources that are difficult to obtain by other means and which improve performance and innovation (Borgatti and Foster 2003). Several regional networks encounter the same conservation issue (e.g., EIDs and increased pressure on declining pteropodids from a variety of factors unite BCA, SEABCRU, CCINSA, ABS; hunting of bats for bushmeat and medicine are concerns for BCA, SEABCRU, ABS) and might benefit from joint-venture approaches or alliances to seek funding for research and conservation action. Global initiatives, such as priority-setting of important areas or sites, would likewise foster bridging ties.

The challenges in establishing and maintaining a global network of networks are essentially those of the regional networks, writ large—limitations on time, resources, communication, and trust. To overcome these constraints, the global network must have a clear identity, mission, and objectives agreed upon by all member networks. Given resource limitations, and the many threats to bats that participant networks deal with within their own regions, member networks must see how involvement benefits not only the global mission but their own. Communication is pivotal to all networks, and at the global scale, there are obvious barriers associated with cultural and linguistic differences, sometimes augmented by insular attitudes. Just as important for communication and expectations is the diversity of the networks themselves; establishing bridging links between networks comprising mostly of researchers and conservation practitioners (RELCOM, SEABCRU), and those made up of NGOs (BatLife Europe), for example, require thought and active fostering of trust among actors. Moreover, clear lines of communication must be established between executives/committees representing societies, and among members at the individual level.

17.5 Recommendations

With the globalization of threats to bats, we recommend the following:

1. The development of a global network of bat researchers and conservationists to respond to such threats and to provide a unified voice for advocacy.
2. That the global network be formed as a federation of regional networks, retaining regional autonomy and identity.

3. The establishment of new networks in regional gaps, specifically East Asia, Central Asia, the Middle East, and the Russian Federation.
4. That existing and planned networks consider social network theory and developing and refining their structure. We recommend that:
   (a) at foundation, networks adopt a centralized structure based around a few well-connected actors;
   (b) as the network matures:
      (i) actively transition to a structure comprising multiple, heterogeneous subgroups differing in knowledge areas and expertise;
      (ii) fill gaps in knowledge, expertise, or geography by developing links with new actors;
      (iii) increase overall membership diversity; and
      (iv) develop ties to peripheral actors with overlapping conservation interests.

Acknowledgements We thank the following sponsors of the networks: Bat Conservation International, Beneficia Family Foundation, British American Tobacco Biodiversity Partnership, The Brown Foundation Inc. of Houston, Disney Worldwide Conservation Fund, The Field Museum, Greater Houston Community Foundation, Island Foundation, National Science Foundation (Grant No. 1051363), Rufford Foundation, USAID, US Forest Service International Program, US Fish & Wildlife Service, Whitley Fund for Nature, WoodTiger Fund. We greatly appreciate input on the structure and functioning of the individual networks from the following: Julia Hanmer, Karen Haysom, Andreas Streit, Sally Walker.

References

Abdul Aziz S, Olival KJ, Bumrungsri S, Richards GC, Racey PA (2016) The conflict between fruit bats and fruit growers: species, legislation and mitigation. In: Voigt CC, Kingston T (eds) Bats in the Anthropocene: conservation of bats in a changing world. Springer International AG, Cham, pp 377–420

Aguirre LF, Barquez (2013) Critical areas for bat conservation: Latin American conservationists build a grand strategy. Bats 31:10–12

Aguirre LF, Aguayo R, Balderrama J, Cortez C, Tarifa T, van Damme PA et al (2009) El método de evaluación del grado de amenaza (MEGA). Libro rojo de la fauna silvestre de vertebrados de Bolivia. Ministerio de Medio Ambiente y Agua, La Paz Bolivia, pp 7–18

Aguirre LF, Nassar JM, Barquez RM, Medellín RA, Navarro L, Rodríguez-Durán A, Rodríguez-Herrera B (2014) De esfuerzos locales a una iniciativa regional: La Red latinoamericano y del Caribe para. Ecol Boliv 49:45–50

Arnett EB, Baerwald EF, Matthews F, Rodrigues L, Rydell J, Voigt CC (2016) Impacts of wind energy development on bats: a global perspective. In: Voigt CC, Kingston T (eds) Bats in the Anthropocene: conservation of bats in a changing world. Springer International AG, Cham, pp 295–317

Aronson E, Patnoe S (2011) Cooperation in the classroom: the jigsaw method, 3rd edn. Pinter & Martin, London

Aslan CE, Pinksy ML, Ryan ME, Souther S, Terrell KA (2014) Cultivating creativity in conservation science. Conserv Biol 28:345–353

Barlow KE, Briggs PA, Haysom KA, Hutson AM, Lechiara NL, Racey PAC, Walsh AL, Langon SD (2015) Citizen science reveals trends in bat populations: the National Bat Monitoring Programme in great Britain. Biol Conserv 182:14–26

BirdLife International (2008) The impact of Important Bird Area directories. In: Presented as part of the BirdLife state of the world's birds website. Available from http://www.birdlife.org/datazone/sowb/casestudy/219. Checked 7 July 2014

Bodin Ö, Crona BI (2009) The role of social networks in natural resource governance: what relational patterns make a difference? Glob Environ Change 19:366–374

Bodin Ö, Prell C (eds) (2011) Social networks and natural resource management: uncovering the fabric of environmental governance. Cambridge University Press, Cambridge

Bodin Ö, Crona B, Ernstson H (2006) Social networks in natural resource management: what is there to learn from a structural perspective. Ecol Soc 11(2): r2. [Online] http://www.ecologyandsociety.org/vol11/iss2/resp2/

Borgatti SP, Foster PC (2003) The network paradigm in organizational research: a review and typology. J Manage 29:991–1013

Breed AC, Field JE, Smith CS, Edmonston J, Meers J (2010) Bats without borders: long-distance movements and implications for disease risk management. EcoHealth 7:204–212

Burke M, Marlow C, Lento T (2010) Social network activity and social well-being. In: Proceedings of the SIGCHI conference on human factors in computing systems. ACM, New York City

Burt R (1992) Structural holes: the social structure of competition. Harvard University Press, Cambridge

Cheruvelil KS, Soranno PA, Weathers KC, Hanson PC, Goring SJ, Filstrup CT, Read EK (2014) Creating and maintaining high-performing collaborative research teams: the importance of diversity and interpersonal skills. Front Ecol Environ 12:31–38

Crona B, Bodin Ö (2006) What you know is who you know? Communication patterns among resource users as a prerequisite for co-management. Ecol Soc 11(2):7. [Online] http://www.ecologyandsociety.org/vol11/iss2/art7/

Ellison NB, Steinfield C, Lampe C (2007) The benefits of Facebook "friends:" Social capital and college students' use of online social network sites. J Comput-Mediat Commun 12:1143–1168

Epstein JH, Olival KJ, Pulliam JRC, Smith C, Westrum J, Hughes T et al (2009) *Pteropus vampyrus*, a hunted migratory species with a multinational home-range and a need for regional management. J Appl Ecol 46:991–1002

Ernstson H, Sörlin S, Elmqvist T (2008) Social movements and ecosystem services–the role of social network structure in protecting and managing urban green areas in Stockholm. Ecol Soc 13(2): 39. [Online] http://www.ecologyandsociety.org/vol13/iss2/art39/

ESA (1973) US Endangered Species Act of 1973, as amended, Pub L No 93-205, 87 Stat 884 (28 Dec 1973). Available at http://www.fws.gov/endangered/esa-library/pdf/ESAall.pdf

Folke C, Hant Olsson P, Norberg J (2005) Adaptive governance of social-ecological systems. Annu Rev Environ Resour 30:441–473

Frick W, Willis C, Puechmaille S (2016) White-nose syndrome: a threat to north American bats. In: Voigt CC, Kingston T (eds) Bats in the Anthropocene: conservation of bats in a changing world. Springer International AG, Cham, pp 245–259

Granovetter M (1973) The strength of weak ties. Am J Sociol 78:1360–1380

Guerrero AM, McAllister RRJ, Corcoran J, Wilson KA (2013) Scale mismatches, conservation planning and the value of social-network analysis. Conserv Biol 27:35–44

Hessels LK (2013) Coordination in the science system: theoretical framework and a case study of an intermediary organization. Minerva 51:319–339

Hutson AM, Mickleburgh SP, Racey PA (comp) (2001) Microchiropteran bats: global status survey and conservation action plan. IUCN/SSC Chiroptera Specialist Group, IUCN, Gland

Ith S, Soisook P, Bumrungsri S, Kingston T, Peuchmaille SJ, Struebig MJ et al (2011) A taxonomic review of *Rhinolophus coelophyllus* Peters 1867 and *R. shameli* Tate 1943 (Chiroptera: Rhinolophidae) in continental Southeast Asia. Acta Chiropterol 13:41–59

IUCN (2012) Guidelines for application of IUCN Red List criteria at regional and national levels: Version 4.0. IUCN, Gland

Jones KE, Pate NG, Levy MA, Storeygard A, Balk D, Gittleman JL, Daszak P (2008) Global trends in emerging infectious diseases. Nature 451:990–993

Kingston T (2010) Research priorities for bat conservation in Southeast Asia: a consensus approach. Biodivers Conserv 19:471–484

Kingston T (2016) Cute, creepy, or crispy—how values, attitudes and norms shape human behavior toward bats. In: Voigt CC, Kingston T (eds) Bats in the Anthropocene: conservation of bats in a changing world. Springer International AG, Cham, pp 571–588

Kingston T, Juliana S, Nurul-Ain E, Hashim R, Zubaid A (2012) The Malaysian bat conservation research unit: from a national model to an international network. Malays Appl Biol 41:1–10

Kunz TH, Racey PA (1998) Bat biology and conservation. Smithsonian Institution, Washington, DC

Lewis J, West A (2009) 'Friending': London-based undergraduates' experience of Facebook. New Media Soc 11:1209–1229

McPherson JM, Smith-Lovin L (1987) Homophily in voluntary orrganizations: status distance and the composition of face-to-face groups. Am Sociol Rev 52:370–379

Meyer CF, Aguiar L, Aguirre LF, Baumgarten J, Clarke FM, Cosson JF et al (2010) Long-term monitoring of tropical bats for anthropogenic impact assessment: gauging the statistical power to detect population change. Biol Conserv 143:2797–2807

Mickleburgh SP, Hutson AM, Racey PA (1992) Old World fruit bats: an action plan for their conservation. IUCN, Gland

Mickleburgh S, Waylen K, Racey PA (2009) Bats as bushmeat: a global review. Oryx 43:217–234

Mildenstein T, Tanshi I, Racey PA (2016) Exploitation of bats for bushmeat and medicine. In: Voigt CC, Kingston T (eds) Bats in the Anthropocene: conservation of bats in a changing world. Springer International AG, Cham, pp 325–363

Mills M, Álvarez-Romero JG, Vance-Borland K, Cohen P, Pressey RL et al (2014) Linking regional planning and local action: towards using social network analysis in systematic conservation planning. Biol Conserv 169:6–13

Newman L, Dale A (2005). Network structure, diversity, and proactive resilience building: a response to Tompkins and Adger. Ecol Soc 10(1):r2. [Online] http://www.ecologyandsociety.org/vol10/iss1/resp2/

Newman L, Dale A (2007) Homophily and agency: creating effective sustainable development networks. Environ Dev Sustain 9:79–90

Olsson P, Folke C, Berkes F (2004) Adaptive comanagement for building resilience in social—ecological systems. Environ Manage 34:75–90

Olsson P, Gunderson LH, Carpenter SR, Ryan P, Lebel L, Folke C, Holling CS (2006) Shooting the rapids: navigating transitions to adaptive governance of social–ecological systems. Ecol Soc 11(1): r18. [online] http://www.ecologyandsociety.org/vol11/iss1/art18/

Popa-Lisseanu AG, Voigt CC (2009) Bats on the move. J Mammal 90:1283–1289

Portes A (1998) Social capital: its origins and applications in modern sociology. Annu Rev Sociol 24:1–24

Prell C, Hubacek K, Quinn C, Reed M (2008) 'Who's in the network?' When stakeholders influence data analysis. Syst Pract Action Res 21:443–458

Prell C, Hubacek K, Reed M (2009) Stakeholder analysis and social analysis in natural resource management. Soc Nat Resour 22:201–518

Provan KG, Kenis P (2008) Modes of network governance: structure, management, and effectiveness. JPART 18:229–252

Putnam R (2000) Bowling alone: the collapse and revival of American community. Simon & Schuster, New York
Racey PA (2013) Bat conservation: past, present and future. In: Adams RA, Pederson SC (eds) Bat evolution, ecology and conservation. Springer, New York, pp 517–531
Racey PA, Hutson AM, Lina PHC (2013) Bat rabies, public health and European bat conservation. Zoonoses Public Health 60:58–68
Reuf M, Aldrich H, Carter N (2003) The structure of founding teams: homophily, strong ties, and isolation among US entrepreneurs. Am Sociol Rev 68:195–222
Richter HV, Cumming GS (2008) First application of satellite telemetry to track African straw-coloured fruit bat migration. J Zool 275:172–176
Sánchez O, Medellín R, Aldama A, Goettsch B, Soberón J, Tambutti M (2007) Método de evaluación del riesgo de extinción de las especies silvestres en México (MER). Instituto Nacional de Ecología, Mexico
Sandström A, Carlsson L (2008) The performance of policy networks: the relation between network structure and network performance. Policy Stud J 36:497–524
Sandström A, Rova C (2010). Adaptive co-management networks: a comparative analysis of two fishery conservation areas in Sweden. Ecol Soc 15(3): 14. [Online] http://www.ecologyandsociety.org/vol15/iss3/art14/
Schneeberger K, Voigt CC (2016) Zoonotic viruses and conservation of bats. In: Voigt CC, Kingston T (eds) Bats in the Anthropocene: conservation of bats in a changing world. Springer International AG, Cham, pp 263–282
Singaravelan N, Marimuthu G, Racey PA (2009) Do all fruit bats deserve to be listed as vermin in the Indian wildlife (protection) and amended acts—a critical review. Oryx 43:608–613
Tsang SM, Cirranello AL, Bates PJJ, Simmons NB (2016) The roles of taxonomy and systematics in bat conservation. In: Voigt CC, Kingston T (eds) Bats in the Anthropocene: conservation of bats in a changing world. Springer International AG, Cham, pp 503–531
Vance-Borland K, Holley J (2011) Conservation stakeholder network mapping, analysis, and weaving. Conserv Lett 4:278–288
Wasserman S, Faust K (1994) Social network analysis—methods and applications. Cambridge University Press, Cambridge
Watts D (2003) Six degrees: the science of a connected age. WW Norton, New York
Wilkinson GS, South JM (2002) Life history, ecology and longevity in bats. Aging Cell 1:124–131
World Wind Energy Association (WWEA) (2014) 2014 Half-year Report http://www.wwindea.org/webimages/WWEA_half_year_report_2014.pdf

Chapter 18
Cute, Creepy, or Crispy—How Values, Attitudes, and Norms Shape Human Behavior Toward Bats

Tigga Kingston

Abstract Bat populations around the world are declining as a consequence of human activities. Bat conservation thus hinges on changing human behavior, but to do so, we must understand the origins and drivers of the behavior. As natural scientists, most bat biologists lack the knowledge and training to implement rigorous studies of the human dimensions of bat conservation, yet such studies are needed to guide successful intervention. As we travel through the Anthropocene, it is critical that bat conservation biologists adopt an interdisciplinary approach and work with researchers from the social sciences who hold these skills and knowledge. To facilitate conversation and collaboration with conservation social scientists, I review the key theoretical and empirical perspectives on human behavior toward wildlife and report on studies of bats in these contexts wherever possible. I also recommend ways in which bat biologists can use some of this knowledge to enhance less structured or opportunistic outreach efforts encountered during our research activities.

18.1 Introduction

Human activities have wrought such intensive and extensive environmental changes to our planet that we now witness the dawn of the Anthropocene—the human epoch. The Anthropocene is not being kind to bats; populations are declining around the world (Voigt and Kingston 2016) in response to land-use change

T. Kingston (✉)
Department of Biological Sciences, Texas Tech University, Lubbock, TX, USA
e-mail: tigga.kingston@ttu.edu

C.C. Voigt and T. Kingston (eds.), *Bats in the Anthropocene: Conservation of Bats in a Changing World*, DOI 10.1007/978-3-319-25220-9_18

and management practices (Law et al. 2015; Meyer et al. 2015; Korine et al. 2015; Williams-Guillén et al. 2016), urbanization and intensification (Altringham and Kerth 2015; Arnett et al. 2015; Jung and Threlfall 2015; Rowse et al. 2016), disturbance and loss of roosts (Furey and Racey 2015; Law et al. 2015; Voigt et al. 2016), and direct exploitation for bushmeat and medicine (Mildenstein et al. 2016). As human populations grow and encroach on remaining bat habitat, human–bat interactions are increasing, often with negative consequences for both parties through disease relationships (Schneeberger and Voigt 2016), occupation of human dwellings (Voigt et al. 2016), and conflict over fruit crops (Abdul Aziz et al. 2015).

The Anthropocene is named for us, and solutions to our environmental problems rest with us, as Mascia et al. (2003) so concisely put it: "Biodiversity conservation is a human endeavor: initiated by humans, designed by humans, and intended to modify human behavior"—(Mascia et al. 2003, p. 650). Bat conservation is no different from any other aspect of biodiversity conservation in this regard; attempts to reduce the many threats to bats ultimately hinge on changing peoples' behavior (Stern 2000; Ehrlich and Kennedy 2005; Schultz 2011; St John et al. 2013; Veríssimo 2013; Clayton and Myers 2015). "People" may range from bat hunters in rural villages to government officials or politicians in administrative centers, but as stakeholders in the issues surrounding bats, they must be motivated to change their actions and decisions (Menon and Lavigne 2006). How do we determine the stakeholders involved and how do we then change people's minds and behavior? The scientific training of most bat biologists leaves us ill-equipped both practically (St John et al. 2010, 2014) and philosophically (Moon and Blackman 2014) and often extraordinarily naïve, when it comes to dealing with people. Surely, if we share our knowledge and "educate" people, they will change their ways. Hunters in Ghana and Indonesia will be so impressed by the importance of bats as pollinators of their favorite fruit, or so fearful of disease risk, that they will stop hunting them. US politicians will mandate turbine cut-in speeds that reduce bat fatalities once they appreciate the critical role that bats play in the suppression of agricultural insect pests. Home owners will learn to live with their seasonal "attic bats" because they are keeping down the summer mosquito population.

Unfortunately, providing people with environmental knowledge alone is rarely enough to promote conservation behavior, and there is an enormous body of research from the social sciences, primarily from social psychology (St John et al. 2010; Teel et al. 2015), addressing the theoretical constructs behind behavior change. These constructs have provided frameworks for empirical assessments of attitudes and behaviors toward the environment and wildlife, and new disciplines such as human dimensions of wildlife (Manfredo 2008; Decker et al. 2012) and conservation psychology (Clayton and Myers 2015) have arisen in recent years, as a growing numbers of social scientists specialize in environmental or biodiversity conservation. Indeed, the Society for Conservation Biology established a Social Science Working Group in 2003 (http://conbio.org/groups/working-groups/social-science), and a recent report from the Group provides an excellent introduction to the conservation social sciences (Bennett and Roth 2015).

I advocate that if we are to be effective in tackling the human dimensions of bat conservation, we need to work collaboratively with scientists who understand and study people in the same depth that we do our bats! But communication across disciplines requires some measure of reciprocal understanding of the theory and practice of each discipline. The goal of this chapter is to facilitate conversation and collaboration with conservation social scientists. As is clear from Bennet and Roth (2015), there are many fields within the broad realm of conservation social science, but my aim is to introduce bat biologists to the core theoretical constructs behind behavior as applied to conservation and to report on empirical studies of human–bat relationships in these frameworks. Arguably, this task should have been left to a social scientist, but I hope that a natural scientist's perspective of the field may help make it accessible to my fellow bat biologists, who share my training, and avoid bias toward particular world views prevailing within the field. Nonetheless, the basic premise of the chapter is as follows:

> very soon it will be unforgiveable to carry out second-rate social science in conservation, just as now it is unacceptable to use shoddy methods to monitor animal abundance (St John et al. 2013, pp. 357–358)

18.2 Theories of Behavior and Behavioral Change

People make behavioral choices based in large part on their values, attitudes, and to conform to societal expectations and pressures. Although early models of behavior assumed linear relationships in which knowledge influences attitude which in turn influences behavior relating to an issue of concern ("deficit" models—Burgess et al. 1998), this has rarely proved to be the case. Although the correlation between attitudes and behavior (Kraus 1995), including pro-environmental behavior, is quite well supported (Iozzi 1989), the relationship between knowledge and attitudes is complex and support variable (Kellert and Westervelt 1984; Kaiser et al. 1999; Kollmuss and Agyeman 2002; Thompson and Mintzes 2002). Providing people with knowledge about bats and logical arguments about the importance of addressing threats to them does not always change attitudes, and if it does, there is no guarantee that the attitude change will affect behavior toward them.

Psychologists came to appreciate that knowledge is just one of many factors influencing attitudes and recognized that external constraints and/or context (Guagnano et al. 1995; Stern 2000) may further influence changes in behavior. These concepts were encapsulated by the work of Martin Fishbein and Icek Ajzen who first added two factors to the simple linear pathway from attitude to behavior in the theory of reasoned action (TRA) (Fishbein and Ajzen 1975; Ajzen and Fishbein 1980). The theory of reasoned action proposes that the effects of attitudes on behavior are indirect and that there is an intermediate predictor of behavior—behavioral intention. Behavioral intention is not only predicted by attitude but also by subjective norms—the perceived social pressure to perform or not perform the

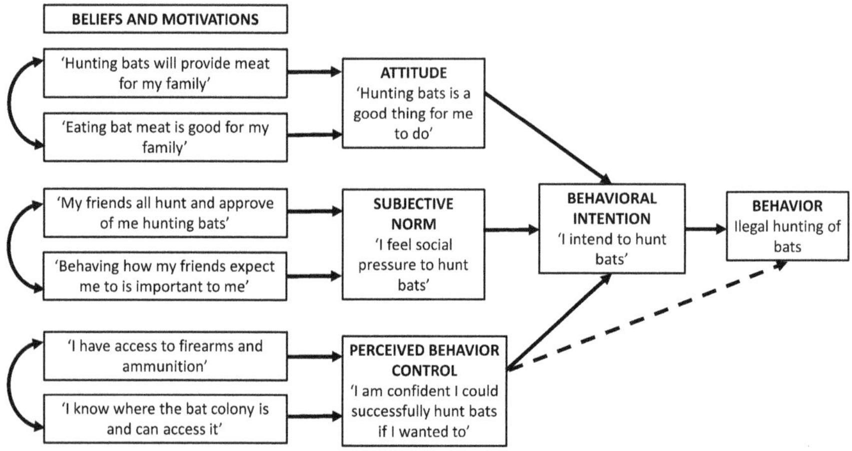

Fig. 18.1 The theory of planned behavior applied to illegal hunting of bats, for example *Pteropus vampyrus* in Sarawak, Malaysia. The strength of the components (attitude, subjective norms, and perceived behavioral control) and the beliefs that underpin them can be measured through interviews or questionnaires that ask respondents their level of agreement with the example statements. This not only provides the overall probability of a behavior, but also identifies the differential influence of the components and thus targets for intervention (adapted from St John et al. 2010, 2014)

behavior. The theory of planned behavior (TPB) added a third factor, perceived behavioral control, to the model. Perceived behavioral control describes whether or not people feel they have the resources, opportunities, or abilities to perform the relevant behavior (Azjen 1991) and can directly influence behavioral intention or the behavior itself (Fig. 18.1). Although the TPB has been a mainstay of attitude-behavior theory and research since its introduction and has received substantial empirical support (Armitage and Conner 2001), the application of the TPB to conservation is more recent (St John et al. 2010, 2013).

A related conceptual framework with a more specific history in environmental conservation and attitudes toward wildlife, particularly in the USA, is the value attitude behavior (VAB) model (Homer and Kahle 1988; Manfredo 2008). The VAB model places values at the base of a cognitive hierarchy of behavior, influencing attitudes and norms through a "value orientation." Values are defined as the set of beliefs held by an individual about what is right and wrong.

The power of these theories for practitioners aiming to induce behavioral change is that the target behavior is broken down into components which may differ in influence (St John et al. 2013) (Fig. 18.1). From a conservation perspective, analysis of the differential influences can help identify the most important barrier to change that can then be the focus of an intervention. So although attitudes are strong predictors of behavioral intention, they are commonly shaped by values, and the significance of subjective norms and perceived behavioral control in the success or otherwise of conservation interventions is becoming increasingly apparent (St John et al. 2013). But just what are these components? What do social scientists mean by values, attitudes, and norms, particularly as they relate to conservation and environmental behaviors?

18.3 Values

18.3.1 Theory

Values are fundamental beliefs about how the world should be, and they express a personal or social preference for an end state of existence or specific mode of conduct (Rokeach 1973). For example, people may value the end states of beauty, peace, wealth, friendship, equality, freedom (Rokeach 1973), and behaviors that can lead to these end states, e.g., self-expression, egalitarianism, belongingness, and humanity toward other living organisms. Values are single beliefs that form slowly in youth over many experiences (Rohan 2000). Consequently, they are stable through time, providing motivational constructs that persist through adulthood (Schwartz 1992), and are thus likely to strongly influence attitudes and guide an individual's processing of information and events.

There is a strong cultural component to values, so values tend to vary less within than they do among different cultures (Kluckhohn and Strodtbeck 1961; Schwartz 1992). Values are thought to be organized into value systems or value orientations (Rokeach 1973), and prioritization of values within these orientations is more individual and appears to explain differences among people in conservation-related attitudes and behaviors within cultures (Teel et al. 2015). Although values of an individual rarely change, they can change across generations as cultural expectations change through time (Manfredo and Teel 2008).

18.3.2 Empirical Values

Given the stability and motivational influence of values, much research has focused on identifying core values or sets of values that influence attitudes toward conservation and wildlife. A central hypothesis guiding this research is that, because of the commonalities of challenges that humans face across cultures, there should be a limited set of universal values (Kluckhohn and Strodtbeck 1961). Kluckhohn and Strodtbeck (1961) identified and tested six dimensions of cultural value orientations, one of which addressed the relationship of individuals and groups with nature. Human–nature relations fell into one of three orientations: mastery, in which humans are seen as superior to nature and have a need and responsibility to attempt to control it; harmony, whereby people work with nature to maintain harmony and balance; and subjugation, in which people cannot and should not exercise control over natural forces but, rather, are subject to the higher power of these forces. The influence of this foundational work persists, with value orientations that affect attitudes and behaviors more specifically toward wildlife variably described as mutualism/harmony/protection orientation versus materialism/domination/mastery/utilization (e.g., Fulton et al. 1996; Manfredo and Teel 2008).

Later influential work by Rokeach (1973) identified at most 36 universal values addressing all aspects of life, but most current conceptual frameworks have their origins in the theoretical structure for life values of Schwartz (1992). Schwartz proposed a typology of ten motivational life value types, comprising 56 value items, clustered along two motivational dimensions: openness to change versus conservation (meaning conservative behavior) and self-enhancement (e.g., materialism, personal ambition) versus self-transcendence (e.g., benevolence, respect for the environment) (Schwartz 1992), and these have proved remarkably consistent across cultures (Schwartz and Sagiv 1995; Schultz et al. 2005). Pro-environmental behaviors tend to correlate positively with self-transcendence values (Stern et al. 1998; Stern 2000).

While values can be hard to influence and change, there has been recent interest in their use in communication strategies intended to motivate conservation behavior (Clayton et al. 2013; Teel et al. 2015). "Deep framing" forges connections between the kind of language used in communication materials and a set of values (Crompton 2010). This approach is central to the "Common Cause" network of NGOs led by WWF-UK (http://valuesandframes.org/) seeking social and environmental change (Crompton 2010). The "Common Cause for Nature" publications comprise a detailed report and a practitioner's guide (Blackmore et al. 2013a, b) commissioned by 13 UK conservation organizations, including the Bat Conservation Trust. The reports focus on the ways in which values can be engaged as part of conservation communication. Schwartz's value topology is adopted, although grouped into "intrinsic" and "extrinsic" motivational clusters, which are broadly equivalent to self-transcendence (self-direction, benevolence, universalism) and self-enhancement (power and achievement), respectively. Blackmore et al. (2013a, b) caution strongly against the use of extrinsic frames that "sell" the conservation issue. They argue that by framing conservation messages in terms of economic or utilitarian value, campaigns appeal to self-interest motivations and may suppress environmental concern. Rather, messaging should appeal to intrinsic values which are more likely to foster environmental concern. This is a pertinent consideration as many bat conservation frames are based on ecosystem services provided by bats, and there are a growing number of studies attaching monetary values to the services (e.g., Cleveland et al. 2006; Wanger et al. 2014).

18.4 Attitudes

18.4.1 Theory

Attitude describes the tendency to think, feel, or act positively or negatively toward objects in our environment (Eagly and Chaiken 1993). Tendency arises because of "an association, in memory, of an evaluation of an object"

(Fazio et al. 1982, p. 341). Whereas values are single beliefs that transcend objects and situations and apply across time, attitudes organize several beliefs around a specific object or situation (Rokeach 1973). In the prevailing multicomponent model of attitude, attitudes are evaluations of an object that comprise three distinct components (Zanna and Rempel 1988; Eagly and Chaiken 1993). The cognitive component encompasses the beliefs and thoughts a person holds about an attitude object and the attributes they associate with it. Whereas bat researchers typically have a positive cognitive response to bats, a member of the public's belief in myths (alternative conceptions) may lead to negative responses and hence attitudes (Prokop and Tunnicliffe 2008; Prokop et al. 2009). The affective component describes the emotions a person feels toward an attitude object. Many people report that bats make them feel scared (e.g., Kahn et al. 2008); they have a negative affective response which can lead to a negative attitude. The behavioral component refers to past behaviors or experiences regarding an attitude object. The multicomponent model of attitude content is informative for educational approaches. As scientists we disdain emotional approaches to research, but this should not bleed into a solely cognitive approach to attitude change. While our training conditions us to address the cognitive component of an attitude, for example by providing information on ecosystem services, or attempting to dispel myths, appealing to affective components and behavioral components may be just as powerful (Pooley and O'Connor 2000) (Sect. 18.4.2.1).

It is also worth noting that an attitude object (bats) may not necessarily hold all three components. For example, a child present at a school visit may hold beliefs about bats and feel positively (or negatively) toward them, but have never encountered them (no behavioral component). Moreover, although associations among components are commonly consistent and even synergist in supporting a particular attitude (Eagly and Chaiken 1993), they can sometimes be inconsistent and even contradictory. This is critical to recognize in the design of conservation messages and interventions. For example, it is possible that someone is aware of and appreciates the ecosystem services that bats provide (positive cognition), but still fears them (negative affect), or has a long history of hunting and eating bats (negative behavior). Thus, appealing to single attitude component will not necessarily lead to a change in attitude, particularly if the other components are stronger. Materials and approaches that are themselves multicomponent may be more effective. For example, the Malaysian Bat Conservation Research Unit produced a comic "Gema's Home" (Benton-Browne and Palmer 2003), a story of an insectivorous bat, Gema (Malay for echo), whose tree roost was being cut down by a local farmer (Mr. Aziz). Gema's distress is palpable, and she appeals to her human friend, a little girl called Nur, for help. Nur and Gema take Mr. Aziz to visit a nectarivorous bat (Polly) and a fruit bat (Fruity), and together they explain the ecosystem services provided by bats and dispel some of the common myths about bats. Mr. Aziz changes his ways and becomes a protector of bats. The cartoon representations and characterizations of the bats are appealing (affective component), and Gema's situation is initially upsetting (affective), but there is explanation of the importance of bats (cognitive). The story is also produced as a shadow puppet show, a traditional performance art in Malaysia, as part of a children's workshop.

Attitudes and attitude components have both valence (positive vs. negative direction of evaluation) and strength. Attitude strength is an important consideration for interventions because strong attitudes are more likely to persist over time, resist change, influence information processing, and predict behavior (Petty and Krosnick 1995; Krosnick and Petty 1995; Holland et al. 2002).

Attitudes are believed to be adaptive, providing a rapid means for processing information and guiding behavior in a complex, data-laden environment and serving four broad functions (Smith et al. 1956; Katz 1960; Maio and Haddock 2014). Awareness of attitude functions is important from a conservation education perspective, because function, like the strength of the components described above, influences susceptibility to attitude change and the kinds of persuasive appeals that might work. First, attitudes can provide an *object-appraisal* function —a summary of the positive and negative attributes of an object to guide how a person should respond to it. Appraisals are commonly based on a utilitarian evaluation—bats provide ecosystem services as agents of pest control, or bats are great bushmeat, but may also derive from a feeling—bats are scary, or bats are cute. Second, attitudes can be used to convey our personal moral values and goals. This *value-expressive* function is related to our self-concept, and, perhaps not surprisingly, attitudes serving this function tend to be central and strong. Attitudes that facilitate relationships with others serve a *social-adjustment* function. Attitudes can also function to protect us against internal conflict (*ego-defensive* or *externalization*) and to defend our self-esteem (for further discussion see Maio and Haddock 2014).

By way of example, let us consider possible attitude functions toward colonies of flying foxes. Attitudes may be based on a utilitarian object-appraisal function in communities who view the bats as a source of bushmeat or income to feed their families (e.g., Kamins et al. 2014). In other communities, such as the Minahasa and Sangir tribes of northern Sulawesi, flying fox consumption may also be associated with a cultural identity (e.g., Sheherazade and Tsang 2015). Now, the attitude function may be value-expressive or social-adjustment. Contrasting attitudes toward the same bats held by biologists may be based on a utilitarian object-appraisal—bats are pollinators and seed dispersers, bats are sources of viruses that may affect human populations, and/or a deeply held belief that bats have a right to exist and not be hunted (value-expressive function). Value-expressive (core moral values and convictions) and object-appraisal functions seem especially predictive of behavior (Fazio 2000) and resistant to change. For example, Kamins et al. (2014) asked Ghanaian bat hunters and vendors what value bats have for people. Four responses were given—no value (14 %), economic value (30 %), meat (30 %), and both meat and money (26 %), reflecting a highly utilitarian object-appraisal function for their attitude toward bats. A subsequent education intervention highlighting the disease risk associated with hunting and butchering bats and the environmental importance of fruit bats had only modest influence—only 45 % of interviewees reported an intention to stop hunting, butchering, or selling bat bushmeat (Kamins et al. 2014).

Because of the adaptive role attitudes play in dealing with the barrage of information we face every day, not only do they influence behavioral intention, but they also influence how we process information about the attitude object. This is important to be aware of in educational or outreach programs. Attitudes influence what information we pay attention to (selective exposure) (Allport 1935; Frey 1986), with preference for information that fits our existing evaluation (Knobloch-Westerwick and Meng 2009); how we evaluate the new information, especially if our existing attitude is strong and hence accessible (Houston and Fazio 1989); and our ability to remember specific information (selective memory) or behaviors. In general, information processing works to minimize cognitive dissonance (Festinger 1957)—the sense of disquiet or mental tension we feel if our behavior or beliefs toward an object are inconsistent. So there is a tendency to select, evaluate, and remember information congruent with our attitudes (otherwise, we have trouble believing in ourselves). In the vernacular, we can think of this as "preaching to the converted" or having our information "fall on deaf ears."

18.4.2 Empirical Attitudes Toward Animals and Factors Affecting Them

The most widely used framework for understanding people's attitudes toward animals (rather than other environmental issues more generally) remains the empirical approach of Stephen Kellert. Kellert led a five-phase report to the US Department of Interior, Fish and Wildlife Service (1977–1983) evaluating the US public's knowledge and attitudes toward animals through surveys of 3945 members of the public (Kellert 1979, 1980; Kellert and Berry 1980; Kellert and Westervelt 1981, 1983). His findings suggested that four major factors influence the US public's attitudes to animals: (i) prior attitude toward, and values of wildlife and nature; (ii) previous experience and knowledge of species or group; (iii) relationship between species and humans, e.g., cultural significance, utility value, or conservation status; and (iv) human perceptions of individual species.

Kellert's work lacked a clear conceptual framework (Manfredo 2008), so below I try to integrate some of the more theoretical perspectives that have since been developed and then examine how we might view these factors from the perspective of bat conservation initiatives and outreach. As detailed below, bats are a mix of good news and bad news when viewed in the context of Kellert's framework.

18.4.2.1 Prior Attitudes and Values of Wildlife and Nature

Kellert developed a typology of attitudes to wildlife (Kellert 1976, 1993, 2002) and identified nine groups, the most common of which were humanistic (primary interest in and strong affection for individual animals, principally pets), moralistic (primary concern for the right and wrong treatment of animals, with strong opposition to exploitation of and cruelty toward animals), utilitarian (primary concern for the practical and material value of animals), and negativistic (primary orientation an active avoidance of animals due to dislike or fear) (Kellert 1980; Kellert and Wilson 1993). In a similar vein, but starting from a theoretical standpoint, Stern and colleagues used Schwartz's work on values (Sect. 18.3.2) to develop a value-based theory of environmental attitudes, describing them as egoistic (reflecting concern about environmental problems for the self), altruistic (concern for the effect on others, such as friends, family, community, and future generations), and biospheric (concern for living things regardless of their value to people) (Stern and Dietz 1994; Stern et al. 1993). Later authors collapse these attitudes to simply anthropocentric (utilization) and biocentric (preservation) (e.g., Milfont and Duckitt 2010). Other motivational frameworks place additional emphasis on the role of emotions (Pooley and O'Connor 2000; Serpell 2004).

Knight (2008) found that people with higher moralistic attitudes report higher levels of support for protection of species (including bats) than those with dominionistic world views. Interestingly, moral reasoning and moralistic attitudes toward animals and nature can develop as early as preschool in children (Kahn 2006), and moral concern and caring can exist alongside a fear orientation (Kahn et al. 2008), the basis for negativistic attitudes. Kahn et al. (2008) interviewed children in four age groups (6–7 years, 9–10 years, 12–13 years, and 15–16 years) as they exited an exhibit of Rodrigues fruit bats (*Pteropus rodricensis*) at Brookfield Zoo (Illinois, USA) and explored caring for bats, fear of bats, and potential moral basis for keeping bats (or not) in captivity. The exhibit presented no barriers between visitors and the bats (glass or mesh), permitting potentially "fearful" encounters as bats swooped by. While just over half the children, especially in the younger age groups, expressed some fear, the same fearful children still cared about bats and the rights of bats. All children gave both anthropocentric and biocentric justifications in response to questions about caring for bats and the rights of bats.

Unfortunately, fear (Prokop et al. 2009) and disgust for bats (Prokop and Tunnicliffe 2008) are widespread. In a study across UK, India, USA, Holland, Korea, Hong Kong, and Japan, bats fell firmly into the "disgust" category falling behind cockroaches, spiders, beetles, maggots, worms, and leeches, and only just beating out wasps, lizards, rats, mice, and slugs (Davey et al. 1998)! Bats are recognized and conceptualized as "bad" animals even among kindergarteners (Kubiatko 2012). Rachman (1977) proposed that fears are learned by children through one of a combination of the following learning pathways: (1) direct conditioning, (2) vicarious learning, and (3) negative verbal information. The power of negative verbal information in engendering fear of *novel* animals has been demonstrated (Field and Lawson 2008) and is especially effective when verbal information comes from a parent (Muris et al. 2010; Remmerswaal et al. 2013). Conversely, there is a reduction in children's fear beliefs when positive information is provided about novel animals (Field and Lawson 2003; Muris et al. 2003; Kelly et al. 2010). However, fear beliefs can be difficult to reverse if they are already well established, rather than invoked toward a novel animal. Williams (2014) sought to reduce fear of bats in US 7- to 9-year-olds with positive verbal information. Although she found a slight change in scores on the Bat Attitude Questionnaire in some children, there were no significant changes in scores on the Fear Belief Questionnaire. Few children will have encountered bats by this age, so direct acquisition of fear through classical conditioning is unlikely. Rather, William's study illustrates just how powerful the indirect negative information coming from the media and culture can be in defining children's fear of bats, and this will be particularly pronounced if conveyed by parents.

If positive information, which is tackling the cognitive component of attitude, is ineffective in changing attitudes, perhaps we would do better to work on the affective component. When an object is paired with an affective sensation, we are tapping into emotion learning, or affective or evaluative conditioning (De Houwer et al. 2001), similar to the classical condition of Pavlov, more familiar to biologists. Although evaluative conditioning is strongest when people have low knowledge about the attitude object, it can still influence attitude change when knowledge or attitudes exist (Olson and Fazio 2006). Bats are frequently paired with scary, negative emotions (e.g., vampires, horror films, haunted houses), so we must work to link positive affect to them. Outreach activities should be fun and participatory: For example, the MBCRU refers to a 3-h children's workshop as the "Malaysian Bat Party" with activities and games that children enjoy. Another approach is to look beyond our rationalist scientific training and promote empathy for bats by levering the universal human tendency to anthropomorphize (project human characteristics onto non-human animals). Anthropomorphism may have been with us since Paleolithic times (Mithen and Boyer 1996), and its use as tool for conservation is receiving growing attention (Tam et al. 2013; Chan 2012; Root-Bernstein et al. 2013). Anthropomorphic bats already prevail in the children's bat literature, led by Jane Cannon's wonderful Stellaluna, although bats in some books lack names and in others look like rodents! From a campaign perspective, probably the earliest example of deliberate anthropomorphic characterization of bats comes from the work of the UK's Mammal Society and the Fauna and Flora Preservation Society to change attitudes to bats when they received full legal protection under Wildlife and Countryside Act in 1981. Artist Guy Troughton deliberately portrayed bats as friendly, fun creatures in books, stickers (Fig. 18.2), mugs, and Christmas cards, and these products were integral to the reversal of public attitudes to bats (Morris 1987).

18.4.2.2 Previous Experience and Knowledge

As nocturnal, volant mammals, people do not experience bats in the way that they might birds and this has consequences for attitudes. Bat sightings are commonly at a distance and fleeting (Sexton and Stewart 2007), while closer encounters may be in a negative or fearful setting, for example as a nuisance in dwellings (Voigt et al. 2016), and/or may prompt fears of disease(Liesener et al. 2006). Bat knowledge is commonly low (e.g., Kingston et al. 2006; Sexton and Stewart 2007; Sheherazade and Tsang 2015) and correlates with attitudes toward bats (Prokop et al. 2009).

Fig. 18.2 Car sticker from the campaign to change the attitudes of the British public to bats c. 1985. Artist Guy Troughton subtly altered the bat to confer greater anthropomorphic appeal (large, soulful eyes and a slightly tremulous smile!) (*Source* Morris 1987)

Moreover, outreach must operate not just from a position of limited or no knowledge, but contend with the abundance of "alternative conceptions" or myths about bats. For example, only 17 % of nearly 200 children (6–16 years) surveyed by Prokop and Tunnicliffe (2008) in Slovakia rejected the idea that bats can tangle in human hair and 36 % asserted that the main diet of bats is blood, a misconception that was still prevalent in undergraduates (Prokop et al. 2009). Not only do alternative conceptions about bats correlate with negative attitudes, but alternative conceptions are depressingly robust and difficult to correct (Mintzes and Wandersee 1998).

It is a rare for a bat biologist to complete a school visit without encountering the "bats lay eggs" question (or assertion!). Viewed through the lens of cognitive psychology, the paucity of knowledge about bats means that many lay people conceptualize them as an exception to the rule "if it has wings, it is a bird." Consequently, they are more likely to use knowledge of the behavior and physiology of birds to reason about bats than they would other mammals (e.g., dogs and hedgehogs) (Davis et al. 2013). Davis et al. asked subjects the likelihood that an internal trait (a protein) and a behavioral trait (a feeding behavior) described in birds or mammals would also be found in bats and dogs. People were significantly more likely to generalize the bird traits to bats than dogs and the mammal traits to dogs than bats. So non-experts automatically assume that the knowledge they have about birds applies to bats and vice versa. This is of conservation concern because bats share little behaviorally or physiologically with birds and respond differently to conservation issues.

18.4.2.3 Relationship Between Species and Humans—Cultural Significance and Utility Value

Conceptions of nature are a social construct created within a historical and cultural context (Clayton and Myers 2015). It is beyond the scope of this chapter to review all the cultural, religious, and symbolic perspectives of bats(see Lawrence

1993), but around the world, bats are commonly associated with aspects of death and sometimes rebirth. For example, bats are believed to be witches in Nigeria (Iroro Tanshi, pers. comm.), spirits of the dead in the Ivory Coast, criminals in Madagascar (Andrade 2009), and souls of the dead searching for rebirth in old Europe. They are deified in Mayan culture, although the bat god Camazotz is thought to represent some kind of giant vampire bat, and is still associated with death, unfortunately. More broadly, while the Bible describes bats as detestable, unclean birds (Leviticus 11:13–20, Deuteronomy 14:11–19), in Shi'a hadith (Nahjul Balagha Sermon 154 or 155 depends on the version), bats are viewed as a testament to "His [Allah's] delicate production, wonderful creation and deep sagacity." Famously, in Chinese, culture bats are viewed as auspicious creatures and symbols of good luck because the word "bat" is a homophone (pronounced the same) of "fortune" in Mandarin Chinese. The Wu Fu, or five lucks, is typically depicted as a ring of five bats signifying the Five Fortunes—longevity, wealth, health and composure, virtue, and a natural death in old age.

On the plus side, bats have great utility to people through the ecosystem services they provide as agents of pest suppression, pollination and seed dispersal, and sources of guano (Kunz et al. 2011). Boyles et al. (2011) estimated that bats may collectively save the US agricultural industry at least $3.7 billion a year by suppressing crop pests, and Wanger et al. (2014) put the value of a single insect-eating species (*Chaerephon plicata*) to rice production in Thailand at over $1 million annually. Such economic evaluations certainly receive substantial press coverage, and it would be interesting to study the influence of this on public attitudes toward bats. Caution is warranted because while featuring ecosystem services can be an effective frame for a campaign, attaching monetary evaluations to wildlife appeals to materialist values which may evoke values and attitudes that are less receptive to conservation (see 18.3.2 above).

18.4.2.4 Human Perceptions of Individual Species

Public support for species' conservation is strongly influenced by human perceptions, predominantly the esthetic appeal of the species (Gunnthorsdottir 2001; Stokes 2007), its similarity to humans (Kellert 1996; Batt 2009), and perceived threat to humans (Knight 2008; Kellert 1996). Unfortunately, to much of the public, bats have little esthetic appeal (Knight 2008) and frequently evoke disgust (Davey et al. 1998; Bjerke and Østdahl 2004) and, despite being mammals, bear very little similarity to humans ("where are its eyes?"). Even well-meaning educational displays may feed rather than extinguish these perceptions, particularly when imagery is at a larger-than-life scale (Fig. 18.3).

Perceptions of the threats of bats to people are becoming a major concern because of the, often alarmist, publicity surrounding their role as reservoir hosts in emerging infectious diseases (Schneeberger and Voigt 2016). This requires careful treatment in education programs because although the likelihood of a bat virus being transmitted to humans is very low, the consequences of infection can

Fig. 18.3 Some portrayals of bats in education settings can have a counterproductive influence on attitudes and perceptions (*Photo* T. Kingston)

be very high, often fatal. In many countries, populations at risk of exposure, such as bat hunters, butchers, consumers, or guano harvesters, have very low rates of risk perception (Harrison et al. 2011; Robertson et al. 2011; Kamins et al. 2014). Educational interventions are needed to reduce target behaviors that increase transmission probabilities, but the challenge from a conservation perspective is to do so without engendering an overall negative attitude toward bats or calls for destruction of populations. Education materials that simultaneously target behaviors and highlight bat ecosystem services are a start (see Appendix 3 of Kamins et al. 2014), but it is unclear how effective these approaches are, and further research on such "mixed messages" is much needed.

18.5 Social Norms

Although the social norm concept has its origins in early twentieth century anthropology and sociology (Hechter and Opp 2001) and was explicitly incorporated into the TRA and TPB (as the subjective norm), recognition of the role of the social context and pressures on people's attitude and behavior toward environmental actions and species protection is more recent (Cialdini et al. 1990; Cialdini 2003; Mascia et al. 2003; Schultz et al. 2007; Goldstein et al. 2008; St John et al. 2013; McDonald et al. 2014).

Social norms are the accepted or implied rules about how members of a social group should, and do, behave (Sherif 1936). Individuals breaking these rules may be sanctioned formally, if the norm is written into law for example, or informally through social disapproval. The more motivated an individual is to identify with a particular social group, the more likely they are to recognize and conform to the group's norms (Deaux 1996; Manfredo 2008), particularly if the norm is central to group identification (Christensen et al. 2004). Social norms are dynamic, and they depend on the person and situation (Ajzen 1971). There are several norm constructs, beginning with the subjective norm of Fishbein and Ajzen (1975) which focuses on beliefs about what important others expect one to do in a given situation. A more operational approach identifies the descriptive norm, which is based on perceptions about what others actually do, and the injunctive norm, perceptions about what others approve of (more akin to the original subjective norm) (Cialdini and Trost 1998). This division is important because appeals in which these conflict can fail to change behavioral intention (Cialdini 2003, McDonald et al. 2014). So if a persuasive appeal is intending to convey disapproval of an action (injunctive norm), but at the same time suggests that many people perform this behavior (descriptive norm), the message is normatively muddled. For example, if a message was to indicate that people should not kill bats (perhaps by hunting, or excluding them from homes) (injunctive norm) but that many people are doing so (descriptive norm), the persuasive appeal is conflicted. If there are high levels of a socially disapproved behavior, it is better to focus on the injunctive norm. Conversely, a descriptive norm approach would be effective in promoting a new behavior, for example building bat houses. In sum, descriptive and injunctive normative messages need to align and whenever possible be used together (Cialdini 2003; Kinzig et al. 2013).

Sociology identifies four basic types of norms: folkways or "customs"; mores—norms of morality including religious doctrines; taboos—behaviors forbidden by culture (which may be enacted into law); and laws—norms that are written down and enforced. The potential of taboos, and the informal institutions that proscribe them, to advance conservation agendas is of growing interest (Colding and Folke 2001), particularly in situations where the influence of external formal institutions is constrained (Jones et al. 2008). Taboos prohibit eating of bats (*Pteropus*) by the Mahafaly and Antandroy people of Madagascar (MacKinnon et al. 2003), while sacred forests provide protection in other parts of the country (Rahaingodrahety et al. 2008). Similarly, sacred groves protect colonies of *Pteropus giganteus* in Tamil Nadu, India (Marimuthu 1988; Tangavelou et al. 2013), and West Kalimantan, Indonesia (Wadley and Colfer 2004). Colonies of *Pteropus* throughout much of Indochina are associated with gardens attached to Buddhist monasteries (pers. obs.), while sacred caves protected by Buddhists provide refuge for diverse insectivorous bat species (e.g., Robinson and Smith 1997). Sacred caves and rocks provide similar protection elsewhere with known examples from Ghana (Hens 2006) and Kenya (Metcalfe et al. 2009).

People do not always adhere to taboos or mores, or practice their nominal religion, especially if there is conflict with utilitarian or cultural use of the animal. For example, although all the Abrahamic religions explicitly prohibit consumption of bats, sales and consumption of flying foxes in North Sulawesi peak around Christian celebrations (Sheherazade and Tsang 2015). Similarly, taboos may not be respected if wildlife resources are scare (Bobo et al. 2015). In addition, bats may be seen as the exception to broader social norms. For example, India's Wildlife Protection Act (1972) schedules bats as vermin, excluding them from

Fig. 18.4 Portrait of *Eidolon helvum*, typical of bat biologists' collections (*Photo* T. Kingston) (*left*) and the author smiling with the same bat (*right*) conveying positive affect that can shape social norms and attitudes toward bats (*Photo* P. Webala)

protection. Nonetheless, appealing to neglected prior norms and taboos may be a point of leverage, but should be done with guidance from local religious/spiritual leaders.

Norms are internalized by three transmission routes (Gintis 2003), vertically from parent to child, obliquely through social institutions (e.g., religion, government, school, media), and horizontally through interactions with peers. Conveyance methods (Cialdini and Trost 1998) include active instruction (stories, myths), passive observation (nonverbal imitation), and inference from behavior around us. Bat researchers can contribute to the oblique transmission route of positive social norms about bats by publicizing their work in the popular scientific press, social media, visiting schools, etc. To be effective, we should be sure to emphasize the wonder of bats, not just our science, and not be afraid to appeal to emotion and anthropomorphic tendencies (Sect. 18.4.2.2). As biologists, when photographing bats we tend to concentrate our efforts on portraiture (head shots), to capture the diversity of bat morphology and diagnostic taxonomic features, or "researchers in action," conveying only a scientific behavioral norm toward bats which often involves trapping of some description. These have their place, but from an outreach perspective intending to lever norms, images of a researcher holding a bat smiling conveys that bats are not a source of fear but happiness (positive affect) and that many people do, and one should, behave positively toward them (Fig. 18.4).

18.6 Assessing Attitudes, Values, and Norms

From the above, it is clear that knowledge of people's values, attitudes, attitude functions, and social norms could be very useful in the design of messages aimed at influencing behavior. A detailed review of methods for measuring these psychosocial constructs is beyond the scope of this chapter, but for conservation purposes, most measures are commonly based on self-reporting (or interviews) through questionnaires with scaled responses. Good questionnaires can gather information on knowledge, values, attitudes, opinions, behaviors, facts, and other information and have been increasingly used in ecology and conservation biology to assess stakeholder opinions and perceptions of and behaviors toward species or issues of conservation concern (White et al. 2005). Although many of us have designed and given questionnaires to or interviewed stakeholders, robust design that can evaluate the values and attitudes behind behaviors and provide an unbiased assessment of the behaviors themselves requires substantial preparation. Central is a solid theoretical understanding of social psychology and psychometrics and design considerations. For example, questionnaires need to quantify and maximize validity (does the questionnaire or "instrument" measure what it intended to) and reliability (does the instrument consistently or accurately measure what it is intended to measure). The procedure for sampling the target population (e.g., random, systematic, comprehensive) needs to be considered as does non-response bias, to name but a few factors.

Measuring behavior through self-reporting or interviews can be particularly tricky if the behavior is illegal or contravenes a social norm (socially disapproved or inappropriate). Respondents may not tell the truth or may skip the question, compromising data validity (King and Bruner 2000). This is key if the prevalence of particular behaviors (such as hunting bats) is the end point of the study and is even more pertinent if the study aims to assess whether attitudes are good predictors of behavior (e.g., St John et al. 2011). Recent applications of sensitive question tools from human health research (e.g., condom use in HIV research) to conservation "rule breaking" provide much higher reporting of illegal activities than conventional approaches (St John et al. 2013; Nuno and St John 2015). If non-sensitive characteristics (attitudes or demographics) can predict sensitive behaviors, then the identification of the target audience for intervention is greatly facilitated (St John et al. 2013).

Scientists are rarely trained in appropriate social science methodologies and indeed may come at human studies from a very different philosophical perspective that can influence our understanding and interpretation of social science data and conclusions (Moon and Blackman 2014). As emphasized in the introduction, we should be collaborating directly with social scientists (Mascia et al. 2003; Sandbrook et al. 2013; St John et al. 2014), but a good introduction to methods is given in Newing (2011).

18.7 Recommendations

There remain few published studies addressing the drivers of human behavior toward bats, yet this is key to their conservation. The primary recommendation for bat biologists directly tackling bat conservation issues is to work with conservation social scientists to fully characterize the human dimension of the problem and identify targets for intervention. Key considerations in the design of interventions and messages are as follows:

- Work to establish the component (value, attitude, norm, perceived behavioral control) of the TPB pathway/VAB framework acting as a barrier to behavioral change. Target the "barrier component" for intervention.
- Recall that many components are structured. Attitudes may be based on cognitive, affective, or behavioral perspectives, and attitude functions serve different roles. If interventions are to resonate with a target audience, it is critical that they not only address the component that is the problem, but that the message matches the content or functions that are the basis for the recipients' attitude. Similarly, social norms can be descriptive or injunctive, and messages should be sure to align with the prevailing norm and avoid conflict between them.
- Work with values and avoid framing messages that appeal to extrinsic or self-enhancement values.
- Remember that much of human behavior is driven by how we feel (affective component of attitudes, social pressure behind norms). Do not be afraid to appeal to emotion and anthropomorphic tendencies in the design of messages and materials.
- Be aware of our influence as scientists on social norms relating to bats and be sure to convey the wonder and positive affect they engender (BATS ARE COOL!!!).

References

Abdul Aziz S, Olival KJ, Bumrungsri S, Richards GC, Racey PA (2016) The conflict between fruit bats and fruit growers: species, legislation and mitigation. In: Voigt CC, Kingston T (eds) Bats in the Anthropocene: conservation of bats in a changing world. Springer International AG, Cham, pp 377–420

Ajzen I (1971) Attitudinal vs. normative messages: an investigation of the differential effects of persuasive communications on behavior. Sociometry 34:263–280

Ajzen I (1991) The theory of planned behavior. Organ Behav Hum Dec 50:179–211

Ajzen I, Fisbein M (1980) Understanding attitudes and predicting social behavior. Prentice-Hall, Englewood Cliffs

Allport GW (1935) Attitudes. In: Murchison C (ed) Handbook of social psychology. Clark University Press, Worcester, MA, pp 798–844

Altringham J, Kerth G (2016) Bats and roads. In: Voigt CC, Kingston T (eds) Bats in the Anthropocene: conservation of bats in a changing world. Springer International AG, Cham, pp 35–58

Andrade E (2009) Bats. In: Asante MK, Mazama A (eds) Encyclopedia of African Religion, vol 1. Sage, Thousand Oaks, pp 113–114

Armitage CJ, Conner M (2001) Efficacy of the theory of planned behaviour: a meta-analytic review. Brit J Soc Psychol 40:471–499

Arnett EB, Baerwald EF, Matthews F, Rodrigues L, Rydell J, Voigt CC (2016) Impacts of wind energy development on bats: a global perspective. In: Voigt CC, Kingston T (eds) Bats in the Anthropocene: conservation of bats in a changing world. Springer International AG, Cham, pp 295–317

Batt S (2009) Human attitudes towards animals in relation to species similarity to humans: a multivariate approach. Biosci Horiz 2:180–190

Bennett NJ, Roth R (eds) (2015) The conservation social sciences: What? How? and Why? Canadian Wildlife Federation and Institute for Resources, Environment and Sustainability, University of British Columbia, Vancouver, BC

Benton-Browne A, Palmer E (2003) Gema's Home. Produced by the Malaysian Bat Conservation Research Unit and available for download http://www.seabcru.org/wp-content/uploads/2013/07/Comic-Gemas-Home.pdf

Bjerke T, Østdahl T (2004) Animal-related attitudes and activities in an urban population. Anthrozoös 17:109–129

Blackmore E, Underhill R, McQuilkin J, Leach R (2013a) Common cause for nature: finding values and frames in the conservation sector. Public Interest Research Centre, Machynlleth, Wales

Blackmore E, Underhill R, McQuilkin J, Leach R, Holmes T (2013b) Common cause for nature: a practical guide to values and frames in conservation. Public Interest Research Centre, Machynlleth, Wales

Bobo KS, Aghomo FFM, Ntumwel BC (2015) Wildlife use and the role of taboos in the conservation of wildlife around the Nkwende Hills Forest Reserve; South-west Cameroon. J Ethnobiol Ethnomed 11:1–24

Boyles JG, Cryan PM, McCracken GF, Kunz TH (2011) Economic importance of bats in agriculture. Science 332:41–42

Burgess J, Harrison CM, Filius P (1998) Environmental communication and the cultural politics of environmental citizenship. Environ Plann A 30:1445–1460

Chan AAH (2012) Anthropomorphism as a conservation tool. Biodiv Conserv 21:1889–1892

Christensen PN, Rothgerber H, Wood W, Matz DC (2004) Social norms and identity relevance: A motivational approach to normative behavior. Pers Soc Psychol B 30:1295–1309

Cialdini RB (2003) Crafting normative messages to protect the environment. Curr Dir Psychol Sci 12:105–109

Cialdini RB, Trost MR (1998) Social influence: social norms, conformity, and compliance. In: Gilbert DT, Fiske ST, Lindzey G (eds) The handbook of social psychology, 4th edn. McGraw Hill, Boston, pp 141–192

Cialdini RB, Reno RR, Kallgren CA (1990) A focus theory of normative conduct: recycling the concept of norms to reduce littering in public places. J Pers Soc Psychol 58:1015–1026

Clayton S, Myers OE (2015) Conservation psychology: understanding and promoting human care for nature, 2nd edn. Wiley-Blackwell, Chichester

Clayton S, Litchfield C, Geller ES (2013) Psychological science, conservation, and environmental sustainability. Front Ecol Environ 11:377–382

Cleveland CJ, Betke M, Federico P, Frank JD, Hallam TG, Horn J et al (2006) Economic value of the pest control service provided by Brazilian free-tailed bats in south-central Texas. Front Ecol Environ 4:238–243

Colding J, Folke C (2001) Social taboos: 'Invisible' systems of local resource management and biological conservation. Ecol Appl 11:584–600

Crompton T (2010) Common Cause: The case for working with our cultural values. WWF, Surrey

Davey GC, McDonald AS, Hirisave U, Prabhu GG, Iwawaki S, Im Jim C et al (1998) A cross-cultural study of animal fears. Behav Res Ther 36:735–750

Davis T, Goldwater MB, Gaylord N, Worthy DA, Otto AR, Glass BD (2013) The cognitive psychology of human-bat interactions: Implications for ecological policy and zoonotic disease transmission. In: Geyer GA (ed) Bats: phylogeny and evolutionary insights, conservation strategies and role in disease transmission. Nova, Hauppauge, pp 1–17

De Houwer J, Thomas S, Baeyens F (2001) Association learning of likes and dislikes: a review of 25 years of research on human evaluative conditioning. Psychol Bull 127:853–869
Deaux K (1996) Social identification. In: Higgins ET, Kruglanski AW (eds) Social psychology: handbook of basic principles. The Guildford Press, New York, pp 777–798
Decker DJ, Riley SJ, Siemer WF (eds) (2012) Human dimensions of wildlife management. John Hopkins University Press, Baltimore
Eagly AH, Chaiken S (1993) The psychology of attitudes. Harcourt, Brace and Jovanovich, San Diego
Ehrlich P, Kennedy D (2005) Millennium assessment of human behaviour. Science 309:562–563
Fazio RH, Chen J, McDonel EC, Sherman SJ (1982) Attitude accessibility, attitude-behavior consistency, and the strength of the object-evaluation association. J Pers Soc Psychol 47:277–286
Fazio RH (2000) Accessible attitudes as tools for object appraisal: their costs and benefits. In: Maio GR, Olson JM (eds) Why we evaluate: functions of attitudes.Erlbaum, Mahwah, pp 1–36
Festinger L (1957) A theory of cognitive dissonance. Standford University Press, Stanford
Field AP, Lawson J (2003) Fear information and the development of fears during childhood: Effects on implicit fear responses and behavioural avoidance. Behav Res Ther 41:1277–1293
Field AP, Lawson J (2008) The verbal information pathway to fear and subsequent causal learning in children. Cogn Emot 22:459–479
Fishbein M, Ajzen I (1975) Belief, attitude, intention, and behavior: an introduction to theory and research. Addison-Wesley, Reading
Frey D (1986) Recent research on selective exposure to information. In: Berkowitz L (ed) Advances in experimental social psychology, vol 19. Academic Press, San Diego, pp 41–80
Fulton DC, Manfredo MJ, Lipscomb J (1996) Wildlife value orientations: a conceptual and measurement approach. Hum Dimens Wildl 1:24–47
Furey NM, Racey PA (2016) Conservation ecology of cave bats. In: Voigt CC, Kingston T (eds) Bats in the Anthropocene: conservation of bats in a changing world. Springer International AG, Cham, pp 463–492
Gintis H (2003) Solving the puzzle of prosociality. Ration Soc 15:155–187
Goldstein NJ, Cialdini RB, Griskevicius V (2008) A room with a viewpoint: Using social norms to motivate environmental conservation in hotels. J Consum Res 35:472–482
Guagnano GA, Stern PC, Dietz T (1995) Influences on attitude-behavior relationships: A natural experiment with curbside recycling. Environ and Behav 27:699–718
Gunnthorsdottir A (2001) Physical attractiveness of an animal species as a decision factor for its preservation. Anthrozoös 14:204–215
Harrison ME, Cheyne SM, Darma F, Ribowo DA, Limin SH, Struebig MJ (2011) Hunting of flying foxes and perception of disease risk in Indonesian Borneo. Biol Conserv 144:2441–2449
Hechter M, Opp KD (2001) Social norms. Russell Sage Foundation, New York
Hens L (2006) Indigenous knowledge and biodiversity conservation and management in Ghana. J Hum Ecol 20:21–30
Holland R, Verplanken B, van Knippenberg A (2002) On the nature of attitude-behavior relations: the strong guide, the weak follow. Eur J Soc Psychol 32:869–876
Homer PM, Kahle LR (1988) A structural equation test of the value-attitude-behavior hierarchy. J Pers Soc Psychol 54:638–647
Houston DA, Fazio RH (1989) Biased processing as a function of attitude accessibility: making objective judgments subjectively. Soc Cognition 7:51–66
Iozzi LA (1989) What research says to the educator: part one, environmental education and the affective domain. J Environ Educ 20:3–9
Jones JPG, Andriamarovololona MM, Hockley N (2008) The importance of taboos and social norms to conservation in Madagascar. Conserv Biol 22:976–986
Jung K, Threlfall CG (2016) Urbanisation and its effects on bats—a global meta-analysis. In: Voigt CC, Kingston T (eds) Bats in the Anthropocene: conservation of bats in a changing world. Springer International AG, Cham, pp 13–28

Kahn PH (2006) Nature and moral development. In: Killen M, Smetana JG (eds) Handbook of moral development. Lawrence Erlbaum Associates, Mahwah, pp 461–480

Kahn PH, Saunders CD, Severson RL, Myers OE, Gill BT (2008) Moral and fearful affiliations with the animal world: children's conceptions of bats. Anthrozoös 21:375–386

Kamins AO, Rowcliffe JM, Ntiamoa-Baidu Y, Cunningham AA, Wood JL, Restif O (2014) Characteristics and risk perceptions of Ghanaians potentially exposed to bat-borne zoonoses through bushmeat. EcoHealth 12:104–120

Katz D (1960) The functional approach to the study of attitudes. Public Opin Quart 24:163–204

Kaiser FG, Wölfing S, Fuhrer U (1999) Enviornmental attitude and ecological behavior. J Environ Psychol 19:1–19

Kellert SR (1976) Perceptions of animals in American society. Trans Nort Am Wildl Nat Res 41:533–546

Kellert SR (1979) Public attitudes toward critical wildlife and natural habitat issues, phase I. United States Department of the Interior Fish and Wildlife Service

Kellert SR (1980) Phase II: Activities of the American public relating to animals. United States Department of the Interior Fish and Wildlife Service

Kellert SR (1993) The biological basis for human values of nature. In: SR Kellert, EO Wilson (eds) The biophilia hypothesis. Island Press, Washington, DC, pp 41–69

Kellert SR (1996) The value of life: biological diversity and human society. Island Press, Washington

Kellert SR (2002) Experiencing nature: affective, cognitive, and evaluative development in children. In: Kahn PH, Kellert SR (eds) Children and nature: psychological, sociocultural, and evolutionary investigations. MIT Press, Cambridge, pp 117–151

Kellert SR, Berry JK (1980) Phase III: Knowledge, affection and basic attitudes toward animals in American society. United States Department of the Interior Fish and Wildlife Service

Kellert SR, Westervelt MO (1981) Trends in animal use and perception in twentieth century America: Phase IV. United States Department of the Interior Fish and Wildlife Service

Kellert SR, Westervelt MO (1983) Children's attitudes, knowledge and behaviors toward animals: Phase V. United States Department of the Interior Fish and Wildlife Service

Kellert SR, Westervelt MO (1984) Children's attitudes, knowledge and behaviors towards animals. Children's Environs Quart 1:8–11

Kellert SR, Wilson EO (1993) The biophilia hypothesis. Island Press, Washington, DC

Kelly VL, Barker H, Field AP, Wilson C, Reynolds S (2010) Can Rachman's indirect pathways be used to unlearn fear? A prospective paradigm to test whether children's fears can be reduced using positive information and modelling a non-anxious response. Behav Res Ther 48:164–170

King MF, Burner GC (2000) Social desirability bias: a neglected aspect of validity testing. Psychol Market 17:79–103

Kingston T, Juliana S, Rakhmad SK, Fletcher CD, Benton-Browne A, Struebig M et al (2006) The Malaysian Bat Conservation Research Unit: Research, capacity building and education in an Old World hotspot. In: Sahir O, Siti Hawa Y, Sivananthan E, Shukor Md. N, Norhayati A, Sharul Anuar Mohd. S (eds) Proceedings of the national seminar on protected areas. Department of Wildlife and National Parks, Malaysia, pp 41–60

Kinzig AP, Ehrlich PR, Alston LJ, Arrow K, Barrett S, Buchman TG et al (2013) Social norms and global environmental challenges: the complex interaction of behaviors, values, and policy. Bioscience 63:164–175

Kluckhohn FR, Strodtbeck FL (1961) Variations in value orientations. Row, Peterson, Oxford

Knight AJ (2008) "Bats, snakes and spiders, Oh my!" How aesthetic and negativistic attitudes, and other concepts predict support for species protection. J Environ Pyschol 28:94–103

Knobloch-Westerwick S, Meng J (2009) Looking the other way: selective exposure to attitude-consistent and counterattitudinal political information. Commun Res 36:426–448

Kollmuss A, Agyeman J (2002) Mind the Gap: why do people act environmentally and what are the barriers to pro-environmental behavior? Environ Educ Res 8:239–260

Korine C, Adams R, Russo D, Fisher-Phelps M, Jacobs D (2016) Bats and water: anthropogenic alterations threaten global bat populations. In: Voigt CC, Kingston T (eds) Bats in the Anthropocene: conservation of bats in a changing world. Springer International AG, Cham, pp 215–233

Kraus SJ (1995) Attitudes and the prediction of behaviour: a meta-analysis of the empirical literature. Pers Soc Psychol B 21:58–75

Krosnick JA, Petty RE (1995) Attitude strength: an overview. In: Petty RE, Krosnick JA (eds) Attitude strength: antecedents and consequences. Erlbaum, Hillsdale, pp 1–24

Kubiatko M (2012) Kindergarten children's perception of animals focusing on the look and fear of animals. Educ Sci: Theory Pract 12:3181–3186

Kunz TH, Braun de Torrez E, Bauer D, Lobova T, Fleming TH (2011) Ecosystem services provided by bats. Ann NY Acad Sci 1223:1–38

Law B, Park K, Lacki M (2016) Insectivorous bats and silviculture: balancing timber production and bat conservation. In: Voigt CC, Kingston T (eds) Bats in the Anthropocene: conservation of bats in a changing world. Springer International AG, Cham, pp 105–141

Lawrence EA (1993) The sacred bee, the filthy pig, and the bat out of hell: Animal symbolism as cognitive biophilia. In: Kellert SR, Wilson EO (eds) The biophilia hypothesis. Island Press, Washington, DC, pp 301–341

Liesener AL, Smith KE, Davis RD, Bender JB, Danila RN, Neitzel DF et al (2006) Circumstances of bat encounters and knowledge of rabies among minnesota residents submitting bats for rabies testing. Vector Borne Zoonotic Dis 6:208–215

MacKinnon JL, Hawkins CE, Racey PA (2003) Pteropodidae, fruit bats, Fanihy, Angavo. In: Goodman SM, Benstead JP (eds) The natural history of Madagascar. University of Chicago Press, Chicago, pp 1299–1302

Maio GR, Haddock G (2014) The psychology of attitudes and attitude change, 2nd edn. Sage, Los Angeles

Manfredo MJ (ed) (2008) Who cares about wildlife? Social science concepts for exploring human-wildlife relationships and conservation issues. Springer, New York

Manfredo MJ, Teel TL (2008) Integrating concepts: demonstration of a multilevel model for exploring the rise of mutualism value orientations in post-industrial society. In: Manfredo MJ (ed) Who cares about wildlife? Social science concepts for exploring human-wildlife relationships and conservation issues. Springer, New York, pp 191–218

Marimuthu G (1988) The sacred flying fox of India. Bats 6:10–11

Mascia M, Brosius JP, Dobson TA, Forbes B, Horowitz L, McKean M, Turner N (2003) Conservation and the social sciences. Conserv Biol 17:649–650

McDonald RI, Fielding KS, Louis WR (2014) Conflicting social norms and community conservation compliance. J Nat Conserv 22:212–216

Menon V, Lavigne D (2006) Attitudes, values and objectives: the real basis of wildlife conservation. In: Lavigne DM (ed) Gaining ground: In pursuit of ecological sustainability. International Fund for Animal Welfare and University of Limerick, Guelph and Limerick, pp 173–189

Metcalfe K, French-Constant R, Gordon I (2009) Sacred sites as hotspots for biodiversity: the Three Sisters Cave complex in coastal Kenya. Oryx 44:118–123

Meyer CFJ, Struebig MJ, Willig MR (2016) Responses of tropical bats to habitat fragmentation, logging, and deforestation. In: Voigt CC, Kingston T (eds) Bats in the Anthropocene: conservation of bats in a changing world. Springer International AG, Cham, pp 63–94

Milfont TL, Duckitt J (2010) The environmental attitudes inventory: a valid and reliable measure to assess the structure of environmental attitudes. J Environ Psychol 30:80–94

Mildenstein T, Tanshi I, Racey PA (2016) Exploitation of bats for bushmeat and medicine. In: Voigt CC, Kingston T (eds) Bats in the Anthropocene: conservation of bats in a changing world. Springer International AG, Cham, pp 325–363

Mintzes JJ, Wandersee JH (1998) Research in science teaching and learning: a human constructivistic view. In: Mintzes JJ, Wandersee JH, Novak JD (eds) Teaching science for understanding. Academic Press, Orlando, pp 59–92

Mithen S, Boyer P (1996) Anthropomorphism and the evolution of cognition. J Roy Anthropol Inst 2:717–721
Moon K, Blackman D (2014) A guide to understanding social science research for natural scientists. Conserv Biol 28:1167–1177
Morris PA (1987) Changing attitudes towards British mammals. Biol J Linn Soc 32:225–233
Muris P, Bodden D, Merckelbach H, Ollendick TH, King N (2003) Fear of the beast: a prospective study on the effects of negative information on childhood fear. Behav Res Ther 41:195–208
Muris P, van Zwol L, Huijding J, Mayer B (2010) Mom told me scary things about this animal: parents installing fear beliefs in their children via the verbal information pathway. Behav Res Ther 48:341–346
Newing H (2011) Conducting research in conservation: social science methods and practice. Routledge, New York
Nuno A, St John FA (2015) How to ask sensitive questions in conservation: a review of specialized questioning techniques. Biol Conserv 189:5–15
Olson MA, Fazio RH (2006) Reducing automatically activated prejudice through implicit evaluative conditioning. Pers Soc Psychol B 32:421–433
Petty RE, Krosnick JA (eds) (1995) Attitude strength: antecedents and consequences. Erlbaum, Hillsdale
Pooley JA, O'Connor M (2000) Environmental education and attitudes: emotions and beliefs are what is needed. Environ Behav 32:711–723
Prokop P, Tunnicliffe SD (2008) "Disgusting" animals: primary school children's attitudes and myths of bats and spiders. Eurasia J Math Sci Tech Ed 4:87–97
Prokop P, Fančovičová J, Kubiatko M (2009) Vampires are still alive: Slovakian students' attitudes towards bats. Anthrozoös 22:19–30
Rachman S (1977) The conditioning theory of fear-acquisition a critical examination. Behav Res Ther 15:375–387
Rahaingodrahety VN, Andriafidison D, Ratsimbazafy JH, Racey PA, Jenkins RK (2008) Three flying fox (Pteropodidae: *Pteropus rufus*) roosts, three conservation challenges in southeastern Madagascar. Madagascar Conserv Develop 3:17–21
Remmerswaal D, Muris P, Huijding J (2013) "Watch out for the gerbils, my child!" The role of maternal information on children's fear in an experimental setting using real animals. Behav Ther 44:317–324
Robertson K, Lumlertdacha B, Franka R, Petersen B, Bhengsri S, Henchaichon S et al (2011) Rabies-related knowledge and practices among persons at risk of bat exposures in Thailand. PLoS Negl Trop Dis 5(6):e1054
Robinson MF, Smith AL (1997) Chiroptera from Loei Province, north-east Thailand. Nat Hist B Siam Soc 45:1–16
Rohan MJ (2000) A rose by any name? The values construct. Pers Soc Psychol Rev 4:255–277
Rokeach M (1973) The nature of human values, vol 438. Free Press, New York
Root-Bernstein M, Douglas L, Smith A, Verissimo D (2013) Anthropomorphized species as tools for conservation: utility beyond prosocial, intelligent and suffering species. Biodivers Conserv 22:1577–1589
Rowse EG, Lewanzik D, Stone EL, Harris S, Jones G (2016) Dark matters: the effects of artificial lighting on bats. In: Voigt CC, Kingston T (eds) Bats in the Anthropocene: conservation of bats in a changing world. Springer International AG, Cham, pp 187–207
Sandbrook C, Adams WM, Buescher B, Vira B (2013) Social research and biodiversity conservation. Conserv Biol 27:1487–1490
Schneeberger K, Voigt CC (2016) Zoonotic viruses and conservation of bats. In: Voigt CC, Kingston T (eds) Bats in the Anthropocene: conservation of bats in a changing world. Springer International AG, Cham, pp 263–282
Schultz PW (2011) Conservation means behaviour. Conserv Biol 25:1080–1083
Schultz PW, Gouveia VV, Cameron LD, Tankha G, Schmuck P, Franěk M (2005) Values and their relationship to environmental concern and conservation behavior. J Cross Cult Psychol 36:457–475

Schultz PW, Nolan JM, Cialdini RB, Goldstein NJ, Griskevicius V (2007) The constructive, destructive, and reconstructive power of social norms. Psychol Sci 18:429–434

Schwartz SH (1992) Universals in the content and structure of values: theoretical advances and empirical tests in 20 countries. Adv Exp Soc Pyschol 25:1–65

Schwartz SH, Sagiv L (1995) Identifying culture-specifics in the content and structure of values. J Cross Cult Psychol 26:92–116

Serpell JA (2004) Factors influencing human attitudes to animals and their welfare. Anim Welfare 13(suppl. 1):145–151

Sexton NR, Stewart SC (2007). Understanding knowledge and perceptions of bats among residents of Fort Collins, Colorado. US Geological Survey

Sheherazade, Tsang SM (2015) Quantifying the bat bushmeat trade in North Sulawesi, Indonesia, with suggestions for conservation action. Global Ecol Conserv 3:324–330

Sherif M (1936) The psychology of social norms. Harper & Brothers, New York

Smith MB, Bruner JS, White RW (1956) Opinions and personality. Wiley, New York

St John FAV, Edwards-Jones G, Jones JPG (2010) Conservation and human behavior: lessons from social science. Wild Res 37:658–667

St John FAV, Keane AM, Edwards-Jones G, Jones L, Yarnell RW, Jones JP (2011) Identifying indicators of illegal behaviour: carnivore killing in human-managed landscapes. Proc R Soc Lond B Bio rspb20111228

St John FAV, Keane AM, Milner-Gulland EJ (2013) Effective conservation depends upon understanding human behaviour. In: MacDonald DW, Willis KJ (eds) Key topics in conservation biology, vol 2. Wiley-Blackwell, Colchester, pp 344–361

St John FAV, Keane AM, Jones JPG, Milner-Gulland EJ (2014) Robust study design is as important on the social as it is on the ecological side of applied ecological research. J Appl Ecol 51:1479–1485

Stern PC (2000) Toward a coherent theory of environmentally significant behavior. J Soc Issues 3:407–424

Stern PC, Dietz T (1994) The value basis of environmental concern. J Soc Issues 50:65–84

Stern PC, Dietz T, Kalof L (1993) Value orientations and environmental concern. Environ Behav 25:322–348

Stern PC, Dietz T, Guagnano GA (1998) A brief inventory of values. Educ Psychol Meas 58:984–1001

Stokes DL (2007) Things we like: human preferences among similar organisms and implications for conservation. Hum Ecol 35:361–369

Tam KP, Lee SL, Chao MM (2013) Saving Mr. Nature: anthropomorphism enhances connectedness to and protectiveness toward nature. J Expt Soc Psychol 49:514–521

Tangavelou AC, Rani PJ, Karthikeyan S (2013) Conservation of Sacred Indian flying fox (Bat) at sacred landscape of Pudukottai district, Tamil Nadu, India. Asian J Conserv Biol 2:178–180

Teel LT, Dietsch AM, Manfredo MJ (2015) A (social) psychology approach in conservation. In: Bennett NJ, Roth R (eds) The conservation social sciences: What? How? and Why? Canadian Wildlife Federation and Institute for Resources, Environment and Sustainability. University of British Columbia, Vancouver, pp 21–25

Thompson TL, Mintzes JJ (2002) Cognitive structure and the affective domain: on knowing and feeling in biology. Int J Sci Educ 24:645–660

Veríssimo D (2013) Influencing human behavior: an underutilised tool for biodiversity management. Conserv Evid 10:29–31

Voigt CC, Kingston T (2016) Bats in the Anthropocene: conservation of bats in a changing world. Springer International AG, Cham

Voigt CC, Phelps KL, Aguirre LF, Schoeman MC, Vanitharani J, Zubaid A (2016) Bats and buildings: the conservation of synanthropic bats. In: Voigt CC, Kingston T (eds) Bats in the Anthropocene: conservation of bats in a changing world. Springer International AG, Cham, pp 427–453

Wadley RL, Colfer CJP (2004) Sacred forest, hunting and conservation in West Kalimantan, Indonesia. Hum Ecol 32:313–338

Wanger TC, Darras K, Bumrungsri S, Tscharntke T, Klein AM (2014) Bat pest control contributes to food security in Thailand. Biol Conserv 171:220–223

White PCL, Jennings NV, Renwick AR, Barker NHL (2005) Questionnaires in ecology: a review of past use and recommendations for best practice. J Appl Ecol 42:421–430

Williams KJ (2014) The effect of positive verbal information on reducing fears about bats in school-aged children. PhD Dissertation, UMI Number 3618417, Walden University, USA

Williams-Guillén K, Olimpi E, Maas B, Taylor PJ, Arlettaz R (2016) Bats in the anthropogenic matrix: challenges and opportunities for the conservation of Chiroptera and their ecosystem services in agricultural landscapes. In: Voigt CC, Kingston T (eds) Bats in the Anthropocene: conservation of bats in a changing world. Springer International AG, Cham, pp 151–178

Zanna MP, Rempel JK (1988) Attitudes: a new look at an old concept. In: BarTal D, Kruglanski AW (eds) The social psychology of knowledge. Cambridge University Press, Cambridge, pp 315–334

Index

C.C. Voigt and T. Kingston (eds.), *Bats in the Anthropocene: Conservation of Bats in a Changing World*, DOI 10.1007/978-3-319-25220-9

N

T

www.ingramcontent.com/pod-product-compliance
Ingram Content Group UK Ltd.
Pitfield, Milton Keynes, MK11 3LW, UK
UKHW022218051026
15624UKWH00001B/215
9783319252186